Lecture Notes in Computer Science 8881

Commenced Publication in 1973
Founding and Former Series Editors:
Gerhard Goos, Juris Hartmanis, and Jan van Leeuwen

More information about this series at http://www.springer.com/series/7407

Zhao Zhang · Lidong Wu
Wen Xu · Ding-Zhu Du (Eds.)

Combinatorial Optimization and Applications

8th International Conference, COCOA 2014
Wailea, Maui, HI, USA, December 19–21, 2014
Proceedings

 Springer

Editors
Zhao Zhang
Zhejiang Normal University
Jinhua, Zhejiang
China

Lidong Wu
University of Texas
Tyler, TX
USA

Wen Xu
Ding-Zhu Du
University of Texas
Dallas, TX
USA

ISSN 0302-9743   ISSN 1611-3349 (electronic)
ISBN 978-3-319-12690-6   ISBN 978-3-319-12691-3   (eBook)
DOI 10.1007/978-3-319-12691-3

Library of Congress Control Number: 2014954574

LNCS Sublibrary: SL1 – Theoretical Computer Science and General Issues

Springer Cham Heidelberg New York Dordrecht London

Printed on acid-free paper

Springer is part of Springer Science+Business Media (www.springer.com)

Preface

The Eighth Annual International Conference on Combinatorial Optimization and Applications (COCOA 2014) was held during December 19–21, 2014, in Maui, Hawaii, USA. COCOA 2014 provides a forum for researchers working in the area of theoretical computer science and combinatorics.

The technical program of the conference includes 56 contributed papers selected by the Program Committee from 133 full submissions received in response to the call for papers. All the papers were peer reviewed by at least three Program Committee members or external reviewers. In addition to selected papers, the conference also included one invited presentation by My T. Thai

The topics cover most aspects of theoretical computer science and combinatorics related to computing, including classic combinatorial optimization, geometric optimization, network optimization, optimization in graphs, applied optimization, complexity, cryptography and game, and miscellaneous. Some of the papers will be selected for publication in special issues of Algorithmic, Theoretical Computer Science (TCS), and Journal of Combinatorial Optimization (JOCO). It is expected that the journal version papers will appear in a more complete form.

The proceeding also includes eight papers selected from a workshop on computational social networks (CSoNet 2014) co-located with COCOA 2014, held during December 19–21, 2014. We appreciate the work by the CSoNet Program Committee that helped with enriching the conference topics.

We thank all the people who made this meeting possible: the authors for submitting papers, the Program Committee members, and external reviewers for volunteering their time to review conference papers. We would like to extend special thanks to the publication, publicity, and local organization chairs for their hard work in making COCOA 2014 a successful event.

September 2014

Xiaodong Hu
Weili Wu

Organization

Committees

General Co-chairs

Xiaodong Hu — Chinese Academy of Sciences, China
Weili Wu — University of Texas at Dallas, USA

Program Chairs

Ding-Zhu Du — University of Texas at Dallas, USA
Thang Dinh — Virginia Commonwealth University, USA
Lidong Wu — University of Texas at Tyler, USA
Zhao Zhang — Zhejiang Normal University, China

Program Committee

Wolfgang Bein — University of Nevada, USA
Sergiy Butenko — Texas A&M University, USA
Gruia Calinescu — Illinois Institute of Technology, USA
Chiuyuan Chen — National Chiao Tong University, Taiwan
Xujin Chen — Chinese Academy of Sciences, China
He Chen — South-Eastern University, China
Zhi-Xiang Chen — University of Texas-Pan American, USA
Zhi-Zhong Chen — Tokyo Denki University, Japan
Yongxi Cheng — Xi'an Jiaotong University, China
Ovidiu Daescu — University of Texas at Dallas, USA
Bhaskar Dasgupta — University of Illinois at Chicago, USA
Vladimir Deineko — The University of Warwick, UK
Ding-Zhu Du — University of Texas at Dallas, USA
Zhenhua Duan — Xidian University, China
Neng Fan — University of Arizona, USA
Juraj Hromkovic — ETH Zentrum, Switzerland
Wenlian Hsu — Academia Sinica, Taiwan
Hejiao Huang — Harbin Institute of Technology, China
Kazuo Iwama — Kyoto University, Japan
Xinwen Jiang — National University of Defense Technology, China
Liying Kang — Shanghai University, China
Naoki Katoh — Kyoto University, Japan
Ilias S. Kotsireas — Wilfrid Laurier University, Canada
Asaf Levin — The Technion, Israel
Minming Li — City University of Hong Kong, China

Wei Wayne Li	Texas Southern University, USA
Zaixin Lu	Marywood University, USA
Mitsunori Ogihara	University of Miami, USA
Suneeta Ramaswami	Rutgers University, USA
Lusheng Wang	City University of Hong Kong, Hong Kong
Weifan Wang	Zhejiang Normal University, China
Lidong Wu	University of Texas at Tyler, USA
Wen Xu	University of Texas at Dallas, USA
Hsu-Chun Yen	National Taiwan University, Taiwan
Louxin Zhang	National University of Singapore, Singapore
Zhao Zhang	Xingjiang University, China
Xiao Zhou	Tohoku University, Japan
Xuding Zhu	Zhejiang Normal University, China

Heterogenous Interdependent Networks: Critical Elements and Cascades Analysis (Keynote)

My T. Thai

Department of Computer and Information Science and Engineering,
University of Florida, Gainesville, FL 32611, USA
mythai@cise.ufl.edu

Abstract. Modern complex networked systems, such as power grids, communication networks, and transportation networks are interdependent in such a way that a failure of an element in one system may cause multiple failures of elements in other systems. This process can propagate back and forth between interdependent systems in a cascading fashion, resulting in a catastrophic widespread failure. Another notable example would be a complex system of social networks where users can have their accounts on many networks simultaneously. Since users can automatically synchronize their posts on all networks, the social contagion is not only propagated between several layers from friends to his colleagues but also cascading across many social networking sites. Therefore, it is urgently call for new models and analytical techniques to understand the dynamics of these cascades and assess the systems vulnerability.

In this talk, we will discuss the cascading behaviors in different networked systems, mathematically quantifying the "depth" and "breadth" of cascades. Based on these models, we present techniques to identify the critical elements whose failures yields the most significant loss of resilience of the interdependent systems.

Contents

Classic Combinatorial Optimization

An Exact Algorithm for Non-preemptive Peak Demand Job Scheduling

Sean Yaw$^{(\boxtimes)}$ and Brendan Mumey

Department of Computer Science, Montana State University, Bozeman, MT, USA
{sean.yaw,mumey}@cs.montana.edu

Abstract. Peak demand scheduling aims to schedule jobs so as to minimize the peak load in the schedule. An important application of this problem comes from scheduling power jobs in the smart grid. Currently, peaks in power demand are due to the aggregation of many jobs being scheduled in an on-demand fashion. Often these have some flexibility in their starting times which can be leveraged to lower the peak demand of a schedule. While the general version of the problem is known to be NP-hard (we observe it is even NP-hard to approximate), we provide an optimal algorithm based on dynamic programming that is fixed-parameter tractable (FPT). Simulation results using household power usage data show that peak power demand can be significantly reduced by allowing some flexibility in job execution times and applying scheduling.

1 Introduction

We consider a variation on the classic job scheduling problem in which the goal is to minimize the peak demand of a schedule. An important scenario where this problem arises is in scheduling power jobs in the smart grid. It is advantageous for utility providers to keep aggregate power demand as flat as possible since this reduces the cost of generation and distribution. In addition, power generation may be performed at a local level in which there is reduced benefit of averaging over a large set of users. Currently, power job requests in the home are not scheduled: power draw begins when the consumer turns on an appliance and ends when the appliance is turned off. This on-demand scheduling can lead to high peaks of power usage when consumers collectively issue many jobs, such as in the evening when numerous household appliances are running. Though many appliances need to be scheduled in an on-demand fashion to provide instant functionality (e.g. television, light bulbs), other appliances may have more flexibility and can be delayed and executed at a later time, when power draw is not so high (e.g. dishwasher, plug-in vehicle charging) [17]. This job flexibility provides the opportunity to flatten out the overall demand by intelligently delaying execution of some jobs.

There are several interesting variations on the peak demand scheduling problem: To begin with, some jobs may be known in advance while others arrive in a dynamic unpredictable fashion. Thus, the problem can be viewed in both offline and online settings. Another variation is whether jobs are preemptable,

© Springer International Publishing Switzerland 2014
Z. Zhang et al. (Eds.): COCOA 2014, LNCS 8881, pp. 3–12, 2014.
DOI: 10.1007/978-3-319-12691-3_1

meaning they can be interrupted for some time, provided they are completed by the deadline. Some loads such as heating an electric water heater or charging a plug-in electric vehicle are preemptable but other loads such as running a dishwasher are not. Finally, some loads may be resizable in the sense that their power requirements may be adjustable between higher or lower power levels in conjunction with different execution times (e.g. battery charging). Each of these variations has an impact on the computational complexity of creating a schedule that minimizes the peak demand.

In this work, we focus on minimizing the peak demand of a schedule of non-preemptable, non-resizable jobs having flexible timelines. We summarize our contributions as follows: We formalize and present two algorithms for the peak demand minimization problem. The first is an optimal FPT algorithm based on dynamic programming and the second is a simple heuristic that is shown to have good results in practice. We leverage current energy disaggregation research to generate real world test scenarios and schedule them using our algorithms.

The rest of this paper is organized as follows. We discuss related work in Sect. 2. We formulate the problem in Sect. 3 and examine its computational complexity. Algorithms are presented in Sect. 4 and simulation results are presented in Sect. 5. The paper is concluded in Sect. 6.

2   Related Work

Several variations on job scheduling for the smart grid have recently been explored. In [11,12], the authors define a job model, derive a lower bound on the optimal schedule cost for the offline, preemptive, non-resizable variation and also provide an optimal online algorithm for the preemptive, non-resizable case, assuming a statistical description of the input. In [7,13], the authors apply linear programming techniques to power job scheduling with an emphasis on pricing and price prediction models with the result of reducing peak loads. The main difference with our work is that we look at specific jobs and aim to schedule them within an acceptable window. These works do not look at specific jobs, but instead aim to reduce overall usage with price incentives. In [16], an approximation algorithm for peak minimization of non-preemptive jobs that have common arrival and completion deadlines was described that has an approximation ratio of 7.82.[1] This approximation ratio was reduced to 4 by an algorithm presented in [19]. Approximation algorithms have also been studied for the special case of peak demand minimization of jobs with unit demand [5,6].

A problem closely related to peak demand scheduling is the much studied problem of job scheduling for a speed scalable CPU [4,14,18]. In this case, the tasks are compute jobs and the processing unit can vary the speed at which the current job executes; higher speed results in more energy use. While the preemptive version of the problem has been well-studied and good offline and

---

[1] While not explicitly stated in [16], the best approximation ratio achieved for the MP algorithm results from minimizing $a + 2 + \frac{2a}{a-1}$, which occurs at $a = \sqrt{2}+1$ and yields an approximation ratio of 7.82.

online algorithms are known, only recently has the non-preemptive version been addressed [1,3]. A version of the speed scaling problem that considers parallelizable jobs has also received recent attention [8]. Scheduling parallelizable compute jobs is similar for power jobs scheduling as both leverage concurrent execution.

Non-preemptive job scheduling is also similar to rectangular strip packing [2,9,15]. The main differences are that, in the general case, jobs are limited in where they can be placed in the strip and once jobs are scheduled, they do not need to remain as intact rectangles. Since job height represents the power required, each segment of a scheduled job will drop to lie on top of the job below it, instead of remaining as an intact rectangle in the strip.

3 Problem Formulation

Job j is defined as a 4-tuple: (a_j, d_j, l_j, h_j), where a_j and d_j are the arrival time and deadline within the time interval $[0, T]$, l_j is the job length, and h_j is the instantaneous resource requirement (demand). Job j is scheduled by assigning it a start time, s_j, and then runs in the closed-open interval $[s_j, s_j + l_j)$ such that $[s_j, s_j + l_j) \subset [a_j, d_j)$. The interval $[a_j, d_j)$ is called the *execution window* of j.

The *demand* at time t is the sum of the heights of jobs that are scheduled during t, i.e.

$$H(t) = \sum_{j:t\in[s_j,s_j+l_j)} h_j.$$

Then, the *peak demand*, $H^{\max}$, of the schedule is the maximum demand of any of the timeslots in $[0, T]$,

$$H^{\max} = \max_{t\in[0,T]} H(t).$$

Definition 1. *The **Peak Demand Minimization Problem (PDM)** is, given a set of n jobs, determine a job schedule, $\mathcal{S} = \langle s_j \rangle$, so as to minimize $H^{\max}$.*

In general, each job can have any valid arrival time and deadline pair which leads to staggered execution windows as shown in Fig. 1. In this paper we consider the general version of this problem as well as the case where jobs have common execution windows (same arrival times and deadlines).

3.1 Computational Complexity

In [16], PDM was shown to be NP-hard via a reduction from the *Subset-Sum* problem. We observe that PDM is even NP-hard to approximate within a ratio of 2 by reducing it from the *Scheduling with Release Times and Deadlines on a Minimum Number of Machines* (SRDM) problem.

Lemma 1. *For all $\epsilon > 0$, PDM is NP-hard to approximate within a factor of $2 - \epsilon$.*

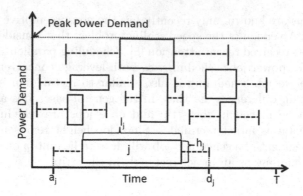

Fig. 1. Non-preemptive power jobs to be scheduled.

Proof. The SRDM problem considers scheduling jobs that consist of release times, deadlines, and lengths in a way that minimizes the number of machines needed to construct a schedule. An instance of SRDM can be reduced to an instance of PDM by retaining each job's release time, deadline, length, and making each job's height to be 1.

Then, any solution to the PDM instance that results in a schedule of height p, will be a schedule for the SRDM instance requiring p machines due to the unit height of all the jobs. Likewise, any solution to the SRDM instance requiring p machines will be a schedule for the PDM instance of height p because that is the minimum number of concurrently executing jobs. It follows that any approximation algorithm for PDM will provide the same approximation ratio for SRDM. Since SRDM cannot be approximated within a factor of $2 - \epsilon$, so the same hardness result applies to PDM [6].   □

4   Algorithms

In this section we present two algorithms for the PDM problem. The first is an optimal, fixed parameter tractable (FPT), algorithm and the second is a simple but effective heuristic algorithm.

4.1   PDM-Exact: An Optimal FPT Algorithm

The PDM-Exact algorithm employs dynamic programming techniques to minimize schedule height for the general, staggered execution window version of the problem. The algorithm searches for the minimum feasible peak power demand by testing whether or not the jobs can be scheduled under a predetermined power threshold τ. The threshold τ is then varied until the minimum value can be determined. Feasibility testing is performed by considering, for each job, every possible configuration of start times of that job and each previous job it overlaps

---

Algorithm 1. PDM-Exact

Step 1   Sort jobs first by deadline and then by arrival time.
        Compute $\mathcal{L}_j = \{i : \text{job } i \text{ overlaps job } j\}$.
        let $\tau_{\max} = 1$

Step 2   **while** (**not** Is-Feasible($\tau_{\max}$))
         $\tau_{\max} = 2 \cdot \tau_{\max}$
        endwhile
        Apply binary search to determine
          $\tau_{\min} = argmin_{\tau \in (\tau_{\max}/2, \tau_{\max}]}$ Is-Feasible(τ)

Step 3   Build schedule $\mathcal{S}$ using the configurations for $\tau_{\min}$:
         select $c \in \mathcal{C}_{n-1}$
         set s_{n-1} according to c
         for each job $j = n-2$ to 0:
            let $c' \in h(c)$ (note: $c' \in \mathcal{C}_j$)
            set s_j according to c'
            let $c = c'$
         endfor

(employing a non-trivial definition of overlap), and then seeing if a set of configurations for each job can be concurrently scheduled. Pseudocode is presented in Algorithm 1 and further explained below.

Job Overlap Lists. Order the jobs by increasing deadline first and then by increasing arrival time. Then determine, for each job j, the set of jobs that j overlaps. We define overlap in a non-trivial manner: Job i is said to *overlap* job j if $i \leq j$ and $\min_{l \geq j}(a_l) < d_i$. We define *job overlap lists* $\mathcal{L}_j$, as follows: $\mathcal{L}_j = \{i : \text{job } i \text{ overlaps job } j\}$. Figure 2 illustrates a non-trivial example of job overlaps.

Lemma 2. *For ordered jobs: $i, ..., k, ..., j$, if job i is in $\mathcal{L}_j$, then job i is also in $\mathcal{L}_k$.*

Proof. This is easy to confirm as $i \in \mathcal{L}_j$ implies that $\min_{l \geq j}(a_l) < d_i$. Since the jobs are ordered by deadline, $\min_{l \geq j}(a_l) < d_i \leq d_k$.   □

Configuration Generation. A value for the power threshold τ is given. For each job j, generate every possible configuration $\langle ..., s_k, ... \rangle$ of start times of

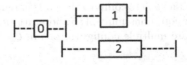

Fig. 2. Illustration of overlap definition: job 0 overlaps job 1, due to job 2's early arrival time. Also note that $\mathcal{L}_0 = \{0\}$, $\mathcal{L}_1 = \{0, 1\}$ and $\mathcal{L}_2 = \{0, 1, 2\}$.

Algorithm 2. Is-Feasible(τ)

Step 1   update $\mathcal{C}_j = \{\langle \ldots, s_k, \ldots \rangle : k \in \mathcal{L}_j, a_k \leq s_k \leq d_k - l_k,$
$\qquad\qquad\qquad \text{maxHeight}(\langle \ldots, s_k, \ldots \rangle) \leq \tau\}.$

Step 2   **for** each job $j = 1$ to $n - 1$:
$\qquad$ **for** each $c' \in \mathcal{C}_j$:
$\qquad\qquad$ Let $h(c') = \{c \in \mathcal{C}_{j-1} : c \sim c'\}$
$\qquad\qquad$ **if** $h(c') = \emptyset$
$\qquad\qquad\qquad$ $\mathcal{C}_j.\text{remove}(c)$
$\qquad\qquad$ **endif**
$\qquad$ **endfor**
$\qquad$ **if** $\mathcal{C}_j = \emptyset$
$\qquad\qquad$ **return false**
$\qquad$ **endif**
$\qquad$ **endfor**
$\qquad$ **return true**

---

jobs k in its overlap list $\mathcal{L}_j$. Define $\text{maxHeight}(\langle \ldots, s_k, \ldots \rangle)$ as the peak demand of the jobs in $\mathcal{L}_j$ given these start times. Each configuration is a tuple consisting of a valid start time for each job k in $\mathcal{L}_j$, such that the peak demand of the configuration is at most τ. The configuration lists are formally defined as:

$$\mathcal{C}_j = \{\langle \ldots, s_k, \ldots \rangle : k \in \mathcal{L}_j, a_k \leq s_k \leq d_k - l_k, \text{maxHeight}(\langle \ldots, s_k, \ldots \rangle) \leq \tau\}$$

Definition 2. *Configuration c in $\mathcal{C}_i$ and c' in $\mathcal{C}_j$ are **compatible** (written $c \sim c'$) if they agree on the starting times of all jointly shared jobs. A configuration, c' in $\mathcal{C}_j$ has a **valid history** if $j = 0$, or there is a c in $\mathcal{C}_{j-1}$ such that $c \sim c'$ and c has a valid history.*

Figure 3 shows an example of compatible configurations.

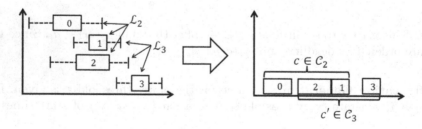

Fig. 3. Configurations in $\mathcal{C}_2$ and $\mathcal{C}_3$ are compatible if all shared jobs $(1, 2)$ have the same start times. Note that since job 0 has multiple valid start times that do not interfere with the other jobs, there are multiple configurations in $\mathcal{C}_2$ that are compatible with $c' \in \mathcal{C}_3$.

Feasibility Testing. Lemma 2 implies that each overlap list $\mathcal{L}_j$ is consecutive run of jobs ending at job j. It follows that a configuration history $c_0 \sim c_1 \sim \cdots \sim c_j$ defines a specific start time for each job 0 to j. For every configuration $c \in \mathcal{C}_j$, ordered by j, $\mathcal{C}_{j-1}$ is searched for valid histories of c. As they are found, pointers are assigned to point from c to its valid history configurations $c' \in \mathcal{C}_{j-1}$. If a valid history for $c \in \mathcal{C}_j$ cannot be found then it is removed from $\mathcal{C}_j$ and cannot serve as part of a valid history for configurations in $\mathcal{C}_{j+1}$. Furthermore, if maxHeight$(c) > \tau$ then c is also removed from $\mathcal{C}_j$. If $\mathcal{C}_j$ is empty at the end, then the peak demand threshold τ is not feasible. If a valid history can be found for at least one configuration of the final job n, then a schedule can be generated with a peak demand at most τ.

Schedule Building. A schedule can be built once the minimum τ is known by beginning with the last job and constructing the valid history in reverse. That configuration includes a pointer to a valid history configuration in the prior configuration list and so on. Proceed through the jobs in reverse, selecting a configuration for job $j - 1$ that belongs to a valid history of the configuration that was chosen for job j.

Theorem 1. *The schedule found by PDM-Exact achieves the optimal peak power demand.*

Proof. Suppose $\mathcal{S}^*$ is an optimal PDM schedule for the given input jobs with peak demand τ^*. Let c_j^* be the configuration for job j that agrees with $\mathcal{S}^*$. Clearly, c_0^* will be an available configuration in $\mathcal{C}_0$ if the current test threshold $\tau \geq \tau^*$. By induction, c_j^* will also be an available configuration in $\mathcal{C}_j$ Thus, τ will reach τ^* and the schedule with the optimal peak demand will be found. $\square$

Fixed Parameter Tractability

Definition 3. *An algorithm is said to be fixed parameter tractable (FPT) if there exists some parameters, $p_1, p_2, ...,$ of the input or output such that the running time of the algorithm is a polynomial function of the input size times some function of the parameters: $f(p_1, p_2, ...) \cdot poly(n)$.*

The complexity of PDM-Exact is driven by the size of the configuration lists $\mathcal{C}_j$, which is dependent on the size of $\mathcal{L}_j$ and the number of possible start times for each job. Thus, a large number of jobs having limited overlap list sizes can still be scheduled efficiently. On the other hand, the worst case is if there is a job that spans the entire timescale, then every configuration of all jobs needs to be considered. If n represents the number of jobs, $m = \max_j(|\mathcal{L}_j|)$, and q is the maximum number of job start times, i.e. $q = \max_j(d_j - a_j - l_j + 1) \leq T$, then the running time of PDM-Exact is $O(\lg \tau^* \cdot q^m \cdot n)$, where τ^* is the optimal peak demand value.

4.2   PDM-Heuristic: A Simple Heuristic Algorithm

As noted above, the general PDM problem is NP-hard to approximate so we consider a heuristic approach that is shown to be effective in practice. Our idea is to schedule jobs that have tight execution windows first and save jobs with more space in their execution windows to be scheduled later. To schedule a job, we simply find the starting time that minimizes the peak demand of the schedule created so far. The complete algorithm is given below as Algorithm 3 (note that $w_j = \frac{l_j}{d_j - a_j} \in (0, 1]$ measures the tightness of job j; values closer to 1 indicate tighter jobs).

Algorithm 3. PDM-Heuristic

Step 1   Sort jobs by decreasing $w_j = \frac{l_j}{d_j - a_j}$ values.

Step 2   Schedule start times as follows:
  forall jobs j (in sorted order)
    $s = a_j$
    for $t = a_j + 1, \ldots, d_j$
      if maxHeight$(\mathcal{S} \cup s_j = t) <$ maxHeight$(\mathcal{S} \cup s_j = s)$
        $s = t$
      endif
    endfor
    $s_j = s$
  endforall

---

5   Experimental Results

Simulations were conducted using both the *PDM-Exact* and *PDM-Heuristic* algorithms, *OnDemand* schedules jobs to start at their arrival time. Jobs were created using appliance specific data from six residences [10]. We identified appliances (e.g. washing machine) likely to have flexible timelines and determined their height, length, and arrival time distributions within a 24 h period. Deadlines were set to be uniformly distributed between the minimum deadline and four times the average length.

We looked at two different scenarios to test the effectiveness of the algorithms. First, we created a simpler scenario in order to test the performance of the PDM-Exact algorithm. Instead of generating jobs as described above, we randomly generated jobs with arrival time 0 to simulate a single peak. Figure 4 shows the average results of running PDM-Exact, PDM-Heuristic and OnDemand on five iterations of five jobs each from this simplified data generation. The average peak power demand was 1.36 for PDM-Exact, 1.44 for PDM-Heuristic, and 2.89 for OnDemand scheduling. As can be seen in the figure, the PDM-Heuristic algorithm achieves a near-optimal schedule.

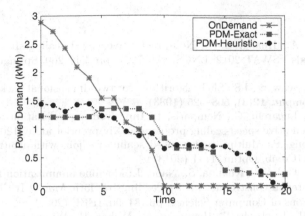

Fig. 4. PDM-Exact vs. PDM-Heuristic

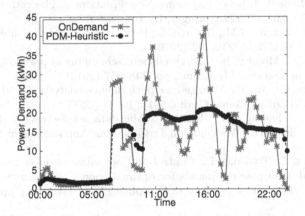

Fig. 5. Domestic power scheduling (24 h scenario)

Next, Fig. 5 shows the average of 1000 iterations of peak power demand versus time of day for both PDM-Heuristic and the OnDemand algorithm (PDM-Exact was unable to schedule a realistic number of these jobs) over a 24 h period. The average peak power demand from OnDemand scheduling is 42.0 kWh while the average peak for the PDM-Heuristic is 20.8 kWh.

6 Conclusions

Emerging smart grid systems will allow for more control over the scheduling of power jobs and thus the PDM problem is timely. In this work we presented a new FPT algorithm for PDM and a simple yet effective heuristic. While approximation algorithms are known for special cases of PDM ([5,6,16,19]), there are none to date for the general case, so this is an interesting problem for future work.

References

1. Antoniadis, A., Huang, C.-C.: Non-preemptive speed scaling. In: Fomin, F.V., Kaski, P. (eds.) SWAT 2012. LNCS, vol. 7357, pp. 249–260. Springer, Heidelberg (2012)
2. Baker, B.S., Schwarz, J.S.: Shelf algorithms for two-dimensional packing problems. SIAM J. Comput. **12**(3), 508–525 (1983)
3. Bampis, E., Lucarelli, G., Nemparis, I.: Improved approximation algorithms for the non-preemptive speed-scaling problem. arXiv preprint arXiv:1209.6481 (2012)
4. Bell, P., Wong, P.: Multiprocessor speed scaling for jobs with arbitrary sizes and deadlines. J. Comb. Optim. 1–11 (2013)
5. Chuzhoy, J., Guha, S., Khanna, S., Naor, J.: Machine minimization for scheduling jobs with interval constraints. In: Proceedings of 45th Annual IEEE Symposium on Foundations of Computer Science, pp. 81–90. IEEE (2004)
6. Cieliebak, M., Erlebach, T., Hennecke, F., Weber, B., Widmayer, P.: Scheduling with release times and deadlines on a minimum number of machines. In: Levy, J.J., Mayr, E., Mitchell, J. (eds.) Exploring New Frontiers of Theoretical Informatics. IFIP, vol. 155, pp. 209–222. Springer, Boston (2004)
7. Conejo, A.J., Morales, J.M., Baringo, L.: Real-time demand response model. IEEE Trans. Smart Grid **1**(3), 236–242 (2010)
8. Fox, K., Im, S., Moseley, B.: Energy efficient scheduling of parallelizable jobs. In: Symposium on Discrete Algorithms, pp. 948–957 (2013)
9. Gu, X., Chen, G., Xu, Y.: Average-case performance analysis of a 2d strip packing algorithm - NFDH. J. Comb. Optim. **9**(1), 19–34 (2005)
10. Kolter, J.Z., Johnson, M.J.: Redd: a public data set for energy disaggregation research. In: SustKDD Workshop on Data Mining Applications in Sustainability (2011)
11. Koutsopoulos, I., Tassiulas, L.: Control and optimization meet the smart power grid: scheduling of power demands for optimal energy management. In: International Conference on Energy-Efficient Computing and Networking, pp. 41–50. ACM (2011)
12. Koutsopoulos, I., Tassiulas, L.: Optimal control policies for power demand scheduling in the smart grid. IEEE J. Sel. Areas Commun. **30**(6), 1049–1060 (2012)
13. Mohsenian-Rad, A.H., Leon-Garcia, A.: Optimal residential load control with price prediction in real-time electricity pricing environments. IEEE Trans. Smart Grid **1**(2), 120–133 (2010)
14. Mu, Z., Li, M.: Dvs scheduling in a line or a star network of processors. J. Comb. Optim. 1–20 (2013)
15. Ortmann, F.G., Ntene, N., van Vuuren, J.H.: New and improved level heuristics for the rectangular strip packing and variable-sized bin packing problems. Eur. J. Oper. Res. **203**(2), 306–315 (2010)
16. Tang, S., Huang, Q., Li, X.Y., Wu, D.: Smoothing the energy consumption: peak demand reduction in smart grid. In: INFOCOM, 2013 Proceedings IEEE, pp. 1133–1141, April 2013
17. Vytelingum, P., Ramchurn, S.D., Voice, T.D., Rogers, A., Jennings, N.R.: Trading agents for the smart electricity grid. In: International Conference on Autonomous Agents and Multiagent Systems, pp. 897–904 (2010)
18. Yao, F., Demers, A., Shenker, S.: A scheduling model for reduced CPU energy. In: Symposium on Foundations of Computer Science, pp. 374–382 (1995)
19. Yaw, S., Mumey, B., McDonald, E., Lemke, J.: Peak demand scheduling in the smart grid. In: IEEE SmartGridComm (2014, to appear)

An Asymptotic Competitive Scheme for Online Bin Packing

Lin Chen, Deshi Ye, and Guochuan Zhang$^{(\boxtimes)}$

College of Computer Science, Zhejiang University, Hangzhou 310027, China
{yedeshi,zgc}@zju.edu.cn

Abstract. We study the online bin packing problem, in which a list of items with integral size between 1 to B arrives one at a time. Each item must be assigned in a bin of capacity B upon its arrival without any information on the next items, and the goal is to minimize the number of used bins. We present an asymptotic competitive scheme, i.e., for any $\epsilon > 0$, the asymptotic competitive ratio is at most $\rho^* + \epsilon$, where ρ^* is the smallest possible asymptotic competitive ratio among all online algorithms.

Keywords: Online algorithms · Competitive scheme · Bin packing

1 Introduction

Bin packing is one of the well-known combinatorial optimization problems in operations research and theoretical computer science. An instance of bin packing consists of a set of items with integral size up to B (a given integer), and the goal is to pack these items into a minimum number of bins of size B. The off-line bin packing problem, where all items are available before packing starts, is NP-hard [7]. In terms of asymptotic performance ratio, a standard measure for bin packing algorithms, de la Vega and Lueker [6] presented an APTAS and Karmakar and Karp [11] improved this result by giving an AFPTAS. Apart from this classical model, one can find many interesting extensions (e.g., [2,17]).

In the scenario of online bin packing, items arrive one by one in a list. Upon arrival of an item it must be irrevocably packed into a bin without knowing the subsequent items. Given an instance I, let $A(I)$ and $OPT(I)$ be the number of bins used by an online algorithm A and the optimal number of bins needed, respectively. The *asymptotic competitive ratio* ρ_A^∞ of algorithm A is the infimum ρ such that the following inequality holds for any instance I, where κ is a constant,

$$A(I) \leq \rho OPT(I) + \kappa.$$

One of the first online bin packing algorithms, *First Fit*, was studied by Ullman and Johnson et al. [9,15]. They proved that the asymptotic competitive

---

Research was supported in part by NSFC(11071215,11271325).

Z. Zhang et al. (Eds.): COCOA 2014, LNCS 8881, pp. 13–24, 2014.
DOI: 10.1007/978-3-319-12691-3_2

ratio of First Fit is 1.7. Then a sequence of improvements was proposed [12, 13, 16] and the currently best known upper bound is 1.58889 [14], while the best known lower bound is 1.54037 [1]. Very recently, the *competitive ratio approximate scheme* was introduced to online parallel machine scheduling problems by Günther et al. [8]. For any given $\epsilon > 0$, there exists an online algorithm $\{A_\epsilon\}$ that achieves a competitive ratio at most of $(1 + \epsilon)$ times the optimal competitive ratio. Motivated by their work, we revisit the online bin packing problem. Following the simplified notion as [4], we use the *competitive scheme* instead of the *competitive ratio approximation scheme* in this paper. Our task is to design an asymptotic competitive scheme for the online bin packing problem. For simplicity, throughout the paper, we use competitive ratio instead of asymptotic competitive ratio.

Our Contribution. Let ρ^* be the competitive ratio of a best possible online algorithm. We show the following result.

Theorem 1. *The online bin packing problem admits an asymptotic competitive scheme $\{A_{\epsilon,\kappa}|\epsilon > 0\}$ satisfying that $A_{\epsilon,\kappa}(I) \leq (\rho^* + O(\epsilon))OPT(I) + \kappa$, where κ and ϵ are constants, and the running time of $A_{\epsilon,\kappa}$ is polynomial if B is fixed.*

General Idea. To prove Theorem 1, we start with the *bounded instances* where the adversary only releases a constant number of items. Indeed, if the adversary only releases C items, then the number of all the possible sequences of items is bounded by B^C, which is also a constant. It is not difficult to imagine that a best possible online algorithm for the bounded instances could be determined. Suppose this algorithm has a competitive ratio of ρ_0, then $\rho^* \geq \rho_0$ since even if we restrict the adversary to release at most C items, no online algorithm has a competitive ratio better than ρ_0. The main technical part is to show that, once C is large enough, we can generalize the algorithm of competitive ratio ρ_0 for bounded instances to an algorithm of competitive ratio $\rho_0 + O(\epsilon)$ for the general instances. To this end, we introduce the notion of *modified instances* as an intermediate. In a modified instance, the adversary can release an arbitrary number of items, however, the item list must conform to a certain pattern. We will show that, an online algorithm for bounded instances could be generalized to an online algorithm for modified instances with a loss of $O(\epsilon)$ in its competitive ratio. Meanwhile, an online algorithm for modified instances could also be generalized to an online algorithm for general instances with a loss of $O(\epsilon)$ in its competitive ratio.

The paper is organized as follows. In Sect. 2, we provide some definitions and notations. In Sect. 3, we show how to derive a best possible algorithm for the bounded instances. It remains to show how the algorithm for bounded instances could be generalized to an algorithm for modified instances, which is further generalized to an algorithm for general instances. The latter part is easier and we address it in Sect. 4, while the former part is presented in Sect. 5.

2   Preliminaries

Given the bin size B, an input of the online bin packing problem is a list
(sequence) of items $(J_1, J_2, \ldots, J_n)$ for $n > 0$, where the i-th item is denoted
by J_i, and we abuse the notation J_i to denote the size of the i-th item, which
is an integer belonging to $\{1, 2, \cdots, B\}$. Given n items as an input, any packing
of these n items into (at most n) bins could be represented by a $(2B)$-tuple
$(r(n), x(n))$, where

- $r(n) = (r_1(n), r_2(n), \ldots, r_B(n))$, where $r_i(n)$ is the number of items of size
  exactly i;
- $x(n) = (x_1(n), x_2(n), \ldots, x_B(n))$, where $x_i(n)$ is the number of bins whose
  free space is exactly $B - i$ for $1 \leq i \leq B$.

Obviously, $\sum_{i=1}^{B} r_i(n) = n$, and the number of bins used is $\sum_{i=1}^{B} x_i(n)$. We call
$(r(n), x(n))$ as a *state* and write $\eta^n = (r(n), x(n))$. If it is clear from context,
we also write $(r(n), x(n))$ as (r, x) for simplicity. Let ST_n be the set of all the
states with n items (i.e., all possible $(r(n), x(n))$'s), and denote its cardinality
as $|ST_n|$. We can thus list these states as $\eta_1^n, \cdots, \eta_{|ST_n|}^n$. Specifically, we will use
η^n to denote an arbitrary state in ST_n. Note that ST_0 consists of a unique state
$\eta_1^0 = (0, 0, \cdots, 0)$.

Given any state $\eta^n = (r(n), x(n))$, we denote as $OPT(r(n))$ the optimal
number of bins used when the items of $r(n)$ are packed. As a consequence, we
define the *instant ratio* of the state η^n as

$$\tilde{\rho}(\eta^n) = \max\{1, (\sum_{i=1}^{B} x_i(n) - \kappa)/OPT(r(n))\}.$$

Specifically, define $\tilde{\rho}(\eta_1^0) = 1$. Here the constant κ in the above definition is the
κ in Theorem 1.

We can interpret an online algorithm for the bin packing problem in terms of
the states. Indeed, when an algorithm is applied to an item list $(J_1, J_2, \ldots, J_n)$,
it returns a list of states $\eta^0 \to \eta^1 \to \cdots \to \eta^n$, where η^i is the state in which the
first i items are packed. Specifically, if the competitive ratio of this algorithm
is ρ, then $\tilde{\rho}(\eta^i) \leq \rho$ for any i, and meanwhile there exists a certain item list
$(J_1^*, J_2^*, \ldots, J_n^*)$ such that $\tilde{\rho}(\eta^n) = \rho$. In this view, the competitive ratio of an
online algorithm is the instant ratio of the worst state it could ever return.

Recall that the Next-Fit algorithm [10] for bin packing has a competitive ratio
of 2 (both in terms of asymptotic competitive ratio and absolute competitive
ratio). Thus $\rho^* \leq 2$ and we focus on states with instant ratio no more than 2.
States with instant ratio larger than 2 are deleted beforehand. Let d be some
constant that will be specified later and $R = ST_d$ for simplicity. For any integer
$k > 0$, we define

$$kR = \{(kr(d), kx(d)) = (kr_1(d), \cdots, kr_B(d), kx_1(d), \cdots, kx_B(d)) | (r(d), x(d)) \in R\}.$$

Obviously, $kR \subset ST_{kd}$. A state $(\hat{r}(kd), \hat{x}(kd)) \in ST_{kd}$ is called a *neighbor*
of $(kr(d), kx(d)) \in kR$ if $|\hat{r}_i(kd) - kr_i(d)| < k$ and $|\hat{x}_i(kd) - kx_i(d)| < k$

for all i. According to this definition, a state in ST_{kd} might be the neighbor of multiple states of kR. To make the notion of 'neighborhood' unique, we define an *assignment* as a mapping that assigns every state in ST_{kd} to be a neighbor of a unique state in kR (which can be achieved by assigning every state in ST_{kd} to an arbitrary one of its neighbors). Given an assignment, all the states in ST_{kd} are divided into $|R|$ disjoint sets, with each containing one state of kR and all its neighbors. Finally we define the *perturbation*. A perturbation is a vector $\Delta = (\Delta(r), \Delta(x))$, where $\Delta(r) = (\Delta_1(r), \cdots, \Delta_B(r))$, $\Delta(x) = (\Delta_1(x), \cdots, \Delta_B(x))$ with each coordinate being an integer. We define $D = ||\Delta||_\infty = \max\{|\Delta_i(r)|, |\Delta_i(x)|\}$, and write $(r', x') = (r, x) + \Delta$ as the normal vector addition. It is not difficult to verify that if $OPT(r) > BD$, then

$$\tilde{\rho}(r', x') \leq \frac{\sum x_i + BD}{OPT(r) - BD}.$$

The above formula is useful in characterizing how a slight perturbation will change the instant ratio of a state.

3   Bounded Instances

We consider bounded instances of bin packing, where the bounded instance refers to the bin packing problem in which no more than C items could be released for some constant C. In this section we will determine the competitive ratio of the best possible online algorithm for the bounded instances via a dynamic programming algorithm. Indeed, a best algorithm for bounded instances could also be simply determined by brute force. However, as it needs to be further generalized, the dynamic programming algorithm will provide additional information on its structure.

We establish a layered graph G, in which there are $|ST_h|$ vertices at the h-th layer, each corresponding to some η_i^h. With a slight abuse of the notation we also use η_i^h to denote its corresponding vertex. For every η_i^h, we construct B vertices, namely $\alpha_{i,j}^h$ for $1 \leq j \leq B$ representing the release of item of size j by the adversary. For simplicity, all the $\alpha_{i,j}^h$ are denoted as vertices of the $(h + 1/2)$-th layer. There are only edges between vertices of the h-th layer and the $(h + 1/2)$-th layer, and between vertices of the $(h + 1/2)$-th layer and the $(h + 1)$-st layer. Indeed, there is an edge between η_i^h and $\alpha_{i,j}^h$ for any h, i and $1 \leq j \leq B$. There is an edge between $\alpha_{i,j}^h$ and η_k^{h+1}, if by packing the item of size j into a certain bin, the state η_i^h is changed to η_k^{h+1}.

Now we can easily associate an online algorithm with a path in the layered graph G. If the adversary releases n items of sizes $J_1, J_2, \cdots, J_n$, and meanwhile the algorithm returns a series of states $\eta_{i_0}^0$ (obviously $i_0 = 1$ since ST_0 contains only one element), $\eta_{i_1}^1$, $\cdots$, $\eta_{i_n}^n$, then associate it with a path in the graph as $\eta_{i_0}^0 \to \alpha_{i_0,j_1}^0 \to \eta_{i_1}^1 \to \cdots \to \alpha_{i_{n-1},j_n}^{n-1} \to \eta_{i_n}^n$.

Meanwhile, any path of length $2n$ that starts at η_1^0 and ends at η_i^n for some i represents the packing of n items by a certain online algorithm. We adopt the idea of [4] to reformulate the problem of finding the best online algorithm for bounded instances into the following problem on a game between the adversary and the packer:

- Initially the game starts at the vertex η_1^0.
- If currently the game arrives at the vertex η_i^h for $h < C$, then the adversary either decides to end the game, or moves the game to some α_{ij}^h that is incident to η_i^h. If the game arrives at the vertex η_i^h for $h = C$, then the game ends.
- If currently the game arrives at the vertex α_{ij}^h, then the packer chooses to move the game to some η_k^{h+1} that is incident to α_{ij}^h.

Again we take the above figure as an example. Starting at η_1^0, if the adversary releases an item of size 1, he moves the game to $\alpha_{1,1}^0$. Then the packer packs this item into one bin, meaning that he moves the game to η_1^1. Now the adversary could either choose to stop the game, or continue to release items. If he releases a new item of size 1, the game arrives at $\alpha_{1,1}^1$.

If the game ends at η_i^h for $h \leq C$, then the utility of the adversary is defined to be $\tilde{\rho}(\eta_i^h)$, while the utility of the packer is defined to be $-\tilde{\rho}(\eta_i^h)$. Starting from η_1^0, if the packer always packs items according to Next-Fit, then obviously the adversary could choose to release a certain list of items such that the game ends at some η_i^h with $\tilde{\rho}(\eta_i^h)$ about 2. If the packer is smart enough, he would resort to an optimum online algorithm (with the competitive ratio of ρ^*) so that no matter how the adversary releases items, he is always able to move the game to some η_i^h with $\tilde{\rho}(\eta_i^h) \leq \rho^*$. Thus, $-\rho^*$ is the largest possible utility the packer could achieve starting at η_1^0, and meanwhile ρ^* is the largest possible utility the adversary could ever achieve. Analogously, we define $\rho(\eta_i^h)$ to be the largest utility the adversary could get by releasing at most $C - h$ additional items, which implies that starting at η_i^h, the optimum "online" algorithm would have a competitive ratio of $\rho(\eta_i^h)$. Now we provide a dynamic programming algorithm to compute the value of $\rho(\eta_i^h)$. Obviously we have $\rho(\eta_i^h) \geq \tilde{\rho}(\eta_i^h)$.

Note that the adversary is no longer able to release items if the current scenario is some $\eta_i^C \in ST_C$. Thus we have $\rho(\eta_i^C) = \tilde{\rho}(\eta_i^C)$.

Let $N(\alpha_{i,j}^h)$ be the set of vertices at the $(h + 1)$-st layer that are incident to the vertex $\alpha_{i,j}^h$, for any i, j and $h \leq C - 1$. Then:

$$\rho(\alpha_{i,j}^h) = \min_k \{\rho(\eta_k^{h+1}) | \eta_k^{h+1} \in N(\alpha_{i,j}^h)\}, \rho(\eta_i^h) = \max\{\tilde{\rho}(\eta_i^h), \max_j \{\rho(\alpha_{i,j}^h)\}\}.$$

We give an explanation for the first equation, and the second one is similar. Suppose currently the game is at η_i^h. The adversary knows that if he releases an item of size j, then all the possible states by packing this new item are given by $\{\eta_k^{h+1} \in N(\alpha_{i,j}^h)\}$. Suppose the adversary is clever enough, who knows that if the game arrives at η_k^{h+1}, the best possible online algorithm, starting at η_k^{h+1}, would have a competitive ratio of $\rho(\eta_k^{h+1})$. Thus, if he chooses to release an item of size j, the largest utility he could get is $\rho(\alpha_{i,j}^h) = \min_k \{\rho(\eta_k^{h+1}) | \eta_k^{h+1} \in N(\alpha_{i,j}^h)\}$ as the "worst case" for him is that the packer chooses to pack items in the way indicated by $\min_k \{\rho(\eta_k^{h+1}) | \eta_k^{h+1} \in N(\alpha_{i,j}^h)\}$.

A best possible online algorithm for bounded instances is described below.

Algorithm 1

1. For a given constant C, construct the graph G and calculate the $\tilde{\rho}(\eta_i^C)$.
2. For all i, let $\rho(\eta_i^C) = \tilde{\rho}(\eta_i^C)$.
3. For $h = C - 1$ down to 1, iteratively calculate, for all i, j,

$$\rho(\alpha_{i,j}^h) = \min_k\{\rho(\eta_k^{h+1}) | \eta_k^{h+1} \in N(\alpha_{i,j}^h)\}, \ \rho(\eta_i^h) = \max\{\tilde{\rho}(\eta_i^h) \max_j\{\rho(\alpha_{i,j}^h)\}\}.$$

4. For the released item of size j when the current state is η_i^h, let $k^* = \operatorname{argmin}_k$ $\{\rho(\eta_k^{h+1}) | \eta_k^{h+1} \in N(\alpha_{i,j}^h)\}$, then we assign this item to the bin such that the state η_i^h is changed to the state $\eta_{k^*}^{h+1}$.

Remark. For simplicity we will call $\rho(\eta_i^h)$ as the *ratio* of the state η_i^h. Furthermore for the ease of analysis we will round up each instant ratio to be its nearest value in $SV = \{1, 1 + \epsilon, 1 + 2\epsilon, \cdots, 2\}$, and as a consequence after computation the ratios of states also belong to SV.

4   From Modified Instances to General Instances

Let A be the best possible online algorithm for bounded instances. As we have mentioned, we need to generalize it to an algorithm for general instances, and the generalization has two steps. First, we generalize it to an algorithm for the modified instances (the definition of a modified instance will be given below). Then, we generalize the algorithm for modified instances to an algorithm for general instances. We deal with the easier part in this section, i.e., roughly speaking, we show that an algorithm of competitive ratio ρ^* for modified instances could be transformed into an algorithm of competitive ratio $\rho^* + O(\epsilon)$ for the general instances.

We give the definition of a modified instance. Let $l = (J_1, \cdots, J_h)$ be any list of h items ($|l| = h$). Given l, we use kl to denote the sequence by duplicating each item of l into k items, i.e., $kl = (J_1', \cdots, J_{kh}')$, where $J_{ki+j}' = J_{i+1}$ for $0 \le i \le h - 1$ and $1 \le j \le k$, or equivalently, $kl = (\underbrace{J_1, \ldots, J_1}_k, \ldots, \underbrace{J_j, \ldots, J_j}_k)$.

Given any integers $k, c > 0$, we say L is a modified instance or a modified list (with respect to (k, c)), if $L = (l_1, kl_2, k^2 l_3, \cdots, k^h l_{h+1})$, where $|l_i| = c$ for $1 \le i \le h$, and $|l_{h+1}| \le c$. The bin packing problem for modified instances is the bin packing problem satisfying the following conditions:

- The items released by the adversary form a modified list.
- The adversary could only stop releasing items at certain times, i.e., he could only do the following:
  • The adversary releases no more than c items, and stops.
  • The adversary releases no more than $c + kc$ items, and stops after he releases the $(c + kj)$-th item ($j \le c$).
  • $\cdots$

- The adversary releases no more than $c + kc + \cdots + k^h c$ items, and stops after he releases the $(c + kc + \cdots + k^h j)$-th item ($j \leq c$).

Theorem 2. *Given any $\epsilon > 0$, if there is an online algorithm of competitive ratio ρ^* on modified instances with respect to any $k > 0$ and $c \geq c(k, B, \epsilon)$ (where $c(k, B, \epsilon)$ is a constant only depending on k, B and ϵ), then there is an algorithm of competitive ratio $\rho^* + \epsilon$ for the general problem.*

Proof. We prove the theorem by modifying the algorithm A of competitive ratio ρ^* for modified instances. Throughout the proof we keep track of two lists, one is the list σ of items released in the general bin packing problem, and the other is the item list σ' of a modified instance which is constructed from σ so that algorithm A could return a feasible packing by taking σ' as an input.

In the following, we construct an algorithm for the general problem with the input σ. If the h-th item in σ arrives, and $h \leq c$, we pack this item according to algorithm A. We have $\sigma = \sigma'$. When the $(c+1)$-st item in σ, say, J_{c+1} releases, we take it as k identical items, one true item and $k-1$ fake items, and pack them according to A. Now we add k copies of J_{c+1} to σ'. Consider the $(c+2)$-nd item. If it is different from J_{c+1}, then again we take it as k identical J_{c+2} and pack them according to A. Meanwhile we add J_{c+2} to σ and k copies of J_{c+2} to σ'. Otherwise, it is the same with J_{c+1}, then we replace one fake J_{c+1} with this item in the current packing. In this case, σ' remains the same. We proceed with the above procedure. Whenever a new item J_n releases, we add it to σ and check if there exists a fake item of the same size in the current solution. If yes, we replace this fake item with this new item and σ' remains the same. Otherwise, we add J_n to σ', and another $k^h - 1$ identical items are released together with it for some h depending on the length of σ'. Now we resort to A to decide a packing for these k^h items, and for J_n, there are $k^h - 1$ fake items now.

Next we check the competitive ratio of the above algorithm. Let μ be the number of bins used, r_i be the number of items of size i according to σ, r_i' be the number of items of size i according to σ'. We use $|\sigma|$ to denote the number of items in the list σ and suppose $c + kc + \cdots + k^{h-1}c < |\sigma'| \leq c + kc + \cdots + k^h c$ for some h. Then it follows that $r_i' - r_i \leq k^h$ according to the construction of σ'.

Since the competitive ratio of A is ρ^*, we have $\mu \leq \rho^* OPT(r_1', \cdots, r_B') + \kappa$. Meanwhile, $OPT(r_1, \cdots, r_B) \geq OPT(r_1', \cdots, r_B') - Bk^h$ as $r_i' - r_i \leq k^h$. Notice that $OPT(r_1', \cdots, r_B') \geq |\sigma'|/B > k^{h-1}c/B$. The competitive ratio for the general bin packing is at most (for simplicity we let $OPT = OPT(r_1', \cdots, r_B')$).

$$\frac{\mu - \kappa}{OPT(r_1, \cdots, r_B)} \leq \frac{\rho^* OPT}{OPT - Bk^h} = \frac{\rho^*}{1 - Bk^h/OPT} \leq \frac{\rho^*}{1 - B^2k/c}.$$

The theorem follows by taking $c > B^2 k/\epsilon^2 = c(k, B, \epsilon)$.

Remark. For ease of analysis, the following sequence is also taken to be a modified instance (with respect to (k, c)): $L = (l_1, kl_2, k^2 l_3, \cdots, k^h l_{h+1})$ where $|l_i| = c$ for $2 \leq i \leq h$, $|l_{h+1}| \leq c$ and $|l_1| \geq c$, i.e., the first part of the list could contain more than c items.

5 From Bounded Instances to Modified Instances

Let ρ_0 be the competitive ratio of the best algorithm for bounded instances (in which the adversary releases at most C items). We show in this section that when C is large enough, we can transform the algorithm into a $(\rho_0 + O(\epsilon))$-competitive algorithm for the bounded instances with respect to (k_0, c_0) for some k_0 and c_0. Combining this result with Theorem 2, Theorem 1 follows directly. The values of C, k_0 and c_0 will be determined at the end of this section.

5.1 Overview of the Technique

We revisit the graph G that contains all the possible states. G is an infinite graph and we can only afford to compute the ratios of states in ST_h for $h \leq C$. Note that once the adversary releases an i-th item with $i \leq C$, the optimal algorithm for bounded instances can refer to the ratio of the current state to decide how to pack this item (the reader may refer to Algorithm 1 in Sect. 3). What if the current state is some η_i^n for $n > C$? A natural idea is to do *state mapping*, i.e., we map η_i^n to some proper $\eta_{i'}^h$ for $h < C$. Once a new item is released, we check $\eta_{i'}^h$ to see how this new item is packed, and then pack it in a similar way for η_i^n.

Modified instances are defined in order that we can carry out the above idea. Roughly speaking, we will specify some constants γ and k such that $k\gamma < C$, and take states of ST_h for $\gamma \leq h \leq k\gamma$ as samples. Consider modified instances with respect to $(k, (k-1)\gamma)$, i.e., the item lists of the form $(l_1, kl_2, \cdots, k^{h-1}l_h, k^h l_{h+1})$, where $|l_1| = k\gamma$, $|l_i| = (k-1)\gamma$ for $2 \leq i \leq h$ and $|l_{h+1}| \leq (k-1)\gamma$.

Suppose the adversary releases at most $k\gamma$ items. Obviously we can run the algorithm for bounded instances as $k\gamma < C$. Otherwise, suppose $k\gamma$ items are released and the current state is some $\eta_i^{k\gamma} = (r, x)$. Then according to our definition of neighborhood, with some slight perturbation Δ_1 we have $\eta_i^{k\gamma} = k\eta_{i'}^{\gamma} + \Delta_1$, where $\eta_{i'}^{\gamma}$ is some state of ST_{γ}. By the definition of modified instances, after $k\gamma$ items the adversary releases k identical items, denoted as kJ_j for simplicity, where J_j is arbitrary. According to the algorithm, for bounded instances, $\eta_{i'}^{\gamma}$ changes to $\eta_{h'}^{\gamma+1}$, when a single item J_j is released. We write $\eta_{h'}^{\gamma+1} = \eta_{i'}^{\gamma} + J_j$ for simplicity. Then we can pack kJ_j in a way such that $k\eta_{i'}^{\gamma} + kJ_j = k\eta_{h'}^{\gamma+1}$, e.g., if a new bin is opened for J_j in $\eta_{i'}^{\gamma}$, then k new bins are opened for kJ_j in $k\eta_{h'}^{\gamma+1}$. Now we have $\eta_i^{k\gamma} + kJ_j = k\eta_{h'}^{\gamma+1} + \Delta_1$, and if the adversary releases the next k identical items, we check $\eta_{h'}^{\gamma+1}$ to see how to pack them.

After the adversary release $k\gamma + k(k-1)\gamma = k^2\gamma$ items, we arrive at some state of $ST_{k^2\gamma}$ which could be expressed as $k\eta_\ell^{k\gamma} + \Delta_1$ for some $\eta_\ell^{k\gamma} \in ST_{k\gamma}$, and again with some slight perturbation Δ_2 we have $\eta_\ell^{k\gamma} = k\eta_{\ell'}^{\gamma} + \Delta_2$. Hence the current state is $k^2\eta_{\ell'}^{\gamma} + k\Delta_2 + \Delta_1$. According to the definition of modified instances, next, the adversary releases k^2 identical items, say, $k^2 J_j$. Now we can again check how a single item J_j is packed for $\eta_{\ell'}^{\gamma}$ and carry on the above arguments. To make the above arguments work, we need to show the following conditions: Given a state η^h, a slight perturbation (changing η^h to $\eta^h + \Delta$) does not change the instant ratio much; and multiplication by an integer k (changing

η^h to $k\eta^h$) does not change the instant ratio much. It results in the following two lemmas (the complete proofs will be given in a full version of the paper).

Lemma 1. *For any integers $k, d > 0$, and for any $(r(d), x(d))$ such that $\sum r_i(d) = d$, we have $kOPT(r(d)) - kB^B \leq OPT(kr(d)) \leq kOPT(r(d))$.*

Note that $OPT(r(d)) \geq d/B$. Take $d = B^{B+2}/\epsilon$ so that $B^B \leq \epsilon OPT(r(d))/B$. Recall that $R = ST_d$. Let $C = 2^\mu d$ for some constant μ. Then we can calculate the ratio of each state of ST_h for $h \leq C$, and an optimal algorithm for bounded instances could be determined. Let ρ_0 be its competitive ratio. Let $q \geq d$. Two states of the q-th layer, say, $\eta_i^q = (r(q), x(q))$ and $\eta_j^q = (r(q)', x(q)')$, are called *near*, if $|r_i(q) - r_i(q)'| \leq q/d$ and $|x_i(q) - x_i(q)'| \leq q/d$. We have

Lemma 2. *For any two near states $\eta_i^q = (r(q), x(q))$ and $\eta_j^q = (r(q)', x(q)')$, $|\rho(r(q), x(q)) - \rho(r(q)', x(q)')| \leq O(\epsilon)$.*

The above lemma implies that the ratio of a state in kR differs at most $O(\epsilon)$ to the ratio of its neighbors for any integer $k > 0$.

5.2   Constructing an Algorithm for Modified Instances

Recall that $R = ST_d$. For any integer $k > 0$, the states in kR are called *principle states* of ST_{kd}. We consider $ST_{2^h d}$ for $h = 1, 2, \cdots, \mu$. There is a vertex for each state of $2^h R$, and as we mention before, all the states of $ST_{2^h d}$ could be partitioned into subgroups where each group consists of a state in $2^h R$ and its neighbors. Since a state not in $2^h R$ might be a neighbor of multiple principle states, as discussed in Sect. 2, we give an assignment so that it becomes a neighbor of a unique principle state.

An assignment is called *compatible*, if according to this assignment, $(r', x') \in ST_{2^h d}$ is a neighbor of $(r, x) \in 2^h R$ implies that $(2^k r', 2^k x') \in ST_{2^{h+k} d}$ is a neighbor of $(2^k r, 2^k x) \in 2^{h+k} R$ for any $k \geq 1$. A compatible assignment could be constructed easily. In the following discussion we take one arbitrary compatible assignment. We use $T(r, x)$ to denote the set of neighbors of any $(r, x) \in kR$ (including (r, x)). Define

$$\rho(T(r, x)) = \min\{\rho(r', x')|(r', x') \in T(r, x)\}.$$

Since for any h the set $ST_{2^h d}$ is always partitioned into $|R|$ subgroups with each group being the set of neighbors of some principle state, we sort the states of $|R|$ in an arbitrary sequence as $\eta_1^d, \eta_2^d, \cdots, \eta_{|R|}^d$, where $\eta_i^d = (r(d), x(d))$ for some $r(d)$ and $x(d)$. We denote as $k\eta_i^d = (kr(d), kx(d))$.

Determining the Parameters. Consider $(\rho(T(2^h \eta_1^d)), \rho(T(2^h \eta_2^d)), \cdots, \rho(T(2^h \eta_{|R|}^d)))$ for $h = h_0, h_0 + 1, \cdots, \mu$, where h_0 is some constant. Each coordinate takes some value of $1 + k\epsilon$ for $0 \leq k \leq 1/\epsilon$ and thus has at most $1/\epsilon + 1$ different possible values. Each vector contains $|R|$ coordinates. There are at most $(1 + 1/\epsilon)^{|R|}$ different vectors. Thus, let $\mu - h_0 = (1 + 1/\epsilon)^{|R|}$. Among the $\mu - h_0 + 1$ vectors, we know that there exist two vectors which are identical. Let them be

$(\rho(T(\xi\eta_1^d)), \cdots, \rho(T(\xi\eta_{|R|}^d)))$ and $(\rho(T(\lambda\xi\eta_1^d)), \cdots, \rho(T(\lambda\xi\eta_{|R|}^d)))$ for some integers λ and ξ. Then $\lambda \leq 2^{\mu-h_0} \leq 2^{(1+\epsilon)|R|}$, which is a constant that depends on $|R|$ and $1/\epsilon$, i.e., B and $1/\epsilon$.

According to the above arguments, we can first apply Theorem 2 with $k = 2^{(1+\epsilon)|R|}$ and determine the parameter $c(k, B, \epsilon)$ that only depends on k, B and ϵ in the theorem. Let it be c_0. Then we take h_0 such that $2^{h_0 d} \geq c_0$ and let $\mu = h_0 + (1 + 1/\epsilon)|R|$. Now we take $C = 2^{\mu}d$ (recall that $d = B^{B+2}/\epsilon$) and compute the ratios of each state and find the two identical vectors from $(T(2^h\eta_1^d), \cdots, T(2^h\eta_{|R|}^d))$ for $h = h_0, h_0+1, \cdots, \mu$. Again we denote the two identical vectors we find out as $(\rho(T(\xi\eta_1^d)), \cdots, \rho(T(\xi\eta_{|R|}^d)))$ and $(\rho(T(\lambda\xi\eta_1^d)), \cdots, \rho(T(\lambda\xi\eta_{|R|}^d)))$.

Let ρ_0 be competitive ratio of the best possible algorithm for bounded instances in which the adversary releases at most C items.

Theorem 3. *There exists an online algorithm whose competitive ratio is $\rho_0 + O(\epsilon)$ for the modified instance with respect to (k, c), where $k = \lambda$ and $c = (\lambda - 1)\xi d$.*

Proof. Before starting the proof, recall that in order to apply Theorem 2, we need to show that $c = (\lambda - 1)\xi d \geq c(k, B, \epsilon) = c(\lambda, B, \epsilon)$. Since $\lambda \leq 2^{(1+\epsilon)|R|}$, we have $c(\lambda, B, \epsilon) \leq c(2^{(1+\epsilon)|R|}, B, \epsilon) = c_0$. Meanwhile, $2^{h_0 d} \geq c_0$, thus $(\lambda - 1)\xi d \geq \xi d \geq 2^{h_0}d \geq c_0 \geq c(\lambda, B, \epsilon)$.

Now we come to the proof of the theorem. Let A be an optimal algorithm for the bounded instance. Now the list of items released by the adversary is of the form $l = (l_1, \lambda l_2, \lambda^2 l_3, \cdots, \lambda^h l_{h+1})$ where $|l_1| = \lambda\xi d$, $|l_i| = (\lambda - 1)\xi d$ for $2 \leq i \leq h$ and $|l_{h+1}| \leq (\lambda - 1)\xi d$. Obviously we can always apply A for the first $\lambda\xi d$ items in the list of items released by the adversary. It remains to show how to pack the list λl_2.

For any $T(\xi\eta_i^d)$, there exists some $(r, x) \in T(\xi\eta_i^d)$ such that $\rho(r, x) = \rho(T(\xi\eta_i^d))$. Denote it as $T_{min}(\xi\eta_i^d)$. Suppose after packing the c items with A, the current state is some state, say, $z_1 \in T(\lambda\xi\eta_h^d)$ for some h. Then

$$\rho(T(\lambda\xi\eta_h^d)) \leq \rho(z_1) \leq \rho_0.$$

Based on the selection of ξ and λ, we have

$$\rho(T_{min}(\xi\eta_i^d)) = \rho(T(\xi\eta_h^d)) = \rho(T(\lambda\xi\eta_h^d)) \leq \rho_0.$$

According to the compatible assignment, $\lambda T_{min}(\xi\eta_i^d) \in T(\lambda\xi\eta_i^d)$, and is near z_1. Let $z_1^* = \lambda T_{min}(\xi\eta_i^d)$, and we let $z_1 = z_1^* + \Delta_1$, where $||\Delta_1||_\infty \leq \lambda\xi$, i.e., the absolute value of each coordinate of Δ_1 is bounded by $\lambda\xi$. Suppose the first λ items of the list are λJ_j. According to A, $z_1^*/\lambda \in T_{min}(\xi\eta_i^d)$ changes to y by adding a single item J_j. Now if we start at the state z_1^* to pack λJ_j, we may view z_1^* as λ copies of z_1^*/λ and we can pack the λ identical items in the same way, i.e., pack items in the way that z_1^* changes to λy. Thus, starting at z_1 to pack these items, we may adopt the same idea as in the proof of Lemma 2. Again, add some dummy bins to alter the state into z_1^* and pack items. It follows that z_1 changes to $\lambda y + \Delta_1$.

We pack items iteratively as the above procedure. Let $z_2 = z_1^*/\lambda + l_2$ according to Algorithm A. Then it can be easily seen that $z_2 \in ST_{\lambda\xi d}$. Meanwhile, the above way of packing cause the current state to be $\lambda z_2 + \Delta_1$. Recall that the algorithm A ensures that $\rho(z_2) \le \rho(z_1^*/\lambda) \le \rho_0$. Suppose $z_2 \in T(\lambda\xi\eta_{h'}^d)$ for some h'. Again we get

$$\rho(T(\xi\eta_{h'}^d)) = \rho(T(\lambda\xi\eta_{h'}^d)) \le \rho_0.$$

Thus there exists some $z_2^* \in T(\lambda\xi\eta_{h'}^d)$ such that $\rho(z_2^*/\lambda) \le \rho_0$. Again z_2 is near z_2^* and we have $z_2 = z_2^* + \Delta_2$ for some $||\Delta_2||_\infty \le \lambda\xi$. Suppose $z_2^*/\lambda + l_3 = z_3$ according to Algorithm A. Starting at $\lambda z_2 + \Delta_1$, the next part of the list is $\lambda^2 l_3$. Thus $\lambda z_2 + \Delta_1 + \lambda^2 l_3 = \lambda^2 z_3 + \lambda\Delta_2 + \Delta_1$.

Iteratively applying the above computation, the final state arrived is $\lambda^h z_{h+1} + \lambda^{h-1}\Delta_h + \cdots + \Delta_1$, where $||\Delta_i||_\infty \le \lambda\xi$, $\rho(z_{h+1}) \le \rho_0$. Due to the final part of the list $\lambda^h l_{h+1}$ that may not be complete, i.e., $|l_{h+1}|$ may not equal to c, z_{h+1} may not be a state of $ST_{\lambda\xi d}$. Nevertheless, z_{h+1} is some state between the $\lambda\xi d$-th layer and ξd-th layer and Algorithm A ensures that $\rho(z_{h+1}) \le \rho_0$.

Compute the instant ratio of $\lambda^h z_{h+1} + \lambda^{h-1}\Delta_h + \cdots + \Delta_1$. Let $z_{h+1} = (r, x)$. Recall Lemma 1. Then $OPT(\lambda^h r) \ge \lambda^h OPT(r) - \lambda^h B^B$.

$$\tilde{\rho}(\lambda^h z_{h+1} + \lambda^{h-1}\Delta_h + \cdots + \Delta_1) \le \frac{\lambda^h \sum x_i + B\lambda\xi \sum_{j=1}^h \lambda^{j-1}}{OPT(\lambda^h r) - B\lambda\xi \sum_{j=1}^h \lambda^{j-1}}$$

$$\le \frac{\lambda^h \rho_0 OPT(r) + B\lambda\xi \cdot 2\lambda^{h-1}}{\lambda^h OPT(r) - \lambda^h B^B - B\lambda\xi \cdot 2\lambda^{h-1}}$$

$$= \frac{\rho_0 OPT(r) + 2B\xi}{OPT(r) - B^B - 2B\xi}$$

Since $z_{h+1} = (r, x)$ is a state between the $\lambda\xi d$-th layer and ξd-th layer, we know that $OPT(r) \ge \xi d/B$. As $d = B^{B+2}/\epsilon$, it follows directly that $\tilde{\rho}(\lambda^h z_{h+1} + \lambda^{h-1}\Delta_h + \cdots + \Delta_1) \le \rho_0 + O(\epsilon)$.

6   Concluding Remarks

In this paper we have designed a competitive scheme for online bin packing such that the competitive ratio of our algorithm is at most of $1 + \epsilon$ times the best possible competitive ratio of any online algorithms, for any given $\epsilon > 0$. Our scheme provided a theoretical approach to narrow the known lower bound 1.54037 [1] and the upper bound 1.58889 [14]. The running time of our scheme is exponential in the bin size B and $1/\epsilon$. If the number of item sizes is a constant, our algorithm runs in polynomial time. But it remains an open problem whether we can design competitive schemes polynomially in both the number of items and $\log B$.

For bin packing, the absolute competitive ratio is another measure for online algorithms in the literature, though it is not as common as the asymptotic competitive ratio. To the knowledge of us, the best known lower bound is 5/3 [3] and the best known upper bound is 1.7 [5] in terms of absolute competitive ratio. Note that the results in this work are also valid even if the performance metric is the absolute competitive ratio. In addition, we claim that the techniques used in this paper can be extended to other variants of bin packing problems, such as the online variable-sized bin packing problem and the online bounded-space bin packing problem.

References

1. Balogh, J., Békési, J., Galambos, G.: New lower bounds for certain classes of bin packing algorithms. In: Jansen, K., Solis-Oba, R. (eds.) WAOA 2010. LNCS, vol. 6534, pp. 25–36. Springer, Heidelberg (2011)
2. Boyar, J., Epstein, L., Favrholdt, L.M., et al.: The maximum resource bin packing problem. Theoret. Comput. Sci. **362**, 127–139 (2006)
3. Brown, D., Baker, B.S., Katseff, H.P.: Lower bounds for on-line two-dimensional packing algorithms. Acta Informatica **18**, 207–225 (1982)
4. Chen, L., Ye, D., Zhang, G.: Approximating the optimal competitive ratio for an ancient online scheduling problem. CoRR, abs/1302.3946v1 (2013)
5. Dośa, G., Sgall, J.: First fit bin packing: A tight analysis. In: Proceedings of the 30th International Symposium on Theoretical Aspects of Computer Science (STACS), pp. 538–549 (2013)
6. Fernandez de La Vega, W., Lueker, G.S.: Bin packing can be solved within $1+ \varepsilon$ in linear time. Combinatorica **1**, 349–355 (1981)
7. Garey, M.R., Johnson, D.S.: Computers and Intractability: A Guide to the theory of of NP-Completeness. Freeman and Company, San Francisco (1979)
8. Günther, E., Maurer, O., Megow, N., Wiese, A.: A new approach to online scheduling: Approximating the optimal competitive ratio. In: Proceedings of the 24th Annual ACM-SIAM Symposium on Discrete Algorithms (SODA), pp. 118–128 (2013)
9. Johnson, D.S., Demers, A., Ullman, J.D., Garey, M.R., Graham, R.L.: Worst-case performance bounds for simple one-dimensional packing algorithms. SIAM J. Comput. **3**, 256–278 (1974)
10. Johnson, D.S.: Fast algorithms for bin packing. J. Comput. Syst. Sci. **8**, 272–314 (1974)
11. Karmarkar, N., Karp, R.: An efficient approximation scheme for the one-dimensional bin-packing problem. In: Proceedings of the 23rd Annual Symposium on Foundations of Computer Science (FOCS), pp. 312–320 (1982)
12. Lee, C.C., Lee, D.T.: A simple online bin packing algorithm. J. ACM **32**, 562–572 (1985)
13. Ramanan, P., Brown, D., Lee, C.C., Lee, D.T.: Online bin packing in linear time. J. Algorithms **10**, 305–326 (1989)
14. Seiden, S.: On the online bin packing problem. J. ACM **49**, 640–671 (2002)
15. Ullman, J.: The performance of a memory allocation algorithm. Technical report. Princeton University. Dept. of Electrical Engineering (1971)
16. Yao, A.C.C.: New algorithms for bin packing. J. ACM **27**, 207–227 (1980)
17. Y. Zhang, F.Y.L. Chin, H.-F. Ting et al. Online algorithms for 1-space bounded 2-dimensional bin packing and square packing. Theortical Comput. Sci. (2014) (to appear)

Randomized Online Algorithms for Set Cover Leasing Problems

Sebastian Abshoff, Christine Markarian$^{(\boxtimes)}$, and Friedhelm Meyer auf der Heide

Computer Science Department, Heinz Nixdorf Institute, University of Paderborn,
Fürstenallee 11, 33102 Paderborn, Germany
{christine.markarian,abshoff,fmadh}@upb.de

Abstract. In the leasing variant of Set Cover presented by Anthony et al.
[1], elements U arrive over time and must be covered by sets from a family
F of subsets of U. Each set can be leased for K different periods of time.
Let $|U| = n$ and $|F| = m$. Leasing a set S for a period k incurs a cost c_S^k
and allows S to cover its elements for the next l_k time steps. The objective
is to minimize the total cost of the sets leased, such that elements arriving
at any time t are covered by sets which contain them and are leased during time t. Anthony et al. [1] gave an optimal $O(\log n)$-approximation for
the problem in the offline setting, unless $\mathcal{P} = \mathcal{NP}$ [22]. In this paper, we
give randomized algorithms for variants of Set Cover Leasing in the online
setting, including a generalization of Online Set Cover with Repetitions
presented by Alon et al. [2], where elements appear multiple times and
must be covered by a different set at each arrival. Our results improve the
$\mathcal{O}(\log^2(mn))$ competitive factor of Online Set Cover with Repetitions [2]
to $\mathcal{O}(\log d \log(dn)) = \mathcal{O}(\log m \log(mn))$, where d is the maximum number
of sets an element belongs to.

Keywords: Set cover · Multicover · Online algorithms · Randomized
algorithms · Leasing

1 Introduction

The *Set Cover* problem is a classical $\mathcal{NP}$-hard problem defined as follows. Given
a set of elements U and a family F of subsets of U. Let $|U| = n$ and $|F| = m$.
Each set $S \in F$ has a non-negative cost $c(S)$. A set cover is a collection of
sets from F whose union is U. The goal is to find a set cover of minimum
cost. SETCOVER has an $\mathcal{O}(\log n)$ approximation ratio [7,8,10,11], which is the
best possible unless $\mathcal{P} = \mathcal{NP}$ [22]. SETCOVER has been studied as a more
general version known as the *Set MultiCover* problem, in which all elements
are required to be covered by (some fixed factor) p different sets and has an
$\mathcal{O}(\log \Delta) = \mathcal{O}(\log n)$ approximation ratio [12,13], where Δ is the maximum
cardinality of the sets.

This work was partially supported by the German Research Foundation (DFG)
within the Collaborative Research Centre "On-The-Fly Computing" (SFB 901) and
the International Graduate School "Dynamic Intelligent Systems".

© Springer International Publishing Switzerland 2014
Z. Zhang et al. (Eds.): COCOA 2014, LNCS 8881, pp. 25–34, 2014.
DOI: 10.1007/978-3-319-12691-3_3

SETCOVER has also been studied in the online setting, where elements arrive over time, as the *Online Set Cover* problem [15, 16]. An online algorithm does not know in advance which elements arrive in the future. Once a set is chosen, it cannot be returned. An online algorithm must cover every element which arrives by a set containing it, while minimizing the total cost of the sets it chooses. To measure the performance of the online algorithm, *competitiveness* is used. The competitive ratio of the algorithm is defined to be the maximum ratio, over all input sequences, of the cost of the online algorithm to that of the optimal offline algorithm. Alon et al. [15] gave an $\mathcal{O}(\log m \log n)$ competitive algorithm for ONLINESETCOVER. This is nearly tight due to a lower bound of $\Omega(\log m \log n/(\log \log m + \log \log n))$ which they give for a wide range of relations among m and n. In the unweighted case where costs are uniform, their $\mathcal{O}(\log n \log d)$ competitive factor was later improved by Buchbinder et al. [19] to $\mathcal{O}(\log(n/Opt) \log d)$, where Opt is the optimal offline solution to the problem and d is the maximum number of sets an element belongs to.

An interesting application to ONLINESETCOVER is in the domain of computer networks. Clients arrive over time and request a service from servers. A server can provide a service to a subset of clients. The job of the online algorithm is to serve the clients which arrive by activating a subset of servers. A cost is associated with activating each server and the goal is to minimize the total cost of activated servers. Beyond that, it could also be that a company does not own servers but leases them in order to serve its clients [21]. Servers can be leased for different durations with different costs. A server can provide a service as long as it is leased. Since the company does not know the clients in advance, it may lease some server for a long time and pay a lot, just to realize in the future that none of the server's clients show up again. This scenario can be formalized as the *Set Cover Leasing* problem. In SETCOVERLEASING, elements arrive over time and must be covered. Each set can be leased for K different periods of time. Leasing a set S for a period k incurs a cost c_S^k and allows S to cover its elements for the next l_k time steps. The objective is to minimize the total cost of the sets leased, such that at any time t an element arrives, it is covered by a set which contains it and is leased at time t. SETCOVERLEASING was introduced by Anthony and Gupta [1] who gave an optimal $\mathcal{O}(\log n)$-approximation for the problem in the offline setting. In this paper, we address the problem in the online setting and give online algorithms for SETCOVERLEASING and many of its variants. SETCOVERLEASING is a generalization of ONLINESETCOVER when $K = 1$ and sets are leased for only an infinite period of time. Moreover, algorithms for ONLINESETCOVER [15, 16] cannot be extended to SETCOVERLEASING because it is difficult to consider the time aspect in them.

1.1   Related Work

A different version of ONLINESETCOVER, where the goal is to pick k sets so as to maximize the elements that are covered, has been studied [16]. The authors [16] gave a randomized $\mathcal{O}(\log m \log(n/k))$ competitive algorithm for the problem, and showed that their bound is optimal for many values of n, m, and k.

More generally, Alon et al. [2] considered the *Online Set Cover Problem with Repetitions* problem within the larger context of admissions control in general networks. In ONLINESETCOVERWITHREPETITIONS, an element may arrive multiple times and must be covered by a different set at each arrival. Alon et al. [2] gave a randomized $\mathcal{O}(\log^2(mn))$ competitive algorithm and a deterministic bi-criteria algorithm for the problem. A very similar problem called *uniform Online Set Multicover*, the online variant of SETMULTICOVER, was studied by Bermana et al. [3], where all arriving elements are requested to be covered by (some fixed factor) p different sets. This is motivated in the client/server scenario, where each client needs to be served by several servers, for the sake of reliability and robustness. Bermana et al. [3] focused on tight analysis of approximability by improving results from [15] up to constant factors and were motivated by applications in systems biology. Their competitive ratios were given in terms of the maximum set size Δ, d, and p. They also gave lower bounds, similar to [15], for unweighted and weighted uniform ONLINESETMULTICOVER and proved $\Omega(\log m/p \log n/p/(\log\log m/p + \log\log n/p))$, and $\Omega(\log m \log n/(\log\log m + \log\log n))$, respectively.

A closely related problem, the *Online Facility Location*, was studied by Alon et al. [4,14]. In ONLINEFACILITYLOCATION, a set of m facilities, each with a setup cost is given. A set of n clients arrive, each with a connection cost to each facility, and must be connected to some facility. An online algorithm must connect arriving clients to facilities, while minimizing the total setup costs and connection costs. ONLINEFACILITYLOCATION is a generalization of ONLINESETCOVER, where the connection cost of a client/element to a facility/set is zero if the set contains the element, and infinity otherwise. *Metric* ONLINEFACILITYLOCATION, where connection costs satisfy the triangle inequality, has also been widely studied. Meyerson [17] presented a randomized $\mathcal{O}(\log n)$ competitive algorithm which was improved to a deterministic $\mathcal{O}(\log n/\log\log n)$ competitive algorithm [18] by Fotakis, who also showed that this bound is optimal.

Along with SETCOVERLEASING, Anthony and Gupta [1] presented two other infrastructure leasing problems: metric *Facility Leasing* and *Steiner Tree Leasing*, the leasing variants of the well known metric Facility Location and Steiner Tree problems. The leasing model was studied in offline and online settings. In the offline setting, Anthony and Gupta [1] gave offline algorithms for the problems which were a result of an interesting relationship between infrastructure leasing problems and stochastic optimization. They gave an $\mathcal{O}(K)$-approximation to metric FACILITYLEASING, which was then improved to a 3-approximation by Nagarajan and Williamson [6]. Nagarajan and Williamson studied the problem in the online setting as well, and gave an $\mathcal{O}(K\log n)$ competitive algorithm [6]. In a previous work [5], we extended the results by Nagarajan and Williamson [6] by removing the dependency on n (and thereby on time). We gave an $\mathcal{O}(l_{max}\log(l_{max}))$ competitive algorithm where l_{max} is the maximum lease length. Moreover, we showed that our algorithm has an $\mathcal{O}(\log^2(l_{max}))$ competitive factor for many natural client arrival patterns. Such patterns include, for example, situations where the number of clients arriving in each time step does not vary too much, or is non-increasing, or is polynomially bounded in l_{max}.

A strongly related problem known as the *Parking Permit Problem* was introduced by Meyerson [20]. Here, each day, depending on the weather, we have to either use the car (if its rainy) or walk (if its sunny). If we take the car, we must have a valid parking permit. There are K different types of parking permits, each with its own duration and cost. The goal is to buy a set of permits in order to cover all rainy days and minimize the total cost of purchases (and without using weather forecasts). Meyerson [20] gave a deterministic $\mathcal{O}(K)$ competitive algorithm and a randomized $\mathcal{O}(\log K)$ competitive algorithm along with matching lower bounds for the PARKINGPERMITPROBLEM.

1.2   Our Contribution

In this paper, we give online algorithms for three variants of SETCOVERLEASING.

- *Non-uniform Online Set Multicover (one lease type, i.e., K = 1):* Non-uniform ONLINESETMULTICOVER is a generalization of uniform ONLINESETMULTICOVER, where each element is requested to be covered by a different number of sets. We give an $\mathcal{O}(\log d \log n) = \mathcal{O}(\log m \log n)$ competitive algorithm for the problem (Theorem 1). This matches the randomized $\Omega(\log m \log n)$ lower bound given by Feige et al. for ONLINESETCOVER [9].
- *Online Set Multicover with Repetitions (one lease type, i.e., K = 1):* ONLINESETMULTICOVERWITHREPETITIONS is a generalization of non-uniform ONLINESETMULTICOVER with the addition that elements may appear several times and must be covered by different sets. We give an $\mathcal{O}(\log d \log(dn)) = \mathcal{O}(\log m \log(mn))$ competitive algorithm for the problem. This improves the $\mathcal{O}(\log^2(mn))$ competitive ratio of ONLINESETCOVERWITHREPETITIONS [2] to $\mathcal{O}(\log d \log(dn)) = \mathcal{O}(\log m \log(mn))$ (Theorem 2).
- *Set Multicover Leasing:* SETMULTICOVERLEASING is the leasing variant of non-uniform ONLINESETMULTICOVER, where sets are leased for K different periods of time and a leased set may cover an element during its lease period only once. We give an $\mathcal{O}(\log(dK) \log n) = \mathcal{O}(\log(mK) \log n)$ competitive algorithm for the problem. This implies a coinciding $\mathcal{O}(\log m \log n)$-competitive ratio for ONLINESETCOVER [15] (K = 1) (Theorem 3) and matches the randomized $\Omega(\log m \log n)$ lower bound by Feige et al. [9].

1.3   Organization of the Paper

The rest of this paper is structured as follows. Section 2 consists of some preliminaries. Section 3 presents a randomized algorithm for SETMULTICOVERLEASING which carries on to ONLINESETMULTICOVERWITHREPETITIONS and ONLINESETMULTICOVER. Section 4 gives the analysis of the algorithm for each of the three problems: SETMULTICOVERLEASING, ONLINESETMULTICOVERWITHREPETITIONS, and ONLINESETMULTICOVER. Section 5 concludes the paper.

2 Preliminaries

Notations: An element j arriving at time t and requesting to be covered by p sets, is denoted as a triplet (j, t, p). Let E be the collection of all these triplets. A set S leased at time t for a period k can cover its elements over the interval $T = [t, t + l_k]$ and incurs a cost c_S^k. We denote such a set as a triplet (S, k, T). Let C be the collection of all these triplets. We refer to a leased set as *active*. Let $N(j, t, p)$ be the triplets $(S, k, T) \in C$ such that $j \in S$ and $t \in T$. We call these triplets *candidates* of (j, t, p). Let $N(S, k, T)$ be the triplets $(j, t, p) \in E$ such that $(S, k, T) \in N(j, t, p)$. We say an element is *p-covered* if at least p of its candidates are active.

Definition (Set Multicover Leasing): Elements in a universe U, $|U| = n$, arrive over time and must be covered by sets in a family F of subsets of U, $|F| = m$. Each arriving element has a number p which specifies the number of different sets it should be covered by. Sets can be leased for K different periods of time and a leased set may cover an element during its lease period only once. Leasing a set S for a period k incurs a cost c_S^k and allows S to cover its elements for the next l_k time steps. The objective is to minimize the total cost of the sets leased, such that elements arriving at any time t are covered by the requested number of sets which contain them and are leased during time t.

Definition (Interval Model): Meyerson [20] assumed the *interval model* for the leases, where leases of the same period may not coincide and showed that this remains competitive for the original problem within a constant factor. We adopt the same model for the sets, where the same sets with the same leases may not coincide. It is easy to see that this maintains the competitiveness within a constant factor as well. Following the interval model, each arriving element may have at most $K \cdot d$ candidates.

3 A Randomized Algorithm

In this section, we present a randomized algorithm for SETMULTICOVERLEAS-ING. Our algorithm maintains a fraction f_{SkT} for each set $(S, k, T) \in C$, initially set to zero and non-decreasing throughout the algorithm. We use two techniques: (1) Randomized Rounding technique and the (2) Greedy technique. When an element $(j, t, p) \in E$ arrives, we first 1-cover it by some candidate $(S, k, T) \in C$. To do this, we increase the fractions of all its candidates until they sum up to 1. Then, using an appropriate Randomized Rounding technique, we select at least one candidate and lease it. To 2-cover (j, t, p), we increase the fractions of all candidates excluding the one leased to 1-cover (j, t, p) (choose arbitrarily any set if there is more than 1), until they sum up to 1. Then, we again use Randomized Rounding to guarantee that a second candidate is leased. We continue with this Greedy technique until (j, t, p) is p-covered. We first describe an algorithm which i-covers any arriving element (j, t, p) (Algorithm 1) and then use it in SET-MULTICOVERLEASING (Algorithm 2). We maintain for each set, $2\lceil \log(n + 1) \rceil$

independent random variables $X_{(SKT)(q)}$, $(1 \leq q \leq 2 \lceil \log(n+1) \rceil)$, distributed uniformly in the interval $[0, 1]$. We define $\mu_{SKT} = min\{X_{(SKT)(q)}\}$.

Algorithm 1. (i-Cover)

An element (j, t, p) arrives. Let Q be the collection of the sets in $N(j, t, p)$ after removing $i - 1$ active candidates.
(i) (fractional) if $\sum_{(S,k,T) \in Q} f_{SkT} < 1$, do the following *increment*.
  While $\sum_{(S,k,T) \in Q} f_{SkT} < 1$;
    $f_{SkT} = f_{SkT} \cdot (1 + 1/c_S^k) + \frac{1}{|Q| \cdot c_S^k}$
(ii) (integer) Lease $(S, k, T) \in Q$ with $f_{SKT} > \mu_{SKT}$.
(iii) If (j, t, p) is not covered by some set in Q (i.e., there is no active set in Q);
    Lease the cheapest $(S, k, T) \in Q$.

Algorithm 2. (Set Multicover Leasing)

An element (j, t, p) arrives. Set $i = 0$.
While($i \leq p$)
  - run **Algorithm 1** (i-Cover)
  - increment i by 1

---

4   Analysis

This section shows that **Algorithm 2** is $\mathcal{O}(\log d \log n)$, $\mathcal{O}(\log d \log(dn))$, and $\mathcal{O}(\log(dK) \log n)$ competitive for ONLINESETMULTICOVER, ONLINESETMULTICOVERWITHREPETITIONS, and SETMULTICOVERLEASING, respectively.

4.1   Online Set Multicover

It is easy to see that **Algorithm 2** constructs a feasible solution to SETMULTICOVERLEASING. As mentioned earlier, SETMULTICOVERLEASING generalizes ONLINESETMULTICOVER when $K = 1$ and sets are leased for an infinite period of time. We show that **Algorithm 2** is an $\mathcal{O}(\log d \log n)$ competitive randomized algorithm for ONLINESETMULTICOVER. We refer to elements and sets as singletons in U and F, respectively. To compute the total expected cost, we first bound the fractional cost (the fractional cost is the sum of all fractions) by $(\mathcal{O}(\log d) = \mathcal{O}(\log m)) \cdot Opt$, where Opt is the cost of an optimal offline solution. Then, we show that the randomized integer solution has an expected cost of at most $\mathcal{O}(\log n)$ times the fractional cost and hence deduce the expected $\mathcal{O}(\log d \log n) = \mathcal{O}(\log m \log n)$ competitive factor of the algorithm. The lemma below bounds the fractional cost.

Lemma 1. *The fractional cost is $\mathcal{O}(\log(d))$ times the optimal offline solution.*

Proof. We show the following two facts to bound the fractional cost.

(i) An increment adds at most two to the fractional cost.
(ii) The total number of increments in the algorithm is $\mathcal{O}\left(\log d\right) \cdot Opt$.

We fix an element j.

Proof of (i). In an increment, the fraction of each candidate $S \in Q$ is increased by $\left(\frac{f_S}{c_S} + \frac{1}{|Q| \cdot c_S}\right)$. Summing up over all candidates in Q, the fractional cost increases, after each increment, by $\sum_{S \in Q} c_S \cdot \left(\frac{f_S}{c_S} + \frac{1}{|Q| \cdot c_S}\right) = \sum_{S \in Q} f_S + 1$. Since an increment is only done if $\sum_{S \in Q} f_S < 1$, each increment adds at most two to the cost of the fractional solution. $\qquad\square$

Proof of (ii). The total number of increments is the sum of the increments for each $1 \le i \le p_{max}$, where p_{max} is the maximum number of times any element wants to be covered by. An increment is only done if $\sum_{S \in Q} f_S < 1$. For any $1 \le i \le p_{max}$ and in any increment the algorithm decides to make, at least one set S_{opt} in the optimal solution is a candidate and therefore increases its fraction (an optimal solution requires p_{max} sets too). The fraction of S_{opt} reaches at least $\frac{1}{|Q|}$ after $c_{S_{opt}}$ increments due to the second part of the increment $\frac{1}{|Q| \cdot c_{S_{opt}}}$. After this, the first part of the increment keeps multiplying the fraction by $(1 + 1/c_{S_{opt}})$ and stops when it becomes larger than 1. Hence, after $\mathcal{O}\left(c_{S_{opt}} \cdot \log |Q|\right)$ increments, $\sum_{S \in Q} f_S$ will be greater than 1. Since $|Q| \le d$, (ii) holds. $\qquad\square$

Since the fractions increase only during an increment, the lemma follows. $\qquad\square$

We now show that the randomized integer solution is at most $\mathcal{O}(\log n)$ times the fractional cost. We fix a set S. The expected cost of choosing S throughout the algorithm is $c_S \cdot Pr(f_S > \mu_S) \le 2 \log(n + 1) \cdot c_S \cdot f_S$. Thus, the total expected cost is upper bounded by $\sum_{S \in F} 2 \log(n+1) \cdot c_S \cdot f_S$. Furthermore, to guarantee a feasible solution, **Algorithm 1** adds the cheapest candidate to the solution if an element j is not covered after randomization (this is a lower bound to the optimal solution Opt). Nevertheless, we show that this only happens with probability at most $1/n^2$ and adds an unnoticed additional cost to the competitive factor.

For a single $1 \le q \le 2\lceil \log(n+1) \rceil$, the probability that j is not covered is $\le \prod_{S \in Q}(1 - f_S) \le e^{-\sum_{S \in Q} f_S} \le 1/e$. The last inequality holds because $\sum_{S \in Q} f_S \ge 1$. Thus the probability that j is not covered, for all $1 \le q \le 2\lceil \log(n+1) \rceil$, is at most $1/n^2$. The additional expected cost is thus upper bounded by $n \cdot 1/n^2 \cdot Opt$. From the analysis above as well as Lemma 1, we can deduce the following theorem.

Theorem 1. *There is an online randomized algorithm for* ONLINESETMULTI-COVER *that is* $\mathcal{O}(\log d \log n) = \mathcal{O}(\log m \log n)$ *competitive.*

4.2   Online Set Multicover with Repetitions

As defined earlier, ONLINESETMULTICOVERWITHREPETITIONS adds to ONLI-NESETMULTICOVER the possibility for elements to appear several times where they must be covered by different sets each time. We modify **Algorithm 1** and the analysis in Sect. 4.1 to show that **Algorithm 2** is an $\mathcal{O}(\log d \log(dn))$ competitive randomized algorithm for ONLINESETMULTICOVERWITHREPETITIONS.

- Modify (**Algorithm 1**): We maintain for each set, $2\lceil\log((dn)+1)\rceil$ independent random variables $X_{(SKT)(q)}$, $(1 \le q \le 2\lceil\log((dn)+1)\rceil)$, distributed uniformly in the interval $[0, 1]$.
- Modify (Analysis (Sect. 4.1)): We show that the randomized integer solution is at most $\mathcal{O}(\log(dn))$ times the fractional cost. We fix a set S. The expected cost of choosing S throughout the algorithm is $c_S \cdot Pr(f_S > \mu_S)$ $\le 2\log((dn)+1) \cdot c_S \cdot f_S$. Thus, the total expected cost is upper bounded by $\sum_{S \in F} 2\log((dn)+1) \cdot c_S \cdot f_S$. Furthermore, to guarantee a feasible solution, **Algorithm 2** adds the cheapest candidate to the solution if an element j is not covered after randomization (this is a lower bound to the optimal solution Opt). Nevertheless, we show that this only happens with a probability at most $1/(dn)^2$ and adds an unnoticed additional cost to the competitive factor. For a single $1 \le q \le 2\lceil\log((dn)+1)\rceil$, the probability that j is not covered is $\le \prod_{S \in Q}(1 - f_S) \le e^{-\sum_{S \in Q} f_S} \le 1/e$. The last inequality holds because $\sum_{S \in Q} f_S \ge 1$. Thus the probability that j is not covered, for all $1 \le q \le 2\lceil\log((dn)+1)\rceil$, is at most $1/(dn)^2$. The additional expected cost is thus upper bounded by $(dn) \cdot 1/(dn)^2 \cdot Opt$ since each element cannot be asked to be covered more than d times.

Hence, we deduce the following theorem.

Theorem 2. *There is an online randomized algorithm for* ONLINESETMULTICOVER *that is* $\mathcal{O}(\log d \log(dn)) = \mathcal{O}(\log m \log(mn))$ *competitive.*

4.3   Set Multicover Leasing

Algorithm 2 gives the first algorithm for the leasing variant of ONLINESETMULTICOVER. We show that **Algorithm 2** is an $\mathcal{O}(\log(dK)\log n)$ competitive algorithm for SETMULTICOVERLEASING. The only part we modify in the analysis of Sect. 4.1 is the bound on the fractional cost.

Modify (**Proof of (ii).**) $|Q| \le (Kd)$ rather than $|Q| \le d$. This holds because of the interval model, where each arriving element may have at most $K \cdot d$ candidates. This results in a fractional cost of $\mathcal{O}(\log(dK))$ times the optimal offline solution (Lemma 1). We thus deduce the following theorem.

Theorem 3. *There is an online randomized algorithm for* SETMULTICOVERLEASING *that is* $\mathcal{O}(\log(dK)\log n) = \mathcal{O}(\log(mK)\log n)$ *competitive.*

5   Conclusion

This paper presents the first algorithm for SETCOVERLEASING. SETCOVERLEASING is an online variant of the classical SETCOVER. Unlike standard online variants, this leasing variant captures a more natural scenario where purchased sets have time durations and expire. Our results are nearly tight in the case where sets can be leased only for an infinite period of time [9,15]. Furthermore, we study online variants of SETCOVERLEASING, including a generalization of ONLINESETCOVERWITHREPETITIONS presented by Alon et al. [2] who gave an $\mathcal{O}(\log^2(mn))$ competitive randomized algorithm for the problem. We improve the $\mathcal{O}(\log^2(mn))$ competitive factor to $\mathcal{O}(\log d \log(dn)) = \mathcal{O}(\log m \log(mn))$. Since any online problem can be studied as a leasing problem, our approach can be of an independent interest for leasing variants of many other problems.

Because any leasing problem generalizes the PARKINGPERMITPROBLEM which has a deterministic $\Omega(K)$ lower bound and a randomized $\Omega(\log K)$ lower bound, $\Omega(K + \log m \log n/(\log \log m + \log \log n))$, and $\Omega(\log K + \log m \log n/(\log \log m + \log \log n))$ are deterministic and randomized lower bounds for SETCOVERLEASING, respectively. The question is whether we can prove stronger lower bounds by combining the lower bounds for ONLINESETCOVER [9,15] and the PARKINGPERMITPROBLEM [20].

All the algorithms for the online leasing variants of the problems studied so far build upon the algorithms for the non-leasing variants of the corresponding problems and the PARKINGPERMITPROBLEM algorithm [20]. This means, we do not know yet the price we have to pay for leasing. In other words, we do not know whether there is a common difficulty among these problems in their leasing variant or the same techniques used for the non-leasing variants could be extended to solve the leasing variants.

References

1. Anthony, B.M., Gupta, A.: Infrastructure leasing problems. In: Fischetti, M., Williamson, D.P. (eds.) IPCO 2007. LNCS, vol. 4513, pp. 424–438. Springer, Heidelberg (2007)
2. Alon, N., Azar, Y., Gutner, S.: Admission control to minimize rejections and online set cover with repetitions. In: 17th ACM Symposium on Parallelism in Algorithms and Architectures (SPAA), Prague, pp. 238–244 (2005)
3. Berman, P., DasGubta, B.: Approximating the online set multicover problems via randomized winnowing. Theor. Comput. Sci. **393**(13), 54–71 (2008)
4. Alon, N., Awerbuch, B., Azar, Y., Buchbinder, N.: A general approach to online network optimization problems. In: 15th Annual ACM-SIAM Symposium on Discrete Algorithms (SODA), New Orleans, LA, pp. 577–586 (2004)
5. Kling, P., Meyer auf der Heide, F., Pietrzyk, P.: An algorithm for online facility leasing. In: Even, G., Halldórsson, M.M. (eds.) SIROCCO 2012. LNCS, vol. 7355, pp. 61–72. Springer, Heidelberg (2012)
6. Nagarajan, C., Williamson, D.P.: Offline and online facility leasing. In: Lodi, A., Panconesi, A., Rinaldi, G. (eds.) IPCO 2008. LNCS, vol. 5035, pp. 303–315. Springer, Heidelberg (2008)
7. Chavatal, V.: A greedy heuristic for the set-covering problem. Math. Oper. Res. **4**(3), 233–235 (1979)
8. Feige, U.: A threshold of ln n for approximating set cover. J. ACM **45**(4), 634–652 (1998)

9. Feige, U., Korman, S.: Personal communication
10. Johnson, D.S.: Approximation algorithms for combinatorial problems. J. Comput. Syst. Sci. **9**, 256–278 (1974)
11. Lovasz, L.: On the ratio of optimal and fractional covers. Discrete Math. **13**, 383–390 (1975)
12. Berman, P., DasGupta, B., Sontag, E.: Randomized approximation algorithms for set multicover problems with applications to reverse engineering of protein and gene networks. Discrete Appl. Math. **155**(67), 733–749 (2007)
13. Vazirani, V.: Approximation Algorithms (2001)
14. Fotakis, D.: A primal-dual algorithm for online non-uniform facility location. J. Discrete Algorithms **5**, 141–148 (2007)
15. Alon, N., Awerbuch, B., Azar, Y., Buchbinder, N., Naor, J.: The online set cover problem. In: 35th Annual ACM Symposium on the Theory of Computation (STOC), San Diego, CA, USA, pp. 100–105 (2003)
16. Awerbuch, B., Azar, Y., Fiat, A., Leighton, T.: Making commitments in the face of uncertainty: how to pick a winner almost every time. In: 28th Annual ACM Symposium on Theory of Computing (STOC), Pennsylvania, USA, pp. 519–530 (1996)
17. Meyerson, A.: Online facility location. In: 42nd Annual IEEE Symposium on Foundations of Computer Science (FOCS), Las Vegas, Nevada, USA, pp. 426–431 (2001)
18. Fotakis, D.: On the competitive ratio for online facility location. In: Baeten, J.C.M., Lenstra, J.K., Parrow, J., Woeginger, G.J. (eds.) ICALP 2003. LNCS, vol. 2719, pp. 637–652. Springer, Heidelberg (2003)
19. Buchbinder, N., Naor, J.S.: Online primal-dual algorithms for covering and packing problems. In: Brodal, G.S., Leonardi, S. (eds.) ESA 2005. LNCS, vol. 3669, pp. 689–701. Springer, Heidelberg (2005)
20. Meyerson, A.: The parking permit problem. In: 46th Annual IEEE Symposium on Foundations of Computer Science (FOCS), Pittsburgh, PA, USA, pp. 274–284 (2005)
21. Malik, S., Huet, F.: Virtual cloud: rent out the rented resources. In: 6th IEEE International Conference for Internet Technology and Secured Transactions (ICITST), Abu Dhabi, UAE, pp. 536–541 (2011)
22. Alon, N., Moshkovitz, D., Safra, S.: Algorithmic construction of sets for k-restrictions. ACM Trans. Algorithms **2**, 153–177 (2006)

Geometric Optimization

Optimizing Squares Covering a Set of Points

Binay Bhattacharya[1], Sandip Das[2], Tsunehiko Kameda[1($\boxtimes$)],
Priya Ranjan Sinha Mahapatra[3], and Zhao Song[4]

[1] School of Computing Science, Simon Fraser University, Burnaby, Canada
{binay,tiko}@sfu.ca
[2] Advanced Computing and Microelectronics Unit, Indian Statistical Institute,
Kolkata, India
sandipdas@isical.ac.in
[3] Department of Computer Science and Engineering, University of Kalyani,
Kalyani, India
priya@klyuniv.ac.in
[4] Department of Computer Science, University of Texas at Austin, Austin, USA
zhaos@utexas.edu

Abstract. We investigate two kinds of optimization problems regarding
points in the 2-dimensional plane that need to be enclosed by squares.
(1) Given a set of n points, find a given number of squares that enclose
all the points, minimizing the size of the largest square used.
(2) Given a set of n points, enclose the maximum number of points, using
a specified number of squares of a fixed size. We provide different techniques to solve the above problems in cases where squares are axis-
parallel or of arbitrary orientation, disjoint or overlapping. All the
algorithms we use run in time that is a low-order polynomial in n.

1 Introduction

Given a set $\mathcal{P}$ of n points, the enclosing problem in computational geometry
is to find the smallest geometrical object of a given type that encloses all the
points in $\mathcal{P}$. There are the problems of finding the minimum enclosing circle
[17], the minimum area triangle [4,11,16], the minimum area rectangle [20], the
minimum bounding box [15], the smallest ellipsoid [21], and the smallest width
annulus [1]. As far as we are aware, there has been little work on finding the
smallest enclosing square(s) with arbitrary orientation. Das *et al.* [6] presented
an algorithm to identify the smallest square of arbitrary orientation, containing
exactly k points in $O(n^2 \log n + kn(n-k)^2 \log n)$ time. It runs in $O(n^2 \log n)$
time when $n = k$.

Katz *et al.* [9] consider the problem of covering the points in $\mathcal{P}$ by two
"constrained" squares whose centers must lie on some specified points. The two
squares may or may not be disjoint. They presented an $O(n \log^2 n)$ time algorithm to find two constrained axis-parallel squares whose union covers $\mathcal{P}$ and
the size of the larger square is minimized. They also presented an $O(n^2 \log^4 n)$

© Springer International Publishing Switzerland 2014
Z. Zhang et al. (Eds.): COCOA 2014, LNCS 8881, pp. 37–52, 2014.
DOI: 10.1007/978-3-319-12691-3_4

time algorithm to find two constrained parallel squares (with arbitrary orientation) whose union covers $\mathcal{P}$ and the size of the larger square is minimized. Jaromczyk and Kowaluk [8] solved the unconstrained version of the above problem in $O(n^2)$ time. They also presented an $O(n^3 \log^2 n)$ time algorithm to find two constrained squares (where each square is allowed to rotate independently) whose union covers $\mathcal{P}$ and the size of the larger square is minimized.

Other researchers have been interested in maximizing the number of points in $\mathcal{P}$ that can be enclosed by unit squares [12,13]. Mahapatra *et al.* proposed an $O(n^2)$ time and space algorithm for computing two axis-parallel unit squares which may be either disjoint or overlapping such that they together cover the maximum number of points [12]. In case the two such squares intersect, the interior of their intersection is not allowed to contain any point of $\mathcal{P}$. They introduced an $O(n^2 \log^2 n)$ time and $O(n \log n)$ space algorithm to find two overlapping axis-parallel unit squares such that they together cover the maximum number of points [13]. They introduced an $O(k^2 n^5)$ time and $O(kn^4)$ space algorithm to find k disjoint axis-parallel unit squares, so as to maximize the number of points covered by them [12].

Kim *et al.* [10] consider two variants of the optimization problem for disjoint *rectangles*: (1) the rectangles are allowed to be oriented freely while being restricted to be parallel to each other, and (2) one rectangle is restricted to be axis-parallel but the other rectangle is allowed to be oriented freely. For both of the problems, they presented $O(n^2 \log n)$ time algorithm using $O(n)$ space. Segal [18] discusses lower bounds on some covering problems by squares, and, in particular, shows that it takes $\Omega(n \log n)$ time to find the smallest square with a given center that encloses all the points.

We assume that no three points in $\mathcal{P}$ lie on the same line. The objective of the first group of problems that we solve is to minimize the size of the largest square used. Table 1 summarizes the time complexities of our algorithms presented in

Table 1. Minimizing the size of the largest square: [*]With a rectangular obstacle, [†]Implied by the two previous results

Number	D/ND/DC	AP/AO	Previous best	Our results	Section
1	–	AO	$O(n^2 \log n)$ [6]	$O(n \log n)$	2.1
1	–	AO[*]	–	$O(n \log n)$	2.2
2	D	AP	$O(n \log n)$ [8]	$O(n)$	2.3
2	ND	AP	–	$O(n)$	2.4
2	DC	AP	$O(n)$ [19]	$O(n)$[†]	–
2	D	AP-AO	$O(n^2 \log n)$ [10]	–	2.5
2	DC	AP-AO	–	$O(n^3 \log n)$	2.6
2	DC	AO	–	$O(n^4 \log n)$	2.7
3	D	AP	–	$O(n \log n)$	2.8
4	D	AP	–	$O(n^2 \log^2 n)$	2.10

Table 2. Maximizing the number of points covered

Number	D/ND/DC	AP/AO	Previous best	Our results	Section
1	–	AO	–	$O(n^3 \log n)$	3.1
2	DC	AP	$O(n^2)$ [12]	–	–
2	ND	AP	$O(n^2 \log^2 n)$ [13]	–	–
2	D	AP	$O(n \log n)$ [12]	–	–
3	D	AP	$O(n^5)$ [12]	$O(n^2 \log n)$	3.2
4	D	AP	$O(n^5)$ [12]	$O(n^3 \log n)$	3.3

this paper. The first column indicates the number of squares used. The acronym AP stands for "axis-parallel," AO for "arbitrary orientation," D for "disjoint," ND for "non-disjoint" (overlapping), and DC for "don't care," which means that the squares involved are not constrained to be D or ND.

The objective of the second group of problems that we solve is to maximize the number of points covered. Only axis-parallel squares are used. Table 2 shows our results.

This paper is organized as follows: In Sect. 2, we discuss the problem of enclosing a set of n points in the plane by k squares for $k =$1, 2, 3, and 4 in such a way that the size of the largest square used is minimized. The algorithms summarized in Table 1 are presented/analyzed in individual subsections. The objective of Sect. 3 is to discuss the problem of enclosing the maximum number of points from among a given set of n points in the plane by k squares for $k = 1, 3$, and 4 in such a way that the size of the largest square used is minimized. Section 4 concludes the paper.

Notation. We use $\mathcal{P}$ to denote a set of n points in the plane. The size of a set S of points is denoted by $|S|$. For a technical reason, we assume that no three points lie on a line. Coordinates in the plane are given by (x, y). For any point $p \in \mathcal{P}$, $p.x$ (resp. $p.y$) denotes its x (resp. y)-coordinate. In general, letter p with a subscript indicates a point in $\mathcal{P}$, and angles are represented by α, β and θ with or without a subscript. The upper case S with or without a subscript indicates a square, and the lower case s with a subscript indicates a side of a square. For a square S, $size(S)$ is the length of its one side.

2   Minimizing the Enclosing Square Size

2.1   One Square, Arbitrary Orientation

Given a convex polygon P,[1] we want to find the enclosing rectangle whose maximum side is minimized. Let $p_1, p_2, \cdots, p_m$ be the vertices of P ordered counter-clockwise (ccw). The angle of the side (p_i, p_{i+1}) from the horizontal

---

[1] This polygon may be the convex hull of the points in $\mathcal{P}$ computed in $O(n \log n)$ time.

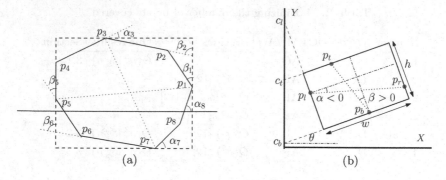

Fig. 1. (a) Angles $\{\alpha_i\}$ and $\{\beta_j\}$; (b) Rotating the enclosing rectangle

(resp. vertical) axis is denoted by α_i (resp. β_i), where $0 \le \alpha_i < 90°$ and $0 \le \beta_i < 90°$. See Fig. 1(a). We now merge $\{\alpha_i\}$ and $\{\beta_i\}$ into a single sorted sequence $\alpha'_1, \alpha'_2, \ldots, \alpha'_m$, from the smallest to the largest. Based on the initial axis-parallel position of the enclosing rectangle R of P, we name its four sides as top, bottom, left and right sides, denoted by s_t, s_b, s_l, s_r, respectively. In Fig. 1(b), the most ccw vertex that touches s_t (resp. s_b, s_l, s_r) is named p_t (resp. p_b, p_l, p_r). It's possible that some of these points may coincide, e.g., $p_t = p_l$. In Fig. 1(a), we have $p_l = p_5, p_r = p_1, p_b = p_7$, and $p_t = p_3$. Let h (resp. w) denote the length of s_l (resp. s_t). We call $\max\{w, h\}$ the *size* of R.

We call $\overrightarrow{p_b p_t}$ (resp. $\overrightarrow{p_l p_r}$) the *bt-spanning vector* (resp. *lr-spanning vector*). Note that if the lr-spanning vector (resp. bt-spanning vector) is parallel to side s_t (resp. s_l) of the enclosing rectangle R, then rotating R ccw makes s_t (resp. s_l) shorter. But if it is tilted ccw from this position, i.e., $\alpha > 0$ (resp. $\beta > 0$), rotating R ccw makes s_t (or s_l) longer. In Fig. 1(b), $\overrightarrow{p_b p_t}$ (resp. $\overrightarrow{p_l p_r}$) are tilted ccw (resp. cw) relative to s_l (resp. s_t), as indicated by $\beta > 0$ (resp. $\alpha < 0$). As we rotate R ccw from $0°$ to $90°$, the sides of R will stretch or shrink. Our approach is to rotate R ccw by degree θ from $0°$, until one of the following two types of events occurs, and repeat.

A. $\theta = \alpha'_i$ for some i $(1 \le i \le m)$.
B. The lr-spanning vector (or bt-spanning vector) becomes parallel to side s_t (or s_l).

Note that there are $O(n)$ events of either type. The minimum size *may be* attained when a type A event occurs. There is another scenario under which the minimum size may be attained. As we observed above, after a type B event occurs, the width or height of the rectangle starts shrinking. Assume that $h < w$. See Fig. 1(b). If $\beta > 0$ (resp. $\alpha < 0$), for example, as in See Fig. 1(b), then as we rotate R ccw, h increases and w decreases. If we reach $h = w$ in this process, R becomes a square, and this square is a candidate for the minimum-size square.

To compute h and w when an event of type A occurs, we represent the four sides of R by linear equations as follows:

$$s_t : \ Y = (\tan\theta)X + c_t \qquad s_b : \ Y = (\tan\theta)X + c_b \qquad (2.1)$$

$$s_l : \ Y = -(\cot\theta)X + c_l \qquad s_r : \ Y = -(\cot\theta)X + c_r \qquad (2.2)$$

The constant c_t in (2.1) can be determined by plugging $X = p_t.x$ and $Y = p_t.y$ in it. The constant c_b in (2.1) can be determined similarly from $p_b.x$ and $p_b.y$. Then we have height $h = (c_t - c_b)\cos\theta$. The width w can be computed similarly.

To detect when an event of type B occurs, we need to keep track of the angles α (resp. β) that the lr-spanning vector (resp. bt-spanning vector) forms with s_t (resp. s_l). They can be easily determined from the coordinates of p_l and p_r (resp. p_t and p_b). Assume that $\beta > 0$ and $\alpha < 0$ as in Fig. 1(b). Angle β may become negative when we increase θ from α'_i to α'_{i+1}. If this is the case, then we know that event of type B has occurred at some angle $\theta = \theta'$, where $\alpha'_i < \theta' \leq \alpha'_{i+1}$. We test if $h = w$ for some angle in the interval (α'_i, θ'), and if so, we record the corresponding R as a candidate for the optimal square. We can do similar processing if $\beta < 0$ and $\alpha > 0$. Each event can be processed in constant time, since we know the sequence of the vertices along polygon P. After rotating R up to $90°$, we pick the snapshot of R found so far with the smallest size. An optimal enclosing square is obtained by expanding the optimal rectangle R into a square.

Theorem 1. *The square of the smallest size that encloses all the points in $\mathcal{P}$ can be computed in $O(n\log n)$ time.*

Proof. We first compute the convex hull of the given set of points in $O(n\log n)$ time, and consider it as polygon P in the above discussion. The processing time for all events is $O(n)$. $\qquad\qquad\square$

2.2   One Square, Arbitrary Orientation, Rectangular Obstacle

We assume that there exists a line that separates the obstacle rectangle from all the points. Suppose we impose the restriction that the square must not intersect a given axis-parallel rectangular object. Figure 2(a) shows the static rectangular obstacle, whose upper left corner is at (x_0, y_0), and a rectangle that is rotated, making sure that it encloses all points, where points p_t, p_b, and p_r are the most ccw vertices, touching s_t, s_b, and s_r of rectangle R, respectively. Let β denote the angle between $\overrightarrow{p_b p_t}$ and s_l, and assume $\beta > 0°$. To see if the lower right corner of the rotating rectangle collides with the obstacle, consider the lines

$$Y = (\tan\theta_1)X + c_b \quad \text{and} \quad Y = (\cot\theta_1)X + c_r \qquad (2.3)$$

that pass through p_b and p_r, respectively. Equation (2.3) have three unknowns c_b, c_r and θ_1. Two constraints on them are given by the fact that the first (resp. second) line in (2.3) must pass through vertex p_b (resp. p_b). The third constraint is that the intersection of these two lines must lie on the line $X = x_0$. If this intersection point lies below y_0, then R hits the obstacle when the rotation angle is θ_1.

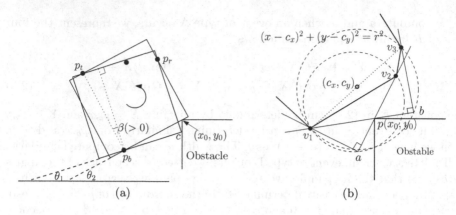

Fig. 2. (a) Rotation angles in (θ_1, θ_2) forbidden; (b) Two positions of the enclosing square that touch the obstacle

To find θ_2 in this case, note that it is the slope of the line connecting p_b and (x_0, y_0). From angle θ_1 to θ_2, we cannot prevent R from intersecting the upper left corner of the obstacle. We call the angle interval (θ_1, θ_2) the *forbidden interval*. There is at most one *forbidden interval*, since we need to rotate R only up to $90°$.

Figure 2(b) shows that the forbidden interval may not be caused by just two vertices of the convex hull P. Suppose the enclosing square touches the convex hull at vertices v_1 and v_2 when it hits the obstacle, rotating ccw. When the enclosing square hits the obstacle, rotating cw, it may be touching vertices v_3 and v_4 that are different from v_1 and v_2. Suppose it is touching v_1 and v_3, as shown in the figure. Let (c_x, c_y) be the mid point between v_1 and v_3, and let r be the radius of the circle with center at (c_x, c_y). Then this circle can be described by the following equation:

$$(x - c_x)^2 + (y - c_y)^2 = r^2. \tag{2.4}$$

The fact that it hits the obstacle can be expressed by $(x_0 - c_x)^2 + (y_0 - c_y)^2 < r^2$, where $p(x_0, y_0)$ is the corner point of the obstacle. Point b, for example, can be computed as the intersection of the line connecting v_1 and p and the circle represented by (2.4). The angle θ_2 in Fig. 2(a) can be computed from the slope of the line connecting v_1 and p.

The outer envelope of the dashed circles, whose diameters are $\overline{v_1 v_2}$ and $\overline{v_1 v_3}$, etc., is called the *angle hull* in [7]. Thus, the enclosing square hits the obstacle, if and only if the corner point of the obstacle is inside the angle hull.

Theorem 2. *Assume that there exists a line that separates the rectangular obstacle from all the points in $\mathcal{P}$. The square of the smallest size enclosing all the points that does not intersect the obstacle can be computed in $O(n \log n)$ time.* $\square$

2.3   Two Disjoint Axis-Parallel Squares

Without loss of generality, we assume that the two squares are divided by a vertical line. We use the *prune-and-search* algorithm as follows.

1. Initialize $\mathcal{P}' = \mathcal{P}$.
2. By a vertical line, separate the points in $\mathcal{P}'$ into two sets $\mathcal{P}'_l$ and $\mathcal{P}'_r$, whose sizes differ by at most one.
3. Compute the smallest and leftmost square S_l and the smallest and rightmost square S_r, containing the points of $\mathcal{P}'_l$ and $\mathcal{P}'_r$, respectively.
4. (a) $size(S_l) = size(S_r)$: Output S_l and S_r and stop.
   (b) Assume without loss of generality that $size(S_l) < size(S_r)$. Identify the four extreme points contained in $\mathcal{P}'_l$, a, b, c, d, along the x and y directions. Update $\mathcal{P}'$ by $\mathcal{P}' = \mathcal{P}' \backslash (\mathcal{P}'_l \backslash \{a, b, c, d\})$.[2] If $|\mathcal{P}'|$ is larger than a preselected constant c then go to Step 2.
5. Use a brute force method to solve the problem and stop.   □

Note that no sorting is required in the above algorithm. Note also that the above algorithm generates two squares when one square with the same maximum size can cover all the points. We thus have

Theorem 3. *One can find two disjoint axis-parallel squares covering all the points in $\mathcal{P}$, such that the size of the larger square is minimized, in $O(n)$ time using $O(n)$ space.*

Proof. Steps 2–4 are repeated $O(\log n)$ times. Each iteration costs linear time in the size of $\mathcal{P}'$. After each iteration (at least) a constant fraction (almost $1/2$) of points are pruned. Therefore, the total cost of the algorithm is linear.   □

2.4   Two Overlapping Axis-Parallel Squares

We assume that the optional solution is not just one square that contains all the points. This case can be easily identified. We first determine the highest, lowest, leftmost and the rightmost points. Based on them, we then identify the

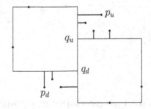

Fig. 3. Expanding two equal sized squares

---

[2] Since the points in $\mathcal{P}'_l \backslash \{a, b, c, d\}$ cannot affect the size of the left square, they can be pruned.

outer sides of the two squares. Without loss of generality, we assume that they are related as shown in Fig. 3. Note that the upper (resp. lower) side of the left (resp. right) square may not touch a point. We expand the two equal sized squares until they touch each other. If there is no point outside the two squares, then the optimal solution can be found by shrinking them to the minimum sizes. So, suppose that there are points outside the two squares. We want to cover them by expanding one or both squares. Let q_u and q_d denote the corner points, as shown in Fig. 3. Consider any point p that lies to the upper right of q_u so that $p.x > q_u.x$ and $p.y > q_u.y$ hold. We compare such points based on the value $\min\{|p.x - q_u.x|, |p.y - q_u.y|\}$, and determine the point p_u that has the maximum value. Similarly, we determine the point p_d that has the maximum value, based on the distance from q_d. Let l be the maximum of the values associated with p_u and p_d. We expand the two squares by length l. Clearly, one of the squares may need to be expanded by less than l. Thus, we have the following result:

Theorem 4. *One can find two overlapping axis-parallel squares covering all the points in $\mathcal{P}$, such that the size of the larger square is minimized, in $O(n)$ time using $O(n)$ space.*   □

2.5 Two Disjoint Squares, One Axis-Parallel, the Other of Arbitrary Orientation

Kim et al. solve the problem for rectangles [10] in $O(n^2 \log n)$ time. We can always convert each rectangle into a square by stretching the shorter side. The complexity remains the same, i.e., $O(n^2 \log n)$.

2.6 Two Squares, One Axis-Parallel, the Other of Arbitrary Orientation, "don't care"

A few possible cases are shown in Fig. 4. Let S denote the axis-parallel square. In Fig. 4(a), the points that are not in S lie to the left of the right side of S. This is not the case for in Fig. 4(b) and (c).

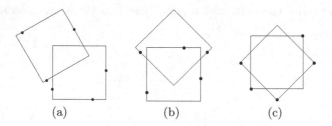

| (a) | (b) | (c) |

Fig. 4. Two overlapping squares

We preprocess n points based on their x-coordinates, ordering them into a sorted sequence, $x_1, x_2, \ldots, x_n$, from the smallest to the largest. This takes $O(n \log n)$ time.

For each of the $O(n^2)$ pairs of points, (p_l, p_b), we draw a vertical (resp. horizontal) line through p_l (resp. p_b). Note that the case $p_l = p_b$ should also be considered. Let o_{p_l, p_b} be the intersection of these two lines. We examine the squares S whose left-bottom corner is at o_{p_l, p_b}. As we expand the size of S, a new point may touch its boundary. There can be $O(n)$ such events, and each event defines a new axis-parallel square S. Clearly for a fixed pair (p_l, p_b), as the size of S increases, the size of the optimal square S' that contains all the points not in S monotonically decreases (or is non-decreasing to be precise). This suggests that we can do binary search on the size of S to find the size that minimizes the maximum of the sizes of S and S'. Thus for a given pair (p_l, p_b), we need to perform $O(n \log n)$ operations to determine the size of S'. This entails the computation of the convex hull for the points outside of S, and finding the min-size square that encloses it. Using a modified Graham scan, it takes $O(n)$ time [14]. We thus have

Theorem 5. *In $O(n^3 \log n)$ time, we can find two (possibly overlapping) squares, one axis-parallel, the other of arbitrary orientation, which jointly enclose all the points in $\mathcal{P}$ such that the size of the larger square is minimized.* □

2.7 Two Squares, Each of Arbitrary Orientation, "don't care"

Consider any two points. The order of their projections on the x-axis can be reversed by rotation the x-axis. So for any given point p_i, there are $n-1$ "critical" angles, and for all points, there are $O(n^2)$ critical angles. Sort them, and let θ_1 and θ_2 be two adjacent critical angles, and consider a group of points. It is possible that, as we rotate the coordinates from θ_1 and θ_2, the height of the enclosing rectangle may become larger than the width, or vice versa. In other words, there may be another critical angle between θ_1 and θ_2. In any case, there are still only $O(n^2)$ critical angles, and for each critical angle, we can apply the algorithm of Sect. 2.6. We need to rotate the "axis-parallel" square S, to make it as small as possible. By performing binary search on the size of the first square, we get

Theorem 6. *In $O(n^4 \log n)$ time, we can find two (possibly overlapping) squares, each of arbitrary orientation, which jointly enclose all the points in $\mathcal{P}$ such that the size of the larger square is minimized.* □

2.8 Three Disjoint Axis-Parallel Squares

Let S_1, S_2 and S_3 be the three squares that will jointly contain all the points in $\mathcal{P}$. Figure 5(a) shows one possible pattern, in which the three squares are aligned horizontally. They can also be aligned vertically. Figure 5(b) shows another possible pattern, where a line separates a square from the other two. There are four variations of the pattern in Fig. 5(b). We need to solve the problem for each of these 6 cases and pick the minimum size. We do binary search on the line L that separates one square from the other two, solving the 2-square problem on the part that contains two squares. This subproblem can be solved in $O(n)$ by the algorithm in Sect. 2.3.

Theorem 7. *In $O(n \log n)$ time, we can find three disjoint axis-parallel squares, which jointly enclose all the points in $\mathcal{P}$ such that the size of the largest square is minimized.*   □

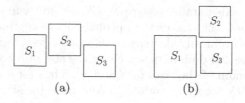

(a)   (b)

Fig. 5. Three disjoint axis-parallel squares

Special case. Here we discuss a special case in Fig. 5(a), and show that we can solve the problem in $O(n)$ time in this special case. Our approach is to use the prune-and-search method as follows. We first partition the set of n points into three parts with an "equal" number of points, $\mathcal{P}_1, \mathcal{P}_2, \mathcal{P}_3$,[3] and carry out the following procedure:

Procedure 1

1. (a) *Solve the 1-square problem on $\mathcal{P}_1$ and the 2-square problem on $\mathcal{P}_2 \cup \mathcal{P}_3$. Let l_1 and l_{23} be the two optimal sizes.*
   (b) *Solve the 1-square problem on $\mathcal{P}_3$ and the 2-square problem on $\mathcal{P}_1 \cup \mathcal{P}_2$. Let l_3 and l_{12} be the two optimal sizes.*
2. (a) *If $l_{23} > l_1$, then $\mathcal{P}_1$ is covered by one square: we can prune $\lceil n/3 \rceil - 4$ points of $\mathcal{P}_1$.*
   (b) *If $l_{12} > l_3$, then $\mathcal{P}_3$ is covered by one square: we can prune at least $\lfloor n/3 \rfloor$ points of $\mathcal{P}_3$.*
   (c) *If $l_1 > l_{23}$ and $l_3 > l_{12}$, then $\mathcal{P}_2$ is covered by one square, and we can prune $\lfloor n/3 \rfloor$ points of $\mathcal{P}_2$.*

Thus after $O(n)$ operations, at least $\lfloor n/3 \rfloor - 4$ points can be pruned. An $O(n)$ algorithm results when the process is repeated.

2.9 Guillotine k-partition

A k-*partition* of a rectangle R is a partition of R into k subrectangles, such that each subrectangle is parallel to a side of R. A *guillotine cut* [5] cuts across a rectangular area all the way from top to bottom or from left to right. Preforming k–1 guillotine cuts on R results in k *sectors*, which we call a *guillotine k-partition*. Thus the original rectangular area with no cut is a guillotine 1-partition. Let us

---

[3] To be exact, $\mathcal{P}_1$ (resp. $\mathcal{P}_2$, $\mathcal{P}_3$) covers the first $\lceil n/3 \rceil$ (resp. next $\lfloor n/3 \rfloor$, the remaining $\lceil n/3 \rceil$ or $\lfloor n/3 \rfloor$) points. They can be determined in linear time [2]. So the difference is at most one.

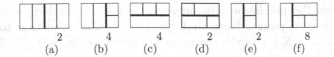

Fig. 6. 4-partitions

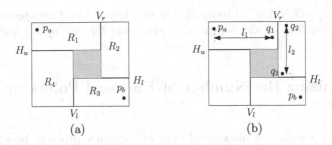

Fig. 7. Non-guillotine 5-partition: (a) p_a and p_b are unique; (b) $l_1 = size(S_1)$ and $l_2 = size(S_2)$

count the number of possible guillotine k-partitions, $N(k)$. It is easy to see that $N(1) = 1$, $N(2) = 2$, and $N(3) = 6$. (See the previous subsection.) Figure 6 shows that $N(4) = 22$. The number below each partition shows the number of variants that are obtained by rotation and flipping (mirror images).

Lemma 1. *For $k \le 4$, every k-partition is a guillotine k-partition.*   □

The above lemma doesn't hold for $k \ge 5$.

2.10   Four Disjoint Axis-Parallel Squares

In Fig. 6, the red segment in each rectangle indicates the first guillotine cut. Note that on either side of this guillotine cut, there are at most 3 sectors, in other words at most three squares need be used to cover them. There are $O(\log n)$ choices for the guillotine cut, and once it is made, we can use the algorithm presented in Sect. 2.4 or 2.8, which runs in $O(n \log n)$ time. It thus appears that we might have an $O(n \log^2 n)$ time algorithm. Unfortunately, however, we have overlooked the pattern shown in Fig. 7(a), where the gray rectangle in the middle does not contain any point. We name the horizontal (resp. vertical) separating lines H_u and H_l (resp. V_l and V_r). Its mirror image is also possible. Let us name the rectangles defined by them clockwise $R_1, \ldots, R_4$, and let S_1 (resp. $S_2, \ldots, S_4$) denote the minimum size square that contains all the points in R_1 (resp. $R_2, \ldots, R_4$). Clearly, there are $O(n^2)$ choices for the vertical positions of H_u and H_l. For a particular choice, the leftmost (resp. rightmost) point p_a (resp. p_b) in R_1 (resp. R_3) are uniquely determined. We perform binary search on V_r to find its position that minimizes $|size(S_1) - size(S_2)|$. This minimizes $\max\{size(S_1), size(S_2)\}$. Using binary search, we now choose V_l at the position that minimizes the maximum size between S_3 and S_4, making sure that the gray area is free of points. Note that the our objective function to be minimized is $\max\limits_{1 \leq i \leq 4} size(S_i)$. Clearly, the more V_r moves to the right, the larger gets $\max\{size(S_3), size(S_4)\}$. Therefore, we should not move V_r to the right, once we reach the point where $\max\{size(S_3), size(S_4)\} \geq \max\{size(S_1), size(S_2)\}$.

Theorem 8. *In $O(n^2 \log^2 n)$ time, we can find four disjoint axis-parallel squares, which jointly enclose all the points in $\mathcal{P}$ such that the size of the largest square is minimized.* $\square$

3 Maximizing the Number of Enclosed Points by Unit Squares

In this section, we discuss the second kind of enclosure problem by square(s). With a given number of unit squares, we want to cover the maximum number of points in $\mathcal{P}$.

3.1 One Unit Square, Arbitrary Orientation

If the square is restricted to be axis-parallel, then it can be done in $O(n \log n)$ time [3, 12]. A unit square is said to be in a *canonical* position if its three sides each touch a point. (A corner point touches two sides.) Clearly, we can restrict ourselves to unit squares in a canonical position.

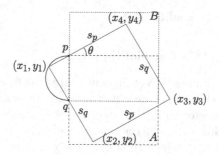

Fig. 8. Rotating the solid square, maintaining contact with points p and q.

Pick two points, p and q, out of $\mathcal{P}$, and consider a square on whose boundary p and q lie. (If $p = q$, then it is a corner point.) See Fig. 8, where the dashed (resp. dotted) square indicates the initial (resp. final) position of the solid square that is being rotated. For all the other points $\in \mathcal{P}$ will enter/leave the solid square no more than a constant number of times, as we rotate the solid square ccw from 0 to 90°. The four sides of the square are named s_p (touching point p), s_q (touching point q), $s_{\bar{p}}$ (opposite to s_p), and $s_{\bar{q}}$ (opposite to s_q). A point $t \in P$ (where $t \neq p$ and $t \neq q$) can touch up to four sides of the solid square under different θ's. We now determine the angle θ, if any, under which t touches each of the four sides. For example, assume that $t \in s_p$. Then we can represent side s_p by the line

$$y - t.y = \frac{t.y - p.y}{t.x - p.x}(x - t.x), \tag{3.1}$$

where $\frac{t.y-p.y}{t.x-p.x} > 0$, which is the angle of rotation when side s_p touches t. From (3.1), we obtain $\theta_p(t) = \arctan(\frac{t.y-p.y}{t.x-p.x})$. Similarly we can obtain the following angles:

1. $t \in s_q$: $\theta_q(t) = \frac{\pi}{2} + \arctan(\frac{t.y-q.y}{t.x-q.x})$.
2. $t \in s_{\overline{p}}$: $\theta_{\overline{p}}(t) = \arctan(\frac{1-t.x+q.x}{t.y-q.y})$.
3. $t \in s_{\overline{q}}$: $\theta_{\overline{q}}(t) = \arctan(\frac{p.y-t.y}{1-t.x+p.x})$.

Finding all these possible angles takes $O(n)$ time and sorting them takes $O(n \log n)$ time. For each "critical" angle, at which a side of the square hits a point, the number of points covered by the square can be updated in constant time. Since there are $O(n^2)$ possible pairs (p, q), we obtain

Theorem 9. *Given a set $\mathcal{P}$ of n points, the location and orientation of a unit square that maximize the number of points covered can be found in $O(n^3 \log n)$ time.* □

3.2 Three Disjoint Axis-Parallel Unit Squares

Lemma 2. [3,12] *An axis-parallel unit square enclosing the maximum number of points from $\mathcal{P}$ can be found in $O(n \log n)$ time using $O(n)$ space.* □

Lemma 3. (Theorem 4 in [3]) *Two disjoint axis-parallel unit squares that jointly enclose the maximum number of points from $\mathcal{P}$, can be found in $O(n \log n)$ time using $O(n)$ space.* □

Using the above two lemmas, we can prove the following result:

Theorem 10. *Given a set $\mathcal{P}$ of n points, three disjoint axis-parallel unit squares that jointly enclose the maximum number of points from $\mathcal{P}$ can be found in $O(n^2 \log n)$ time and $O(n)$ space.*

Proof. Construct an array A_x (resp. A_y) containing the points of $\mathcal{P}$ in the ascending order of their x-coordinates (resp. y-coordinates). This takes $O(n \log n)$ time. Observe that among the three squares, one square is separated from other two by a horizontal or vertical line. See Fig. 5.

We consider one of the six cases we mentioned regarding Fig. 5, because the remaining cases are symmetric. We thus assume that the separating line is vertical. For an array index i, let $\mathcal{P}_i \subset \mathcal{P}$ (resp. $\mathcal{P}'_i \subset \mathcal{P}$) be the set of points in A_x with index not larger (resp. larger) than i. We apply Lemma 2 to find a square that encloses the maximum number of points from $\mathcal{P}_i$, and apply Lemma 3 to find a pair of disjoint squares enclosing the maximum number of points from

$\mathcal{P}'_i$. We check every i in order to maximize the total number of points covered. Thus, in $O(n^2 \log n)$ time, we can find the maximum number found so far. Now we repeat this process for the other five cases mentioned above, and pick the maximum among them.   □

3.3 Four Disjoint Axis-Parallel Unit Squares

In addition to guillotine 4-partitions, we must consider the 5-partition, illustrated in Fig. 7. Here we concentrate on this problem. We need to check $O(n^2)$ pairs of H_u and H_l. In order to avoid checking $O(n^2)$ pairs of V_l and V_r, we now perform some preprocessing. Let $A_x = \langle x_1, \ldots, x_n \rangle$ (resp. $A_y = \langle y_1, \ldots, y_n \rangle$) be the sorted array of x-coordinates (resp. y-coordinates) of the points of $\mathcal{P}$ that we constructed earlier. In preprocessing, for each pair (x_i, y_j), we want to find a unit square $s_{i,j}$ which lies in the lower left quadrant of the vertical (resp. horizontal) line $x = x_i$ (resp. $y = y_j$) and contains the largest number of points. When $i = j = 1$, there is a unique such unit square, whose upper right point is $(x_1 y_1)$. In constant time, we can find the unit square $s_{i,n}$, since we assume that at most two points have the same x-coordinate. Suppose, we have determined

$$\{s_{k,l} \mid 1 \le k \le i-1; 1 \le l \le n\} \cup \{s_{k,l} \mid 1 \le k \le n; 1 \le l \le j-1\}.$$

In order to find $\{s_{k,l} \mid 1 \le k \le i; 1 \le l \le n\}$, note that there are at most two points with x-coordinate equal to x_i. For each of them, we check the number of points enclosed in the unit squares that lie between two vertical lines $x = x_{k-1}$ and $x = x_k$ and below $y = y_l$ for each y_l $(1 \le l \le n)$.

Lemma 4. [12] *Let $s_{i,j}$ be a unit square which lies in the lower left quadrant of the vertical (resp. horizontal) line $x = x_i$ (resp. $y = y_j$) and contains the largest number of points. We can compute the set $\{s_{i,j} \mid 1 \le i \le n; 1 \le j \le n\}$ in $O(n^2)$ time.*   □

Define 2-dimensional array $S = [s_{i,j}]_{1 \le i \le n; 1 \le j \le n}$. Similarly, we compute three other 2-dimensional arrays $P = [p_{i,j}]_{1 \le i \le n; 1 \le j \le n}$, $Q = [q_{i,j}]_{1 \le i \le n; 1 \le j \le n}$, and $R = [r_{i,j}]_{1 \le i \le n; 1 \le j \le n}$ that store a unit square that contains the largest number points in the upper-left, upper-right, and lower-right quadrant of point (x_i, y_j), respectively, in $O(n^2)$ time. For fixed H_u, H_l, and V_r, we can use $P[\cdot]$ (resp. $Q[\cdot]$) to find the unit square that contains the largest number of points in the upper left (resp. upper right) of the partition defined by $y = H_u$, $y = H_l$, and $x = V_r$ in constant time.

How about the unit squares in the lower left and lower right of the partition defined by $y = H_u$, $y = H_l$, and $x = V_r$? Ignoring V_r for now, assume that $y = H_u$, $y = H_l$ are fixed. We set $V_l = x_1, x_2, \ldots, x_n$, and consult $S[\cdot]$ (resp. $R[\cdot]$) to find the optimal unit squares in the lower left (resp. lower right) of the partition defined by $y = H_u$, $y = H_l$, and $x = V_l$. We put this pair of optimal squares in a 1-dimensional array $A[\cdot]$. This takes $O(n)$ time.

When $x = V_r$ is fixed, the range of V_l is also fixed, since the rectangular area in the middle must contain no point. As we advance V_r to the right, one point at a time, we update the leftmost and rightmost positions of V_l, which monotonically move to the right. We also keep track of the optimal pair in array $A[\cdot]$ in the current range of V_l. This takes amortized constant time per move. Therefore, the total time required is $O(n^3)$.

We now consider the cases shown in Fig. 6. There are $O(n)$ choices for the thick separating line, on each side of which we place at most 3 squares. For a given separating line we can solve the optimal squares in $O(n^2 \log n)$ time by Theorem 10. We thus have

Theorem 11. *Given a set $\mathcal{P}$ of n points, four disjoint axis-parallel unit squares that jointly enclose the maximum number of points from $\mathcal{P}$ can be found in $O(n^3 \log n)$ time.* □

4 Conclusion

We have presented a number of results which are summarized in Tables 1 and 2. Some of them are improvements over existing algorithms, while others are new problems that have not been previously considered. The obvious open problems are to consider the case where the number of squares is more than four.

References

1. Agarwal, P.K., Sharir, M.: Planar geometric locations problems. Algorithmica **11**, 185–195 (1994)
2. Blum, M., Floyd, R., Pratt, V., Rivest, R., Tarjan, R.: Time bounds for selection. J. Comput. Syst. Sci. **7**(4), 448–461 (1973)
3. Cabello, S., Díaz-Báñez, J.M., Seara, C., Sellares, J.A., Urrutia, J., Ventura, I.: Covering point sets with two disjoint disks or squares. Comput. Geom. **40**(3), 195–206 (2008)
4. Chandran, S., Mount, D.: A parallel algorithm for enclosed and enclosing triangles. Int. J. Comput. Geom. Appl. **2**, 191–214 (1992)
5. Christofides, N., Hadjiconstantinou, E.: An exact algorithm for orthogonal 2-D cutting problems using guillotine cuts. Eur. J. Oper. Res. **83**(1), 21–38 (1995)
6. Das, S., Goswami, P.P., Nandy, S.C.: Smallest k-point enclosing rectangle and square of arbitrary orientation. Inform. Process. Letts. **94**, 259–266 (2005)
7. Hoffmann, C.M.: Robustness in geometric computations. J. Comput. Inf. Sci Eng. **1**(2), 143–155 (2001)
8. Jaromczyk, J.W., Kowaluk, M.: Orientation independent covering of point sets R^2 with pairs of rectangles or optimal squares. In: Proceedings of the European Workshop on Computational Geometry. LNCS, vol. 871, pp. 71–78. Springer, Heidelberg (1996)
9. Katz, M.J., Kedem, K., Segal, M.: Discrete rectilinear 2-center problems. Comput. Geom. **15**(4), 203–214 (2000)
10. Kim, S.S., Bae, S.W., Ahn, H.K.: Covering a point set by two disjoint rectangles. Int. J. Comput. Geom. Appl. **21**(3), 313–330 (2011)
11. Klee, V., Laskowski, M.: Finding the smallest triangles containing a given convex polygon. J. Algorithms **6**, 359–375 (1985)

12. Mahapatra, P.R.S., Goswami, P.P., Das, S.: Covering points by isothetic unit squares. In: Proceedings of the Canadian Conference on Computational Geometry (CCCG), pp. 169–172 (2007)
13. Mahapatra, P.R.S., Goswami, P.P., Das, S.: Maximal covering by two isothetic unit squares. In: Proceedings of the Canadian Conference on Computational Geometry (CCCG), pp. 103–106 (2008)
14. Melkman, A.A.: On-line construction of the convex hull of a simple polyline. Inform. Process. Letts. **25**(1), 11–12 (1987)
15. O'Rourke, J.: Finding minimal enclosing boxes. Int. J. Comput. Inf. Sci. **14**, 183–199 (1985)
16. O'Rourke, J., Aggarwal, A., Maddila, S., Baldwin, M.: An optimal algorithm for finding minimal enclosing triangles. J. Algorithms **7**, 258–269 (1986)
17. Preparata, F., Shamos, M.: Computational Geometry: An Introduction. Springer, Berlin (1990)
18. Segal, M.: Lower bounds for covering problems. J. Math. Model. Algorithms **1**, 17–29 (2002)
19. Sharir, M., Welzl, E.: Rectilinear and polygonal p-piercing and p-center problems. In: ACM Symposium on Computational Geometry, pp. 122–132 (1996)
20. Toussaint, G.: Solving geometric problems with the rotating calipers. In: Proceedings of the IEEE MELECON (1983)
21. Welzl, E.: Smallest enclosing disks (balls and ellipses). In: Maurer, H.A. (ed.) New Results and Trends in Computer Science. LNCS, vol. 555, pp. 359–370. Springer, Heidelberg (1991)

Algorithms for Fair Partitioning
of Convex Polygons

Bogdan Armaselu[1][(✉)] and Ovidiu Daescu[2]

[1] Department of Computer Science, The University of Texas at Dallas,
Richardson, TX, USA
bxa120530@utdallas.edu
[2] Department of Computer Science, The University of Texas at Dallas,
Richardson, TX, USA
daescu@utdallas.edu

Abstract. In this paper we study the problem of partitioning a convex
polygon P with n vertices into m polygons of equal area and perimeter
and give efficient algorithms for $m = 2$ and for the more general case
when m is a power of 2. While it was known such a partition exists, no
algorithmic results were published so far. Our algorithm for $m = 2$ is
optimal and runs in $O(n)$ time, while the algorithm for $m = 2^k$, where
$k \geq 2$ is an integer, runs in $O(n(2n)^{k-1})$ time. The algorithms have been
implemented and tested on randomly generated convex polygons.

1 Introduction

The problem we address in this paper, called *Fair Partitioning of Convex Polygons*, is a geometric optimization problem that is stated as follows: **Given a
convex polygon P with n vertices, partition P into m disjoint polygons
of equal area and perimeter**.

The problem is related to the Cake Cutting Problem [1], in which, given a
convex polygon, one is supposed to find a partitioning in which all the sides have
equal area (but perimeters can be different). It finds applications in important
domains, such as electrical engineering, convex optimizations, and special dis-
cretizations of spatial domains. In electrical engineering, for instance, consider
a given (convex) surface that is crossed by an electric current. The goal is to
split the surface into a given number of smaller surfaces such that the contour of
each of them has the same electric field, and the rate of change of the magnetic
field in time over each of them is the same. By applying Maxwell's equations,
this problem can be reduced to the Fair Partitioning problem. Likewise, for an
application in convex optimization, consider a country with convex borders that
is to be divided into administrative regions (say, counties) that should be bor-
dered with fences. If the goal is for each county to have an equal share, each one
should receive the same area and the same border perimeter.

Daescu's research has been partially supported by NSF award CNS1035460.

1.1   Definitions, Notations, and Related Work

Definition 1. *Let L be a line in the plane. The **left side** of L is the side that contains a point q with x-coordinate smaller than the x-coordinate of its projection on L, if L is not horizontal, and the side above L, if L is horizontal. The other side of L is called the right side.*

Definition 2. *The **area bisector** of a polygon P is a line L that splits P into two sides of equal areas.*

Definition 3. *Let L be an area bisector of P and $U(x, y)$ its upper intersection with P. We define the **left perimeter function** as follows: $LPF_P(x, y)$ is the perimeter of the portion of P that lies to the left side of L. We also write $LPF_P(L) = LPF_P(x, y)$.*

The cake-cutting problem, in which one is to find a bisecting line of a polygon, was studied by Steinhaus in 1948 [1]. This problem has extensions and relaxations. In one of them, there are m different measures μ_k of area on the polygon P, and each resulting sub-polygon P_k must have $\mu_k(P_k) \geq 1/m$, given that $\mu_k(P) = 1$. In other words, if we are to split a convex polygonal cake to m people, everyone should believe they have received at least $1/m$ of it, according to their own measure [7].

The problem of fair partitioning of convex polygons was introduced in 2006 by Nandakumar and Rao [2,3]. They conjectured that for any $m > 1$, a polygon can be convex fair partitioned, that is, all resulting m sides are convex polygons. They also presented a proof of exiatence of a fair partition for $m = 2$ and $m = 2^k$, for any integer $k > 1$. In 2009, Barany et al. [4] gave a proof for $m = 3$, using a more complicated approach based on equivariant topologies. In 2010, Hubard and Aronov [5] proved the conjecture for $m > p^k$, for any prime p and $k > 0$ [5].

In the following, we are going to briefly present the results in [2,3,5,6].

Theorem 1 *[3]. Given a convex polygon P, there is a convex fair partitioning of P into 2 sides of equal area and perimeter.*

To prove that, they start with an area bisector L. Such a line always exists due to continuity of the area of one side of L inside P. Then, they argue that L can be "rotated" by changing its intersections with the polygon while keeping it an area bisector, until L gives a fair partitioning of P. This eventually happens due to continuity of the left perimeter function $LPF_{P,L}$.

See Fig. 1 for an illustration.

Theorem 2 *[3]. Given a convex polygon P, there is a convex fair partitioning of P into 2^k sides of equal area and perimeter, for any $k > 1$.*

To prove that, they follow the same approach as for $m = 2$, except that, after finding an area bisector L, one has to recursively take each side of L and find a fair partitioning of it, then keep updating L and then recursively do the same operation on each side, until to resulting perimeters are equal.

Figure 2 illustrates this proof for $m = 4$.

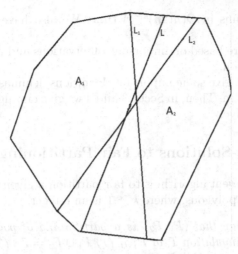

Fig. 1. Line L splits the polygon into two sides of equal areas A_1 and A_2. Lines L_1, L_2 also split the polygon into two sides of equal areas, but their perimeters may differ for L_1, L_2, as well as for L

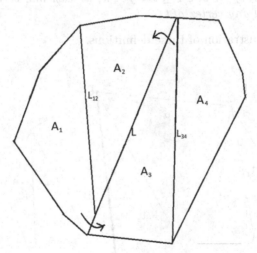

Fig. 2. After an area bisector L of the polygon P is found, the fair partitionings (A_1, A_2) of the left side, and (A_3, A_4) of the right side, are computed. If the perimeters of all sides are not equal, L is rotated accordingly, by keeping its sides of equal areas, and changing its intersections with P

1.2   Our Results

We present an $O(n)$ time, optimal algorithm to find a fair partitioning of P into $m = 2$ disjoint polygons. Note that, while a mathematical proof of existence of such partitioning was given in [3], this is the first algorithm to find such a partitioning. We have implemented this algorithm. For $m = 2^k$, $k > 1$, we give an algorithm that runs in $O(n(2n)^{k-1})$ time. We also have implemented this algorithm.

Our algorithms are based on important observations and properties outlined in the following section.

In Sect. 2, we first give some important definitions, lemmas, and observations concerning the problem. Then, in Sects. 3 and 4 we give our algorithms for $m = 2$ and $m = 2^k$, respectively.

2 Algorithmic Solutions to Fair Partitioning

In this section we present algorithms to fair partition a given convex polygon P into $m = 2^k$ disjoint polygons, where $k \geq 1$ is an integer.

Definition 4. *We say that (P_1, P_2) is a partitioning of polygon P by edge e with respect to a triangulation T of P, if (1) $P_1 \cup P_2 = P$; (2) The border edge between P_1 and P_2 is e, and (3) e does not cross any triangle of T.*

Definition 5. *We say that a triangulation $T = t_1, \ldots t_k$ is a **left-to-right triangulation** of P if for any $1 \leq i < j < k$, at least one vertex of t_i is to the left of (or equal to) any vertex of t_j.*

See Fig. 3 for an illustration of these definitions.

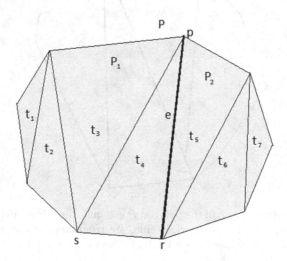

Fig. 3. A partitioning of P into (P_1, P_2) by edge e, with respect to left-to-right triangulation T

For any two points p and q we denote by $p - q$ the vector from q to p, and for any vector $\vec{v}$ and real number α, we denote by $p + \alpha\,\vec{v}$ the point where one arrives by starting from p and moving in the direction of $\vec{v}$ α times. Note that, if p, q, and r area colinear then $p - q = pq/pr \cdot (p - r)$.

Lemma 3. *Let T be a left-to-right triangulation of P, and let (P_1, P_2) be a partitioning of P by a non-horizontal line segment e with respect to T such that $area(P_1) >= area(P_2)$ and $area(P_1 \setminus t_1) < area(P_2 \cup t_1)$, where $t_1 \in T$ is the triangle just right of e that has e as its sides. Let $t_1 = \triangle prs$, where p, r, and s are vertices of P and rs is an edge of P. Also, let $D_1 = area(P_1 \setminus t_1) - area(P_2 \cup t_1)$. Then, by setting $q = D_1/(2area(t_1))(r - s) + s$, line $L = pq$ partitions P into (P_1', P_2'), where p is a vertex of P and q is on edge rs (can be an endpoint), such that $area(P_1') = area(P_2')$, i.e., L is an area bisector of P.*

Figure 4 supports the proof of this lemma.

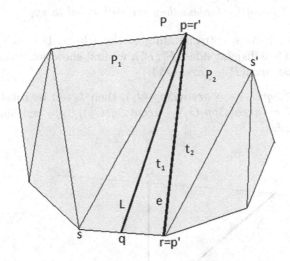

Fig. 4. Line e splits P into P_1 and P_2. Triangle t_1 is the right-most triangle of P_1, t_2 is the left-most triangle of P_2. Line L is an area bisector of P.

Proof. We need to show that $area(P_1') = area(P_2')$. Note that $P_1' = P_1 - \triangle prq$ and $P_2' = P_2 \cup \triangle prq$. Hence, $area(P_1') = area(P_1) - area(\triangle prq) = area(P_1) - area(\triangle prs) \cdot rq/rs = area(P_1) - area(t_1)(rs - qs)/rs$. Because $q = D_1/(2area(t_1))(r - s) + s$, we get $area(P_1') = area(P_1) - area(t_1) \cdot (rs - D_1/(2area(t_1)) \cdot rs)/rs = area(P_1) - (area(t_1) - D_1/2) = area(P_1) - area(t_1) + D_1/2$. Since $D_1 = area(P_1 \setminus t_1) - area(P_2 \cup t_1)$, we have $D_1 = area(P_2) - area(P_1) + 2area(t_1)$, thus $area(P_1') = area(P_1)/2 + area(P_2)/2 = area(P)/2 = area(P_2')$. ∎

We have showed how to find a area bisector. Now we show how to find a partitioning into polygons of equal area and perimeter.

Definition 6. *A line L that crosses a polygon P is said to be rotated into another line L' with respect to P, if each side of L has the same area as the corresponding side of L' (but possibly different perimeters).*

Let T be a left-to-right triangulation of P, and $L = pq$ be an area bisector of P that splits it into sides (P_1, P_2) with p a vertex on P, q on rs, and rs an edge of P. Let pp_1, pp_2 be edges of P such that lines p_1r and p_2s do not cross L. Note that, if $n = 3$, we have $p_1 = r, p_2 = s$. Also, let $O_1 = L \cap sp_1, O_2 = L \cap rp_2$.

Lemma 4. *There are 4 possible lines L_1, L_2, L_3, L_4 into which L can be rotated with respect to P, keeping endpoints on the same edges, that have a vertex of P and a point on an edge of P as endpoints. Specifically, L_1 has an endpoint at r and one on edge pp_2, L_2 has an endpoint at s and one on edge pp_1, L_3 has an endpoint at p_1 and one on edge qs, L_4 has an endpoint at p_2 and one on edge qr. In addition, L_1 and L_4 cannot co-exist (unless they are both equal to rp_2), and neither can L_2 and L_3 (unless they are both equal to sp_1).*

Proof. There are 4 cases, corresponding to the 4 lines. For each of these lines that splits P into two disjoint sides (P_1', P_2'), we first show how to compute them, then we show that $area(P_1') = area(P_1)$.

Case 1. If $area(\Delta qrO_2) <= area(\Delta pp_2O_2)$, then L can be rotated into a line $L_1 = rp'$, with $p' = (area(\Delta pp_2O_2) - area(\Delta qrO_2))/(pp_2 \cdot rp_2 \sin \angle rp_2p) \cdot (p - p_2) + p_2$. See Fig. 5.

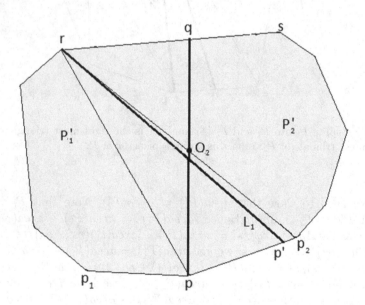

Fig. 5. Line L can be rotated into a line $L_1 = rp'$, with $p' \in pp_2$.

Let $O = pq \cap rp'$. Then $P_1' = P_1 \cup \Delta pp'O - \Delta qrO$, therefore $area(P_1')$ $= area(P_1) + area(\Delta pp'O) - area(\Delta qrO) = area(P_1) + (area(\Delta pp_2O_2) - area(p_2p'OO_2)) - (area(\Delta qrO_2) + area(\Delta rOO_2)) = area(P_1) + (area(\Delta pp_2O_2) - area(\Delta qrO_2)) - (area(p_2p'OO_2) + area(\Delta rOO_2)) = area(P_1) + p_2p'/pp_2 \cdot (pp_2 \cdot$

$rp_2 \sin \angle rp_2 p) - area(\Delta rp_2 p') = area(P_1) + p_2 p' \cdot rp_2 \sin \angle rp_2 p' - area(rp_2 p') = area(P_1)$.

Case 2. If $area(\Delta qsO_1) <= area(\Delta pp_1 O_1)$, then L can be rotated into a line $L_2 = sp''$, with $p'' = (area(\Delta pp_1 O_1) - area(\Delta qsO_1))/(pp_1 \cdot sp_1 \sin \angle sp_1 p) \cdot (p - p_1) + p_1$. See Fig. 6.

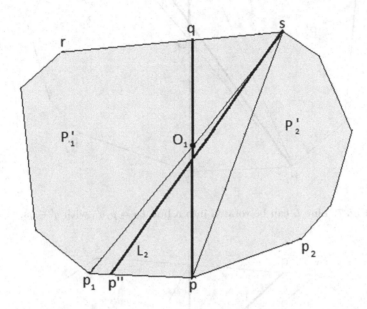

Fig. 6. Line L can be rotated into a line $L_2 = sp''$, with $p'' \in pp_1$.

Let $O = pq \cap sp''$. Then $P_1' = P_1 \cup \Delta qsO - \Delta pp''O$, therefore $area(P_1') = area(P_1) + area(\Delta qsO) - area(\Delta pp''O) = area(P_1) + (area(\Delta qsO_1) + area(\Delta sOO_1)) - (area(\Delta pp_1 O_1) - area(p_1 p''OO_1)) = area(P_1) + (area(\Delta qsO_1) - area(\Delta pp_1 O_1)) + area(\Delta sOO_1) + area(p_1 p''OO_1) = area(P_1) - p_1 p''/pp_1 \cdot (pp_1 \cdot sp_1 \sin \angle sp_1 p) + area(\Delta sp_1 p'') = area(P_1) - p_1 p'' \cdot sp_1 \sin \angle sp_1 p'' + area(\Delta sp_1 p'') = area(P_1)$.

Case 3. If $area(\Delta qsO_1) >= area(\Delta pp_1 O_1)$, then L can be rotated into a line $L_3 = p_1 q'$, with $q' = (area(\Delta qsO_1) - area(\Delta pp_1 O_1))/(sq \cdot sp_1 \sin \angle qsp_1) \cdot (q - s) + s$. See Fig. 7.

Let $O = pq \cap p_1 q'$. Then $P_1' = P_1 \cup \Delta qq'O - \Delta pp_1 O$, therefore $area(P_1') = area(P_1) + area(\Delta qq'O) - area(\Delta pp_1 O) = area(P_1) + (area(\Delta qsO_1) - area(sq'OO_1)) - (area(\Delta pp_1 O_1) + area(\Delta p_1 OO_1)) = area(P_1) + (area(\Delta qsO_1) - area(\Delta pp_1 O_1)) - area(sq'OO_1) - area(\Delta p_1 OO_1) = area(P_1) + sq'/sq \cdot (sqsp_1 \sin \angle qsp_1) - area(\Delta sq'p_1) = area(P_1) + sq' \cdot sp_1 \sin \angle q'sp_1 - area(\Delta sq'p_1) = area(P_1)$.

Case 4. If $area(\Delta qrO_2) >= area(\Delta pp_2 O_2)$, then L can be rotated into a line $L_4 = p_2 q''$, with $q'' = (area(\Delta qrO_2) - area(\Delta pp_2 O_2))/(rq \cdot rp_2 \sin \angle qrp_2) \cdot (q - r) + r$. See Fig. 8.

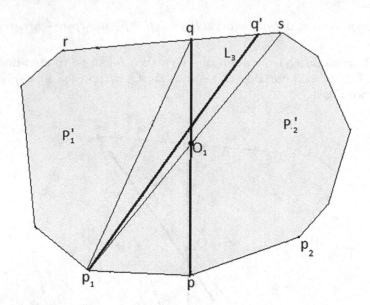

Fig. 7. Line L can be rotated into a line $L_3 = p_1q'$, with $q' \in qs$.

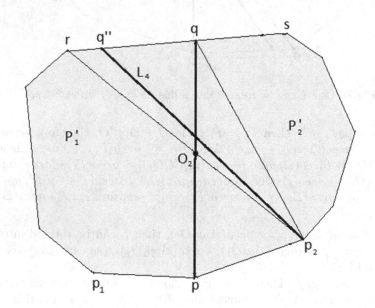

Fig. 8. Line L can be rotated into a line $L_4 = p_2q''$, with $q'' \in qr$.

Let $O = pq \cap p_2q''$. Then $P_1' = P_1 \cup \Delta pp_2O - \Delta qq''O$, therefore $area(P_1') = area(P_1)+area(\Delta pp_2O)-area(\Delta qq''O)=area(P_1')=area(P_1)+(area(\Delta pp_2O_2)+ area(\Delta pOO_2)) - (area(\Delta qrO_2) - area(rq''OO_2)) = area(P_1) + (area(\Delta pp_2O_2) - area(\Delta qrO_2)) + (area(\Delta pOO_2 + area(rq''OO_2)) = area(P_1) - rq''/rq \cdot (rqrp_2$

$\sin \angle qrp_2) + area(\Delta rq''p_2) = area(P_1) + rq'' \cdot rp_2 \sin \angle q''rp_2 + area(\Delta rq''p_2) = area(P_1)$. ∎

Note that either L_2 or L_3 exist (unless they are both equal to sp_1) and either L_1 or L_4 exist (unless they are both equal to rp_2). In the following, we denote by L'_1 the line among L_1 and L_4 that exists, and by L'_2 the line among L_2 and L_3 that exists.

Let p, r, s, p_1, p_2 be points on the edges of P (can be vertices of P and can have $r = p_1, s = p_2$) and $p' \in pp_2$, $p'' \in pp_1$, $q, q', q'' \in rs$. Let L, L'_1, L'_2 be area bisectors of P into (P_1, P_2), (P_{11}, P_{12}), (P_{21}, P_{22}) respectively, with $L = pq, L'_1 = rp', L'_2 = sp''$. Let $Dp = perimeter(P_1) - perimeter(P_2)$, $Dp_1 = perimeter(P_{11}) - perimeter(P_{12})$, $Dp_2 = perimeter(P_{21}) - perimeter(P_{22})$.

Lemma 5. *There exists a line L' in the cone formed by L'_1 and L'_2, that partitions P into (P'_1, P'_2) such that $area(P''_1) = area(P''_2)$ and $|D'p|$ is minimized, where $D'p = perimeter(P''_1) - perimeter(P''_2)$, and $|x|$ denotes the modulus of the real number x.*

Proof. There are 4 cases. In each of them, we show how to find the two endpoints of L', one on rs and one on either pp_1 or pp_2. Then we show that either $D'p = 0$ or $|D'p| < \min |Dp_1|, |Dp_2|$ holds.

Case 1. If $Dp = 0$, then $a = p, b = q$, and we have $P''_1 = P_1, P''_2 = P_2$. Therefore, $D'p = perimeter(P''_1) - perimeter(P''_2) = perimeter(P_1) - perimeter(P_2) = 0$.

Case 2. If $DpDp_1 < 0$, then $a = Dp/(Dp-Dp_1)\cdot(p'-p)+p$, $b = Dp/(Dp-Dp_1)\cdot (r-q)+q$. We have $D'p = perimeter(P''_1) - perimeter(P''_2) = (perimeter(P_1) + pa - qb + ab - pq) - (perimeter(P_2) - pa + qb + ab - pq) = (perimeter(P_1) - perimeter(P_2)) + 2(pa - qb) = Dp + 2(pa - qb)Dp + 2Dp/(Dp-Dp_1)\cdot(pp'-qr) = Dp + 2Dp/(Dp - (Dp + 2(pp' - qr))) \cdot (pp' - qr) = 0$. See Fig. 9.

Case 3. If $DpDp_2 < 0$, then $a = Dp/(Dp-Dp_2)\cdot(p''-p)+p$, $b = Dp/(Dp-Dp_2)\cdot (s-q)+q$. We have $D'p = (perimeter(P_1)-pa+qb+ab-pq)-(perimeter(P_2)+pa - qb + ab - pq) = Dp - 2(pa - qb) = Dp + 2Dp/(Dp-Dp_2) \cdot (qs - pp'') = Dp + 2Dp/(Dp - (Dp + 2(qs - pp''))) \cdot (qs - pp'') = 0$.

Case 4. Otherwise, we set $L' = L'_i$ such that $D'p = Dp_i = min(Dp_j, j = 1, 2)$. It is obvious that $D'p = Dp_i = min(Dp_j, j = 1 \ldots 4)$. ∎

Note. Let $f(L) = perimeter(left(L)) - perimeter(right(L))$ (can be either Dp_1 or Dp_2). In some cases, when we do not find a line L' for which $f(L') = 0$, we take the one that minimizes $|f(L')|$. At the next step, we may find that the line L'' that minimizes $|f(L'')|$ may be already taken (may be one of the initial lines L_1, L_2). To avoid this, we maintain a stack Λ of lines already considered. At each step, if we don't find a fair partitioning, we push L onto Λ, and then update L to be the line $\in L_1, L_2$ that exists, has the smallest $|f(L)|$ and is not among the top two lines in Λ. Then, we repeat the process described in the proof of Lemma 5 using the new area bisector L.

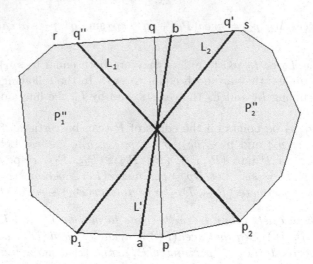

Fig. 9. $L' = ab$ is within the cone generated by L'_1, L'_2. We have $DpDp_1 < 0$, so a is on pp_1.

3   Algorithm for $m = 2$

We first give the pseudocode and then prove the correctness of our algorithm.

```
Procedure fairParitioning2(P):
input: convex polygon P
output: (L, P₁, P₂), where L is a fair partitioning line and P₁,
P₂ are the resulting polygons.
Let (L, P₁, P₂, Per₁, Per₂) = areaBisector(P)
Let Λ = {L}
While Per₁ ≠ Per₂ do:
      (L″, P₁″, P₂″) = update(L, P, Λ)
      Set P₁ = P₁″, P₂ = P₂″, L = L″.
      Let Per₁ = perimeter(P₁), Per₂ = perimeter(P₂)
End While;
Return (L, P₁, P₂).
End.
```

```
Procedure areaBisector(P):
input: convex polygon P
output: (L, P₁, P₂, Per₁, Per₂), where L is a area bisector of P, P₁,
P₂ are its sides, and Per₁, Per₂ are their corresponding perimeters
Let T = triangulate − left − to − right(P)
Let P₁ = ∅, P₂ = P, Per₁ = 0, Per₂ = perimeter(P)
For all triangles t ∈ T do
      if area(P₁ ∪ t) > area(P₂ − t) then
```

Let $t = \Delta prs$, where rs is an edge of P, $D_1 = area(P_1 \cup t_1) - area(P_2 \setminus t_1)$, $q = D_1/(2 area(t_1))(r-s)+s$

Let $L = pq$

$P_1 = P_1 \cup t$, $P_2 = P_2 \setminus t$

if t is the 1st triangle of T then $Per_1 = Per_1 + rs + rp$, $Per_2 = Per_2 - rs - rp$

else $Per_1 = Per_1 + rs$, $P_2 = P_2 - rs$

End For;

Return $(L, P_1, P_2, Per_1, Per_2)$

End.

Procedure $update(L, P, \Lambda)$:

input: convex polygon P, its area bisector L and set of previously considered area bisectors Λ

output: (L', P_1', P_2'), where L' is a area bisector and P_1, P_2 are the resulting polygons, such that $|perimeter(P_1') - perimeter(P_2')| < |perimeter(P_1) - perimeter(P_2)|$, where P_1, P_2 are the sides of L

Let $L = pq$, rs an edge of P such that q is on rs and pp_1, pp_2 be edges of P such that p_1r, p_2s do not cross L. Also, let $O_1 = L \cap sp_1$, $O_2 = L \cap rp_2$.

If $area(\Delta qrO_2) \leq area(\Delta pp_2O_2)$, then set $L_1 = rp'$, with $p'' = (area(\Delta pp_2O_2) - area(\Delta qrO_2))/(pp_2 \cdot rp_2 \sin \angle rp_2p) \cdot (p - p_2) + p_2$;

Else If $area(\Delta qsO_1) \leq area(\Delta pp_1O_1)$, then set $L_2 = sp'$, with $p' = (area(\Delta pp_1O_1) - area(\Delta qsO_1))/(pp_1 \cdot sp_1 \sin \angle sp_1p) \cdot (p - p_1) + p_1$.

Else If $area(\Delta qsO_1) \geq area(\Delta pp_1O_1)$, then set $L_3 = p_1q'$, with $q' = (area(\Delta qsO_1) - area(\Delta pp_1O_1))/(sq \cdot sp_1 \sin \angle rsp_1) \cdot (q - s) + s$;

Else If $area(\Delta qrO_2) \geq area(\Delta pp_2O_2)$, then set $L_4 = p_2q'$, with $q'' = (area(\Delta qrO_2) - area(\Delta pp_2O_2)/(rq \cdot rp_2 \sin \angle srp_2) \cdot (q - r) + r$;

Let $Dp = perimeter(P_1') - perimeter(P_2')$

For $i = 1...2$ do $Dp_i = perimeter(P_{i1}') - perimeter(P_{i2}')$.

Let $L_1' \in \{L_1, L_2\}$ that exists, $L_2' \in \{L_3, L_4\}$ that exists.

Let $L' = ab$, where

If $Dp = 0$, then $a = p, b = q$, and we are done.

Else If $DpDp_1 < 0$, then $a = Dp/(Dp - Dp_1) \cdot (p' - p) + p$, $b = Dp/(Dp - Dp_1) \cdot (r - q) + q$

Else If $DpDp_2 < 0$, then $a = Dp/(Dp - Dp_2) \cdot (p'' - p) + p$, $b = Dp/(Dp - Dp_2) \cdot (s - q) + q$.

Else, $L' = L_i$ such that: $L_i \notin \Lambda, Dp_i = min(Dp_j, j = 1, 2)$, where P_1'' and P_2'' are partitions of P by line L'.

$\Lambda = \Lambda \cup \{L'\}$

Return (L, P_1'', P_2'').

End.

Theorem 6. *Procedure $fairPartitioning2$ correctly computes a fair partitioning of P into two parts in $O(n)$ time.*

Proof. Since P is convex, it can be triangulated in $O(n)$ time. After setting $L = pq$, by Lemma 3, L is an area bisector. Hence, procedure $areaBisector(P)$ computes an area bisector L of P in $O(n)$ time. By Lemma 4, there are 4 ways in which L can be rotated into L'. By selecting the one that satisfies the given 3 criteria, then (by Lemma 5), we find a line L' that splits P into (P_1'', P_2''), to minimize $D'p = perimeter(P_1'') - perimeter(P_2'')$. If $D'p > 0$, by our observations, we continue by rotating L', otherwise we are done. Also, when computing areas (perimeters) of sides of lines, we need only update the previously computed areas (perimeters), by adding or subtracting the area (length) of the triangle (segment) that is added (removed). Hence, procedure $update(L, P, \Lambda)$ correctly updates line L such that the difference of the perimeters of its sides is reduced. The procedure takes constant time and it may be called $O(n)$ times, since the while-loop proceeds in steps that are based on the vertices of P. The total running time is thus $O(n)$. ∎

4 Algorithm for $m = 2^k$

In order to hold all the needed information about the fair partitioning of P, when computing a fair partitioning line L, we associate L with a hierarchical data structure, and call it a Partition Line. The data structure contains, in addition to the line, the sides of the line as polygons, along with their fair partitioning line.

```
Procedure fairParitioning2^k(P, k):
input: convex polygon P
output: (L, P_1, P_2), where L is the a fair Partition Line of P
and P_1, P_2 are its sides
   Let (L, P_1, P_2, Per_1, Per_2) = areaBisector(P)
   Let Λ = {}
   Let (L_1, P_11, P_12) = fairParitioning2^k(P_1, k - 1)
   Let (L_2, P_21, P_22) = fairParitioning2^k(P_2, k - 1)
   While Per_1 ≠ Per_2) do:
        (L', P_1', P_2') = update(L, P, Λ)
        Set P_1 = P_1', P_2 = P_2', L = L'.
        Update (L_1, P_11, P_12) = fairParitioning2^k(A, k - 1)
        Update (L_2, P_21, P_22) = fairParitioning2^k(B, k - 1)
        Set Per_1 = perimeter(P_11), Per_2 = perimeter(P_21)
   End While;
   Return (L, P_1, P_2)
End.
```

Theorem 7. *Procedure $fairPartitioning2^k$ correctly computes a fair partitioning of P into 2^k polygons. Its running time is $O(n(2n)^{k-1})$.*

Proof. We prove this by induction. We already proved before that the algorithm for $m = 2$ computes a fair partitioning of P. Assume for $m = 2^{k-1}$ that

$(L_1, P_{11}, P_{12}), (L_2, P_{21}, P_{22})$ are fair partitionings. Note that all areas are equal. The perimeters may not be equal. Let $D'_e = perimeter(P_1) - perimeter(P_2)$ and assume $D'_e \neq 0$. Also let $L = qv$. Similarly to the proof of Theorem 6, there exists $L' = ab$ that crosses P such that aq and bv are edges on P, and either $|D'_e + aq - bv| < |D'_e|$ or $|D'_e - aq + bv| < |D'_e|$. Calling the algorithm recursively on each side A', B', we get new fair partitionings for each side, $(L'_1, P'_{11}, P'_{12}), (L'_2, P'_{21}, P'_{22})$. Repeating this iteration, we will eventually get a line L' with $D'_e = 0$, i.e. $perimeter(P'_1) = perimeter(P'_2)$. Since $perimeter(P'_{11}) = perimeter(P'_{12})$ and $perimeter(P'_{21}) = perimeter(P'_{22})$, it follows that $perimeter(P'_{11}) = perimeter(P'_{12}) = perimeter(P'_{21}) = perimeter(P'_{22})$. That is, (L', P'_1, P'_2) is a fair partitioning of P.

As for the running time, note that we recursively call the algorithm on each side $O(n)$ times to update them. We also use additional $O(n)$ time for calling $areaBisector(P)$ once and $update(L, P, \Lambda)$. The running time is $T(n,m) = O(n + 2nT(n, m/2)) = n + 2nT(n, m/2) = n + 2n(n + 2nT(n, m/4)) = n + 2n^2 + 4n^2T(n, m/4) = ... = n + 2n^2 + 4n^3 + ... + 2^{\log m - 1}n^{\log m - 1}T(n, 2) = n + 2n^2 + ... + 2^{\log m - 1}n^{\log m} = n(1 + 2n + (2n)^2 + ... + (2n)^{\log m - 1})$, which solves to $T(n,m) = n((2n)^{\log m} - 1)/(2n - 1) = O(n/(2n - 1)2^{\log m}n^{logm}) = O(m/2n^{\log m})$ or $T(n, k) = O(n(2n)^{k-1})$.   ∎

5 Conclusions

We have designed optimal algorithms to fair partition a convex polygon into m parts, for $m = 2^k, k \geq 1$. We have also implemented our algorithms in JAVA, using the toolkits java.swing for the user interface, and java.awt geometry classes and methods (for points, lines, and other geometric objects). Due to the space limits we leave discussion of implementation and experimental results for the full version of the paper.

References

1. Steinhaus, H.: The problem of fair division. Econometrica **16**, 101–104 (1948)
2. Nandakumar, R.: Cutting Shapes (2006). http://nandacumar.blogspot.com/2006/09/cutting-shapes.html
3. Nandakumar, R., Rao, N.R.: Fair Partitioning of Polygons: An Introduction (2010)
4. Barany, I., Blagojevic, P., Szucs, A.: Equipartitioning by a convex 3-fan (2009)
5. Hubard, A., Aronov, B.: Convex Equipartitions of Volume and Surface Area (2011)
6. Karasev, R.N.: Equipartition of Several Measures (2010)
7. Steinhaus, H.: Mathematical Snapshots, 3rd edn. Dover, New York (1999)

A Quasi-polynomial Time Approximation Scheme for Euclidean CVRPTW

Liang Song[1,2], Hejiao Huang[1,2]([✉]), and Hongwei Du[1,2]

[1] Shenzhen Graduate School, Harbin Institute of Technology, Shenzhen, China
[2] Shenzhen Key Laboratory of Internet Information Collaboration, Shenzhen, China
{songliang,hjhuang,hwdu}@hitsz.edu.cn
http://carc.hitsz.edu.cn/MICCRC/index.html

Abstract. The capacitated vehicle routing problem with time windows (CVRPTW) is a variant of the classical vehicle routing problem. In a category of CVRPTW, each customer has same unit-demand and must be served within a time window from a finite set of consecutive time windows. This paper gives a quasi-polynomial time approximation scheme (Q-PTAS) for this category of CVRPTW under the Euclidean setting. With a reasonable vehicle speed requirement, our algorithm could generate a set of routes of the length of $(1 + O(\epsilon))OPT$ on expectation.

Keywords: Modern logistics · CVRPTW · Approximation algorithm

1 Introduction

Because of the economic significance in modern logistics, the classical capacitated vehicle routing problem (CVRP) [1] again attracts the researchers in the area of math, computer science and so on. The capacitated vehicle routing problem with time windows (CVRPTW), which is of great significance in real logistics environment, has been a classical variant of CVRP and could be specified by more constraints. For example, some e-commercial companies give their customers some options to designate which time windows they would like to receive goods, such as morning, afternoon or evening. Whether goods are delivered within customers' designated time windows plays an important role in customers' satisfaction. So, the goal of e-commercial or logistics companies is to find the routing plans which serve their customers on time and minimize the total routing length.

CVRP is an NP-hard problem [2], so is the CVRPTW [3–5] and its variants. Therefore, most papers solving on CVRPTW adopt the algorithms of following categories: metaheuristic algorithms, math program algorithms and combinatorial algorithms. Metaheuristic algorithms [6] could solve the large scale

---

Hejiao Huang: This work was financially supported by National Natural Science Foundation of China with Grants No. 11071271, No. 11371004, No. 61100191 and No. 61370216, and Shenzhen Strategic Emerging Industries Program with Grants No. ZDSY20120613125016389, No. JCYJ20120613151201451 and No. JCYJ20130329153215152.

Z. Zhang et al. (Eds.): COCOA 2014, LNCS 8881, pp. 66–73, 2014.
DOI: 10.1007/978-3-319-12691-3_6

problems and adapts to many environments. For example, Toklu et al. [7] study CVRPTW under travel time uncertainty by the ant colony system. However, the approximation ratio of metaheuristic algorithms cannot be proved, and hence the computational results vary a lot according to different logistics instances. The algorithms based on the math program, especially the linear program [8,9], could solve CVRP in the moderate scale. Though the approximation ratio is still not guaranteed, this kind of algorithms could solve the problems within tolerance. In Sousaa et al. [10], the mixed linear programm is adopted to solve real logistics instances of CVRPTW with other objectives together.

Theoretically, combinatorial algorithms generate the best solutions for NP-hard problem. Arora [11,12] gave a polynomial time approximation scheme (PTAS) for TSP, which is very relevant problem of CVRPTW. Based on Arora's work, Das [13,14] gave a quasi-polynomial time approximation scheme (Q-PTAS) for the CVRP with Unit-demand constraint and without vehicle number constraint. However, their work does not extend to CVRPTW.

To solve the weakness of the above work, this paper formulates the CVRPTW in a graph model and give its quasi-polynomial time approximation scheme. We focus on the CVRPTW that the vehicle number constraint is ignored and the customers' demands are restrict to Unit. With a reasonable vehicle speed requirement, our algorithm could generate a set of routes of $(1 + O(\epsilon))OPT$ on the probability of expectation. The remaining part of this paper is organized as below. Section 2 describes the CVRPTW considered in this paper. In Sects. 3 and 4, the algorithm and its proof are given in detail, respectively. Conclusions and future work are listed in Sect. 5.

2   Problem Description

We consider the CVRPTW in the following complete graph under the Euclidean setting. There are n customers and a depot, and each customer has a unit demand. Each vehicle has the same capacity of Q and a speed requirement given in expression (1) as below, in which $Length_{\max}$ is maximum length that a vehicle could travel at it's maximum speed, and $Length_Q$ is maximum length for serving Q customers out of n.

$$Length_{\max} \geq Length_Q. \tag{1}$$

There is a constant number w of consecutive time windows for customers to choose. The objective is to findF a set of routes of the minimum total length, which start and end at depot and each customer is served with its time window. An example of $w = 2$ is illustrated in Fig. 1.

Let OPT be the total length of the optimum solution of CVRPTW described above. Let OPT^L be the optimum solution of the CVRPTW after randomized dissection. Let OPT^{DP} be the total length of the optimum solution from the Dynamic Program described in Sect. 3. Let OPT' be the total length of the optimum solution in Das [13]. Then, we give the following two definitions.

Definition 1 (Structured Solution). Let D be a randomized dissection. An feasible solution S is called a structured solution if it satisfies the following conditions:

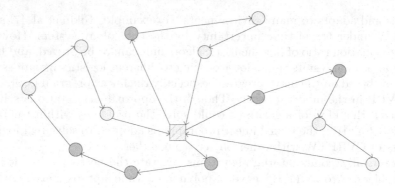

Fig. 1. An example of CVRPTW with 2 time windows, in which gray and white customers must be served in the morning and afternoon, respectively.

- Each route crosses any box of D at portals and within $r = O(1/\epsilon)$ times.
- Each dissection box contains only rounded route segments if the number of its route segments is greater than $\gamma = \lceil \log^4 n/\epsilon^4 \rceil$.
- The round route segments is remembered by a largest threshold t_i from the set of $\{t_i = i | i \in [1, 1/\epsilon]\} \cup \{t_i = t_{i-1}(1 + \epsilon/\log n) | i \in [1/\epsilon, \tau]\} \cup \{t_\tau = Q\}$, where $\tau = \lceil log_{1+\epsilon/\log n}(Q\epsilon) \rceil + 1/\epsilon$, rather than its exact number x of served customers.

Definition 2 (Extended Objective Function). Let $S = \{s_i\}$ be a set of routes. For every level ℓ, let $c(s_i, \ell)$ be the times that s_i crosses level ℓ boxes, and let d_ℓ be the length of a level ℓ box. The Extended Objective Function is

$$F(S) = \sum_i length(s_i) + \frac{\varepsilon}{\log n} \sum_{level\ell} \sum_i c(s_i, \ell) \cdot d_\ell \qquad (2)$$

The first term of the Extended Objective Function is the total length of all the routes in the solution. The second term is how many times that the routes in the solution cross all the dissection boxes, which is multiplied by a coefficient. The Dynamic Program in Sect. 3 will use the Extended Objective Function as the objective function to generate OPT^{DP}.

3   The Quasi-PTAS for CVRPTW

The Algorithm 1 consists of 4 phases, i.e., preprocessing, Dynamic Program, route construction and route partitioning. The preprocessing phase perturbs an instance into a grid, then the subsequent phases run only on the grid. The Dynamic Program runs in bottom-up order to compute the minimum length of total routes. The preprocessing is described in Sect. 3.1, and the Dynamic Program is described in Sect. 3.2. The route construction and route partitioning phases are similar to those in Das [13]. The route construction phase just constructs the routes corresponding to the OPT^{DP} generated in the Dynamic

Program, and randomly dropped customers on a route when the number of its customers is greater than Q. In the route partitioning phase, all the dropped customers are served. A 3-approximation algorithm is used to construct a single TSP route for all the dropped customers firstly, which is proposed in Haimovich and Kan [15]. Then, a new routes is generated each time the customer number is up to Q. The final output consists of the routes generated from the route construction and partitioning phases.

Algorithm 1. Main Algorithm

   Input: Graph, Time Windows, Vehicle Capacity
   Output: a feasible solution S with minimum total length
 1 Preprocessing: generate randomized dissection grid D
 2 Dynamic Program: find the minimum Structured Solution of OPT^L
 3 route construction: generate the routes corresponding to the Dynamic Program
 4 route Partitioning: generate the routes for the dropped customers.

3.1 Preprocessing

Arora [11,12] and Das [13] adopt the preprocessing to solve TSP and CVRP, respectively. We also adopt the prepocessing as the first phase of algorithm, and this is main reason why the algorithm generate its solution of $(1 + O(\epsilon))OPT$ on the probability of expectation.

Perturbation. Firstly, we define the bounding box as the smallest box whose side length L is a power of 2 and contains all the n customers and the depot. Let d denote the maximum distance between any customers or the depot. Then Place a grid of granularity $d\epsilon/n$ inside the bounding box and move customers and the depot to the center of the grid box they lie in. We only solve the problem on the perturbed instance, because a solution for the perturbed instance can be extended into a solution for the original instance by taking detours from the grid centers to the original locations of customers and the depot. The total cost of detours is at most $n \cdot \sqrt{2}d\epsilon/n$. And because it's obvious that $2d \leq OPT$, thus the total cost of detours is at most ϵOPT. Finally, scale all the distances by $n/(\epsilon d)$ so that all coordinates become integral and the minimum distance between any two grid centers is 1. After scaling the maximum distance between customers and depot and hence the side length of the bounding box is $L = O(n)$. An example of the result after perturbation is give in Fig. 2.

Randomized Dissection. Obtain a dissection by recursively partitioning the bounding box into 4 smaller boxes of equal size, using one horizontal and one vertical dissection line, until the smallest boxes are of size 1×1. The dissection can be viewed as a quad-tree with the bounding box as its root and the smallest boxes as the leaves. The bounding box has level 0, the 4 boxes created by the first dissection have level 1, the 16 boxes of the second dissection have level 2, and so on. Since $L = O(n)$, the level of the smallest boxes will be $\ell_{max} = O(logL)$. The horizontal and vertical dissection lines are also assigned levels. The boundary of

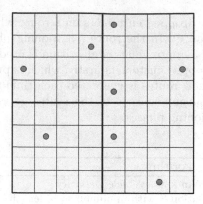

Fig. 2. The randomized dissection D.

the bounding box has level 0, the horizontal and 2^ℓ vertical lines that form level i boxes by partitioning the level boxes are each assigned level ℓ. A randomized dissection of the bounding box is obtained by randomly choosing integers a, b, and shifting the x coordinates of all horizontal dissection lines by a and all vertical dissection lines by b and reducing modulo L.

Portals. Place portals along the dissection lines as follows. Let $m = O(log L/\epsilon)$. Place $2m$ portals equidistant apart on each level ℓ dissection line for all $\ell \leq \ell_{max}$. It's obvious that there will be at most a $4\,m$ portals along the boundary of any dissection box b. As m and L are powers of 2, portals at lower level boxes will also be portals in higher level boxes. We will compute a route that always enters and exits boxes at portals.

3.2 Dynamic Program

For each dissection box b, we construct a rounded configuration $r_{(p,q,i,d,tw)}$, which means the number of the route segments start at p, end at q, cover t_i customers and uses the time window tw. Similarly, we construct an unrounded configuration $u_{(p,q,j,d,tw)}$, which means the exact number of customers covered by the route segments, each of which start at p, end at q, ordered at j and uses the time window tw. Each configuration corresponds to a minimum value $L^b[C]$ by computing the Extended Objective Function defined in Eq. (2). The Dynamic Program runs bottom-up from leaves to the root of the randomized dissection D, and generates the minimum $L^b[C]$ at root which is equal to OPT^{DP}.

Inductively, let b be a box at level ℓ and let b_1, b_2, b_3, b_4 be the children of b at level $\ell + 1$. As every route segment in children is structured, a route segment in b crosses the boundaries of boxes b_1, b_2, b_3, b_4 inside b, at most $4r$ times. Thus each segment in b is the concatenation of at most $4r + 1$ pieces, which could be described by a concatenation profile $\Phi = (p_1, q_1, k_1, tw_1), (p_2, q_2, k_2, tw_2), ..., (p_v, q_v, k_v, tw_v)$ where k is either a threshold index or a sequence number. Especially, Φ is feasible only if each pair of pieces $\Phi(i, j)$ satisfies that tw_i is rowed ahead of tw_j in the time window set, and an example is illustrated in Fig. 3.

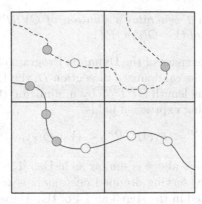

Fig. 3. Two routes are shown in this figure. The one in solid line is feasible because it serves all the gray customers (time window = morning) and then serves the white customers (time window = afternoon). The one in dashed line is not feasible because its white customers are between gray customers.

If b has more than γ route segments, then they must be rounded. Let φ denote the number of different concatenation and n_i the number of route segments of Φ_i, then, the Interface $\Gamma = (n_i)_{i \leq \varphi}$ enumerates all the possiblities that could be combined by the 4 children of box b. Finally, by pure enumeration of Γ, for each $Lb[C]$ of b could be computed.

4 Running Time and Approximation Ratio

Theorem 1. *The Algorithm 1 runs within* $n^{\log^{O(1/\epsilon)} n}$.

Proof. The running time of Algorithm 1 is dominated by the Dynamic Program, which is further dominated by its size of state space. For any configuration of box b, there are $O(W(\tau + \gamma)\log L)$ entries. Here, $W = 2^w$ is constant and w is the number of time windows. Each entry is an integer less than n, thus the total number of configurations for box b is $n^{O(W(\tau+\gamma)\log L)} = n^{O(\log^6 L)}$. As there are $O(n^2)$ dissection boxes, the Dynamic Program has size of $n^{O(\log^6 L)}$ overall.

For the number of concatenation profiles, each Φ has a list of $O(r)$ tuples. Each tuple has $O(\log L)$ choices of portals p, q, $(\tau + \gamma)$ choices of m and W choices of route types. So there are $O(W(\tau + \gamma)\log L) = O(\log^6 L)$ possibilities for each tuple. Thus there are $= (\log^6 L)^{O(r)} = (\log^6 L)^{O(1/\epsilon)}$ possible values of Φ.

For the number of interfaces of a box b. As each n_i is an integer less than n and $L = O(n)$, the total number is $n^{\log^{O(1/\epsilon)} n}$, thus we have a quasi-polynomial number of possibilities for the interfaces of box b.

Checking for consistency of an interface takes time polynomial time. An interface could be transformed in polynomial time to 4 configuration (C_1, C_2, C_3, C_4) of the 4 children of b, and checking the feasibility of (C_1, C_2, C_3, C_4) takes $O(1)$ time. Therefore, the computation of each $L_b[C]$ of box b still takes time of $n^{\log^{O(1/\epsilon)} n}$.

Theorem 2. *Algorithm 1 generates a solution of CVRPTW, which is defined in Sect. 2, of the length of* $(1 + O(\epsilon))OPT$.

Proof. Because the state space of the Dynamic Program enumerates all the feasible route segments on the randomized dissection D, the Dynamic Program will output the solution of the length OPT^{DP} by minimizing the extended objective function, and the following expression holds.

$$OPT^{DP} \leq F(OPT^{DP}) \leq (1 + O(\epsilon))OPT^{L} \tag{3}$$

The proof of the expression above is similar as in Das [13], so, we only prove the total length of the routes serving dropped customers is within $O(\epsilon)OPT$. Note that each routes generated in the step 3 and 4 of the Algorithm 1 serves at most Q customers. So, by our assumption of vehicle speed formulated in expression (1), all the routes in the solution are feasible to the time windows.

By [15], the total length of the routes serving dropped customers $Length_{dropped}$ $= Rad(DR) + 2TSP(DR \cup depot)$, where $Rad(DR) = \sum_{i \in DR}(2/Q)distance_{(i,depot)}$ and $DR = droppedcustomers$. Das [13] has proved $Rad(DR) + 2TSP(DR \cup depot) \leq O(\epsilon)OPT'$, and $OPT' \leq OPT$ always holds, therefore we have the following expression.

$$Length_{dropped} \leq O(\epsilon)OPT \tag{4}$$

Finally, by expression (3) and (4), the total length of the routes generated from Algorithm 1 is within $(1 + O(\epsilon))OPT$.

5   Conclusion and Future Work

We give a quasi-polynomial time approximation scheme (Q-PTAS) for the CVRPTW under the Euclidean setting. With a reasonable vehicle speed requirement, our algorithm could generate a set of routes of $(1 + O(\epsilon))OPT$.

The running time of quasi-polynomial time algorithm is not practical for running, so the running time may be reduced by some other methods. Still, the classical CVRPTW consider vehicle number constraint, which is ignored in our work. Finally, the algorithm has the vehicle speed requirement, and it will have more general application in logistics if the speed requirement could be relaxed.

References

1. Dantzig, G.B., Ramser, J.H.: The truck dispatching problem. Manag. Sci. **6**, 80–91 (1959)
2. Toth, P., Vigo, D.: The Vehicle Routing Problem. Society for Industrial and Applied Mathematics, Philadelphia (2001)
3. Bao, X., Liu, Z.: Approximation algorithms for single vehicle scheduling problems with release and service times on a tree or cycle. Theor. Comput. Sci. **434**, 1–10 (2012)
4. Nagamochi, H., Ohnishi, T.: Approximating a vehicle scheduling problem with time windows and handling times. Theor. Comput. Sci. **393**, 133–146 (2008)

5. Karuno, Y., Nagamochi, H.: An approximability result of the multi-vehicle scheduling problem on a path with release and handling times. Theor. Comput. Sci. **312**, 267–280 (2004)
6. Lacomme, P., Prins, C., Ramdane-Chérif, W.: Competitive memetic algorithms for arc routing problems. Ann. Oper. Res. **131**, 159–185 (2004)
7. Toklu, N.E., Gambardella, L.M., Montemanni, R.: A multiple ant colony system for a vehicle routing problem with time windows and uncertain travel times. J. Traffic Logist. Eng. **2**, 52–58 (2014)
8. Baldacci, R., Mingozzi, A., Roberti, R., Calvo, R.W.: An exact algorithm for the two-echelon capacitated vehicle routing problem. Oper. Res. **61**, 298–314 (2013)
9. Baldacci, R., Mingozzi, A., Roberti, R.: New route relaxation and pricing strategies for the vehicle routing problem. Oper. Res. **59**, 1269–1283 (2011)
10. Sousaa, J.C., Biswasa, H.A., Britob, R., Silveirab, A.: A multi objective approach to solve capacitated vehicle routing problems with time windows using mixed integer linear programming. Int. J. Adv. Sci. Technol. **28**, 1–8 (2011)
11. Arora, S.: Approximation schemes for NP-hard geometric optimization problems: a survey. Math. Program. **97**, 43–69 (2003)
12. Arora, S.: Polynomial time approximation schemes for euclidean traveling salesman and other geometric problems. J. ACM **45**, 753–782 (1998)
13. Das, A., Mathieu, C.: A quasi-polynomial time approximation scheme for Euclidean capacitated vehicle routing. In: The Twenty First Annual ACM-SIAM Symposium on Discrete Algorithms. Society for Industrial and Applied Mathematics, Philadelphia (2009)
14. Das, A.: Approximation schemes for euclidean vehicle routing problems. Dissertation, Brown University, Providence, Rhode Island, USA (2011)
15. Haimovich, M., Rinnooy Kan, A.H.G.: Bounds and heuristic for capacitated routing problems. Math. Oper. Res. **10**, 527–542 (1985)

On-Line Strategies for Evacuating from a Convex Region in the Plane

Qi Wei[1,2(✉)], Xuehou Tan[1,3], Bo Jiang[1], and Lijuan Wang[1,2]

[1] School of Information Science and Technology, Dalian Maritime University,
Linghai Road 1, Dalian, China
qwei2009@hotmail.com
[2] School of Information Science and Technology,
Dalian Institute of Science and Technology, Bingang Road 999-26,
Dalian, China
[3] School of Information Science and Technology, Tokai University,
4-1-1 Kitakaname, Hiratsuka 259-1292, Japan

Abstract. This paper considers the problem faced by a group of evacuees who must leave from an affected area as quickly as possible. We seek the strategies that achieve a bounded ratio of evacuation path length without any boundary information to that with. We restrict the affected area to a convex region in the plane, and present a 19.64-competitive strategy. It can be considered as the planar generalization of the *doubling strategy*. Also, we give a 21-competitive strategy in the grid network.

Keywords: Computational geometry · Evacuation strategy · Competitive analysis · Convex region · Grid network

1 Introduction

Motivated by the relations to the well-known searching problem and evacuation problem, much attention has recently been devoted to the problem of how to evacuate from an affected area efficiently when an emergency occurs. In this paper, suppose that there are some people in the affected area, and they don't know any boundary information of the affected area. We seek strategies for evacuating the affected people from the dangerous region as quickly as possible.

The quality of an on-line strategy is usually measured by a competitive ratio, which is defined as follows. Let P denote an affected area which is restricted to a convex region in the plane. The evacuees are modeled as the points inside P, whose initial positions are all same. When a strategy SWI without any boundary information of P is used to evacuate from P, we denote by $|SWI(P)|$ the tour length (cost) of the evacuation group to evacuate from P by SWI. Let $|OPT(P)|$ denote the tour length (cost) required to evacuate from P in the case that the evacuees know the boundary information of P, i.e., $|OPT(P)|$ is the shortest distance between the evacuees and the boundary of P. Then, the competitive ratio r is defined as follows.

$$r = \sup \frac{|SWI(\mathrm{P})|}{|OPT(\mathrm{P})|}$$

© Springer International Publishing Switzerland 2014
Z. Zhang et al. (Eds.): COCOA 2014, LNCS 8881, pp. 74–85, 2014.
DOI: 10.1007/978-3-319-12691-3_7

Previous work: The evacuation problem has been extensively studied. Chen et al. [1] used the methods of system simulation to compare the evacuation efficiency of three different networks, which include the grid network that we also consider in this paper. Lu et al. [2] and Shekhar et al. [3] studied the shortest path algorithm of evacuation with the consideration of capacity constraints and the increasing number of people in time and space. Berman [4] surveyed the problems of on-line searching and navigation. Burgard [5] considered the evacuate ratio with the cost of strategy as the time spent, instead of path lengths.

The previous studies mainly focus on details of evacuation such as flow and other constraints, so as to analyze the strategy under complete information on the boundary. In a recent work [6], Xu et al. considered a new situation in which the evacuees don't know the boundary information of the affected area. This really occurs in emergency, as the affected region is not known to the evacuees in some cases. For the convex region in the plane, Xu et al. [6] gave an evacuation strategy for k ($k \geq 3$) groups of the evacuees with a competitive ratio of $3/\cos(\pi/k)$, provided that the evacuees can communicate with each other during the evacuation. But, for $k = 1$ or 2, the problem is unsolved. For the case $k = 3$, Wei et al. improved it to $2 + 2\sqrt{3}$ [7]. The evacuation strategies for grid networks are also studied in [6]. But, their strategies rely on a more parameter R, which is the radius of the largest inner circle of the convex region.

Our work: In the strategy of Xu et al. [6], it is required that the communication among different groups be always available. Note also that the strategies of Xu et al. do not work in the case that $k = 1$ or 2. In this paper, we further release the communication restriction, and consider this problem for one group evacuation. Our model is clearly more practical. We present a 19.64-competitive strategy for one group evacuation in the plane, and a 21-competitive strategy in grid network. In general plane, the evacuees basically move on a sequence of the semicircles, whose centers are all the same, such that the radius of a semicircle is twice that of the previous one in the sequence. Two consecutive semicircles are simply connected by a line segment. It can be considered as the planar generalization of the *doubling strategy* [8]. In grid network, we adjust our strategy to the property of the grid. Our competitive ratio for grid network is independent of the parameter R, which thus gives a significant improvement upon the previous work [6].

The rest of this paper is organized as follows. In Sect. 2, we give some basic definitions relevant to this paper. In Sect. 3, we present efficient strategy and analyze the competitive ratio in general plane. In Sect. 4, we study the evacuation problem for one group in grid network. In Sect. 5, we conclude the paper with a discussion of further research and open questions.

2 Preliminaries

In this paper, we define a convex polygon as a closed polygonal chain with all interior angles equal and less than 180°. Let P be a convex polygon and O be the origin. The evacuees starting at O in P don't know any boundary information and their locations. The evacuees converge to a group G to evacuate, and their goal is to leave from P as soon as possible.

Successful evacuation of the evacuees requires that the group G has reached the boundary of the affected area. The cost of the strategy is the tour length that G walks during the evacuation. Let $S(a, b)$ denote the path from a to b on the strategy path. The performance of a strategy is measured by a competitive ratio r which is defined as above. Our objective is to minimize the competitive ratio r.

We make some assumptions about the evacuation in this paper: First, the evacuees starting at O in P don't know any boundary information of P and their location. Second, the evacuees move at the unit speed during the evacuation. Third, when we consider in grid network, the network consists of several grid units with edge length of 1. Evacuees travel along the edges of network and cannot stay on the edge but the node of the network. The origin O is a node in P.

3   Scenario 1: General Plane

In this section, we study the evacuation problem in the situation of general plane. In reality, the evacuees have several choices for grouping in the evacuation. One is that all the evacuees converge to a group to evacuate from the affected area. The other is that the evacuees are divided into several groups, and choose different routes to escape from the affected area with information sharing among them. The evacuation strategy of n groups ($n \geq 3$) had been studied in [6], provided that the communication among different groups be always available. Now, we study the strategy for one group evacuation which is more practical in the actual situation.

In this situation, only one group of the evacuees is required to escape from the affected area. The main idea of our strategy is to iteratively follow a sequence of semicircles, whose centers are all the same, such that the radius of a semicircle is twice that of the previous one in the sequence. The very first radius is assumed to be on the horizontal line L and of length one, which is assumed to be small, as compared to $|OPT(P)|$. Two consecutive semicircles are simply connected by a line segment. An instance of our strategy is shown in Fig. 1, where O denotes the starting position of the evacuees' group, and A_i, B_i the points of L intersecting the ith half-circle in our strategy. OA_1 is the first radius, $|OA_1| = 1$. OA_2 is the second radius, $|OA_2| = 2$. OA_3 is the third radius, $|OA_3| = 4$, and so on. $|OPT(P)|$ is $|OI|$.

Spiral Evacuation Strategy in General Plane (SEP)

Step 1: Make a directed line L which pass through O and its direction is arbitrary, as shown in Fig. 1. Let $Z = \{Z_1, Z_2\} = \{+L, -L\}$ denote the direction set defined by L, where $+L$ is the direction of L, and $-L$ is the direction opposite to $+L$. Let j denote the sequence number of Z and k denote the sequence number of the set of half-circles. At the beginning, let $j = 1$, $k = 1$ (j, $k \in \mathbb{N}$) and G starts at point O. Here, G denotes the group of the evacuees.

Step 2: Make a segment $\overline{OA_1}$ with length 1 on L towards Z_1. G walks along $\overline{OA_1}$. Make a semi-circle with radius $r_1 = |\overline{OA_1}|$, which intersects L with B_1. G walks along with $\widehat{A_1B_1}$ counterclockwise. $j = j + 1$, $k = k + 1$.

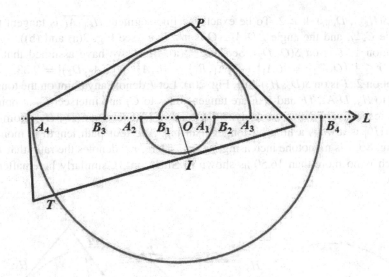

Fig. 1. The strategy SEP

Step 3: While $k \geq 2$, do the followings:

(1) Make a segment $\overline{B_{k-1}A_k}$ with length $|r_{k-1}|$ on L towards Z_j. G walks along $\overline{B_{k-1}A_k}$. Make a semicircle with radius $r_k = |\overline{OA_k}|$, which intersects L with B_k. G walks along with $\widehat{A_k B_k}$ counterclockwise.

(2) if (G reach the boundary of P) stop.

(3) $j = j + 1$, $k = k + 1$.

(4) if ($j > 2$) let $j = 1$.

Theorem 1: In general plane, the evacuate ratio of **SEP** is no more than 19.64.

Proof. As described above, the tour path of SEP consists of several semicircles and line segments. Let $C = \{ C_1, C_2, \ldots C_k \}$ and $r = \{ r_1, r_2, \ldots r_k \}$ denote the semicircle set and its radius set, which are used in SEP, respectively. Let $d = \{ d_1, d_2, \ldots d_k \}$ denote the line segment set used in SEP. Then, $|C_k|$, $|r_k|$ and $|d_k|$ denote the perimeter of C_k, the length of r_k and d_k, respectively.

First, consider the relationship between $|C_k|$, $|r_k|$ and $|d_k|$. With the property of SEP path, we have $r_1 = 1$, $C_1 = \pi$, $d_1 = 1$. And $|r_k| = 2|r_{k-1}|$, $|C_k| = \pi^*|r_k|$, $|C_k| = 2|C_{k-1}|$, $|d_k| = |r_{k-1}|$ when $k \geq 2$.

Then, turn to the competitive ratio of SEP. Let T denote the first intersection point between P and the tour path. To get the up bound of the ratio, the shortest distance from O to the boundary of P must be as small as possible. So, the boundary of P which has the shortest distance to O must approaches but does not meet the tour path at any point except T, see Fig. 1. For T may be on different locations of the tour path, we distinguish the following situations. Denote by D_k and H_k the points on the semicircle C_k such that $|OPT(P)|$ is $|r_{k-1}|$ when T is on $S(D_k, H_{k+1})$ and $|OPT(P)|$ is monotone increasing when

T is on $S(H_{k+1}, D_{k+1})$, $k \geq 2$. To be exact, the line segment $\overline{H_{k+1}A_k}$ is tangent to the semicircle C_{k-1}, and the angle $\angle D_k A_{k-1} O$ is $\pi/2$ (e.g., see Fig. 2(a) and (b)).

Situation 1. T is on $S(O, D_2)$. See, Fig. 2(a). Since we have assumed that $|OPT$ $(P)| \geq 1$, $r \leq |S(O, D_2)| = |OA_1| + |S(A_1, B_1)| + |B_1 A_2| + |S(A_2, D_2)| = 9.33$.

Situation 2. T is on $S(D_2, H_3)$. See, Fig. 2(a). Let T denote any point on the tour path from D_2 to H_3. $D_2 A_1$, TF and $H_3 E$ are tangent lines to C_1 and intersect C_1 at point A_1, F and E, respectively. Now, we analyze the trend of r. If T is on $S(D_2, H_3)$, from D_2 to H_3, $|OPT(P)|$ is always a little bit longer than $|r_1|$. The tour path length is monotone increasing. So, r is monotone increasing, $r \leq r_{H_3}$. Here, r_{H_3} denotes the ratio that T is at H_3 (which is no more than 16.50 as shown in Situation 3), similarly hereinafter. See Fig. 2 (a).

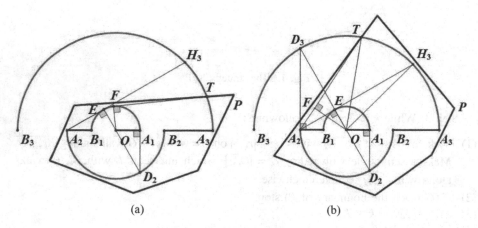

Fig. 2. The proof of Theorem 1. (a) Situation 2. T is on $S(D_2, H_3)$. (b) Situation 3. T is on $S(H_3, D_3)$.

Situation 3. T is on $S(H_3, D_3)$. See, Fig. 2(b). Let T denote any point on the tour path from H_3 to D_3. $H_3 E$ is tangent line to C_1 and intersect C_1 at point E. $D_3 A_2$ is tangent line to C_2 and intersect C_2 at point A_2. Now, we analyze the trend of r. If T is on $S(H_3, D_3)$, from H_3 to D_3, $|OPT(P)|$ is always a little bit longer than $|OF|$. Obviously, $|OF|$ is monotone increasing from $|OE|$ to $|OA_2|$, see Fig. 2(b). So, both $|OPT(P)|$ and the tour path length are monotone increasing.

Let x and y denote the radian of $\angle TA_2 H_3$ and $\angle H_3 OT$, respectively. When T moves from H_3 to D_3, $|OPT(P)|$ changes from $|OE|$ to $|OA_2|$ synchronously. In this process, $0 \leq x \leq \angle D_3 A_2 E$ and $0 \leq y \leq \angle H_3 O D_3$. In $\triangle D_3 A_2 O$, $\sin \angle A_2 D_3 O = |r_2|/|r_3| = 0.5 \doteq 0.5$. So, $\angle A_2 D_3 O = \pi/6$, $\angle D_3 A_2 E = \pi/3$. In $\triangle EOH_3$, $\cos \angle EOH_3 = |r_1|/|r_3| = 0.25$.

So, $\angle H_3 O D_3 = \arccos \frac{1}{4}$. $0 \leq x \leq \frac{\pi}{3}$, $0 \leq y \leq \arccos \frac{1}{4}$.

In $\triangle A_2 OT$, $\frac{|OT|}{\sin \angle OA_2 T} = \frac{|OA_2|}{\sin \angle A_2 TO}$. $\angle A_2 TO = \arcsin \frac{|r_2| \sin(\angle H_3 A_2 T + \angle EA_2 O)}{|r_3|}$.

$$y = \angle H_3OD_3 - \angle TOD_3 = \angle H_3OD_3 - (\angle TOF - \angle FOD_3)$$

$$= \angle H_3OD_3 - (\pi/2 - \angle A_2TO - x) = \arccos\frac{1}{4} - (\frac{\pi}{2} - \arcsin(\frac{1}{2}\sin(\frac{\pi}{6} + x)) - x)$$

$$= x + \arcsin(\frac{1}{2}\sin(\frac{\pi}{6} + x)) - 0.25$$

$$|S(O, H_3)| = |r_1| + \pi|r_1| + |r_1| + \pi|r_2| + |r_2| + \angle A_3OH_3 * |r_3|$$
$$= 13.42 + (\pi - \angle A_2OD_3 - \angle H_3OD_3) * 4 = 16.50$$

$$r = \frac{|SWI(P)|}{|OPT(P)|} = \frac{|S(O, H_3)| + |\widehat{H_3T}|}{|OF|} = \frac{16.5 + r_3y}{r_2\sin\angle TA_2O} = \frac{8.25 + 2y}{\sin(x + \frac{\pi}{6})}.$$

$$r'(x) = \frac{2y'\sin(x + \frac{\pi}{6}) - (8.25 + 2y)\cos(x + \frac{\pi}{6})}{\sin^2(x + \frac{\pi}{6})}$$

Let $t(x) = 2y'\sin(x + \frac{\pi}{6}) - (8.25 + 2y)\cos(x + \frac{\pi}{6})$,

$$r''(x) = \frac{t'\sin(x + \frac{\pi}{6}) - 2t\cos(x + \frac{\pi}{6})}{\sin^3(x + \frac{\pi}{6})}. \qquad t'(x) = (2y'' + 8.25 + 2y)\sin(x + \frac{\pi}{6})$$

$$y'(x) = 1 + \frac{\cos(x + \frac{\pi}{6})}{\sqrt{3 + \cos^2(x + \frac{\pi}{6})}} \le 1.45 \quad (0 \le x \le \frac{\pi}{3})$$

$$y''(x) = \frac{-3\sin(x + \frac{\pi}{6})}{(3 + \cos^2(x + \frac{\pi}{6}))^{\frac{3}{2}}} \ge -0.58 \quad (0 \le x \le \frac{\pi}{3}).$$

$$t'\sin(x + \frac{\pi}{6}) - 2t\cos(x + \frac{\pi}{6})$$
$$= 2y''\sin^2(x + \frac{\pi}{6}) + (8.25 + 2y)\cos^2(x + \frac{\pi}{6}) + 8.25 + 2y - 2y'\sin(2x + \frac{\pi}{3}) > 0$$

So, $r''(x) > 0$, $r(x)$ $(0 \le x \le \frac{\pi}{3})$ is a convex function.

$$r_{max}(x) = \max\{r(0), r(\frac{\pi}{3})\} = r(0) = r_{H_3} = \frac{|SWI(P)|}{|OPT(P)|} \le \frac{|S(O, H_3)|}{|r_1|} = 16.50$$

So, when T is on $S(H_3, D_3)$, $r \le 16.50$.

Situation 4. T is on $S(D_k, D_{k+1})$, $k \ge 2$ (k denotes the sequence number of the semicircles, D_k is on C_k). As discussed in Situations 2 and 3, we have $r \le r_{H_3} = 16.50$ when k = 2, i.e., T is on $S(D_2, D_3)$. Now, we consider the situation that T is on $S(D_k, D_{k+1})$, $k \ge 3$. The tour path of strategy SEP is cyclical, see Fig. 3 So, as proved in Situations 2 and 3, we have $r \le r_{H_{k+1}}$ when T is on $S(D_k, D_{k+1})$, $k \ge 3$.

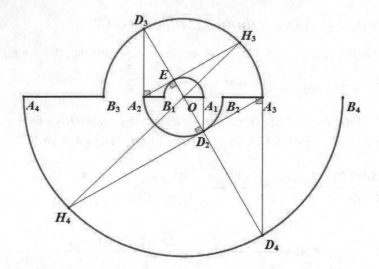

Fig. 3. The proof of Theorem 1. Situation 4. T is on $S(D_k, D_{k+1})$, $k \geq 2$.

All situations have been enumerated above, in term of the intersection point T. As discussed above, we have $r \leq \max\{r_{D_2}, r_{H_3}, r_{H_4}, \ldots r_{H_{k+1}}\}, k \geq 2$, during the whole SEP tour path.

$$
\begin{aligned}
r_{H_{k+1}} &= \frac{|SWI(P)|}{|OPT(P)|} \leq \frac{|S(O, H_{k+1})|}{|r_{k-1}|} = \frac{\sum_{n=1}^{k}(|d_n + C_n + d_{k+1}S(A_{k+1}, H_{k+1})|)}{|r_{k-1}|} \\
&= \frac{|r_1| + \pi|r_1| + |r_1| + \pi|r_2| + \ldots + |r_{k-1}| + \pi|r_k| + |r_k| + \angle A_{k+1}OH_{k+1} * |r_{k+1}|}{|r_{k-1}|} \\
&= \frac{1}{2^{(k-2)}} + \frac{\pi}{2^{(k-2)}} + \frac{1}{2^{(k-2)}} + \frac{\pi}{2^{(k-3)}} + \ldots + \frac{1}{2^{(k-1-(k-1))}} + \frac{\pi}{2^{(k-1-k)}} + \frac{1}{2^{(k-1-k)}} \\
&\quad + 4(\pi - \angle A_kOD_{k+1} - \angle H_{k+1}OD_{k+1}) \\
&= \frac{1}{2^{(k-2)}} + (\frac{1}{2^{(k-2)}} + \frac{1}{2^{(k-3)}} + \ldots + \frac{1}{2^{(k-1-k)}})(\pi + 1) + 4(\frac{2}{3}\pi - \arccos\frac{1}{4}) \\
&= \frac{1}{2^{(k-2)}} + \frac{2(1 - (\frac{1}{2})^k)(\pi + 1)}{1 - \frac{1}{2}} + 3.08 \\
&= \frac{1}{2^{(k-2)}} + (4 - \frac{1}{2^{(k-2)}})(\pi + 1) + 3.08 \\
&= -\frac{\pi}{2^{(k-2)}} + 4 + 4\pi + 3.08 < 4 + 4\pi + 3.08 \\
&= 19.64 \quad (k \geq 2)
\end{aligned}
$$

The proof is thus complete.

4 Scenario 2: Grid Network

Most urban cities have the structure of road network like a grid one [9], See Fig. 4. In this section, we study the evacuation problem in the situation of grid network. We establish a rectangular coordinate system [10, 11] coincides with grid network with O as the origin. The evacuees start from O, and walk along the edges of the network, see Fig. 5. Let $P(x, y)$ denote coordinates of a point in grid network.

Fig. 4. Part of the traffic network of Beijing and Dalian, China

Definition 1 [6]: For any points $P(x_1, y_1)$ and $P(x_2, y_2)$ in grid network, $L = |x_1 - x_2| + |y_1 - y_2|$ is called the evacuation path length.

The main idea of our strategy is to iteratively follow a sequence of semi squares, whose centers are all the same, such that the edge of a semi square is twice that of the previous one in the sequence. The edge length of the first square is two, half of which is assumed to be small, as compared to $|OPT(P)|$. Two consecutive semi squares are simply connected by a line segment. An instance of our strategy is shown in Fig. 5, where O denotes the starting position of the evacuees' group, and A_i, B_i the points of x axis intersecting the ith half-square in our strategy. $|OPT(P)|$ is $|OI|$.

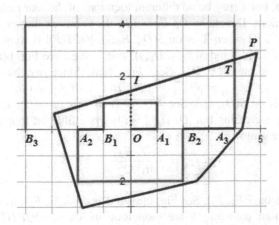

Fig. 5. The strategy SEG

Spiral Evacuation Strategy in Grid Network (SEG)

Step 1: Let $Z = \{Z_1, Z_2\} = \{+x, -x\}$ denote the direction set in grid network. Let i denote edge length of the square, j denote the sequence number of Z and k denote the sequence number of the set of half squares. At the beginning, let $i_1 = 2$, $j = 1$, $k = 1$ $(j, k \in N)$ and G starts at point O.

Step 2: Make a segment $\overline{OA_1}$ with length 1 on axis x towards Z_1. G walks along $\overline{OA_1}$. Make a semi square with center O and side length i_1, which intersects axis x with B_1. G walks along with $S(A_1, B_1)$ counterclockwise. $j = j + 1$, $k = k + 1$.

Step 3: While $k \geq 2$, do the followings:

(1) Make a segment $\overline{B_{k-1}A_k}$ with length $|i_{k-1}|/2$ on L towards Z_j. G walks along $\overline{B_{k-1}A_k}$. Make a semi square with center O and side length i_k, which intersects axis x with B_k. G walks along with $S(A_k, B_k)$ counterclockwise.
(2) if (G reach the boundary of P) stop.
(3) $j = j + 1$, $k = k + 1$.
(4) *if ($j > 2$) let $j = 1$.*

Theorem 2: In grid network, the evacuate ratio of **SEG** is no more than 21.

Proof. As described above, the tour path of SEG consists of several semi squares and line segments. Let $C = \{C_1, C_2, \dots C_k\}$ and $i = \{i_1, i_2, \dots i_k\}$ denote the semi square set and its edge set which are used in SEG, respectively. Let $d = \{d_1, d_2, \dots d_k\}$ denote the line segment set used in SEG. Then, $|C_k|$, $|i_k|$ and $|d_k|$ denote the perimeter of C_k, the length of i_k and d_k, respectively.

First, consider the relationship between $|C_k|$, $|i_k|$ and $|d_k|$. With the property of SEG path, we have $i_1 = 2$, $C_1 = 4$, $d_1 = 1$. And $|i_k| = 2|i_{k-1}|$, $|C_k| = 2|i_k|$, $|C_k| = 2|C_{k-1}|$, $|d_k| = |i_{k-1}|/2$ when $k \geq 2$.

Then, turn to the competitive ratio of SEG. Let T denote the first intersection point between P and the tour path. To get the up bound of the ratio, the shortest distance from O to the boundary of P must be as small as possible. So, the boundary of P which has the shortest distance to O must approaches but does not meet the tour path at any point except T, see Fig. 5. For T may be on different locations of the tour path, we distinguish the following situations. Denote by D_k, E_k and F_k the points on the semi square C_k such that $|OPT(P)|$ is $|i_{k-1}|/2$ when T is on $S(D_k, E_{k+1})$, $|OPT(P)|$ is monotone increasing when T is on $S(E_{k+1}, F_{k+1})$ and $S(F_{k+1}, D_{k+1})$, $k \geq 2$ (e.g., see Fig. 6(a) and (b)).

Situation 1. T is on $S(O, D_2)$. See, Fig. 6(a). Since we have assumed that $|OPT(P)| \geq 1$, $r \leq |S(O, D_2)| = |OA_1| + |S(A_1, B_1)| + |B_1A_2| + |S(A_2, D_2)| = 9$.

Situation 2. T is on $S(D_2, E_3)$. See, Fig. 6(a). If T is on $S(D_2, E_3)$, the tour path length is monotone increasing but $|OPT(P)|$ is always a little bit longer than $|i_1|/2$. So, r is monotone increasing.

$$r \leq r_{E_3} = \frac{|S(O, E_3)|}{|i_1|/2} = 17.$$

Situation 3. T is on $S(E_3, F_3)$. See Fig. 6(a). If T is on $S(E_3, F_3)$, from E_3 to F_3, both $|OPT(P)|$ and the tour path length are monotone increasing. $|OPT(P)|$ is a little bit longer than $|OI|$.

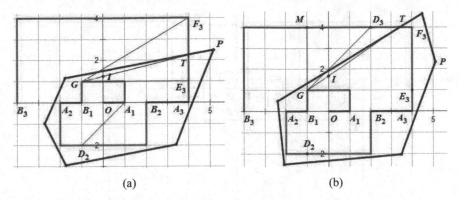

(a)   (b)

Fig. 6. The proof of Theorem 2. (a) Situations 2 and 3. T is on $S(D_2, F_3)$. (b) Situation 4. T is on $S(F_3, D_3)$.

Let x denote the radian of $\angle TGE_3$.

$$|OI| = |i_1|/2 + |i_1|/2 {}^* \tan x = 1 + \tan x, \quad |TE_3| = (|i_1|/2 + |i_3|/2) {}^* \tan x = 5 \tan x$$

$$r = \frac{|SWI(P)|}{|OPT(P)|} \leq \frac{|S(O,E_3)| + |E_3 T|}{|OI|} = \frac{17 + 5 \tan x}{1 + \tan x} = 5 + \frac{12}{1 + \tan x}$$

In this situation, $0 \leq x \leq \angle F_3 GE_3 < \pi/4$. $\tan(x)$ is monotone increasing, so, r is monotone decreasing. Thus, $r \leq r_{E_3} = 17$.

Situation 4. T is on $S(F_3, D_3)$. See Fig. 6(b). If T is on $S(F_3, D_3)$, from F_3 to D_3, $|OPT(P)|$ is always a little bit longer than $|OI|$. Obviously, it is monotone increasing. And the tour path length is monotone increasing too.

Let x denote the radian of $\angle GTM$. $|OI| = |i_1|/2 + |i_1|/2 {}^* \tan x = 1 + \tan x$.

$$|TF_3| = |MF_3| - |MT| = (|i_1|/2 + |i_3|/2) - \frac{|GM|}{\tan x}$$

$$= (|i_1|/2 + |i_3|/2) - \frac{|i_3|/2 - |i_1|/2}{\tan x} = 5 - \frac{3}{\tan x}$$

$$r = \frac{|SWI(P)|}{|OPT(P)|} \leq \frac{|S(O,F_3)| + |F_3 T|}{|OI|} = \frac{20 + 5 - \dfrac{3}{\tan x}}{1 + \tan x} = \frac{25 - \dfrac{3}{\tan x}}{1 + \tan x}$$

$$r'(x) = \frac{\dfrac{3}{\sin^2 x} + \dfrac{6}{\sin x \cos x} - \dfrac{25}{\cos^2 x}}{(1 + \tan x)^2}. \quad \text{Let} \quad w = \frac{3}{\sin^2 x} + \frac{6}{\sin x \cos x} - \frac{25}{\cos^2 x}.$$

In this situation, $\angle GF_3 M \leq x \leq \angle GD_3 M = \dfrac{\pi}{4}$, $\sin x \leq \cos x$.

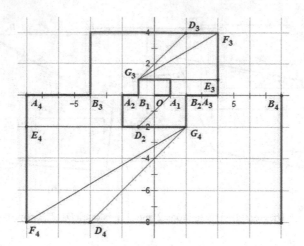

Fig. 7. The proof of Theorem 2. Situation 5. T is on $S(D_k, D_{k+1})$, $k \geq 2$.

$$w \leq \frac{3}{\sin^2 x} + \frac{6}{\sin^2 x} - \frac{25}{\cos^2 x} = \frac{9}{\sin^2 x} - \frac{25}{\cos^2 x} = \frac{9 - 34\sin^2 x}{\sin^2 x \cos^2 x} < 0.$$

So, $r'(x) < 0$, r is monotone decreasing. $r \leq r_{F_3} = 12.5$.

Situation 5. T is on $S(D_k, D_{k+1})$, $k \geq 2$ (k denotes the sequence number of the semi squares, D_k is on C_k). As discussed in situation 2, 3 and 4, we have $r \leq r_{E_3} = 17$ when $k = 2$, i.e., T is on $S(D_2, D_3)$. Now, we consider the situation that T is on $S(D_k, D_{k+1})$, $k \geq 3$. The tour path of strategy SEG is cyclical, see Fig. 7. So, as proved in situations 2, 3 and 4, we have $r \leq r_{E_{k+1}}$ when T is on $S(D_k, D_{k+1})$, $k \geq 3$.

All situations have been enumerated above, in term of the intersection point T. As discussed above, we have $r \leq \max\{r_{D_2}, r_{E_3}, r_{E_4}, \dots r_{E_{k+1}}\}, k \geq 2$, during the whole SEG tour path.

$$r_{E_{k+1}} = \frac{|SWI(P)|}{|OPT(P)|} \leq \frac{|S(O, E_{k+1})|}{|i_{k-1}|/2} = \frac{\sum_{n=1}^{k}(|d_n| + |C_n|) + |d_{k+1}| + |S(A_{k+1}, E_{k+1})|}{|i_{k-1}|/2}$$

$$= \frac{|i_1|/2 + 2|i_1| + |i_1|/2 + 2|i_2| + \dots + |i_{k-1}|/2 + 2|i_k| + |i_k|/2 + |i_{k-1}|/2}{|i_{k-1}|/2}$$

$$= \frac{1}{2^{(k-2)}} + \frac{4}{2^{(k-2)}} + \frac{1}{2^{(k-2)}} + \frac{4}{2^{(k-3)}} + \dots + \frac{1}{2^{(k-1-(k-1))}} + \frac{4}{2^{k-1-k}} + \frac{1}{2^{k-1-k}} + 1$$

$$= \frac{1}{2^{(k-2)}} + 5(\frac{1}{2^{(k-2)}} + \frac{1}{2^{(k-3)}} + \dots + \frac{1}{2^0} + \frac{1}{2^{-1}}) + 1$$

$$= \frac{1}{2^{(k-2)}} + 20(1 - \frac{1}{2^k}) + 1$$

$$= 21 - \frac{16}{2^k}$$

$$< 21 \quad (k \geq 2)$$

The proof is thus complete.

5 Concluding Remarks

In this paper, we study the problem of finding efficient strategies for evacuating from a convex region as soon as possible without any boundary information. We consider the problem for one group evacuation which is more practical. We present a 19.64-competitive strategy in the general plane, and a 21-competitive strategy in the grid network.

We pose several open questions for further research. First, suppose that all the evacuees are not located at one point but different points in a convex region, how to design an efficient strategy to solve the evacuation problem. Second, it is an interesting work to further improve our 19.64-competitive strategy. Since the *doubling strategy* gives the optimal 9-competitive ratio in one dimensional case [8], our strategy may be rather close to the optimal.

Acknowledgements. This work was supported in part by National Natural Science Foundation of China (No. 61173034) and the General Project of Liaoning Province Science and Research (No. L2012487).

References

1. Chen, X., Zhan, F.B.: Agent-based simulation of evacuation strategies under different road network structures. J. Oper. Res. Soc. **59**, 25–33 (2008)
2. Lu, Q., George, B., Shekhar, S.: Capacity constrained routing algorithms for evacuation planning: a summary of results. In: Medeiros, C.B., Egenhofer, M., Bertino, E. (eds.) SSTD 2005. LNCS, vol. 3633, pp. 291–307. Springer, Heidelberg (2005)
3. Shekhar, S., Yang, K., Gunturi, V.M.V., Manikonda, L., et al.: Experiences with evacuation route planning algorithms. Int. J. Geogr. Inf. Sci. **26**(12), 2253–2265 (2012)
4. Berman, P.: On-line searching and navigation. In: Fiat, A. (ed.) Online Algorithms 1996. LNCS, vol. 1442, pp. 232–241. Springer, Heidelberg (1998)
5. Burgard, W., Moors, M., Fox, D., Simmons, R., Thrun, S.: Collaborative multirobot exploration. In: Proceedings 2000 ICRA. Millennium Conference, IEEE International Conference on Robotics and Automation. Symposia Proceedings, vol. 1, pp. 476–481 (2000)
6. Xu, Y., Qin, L.: Strategies of groups evacuation from a convex region in the plane. In: Fellows, M., Tan, X., Zhu, B. (eds.) FAW-AAIM 2013. LNCS, vol. 7924, pp. 250–260. Springer, Heidelberg (2013)
7. Wei, Q., Wang, L.J., Jiang, B.: Tactics for evacuating from an affected area. Proceedings 2013 ICCCI. IJMLC **3**(5), 435–439 (2013)
8. BaezaYates, R.A., Culberson, J.C., Rawlins, G.J.: Searching in the plane. Inf. Comput. **106** (2), 234–252 (1993)
9. Xu, Y.F., et al.: The canadian traveler problem and its competitive analysis. J. Comb. Optim. **18**(2), 17–28 (2009)
10. Du, H., Xu, Y.F.: An approximation algorithm for k-center problem on a convex polygon. J. Comb. Optim. **27**(3), 504–518 (2014)
11. Concalves, J.F., Resende, M.G.C.: A parallel multi-population genetic algorithm for a constrained two-dimensional orthogonal packing problem. J. Comb. Optim. **22**(2), 180–201 (2010)

Rectilinear Duals Using Monotone Staircase Polygons

Yi-Jun Chang and Hsu-Chun Yen[✉]

Department of Electrical Engineering, National Taiwan University, Taipei 106,
Taiwan, Republic of China
yen@cc.ee.ntu.edu.tw

Abstract. A rectilinear dual of a plane graph refers to a partition
of a rectangular area into nonoverlapping rectilinear polygonal modules, where each module corresponds to a vertex such that two modules have side-contact iff their corresponding vertices are adjacent. It is
known that 8-sided rectilinear polygons are sufficient and necessary to
construct rectilinear duals of maximal plane graphs. The result stands
even if modules are restricted to T-shape polygons. We show that the
optimum polygonal complexity of T-free rectilinear duals is exactly 12.
It justifies the intuition that T-shape is more powerful than other
8-sided modules. Our construction of 12-sided T-free rectilinear duals
only requires monotone staircase modules. We also consider the issue of
area-universality, and show that monotone staircase modules are not sufficient to construct area-universal rectilinear duals in general even when
an unbounded polygonal complexity is allowed; however, eight sides are
sufficient for Hamiltonian plane graphs. This line of research regarding
monotone staircase modules is also motivated by the so-called monotone
staircase cuts in VLSI floorplanning. We feel that our results provide a
new insight towards a comprehensive understanding of modules in rectilinear duals.

Keywords: Floorplanning · Rectilinear polygon · Plane graph · Rectilinear dual

1 Introduction

Given a graph with nodes and edges representing circuit components and inter-
connections, respectively, *floor-planning* in VLSI chip design refers to the partition of a rectangular chip area into nonoverlapping rectilinear polygonal modules
in such a way that modules of adjacent nodes share a common boundary. If we
further require that each rectilinear polygon be a rectangle, such a floorplan
is called a *rectangular dual*. As they play key roles in VLSI physical design,
rectangular duals of planar graphs have been studied extensively over the years

---

Research supported in part by National Science Council of Taiwan under Grants
NSC-100-2221-E-002-132-MY3 and NSC-100-2923-E-002-001-MY3.

Z. Zhang et al. (Eds.): COCOA 2014, LNCS 8881, pp. 86–100, 2014.
DOI: 10.1007/978-3-319-12691-3_8

from both theoretical and practical viewpoints. A necessary and sufficient condition for the existence of a rectangular dual for a planar graph was established by Kozminski and Kinnen [9]. A floorplan is said to be *sliceable* if it can be obtained by recursively cutting a rectangle into two parts by a horizontal or a vertical line. Sliceable floorplans enjoy certain nice properties, facilitating global routing by taking advantage of the hierarchical structure of partitioning by the cut lines, for instance. As a generalization of sliceability in floorplanning, *monotone staircase cuts* have been proposed (see, e.g., [7,11]), which are able to yield a richer set of floorplan structures while retaining certain attractive properties enjoyed by sliceable floorplans.

Rectilinear duals represent a generalization of rectangular duals, in which vertices are represented by interior-disjoint rectilinear polygons such that edges correspond to side-contact of polygons. A *monotone staircase polygon* is a polygon formed by two monotonically rising staircases, each of which is a sequence of alternatingly horizontal and vertical line segments from the bottom-left corner to the top-right corner of the polygon. A *staircase polygon* is a rectilinear polygon resulting from rotating a monotone staircase polygon 90°, 180°, or 270°. Figure 1(2) is a rectilinear dual of the plane graph given in Fig. 1(1) using monotone staircase polygons. In fact, floorplans using monotone staircase polygons are exactly those that can be obtained using monotone staircase cuts. See Fig. 1(2)–(3), for instance. Note that staircase polygons are *orthogonally convex*, in the sense that for any horizontal or vertical line, if two points on the line are inside a polygonal region, then the entire line segment between these two points is also inside the polygonal region. The reader is referred to [5,6,8] for more about sliceability, rectangular and rectilinear duals of plane graphs.

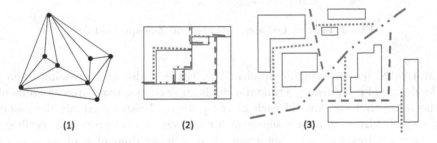

(1)   (2)   (3)

Fig. 1. A floorplan constructed by monotone staircase cuts.

The study of rectilinear duals using monotone staircase polygons is also of interest from the graph drawing perspective. In the field of graph drawing, there is a large amount of work investigating contact representations of graphs. In such drawing styles, vertices are represented by geometric objects such that edges correspond to certain contacts between those objects. The so-called *polygonal complexity* of a rectilinear dual refers to the number of sides in any of the rectilinear polygons found in the rectilinear dual. Yeap and Sarrafzadeh [13] showed that

every maximal plane graph admits a rectilinear dual using polygons of at most
eight sides, which matches the lower bound. Liao *et al.* [10] later improved the
above result by showing that it suffices to use only I-shape, L-shape, and T-shape
modules, whereas in [13], Z-shape modules are also required. See Fig. 2 for these
four types of modules. In fact, I-shape and L-shape modules are degenerated cases
of T-shape modules, as an I (resp., L) can be obtained from a T by chopping off
two ends (resp., one end) of the horizontal segment of the T. As a result, [10] suggests that T-shape modules are sufficient in constructing rectilinear duals of maximal plane graphs. See also [1–3]. It is not difficult to see that I-shape and L-shape
modules are also degenerated cases of Z-shape modules; however, T-shape and
Z-shape modules are incomparable to each other. Note that a Z-shape is an 8-
sided staircase polygon. If we take into account the ⊔-shape, another 8-sided
rectilinear polygon incomparable with T-shape and Z-shape, floorplans using
staircase polygons are exactly those free from T-shape and ⊔-shape modules (and
their generalizations), which can also be characterized as those using orthogonally convex polygons without T-shape modules (and their generalizations).

In view of the above, it is therefore of interest and importance to study
rectilinear duals of plane graphs using staircase polygons. Along this line of
research, a natural question to ask (in view of [10]) is whether Z-shape modules
and their degenerated cases are sufficient in constructing a rectilinear dual of
any maximal plane graph. As it turns out, we are able to answer the question in
the negative, suggesting that T-shape is, in a sense, more powerful than Z-shape
from the viewpoint of polygonal complexity in rectilinear duals.

Fig. 2. I-shape, L-shape, T-shape and Z-shape modules

Motivated by cartographic applications, there is also an increasing interest
in the design of rectilinear/rectangular duals subject to a given area assignment,
in the sense that the area of each of the polygonal regions equals the associated weight. Given an area assignment for vertices, a *cartogram* is a rectilinear
dual that realizes such an assignment. A rectilinear dual of a plane graph is
area-universal iff every possible area assignment can be realized by a combinatorially equivalent rectilinear dual. For rectangular duals, it is known that
area-universality can be characterized by one-sidedness [4]; however, no simple
characterizations are known for rectilinear duals. Regarding the polygonal complexity of area-universal rectilinear duals for maximal plane graphs, it is known
that eight is sufficient, which is also optimal [1]. In this paper, we also investigate the issue of area-universality of rectilinear duals using monotone staircase
polygons.

Our contributions in this paper include the following:

1. We define a partial order on the structure of rectilinear polygons that captures the intuitive idea of degeneracies of polygons naturally. Such a partial order provides a foundation for our subsequent discussions on rectilinear duals using monotone staircase polygons (which are T-free and orthogonally convex).
2. An algorithm is devised for constructing rectilinear duals of maximal plane graphs using 12-sided monotone staircase polygons. The bound (i.e., 12 sides) is also shown to be tight.
3. Regarding area-universality, we prove that there exists a maximal plane graph that does not admit any monotone staircase area-universal rectilinear dual; however, for Hamiltonian maximal plane graphs, eight sides suffice.

2 Preliminaries

A graph $G = (V, E)$ is *planar* iff it can be drawn in the plane without edge crossings. In a planar drawing of a graph, the *outer face* is the unbounded one; a face is called *inner* iff it is not outer. A *plane graph* is a planar graph with a fixed combinatorial embedding and a prescribed outer face. A cycle C divides a plane graph G into two regions. The one that is inside (resp., outside) cycle C is called the *interior region* (resp., *outer region*) of C. We write $G(C)$ to denote the subgraph of G containing exactly C and vertices and edges residing in its interior region. When H is a connected subgraph of G with C as its boundary cycle, $G(H)$ is defined to be $G(C)$.

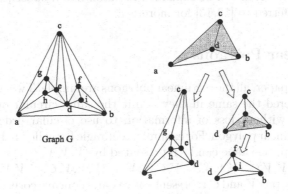

Fig. 3. An example of a separation-tree.

Let $\triangle$ be a triangle (a cycle of length 3). We call $\triangle$ a *separating triangle* (also known as a complex triangle) iff $G(\triangle) \neq \triangle$ in any planar embedding G. $G_\triangle$ is defined to be the induced subgraph of the set of vertices $\{v \in V(G(\triangle)) \mid$ for any triangle $\triangle' \neq \triangle$ in $G(\triangle)$, v does not reside in the interior region of $\triangle'\}$. For graph G depicted in Fig. 3, $G_{\{a,b,c\}}$ is the subgraph induced by $\{a, b, c, d\}$.

The *separation-tree* of a maximal plane graph G is defined to be the unique rooted tree whose vertices are separating triangles and the boundary triangle in G, with $\triangle$ being a descendant of $\triangle'$ iff $\triangle$ is contained in $G(\triangle')$. See Fig. 3. The reader is referred to [12] for more about the above notations and definitions.

The *contraction* of a triangle $\triangle$ is an operation that replaces $G(\triangle)$ with $\triangle$; the *un-contraction* of a (previously contracted) triangle $\triangle$ is an operation that replaces $\triangle$ with $G_\triangle$. The descendants of $\triangle$ remain contracted when we un-contract $\triangle$. For convenience purpose, we write region(S), $S \subseteq V$, to denote the union of region of x, $x \in S$, in the rectilinear dual of G.

An internally triangulated plane graph G admits a rectangular dual iff we can augment G with four vertices $\{N, E, S, W\}$ such that (1) the new outer face is the quadrangle $\{N, E, S, W\}$ and (2) the resulting graph is internally triangulated and contains no separating triangle [9]. The tight connection between separating triangles and rectangular duals makes separation-trees particularly useful in constructing rectilinear duals. Here we sketch a general framework of building rectilinear

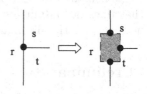

Fig. 4. Illustration of inserting sub-drawing.

duals based on separation-trees: (1) Let $\triangle_1, \triangle_2, \ldots, \triangle_k$ be a level-order traversal of the separation-tree. We let $G' = \triangle_1$ (the boundary triangle). (2) Construct a rectangular dual of G' as the initial drawing. (3) For $i = 1$ to k, we un-contract $\triangle_i$, and plug-in the rectangular drawing of $G_{\triangle_i} \setminus \triangle_i$ to the current drawing. See Fig. 4 for a conceptual illustration of inserting rectangular dual of $G_{\triangle_i} \setminus \triangle_i$ when $\triangle_i = \{s, t, r\}$ is un-contracted. Note that the exact location at which the rectangular dual is inserted will be explained in detail in our subsequent discussion. The reader is referred to [12,13] for more.

3   Rectilinear Polygons

Throughout the paper, all the rectilinear polygons are simple. Two rectilinear polygons are considered the same iff they admit the same circular order of angles. Therefore, it is without loss of information to use circular order of angles to represent a rectilinear polygon. For example, rectangle (or called I-shape), L-shape, T-shape, W-shape, ⊔-shape can be represented by (V, V, V, V), (V, V, V, C, V, V), (V, V, C, V, V, C, V, V), (V, V, V, C, V, C, V, V), (V, V, V, C, C, V, V, V), respectively, where the letters V and C represent convex and concave corners, respectively. Given a sequence P of Cs and Vs, we let $\sharp_C(P)$ and $\sharp_V(P)$ denote the numbers of concave and convex corners, respectively.

Here we define a partial order "$\preceq$" on rectilinear polygons as follows:

Definition 1. *Let P and Q be two rectilinear polygons. $P \preceq Q$ iff Q can be obtained by iteratively inserting (C, V) or (V, C) to P.*

Let R be a rectilinear dual, we call it *Q-free* iff for each module of shape P used in R, we have $Q \npreceq P$. We remark that the partial order "$\preceq$" actually reflects the intuitive idea of degeneracy in the way that $P \preceq Q$ indicates that Q can degenerate to P. Therefore, the notion of "Q-freeness" captures the idea of "Q is not a degenerated form of any rectilinear region in the drawing". Here we define a class of rectilinear polygons called *staircase* as follows:

Definition 2. *Let P be a rectilinear polygon. P is monotone staircase iff $P = (a, S_1, b, S_2)$ clockwise, where $a = b = V$ and both S_1 and S_2 consist of Cs and Vs appearing alternatively and both start and end with V, where the two points a, b are exactly at the most south-western (i.e., lower left-hand) and the most north-eastern (i.e., upper right-hand) corners.*

In words, a monotone staircase polygon is a polygon formed by two monotonically rising staircases, each of which is a sequence of alternatingly horizontal and vertical line segments from the bottom-left corner to the top-right corner of the polygon. A *staircase polygon* is a rectilinear polygon resulting from rotating a monotone staircase polygon $90°$, $180°$, or $270°$. The following facts are easy to

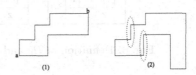

Fig. 5. Some examples.

observe, and may be explicitly or implicitly applied in the discussion throughout the paper.

Fact 1. $\sharp_V(P) - \sharp_C(P) = 4$ *in any rectilinear polygon P.*

Fact 2. *A rectilinear polygon is orthogonally convex iff it does not contain consecutive concave corners.*

Fact 3. *A rectilinear polygon P is staircase iff $\mathsf{T} \npreceq P$ and P is orthogonally convex.*

Fact 4. *A rectilinear polygon P satisfies $\mathsf{T} \preceq P$ iff $P = (S_1, a, S_2, b, S_3, S_4)$ such that $a = b = V$ and $\sharp_C(S_2) = \sharp_V(S_2)$, $\sharp_C(S_1) - \sharp_V(S_1) = \sharp_C(S_3) - \sharp_V(S_3) = 1$.*

Figure 5(1) is an example of a staircase module. The polygon in Fig. 5(2) is degeneratable to T-shape by removing the two pairs of corners (circled in the picture); the reader can verify Fact 4 by considering its representation $(C, V, V, V, C, C, V, V, V, V, C, V) = (S_1 = (C), a = V, S_2 = (), b = V, S_3 = (V, C, C), S_4 = (V, V, V, V, C, V))$.

4   Lower Bound of Polygonal Complexity

In this section, we prove that T-free rectilinear duals of maximal plane graphs must have polygonal complexity of at least 12, which is higher than the 8 in the general case when T-shape modules are allowed. We define the plane graph H_0 in Fig. 6, which is a key structure that is behind the higher polygonal complexity of T-free rectilinear duals.

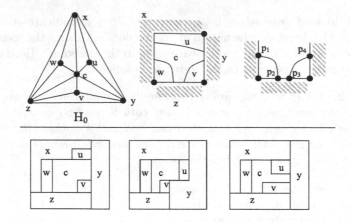

Fig. 6. Definition of H_0 and illustration of the proof of Lemma 1.

Lemma 1. *Let H be a subgraph of a maximal plane graph G such that H is isomorphic to H_0 (with x, y and z as the three vertices on the outer cycle) and $G(H) = H$. For any $\top$-free rectilinear dual of G, there must be at least two concave corners in regions associated with x, y, and z which are located along the border between region($\{x, y, z\}$) and region($\{u, v, w, c\}$).*

Proof. Clearly there must be at least one concave corner. If there is only one such concave corner, without loss of generality, we let x be the one containing the concave corner. Now, the boundary of the region($\{u, v, w, c\}$) must be a rectangle, illustrated in Fig. 6. Note that since module c in Fig. 6 must touch x and x borders the rectangular region along its west and north sides, by symmetry, we can assume that c touches the west boundary without loss of generality. We identify four points p_1, p_2, p_3 and p_4 on the boundary of c as illustrated in Fig. 6. By setting $a = p_2, b = p_3$, $S_1 =$ the segment between p_1 and p_2, and $S_3 =$ the segment between p_2 and p_3, it is easy to see that if all the regions are drawn rectilinearly, such assignments must satisfy the statement of Fact 4, and hence we must have $\top \preceq c$, which is a contradiction. $\qquad\square$

Theorem 1. *There exists a maximal plane graph G such that every $\top$-free rectilinear dual of G must have polygonal complexity of at least 12.*

Proof. Let $G_0 = (V, E)$ be an n-node maximal plane graph with $n \geq 10$. We replace each inner face of G_0 with a copy of H_0 by adding new vertices and edges. Let the resulting graph be G_1, and we let R be any $\top$-free rectilinear dual of G_1. According to Lemma 1, since the number of inner faces in G_0 is $2n - 5$, the number of concave corners in regions associated with vertices of V must be at least $2 \times (2n - 5) = 4n - 10 > 3n = 3|V|$. Therefore, there must be a region in R containing at least 4 concave corners. By Fact 1, such a region has at least $4 + (4 + 4) = 12$ corners. $\qquad\square$

In the lower part of Fig. 6, we give three examples of rectilinear duals of H_0. The left two drawings are both $\top$-free and contain two concave corners in regions associated with x, y, and z; the rightmost one contains only one concave corner but the region associated with c is a T-shape. These two T-free drawings serve as prototypical concepts of the algorithm presented in the next section.

5   Construction of 12-Sided T-free Rectilinear Duals

In this section, we present an algorithm to construct 12-sided T-free rectilinear duals for maximal plane graphs. Our construction uses only monotone staircase modules, which are orthogonally convex ones that cannot degenerate to T-shape modules (Fact 3).

Our algorithm is an inductive approach based on the separation-trees described in Sect. 2. What remains to be done is to make sure that when inserting the rectangular dual of $G_\triangle \setminus \triangle$ to the current rectilinear dual during the course of the construction, (1) every region preserves the shape of a monotone staircase and (2) the total number of concave corners on the boundary of each region is at most 4.

5.1   Un-contracting a Separating Triangle

When we un-contract a triangle $\triangle = \{x, y, z\}$, a rectangular space is allocated to accommodate the details of $G_\triangle \setminus \triangle$, which in turn imposes a concave corner to one of the border of region($\{x, y, z\}$). Without loss of generality, we assume such a concave corner to be on the boundary of region x. As observed in Sect. 4, one concave corner in region($\{x, y, z\}$) may not be enough in some cases. In order to enforce the staircase constraint, we further annotate one of its four sides as *"allowed to add a concave corner"*, which is indicated by an arrow in the illustration. See Fig. 7.

Since region x is monotone staircase, x borders either the entire west and north boundary or the entire east and south boundary of the rectangular space. Therefore, there are eight cases in total since the arrow can point to any one of the four sides of the rectangular boundary. It is sufficient to consider the following two cases: (1) region x borders the west and north sides, and the arrow points to the north side; and (2) region x borders the west and north sides, and the arrow points to the east side, as the remaining cases are symmetric (by flipping the entire drawing around the north-west to south-east line or the north-east to south-west line). Now, we fix region y to be the one that borders the east side of the rectangular space.

A key in our un-contracting process is to identify three special vertices associated with each separating triangle. Consider Fig. 7. Let u, v, and w be the three vertices in $G_\triangle \setminus \{x, y, z\}$ such that u, v, and w are adjacent to $\{x, y\}, \{y, z\}, \{x, z\}$, respectively. It is easy to see that u, v, and w are uniquely determined; otherwise, there must be a separating triangle in $G_\triangle$, which contradicts its definition. Unless $|V(G_\triangle)| = 4$ (in this case, $u = v = w$), u, v and w must be different from each other (otherwise, a separating triangle can be found in $G_\triangle$). When the rectangular dual of $G_\triangle \setminus \{x, y, z\}$ is constructed to fill the rectangular space, we may further assume the rectangular dual to have region w adjacent to the entire west side. Such a drawing must exist since there is no separating triangle inside the quadrangle $\{x, y, z, w\}$.

If we consider the children of the node associated with $\triangle = \{x, y, z\}$ in the separation tree, there are two types of separating triangles:

1. Separating triangles associated with triangles $\{x, y, u\}$, $\{y, z, v\}$, and $\{x, z, w\}$ possibly. (Note that some of these three triangles may not be separating triangles.)
2. Separating triangles in the subgraph surrounded by vertices $\{x, u, y, v, z, w\}$. See Fig. 7.

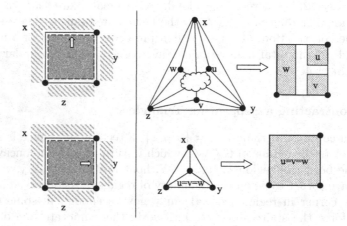

Fig. 7. The left part shows the two cases explained in Sect. 5.1. The right part shows how the rectangular regions associated with u, v, and w are located in the rectangular space for the separating triangle $\{x, y, z\}$.

We first consider Type 1 separating triangles, i.e., those associated with triangles $\{x, y, u\}$, $\{y, z, v\}$, and $\{x, z, w\}$ possibly. Depending on whether the arrow points to the north side or the east side of the rectangular space for $G_{\{x,y,z\}}$, the solutions are depicted as those regions surrounded by dashed boundaries in the rightmost figures in the upper and lower parts of Fig. 8, respectively. Note that in each of the cases, these three regions correspond to the regions allocated for Type 1 separating triangles, with some of which possibly be void if they are not separating triangles. Special attention should be given to the directions of the arrows in those regions. The checkered regions represent those allocated for the special vertices u, v, and w. The white spaces in Fig. 8 are parts of rectangular duals of the subgraph surrounded by vertices $\{x, u, y, v, z, w\}$, and the white dots indicate points at which there may be separating triangles when Type 2 separating triangles are included.

Now we consider Type 2 separating triangles. Recall from Fig. 4 that when a separating triangle is un-contracted, a rectangular region is inserted at the juncture of the three regions associated with the three vertices of the separating triangle. Depending on the orientation of the three regions, there are four cases as illustrated in the upper part of Fig. 9. Attention should be given to the arrows in those regions, which indicate the sides where additional concave corners are possible during the course of future un-contraction. Although Fig. 9 shows the general rule for allocating spaces to accommodate Type 2 separating triangles, there is an exception. Consider the white point incident to both u and x in the upper illustration of Fig. 8 (see also the lower illustration of Fig. 9). One can see from Fig. 8 that the u module stretches upwards in order to make room for the separating triangle $\{x, u, y\}$ (i.e., the top-most red region in the upper illustration of Fig. 8). As a result, we apply rule (3) in Fig. 9, as opposed to rule (4), as illustrated in the lower illustration of Fig. 9. Such special care prevents the creation of an additional (undesired) concave corner to region($\{x, y, z\}$). A similar situation occurs at the juncture of regions u and y in the lower illustration of Fig. 8.

It is clear from the above that all operations preserve monotone staircase shape. What remains to do is to count the number of concave corners for each region. In Fig. 9, when we make a rectangular space at point p, if p is a non-corner of region s, no concave corner is imposed on s. Therefore, for any vertex s not belonging to $\{u, v, w\}$, the number of concave corners imposed on s is at most four since a rectangle has four corners.

For the special vertices u, v, w, if $s = u = v = w$, it is easy to see that we also impose at most four concave corners on s. In Fig. 8, we make one concave corner and three arrows (which may potentially become concave corners) to s. For the case that u, v, w are distinct vertices, the results are summarized in the following:

u: 0 concave corner, 1 arrow, and 3 white dots; the total amount is 4.
v: 1 concave corner, 1 arrow, and 3 white dots; the total amount is 5.
w: 0 concave corner, 1 arrow, and 2 white dots; the total amount is 3.

So far, our algorithm can compute a monotone staircase rectilinear dual that uses modules of at most 14 ($=2 \times 5 + 4$) sides, as the number of sides $= 2 \times$ (the number of concave corners) $+ 4$. To lower the polygonal complexity from 14 to 12, our approach is to transfer one concave corner from v to w. Our solution is presented in the following sub-section.

5.2   Transferring Concave Corners

Let $S = V(G_\triangle) \setminus \{x, y, z\}$, and given a rectangular dual R_0 of $G_\triangle \setminus \{x, y, z\}$ that satisfies the conditions described in Sect. 5.1, i.e., (1) the west and north boundaries of R_0 are adjacent to x, the east and south boundaries of R_0 are adjacent to y and z, respectively, and (2) w (the unique vertex adjacent to x and z) borders the entire west boundary of R_0, we define a relation "$\leftarrow$" on S:

Definition 3. *Given S and R_0, "$\leftarrow$" is a relation on S such that: $s \leftarrow s'$ iff (1) the west side of s is more west than (i.e., on the left-hand side of) the west*

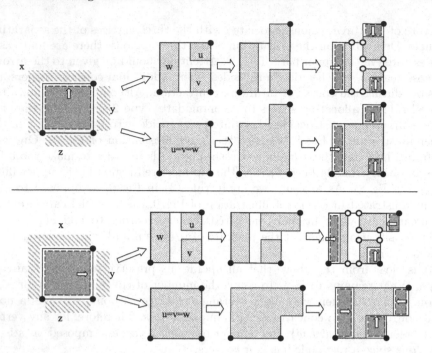

Fig. 8. Illustration of un-contracting Type 1 triangles.

side of s', and (2) there is a point p in R_0 such that p is a $180°$ corner in s and a $90°$ corner in s'.

Regarding the separating triangle $\{x, y, z\}$ discussed in Sect. 5, the following lemmas are easy to observe. Lemma 2 directly follows from the fact that w is the unique vertex adjacent to x and z.

Lemma 2. w is the only one that touches both the north boundary and the south boundary of R_0.

Lemma 3. There exists a path $v = s_1, s_2, \ldots, s_k = w$ in S such that $s_{i+1} \leftarrow s_i$ for $1 \leq i \leq k - 1$.

Let $s_1, s_2, \ldots, s_k$ be the path that satisfies Lemma 3. Our concave corner transfer algorithm works as follows: For $i = 1$ to k, if there is a separating triangle $\triangle' = \{s_i, s_{i+1}, t\}$ for some $t \in S$, we re-build the rectangular space for $\triangle'$ as depicted in Fig. 10, which is capable of "shifting" 1 concave corner from s_i to s_{i+1}. Note that in the first two illustrations in Fig. 10, the concave corner in s_i is explicit, whereas in the last two illustrations, the concave corner in s_i is implicitly indicated by the arrow. The procedure terminates if there is no such triangle; in this case, the number of concave corners in s_i must be smaller than four before the execution of this algorithm. Therefore, all the regions must have at most four concave corners after the concave corner transfer algorithm ends.

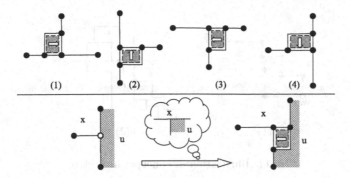

Fig. 9. Illustration of un-contracting Type 2 triangles.

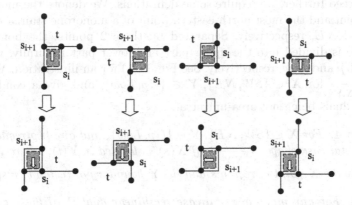

Fig. 10. Illustration of transferring concave corners.

It is easy to see that the algorithm presented in this section for constructing monotone staircase rectilinear duals can be implemented in linear time.

Theorem 2. T-*free rectilinear duals for maximal plane graphs have polygonal complexity of at most 12. Moreover, there is a linear time algorithm that constructs monotone staircase rectilinear duals for maximal plane graphs.*

6   Area-Universal Drawing

Motivated by the applications to cartogram design and floorplanning, for instance, there has been an increasing interest in the study of area-universality of rectilinear duals. A rectilinear dual is *area-universal* iff every possible area assignment can be realized by a combinatorially equivalent one. In view of our earlier discussion, it is natural to investigate how or whether area-universal rectilinear duals can be constructed in the absence of T-shape modules. Our first result shows that restricting modules to monotone staircases is insufficient to construct area-universal rectilinear duals for maximal plane graphs.

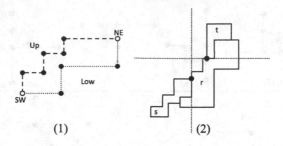

Fig. 11. Illustration of concepts in Sect. 6.

To proceed further, we require some definitions. We denote the most south-western point and the most north-eastern point of a monotone staircase module as SW and NE, respectively. Separated by these 2 points, the boundary of the module is divided into the upper part and lower part naturally, which we denote as Up and Low, respectively. See Fig. 11(1) for an illustration. We define relations $\xrightarrow[X,Y]{}$ for $X \in \{SW, NE\}$, $Y \in \{Up, Low\}$, and give a condition for rectilinear duals being not area-universal.

Definition 4. *For $X \in \{SW, NE\}$, $Y \in \{Up, Low\}$, and any two regions s, t of monotone staircase shape, $s \xrightarrow[X,Y]{} t$ iff $X(s)$ is located in $Y(t)$, where $X(s)$ and $Y(t)$ denote the X point of region s and the Y boundary of region t, respectively.*

Lemma 4. *For any monotone staircase rectilinear dual R, if there exist three regions r, s, t and $Y \in \{Up, Low\}$ such that $s \xrightarrow[NE,Y]{} r$, $t \xrightarrow[SW,Y]{} r$, and $NE(s)$ is more south-west than $SW(t)$, then R is not area-universal.*

Proof. (Sketch) In Fig. 11(2), s, t must be confined in quadrant I, III, respectively, making the sum of areas of s and t to never exceed 50 % of the enclosing rectangular region. $\qquad\square$

Similar to what we have done in Sect. 4, let G be a maximal plane graph and H be a sub-graph of G such that H is isomorphic to H_0 and $G(H) = H$. We consider a monotone staircase rectilinear dual R of G. It is easy to observe that the border between 2 monotone staircase regions cannot intersect with both Up and Low of one of them (otherwise, the other cannot be monotone staircase). Therefore, by the pigeonhole principle, in H we have that two of $\{u, v, w\}$ border c in one of $\{Up(s), Low(s)\}$. We denote these two vertices as s_1, s_2, and the border between s_1, c is more south-west than that of s_2, c. Let $r \in \{x, y, z\}$ be the unique vertex adjacent to both s_1 and s_2, the next lemma reveals a relationship between s_1, s_2, c, and r.

Lemma 5. *If we require R to be area-universal, exactly one of the following must be satisfied for H:*

1. $SW(r)$ is located in $Up(c)$, $s_1 \xrightarrow[NE,Up]{} r$, and $s_2 \xrightarrow[NE,Low]{} r$.
2. $NE(r)$ is located in $Up(c)$, $s_1 \xrightarrow[SW,Low]{} r$, and $s_2 \xrightarrow[SW,Up]{} r$.
3. $SW(r)$ is located in $Low(c)$, $s_1 \xrightarrow[NE,Low]{} r$, and $s_2 \xrightarrow[NE,Up]{} r$.
4. $NE(r)$ is located in $Low(c)$, $s_1 \xrightarrow[SW,Up]{} r$, and $s_2 \xrightarrow[SW,Low]{} r$.

Theorem 3. *There exists a maximal plane graph G such that every monotone staircase rectilinear dual of G is not area-universal.*

Proofs of Lemma 5 and Theorem 3 are omitted due to space limitation. Intuitively speaking, Lemma 5 says that, when the structure depicted in Fig. 11(2) is forbidden, the relative positions of r, s_1, s_2 must obey certain constraints. The idea behind the proof of Theorem 3 is similar to that of Theorem 1. In particular, we are able to show that, for the graph G_1 defined in the proof of Theorem 1, the structure depicted in Fig. 11(2) is unavoidable.

Such an impossibility result motivates us to consider sub-classes of maximal plane graphs. It is easy to observe that the above theorem still holds even restricting to plane 3-trees since replacing a triangle with H_0 preserves the property of being a plane 3-tree. Beside plane 3-trees, Hamiltonian maximal plane graphs, which subsume maximal outer plane graphs and 4-connected plane graphs, are another important sub-class of maximal plane graphs. Contrasting the above results, the following theorem can be shown easily by modifying the ⊔-shape cartogram drawing algorithm described in [1] (by reversing the construction order of the right part, we can get a monotone Z-shape drawing).

Theorem 4. *Hamiltonian maximal plane graphs admit 8-sided area-universal monotone staircase rectilinear duals.*

7 Conclusion

We formalized the concept of "freedom from certain module type" in the study rectilinear duals, and presented a linear-time algorithm to construct T-free rectilinear duals for maximal plane graphs with optimal polygonal complexity 12. Our construction uses only monotone staircase modules. Towards a more comprehensive understanding of modules in rectilinear duals, a natural direction for future research is to investigate other kinds of previously unstudied restrictions to modules.

We also showed that monotone staircase modules are insufficient for constructing area-universal rectilinear duals of maximal plane graphs, and proved that for Hamiltonian plane graphs, eight sides are sufficient. As all existing constructions for area-universal rectilinear dual of maximal plane graphs are not T-free, whether T-free modules are sufficient to construct such drawings remains an interesting open problem.

References

1. Alam, J.M., Biedl, T., Felsner, S., Kaufmann, S.G., Ueckert, T.: Computing cartograms with optimal complexity. Discrete Comput. Geom. **50**(3), 784–810 (2013)
2. Alam, J.M., Biedl, T., Felsner, S., Gerasch, A., Kaufmann, M., Kobourov, S.G.: Linear-time algorithms for hole-free rectilinear proportional contact graph representations. Algorithmica **67**(1), 3–22 (2013)
3. Biedl, T., Ruiz Velázquez, L.E.: Orthogonal cartograms with few corners per face. In: Dehne, F., Iacono, J., Sack, J.-R. (eds.) WADS 2011. LNCS, vol. 6844, pp. 98–109. Springer, Heidelberg (2011)
4. Eppstein, D., Mumford, E., Speckmann, B., Verbeek, K.: Area-universal rectangular layouts. SIAM J. Comput. **41**(3), 537–564 (2012)
5. He, B.D.: A simple optimal binary representation of mosaic floorplans and Baxter permutations. Theor. Comput. Sci. **532**(1), 40–50 (2014)
6. Kant, G., He, X.: Regular edge labeling of 4-connected plane graphs and its applications in graph drawing problems. Theor. Comput. Sci. **172**(1–2), 175–193 (1997)
7. Kar, B., Sur-Kolay, S., Rangarajan, S.H., Mandal, C.R.: A faster hierarchical balanced bipartitioner for VLSI floorplans using monotone staircase cuts. In: Rahaman, H., Chattopadhyay, S., Chattopadhyay, S. (eds.) VDAT 2012. LNCS, vol. 7373, pp. 327–336. Springer, Heidelberg (2012)
8. Kawaguchi, A., Nagamochi, H.: Drawing slicing graphs with face areas. Theor. Comput. Sci. **410**(11), 1061–1072 (2009)
9. Kozminski, K., Kinnen, E.: Rectangular dual of planar graphs. Networks **15**, 145–157 (1985)
10. Liao, C., Lu, H., Yen, H.: Compact floor-planning via orderly spanning trees. J. Algorithms **48**(2), 441–451 (2003)
11. Majumder, S., Sur-Kolay, S., Bhattacharya, B., Das, S.: Hierarchical partitioning of VLSI floorplans by staircases. ACM Trans. Des. Automat. Elect. Syst. **12**(1), 141–159 (2007)
12. Ueckerdt, T.: Geometric Representations of Graphs with Low Polygonal Complexity. Ph.D. thesis, Technische Universitat Berlin (2011)
13. Yeap, K., Sarrafzadeh, M.: Floor-planning by graph dualization: 2-concave rectilinear modules. SIAM J. Comput. **22**, 500–526 (1993)

Optimal Strategy for Walking in Streets with Minimum Number of Turns for a Simple Robot

Azadeh Tabatabaei[1]([⊠]) and Mohammad Ghodsi[2,3]

[1] Department of Computer Engineering,
Sharif University of Technology, Tehran, Iran
atabatabaei@ce.sharif.edu
[2] Sharif University of Technology, Tehran, Iran
[3] School of Computer Science,
Institute for Research in Fundamental Sciences (IPM), Tehran, Iran
ghodsi@sharif.edu

Abstract. We consider the problem of walking a simple robot in an unknown street. The robot that cannot infer any geometric properties of the street traverses the environment to reach a target t, starting from a point s. The robot has a minimal sensing capability that can only report the discontinuities in the depth information (gaps), and location of the target point once it enters in its visibility region. Also, the robot can only move towards the gaps while moving along straight lines is cheap, but rotation is expensive for the robot. We maintain the location of some gaps in a tree data structure of constant size. The tree is dynamically updated during the movement. Using the data structure, we present an *online* strategy that generates a search path for the robot with optimal number of turns.

Keywords: Computational geometry · Minimum link path · Simple robot · Street polygon · Unknown environment

1 Introduction

Due to many real life applications, path planning in unknown environments is considered as a fundamental problem in robotics, computational geometry and online algorithms [2,12,17]. A robot based on the information gathered from its tactile sensors moves in the environment until it achieves its target. Neither the geometric map of the environment nor the location of the target point are known to the robot. The volume of information provided to the robot depends on the strength of its sensors. Employing a simple robot with a simple sensing model has many advantages such as: low cost of hardware, being applicable to many situations, and being robust against sensing uncertainty and noise [3–5,20].

Here a simple robot with an abstract sensor that can only detect the order of discontinuities in the depth information (or gaps) in its visibility region is

© Springer International Publishing Switzerland 2014
Z. Zhang et al. (Eds.): COCOA 2014, LNCS 8881, pp. 101–112, 2014.
DOI: 10.1007/978-3-319-12691-3_9

considered. Each discontinuity represents a portion of the environment that is not visible to the robot (Fig. 1). A label of L or R is assigned to each gap g depending on which side of the gap the hidden region is. Also, the robot recognizes a target point t as it enters in its omnidirectional and unbounded field of view. Moving along straight lines is cheap, but rotation is expensive for the robot. The robot using the information gathered through the sensor starts navigating a street environment from a start point s to reach a target t. A street is a simple polygon P with two vertices s and t such that the counter-clockwise polygonal chain R_{chain} from s to t and the clockwise one L_{chain} from s to t are mutually weakly visible. In other words each point on the left chain is visible from at least one point on the right chain and vice versa [8], (Fig. 1.a). Note that minimizing the number of turns is an essential criterion in path planning for such robot. This problem is also known as the shortest path problem in the link metric, in some literatures [10, 13].

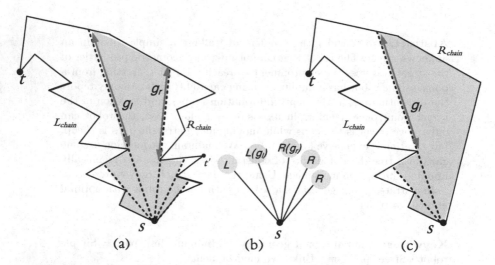

Fig. 1. (a) A street in which L_{chain} is the left chain and R_{chain} is the right chain. The colored region is the visibility polygon of the point robot at the start point s. (b) The position of discontinuities in the depth information detected by the sensor at the start point in (a). (c) A street that has only left gaps at the start point.

The simple robot system applied in this paper was first presented by Lavalle *et al.* [19]. They proposed Gap Navigation Tree (GNT) as a mean to maintain location of the gaps sensed by the robot for navigating the unknown scene. The topological changes of the robot's visibility region, along a path, modify the position of the gaps. Once the GNT is completed, it encodes the shortest path from the current position of the robot to any place in a simply connected environment. Also, it is proved in [18] that, using this data structure, locally optimal searching can be achieved. An online algorithm for the well-known visibility problem pursuit-evasion in an unknown simply connected environment is proposed by Guilamo *et al.* [4,14] using GNT. An optimal search algorithm is offered for a disc robot, using GNT, in a simply connected environment [11]. A competitive strategy is presented for walking in streets for a point robot that is equipped with the gap sensor in [16] such that the generated path is at most 11 times longer the shortest path.

Other minimal sensing models have been introduced by other researchers. Suri *et al.* [15] offered a robot system in which the robot can only sense the combinatorial (non-metric) properties of its surroundings. The sensor detects the vertices of the polygon in its visibility region, and can distinguish if there is an edge between consecutive vertices of the region. Their robot is equipped with a compass in [1]. Katsev *et al.* [7] introduced a simple robot that performs wall-following motions and can traverse the interior of the scene only by following a direction that is parallel to an edge of the environment. Despite of the minimal capabilities, all of them shown that their robot can provide many geometric reasoning and executes many non-trivial tasks such as counting vertices, solving pursuit-evasion problems and mapping a polygon.

The problem of walking in unknown streets with minimum number of turns was first studied in [9]. They presented an optimal online algorithm for a robot with an on-board vision system based on the map that was provided by the robot of its visibility region. In other words their robot has access to the map of its visibility region in the street. In this research, our robot is a simple robot that cannot realize any geometric properties of the surroundings such as coordinates, angles or distances in contrast with the robot system applied in [9] for this problem. Our simple robot traverses interior of the street detecting the positions of the gaps while it can only move towards them. Even with such a confined sensing model, we show that the robot can traverse an optimal path as well as the robot employed in [9]. We present a tree data structure of constant size that maintains the essential data for leading the robot. The tree dynamically changes during the movement.

2   The Sensing Model and Motion Primitive

A point robot starts exploring an un known street until the target t is achieved, starting from s. The robot is equipped with a sensor that detects each discontinuities in depth information that referred as gaps. The sensor reports a cyclically ordered location of the gaps in its visibility region, (Fig. 1). Also, the robot allocates a label of L or R (left or right) to each gaps. Each label shows direction of the hidden region behind a gap relative to the robot's heading [18]. The robot can only track the gaps and records their topological changes. These changes are: appearance, disappearance, merging, and splitting of gaps. The appearance and disappearance events happen when the robot crosses the inflection rays (Fig. 2). Each appearance event generates a gap that corresponds to a portion of the environment that was so far visible, and now is invisible. A gap that is generated by an appearance event, during the movement, is called a primitive gap and the other gaps are non-primitive gaps. The merge and split events occur when the robot crosses the bitangent complements (Fig. 2).

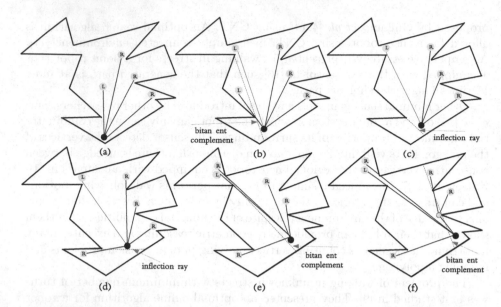

Fig. 2. Illustration of the dynamically changes of the gaps as the robot moves towards a gap. The dark circle denotes the location of the robot, and squares and other circles display primitive and non-primitive gaps respectively. (a) Existing gaps at the beginning. (b) A split event. (c) A disappearance event. (d) An appearance event. (e) Another split event. (f) A merge event.

In order to cover entire region, the robot follows the non-primitive gaps; the region hidden behind the primitive gaps has already been covered. The robot moves along a straight line towards non-primitive gaps, and may rotate as a critical event occurs. Also, the robot makes a turn when a wall of the environment is hit. As the target enters in the robot's visibility region, the robot orients its heading with the target, and walks towards it. Note that moving along straight lines is cheap, but rotation is expensive. At the point in which there is no non-primitive gap the entire environment has been observed by the robot.

3   Algorithm

Here we explain our strategy for leading the robot in a street to reach the target t, starting from s, such that the number of turns in the robot's search path is as small as possible. At each point of the search path, the sensor detects the target t or reports a set of gaps with the label of L or R (l-gap and r-gap for abbreviation).

Definition 1. *[16] In the set of non-primitive l-gaps, the gap which is in the right side of the others is called most advanced left gap and is denoted by g_l. Analogously, in the set of non-primitive r-gaps, the gap which is in left side of the others is called the most advanced right gap and is denoted by g_r (Fig. 1.b).*

We use the following property of streets in searching the target t.

Lemma 1. *[8, 16] On any point of the robot search path, the target t is hidden behind one of the most advanced gaps unless t is visible to robot.*

Proof. Let the target be hidden behind another gap, for example t' in Fig. 1.a is behind an r-gap. Then, the points that are immediately behind g_r are not visible from any point on the R_{chain} that connects s to t'. This is a contradiction with the definition of street.

From Lemma 1, the target is constantly behind the most advanced gaps. So, at the start point s, there is at least one of g_l and g_r unless t is visible from s. If only one of them exists, the robot moves towards the gap in order to cover the region that is behind it (Fig. 1.c). The case in which both of advanced gaps exist is called a funnel case [6,16]. When a funnel situation arises at the start point, the robot moves towards one of the most advanced gaps, for example g_l, to cover the region hidden behind it (Fig. 1.a). Observe that moving along an advanced gap in a funnel situation also decreases the region that is behind the other advanced gap. As the robot traverses interior of the street, g_l and g_r dynamically change. During the traversing, we hold the essential data for memorizing the location of two gaps in a tree. The tree has only two branches that address paths towards g_l and g_r. At the beginning, the start point is root of the tree and g_l and g_r are its children that are circularly ordered around the root. Always the robot's location is root of the tree. The critical events (appearance, disappearance, merge, and split) that change the structure of the robot's visibility region, dynamically change branches of the tree, g_l and g_r, (Fig. 3). The tree data structure is updated as follows:

1. When the robot crosses a bitangent complement of g_r/g_l and another r-gap/l-gap, then g_r/g_l splits and will be replaced by the r-gap/l-gap, (point 1 in Fig. 3).
2. When the robot crosses a bitangent complement of g_r/g_l and an l-gap/r-gap, then g_r/g_l splits into two gaps. g_l/g_r will be replaced by the l-gap/r-gap, (point 2 in Fig. 3 and point 2 in Fig. 4.a).
3. Each appearance event generates a gap that hides a portion of the street that already was visible. Such gap is a primitive gap. So, the data structure does not update by arising an appearance event, (point 3 in Fig. 3). A primitive gap changes the data structure when a most advanced gap merges with the gap; the most advanced gap will be a child of the primitive gap in the data structure, (point 4 in Fig. 3).
4. When the robot crosses over an inflection ray, each of g_l or g_r which is adjacent to the ray, disappears and is eliminated, (point 5 in Fig. 3 and point 1 in Fig. 4.a).

The dynamically changes of the data structure, as the robot traverses the street, are illustrated in Fig. 3.

Our main idea for reducing the number of turns (links) of the search path is maximum use of a selected direction. In other words the robot continues moving

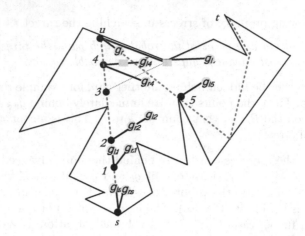

Fig. 3. g_{rs} and g_{ls} are the most advanced gaps at the start point s. g_{ri} and g_{li} are the most advanced gaps at point i. The dotted path that connects s to t is the robot search path. Both of the two most advanced gaps are active at the start point. At point 1, event (1) arises that updates g_r. At point 2, event (2) arises that updates g_l and sets it as an inactive gap. At point 4, event (3) arises. At point 5, the right most advanced gap disappears.

along a selected direction as long as at least one of the hidden region behind the two gaps decreases. In our strategy, each decision for turning is only based on the critical events (appearance, disappearance, merge, and split) which change the robot's visibility region while in the presented strategy in [9] the robot based on the available map of its visibility region selects a direction to move, and makes its decisions for turning. We refer to a most advanced gap as an *active* gap if further movement along the selected direction allows the robot to see more of the hidden region behind the gap.

At the start point, if the funnel situation arises, the robot moves towards g_l to cover the hidden region behind it. By this movement the hidden region behind g_r also decreases. So, by the definition, g_l and g_r are active (Fig. 3). If the other situation arises, in which there exists only on advanced gap, the robot moves towards it to cover the region behind this gap. So, the gap is also an active gap, by the definition. During the movement a gap may switches from being active to being inactive and vice versa. Some of the above events which update g_l and g_r may exchange an advanced gap from being active to being inactive and vice versa. In the following, the events that exchange a gap from being active to being inactive and vice versa is explained in both conditions, funnel condition and the other condition.

– In a funnel condition, the active most advanced gaps may switch from being active to being inactive as soon as the condition ends. It ends if one of these events occurs: (i) The robot enters a point in which the two most advanced gaps are collinear, (point 1 in Fig. 4.b). Further movement along the current direction constructs a new funnel case. One of the most advanced gaps of this funnel is a most advanced gap of the previous funnel, so this gap remains active. The other most advanced gap of new funnel is revealed via splitting the gap of the previous funnel. Further movement along the current direction does not allow the robot to see more of the hidden region behind the gap. So, the gap is set as inactive gap, (point * and gap * in Fig. 4.b). (ii) The robot enters a point in which one of the two active gaps disappears. Note that at this point, only one of the most advanced gaps exists and it is active, (point 1 in Fig. 4.a).

– When there exist only one of the most advanced gaps, for example only g_l exists, a funnel situation starts via splitting g_l into g_l and an r-gap. The most advanced gap of this funnel are g_l and the r-gap. So the r-gap is current g_r. Further movement along the current direction does not allow the robot to see more of the hidden region behind g_r. So, this gap is set as inactive gap. g_l remains active, (point 2 in Fig. 4.a). The situation in which only g_r exists is symmetric, and an inactive g_l may be generated analogously.

The robot continues to walk along the selected direction and makes a turn as often as each of the following conditions occurs.

1. The robot hits a point u on a wall such that it cannot proceed further.
   Recall that the counter-clockwise polygonal chain from s to u or the clockwise one from s to u or both are weakly visible from each simple path that connects s to u [9]. So, by arising a disappearance event or a split event, the most advanced gap that lies on the chain has become inactive before reaching the hit point u. At the hit point, the robot moves towards the gap that has not become inactive. Also at the turn point, the existing most advanced gaps will be set as active gaps again, (point u in Fig. 3).
2. The robot achieves a point in which none of the most advanced gaps are active. Further movement along the current direction does not allow the robot to see more hidden region behind the gaps. One of the situations below has arisen.
   – The existing active most advanced gap merges with an inactive gap (point 3 in Fig. 4.a and point 2 in Fig. 4.b). At this point, the robot turns towards the merged gap and the gap will be set as an active gap.
   – The existing active most advanced gap disappears, (point 2 in Fig. 5). At this point, the robot turns towards the existing advanced gap and the gap will be set as an active gap.

4   Analysis of the Algorithm

In this section, we enumerate the number of the links of the generated path by our strategy for the robot to reach the target t starting from s. Also we prove a competitive ratio for our strategy. Assume that $SP(s,t)$ is the Euclidean shortest path from s to t. An edge $u_i u_j$ of $SP(s,t)$ is called an eave if u_{i-1} and u_{j+1} lie on the different sides of the line that connects u_i to u_j [10] (Fig. 5). First we consider a simple case in which the shortest path between s and t has no eave; $SP(s,t)$ has only right turns or only left turns.

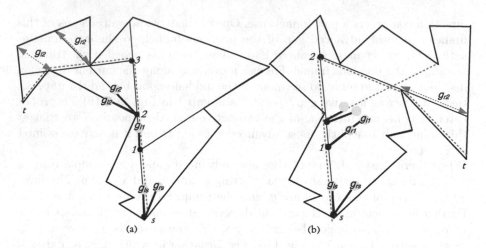

Fig. 4. The dotted path is the robot search path from s to t. (a) A disappearance event occurs at point 1. At point 2, event (1) arises that updates g_l, also an event (2) arises that updates g_r and sets g_r as an inactive gap. At point 3, the active gap g_l merges with the inactive gap g_r. (b) At point 2 the active gap g_r merges with a non-primitive gap.

Lemma 2. *Consider the case that $SP(s,t)$ has only left turns and t is not in the robot's visibility region at the start point s. Our strategy generates a path for the robot to reach the target t with an optimal number of links.*

Proof. Since t is not visible, the robot moves towards g_l. If the target is visible from a point along this direction, by the strategy, the robot does not turn before reaching the point. So, the robot turns only at the point and moves towards t. In the case when t is not visible from any point along the direction, the robot can make a turn when either condition (1) or condition (2) for turning occurs, (Fig. 6.a). Here the key observation is that at each of the turn points, $z_1, z_2, ..., z_n$, both of the left tangent and the right tangent to $SP(s,t)$ lie inside the street (Fig. 6.a). Any minimum link path between s and t must intersect the left tangents to $SP(s,t)$ from $z_1, z_2, ..., z_n$ [10]. Whereas no link can intersect more than one of the left tangents, the path is a minimum link path.

Lemma 3. *If $SP(s,t)$ has only right turns and t is not in the robot's visibility region at the start point s, then our strategy generates a path for the robot to reach the target t that has at most one link more than the optimal path.*

Proof. At the start point s there is two situations: In the first situation, some l-gap exist at the start point. So, the robot moves towards g_l. The case that t is visible from a point on this direction is similar to the corresponding case in the proof of Lemma 2. So, the robot moves from the point towards t. When t is not visible from any point along the direction, g_l becomes inactive at some point x. So either case (1) or case (2) for turning arises, the robot makes a right turn

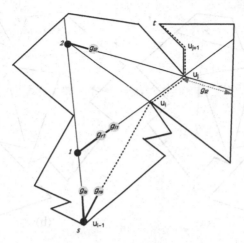

Fig. 5. At the start point both of g_l and g_r are active. At point 1, g_l becomes inactive. At point 2, g_r becomes inactive, as well as g_l. So, the robot makes a turn at this point. The dotted path is the shortest path from s to t and $u_i u_j$ is an eave.

towards the current g_r. Similarly at next turn points the robot makes a right turn. A similar argument to the proof of Lemma 2 shows that the generated path from the first turn point to t has minimum number of links. Then the robot achieves t with at most one additional turn than the optimal (Figs. 4.b and 6.b). In the second situation, no l-gap exists at the start point. The robot moves towards g_r. This condition is symmetric to the situation of Lemma 2 and the number of the robot's turns to achieve the goal is optimal. Hence the claim is proved at the both first and second situations.

Now we consider the general case in which $SP(s,t)$ has both left and right turns. Following property of a street is used as a guideline for obtaining the competitive ratio of our algorithm.

Lemma 4. *[10]   There exists a minimum link path that contains all eaves of $SP(s,t)$.*

In the general case, there are some eaves in the path. Let $u_i u_j$ be the first eave of the path. The shortest path from s to u_i has only right turns or only left turns. If we extend the eave to the street boundary of both side, by our strategy, the robot achieves a point x_i on the first extension with at most one link more than the optimal path, see Fig. 7.

Assume that the $SP(s,t)$ makes left turns from s to u_i and makes a right turn at u_j. The robot, after passing through the point x_i, turns left as soon as each of the conditions (1) or (2) for turning arises and crosses the eave. by our strategy, g_l becomes inactive before the robot achieves the next turn point. So, the robot turns towards the current g_r and crosses the other extension of the eave, x_j. Thus the robot traverses from first extension of an eave to the other

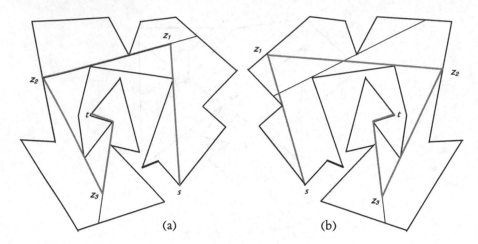

Fig. 6. The bold path is the robot search path. The left and right tangent from each turn point z_i lies inside the street. (a) Shortest path from s to t has only left turns. (a) Shortest path from s to t has only right turns.

extension of the eave with at most two links. Above discussion and Lemma 4 prove the main result of this paper.

Theorem 1. *The robot achieves the target t, starting from s, with at most $m + 1 + e$ links where m is the link distance between s and t and e is the number of eaves in $SP(s, t)$. Also, the generated path by our strategy is optimal.*

Proof. From Lemma 3, the robot may make an additional turn if $SP(s, t)$ turns right at the start point to reach a point on the extension of the first eave. Above discussion shows that the robot makes an additional turn for traversing from first extension to the other extension for every eaves. Since shortest path between two consecutive eave has only right/left turns, the arguments used in the proof of Lemma 2 proves that robot traverses the distance with optimal number of links. So, the robot achieves the target using at most $m + 1 + e$ links. The number of links is equal to the number of the links of the optimal path proposed in [9] for a robot with on-board vision system. Hence this result is optimal for our robot with the minimal sensing capability.

Theorem 2. *Our online strategy terminates while a search path from s to t with the competitive ratio of $2 - 1/m$ is generated for the simple robot, using a constant size memory space.*

Proof. In order to find the target, the robot moves towards the most advanced gaps. So all things that we maintain, during the traversing, are the location of the two gaps in the tree data structure of constant size. Although new most advanced gaps may reveal via splitting during the movement, the number of such events are finite; each corresponds to a crossing over a bitangent complement. The robot is one step closer to reach the target as soon as a most advanced gap

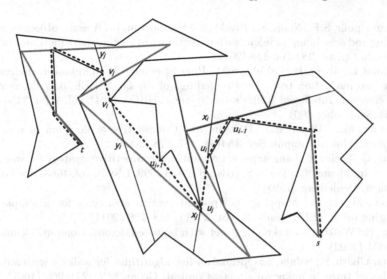

Fig. 7. The general case in which shortest path from s to t has both left and right turns. Bold path is the robot search path.

disappears. So, the strategy should terminate while the target is achieved. From Lemma 4, $e \leq m - 2$. Then, the number of links of the path is at most $2m - 1$, from Theorem 1. Hence the competitive ratio is $2 - 1/m$.

5 Conclusions

In this paper, we considered the problem of walking in streets for a point robot that has a minimal sensing capability. The robot can only detect the gaps and the target in the street. Also, moving along straight lines is cheap, but rotation is expensive for the robot. We proposed an online search strategy that generates a search path for such robot, using a tree data structure of constant size. The robot, starting from s, traverses the street to reach the target t with at most $2m - 1$ links, where m is the link distance between s and t. In other words our strategy has optimal competitive ratio of $2 - 1/m$. Proposing a competitive search strategy for the simple robot in more general classes of polygons is an attractive open problem.

References

1. Disser, Y., Ghosh, S.K., Mihalk, M., Widmayer, P.: Mapping a polygon with holes using a compass. Theor. Comput. Sci. (in press, corrected proof) (Available online 18 December 2013)
2. Fekete, S.P., Mitchell, J.S.B., Schmidt, C.: Minimum covering with travel cost. J. Comb. Optim. **24**, 32–51 (2003)

3. Ghannadpour, S.F., Noori, S., Tavakkoli-Moghaddam, R.: A multi-objective vehicle routing and scheduling problem with uncertainty in customers request and priority. J. Comb. Optim. **28**, 414–446 (2012)
4. Guilamo, L., Tovar, B., LaValle, S.M.: Pursuit-evasion in an unknown environment using gap navigation trees. In: Proceedings of the 2004 IEEE/RSJ International Conference on Intelligent Robot's and Systems, (IROS 2004), vol. 4, pp. 3456–3462. IEEE, September 2004
5. Hammar, M., Nilsson, B.J., Persson, M.: Competitive exploration of rectilinear polygons. Theor. Comput. Sci. **354**(3), 367–378 (2006)
6. Icking, C., Klein, R., Langetepe, E.: An optimal competitive strategy for walking in streets. In: Meinel, C., Tison, S. (eds.) STACS 1999. LNCS, vol. 1563, pp. 110–120. Springer, Heidelberg (1999)
7. Katsev, M., et al.: Mapping and pursuit-evasion strategies for a simple wall-following robot. IEEE Trans. Robot. **27**(1), 113–128 (2011)
8. Klein, R.: Walking an unknown street with bounded detour. Comput. Geom. **1**(6), 325–351 (1992)
9. Kumar Ghosh, S., Saluja, S.: Optimal on-line algorithms for walking with minimum number of turns in unknown streets. Comput. Geom. **8**(5), 241–266 (1997)
10. Kumar Ghosh, S.: Computing the visibility polygon from a convex set and related problems. J. Algorithms **12**(1), 75–95 (1991)
11. Lopez-Padilla, R., Murrieta-Cid, R., LaValle, S.M.: Optimal gap navigation for a disc robot. In: Frazzoli, E., Lozano-Perez, T., Roy, N., Rus, D. (eds.) Algorithmic Foundations of Robotics X. STAR, vol. 86, pp. 123–138. Springer, Heidelberg (2013)
12. Lpez-Ortiz, A., Schuierer, S.: On-line parallel heuristics, processor scheduling and robot searching under the competitive framework. Theor. Comput. Sci. **310**(1), 527–537 (2004)
13. Mitchell, J.S., Rote, G., Woeginger, G.: Minimum-link paths among obstacles in the plane. Algorithmica **8**(1–6), 431–459 (1998)
14. Sachs, S., LaValle, S.M., Rajko, S.: Visibility-based pursuit-evasion in an unknown planar environment. Int. J. Robot. Res. **23**(1), 3–26 (2004)
15. Suri, S., Vicari, E., Widmayer, P.: Simple robots with minimal sensing: from local visibility to global geometry. Int. J. Robot. Res. **27**(9), 1055–1067 (2008)
16. Tabatabaei, A., Ghodsi, M.: Walking in streets with minimal sensing. In: Widmayer, P., Xu, Y., Zhu, B. (eds.) COCOA 2013. LNCS, vol. 8287, pp. 361–372. Springer, Heidelberg (2013)
17. Tan, X., Bo, J.: Minimization of the maximum distance between the two guards patrolling a polygonal region. Theor. Comput. Sci. **532**, 73–79 (2014)
18. Tovar, B., Murrieta-Cid, R., LaValle, S.M.: Distance-optimal navigation in an unknown environment without sensing distances. IEEE Trans. Robot. **23**(3), 506–518 (2007)
19. Tovar, B., LaValle, S.M., Murrieta, R.: Optimal navigation and object finding without geometric maps or localization. In Proceedings of the IEEE International Conference on Robotics and Automation, ICRA'03, vol. 1, pp. 464–470. IEEE, September 2003
20. Xu, Y., et al.: The canadian traveller problem and its competitive analysis. J. Comb. Optim. **18**(2), 195–205 (2009)

Guarding Monotone Art Galleries
with Sliding Cameras in Linear Time

Mark de Berg[1], Stephane Durocher[2], and Saeed Mehrabi[2(✉)]

[1] Department of Mathematics and Computer Science, TU Eindhoven,
Eindhoven, The Netherlands
mdberg@win.tue.nl
[2] Department of Computer Science, University of Manitoba, Winnipeg, Canada
{durocher,mehrabi}@cs.umanitoba.ca

Abstract. A sliding camera in an orthogonal polygon P is a point guard g that travels back and forth along an orthogonal line segment s inside P. A point p in P is guarded by g if and only if there exists a point q on s such that line segment pq is normal to s and contained in P. In the minimum sliding cameras (MSC) problem, the objective is to guard P with the minimum number of sliding cameras. We give a linear-time dynamic programming algorithm for the MSC problem on x-monotone orthogonal polygons, improving the 2-approximation algorithm of Katz and Morgenstern (2011). More generally, our algorithm can be used to solve the MSC problem in linear time on simple orthogonal polygons P for which the dual graph induced by the vertical decomposition of P is a path. Our results provide the first polynomial-time exact algorithms for the MSC problem on a non-trivial subclass of orthogonal polygons.

1 Introduction

Art-gallery problems, introduced by Klee in 1973 [10], are one of the most widely studied problems in computational geometry. In the standard version of the problem one is given a simple polygon P in the plane that needs to be guarded by a set of point guards. In other words, one wants to find a set of point guards such that every point in P is seen by at least one of the guards, where a guard g sees a point p if and only if the segment gp is contained in P. Chvátal [1] proved that $\lfloor n/3 \rfloor$ point guards are always sufficient and sometimes necessary to guard a simple polygon with n vertices. Lee and Lin [9] showed that finding the minimum number of point guards needed to guard an arbitrary polygon is NP-hard for arbitrary polygons. The art-gallery problem is also NP-hard for orthogonal polygons [11] and it even remains NP-hard for monotone polygons [8].

Mark de Berg—Supported by the Netherlands Organisation for Scientific Research (NWO) under project 024.002.003.

Stephane Durocher—Supported in part by the Natural Sciences and Engineering Research Council of Canada (NSERC).

Saeed Mehrabi—Supported in part by a Manitoba Graduate Scholarship (MGS).

Z. Zhang et al. (Eds.): COCOA 2014, LNCS 8881, pp. 113–125, 2014.
DOI: 10.1007/978-3-319-12691-3_10

Eidenbenz [4] proved that the art-gallery problem is APX-hard on simple polygons and Ghosh [5] gave an $O(\log n)$-approximation algorithm that runs in $O(n^4)$
time on simple polygons. Krohn and Nilsson [8] gave a constant-factor approximation algorithm on monotone polygons. They also gave a polynomial-time
algorithm for the orthogonal art-gallery problem that computes a solution of
size $O(\text{OPT}^2)$, where OPT is the cardinality of an optimal solution.

Many variants of the art-gallery problem have been studied. The version in
which we are interested was introduced recently by Katz and Morgenstern [6],
and it concerns *sliding cameras* in orthogonal polygons. A sliding camera in an
orthogonal polygon P is a point guard that travels back and forth along an
orthogonal line segment $s \subset P$. The camera can see a point $p \in P$ if and only if
there is a point $q \in s$ such that the line segment pq is normal to s and contained
in P. The *minimum sliding cameras (MSC) problem* is to guard P with the
minimum number of sliding cameras.

Katz and Morgenstern first considered a restricted version of the MSC problem in which only vertical cameras are allowed. They solved this restricted version in polynomial time for simple orthogonal polygons. For the unrestricted
version, where both vertical and horizontal cameras are allowed, they gave a 2-
approximation algorithm for x-monotone orthogonal polygons. (An orthogonal
polygon P is x-monotone if the intersection of P with any every vertical line is
connected.) Later Durocher *et al.* [2] gave an $O(n^{2.5})$-time (3.5)-approximation
algorithm for the MSC problem on simple orthogonal polygons. Durocher and
Mehrabi [3] showed that the MSC problem is NP-hard when the polygon P is
allowed to have holes. Durocher and Mehrabi also considered a variant of the
problem, called the MLSC problem, in which the objective is to minimize the
sum of the lengths of line segments along which cameras travel, and proved
that the MLSC problem is polynomial-time solvable even on orthogonal polygons with holes. Seddighin [12] proved that the MLSC problem is NP-hard under
k-visibility for any fixed $k \geq 2$.

Our main interest is in the standard MSC problem, where the objective is
to minimize the number of cameras. As discussed above, the complexity of the
MSC problem on simple orthogonal polygons remains unknown. Indeed, even for
x-monotone orthogonal polygons there is only an approximation algorithm for
the problem. Recall that the classical art-gallery problem is NP-hard on simple
orthogonal polygons [11], simple monotone polygons [8] and even on terrains [7].

Our results. In this paper, we give a linear-time dynamic programming algorithm for the MSC problem on orthogonal x-monotone polygons P. This not only
improves the 2-approximation algorithm of Katz and Morgenstern [6], but also
provides, to the best of our knowledge, the first polynomial-time algorithm for
the MSC problem on a non-trivial subclass of orthogonal polygons. We also show
how to extend this result to so-called *orthogonal path polygons*. These are orthogonal polygons for which the dual graph induced by the vertical decomposition of P
is a path. (The vertical decomposition of an orthogonal polygon P is the decomposition of P into rectangles obtained by extending the vertical edge incident to
every reflex vertex of P inward until it hits the boundary of P. The dual graph of

the vertical decomposition is the graph that has a node for each rectangle in the decomposition and an edge between two nodes if and only if their corresponding rectangles are adjacent.) Observe that the class of orthogonal monotone polygons is a subclass of orthogonal path polygons.

2   Preliminaries

For a simple orthogonal and x-monotone polygon P, the leftmost and rightmost vertical edges of P are unique and we denote them by leftEdge(P) and rightEdge(P), respectively. For a sliding camera s in P, we define the *visibility polygon* of s as the maximal subpolygon $P(s)$ of P such that every point in $P(s)$ is guarded by s.

Let $V_P = \{e_1 = \text{leftEdge}(P), e_2, \ldots, e_m = \text{rightEdge}(P)\}$, for some $m > 0$, be the set of vertical edges of P ordered from left to right. For simplicity we assume that every two vertical edges in V_P have distinct x-coordinates, but it is easy to adapt the algorithm to handle degenerate cases. Let P_i^+ (resp., P_i^-), for some $1 \leq i \leq m$, denote the subpolygon of P that lies to the right (resp., to the left) of the vertical line through e_i.

For an orthogonal line segment s in P, we denote the left endpoint and the right endpoint of s by left(s) and right(s), respectively. If s is vertical, we define its left and right endpoints to be its upper and lower endpoints, respectively. We denote the x-coordinate of a point p by $x(p)$. Let s_i and s_j be two horizontal line segments in P. We define the *overlap region of s_i and s_j* as the set of points in P that are visible to both s_i and s_j; if $P(s_i) \cap P(s_j)$ is a line or a point (i.e., it has measure zero), then we consider the overlap region of s_i and s_j to be empty. We first show that we can restrict our attention to solutions that are in some suitable canonical form.

Canonical Form. A *feasible solution* to the MSC problem is a set M of sliding cameras that guards the entire polygon P. We say that a feasible solution M is in *canonical form* if and only if the following properties hold:

 (i) Every vertical line segment in M is *vertically maximal*, meaning that it extends as far upwards and downwards as possible.
 (ii) No vertical line segment in M intersects the interior or passes through the right endpoint of any horizontal line segment in M.
 (iii) The overlap region of s_i and s_j is empty, for every two horizontal line segments $s_i, s_j \in M$ such that $s_i \neq s_j$.
 (iv) Every horizontal line segment $s \in M$ is *rightward maximal*, meaning that s extends at least as far to the right as any horizontal line segment $s' \subset P$ starting at the same x-coordinate, that is, with $x(\text{left}(s)) = x(\text{left}(s'))$.
 (v) Let $s_1, \ldots, s_k$ be the sequence of line segments in M ordered from left to right according to their left endpoint, where in case of ties vertical line segments come before horizontal line segments, and let $M_i := \{s_1, \ldots, s_i\}$. Then, M_i guards every point of P that is to the left of the vertical line $x = x(right(s_i))$.

Lemma 1. *For any x-monotone orthogonal polygon P, there exists an optimal solution M for the MSC problem on P that is in canonical form.*

Proof. Consider the sequence $s_1, \ldots, s_k$ of line segments in M ordered from left to right according to their left endpoint, where in case of ties vertical line segments come before horizontal line segments. This ordering is well defined, because an optimal solution will never have two vertical line segments with the same x-coordinates or two horizontal line segments whose left endpoints have the same x-coordinates. We now show how to modify the line segments in M to get an optimal solution in canonical form. Without loss of generality, we assume that all vertical segments in M are already vertically maximal.

We first modify M so that if s_1 is horizontal, then left(s_1) lies on leftEdge(P), the leftmost vertical edge of P. Assume this is not the case. Then, leftEdge(P) is seen by a vertical line segment s_j, for some $j > 1$. We now replace s_1 and s_j by two horizontal line segments, as follows. The first line segment is a rightward maximal line segment s starting on leftEdge(P)—note that s must intersect s_j—and the second horizontal line segment is a rightward maximal line segment s' with $x(\text{left}(s')) = x(\text{right}(s))$. Clearly replacing s_1, s_j by s, s' gives another optimal solution. With a slight abuse of notation we let M denote this new optimal solution, and we let $s_1, \ldots, s_k$ denote the ordered set of line segments in the new solution. Note that we now have that if s_1 is horizontal, then it starts at leftEdge(P) and it is rightward maximal.

Next we turn M into an optimal solution in canonical form. To this end we go over the line segments in order. When we handle line segment s_i we will replace s_i by a line segment s_i', but we will not modify any other line segment. Let $M_i := \{s_1', \ldots, s_i'\}$. We maintain the following invariant:

> *Invariant:* After handling s_i, the modified set M is still an optimal solution. Moreover, M_i has all the required properties: (i) all vertical line segments in M_i are vertically maximal, (ii) no vertical line segment in M_i intersects the interior or passes through the right endpoint of any horizontal line segment in M_i, (iii) the overlap region of any two horizontal line segments in M_i is empty, (iv) every horizontal line segment in M_i is rightward maximal, and (v) M_i guards everything to the left of the vertical line $x = x(\text{right}(s_i))$.

Handling s_1 is trivial: we simply set $s_1' := s_1$. If s_1 is vertical then this clearly establishes the invariant—note that no line segment s_j with $j > 1$ can see anything to the left of s_1 that is not also seen by s_1 (since s_1 is vertically maximal), which implies that s_1 must see everything to its left. If s_1 is horizontal then the invariant holds as well, since we already made sure that s_1 is rightward maximal if it is horizontal. Now suppose the invariant holds after we have handled s_{i-1}, and consider s_i. There are two cases.

- If s_i is vertical, then we proceed as follows. Observe that M_i must guard everything to the left of s_i, since M is feasible and no line segment s_j with $j > i$ can see a point to the left of s_i that is not seen by s_i. Since s_i is vertical, the fact that M_{i-1} satisfies the invariant immediately implies that M_i has properties (iii) and (iv). So the only problem is that s_i may intersect the interior or may pass through the right endpoint of some line segment s'_j with $j < i$. If this is not the case we simply set $s'_i := s_i$, otherwise we replace s_i by a rightward maximal line segment s'_i with $x(\text{left}(s'_i)) = x(\text{right}(s'_i))$.

After the replacement, M is still a feasible (and, hence, optimal) solution. Indeed, everything to the left of the vertical line $x = x(\text{right}(s'_j))$ is guarded by M_j and to the right of the vertical line $x = x(\text{right}(s'_j))$, the new line segment s'_i sees at least as much as s_i. By the same argument, M_i guards everything to the left of the vertical line $x = x(\text{right}(s'_i))$ and, therefore, the property (v) holds. Finally, M_i has properties (i) and (ii) because M_{i-1} had those properties and the new line segment s'_i is horizontal, and M_i has properties (iii) and (iv) by construction.

There is one subtlety that we must address. Namely, we have to show that after replacing s_i by s'_i the order of the line segments does not change. In other words, we must show that s'_i is still the i-th line segment in the order. (Otherwise we would have to argue about a different set M_i.) Obviously left(s'_i) lies to the right of left(s'_j) for all $j < i$. Moreover, there cannot be any line segment s_k with $k > i$ such that left(s_k) lies in between s_i and left(s'_i). Indeed, such a line segment could be omitted, contradicting the optimality of M.

- If s_i is horizontal, we proceed as follows. Obviously, the only properties that may be violated are properties (iii) and (iv). It might be the case that a vertical line segment $s_j \in M$ intersects the interior or passes through the right endpoint of s_i (thus violating the property (ii)), but this may happen only if $j > i$ and, therefore, the invariant is still maintained for M_i; if such line segment s_j exists, then the set M will be modified when we later handle s_j. If M_i only violates property (iv), then we replace s_i by a rightward maximal line segment s'_i with $x(\text{left}(s'_i)) = x(\text{left}(s_i))$. If s_i violates property (iii), then let $s'_j \in M_{i-1}$ be the horizontal line segment that has an overlap with s_i. We now replace s_i by a rightward maximal line segment s'_i with $x(\text{left}(s'_i)) = x(\text{right}(s'_j))$.

  Since s'_i sees at least as much as s_i (except possibly for points that were already seen by s'_j), the new solution is still feasible. Moreover, M_i sees everything to the left of the vertical line $x = x(\text{right}(s'_i))$. Therefore, since M_{i-1} satisfies the invariant and because of the way s'_i is constructed, M_i has all the properties (i)–(v). Finally, the new line segment s'_i is still the i-th line segment in the order, as can be verified in the same way as before.

After handling the last line segment s_k in M, the set M_k is an optimal solution in canonical form, thus proving the lemma.   □

3 A Dynamic Programming Algorithm

In this section, we present the linear-time exact algorithm for the MSC problem on orthogonal and x-monotone polygons. Our algorithm is based on a dynamic programming approach.

3.1   The Recursive Structure

Let P be an orthogonal x-monotone polygon with n vertices. Below we discuss the recursive structure of the MSC problem on P and we define the subproblems we use in our dynamic programming algorithm.

Let $M_{\mathrm{OPT}} = \{s_1, \ldots, s_k\}$ be an optimal solution for the MSC problem on P that is in canonical form, where the segments are numbered from left to right. Consider a segment $s_j \in M_{\mathrm{OPT}}$. By property (v) of the canonical form, no segment $s_{j'} \in M_{\mathrm{OPT}}$ with $j > j$ is needed to guard anything to the left of $\mathrm{right}(s_j)$. Hence, after having selected $s_1, \ldots, s_j$, the subproblem we are left with is to guard P_i^+, where i is such that $\mathrm{right}(s_j)$ lies on the line containing the vertical edge e_i. Note that when s_j is vertical, we already guarded a part of P_i^+, and we have to take this into account in our subproblem. Hence, we define two types of subproblems.

Type A. Given $1 \leq i \leq k$, guard P_i^+ with the minimum number of sliding cameras.

Type B. Given $1 \leq i \leq k$, guard P_i^+ with the minimum number of sliding cameras, under the assumption that the subregion of P_i^+ that is visible from $\mathrm{leftEdge}(P_i^+)$ has already been guarded.

We denote the number of guards needed in an optimal solution of Type A on the polygon P_i^+ by $A[i]$ and the number of guards needed in an optimal solution of Type B on the polygon P_i^+ by $B[i]$. Note that the minimum number of cameras needed to guard the entire polygon P is $A[1]$. In the sequel we show how to compute the values $A[i]$ and $B[i]$; computing the actual solution can then be done in a standard manner.

3.2   Solving the Subproblems

We now give the recursive formulas on which our dynamic programming algorithm is based. Recall that the vertical edges of P are number $e_1, \ldots, e_k$ from left to right. We denote the vertical line containing e_i by $\ell(e_i)$. The following lemma gives the recursive formula for solving the subproblem of Type A on P_i^+.

Lemma 2. *Let s be a rightward maximal line segment whose left endpoint lies on $\ell(e_i)$, and let e_{i_1} be the vertical edge of P on which $\mathrm{right}(s)$ lies. Furthermore, let e_{i_2} be the rightmost vertical edge of P such that s', the vertically maximal segment aligned with e_{i_2}, guards everything of P_i^+ lying to the left of e_{i_2}. See Fig. 1 for an illustration. Then,*

$$A[i] = \begin{cases} 0 & if \ i = k \\ \min\left(A[i_1], B[i_2]\right) + 1 & if \ i < k \end{cases}$$

Proof. Trivially $A[i] = 0$ for $i = k$, so assume $i < k$.

Consider the first segment s^* of an optimal solution for P_i^+ that is in canonical form. By property (v), we know that s^* must guard $\mathrm{leftEdge}(P_i^+)$. Hence, if it is horizontal, it must start at $\mathrm{leftEdge}(P_i^+)$. By property (iv), segment s^* is

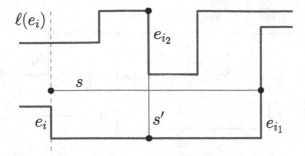

Fig. 1. An illustration of the two cases in solving subproblem of Type A on P_i^+.

rightward maximal. Hence, if the first segment is horizontal then the segment s is the correct choice. After choosing s, we have to guard everything to the right of s. Note that properties (ii) and (iii) imply that the next segment to be chosen lies in $P_{i_1}^+$ (see Fig. 1). Hence, if we decide to pick segment s then we are indeed left with solving the subproblem of Type A on $P_{i_1}^+$. Thus in this case $A[i] = A[i_1] + 1$.

The other option is that the first segment s^* is vertical. Again, by property (v) we know that s^* must guard everything between leftEdge(P_i^+) and s^*. But then it is obviously best to choose s^* as far to the right as possible. Hence, s' is the correct choice. Now the subproblem we are left with is of Type B and on $P_{i_2}^+$ (see Fig. 1), so we have $A[i] = B[i_2] + 1$.

The best way to solve subproblem of Type A on P_i^+ is the best of these two options, which proves the lemma. $\square$

For the subproblems of Type B we have a similar lemma.

Lemma 3. *Let $e_{i'}$ be the leftmost vertical edge in P_i^+ that is not seen by leftEdge(P_i^+), let s be a rightward maximal line segment whose left endpoint lies on $\ell(e_{i'})$, and let e_{i_1} be the vertical edge of P on which right(s) lies. Furthermore, let e_{i_2} be the rightmost vertical edge of P such that s', the vertically maximal segment aligned with e_{i_2}, together with leftEdge(P_i^+) guards everything of P_i^+ lying to the left of e_{i_2}. See Fig. 2 for an illustration. Then,*

$$B[i] = \begin{cases} 0 & if \ i = k \\ \min\left(A[i_1], B[i_2]\right) + 1 & if \ i < k \end{cases}$$

Proof. Trivially $B[i] = 0$ for $i = k$, so assume $i < k$.

Consider the first segment s^* of an optimal solution for P_i^+ that is in canonical form. First, suppose that s^* is horizontal. Obviously it is best to make s^* extend to the right as much as possible, which means left(s^*) should be to the right as far as possible. However, left(s^*) cannot go beyond $e_{i'}$ by property (v). By property (iv), segment s^* is rightward maximal. Hence, if the first segment is horizontal then the segment s is the correct choice. After choosing s, we have

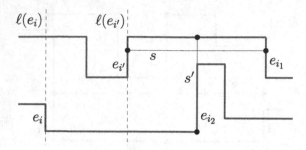

Fig. 2. An illustration of the two cases in solving subproblem of Type B on P_i^+.

to guard everything to the right of s. Note that properties (ii) and (iii) imply that the next segment to be chosen lies in $P_{i_1}^+$. Moreover, since s is rightward maximal and starts to the right of leftEdge(P_i^+), the edge leftEdge(P_i^+) cannot see anything to the right of right(s). Hence, if we decide to pick segment s then we are indeed left with solving the subproblem of Type A on $P_{i_1}^+$ (see Fig. 2). Thus, in this case $A[i] = A[i_1] + 1$.

The other option is that the first segment s^* is vertical. Again, by property (v) we know that s^*, together with leftEdge(P_i^+), must guard everything between leftEdge(P_i^+) and s^*. But then it is best to choose s^* as far to the right as possible. Hence, s' is the correct choice. Now the subproblem we are left with is of Type B and on $P_{i_2}^+$ (see Fig. 2), so we have $A[i] = B[i_2] + 1$.

The best way to solve subproblem of Type A on P_i^+ is the best of these two options, which proves the lemma. □

3.3   Algorithmic Details

In this section, we analyze the algorithm and describe how it can be implemented in linear time. To compute the optimal solution for guarding P_i^+, we need to solve two subproblems; that is, we need to solve a subproblem of Type A and a subproblem of Type B for P_i^+. To solve the subproblem of Type A for P_i^+, we need to solve two subproblems: one is of Type A for which we need to find the vertical edge e_{i_1} described in Lemma 2, and the other one is of Type B for which we need to find the vertical edge e_{i_2} described in Lemma 2. Similarly, to solve the subproblem of Type B for P_i^+, we need to solve two subproblems: one is of Type A for which we need to find the vertical edge e_{i_1} described in Lemma 3, and the other one is of Type B for which we need to find the vertical edge e_{i_2} described in Lemma 3. Therefore, each vertical edge $e_i \in V_P$ is associated with at most four other vertical edges of P; we call these four edges the *associated edges* of e_i. In the following, we show how the associated edges can be computed in $O(n)$ time for all the vertical edges in V_P.

Lemma 4. *The associated edges of all the vertical edges in V_P can be computed in $O(n)$ time.*

Proof. We show that each type of associated edge can be computed in linear time for all the vertical edges in V_P. The lemma then follows from the fact that there are four types of associated edges. We first give some definitions. A reflex vertex v of P is called *right* reflex (resp. *left* reflex) if the interior of P lies to the right (resp., to the left) of the vertical edge incident to v. Moreover, for a reflex vertex v_i of P, we denote the vertical edge incident to v_i by e_i and the maximal vertical line segment in P aligned with e_i by L_i. In the following, we assume that the sequence of the reflex vertices of P ordered from right to left is given.

Step 1: the associated edge e_{i_1} described in Lemma 2. To compute this associated edge, we use a vertical line sweeping P from right to left; the sweep line halts at each reflex vertex of P. Let UQ and LQ be two double-ended queues that store the reflex vertices, respectively, on the upper chain and lower chain of P. By one exception, we assume that UQ (resp., LQ) contains initially the upper vertex (resp., the lower vertex) of rightEdge(P). Reflex vertices are added to the end of the queues, but they might be removed from either the front or the end of the queues. The vertices are removed from a queue depending on whether the vertex v_i at which the sweep line is currently halted lies on the upper chain or on the lower chain of P and also depending on where v_i lies on the chain relative to the previously visited vertices. We maintain the following invariant:

> *Invariant:* When the sweep line halts at the reflex vertex v_i, then (i) the queue UQ stores a reflex vertex v_j of the upper chain if and only if v_j lies to the right of L_i and L_i can see at least one point on L_j; the part of L_j that is visible to L_i is also stored. The vertices in UQ are sorted from right to left by their x-coordinate, and (ii) the queue LQ stores a reflex vertex $v_{j'}$ of the lower chain if and only if $v_{j'}$ lies to the right of L_i and L_i can see at least one point on $L_{j'}$; the part of L_j that is visible to L_i is also stored. The vertices in LQ are sorted from right to left by their x-coordinate.

Consider v_i, the vertex at which the sweep line is currently halted, and suppose that v_x and v_y are the vertices at the front of the two queues. First, we maintain the invariant. To this end, if the part of L_j that is visible to L_i is empty, then we remove v_j from UQ for all v_j in UQ. Similarly, if the part of $L_{j'}$ that is visible to L_i is empty, then we remove $v_{j'}$ from LQ for all $v_{j'}$ in LQ. Next, we set the associated edge for e_i to e_x or e_y whichever is further to the right from e_i. The vertex v_i is then added to the appropriate queue. See Fig. 3 for an example. Since every reflex vertex of P is added to a queue at most once, this step can be completed in $O(n)$ time.

Step 2: the associated edge e_{i_2} described in Lemma 2. To compute this associated edge, consider the reflex vertices of P from right to left sorted by their x-coordinate. Then, the associated edge e_{i_2} described in Lemma 2 for a vertical edge e_i is the edge $e \in V_P$ such that the reflex vertex v incident to e is the leftmost left reflex vertex of P such that $x(v) > x(v_i)$; such vertex v and,

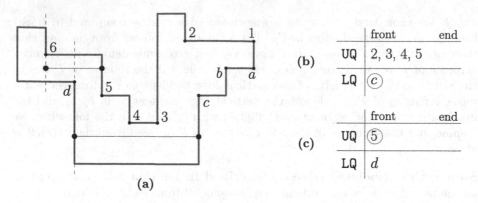

(a)

(b)

	front	end
UQ	2, 3, 4, 5	
LQ	ⓒ	

(c)

	front	end
UQ	⑤	
LQ	d	

Fig. 3. An illustration of the sweep line algorithm. (a) An orthogonal x-monotone polygon P with its reflex vertices on the upper and lower chains labeled from right to left. (b) The status of queues UQ and LQ when the sweep line halts at vertex d and the invariant is maintained: the associated edge e_{i_1} for the vertical edge incident to d is set to the vertical edge incident to vertex c and vertex d is then added to LQ. (c) The status of queues UQ and LQ when the sweep line halts at vertex 6 and the invariant is maintained: the associated edge e_{i_1} for the vertical edge incident to 6 is set to the vertical edge incident to vertex 5 and vertex 6 is then added to UQ.

therefore, its incident vertical edge e can be computed in linear time for all the vertical edges in V_P.

Step 3: the associated edge e_{i_1} described in Lemma 3. To compute this associated edge for an edge $e_i \in V_P$, we first need to compute the vertical edge $e_{i'}$ of P. The edge $e_{i'}$ for e_i is the edge $e \in V_P$ such that the reflex vertex v incident to e is the leftmost right reflex vertex of P such that $x(v) > x(v_i)$. Then, the edge e_{i_1} for e_i is exactly the associated edge that we have already computed for $e_{i'}$ in Step 1. Both vertical edges $e_{i'}$ and e_{i_1} can be computed in linear time for all the vertical edges in V_P.

Step 4: the associated edge e_{i_2} described in Lemma 3. To find this associated edge for a vertical edge e_i, we first find the leftmost right reflex vertex v_j such that $x(v_j) > x(v_i)$; observe that every point of P that lies between L_i and L_j (i.e., the maximal vertical line segments in P aligned with e_i and e_j, respectively) is guarded by leftEdge(P_i^+). Therefore, the associated edge e_{i_2} for e_i is in fact the vertical edge that is furthest to the right from L_j such that every point between L_j and L_{i_2} is guarded by L_{i_2}. But, L_{i_2} is aligned with exactly the associated edge that we have already computed for e_j in Step 2. Therefore, to compute the associated edge e_{i_2} for a vertical edge e_i, we first find the leftmost right reflex vertex v_j to the right of e_i and then return the associated edge computed in Step 2 for e_j. Both vertical edges v_j and e_{i_2} can be computed in $O(n)$ time for all the vertical edges in V_P.

Therefore, we can compute all the four associated edges in $O(n)$ time for all the vertical edges in V_P. This completes the proof of the lemma.   □

By Lemma 4, we first compute the associated edges of all the vertical edges of P in $O(n)$ time. Then, we consider the vertical edges of P in order from right to left and compute the optimal solution for guarding P_i^+ in $O(1)$ time by computing $A[i]$ and $B[i]$ as described, respectively, in Lemma 2 and Lemma 3. Finally, $A[1]$ is returned as the optimal solution for the MSC problem on P. Therefore, we have the main result of this section:

Theorem 1. *There exists an algorithm that solves the MSC problem on any simple orthogonal and x-monotone polygon with n vertices in $O(n)$ time.*

4   Orthogonal Path Polygons

In this section, we show that the dynamic programming algorithm given in Sect. 3 can be used to solve the MSC problem on any orthogonal path polygon P with n vertices in $O(n)$ time; that is, we show that the MSC problem can be solved in $O(n)$ time on any simple orthogonal polygon P for which the dual graph $G(P)$ is a path. To this end, we first describe the structure of P and then will show that P can be converted into an x-monotone polygon by *unfolding*.

Let P be an orthogonal path polygon with n vertices. If P is x-monotone, then we solve the MSC problem on P in linear time by Theorem 1. If polygon P is not x-monotone, then we first partition P into x-monotone subpolygons as follows. Since polygon P is not x-monotone, it must have a vertical edge e whose both endpoints are reflex vertices of P. Parti-

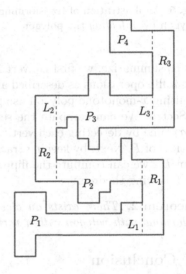

Fig. 4. An example of an orthogonal path polygon P that is not x-monotone along with an illustration of partitioning P into x-monotone subpolygons.

tion P into three subregions by the maximal vertical line segment L that is aligned with e. The subregions induced by L are a rectangle R and two subregions P_L and P_U that are connected to lower and upper parts of one of the sides of R, respectively. Partition P_L and P_R recursively until the subregions induced by the partitions become x-monotone; see Fig. 4 for an illustration. Let $P_1, P_2, \ldots, P_k$ be the set of x-monotone subpolygons of P from bottom to top. Moreover, let L_i, for all $1 \leq i < k$, be the maximal line segment by which we perform the partition and let R_i, for all $1 \leq i < k$, be the corresponding rectangle. Now, for each rectangle R_i in order, we unfold P by flipping the subregion $P_{i+1} \cup P_{i+2} \cup \cdots \cup P_k$ across the line through L_i such that R_{i+1} lies to the same side of L_i as R_i lies. The i-th flip ensures that the subregion $P_1 \cup P_2 \cup \cdots \cup P_{i+1}$ of P is an x-monotone polygon. Therefore, polygon P is converted to an x-monotone polygon after the last flip. See Fig. 5 for an illustration.

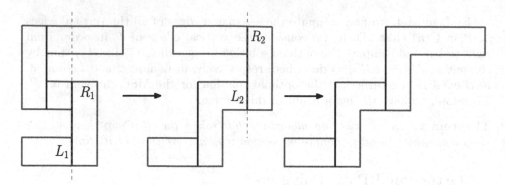

Fig. 5. An illustration of transforming a non-x-monotone polygon into an x-monotone polygon by *unfolding* the polygon.

To summarize, we first convert P into an x-monotone polygon using at most $k < n$ flip operations as described above and then solve the MSC problem on the resulting x-monotone polygon using the dynamic programming algorithm given in Sect. 3. We can compute the set of line segments L_i, for all $1 \leq i < k$, in $O(n)$ time by detecting each vertical edge of P whose both endpoints are reflex vertices of P. Next, by keeping track of the lower and upper chains of P starting from L_1, we can compute the flipped polygon in $O(n)$ time. Therefore, we have the following theorem:

Theorem 2. *There exists an algorithm that solves the MSC problem on any orthogonal path polygon with n vertices in $O(n)$ time.*

5   Conclusion

In this paper, we gave a linear-time exact dynamic programming algorithm for the problem of guarding a simple orthogonal and x-monotone polygon with the minimum number of sliding cameras (i.e., the MSC problem). This improves the 2-approximation algorithm of Katz and Morgenstern [6]. Moreover, we showed that our dynamic program can be used to solve the MSC problem on orthogonal polygons for which the dual graph induced by the vertical decomposition of P is a path (i.e., orthogonal path polygons). However, the complexity of the MSC problem on any simple orthogonal polygon remains open.

References

1. Chvátal, V.: A combinatorial theorem in plane geometry. J. Comb. Theor. Ser. B **18**, 39–41 (1975)
2. Durocher, S., Filtser, O., Fraser, R., Mehrabi, A.D., Mehrabi, S.: A (7/2)-approximation algorithm for guarding orthogonal art galleries with sliding cameras. In: Pardo, A., Viola, A. (eds.) LATIN 2014. LNCS, vol. 8392, pp. 294–305. Springer, Heidelberg (2014)

3. Durocher, S., Mehrabi, S.: Guarding orthogonal art galleries using sliding cameras: algorithmic and hardness results. In: Chatterjee, K., Sgall, J. (eds.) MFCS 2013. LNCS, vol. 8087, pp. 314–324. Springer, Heidelberg (2013)
4. Eidenbenz, S.: Inapproximability results for guarding polygons without holes. In: Chwa, K.-Y., Ibarra, O.H. (eds.) ISAAC 1998. LNCS, vol. 1533, pp. 427–436. Springer, Heidelberg (1998)
5. Ghosh, S.K.: Approximation algorithms for art gallery problems in polygons. Disc. App. Math. **158**(6), 718–722 (2010)
6. Katz, M.J., Morgenstern, G.: Guarding orthogonal art galleries with sliding cameras. Int. J. Comp. Geom. App. **21**(2), 241–250 (2011)
7. King, J., Krohn, E.: Terrain guarding is NP-hard. In: SODA, pp. 1580–1593 (2010)
8. Krohn, E., Nilsson, B.J.: Approximate guarding of monotone and rectilinear polygons. Algorithmica **66**(3), 564–594 (2013)
9. Lee, D.T., Lin, A.K.: Computational complexity of art gallery problems. IEEE Trans. Inf. Theor. **32**(2), 276–282 (1986)
10. O'Rourke, J.: Art Gallery Theorems and Algorithms. Oxford University Press Inc, New York (1987)
11. Schuchardt, D., Hecker, H.-D.: Two NP-hard art-gallery problems for ortho-polygons. Math. Log. Quart. **41**(2), 261–267 (1995)
12. Seddighin, S.: Guarding polygons with sliding cameras. Master's thesis, Sharif University of Technology (2014)

Network Optimization

Information Gathering in Ad-Hoc Radio Networks with Tree Topology

Marek Chrobak[1(✉)], Kevin Costello[2],
Leszek Gasieniec[3], and Darek R. Kowalski[3]

[1] Department of Computer Science,
University of California at Riverside, Riverside, USA
marek@cs.ucr.edu
[2] Department of Mathematics, University of California at Riverside, Riverside, USA
[3] Department of Computer Science, University of Liverpool, Liverpool, UK

Abstract. We study information gathering in ad-hoc radio networks without collision detection, focussing on the case when the network forms a tree with edges directed towards the root. Initially, each node has a piece of information that we refer to as a *rumor*. The goal is to deliver all rumors to the root of the tree as quickly as possible. The protocol must complete this task even if the tree topology is unknown. In the deterministic case, assuming that the nodes are labeled with small integers, we give an $O(n)$-time protocol that uses unbounded messages, and an $O(n \log n)$-time protocol using bounded messages. We also consider fire-and-forward protocols, in which a node can only transmit its own rumor or the rumor received in the previous step. We give a deterministic fire-and-forward protocol with running time $O(n^{1.5})$, and we show that it is asymptotically optimal. We then study randomized algorithms where the nodes are not labelled. In this model, we give an $O(n \log n)$-time protocol and we prove that this bound is asymptotically optimal.

1 Introduction

We consider the problem of information gathering in ad-hoc radio networks, where initially each node has a piece of information called a *rumor*, and all these rumors need to be delivered to a designated target node as quickly as possible. A radio network is defined as a directed graph G with n vertices. At each time step any node v of G may attempt to transmit a message. This message is sent immediately to all out-neighbors of v. However, an out-neighbor u of v will receive this message only if no *collision* occurs, that is if no other in-neighbor of u transmits at this step. We do not assume any collision detection mechanism, so neither u nor any other node knows whether a collision occurred.

Another feature of our model is that the topology of G is unknown. We are interested in distributed protocols, where the computation at a node v depends only on the label of v and the information gathered from the received messages.

Research supported by grants CCF-1217314 (NSF) and H98230-13-1-0228 (NSA).

© Springer International Publishing Switzerland 2014
Z. Zhang et al. (Eds.): COCOA 2014, LNCS 8881, pp. 129–145, 2014.
DOI: 10.1007/978-3-319-12691-3_11

The protocol needs to complete its task within the allotted time, independent of the topology of G. Randomized protocols typically do not use the labels, and thus they work even if the nodes are indistinguishable from each other.

The two primitives for information dissemination in ad-hoc radio networks that have been most extensively studied are *broadcasting* and *gossiping*. The *broadcasting problem* is the one-to-all dissemination problem, where initially only one node has a rumor that needs to be delivered to all nodes in the network. Assuming that the nodes of G are labelled with consecutive integers $0, 1, ..., n - 1$, the fastest known deterministic algorithms for broadcasting run in time $O(n \log n \log \log n)$ [12] or $O(n \log^2 D)$ [11], where D is the diameter of G. The best lower bound on the running time in this model is $\Omega(n \log D)$ [10]. (See also [4,5,8,18] for earlier work.) Allowing randomization, broadcasting can be accomplished in time $O(D \log(n/D) + \log^2 n)$ with high probability [11,19], even if the nodes are not labelled. This matches the lower bounds in [2,20].

The *gossiping problem* is the all-to-all dissemination problem. Here, each node starts with its own rumor and the goal is to deliver all rumors to each node. There is no restriction on the size of messages; in particular, different rumors can be transmitted together in a single message. With randomization, gossiping can be solved in expected time $O(n \log^2 n)$ [11] (see [9,21] for earlier work), even if the nodes are not labelled. In contrast, for deterministic algorithms, with nodes labelled $0, 1, ..., n - 1$, the fastest known gossiping algorithm runs in time $O(n^{4/3} \log^4 n)$ [16], following earlier progress in [8,25]. (See also [15] for more information.) For graphs with arbitrary diameter, the best known lower bound is $\Omega(n \log n)$, the same as for broadcasting. Reducing the gap between lower and upper bounds for deterministic gossiping to a poly-logarithmic factor remains a central open problem in the study of radio networks with unknown topology.

Our work has been inspired by this open problem. It is easy to see that for arbitrary directed graphs gossiping is equivalent to information gathering, in the following sense. On one hand, trivially, any protocol for gossiping also solves the problem of gathering. On the other hand, we can apply a gathering protocol and follow it with a protocol that broadcasts all information from the target node r; these two protocols combined solve the problem of gossiping.

Our Results. To gain better insight into information gathering in radio networks, we focus on tree topologies. Thus we assume that our graph is a tree $\mathcal{T}$ with root r and with all edges directed towards r. A gathering protocol knows that the network is a tree, but it does not know its topology. We consider several variants of this problem, providing the following results:

(1) We first study deterministic algorithms, under the assumption that the nodes of $\mathcal{T}$ are labelled $0, ..., n - 1$. In Sect. 4 we examine the model without any bound on the message size, for which we give an optimal, $O(n)$-time protocol.
(2) Next, in Sect. 5, we consider the model with bounded messages, where a message may contain only one rumor, for which we give an $O(n \log n)$ protocol.
(3) In Sect. 6 we introduce a more restrictive model of *fire-and-forward* protocols, in which a node can only transmit either its own rumor or the rumor

received in the previous step. We give a deterministic fire-and-forward protocol with running time $O(n^{1.5})$ and we show a matching lower bound of $\Omega(n^{1.5})$.

(4) We then consider randomized algorithms (Sect. 7), that do not use node labels. In this model, we give an $O(n \log n)$-time gathering protocol and we prove a matching lower bound of $\Omega(n \log n)$. If the tree is a star, we show that our lower bound is in fact optimal with respect to the leading constant.

Our algorithms for deterministic protocols easily extend to the model with labels drawn from a set $0, 1, ..., L$ where $L = O(n)$, without affecting the running times. If L is arbitrary, our algorithms for bounded and unbounded messages can be implemented in time, respectively, $O(n^2 \log L)$ and $O(n^2 \log n \log L)$.

We remark that some protocols for radio networks use forms of information gathering on trees as a sub-routine; see for example [3,6,17]. However, these solutions typically focus on undirected graphs, which allow feedback, and on a relaxed variant of information gathering where the goal is to gather only a fraction of rumors in the root. In contrast, we study directed trees without any feedback mechanism, and we require all rumors to be collected at the root.

Due to space limitations, some proofs are omitted in this extended abstract. The missing proofs are provided in [7].

2 Preliminaries

We define a radio network as a directed graph $G = (V, E)$ with n nodes, with each node assigned a different label from the set $[n] = \{0, 1, ..., n-1\}$. Denote by $\mathsf{label}(v)$ the label assigned to a node $v \in V$. One node r is distinguished as the *target* node, and we assume that r is reachable from all other nodes. Initially, at time 0, each node v has some piece of information that we will refer to as *rumor* and we will denote it by ρ_v. The objective is to deliver all rumors ρ_v to r as quickly as possible, according to the rules described below.

The time is discrete, namely it consists of time steps numbered with non-negative integers $0, 1, 2, ...$ At any step, a node v may be either in the *transmit state* or the *receive state*. A gathering protocol $\mathcal{A}$ determines, for each node v and each time step t, whether v is in the transmit or receive state at time t. If v is in the transmitting state, then $\mathcal{A}$ also determines what message is transmitted by v, if any. This specification of $\mathcal{A}$ may depend only on the label of v, time t, and on the content of all messages received by v until time t.

All nodes start executing the protocol simultaneously at time 0. If a node v transmits at a time t, the transmitted message is sent immediately to all out-neighbors of v, that is to all u such that (v, u) is an edge. If (v, u) and (v', u) are edges and both v, v' transmit at time t then a *collision* at u occurs and u does not receive a message. We do not assume any feedback from the transmission channel or any collision detection features, so, in case of a collision, neither the sender nor any node within its range knows that a collision occurred.

Throughout the paper, we will focus on the case when the graph is a tree, denote by $\mathcal{T}$, with root r and with all edges directed towards the root r.

The running time of a deterministic gathering protocol $\mathcal{A}$ is defined as the minimum time $T(n)$ such that, for any tree $\mathcal{T}$ with root r and n nodes, any assignment of labels from $[n]$ to the nodes of $\mathcal{T}$, and any node v, the rumor ρ_v of v is delivered to r no later than at step $T(n)$. In case of randomized protocols, we use the expectation of their running time $T(n)$, which is now a random variable, or we show that $T(n)$ does not exceed a desired time bound with high probability.

We consider three types of gathering protocols. In the model with *unbounded messages* a node can transmit arbitrary information in a single step. In particular, multiple rumors can be aggregated into a single message. In the model with *bounded messages*, no aggregation of rumors is allowed. Each message consists of at most one rumor and $O(\log n)$ bits of additional information. Our third model is called *fire-and-forward*. In a fire-and-forward protocol, a node can either transmit its own rumor or the rumor received in the previous step, if any. Thus a message originating from a node travels towards the root one hop at a time, until either it vanishes or it successfully reaches the root.

For illustration, consider a protocol ROUNDROBIN (known the literature), where all nodes transmit in a cyclic order, one at a time. The running time is $O(n^2)$, because in any consecutive n steps each rumor will decrease its distance to the root. For information gathering in trees, ROUNDROBIN can be adapted to use only bounded messages. At any round, if a node v has the rumor ρ_u with label$(u) = t \bmod n$ then v transmits ρ_u. Only one child of a node can have ρ_u, so no collisions will occur, and after at most n^2 steps r will receive all rumors.

3   Some Structure Properties of Trees

The running times of our algorithms in Sects. 4 and 5 depend on the distribution of high-degree nodes. To capture the structure of this distribution we define the concept of γ-depth which measures how "bushy" the tree is.

γ-**Depth of trees.** Let $\mathcal{T}$ be the given tree network with root r and n nodes. Fix an integer γ with $2 \le \gamma \le n-1$. We define the γ-*height* of each node v of $\mathcal{T}$, denoted $height_\gamma(v)$, as follows. If v is a leaf then $height_\gamma(v) = 0$. If v is an internal node then let g be the maximum γ-height of a child of v. If at least γ children of v have γ-height equal g then $height_\gamma(v) = g + 1$; otherwise $height_\gamma(v) = g$. (For $\gamma = 2$, our definition of 2-height is equivalent to Strahler numbers; see [22,24].) We then define the γ-*depth of* $\mathcal{T}$ as $D_\gamma(\mathcal{T}) = height_\gamma(r)$.

In our proofs, we may also consider trees other than the input tree $\mathcal{T}$. If v is a node of $\mathcal{T}$ then $\mathcal{T}_v$ will denote the subtree of $\mathcal{T}$ rooted at v and containing all descendants of v. If $\mathcal{H}$ is any tree and v is a node of $\mathcal{H}$ then, to avoid ambiguity, we will write $height_\gamma(v, \mathcal{H})$ for the γ-height of v with respect to $\mathcal{H}$. Note that if $\mathcal{H}$ is a subtree of $\mathcal{T}$ and $v \in \mathcal{H}$ then, trivially, $height_\gamma(v, \mathcal{H}) \le height_\gamma(v)$.

By definition, the 1-height of a node is the same as its height, namely the longest distance from this node to a leaf in its subtree. For a tree, its 1-depth is equal to its depth. Figure 1 shows an example of a tree whose depth equals 4, 2-depth equals 3, and 3-depth equals 1.

The lemma below follows by simple induction on the depth of $\mathcal{T}$ (see [24] for the special case $\gamma = 2$).

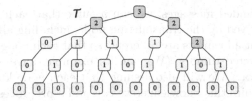

Fig. 1. An example illustrating the concept of γ-depth of trees, for $\gamma = 1, 2, 3$. The depth of this tree T is 4. The number in each node is its 2-height; thus the 2-depth of this tree is 3. All light-shaded nodes have 3-height equal 0 and the four dark-shaded nodes have 3-height equal 1, so the 3-depth of this tree is 1.

Lemma 1. $D_\gamma(T) \leq \log_\gamma n$.

We will be particularly interested in subtrees of T consisting of the nodes whose γ-height is above a given threshold. Specifically, for $h = 0, 1, ..., D_\gamma(T)$, let $T^{\gamma,h}$ be the subtree of T induced by the nodes whose γ-height is at least h Note that $T^{\gamma,h}$ is indeed a subtree of T rooted at r.

For any h, $T - T^{\gamma,h}$ is a collection of subtrees of type T_v, where v is a node of γ-height less than h whose parent is in $T^{\gamma,h}$. When $h = 1$, such subtrees contain only nodes of γ-height equal 0, which implies that they all have degree less than γ. In particular, for $\gamma = 2$, each such subtree T_v is a path from a leaf to v.

Lemma 2. *For any node $v \in T^{\gamma,h}$ we have $height_\gamma(v, T^{\gamma,h}) = height_\gamma(v) - h$. Thus, in particular, we also have $D_\gamma(T^{\gamma,h}) = D_\gamma(T) - h$.*

Proof. It is sufficient to prove the lemma for the case $h = 1$, which can be then extended to arbitrary values of h by induction. So let $h = 1$ and $T' = T^{\gamma,1}$. The proof is by induction on the height of v in T'. If v is a leaf of T' then $height_\gamma(v, T') = 0$, by definition. All children of v in T must have γ-height equal 0, so $height_\gamma(v) = 1$ and thus the lemma holds for v.

Suppose now that v is not a leaf of T' and that the lemma holds for all children of v. This means that for each child u of v in T, either $height_\gamma(u) = 0$ (that is, $u \notin T'$) or $height_\gamma(u, T') = height_\gamma(u) - 1$. Let $height_\gamma(v) = f$. If v has a child with γ-height equal f then there are fewer than γ such children. By induction, these children will have γ-height in T' equal $f - 1$, and each other child that remains in T' has γ-height in T' smaller than $f - 1$. So $height_\gamma(v, T') = f - 1$. If all children of v have γ-height smaller than f then $f \geq 2$ (for otherwise v would have to be a leaf of T') and v must have more than γ children with γ-height $f - 1$. These children will be in T' and will have γ-height in T' equal $f - 2$. So $height_\gamma(v, T') = f - 1$ in this case as well, completing the proof.

4   Deterministic Algorithms with Aggregation

We now prove that using unbounded-size messages we can complete information gathering in time $O(n)$, which is optimal even for paths or star graphs.

Since we use unbounded messages, we can assume that each message contains all information received by the transmitting node, including all received rumors. We also assume that all rumors are different, so that each node can keep track of the number of collected rumors. (We can, for example, have each node append its label to its rumor.) We will also assume that each node knows the labels of its children. To acquire this knowledge, we can add a preprocessing phase where nodes with labels $0, ..., n - 1$ transmit, one at a time, in this order. Thus after n steps each node will receive the messages from its children.

Simple Algorithm. We now present an algorithm for information gathering on trees that runs in time $O(n \log n)$. In essence, any node waits until it receives the messages from its children, then for $2n$ steps it alternates ROUNDROBIN steps with steps when it always attempts to transmit.

Algorithm UNBDTREE1. We divide the time steps into *rounds*, where round s consists of two consecutive steps $2s$ and $2s + 1$, which we call, respectively, the RR-step and the All-step of round s.

For each node v we define its *activation round*, denoted α_v, as follows. If v is a leaf then $\alpha_v = 0$. For any other node v, α_v is the first round such that v has received messages from all its children when this round is about to start.

For each round $s = \alpha_v, \alpha_v + 1, ..., \alpha_v + n - 1$, v transmits in the All-step of round s, and if $\mathsf{label}(v) = s \bmod n$ then it also transmits in the RR-step of round s. In all other steps, v stays in the receiving state.

Analysis. For a node v, we say that v is *dormant* in rounds $0, ..., \alpha_v - 1$, v is *active* in rounds $\alpha_v, ..., \alpha_v + n - 1$, and that v is *retired* in every round thereafter. Since v will make at least one RR-transmission when it is active, v will successfully transmit its message to its parent before retiring, and before this parent is activated. Therefore, by a simple inductive argument, Algorithm UNBDTREE1 is correct, namely that eventually r will receive all rumors from $\mathcal{T}$. This argument shows in fact that, at any round, Algorithm UNBDTREE1 satisfies the following invariants: (i) any path from a leaf to r consists of a segment of retired nodes, followed by a segment of active nodes, which is then followed by a segment of dormant nodes; and (ii) any dormant node has at least one active descendant.

Lemma 3. *Let $d = D_2(\mathcal{T})$. For any $h = 0, 1, ..., d$ and any node v with $\mathsf{height}_2(v) = h$, v gets activated no later than in round $2nh$, that is $\alpha_v \leq 2nh$.*

Proof. The proof is by induction on h. By the algorithm, the lemma trivially holds for $h = 0$. Suppose that the lemma holds for $h - 1$ and consider a node v with $\mathsf{height}_2(v) = h$. To reduce clutter, denote $\mathcal{Z} = \mathcal{T}^{2,h}$. From Lemma 2, we have that $\mathsf{height}_2(v, \mathcal{Z}) = 0$, which implies that $\mathcal{Z}_v$ is a path from a leaf of $\mathcal{Z}$ to v. Let $\mathcal{Z}_v = v_1, v_2, ..., v_q$ be this path, where v_1 is a leaf of $\mathcal{Z}$ and $v_q = v$.

We now consider round $s = 2n(h - 1) + n$. The nodes in $\mathcal{T} - \mathcal{Z}$ have 2-height at most $h - 1$, so, by the inductive assumption, they are activated no later than in round $2n(h-1)$, and therefore in round s they are already retired. If $\alpha_v \leq s$ then $\alpha_v \leq 2nh$, and we are done. Otherwise, v is dormant in round s. Then, by invariant (ii) above, at least one node in $\mathcal{Z}_v$ must be active. Choose the largest p for which v_p is active in round s. In round s and later, all children of the nodes $v_p, v_{p+1}, ..., v_q$ that are not on $\mathcal{Z}_v$ do not transmit, since they are already retired. This implies that for each $\ell = 0, ..., q-p-1$, node $v_{p+\ell+1}$ will get activated in round $s + \ell + 1$ as a result of the All-transmission from node $v_{p+\ell}$. In particular, we obtain that $\alpha_v \leq s + q - p \leq 2nh$, completing the proof of the lemma.

We have $height_2(r) = d$ and $d = O(\log n)$, by Lemma 1. Applying Lemma 3, this implies that $\alpha_r \leq 2nd = O(n \log n)$, so the overall running time is $O(n \log n)$.

Theorem 1. *For any tree with n nodes and any assignment of labels, Algorithm* UNBDTREE1 *completes information gathering in time $O(n \log n)$.*

Linear-Time Algorithm. We show how to improve the running time of information gathering in trees to linear time, assuming unbounded messages. The basic idea is to use strong k-selective families to speed up the computation.

Recall that a *strong k-selective family*, where $1 \leq k \leq n$, is a collection $F_0, F_1, ..., F_{m-1} \subseteq [n]$ of sets such that for any set $X \subseteq [n]$ with $|X| \leq k$ and any $x \in X$, there is j for which $F_j \cap X = \{x\}$. For any $k = 1, 2, ..., n$, there is a strong k-selective family with $m = O(k^2 \log n)$ sets [10,13].

In essence, the strong k-selective family is used to speed up information dissemination through low-degree nodes. To achieve linear time, we will interleave the steps using the selective family with ROUNDROBIN (to deal with high-degree nodes) and steps where all active nodes transmit (to deal with long paths).

Below, we fix parameters $\kappa = \lceil n^{1/3} \rceil$ and $m = O(\kappa^2 \log n)$, the size of a strong κ-selective family $F_0, ..., F_{m-1}$. Without loss of generality, we assume $m \leq n$.

Algorithm UNBDTREE2. We divide the steps into rounds, where each round s consists of three consecutive steps $3s$, $3s + 1$, and $3s + 2$, that we will call the RR-step, All-step, and Sel-step of round s, respectively.

For each node v we define its *activation round*, denoted α_v, as follows. If v is a leaf then $\alpha_v = 0$. For any other node v, α_v is the first round such that before this round starts v has received all messages from its children.

In each round $s = \alpha_v, \alpha_v + 1, ..., \alpha_v + m - 1$, v transmits in the All-step of round s, and if $label(v) \in F_{s \bmod m}$ then v also transmits in the Sel-step of round s. In each round $s = \alpha_v, \alpha_v + 1, ..., \alpha_v + n - 1$, if $label(v) = s \bmod n$ then v transmits in the RR-step of round s. If v does not transmit according to the above rules then v stays in the receiving state.

Analysis. Similar to Algorithm UNBDTREE1, in Algorithm UNBDTREE2 each node v goes through three stages. We call v *dormant* in rounds $0, 1, ..., \alpha_v - 1$, *active* in rounds $\alpha_v, \alpha_v + 1, ..., \alpha_v + n - 1$, and retired thereafter. We will also refer to v as being *semi-retired* in rounds $\alpha_v + m, \alpha_v + m + 1, ..., \alpha_v + n - 1$ (when it is still active, but only uses RR-transmissions). Assuming that v gets activated in some round, since v makes at least one RR-transmission when it is active, it will successfully transmit its message to its parent before retiring, and before its parent gets activated. By induction on the depth of $\mathcal{T}$, each node will eventually get activated, proving correctness. By a similar argument, Algorithm UNBDTREE2 satisfies the following two invariants in each round: (i) Any path from a leaf to r consists of a segment of retired nodes, followed by a segment of active nodes (among the active nodes, the semi-retired nodes precede those that are not semi-retired), which is then followed by a segment of dormant nodes. (ii) Any dormant node has at least one active descendant.

It remains to show that the running time of Algorithm UNBDTREE2 is $O(n)$. The idea is to show that Sel- and All-steps disseminate information in linear time through subtrees where all node degrees are less than κ. The process can stall, however, if all active nodes have parents of degree larger than κ. In this case, a complete cycle of ROUNDROBIN will transmit the messages from these nodes to their parents. We show, using Lemma 1, that such stalling can occur at most the total of 3 times. So the overall running time will be still $O(n)$.

To formalize this argument, let $\bar{d} = D_\kappa(\mathcal{T})$. From Lemma 1, we have $\bar{d} \leq 3$. We fix some $g \in \{0, 1, 2, 3\}$, a node w with $height_\kappa(w) = g$, and we let $\mathcal{Y} = \mathcal{T}_w^{\kappa, g}$. Thus $\mathcal{Y}$ consists of the descendants of w whose κ-height in $\mathcal{T}$ is exactly g, or, equivalently (by Lemma 2), the descendants of w in $\mathcal{T}^{\kappa, g}$ whose κ-height in $\mathcal{T}^{\kappa, g}$ is equal 0. So all nodes in $\mathcal{Y}$ have degree smaller than κ. We also fix $\bar{s}$ to be the first round when all nodes in $\mathcal{T} - \mathcal{T}^{\kappa, g}$ are active or already retired. In particular, for $g = 0$ we have $\bar{s} = 0$. Our goal now is to show that w will get activated in at most $O(n)$ rounds after round $\bar{s}$.

Lemma 4. $\alpha_w \leq \bar{s} + O(n)$.

Proof. Let $d = D_2(\mathcal{Y})$. By Lemma 1, $d = O(\log |\mathcal{Y}|) = O(\log n)$. For $h = 0, ..., d$, let l_h be the number of nodes $u \in \mathcal{Y}$ with $height_2(u, \mathcal{Y}) = h$. The overall idea of the proof is similar to the analysis of Algorithm UNBDTREE1. The difference is that now, since all degrees in $\mathcal{Y}$ are less than κ, the number of rounds required to advance through the h-th layer of $\mathcal{Y}$, consisting of nodes of 2-height equal h, can be bounded by $O(m + l_h)$, while before this bound was $O(n)$. Adding up the bounds for all layers, all terms $O(l_h)$ amortize to $O(n)$, and the terms $O(m)$ up to $O(md) = O(n^{2/3} \log^2 n) = O(n)$ as well. We now fill in the details.

Claim A: Let v be a node in $\mathcal{Y}$ with $height_2(v, \mathcal{Y}) = h$. Then the activation round of v satisfies $\alpha_v \leq s_h$, where $s_h = \bar{s} + 2n + \sum_{i < h} l_i + hm$.

First, we observe that Claim A implies the lemma. This is because for $v = w$ we get the bound $\alpha_w \leq \bar{s} + 2n + \sum_{i \leq d} l_i + dm \leq \bar{s} + 2n + n + O(\log n) \cdot O(n^{2/3} \log n) = \bar{s} + O(n)$, as needed. Thus, to complete the proof, it remains to justify Claim A. We proceed by induction on h.

Consider first the base case, when $h = 0$. We focus on the computation in the subtree $\mathcal{Y}_v$, which (for $h = 0$) is simply a path $v_1, v_2, ..., v_q = v$, from a leaf v_1 of $\mathcal{Y}$ to v. In round $\bar{s} + n$ all the nodes in $\mathcal{T} - \mathcal{Y}$ must be already retired. If v is active in round $\bar{s} + n$, we are done, because $\bar{s} + n \leq s_0$. If v is dormant, at least one node in $\mathcal{Y}_v$ must be active (see the invariant (ii)), so choose p to be the maximum index for which v_p is active. Since we have no interference from outside $\mathcal{Y}_v$, using a simple inductive argument, v will be activated in $q - p$ rounds using All-transmissions. Also, $q-p \leq l_0$, and therefore $\alpha_v \leq \bar{s}+n+q-p \leq \bar{s}+2n+l_0 = s_0$, which is the bound from Claim A for $h = 0$.

In the inductive step, fix some $h > 0$, and assume that Claim A holds for $h - 1$. Denoting $\mathcal{Z} = \mathcal{Y}^{2,h}$, we consider the computation in $\mathcal{Z}_v$, the subtree of $\mathcal{Z}$ rooted at v. $\mathcal{Z}_v$ is a path $v_1, v_2, ..., v_q = v$ from a leaf v_1 of $\mathcal{Z}$ to v. The argument is similar to the base case. There are two twists, however. One, we need to show that v_1 will get activated no later than at time $s_{h-1} + m$; that is, after delay of only m, not $O(n)$. Two, the children of the nodes on $\mathcal{Z}_v$ that are not on $\mathcal{Z}_v$ are not guaranteed to be retired anymore. However, they are semi-retired, which is good enough for our purpose.

Consider v_1. We want to show first that v_1 will get activated no later than at time $s_{h-1} + m$. All children of v_1 can be grouped into three types. The first type consists of the children of v_1 in $\mathcal{T} - \mathcal{Y}$. These are activated no later than in round $\bar{s}$, so they are retired no later than in round $\bar{s} + n$. All other children of v_1 are in $\mathcal{Y} - \mathcal{Z}$. Among those, the type-2 children are those that were activated before round $\bar{s} + n$, and the type-3 children are those that were activated at or after round $\bar{s} + n$. Clearly, v_1 will receive the messages from its children of type 1 and 2, using RR-transmissions, no later than in round $\bar{s}+2n$. The children of v_1 of type 3 activate no earlier than in round $\bar{s} + n$. Also, since they are in $\mathcal{Y} - \mathcal{Z}$, their 2-height in $\mathcal{Y}$ is strictly less than h, so they activate no later than in round s_{h-1}, by induction. (Note that $s_{h-1} \geq \bar{s}+2n$.) Thus each child u of v_1 in $\mathcal{Y}$ of type 3 will complete all its Sel-transmissions, that include the complete κ-selector, between rounds $\bar{s} + n$ and $s_{h-1} + m - 1$ (inclusive). In these rounds all children of v_1 that are not in $\mathcal{Y}$ are retired, so fewer than κ children of v_1 are active in these rounds. This implies that the message of u will be received by v_1. Putting it all together, v_1 will receive messages from all its children before round $s_{h-1} + m$, and thus it will be activated no later than in round $s_{h-1} + m$.

From the paragraph above, we obtain that in round $s_{h-1} + m$ either there is an active node in $\mathcal{Z}_v$ or all nodes in $\mathcal{Z}_v$ are already retired. The remainder of the argument is similar to the base case. If v itself is active or retired in round $s_{h-1} + m$ then we are done, because $s_{h-1} + m \leq s_h$. So suppose that v is still dormant in round $s_{h-1} + m$. Choose p to be the largest index for which v_p is active in this round. All children of the nodes on $\mathcal{Z}_v$ that are not on $\mathcal{Z}_v$ are either retired or semi-retired. Therefore, since there is no interference, v will get activated in $q - p$ additional rounds using All-transmissions. So $\alpha_v \leq s_{h-1} + m + q - p \leq s_{h-1} + m + l_h = s_h$, completing the inductive step, the proof of Claim A, and the lemma.

From Lemma 4, all nodes in $\mathcal{T}$ with κ-height equal 0 will get activated in at most $O(n)$ rounds. For $g = 1, 2, 3$, all nodes with κ-height equal g will activate no later than $O(n)$ rounds after the last node with κ-height less than g is activated. This implies that all nodes in $\mathcal{T}$ will be activated within $O(n)$ rounds.

Theorem 2. *For any tree with n nodes and any assignment of labels, Algorithm UNBDTREE2 completes information gathering in time $O(n)$.*

5   Deterministic Algorithms Without Aggregation

In this section we consider deterministic information gathering without aggregation, where each message can contain at most one rumor, plus additional $O(\log n)$ bits of information. In this model, we give an algorithm with running time $O(n \log n)$. For simplicity, assume here that we are allowed to receive and transmit at the same time. We later explain how to remove this assumption.

Algorithm BNDDTREE. First, we use a modification of Algorithm UNBDTREE2 to compute the 2-height of each node v. In this modified algorithm, the message from each node contains its 2-height. When v receives such messages from its children, it can compute its own 2-height.

Let $\ell = \lceil \log n \rceil$. We divide the computation into $\ell + 1$ phases. Phase h, for $h = 0, ..., \ell$, consists of steps $3nh, ..., 3n(h+1)-1$. In phase h, only the nodes of 2-height equal h participate in the computation. Specifically, consider a node v with $height_2(v) = h$. We have two stages: (**Stage All:**) In each step $t = 3nh, ..., 3nh + 2n - 1$, if v contains any rumor ρ_u that it still has not transmitted, v transmits ρ_u. (**Stage RR:**) In each step $t = 3nh + 2n + u$, for $u = 0, ..., n - 1$, if v has rumor ρ_u, then v transmits ρ_u. In other steps, v is in the receiving state.

Analysis. By Lemma 1, the number of phases is $O(\log n)$, so the algorithm makes $O(n \log n)$ steps. The lemma below will thus complete the analysis.

Lemma 5. *At the beginning of phase h, every node v has rumors from all its descendants in $\mathcal{T} - \mathcal{T}^{2,h}$, namely the descendants whose 2-height is strictly smaller than h. (In particular, if $height_2(v) < h$ then v has all rumors from $\mathcal{T}_v$.)*

Proof. The proof is by induction on h. The lemma trivially holds when $h = 0$. Assume that the claim holds for some $h < \ell$, and consider phase h. We want to show that each node v has rumors from all descendants in $\mathcal{T} - \mathcal{T}^{2,h+1}$. By the inductive assumption, v has all rumors from its descendants in $\mathcal{T} - \mathcal{T}^{2,h}$. So if v does not have any descendants of height h then we are done.

It thus remains to prove that if v has a child u with $height_2(u) = h$ then right after phase h all rumors from $\mathcal{T}_u$ will also be in v. (Of course, this case applies only if $height_2(v) \geq h$.) The subtree $\mathcal{T}_u^{2,h}$, namely the subtree consisting of the descendants of u with 2-height equal h, is a path $\mathcal{P} = u_1, u_2, ..., u_q = u$, where u_1 is a leaf of $\mathcal{T}^{2,h}$. We show that, thanks to pipelining, all rumors that are in $\mathcal{P}$ when phase h starts will reach u during Stage All.

In phase h, none of the children of the nodes in $\mathcal{P}$ transmits, except possibly for the one that is also on $\mathcal{P}$. For any step $3nh + s$, $s = 0, 1, ..., 2n - 1$, and for $i = 1, 2, ..., q-1$, we define $\phi_{s,i}$ to be the number of rumors in u_i that are still not transmitted, and we let $\Phi_s = \sum_{i=a_s}^{q-1} \max(\phi_{s,i}, 1)$, where a_s is the smallest index for which $\phi_{s,a_s} \neq 0$. We claim that as long as $\Phi_s > 0$, its value will decrease in step s. Indeed, for $i < q$, each node v_i with $\phi_{s,i} > 0$ will transmit a new rumor to v_{i+1}. Since $\phi_{s,i} = 0$ for $i < a_s$, node u_{a_s} will not receive any new rumors. We have $\phi_{s,a_s} > 0$, by the choice of a_s. If $\phi_{s,a_s} > 1$ then $\max(\phi_{s,a_s}, 1)$ will decrease by 1. If $\phi_{s,a_s} = 1$ then the index a_s itself will increase. In either case, u_{a_s}'s contribution to Φ_s will decrease by 1. For $i > a_s$, even if u_i receives a new rumor from its child, the term $\max(\phi_{s,i}, 1)$ cannot increase, because if $\phi_{s,i} > 0$ then u_i transmits a new rumor to u_{i+1}, and if $\phi_{s,i} = 0$ then this term is 1 anyway. Therefore, overall, Φ_s will decrease by at least 1.

Since Φ_s strictly decreases in each step, and its initial value is at most $q + n \leq 2n$, Φ_s will become 0 in at most $2n$ steps. In other words, in $2n$ steps u will receive all rumors from $\mathcal{P}$, and thus all rumors from $\mathcal{T}_u$.

In Stage RR, u will transmit all collected rumors to v, without collisions. As a result, at the beginning of the next phase v will contain all rumors from $\mathcal{T}_u$, completing the proof of the inductive step.

The assumption that we can transmit and receive at the same time can be eliminated. The idea is that along paths where all nodes have the same 2-height, we can synchronize the computation by having even and odd nodes along this path transmit at different times. See [7] for a complete proof.

Theorem 3. *For any tree with n nodes and any assignment of labels, Algorithm* BNDDTREE *completes information gathering in time $O(n \log n)$.*

6 Deterministic Fire-and-Forward Protocols

We now consider a very simple type of protocols that we call *fire-and-forward* protocols. For convenience, in this model we allow nodes to receive and transmit messages at the same step. In a fire-and-forward protocol, at any time t, any node v can either be idle or make one of two types of transmissions:

Fire: v can transmit its own rumor, or
Forward: v can transmit the rumor received in step $t - 1$, if any.

In Sect. 7 we give a randomized fire-and-forward protocol that works in expected time $O(n \log n)$. This raises the question whether this running time can be achieved by a deterministic fire-and-forward protocol. (Time $O(n^2)$ is trivial: release all rumors one at a time, spaced at intervals of length n.) We now show that this can be improved to $O(n^{1.5})$ and that this bound is optimal.

The key property of fire-and-forward protocols is that any rumor, once fired, moves up the tree one hop per step, unless either it collides, or is dropped, or it reaches the root. If rumors fired from two nodes collide at all, they will collide at their lowest common ancestor. (We extend the definition of collision to include the situation when a node attempts to fire right after receiving a rumor, when nothing is transmitted). This happens only when the difference in times between these two firings is equal to the difference of their depths in the tree.

Any fire-and-forward protocol $\mathcal{A}$ can be made *oblivious*, in the sense that the decision whether to fire or not depends only on the label of the node and the current time. Let $T(n)$ be the running time of $\mathcal{A}$. Imagine that we run $\mathcal{A}$ on the tree $\mathcal{T}'$ obtained from $\mathcal{T}$ by adding a leaf to any node v and giving it the label of v. Label the original nodes with the remaining labels. This at most doubles the number of nodes, so $\mathcal{A}$ will complete in time $O(T(n))$ on $\mathcal{T}'$. In the execution of $\mathcal{A}$ on $\mathcal{T}'$ the leaves receive no information and all rumors from the leaves will reach the root. This implies that if we apply $\mathcal{A}$ on $\mathcal{T}$ and ignore all information received during the computation, the rumors will also reach the root.

An $O(n^{1.5})$ Upper Bound. We now present our $O(n^{1.5})$-time fire-and-forward protocol. As explained earlier, this protocol should specify a set of firing times for each label, so that for any mapping $[n] \rightarrow [n]$, that maps each label to the depth of the node with this label, each node will have at least one firing time for which there will not be a collision along the path to the root. We want each of these firing times to be at most $O(n^{1.5})$. To this end, we will partition all labels into batches, each of size roughly $\sqrt{n}$, and show that for any batch we can define such collision-avoiding firing times from an interval of length $O(n)$. Since we have about $\sqrt{n}$ batches, this will give us running time $O(n^{1.5})$.

Our construction is based on a concept of dispersers, defined below, which are reminiscent of various *rulers* studied in number theory, for example Sidon sequences. The particular construction we give in the paper is, in a sense, a multiple set extension of a Sidon-set construction by Erdös and Turán [14].

We now give the details. For $z \in \mathbb{Z}$ and $X \subseteq \mathbb{Z}$, let $X + z = \{x + z : x \in X\}$. Let also s be a positive integer. A set family $D_1, ..., D_m \subseteq [s]$ is called an (n, m, s)-*disperser* if for each function $\delta : \{1, ..., m\} \rightarrow [n]$ and each j we have $D_j + \delta(j) \not\subseteq \bigcup_{i \neq j}(D_i + \delta(i))$. The intuition is that D_j represents the set of firing times of node j and $\delta(j)$ represents j's depth in the tree. Then the disperser condition says that some firing in D_j will not collide with firings of other nodes.

Lemma 6. *There exists an (n, m, s)-disperser with $m = \Omega(\sqrt{n})$ and $s = O(n)$.*

Proof. Let p be the smallest prime such that $p^2 \geq n$. For each $a = 1, 2, ..., p - 1$ and $x \in [p]$ define $d_a(x) = (ax \bmod p) + 2p \cdot (ax^2 \bmod p)$. We claim that for any $a \neq b$ and any $t \in \mathbb{Z}$ the equation $d_a(x) - d_b(y) = t$ has at most two solutions $(x, y) \in [p]^2$. For the proof, fix a, b, t and one solution $(x, y) \in [p]^2$. Suppose that $(u, v) \in [p]^2$ is a different solution. Thus we have $d_a(x) - d_b(y) = d_a(u) - d_b(v)$. After substituting and rearranging, this can be written as

$$(ax \bmod p) - (by \bmod p) - (au \bmod p) + (bv \bmod p)$$
$$= 2p[-(ax^2 \bmod p) + (by^2 \bmod p) + (au^2 \bmod p) - (bv^2 \bmod p)].$$

The expression on the left-hand side is strictly between $-2p$ and $2p$, so both sides must be equal 0. This implies that

$$ax - au \equiv by - bv \pmod{p} \quad \text{and} \tag{1}$$
$$ax^2 - au^2 \equiv by^2 - bv^2 \pmod{p} \tag{2}$$

From Eq. (1), the assumption that $(x, y) \neq (u, v)$ implies that $x \neq u$ and $y \neq v$. We can then divide the two equations, getting

$$x + u \equiv y + v \pmod{p}. \tag{3}$$

With addition and multiplication modulo p, $\mathbb{Z}_p$ is a field. Therefore for any x and y, and any $a \neq b$, Eqs. (1) and (3) uniquely determine u and v, completing the proof of the claim.

Now, let $m = (p - 1)/2$ and $s = 2p^2 + p$. By Bertrand's postulate we have $\sqrt{n} \leq p < 2\sqrt{n}$, which implies that $m = \Omega(\sqrt{n})$ and $s = O(n)$. For each $i = 1, 2, ..., m$, define $D_i = \{d_i(x) : x \in [p]\}$. It is sufficient to show that the sets $D_1, D_2, ..., D_m$ satisfy the condition of the (n, m, s)-disperser.

The definition of the sets D_i implies that $D_i \subseteq [s]$ for each i. Fix some δ and j from the definition of dispersers. It remains to verify that $D_j + \delta(j) \not\subseteq \bigcup_{i \neq j}(D_i + \delta(i))$. For $x \in [p]$ and $i \in \{1, 2, ..., m\}$, we say that i *kills* x if $d_j(x) + \delta(j) \in D_i + \delta(i)$. Our earlier claim implies that any $i \neq j$ kills at most two values in $[p]$. Thus all indices $i \neq j$ kill at most $2(m - 1) = p - 3$ integers in $[p]$, which implies that there is some $x \in [p]$ that is not killed by any i. For this x, we will have $d_j(x) + \delta(j) \notin \bigcup_{i \neq j}(D_i + \delta(i))$, completing the proof that $D_1, ..., D_m$ is indeed an (n, m, s)-disperser.

Algorithm MLSDTREE. Let $D_1, D_2, ..., D_m$ be the (n, m, s)-disperser from Lemma 6. We partition all labels (and thus also the corresponding nodes) arbitrarily into batches $B_1, B_2, ..., B_l$, for $l = \lceil n/m \rceil$, with each batch B_i having m nodes (except the last batch, that could be smaller). Order the nodes in each batch arbitrarily, for example according to increasing labels.

The algorithm has l phases. Each phase $q = 1, 2, ..., l$ consists of $s' = s + n$ steps in the time interval $[s'(q - 1), s'q - 1]$. In phase q, the algorithm transmits rumors from batch B_q, by having the j-th node in B_q fire at each time $s'(q-1)+\tau$, for $\tau \in D_j$. Note that in the last n steps of each phase none of the nodes fires.

Analysis. We now show that Algorithm MLSDTREE correctly performs gathering in any n-node tree in time $O(n^{1.5})$. Since $m = \Omega(\sqrt{n})$, we have $l = O(\sqrt{n})$. Also, $s' = O(n)$, so the total run time of the protocol is $O(n^{1.5})$.

It remains to show that during each phase q each node in B_q will have at least one firing that will send its rumor to the root r without collisions. Fix some tree $\mathcal{T}$ and let $\delta(j) \in [n]$ be the depth of the jth node in batch B_q. For any batch B_q and any $v \in B_q$, if v is the jth node in B_q then v will fire at times $s'(q - 1) + \tau$, for $\tau \in D_j$. From the definition of dispersers, there is $\tau \in D_j$ such that $\tau + \delta(j) - \delta(i) \notin D_i$ for each $i \neq j$. This means that the firing of v at time $s'(q-1) + \tau$ will not collide with any firing of other nodes in batch B_q. Since the batches are separated by empty intervals of length n, this firing will not collide with any firing in other batches. So v's rumor will reach r, implying

Theorem 4. *There is a fire-and-forward protocol for information gathering in trees with running time $O(n^{1.5})$.*

It is easy to remove the assumption that nodes are allowed to receive and transmit at the same time, by incorporating the extended definition of collisions into the construction from Lemma 6. The details will appear in the full paper.

An $\Omega(n^{1.5})$ Lower Bound. We match our lower bound by showing that any fire-and-forward protocol needs time $\Omega(n^{1.5})$. The basic idea of the proof is that the firings from the leaves are independent of the computation in the rest of the tree. We use this property to show that for any fire-and-forward protocol $\mathcal{A}$ with running time $o(n^{1.5})$ we can construct a caterpillar tree on which at least one leaf will fail. The construction is based on a counting argument and analyzing the structure of the bipartite graphs representing collisions between firings (see [7]).

Theorem 5. *If $\mathcal{A}$ is a deterministic fire-and-forward protocol for information gathering in trees, then the running time of $\mathcal{A}$ is $\Omega(n^{1.5})$.*

7   Randomized Algorithms

Upper Bound of $O(n \log n)$. We now show a randomized algorithm with expected running time $O(n \log n)$ that does not use any labels. Our algorithm also does not use any aggregation; each message consists only of one rumor. In the description of the algorithm we assume that the number n of nodes is known. Using a standard doubling trick, the algorithm can be extended to one that does not depend on n. We present the algorithm as a fire-and-forward algorithm (see the previous section). In particular, for simplicity, we assume for now that at each step a node can listen and transmit at the same time.

Algorithm RTREE. At any time t, each node $v \neq r$, independently of other nodes, decides to fire with probability $1/n$. If v decides to fire and no rumor arrived at v at step $t - 1$, v fires. If v decides not to fire and v received some rumor in step $t - 1$, then v forwards this rumor in step t. Otherwise, v is idle.

Analysis. We start with the following lemma.

Lemma 7. *At each step $t \geq n$, for each node $z \neq r$, the probability that r receives rumor ρ_z at step t is at least $\frac{1}{n}(1 - \frac{1}{n})^{n-1}$. Further, for different t these events (receiving ρ_z by r) are independent.*

Proof. To prove the lemma, it helps to view the computation in a slightly different, but equivalent way. Imagine that we ignore collisions, and we allow each message to consist of a set of rumors. If some children of v transmit at a time t, then v receives all rumors in the transmitted messages, and at time $t + 1$ it transmits a message containing all these rumors, possibly adding its own rumor, if v decides to fire at that step. We will refer to these messages as *virtual messages*. In this interpretation, if r receives a virtual message that is a singleton set ρ_z, at some time t, then in Algorithm RTREE this rumor ρ_z will be received by r at time t. (Note that the converse is not true.)

Fix a time t and some $z \in \mathcal{T} - \{r\}$. By the above paragraph, it is sufficient to show that the probability that at time t the virtual message reaching r is the singleton $\{z\}$ is equal $\frac{1}{n}(1 - \frac{1}{n})^{n-1}$. This event will occur if and only if: (i) At time $t - depth(z)$, z decided to fire, and (ii) For each $u \in \mathcal{T} - \{z,r\}$, u did not decide to fire at time $t - depth(u)$. By the algorithm, all these events are independent, so the probability of this combined event is exactly $\frac{1}{n}(1 - \frac{1}{n})^{n-1}$.

By Lemma 7, for any step $t \geq n$, the probability of any given rumor ρ_z reaching r in step t is at least as large as the probability of collecting a given coupon in the coupon collector problem. We thus obtain the following theorem.

Theorem 6. *Algorithm RTREE has expected running time $O(n \log n)$. In fact, it will complete gathering in time $O(n \log n)$ with probability $1 - o(1)$.*

Lower Bound of $\Omega(n \log n)$. We now show that Algorithm RTREE is within a constant factor of optimal. Actually we will show something a bit stronger, namely that there is a constant c such that any label-less algorithm with running time less than $cn \ln n$ will almost surely have some rumors fail to reach the root on the tree that is a star graph (consisting of the root with n children that are also the leaves in the tree). In this tree, at each time step t, each leaf v transmits with a probability that can depend only on t and on the set of previous times at which v attempted to transmit. Note that the actions of v at different steps may not be independent. Allowing some dependence, in fact, can help reduce the running time, although only by a constant factor (see Theorem 8).

For the star graph, we can equivalently think of a label-less algorithm running in time T as a probability distribution over all subsets of $\{0, 1, \ldots, T-1\}$ representing the sets of transmission times of each node. Each node v independently picks a subset S_v according to the distribution, and transmits only at the times in S_v. The label-less requirement is equivalent to the requirement that the S_v are identically distributed. Node v succeeds in transmitting if there is a time t such that $t \in S_v$, but $t \notin S_w$ for any $w \neq v$.

Theorem 7. *If $\mathcal{R}$ is a randomized protocol for information gathering on trees then the expected running time of $\mathcal{R}$ is $\Omega(n \ln n)$. More specifically, if n is large enough and we run $\mathcal{R}$ on the n-node star graph for $T \leq cn \ln n$ steps, where $c < \frac{1}{\ln^2 2}$, then there will almost surely be a rumor that fails to reach the root.*

The proof appears in [7]. We also show in [7] that our lower bound is in fact tight, in the sense that the value of the constant c in the above theorem is best possible for star graphs.

Theorem 8. *If $T = cn \ln n$, where $c > \frac{1}{\ln^2 2}$, then there is a protocol which succeeds on the star graph in time T with probability $1 - o(1)$.*

References

1. Alon, A., Spencer, J.: The Probabilistic Method. Wiley, Hoboken (2008)
2. Alon, N., Bar-Noy, A., Linial, N., Peleg, D.: A lower bound for radio broadcast. J. Comput. Syst. Sci. **43**(2), 290–298 (1991)
3. Bar-Yehuda, R., Israeli, A., Itai, A.: Multiple communication in multihop radio networks. SIAM J. Comput. **22**(4), 875–887 (1993)
4. Bruschi, D., Del Pinto, M.: Lower bounds for the broadcast problem in mobile radio networks. Distrib. Comput. **10**(3), 129–135 (1997)
5. Chlebus, B.S., Gasieniec, L., Gibbons, A., Pelc, A., Rytter, W.: Deterministic broadcasting in ad hoc radio networks. Distrib. Comput. **15**(1), 27–38 (2002)
6. Chlebus, B.S., Kowalski, D.R., Radzik, T.: Many-to-many communication in radio networks. Algorithmica **54**(1), 118–139 (2009)
7. Chrobak, M., Costello, K., Gasieniec, L., Kowalski, D.R.: Information gathering in ad-hoc radio networks with tree topology (2014). arXiv 1407.1521
8. Chrobak, M., Gasieniec, L., Rytter, W.: Fast broadcasting and gossiping in radio networks. J. Algorithms **43**(2), 177–189 (2002)
9. Chrobak, M., Gasieniec, L., Rytter, W.: A randomized algorithm for gossiping in radio networks. Networks **43**(2), 119–124 (2004)
10. Clementi, A.E.F., Monti, A., Silvestri, R.: Distributed broadcast in radio networks of unknown topology. Theor. Comput. Sci. **302**(1–3), 337–364 (2003)
11. Czumaj, A., Rytter, W.: Broadcasting algorithms in radio networks with unknown topology. J. Algorithms **60**(2), 115–143 (2006)
12. De Marco, G.: Distributed broadcast in unknown radio networks. In: Proceedings of the 19th Annual ACM-SIAM Symposium on Discrete Algorithms (SODA'08), pp. 208–217 (2008)
13. Erdős, P., Frankl, P., Füredi, Z.: Families of finite sets in which no set is covered by the union of r others. Isr. J. Math. **51**(1–2), 79–89 (1985)
14. Erdős, P., Turán, P.: On a problem of Sidon in additive number theory and on some related problems. J. London Math. Soc. **16**, 212–215 (1941)
15. Gąsieniec, L.: On efficient gossiping in radio networks. In: Kutten, S., Žerovnik, J. (eds.) SIROCCO 2009. LNCS, vol. 5869, pp. 2–14. Springer, Heidelberg (2010)
16. Gasieniec, L., Radzik, T., Xin, Q.: Faster deterministic gossiping in directed ad hoc radio networks. In: Hagerup, T., Katajainen, J. (eds.) SWAT 2004. LNCS, vol. 3111, pp. 397–407. Springer, Heidelberg (2004)
17. Khabbazian, M., Kowalski, D.R.: Time-efficient randomized multiple-message broadcast in radio networks. In: Proceedings of the 30th Annual Symposium on Principles of Distributed Computing, PODC'11, pp. 373–380 (2011)
18. Kowalski, D.R., Pelc, A.: Faster deterministic broadcasting in ad hoc radio networks. SIAM J. Discrete Math. **18**(2), 332–346 (2004)
19. Kowalski, D.R., Pelc, A.: Broadcasting in undirected ad hoc radio networks. Distrib. Comput. **18**(1), 43–57 (2005)
20. Kushilevitz, E., Mansour, Y.: An $\Omega(D \log(N/D))$ lower bound for broadcast in radio networks. SIAM J. Comput. **27**(3), 702–712 (1998)
21. Liu, D., Prabhakaran, M.: On randomized broadcasting and gossiping in radio networks. In: Ibarra, O.H., Zhang, L. (eds.) COCOON 2002. LNCS, vol. 2387, pp. 340–349. Springer, Heidelberg (2002)
22. Strahler, A.: Hypsometric (area-altitude) analysis of erosional topology. Bull. Geol. Soc. Amer. **63**, 117–1142 (1952)

23. Talagrand, M.: Concentration of measure and isoperimetric inequalities in product spaces. Publl. Math. de l'I.H.E.S. **81**, 73–205 (1995)
24. Viennot, X.: A Strahler bijection between Dyck paths and planar trees. Discrete Math. **246**, 317–329 (2003)
25. An, Y.X.: $O(n^{1.5})$ deterministic gossiping algorithm for radio networks. Algorithmica **36**(1), 93–96 (2003)

Improved Algorithms for Computing Minmax Regret 1-Sink and 2-Sink on Path Network

Binay Bhattacharya and Tsunehiko Kameda[(⊠)]

School of Computing Science, Simon Fraser University, Burnaby, Canada
{binay,tiko}@sfu.ca

Abstract. Suppose that in an emergency, such as an earthquake or fire, a number of people need to be evacuated to a safe "sink" from every vertex of a network. The k-sink problem seeks to minimize the evacuation time of all the evacuees to the sinks. In the minmax regret version of this problem, the exact number of evacuees at each vertex is unknown, but only an interval of possible numbers is given. We want to minimize the evacuation time in the worst case, where the actual numbers of evacuees are most unfavorable to the chosen sink locations. We present an optimal $O(n)$ time algorithm for finding the minmax regret 1-sink on a path network, improving the previously best time complexity of $O(n \log n)$ [6,12]. Some ideas we conceived for the new algorithm have other useful applications. For example, we demonstrate that it leads to an algorithm for computing the minmax regret 2-sink in path networks in $O(n \log^4 n)$ time, which is a fairly significant improvement over an $O(n^2 \log^2 n)$ time algorithm [1]. Moreover, the two sinks that our algorithm finds are not restricted to vertices.

1 Introduction

Investigation of evacuation problems dates back many years [5,10]. The objective is to evacuate all evacuees to some sinks as quickly as possible. The problem can be modeled by a network whose vertices represent the places where the evacuees are located and the edges represent possible evacuation routes. Associated with each edge is the transit time across the edge and its capacity in terms of the number of people who can traverse it per unit time [5]. Mamada et al. [11] solved this "classical" 1-sink problem for a tree network in $O(n \log^2 n)$ time, which has recently been improved to $O(n \log n)$ time by Higashikawa et al. [7] for the special case of uniform edge capacity.

The concept of *regret* was introduced by Kouvelis and Yu [8], to model the situations where optimization is required when the exact values (such as the number of evacuees at the vertices) are unknown. Their model only assumes that the upper and lower bounds of those values are known. The objective is to find a solution which is as good as any other solution in the worst case, where the actual values are the most unfavorable.

Recently, motivated by the 2011 earthquake in Japan, Cheng et al. [4] applied minmax regret optimization to the 1-sink problem, and proposed an $O(n \log^2 n)$

© Springer International Publishing Switzerland 2014
Z. Zhang et al. (Eds.): COCOA 2014, LNCS 8881, pp. 146–160, 2014.
DOI: 10.1007/978-3-319-12691-3_12

time algorithm for a path network with uniform edge capacity. There has been a flurry of research activities on this problem since then. The initial result was soon improved to $O(n \log n)$, independently by two different groups in [6,12]. Higashikawa et al. proposed an $O(n^2 \log^2 n)$ time algorithm for a tree network [7]. In this paper we further improve the result on the path network to $O(n)$, by adapting some of the tools we developed for the minmax-reget 1-center problem [2]. Our model, as well as those in [7,12] assume that all the edges have the same capacity. Li et al. [9] and Arumugam et al. [1] investigate the minmax regret k-sink problem on a path network with uniform edge capacity. The two algorithms in [1] run in $O(kn^3 \log n)$ and $O(kn^2 \log^k n)$ time, respectively, for general k.

The first contribution of this paper is an optimal algorithm for computing the minmax regret 1-sink for path networks. It is based on the discovery that the optimal sinks under the "non-dominated" scenarios are related to each other in a way that makes their computation efficient. Similarly, we take advantage of the fact that "regrets" can be compared locally. The second contribution is a sub-quadratic time algorithm for computing the minmax regret 2-sink for path networks. We achieve this by preprocessing data, investing $O(n)$ time, so that the 2-sinks under many scenarios can be found in sub-linear time.

This paper is organized as follows. In the next section, we define our model and introduce terms that are used throughout the paper. Section 3 presents our major result, i.e., an optimal algorithm for path networks. We then present in Sect. 4 an $O(n \log^4 n)$ time algorithm for the minmax regret 2-sink problem for path networks.

2 Preliminaries

2.1 Model

Let $P(V, E)$ be a path network, whose vertices $v_1, v_2, \ldots, v_n$ are arranged in this order from left to right. We assume that each edge has a fixed non-negative length. By $x \in P$, we mean that point x lies on either an edge or a vertex of P. For $a, b \in P$, let $d(a, b)$ denote the distance between a and b. The subpath from vertex v_i to v_j, including v_i and v_j, is denoted by $P[v_i, v_j]$. The integral weight, representing the number of evacuees, of vertex v is known to be in an interval $[\underline{w}(v), \overline{w}(v)]$, where $0 \leq \underline{w}(v) \leq \overline{w}(v)$. A particular realization of the weights of the vertices is called a *scenario*. Define the Cartesian product, $\mathcal{S} \triangleq \prod_{v \in V} [\underline{w}(v), \overline{w}(v)]$. Under a scenario $s \in \mathcal{S}$, let $w^s(v) \in [\underline{w}(v), \overline{w}(v)]$ denote the number of evacuees at vertex v. The scenario under which all vertices have their minimum (resp. maximum) weights is denoted by s_0 (resp. s_M). As in [4,6] we assume that each edge has a unit capacity, i.e., 1 person per unit time.

Let $\boldsymbol{x} = (x_1, x_2)$, where $x_1, x_2 \in P$, for the 2-sink problem and $\boldsymbol{x} = x$, where $x \in P$, for the 1-sink problem. The time needed for all evacuees to reach the sink(s), assuming that evacuation from all vertices start at the same time, is called the *evacuation time*. Under a scenario $s \in \mathcal{S}$, we define our objective function [4] for x under s by

$$\Theta(x, s) \triangleq \text{evacuation time if } x \text{ represents sink(s).} \qquad (1)$$

Let $c(s)$ denote the *optimal 1-sink*, i.e., $c(s) \triangleq \text{argmin}_{x \in P}\Theta(x, s)$, or the *optimal 2-sink*, i.e., $c(s) \triangleq \text{argmin}_{x \in P \times P}\Theta(x, s)$. We now define the *regret* [8] under s by

$$R^s(x) \triangleq \Theta(x, s) - \Theta(c(s), s). \qquad (2)$$

A scenario s is *dominated at* "point" x by another scenario s', if $R^s(x) \leq R^{s'}(x)$ holds. Finally, the *maximum regret* at x is defined by

$$R_{max}(x) \triangleq \max_{s \in S} R^s(x). \qquad (3)$$

The scenario $\text{argmax}_{s \in S}R^s(x)$ is called the *worst-case scenario* for x [8]. Our objective is to find the "point" x, called the *minmax regret sink*, that minimizes $R_{max}(x)$.

Throughout the paper the time for an evacuee to travel a unit distance is assumed to be a constant, denoted by τ.

3   Minmax Regret 1-Sink

For a point $x \in P$, let $V^L(x)$ (resp. $V^R(x)$) denote the set of vertices that lie to the left (resp. to the right) of x. Under a scenario $s \in S$, we can express the evacuation time (1) as follows, as in [4].

$$\Theta(x, s) = \max\{ \max_{v_j \in V^L(x)} \{d(v_j, x)\tau + \sum_{1 \leq i \leq j} w^s(v_i)\},$$

$$\max_{v_j \in V^R(x)} \{d(v_j, x)\tau + \sum_{j \leq i \leq n} w^s(v_i)\}\}. \qquad (4)$$

The first (resp. second) line above is the time it takes for all the evacuees at the vertices in $V^L(x)$ (resp. $V^R(x)$) to evacuate to x.

For $i = 1, 2, \ldots, n$, define the vertex set $V_i \triangleq \{v_1, \ldots, v_i\} \subseteq V$. Given a point $x \in P$ and a scenario $s \in S$, let vertex v_i satisfy $\Theta(x, s) = d(v_i, x)\tau + \sum_{v \in V_i} w^s(v)$ or $\Theta(x, s) = d(v_i, x)\tau + \sum_{v \in V \setminus V_i} w^s(v)$. Such a vertex v_i is called a *critical vertex* for x under s. For $i = 1, 2, \ldots, n$, let $\overline{s}_i$ (resp. $\underline{s}_i$) denote the scenario under which all the vertices in V_i have the maximum (resp. minimum) weights and the rest of the vertices have the minimum (resp. maximum) weights.[1] Thus, a consecutive vertices from one end of the path have the maximum weights, and the rest, if any, have the minimum weights. We call them the *bipartite* scenarios [3]. Scenario $\underline{s}_n$ (resp. $\overline{s}_n$) is the scenario s_0 (resp. s_M) that we introduced earlier. Let $S^* \triangleq \{\overline{s}_i, \underline{s}_i \mid 1 \leq i \leq n\}$. For a scenario $\overline{s}_i \in S^*$, we call the subpath consisting of

---

[1] They are called *left-dominant* (resp. *right-dominant*) scenarios in [4].

vertices in V_i (resp. $V \backslash V_i$) the *max-weighted* (resp. *min-weighted*) subpath under $\bar{s}_i$. Similarly, for a scenario $\underline{s}_i \in \mathcal{S}^*$, we call the subpath consisting of vertices in V_i (resp. $V \backslash V_i$) the *min-weighted* (resp. *max-weighted*) subpath under $\underline{s}_i$.

It is known [4] that any $s \notin \mathcal{S}^*$ can be converted to a bipartite scenario $s' \in \mathcal{S}^*$ without decreasing the regret $R^s(x)$, so that s is dominated by s' at x.

Theorem 1 [4]. *For any $x \in P$, there is a bipartite scenario that is a worst-case scenario for x. Hence $R_{max}(x) \triangleq \max\limits_{s \in \mathcal{S}^*} R^s(x)$.* □

3.1 Approach

With the concepts introduced so far, we can state our algorithm.

Algorithm 1. `Minmax-Regret 1-Sink`$(P(V, E))$

1. *For every bipartite scenario s,*
    (a) *Determine the optimal sink $c(s)$ under it.*
    (b) *Compute the regret, $R^s(x) = \Theta(x, s) - \Theta(c(s), s)$.*
2. *Find the lowest point in $R_{max}(x) = \max\limits_{s \in \mathcal{S}^*} R^s(x)$.* □

In what follows we shall first identify non-dominated scenarios, and then show how to compute the optimal sink under each such scenario. Finally we carry out Step 2. We shall show that the total time required is $O(n)$.

3.2 Useful Properties

Theorem 1 follows from the following lemma, which is independently useful. It identifies a scenario that is not dominated at x.

Lemma 1. *Let v_i be a critical vertex for $x \in P$ under scenario s, such that v_i lies to the left of x.*

(a) *$R^s(x)$ does not decrease when the weight of any vertex $v \in P[v_{i+1}, v_n]$ is decreased by $\delta = w^s(v) - \underline{w}(v)$.*
(b) *$R^s(x)$ does not decrease when the weight of any vertex $v \in P[v_1, v_i]$ is increased by $\delta = \overline{w}(v) - w^s(v)$.*

Proof. (a) Clearly, reducing the weight of v has no effect on the first term of (2), $\Theta(x, s)$, because v_i continues to be critical. Also, decreasing the weight of any vertex cannot increase $\Theta(c(s), s)$. Therefore, $R^s(x)$ cannot decrease.

(b) Let s' be the scenario that results from the change. Then we have $\Theta(x, s') - \Theta(x, s) = \delta$. We claim that $\Theta(c(s'), s') - \Theta(c(s), s) \leq \delta$, so that $R^{s'}(x) \geq R^s(x)$ holds. To prove this claim, let v_h and v_k be critical vertices for $c(s)$ such that $c(s)$ lies on $P[v_h, v_k]$. Then by definition we have

$$d(v_h, c(s))\tau + \sum_{v \in V_h} w^s(v) = d(v_k, c(s))\tau + \sum_{v \in V \backslash V_{k-1}} w^s(v),$$

if $c(s)$ is not at a vertex.[2]

For $i = 1, 2, \ldots, n$, let us define the evacuation time to a facility at point x under scenario s for all the evacuees at the vertices in V_i or $V \backslash V_{i-1}$, respectively.

$$\overrightarrow{f}_i^s(x) \triangleq d(x, v_i)\tau + \sum_{v \in V_i} w^s(v) \text{ for } x \in P(v_i, v_n] \tag{5}$$

$$\overleftarrow{f}_i^s(x) \triangleq d(x, v_i)\tau + \sum_{v \in V \backslash V_{i-1}} w^s(v) \text{ for } x \in P[v_1, v_i), \tag{6}$$

where $P[v_1, v_i)$ and $P[v_1, v_i)$ do not include v_i. Function $\overrightarrow{f}_i^s(x)$ (resp. $\overleftarrow{f}_i^s(x)$) increases (resp. decreases) linearly as x moves from v_i to v_n (resp. v_1 to v_i). See Fig. 1. If we increase the weight of some vertex $v \in V_h$ by δ, resulting in scenario s', the critical vertex for $c(s')$ may change from v_h to $v_{h'}$. In Fig. 1(a) it is seen that function $\overrightarrow{f}_{h'}^{s'}(x)$ is shifted up by δ from $\overrightarrow{f}_h^s(x)$ for x to the right of v_h. Note that the height difference between the intersection points b and a in the figure is no more than δ. Therefore, $\Theta(c(s'), s') - \Theta(c(s), s) \leq \delta$, which is no more than the amount of increase in $\Theta(x, s)$ we thus have $R^{s'}(x) \geq R^s(x)$. It is possible that $\overrightarrow{f}_h^{s'}(x)$ intersects $\overleftarrow{f}_l^s(x)$ ($l \neq k$), as shown in Fig. 1(b). It is easy to see that $\Theta(c(s'), s') - \Theta(c(s), s) \leq \delta$ in such a case as well. $\qquad \square$

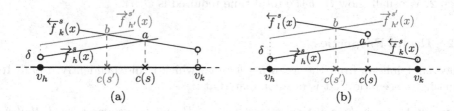

Fig. 1. (a) $\overrightarrow{f}_h^s(c(s)) = \overleftarrow{f}_k^s(c(s))$; (b) $\overrightarrow{f}_h^s(c(s)) > \overleftarrow{f}_k^s(c(s))$.

The symmetric case to the above lemma, where v_i lies to the right of x, can be proved analogously. The following corollary is immediate from these results, and Theorem 1 follows from it.

Corollary 1. *Let v_i be a critical vertex for $x \in P$ under scenario s, such that v_i lies to the left (resp. right) of x. Then s is dominated by $\bar{s}_i$ (resp. $\underline{s}_i$) at x.* $\square$

Based on Lemma 1, we introduce several cost functions (evacuation times) under bipartite scenarios.

$$\underline{f}_i^L(x) \triangleq \begin{cases} 0 & \text{for } x \in P[v_1, v_i] \\ \overrightarrow{f}_i^{s_0}(x) & \text{for } x \in P(v_i, v_n] \end{cases} \tag{7}$$

---

[2] If $c(s)$ is at a vertex, the situation looks like Fig. 1(b), and the equality doesn't hold.

$$\overline{f}_i^L(x) \triangleq \begin{cases} 0 & \text{for } x \in P[v_1, v_i] \\ \overrightarrow{f}_i^{\overline{s}_i}(x) & \text{for } x \in P(v_i, v_n] \end{cases} \tag{8}$$

$$\underline{f}_i^R(x) \triangleq \begin{cases} \overleftarrow{f}_i^{s_o}(x) & \text{for } x \in P[v_1, v_i) \\ 0 & \text{for } x \in P[v_i, v_n] \end{cases} \tag{9}$$

$$\overline{f}_i^R(x) \triangleq \begin{cases} \overleftarrow{f}_i^{\overline{s}_i}(x) & \text{for } x \in P[v_1, v_i) \\ 0 & \text{for } x \in P[v_i, v_n]. \end{cases} \tag{10}$$

Intuitively, $\underline{f}_i^L(x)$ and $\overline{f}_i^L(x)$ (resp. $\underline{f}_i^R(x)$ and $\overline{f}_i^R(x)$) represent the evacuation times of evacuees at the vertices in V_i (resp. $V \setminus V_{i-1}$) if the sink is located at x, provided v_i is a critical vertex. We now define upper envelopes for the evacuation times under s,

$$\Theta_L(x, s) \triangleq \max\{\overrightarrow{f}_i^s(x) \mid 1 \leq i \leq n\}, \tag{11}$$

which is monotonically increasing as x moves from v_1 to v_n, and

$$\Theta_R(x, s) \triangleq \max\{\overleftarrow{f}_i^s(x) \mid 1 \leq i \leq n\}, \tag{12}$$

which is monotonically decreasing as x moves from v_1 to v_n.[3] Let

$$\Theta(x, s) \triangleq \max\{\Theta_L(x, s), \Theta_R(x, s)\}.$$

Clearly, $\Theta(x, s)$ is a unimodal function. It is easy to see that $c(s)$ lies at the lowest point of $\Theta(x, s)$, which is at the intersection of $\Theta_L(x, s)$ and $\Theta_R(x, s)$ [4].

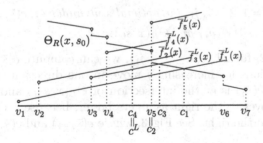

Fig. 2. Intersection of $\overline{f}_j^L(x)$ and $\Theta_R(x, s_0)$

Example 1. Figure 2 shows $\{\overline{f}_i^L(x) \mid 1 \leq i \leq 5\}$ and $\Theta_R(x, s_0)$. Each $\overline{f}_i^L(x)$ is represented by a left-open half line. Function $\overline{f}_i^L(x)$ $(1 \leq i \leq 4)$ intersects $\Theta_R(x, s_0)$ at c_i, and c_4 is the leftmost intersection in this example, which we name c^L. Note that function $\overline{f}_5^L(x)$ does not intersect $\Theta_R(x, s_0)$. Of course, this does not mean that $c(\overline{s}_5)$ does not exist. It only means that $c(\overline{s}_5)$ is not

---

[3] $\Theta_L(x, s)$ and $\Theta_R(x, s)$ are in general discontinuous.

at the intersection of $\overline{f}_5^L(x)$ and $\Theta_R(x, s_0)$. Instead, it is at the intersection of $\overline{f}_4^L(x)$ and $\Theta_R(x, \overline{s}_5)$, namely, v_4 is the critical vertex for $c(\overline{s}_5)$.[4] Note also that $c(\overline{s}_3) = c_2 \neq c_3$. This is because, under $\overline{s}_3$, the critical vertex for point x to the right of v_3 is v_2, not v_3.   □

Let j_1 be the largest index such that $\overline{f}_{j_1}^L(x)$ and $\Theta_R(x, s_0)$ intersect, and let c^L be at the intersection $\overline{f}_{j_0}^L(x)$ and $\Theta_R(x, s_0)$, where $j_1 \geq j_0$. It is possible that $j_1 > j_0$ holds. This would be the case in Example 1, if v_4 didn't exist.

To find the correct $c(\overline{s}_j)$ for $j \leq j_1$ even when $c(\overline{s}_j) \neq c_j$, we perform some preprocessing. Define $\overline{W}^L(v_j)$ by

$$\overline{W}^L(v_j) \triangleq \begin{cases} \overline{w}(v_1) & \text{for } j = 1 \\ \max\{d(v_j, v_{j-1})\tau + \overline{W}^L(v_{j-1}), \sum_{i=1}^{j} \overline{w}(v_i)\} & \text{for } j \geq 2. \end{cases} \quad (13)$$

Clearly, $\overline{W}^L(v_j)$ can be computed in $O(n)$ time for all j. In Example 1, we have $\overline{W}^L(v_3) = \overline{f}_2^L(v_3)$, and $c(\overline{s}_3)$ $(= c_2)$ can be found at the intersection of $d(x, v_3)\tau + \overline{W}^L(v_3)$ and $\Theta_R(x, s_0)$.

Note that all the information about the upper envelope of $\{\overline{f}_j^L(x) \mid j = 1, 2, \ldots\}$ is contained in $\{\overline{W}^L(v_j) \mid j = 1, 2, \ldots\}$. Using them, we can easily compute the highest intersection between $\{\overline{f}_j^L(x) \mid j = 1, 2, \ldots\}$ and $\Theta_R(x, s_0)$. Here we are making use of the fact the all these functions have the same slope τ.

Lemma 2. *For $j = 1, 2, \ldots, j_1$, the optimal sink under $\overline{s}_j$, $c(\overline{s}_j)$, is at the intersection of $d(x, v_j)\tau + \overline{W}^L(v_j)$ and $\Theta_R(x, s_0)$.*   □

Based on Lemma 2, for all j ($1 \leq j \leq j_1$), we can compute $c(\overline{s}_j)$ in $O(n)$ time, since the intersections for increasing j move in one direction (to the left). For $j > j_1$, note that $c(\overline{s}_j)$ is at the intersection of $\Theta_L(x, s_M)$ and $\Theta_R(x, \overline{s}_j)$. It is seen that $c(\overline{s}_j)$ moves to the right as j gets larger, because there are more and more evacuees from the right. See Fig. 3, where $c(\overline{s}_{j-1})$ and $c(\overline{s}_j)$ are compared.

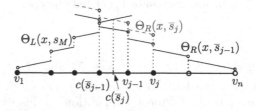

Fig. 3. Computing $\{c(\overline{s}_j) \mid j_1 < j \leq n\}$

---

[4] See the discussion in the paragraph after Lemma 2.

If we flip Fig. 3 horizontally, we get a figure similar to Fig. 2. Therefore, using the method that computed $\{c(\overline{s}_j) \mid 1 \leq j \leq j_1\}$, we can also compute $\{c(\overline{s}_j) \mid j_1 < j \leq n\}$ in $O(n)$ time. We can compute $\{c(\underline{s}_j) \mid 1 \leq j \leq n\}$ similarly. We thus have

Lemma 3. *1. For all j ($1 \leq j \leq n$), optimal sinks $c(\overline{s}_j)$ can be computed in $O(n)$ time.*
2. For all j ($1 \leq j \leq n$), optimal sinks $c(\underline{s}_j)$ can be computed in $O(n)$ time. $\square$

From the above discussion, we also have

Lemma 4. *1. There is an index j_1 such that for j ($1 \leq j \leq j_1 - 1$), optimal sink $c(\overline{s}_j)$ does not lie to the left of $c(\overline{s}_{j+1})$, and for j ($j_1 \leq j \leq n - 1$), optimal sink $c(\overline{s}_j)$ does not lie to the right of $c(\overline{s}_{j+1})$.*
2. There is an index j_1' such that for j ($j_1' \leq j \leq n$), optimal sink $c(\underline{s}_j)$ does not lie to the right of $c(\overline{s}_{j-1})$, and for j ($2 \leq j \leq j_1'$), optimal sink $c(\underline{s}_j)$ does not lie to the left of $c(\overline{s}_{j-1})$. $\square$

3.3 Regret

Now that we know how to compute $\{c(\overline{s}_j), c(\underline{s}_j) \mid j = 1, \ldots, n\}$, we can determine $\{R^{\overline{s}_j}(x), R^{\underline{s}_j}(x) \mid j = 1, \ldots, n\}$ from them by (2). See Fig. 4. The remaining issue is how to perform Step 2 of our approach stated in Sect. 3.1, i.e., finding the lowest point in the upper envelope $R_{max}(x) = \max_{1 \leq j \leq n} \{R^{\overline{s}_j}(x), R^{\underline{s}_j}(x)\}$. From Fig. 4 it is clear that for this purpose we need to use only the part of $R^{\overline{s}_j}(x)$ that is to the right of $c(\overline{s}_j)$ and the part of $R^{\underline{s}_j}(x)$ that is to the left of $c(\underline{s}_j)$. In addition to the useful properties stated in Lemma 4, we have the following nice property:

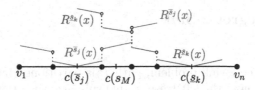

Fig. 4. $R^{\overline{s}_j}(x)$ and $R^{\underline{s}_k}(x)$

Lemma 5. *1. For all j ($1 \leq j \leq n-1$), if $R^{\overline{s}_{j+1}}(x)$ crosses $R^{\overline{s}_j}(x)$ to the right of $c(\overline{s}_{j+1})$, then it does so once.*
2. For all j ($2 \leq j \leq n$), if $R^{\overline{s}_{j-1}}(x)$ crosses $R^{\underline{s}_j}(x)$ to the left of $c(\overline{s}_{j-1})$, then it does so once.

Proof. (a) As in Lemma 4, we consider two cases, depending on the index j. For each j ($1 \leq j \leq j_1$), regret $R^{\overline{s}_j}(x)$ to the right of $c(\overline{s}_j)$ is $d(x, c(\overline{s}_j))\tau$ plus a constant (independent of j) "jump" by $\underline{w}(v_k)$ at each vertex v_k. Thus

$R^{\bar{s}_j}(x)$ to the right of $c(\bar{s}_j)$ for different j's are parallel to each other and don't cross each other. Moreover, we have $R^{\bar{s}_{j_0}}(x) \geq R^{\bar{s}_j}(x)$ to the right of $c(\bar{s}_j)$ for j $(1 \leq j \leq j_1)$. If $j > j_1$, it is clear from Fig. 5, where v_{j+1} has the maximum (resp. minimum) weight under $\bar{s}_{j+1}$ (resp. $\bar{s}_j$), that the "jump" amounts at v_{j+1} are different for $R^{\bar{s}_{j+1}}(x)$ and $R^{\bar{s}_j}(x)$. All other vertices have the same weight under $\bar{s}_{j+1}$ and $\bar{s}_j$. If they cross each other, namely, if $R^{\bar{s}_{j+1}}(v_{j+1}) > R^{\bar{s}_j}(v_{j+1})$ then we have $R^{\bar{s}_{j+1}}(x) > R^{\bar{s}_j}(x)$ for all x to the right of v_{j+1}.

(b) Symmetric to (a).   □

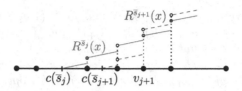

Fig. 5. $R^{\bar{s}_{j+1}}(x)$ may cross $R^{\bar{s}_j}(x)$ from below.

Based on Lemma 5, it is easy to show

Lemma 6. *The upper envelopes,* $\max_{1\leq j\leq n}\{R^{\bar{s}_j}(x)\}$ *and* $\max_{1\leq j\leq n}\{R^{\underline{s}_j}(x)\}$*, can be computed in* $O(n)$ *time.*   □

Thus the lowest point of $R_{max}(x) = \max\{\max_{1\leq j\leq n}\{R^{\bar{s}_j}(x), \max_{1\leq j\leq n}\{R^{\underline{s}_j}(x)\}\}$ can also be computed in $O(n)$ time in Step 2 of Algorithm **Minmax-Regret 1-Sink**$(P(V,E))$. Together with Lemma 3, we have

Theorem 2. *The minmax-regret 1-sink on a path can be computed in* $O(n)$ *time.*   □

4   Minmax Regret 2-Sink

4.1   Approach

In any solution to the 2-sink problem, path P is partitioned into two subpaths. Let P_1 (resp. P_2) denote the left (resp. right) subpath, such that all evacuees at the vertices in P_1 (resp. P_2) evacuate to the sink in P_1 (resp. P_2). Let $\tilde{S}$ denote the set of all *consecutive dominant* scenarios [9]. This means that under each scenario in $\tilde{S}$, a contiguous group of vertices are max-weighted and the rest of vertices are min-weighted. Clearly, a scenario under which only one subpath, either P_1 or P_2, has a bipartite scenario in that subpath belongs to $\tilde{S}$. Let $\Theta^{P_1}(x,s)$ (resp. $\Theta^{P_2}(x,s)$) denote the "local" evacuation time under scenario s to point x in P_1 (resp. P_2) from the vertices of P_1 (resp. P_2). The major difference between the minmax regret 1-sink and 2-sink problems is that in the 2-sink problem, the regret is the difference between $\Theta^{P_1}(x,s)$ (resp. $\Theta^{P_2}(x,s)$), which is local, and the 2-sink evacuation time, which is global, namely over the entire path P. Arumugam et al. proved

Lemma 7 [1]. *Any scenario that is not consecutive dominant is dominated by a consecutive dominant scenario at all points.*[5] □

For a fixed partition $\{P_1, P_2\}$, there are $O(n)$ consecutive dominant scenarios for P_1 and $O(n)$ consecutive dominant scenarios for P_2, but we will examine only $O(\log n)$ partitions. Thus, although $\tilde{S}$ consists of $O(n^2)$ scenarios, our algorithm will examine only $O(n \log n)$ scenarios in it. Given a partition $\{P_1, P_2\}$, let $\tilde{S}(P_1/P_2) \subset \tilde{S}$ denote the set of all scenarios under which the sub-scenarios in P_1 are bipartite and all vertices on P_2 are min-weighted. Similarly, let $\tilde{S}(P_2/P_1) \subset \tilde{S}$ denote the set of all scenarios under which the sub-scenarios in P_2 are bipartite and all vertices on P_1 are min-weighted.

We present our algorithm at two levels. At the lower level, we compute the minmax regret solution under the scenarios in $\tilde{S}(P_1/P_2)$ and $\tilde{S}(P_2/P_1)$ for a given partition $\{P_1, P_2\}$.

Procedure 1. `Local-minmax`(P_1/P_2)

1. *Under each scenario $s \in \tilde{S}(P_1/P_2)$,*
   (a) Compute the optimal 2-sink in P, and determine the evacuation time t^s.
   (b) Determine the 1-sink under s locally within P_1.
2. *Determine the 1-sink under s_0 locally within P_2.*
3. *For each scenario $s \in \tilde{S}(P_1/P_2)$, let $R_1^s(x) = \Theta^{P_1}(x, s) - t^s$ and $R_2^s(x) = \Theta^{P_2}(x, s) - t^s$. Compute the upper envelopes $R_1^*(x) = \max_s R_1^s(x)$ and $R_2^*(x) = \max_s R_2^s(x)$.*
4. *Determine the evacuation time t_{11} (resp. t_{12}) at the lowest point of $R_1^*(x)$ (resp. $R_2^*(x)$), and return $t_1 = \max\{t_{11}, t_{12}\}$.* □

For the scenarios in $\tilde{S}(P_2/P_1)$, we define `Local-minmax`(P_2/P_1), which is obtained from `Local-minmax`(P_1/P_2) by interchanging the indices 1 and 2. From Property 3 in [1] it follows that

Lemma 8 [1]. *The minmax regret is unimodal with respect to the position of the dividing edge that separates P_1 and P_2.* □

Lemma 8 implies that if $t_1 \geq t_2$ (resp. $t_1 < t_2$), we should move the dividing edge to the left (resp. right), where t_1 (resp. t_2) is returned by Procedure `Local-minmax`(P_1/P_2) (resp. `Local-minmax`(P_2/P_1)). Our algorithm will invoke `Local-minmax`(P_1/P_2) and `Local-minmax`(P_2/P_1) $O(\log n)$ times, using binary search.

Algorithm 2. `Minmax-Regret 2-Sink`$(P(V, E))$

1. *Compute the optimal 2-sink under s_0. Let v_k be the rightmost vertex of the left subpath corresponding to the optimal 2-sink. Initialize $P_1 = P[v_1, v_k]$ and $P_2 = P[v_{k+1}, v_n]$.*
2. *Execute `Local-minmax`(P_1/P_2) (resp. `Local-minmax`(P_2/P_1)), which returns the evacuation time t_1 (resp. t_2).*

---

[5] Not necessarily by the same consecutive dominant scenario at every point.

3. *If $t_1 \geq t_2$ (resp. $t_1 < t_2$), move the partitioning edge to the left (resp. right), using binary search on the edges to partition P. If P_1 changes, then repeat this step with the new P_1.*

4. *If $t_1 \geq t_2$ (resp. $t_1 < t_2$), then the t_1 (resp. t_2) is the evacuation time corresponding to the minmax regret 2-sink.*   □

Here is a rough complexity analysis. We will show below (Lemma 10) that Step 1(a) of `Local-minmax`(P_1/P_2) and `Local-minmax`(P_2/P_1), which is the most time-consuming, takes $O(\log^3 n)$ time per scenario, and it must be repeated $O(n)$ times, totaling in $O(n\log^3 n)$ per partition $\{P_1, P_2\}$. Since we test $O(\log n)$ partitions in Algorithm 2, the total time for all partitions is $O(n\log^4 n)$. Everything else, such as the preprocessing we will discuss in the next subsection, takes less time.

4.2   Upper Envelope Tree

For the left-dominant scenarios, we can compute $\overline{W}^L(v_j)$ by (13), as in Sect. 3.2. But we need $\Theta_R(x, s_0)$ for the left subpath P_1. We have a problem here, since the right end of P_1 is not fixed. We need a data structure that makes it easy to compute this upper envelope when the right boundary of P_1 varies.

For this purpose, we introduce the *upper envelope tree* T^s built over the vertices in V for scenario s. As we show later, we will be using only two such trees, T^{s_0} and T^{s_M}. T^s is a balanced binary tree with the vertices of P as its leaf nodes. See Fig. 6.

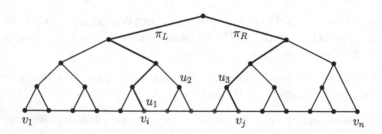

Fig. 6. Upper envelope tree T^s

Let $v_L(u)$ (resp. $v_R(u)$) denote the leftmost (resp. rightmost) vertex on P that belongs to subtree $T^s(u)$ rooted at u. With each node u of T^s, we associate function $E_u^L(x)$ (resp. $E_u^R(x)$), which is the upper envelope for the evacuation times to point x that lies to the right (resp. left) of $v_R(u)$ (resp. $v_L(u)$) for all the evacuees at the vertices on $P[v_L(u), v_R(u)]$. They have the following forms:

$$E_u^L(x) = d(v_R(u), x)\tau + W^L(u) \text{ for } x \in P[v_R(u), v_n]$$
$$E_u^R(x) = d(v_L(u), x)\tau + W^R(u) \text{ for } x \in P[v_1, v_L(u)] \tag{14}$$

Note that it is not necessary to remember the first term $d(v_R(u), x)\tau$ (resp. $d(v_L(u), x)\tau$) in $E_u^L(x)$ (resp. $E_u^R(x)$), since they can be easily reconstructed in constant time. The following algorithm computes $W^L(u)$ and $W^R(u)$ iteratively for all nodes u of T^s.

Algorithm 3. Upper-Envelope-Tree(T, s)

1. *For each vertex $v_i \in V$, let $W^L(v_i) = W^R(v_i) = w^s(v_i)$.*
2. *Let u_l (resp. u_r) denote the left (resp. right) child of node u of T^s. Construct $W^L(u)$ from $W^L(u_l)$ and $W^L(u_r)$ by[6]*

$$W^L(u) = \max\{d(v_R(u), v_R(u_l))\tau + W^L(u_l), W^L(u_r)\}. \tag{15}$$

3. *Construct $W^R(u)$ from $W^R(u_l)$ and $W^R(u_r)$ by*

$$W^R(u) = \max\{d(v_L(u), v_L(u_r))\tau + W^R(u_r), W^R(u_l)\}. \tag{16}$$

$\square$

Lemma 9. *Algorithm* Upper-Envelope Tree(T, s) *computes* $W^L(u)$ *and* $W^R(u)$ *for all non-leaf nodes u of T^s in $O(n)$ time and $O(n)$ space.* $\square$

By invoking Upper-Envelope Tree(T, s) with $s = s_0$ and $s = s_M$, we construct T^{s_0} and T^{s_M}, respectively, in $O(n)$ time.

4.3 Finding the 2-Sink Under a Consecutive Dominant Scenario

Recall Step 1(a) of Procedure Local-minmax(P_1/P_2) and Local-minmax(P_2/P_1) in Sect. 4.1. This step needs to be executed for many consecutive dominant scenarios. Some example scenarios are given in Fig. 7, where the vertices on $P[v_h, v_k]$, shown as hollow circles, are max-weighted, and the rest, shown as filled circles, are min-weighted. We thus have $P_1 = P[v_1, v_k]$ and $P_2 = P[v_{k+1}, v_n]$. In carrying out Step 1(a), we perform nested binary search. The vertical line in Fig. 7 shows the dividing line between two subpaths that are assumed to correspond to the optimal partition $\{P_1', P_2'\}$ for the optimal 2-sink for P under the given consecutive dominant scenario. The optimal 1-sink for the left subpath P_1' is assumed to lie on the edge $e = (v_{i-1}, v_i)$. To verify if it is indeed where the true optimal 1-sink for the left subpath lies, we need to compute the evacuation time to point $x \in e$ for the evacuees at the vertices on $P[v_i, v_j]$, as well as the evacuation time to x for the evacuees at the vertices on $P[v_1, v_{i-1}]$. We now illustrate how to compute the former evacuation time, making use of the upper envelope tree T^{s_M}.

Example 2. Let us consider the case shown in Fig. 7(a), where v_h, v_i, v_j and v_k appear in this order. Assume that the upper envelope tree in Fig. 6 is T^{s_M}, and let π_L (resp. π_R) denote the path from leaf v_i (resp. v_j) to the root of

---

[6] Note that $v_R(u) = v_R(u_r)$.

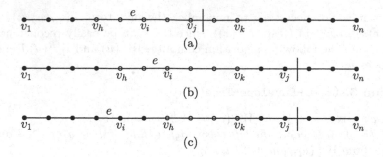

Fig. 7. Right-dominant scenarios in P_1: Hollow vertices are max-weighted.

T^{sM}. We make use of the evacuation time stored at the root of each maximal subtree all of whose leaf vertices belong to $P[v_i, v_j]$. They can be found tracing π_L and π_R, say, from the root downwards, and there are $O(\log n)$ of them. $T^{sM}(u_1)$, $T^{sM}(u_2)$, and $T^{sM}(u_3)$ are such maximal subtrees in this example. Let $\overline{W}[v_i, v_j] = \sum_{i \leq k \leq j} \overline{w}(v_k)$.[7] It is easy to see that the evacuation time to $x \in (v_{i-1}, v_i)$ for the evacuees at the leaf vertices of these subtrees is given by

$$\overline{\tau}_{P[v_i,v_j]}(x) = \max\{d(x, v_i)\tau + \overline{W}[v_i, v_j], d(x, v_L(u_2))\tau + W^L(u_2) + \overline{W}[v_L(u_3), v_j],$$
$$d(x, v_L(u_3))\tau + W^L(u_3)\}. \tag{17}$$

This is essentially the same as (4).   □

The important points in the above example are that (i) the maximum in (17) is taken over $O(\log n)$ items, and (ii) each item is a sum of no more than three terms. Thus the above maximum can be computed in $O(\log n)$ time. The cases shown in Fig. 7(b) and (c) are more complicated. We need to make use of both T^{sM} and T^{so} in those cases, but the idea illustrated in the above example can be generalized.

Lemma 10. *Step 1(a) of* `Local-minmax`(P_1/P_2) *can be carried out in* $O(\log^3 n)$ *time. Similarly for Step 1(a) of* `Local-minmax`(P_2/P_1).

Proof. In the case shown in Fig. 7(b), to the right of edge e, the vertices on $P[v_i, v_k]$ (resp. $P[v_{k+1}, v_j]$) are max-weighted (resp. min-weighted). For point $x \in e = (v_{i-1}, v_i)$, we want to compute the evacuation time for the evacuees at the vertices on $P[v_i, v_j]$. For the evacuees on $P[v_{k+1}, v_j]$, we collect evacuation time data from T^{so} and for the evacuees on $P[v_i, v_k]$, we collect evacuation time data from T^{sM}.[8] The total amount of data collected is still $O(\log n)$. This is also true of the case shown in Fig. 7(c).

Consider the subpath to the left of the vertical line in each of Fig. 7(a), (b) and (c). We can compute the evacuation times to point $x \in e$ from the left and

---

[7] We can precompute $\overline{W}[v_i] = \sum_{1 \leq k \leq i} \overline{w}(v_k)$ for all i in $O(n)$ time. Then $\overline{W}[v_i, v_j] = \overline{W}[v_j] - \overline{W}[v_{i-1}]$ can be found in constant time.

[8] It must be adjusted by $\underline{W}[v_{k+1}, v_l]$.

right parts of the subpath, using the method described above. If they intersect at some $x \in e$, then x is the 1-sink for this subpath. Note that they are not equal if they intersect at an end vertex of e (see $\overline{f}_2^L(x)$ in Fig. 2). Otherwise, we perform binary search to find the appropriate new edge e to test. We need to move the separating edge e $O(\log n)$ times, and each time we spend $O(\log n)$ time to compute evacuation times, using T^{sM} and T^{so}. Thus the total time per choice of the vertical line is $O(\log^2 n)$. We need to examine $O(\log n)$ edges for the vertical line, so that the total time needed to find the 2-sink under a given scenario is $O(\log^3 n)$. □

Let us review Algorithm `Minmax-Regret 2-Sink`$(P(V, E))$. We partition path P into two subpaths, $\{P_1, P_2\}$, by removing an edge, and want to test if this partition corresponds to the minmax regret solution. We consider $O(\log n)$ such partitions, using binary search, and for each partition, $\{P_1, P_2\}$, we examine $O(n)$ left and right dominant scenarios for P_1 and P_2. It is shown in [1] that when some vertices in P_1 (resp. P_2) are max-weighted, then we may assume that all the vertices in P_2 (resp. P_1) are min-weighted. For all left and right dominant scenarios for P_1, we run Procedure `Local-minmax`(P_1/P_2) to compute the minmax regret evacuation time t_1. Similarly, for all left and right dominant scenarios for P_2, we run Procedure `Local-minmax`(P_2/P_1) to compute the minmax regret evacuation time t_2. Based on t_1 and t_2, Step 3 of Algorithm 2 decides which way to move the partitioning edge. We already discussed the time complexity at the end of Sect. 4.1.

Theorem 3. *The minmax regret 2-sink on a path network can be computed in* $O(n \log^4 n)$ *time.* □

5   Conclusion

The first contribution of this paper is an optimal algorithm for finding the minmax regret 1-sink for a path network. We achieved it by making the computation of the optimal sinks under left-dominant and right-dominant scenarios more efficient than the previously used methods [6,12]. We also improved the method for finding the lowest point in the upper envelope of the regret functions.

The second contribution is an $O(n \log^4 n)$ time algorithm for computing the minmax regret 2-sink for a path network, which is the first sub-quadratic result and is a significant improvement over the $O(n^2 \log^2 n)$ time algorithm in [1]. This sub-quadratic time complexity is somewhat surprising in view of the fact that $|\tilde{\mathcal{S}}| = O(n^2)$. We believe that we can generalize our approach to finding the minmax regret k-sink for path networks in sub-quadratic time in n, when k is a constant, improving the results in [1] for small k.

Using our 1-sink algorithm, it is easy to solve the minmax regret 1-sink problem on a cycle network in $O(n^2)$ time. Some open problems are if the algorithms for the minmax regret 1-sink (resp. k-sink) problem for tree networks in [7] (resp. [1]) can be improved.

References

1. Arumugam, G.P., Augustine, J., Golin, M., Srikanthan, P.: A polynomial time algorithm for minimax-regret evacuation on a dynamic path. arXiv:1404,5448v1 [cs.DS] 22 April 2014 165 (2014)
2. Bhattacharya, B., Kameda, T., Song, Z.: Improved minmax regret 1-center algorithms for cactus networks with c cycles. In: Pardo, A., Viola, A. (eds.) LATIN 2014. LNCS, vol. 8392, pp. 330–341. Springer, Heidelberg (2014)
3. Bhattacharya, B., Kameda, T., Song, Z.: Linear time algorithm for finding minmax regret 1-median on a tree with positive vertex weights. Algorithmica **70**, 2–21 (2014)
4. Cheng, S.-W., Higashikawa, Y., Katoh, N., Ni, G., Su, B., Xu, Y.: Minimax regret 1-sink location problems in dynamic path networks. In: Chan, T.-H.H., Lau, L.C., Trevisan, L. (eds.) TAMC 2013. LNCS, vol. 7876, pp. 121–132. Springer, Heidelberg (2013)
5. Hamacher, H., Tjandra, S.: Mathematical modelling of evacuation problems: a state of the art. In: Schreckenberg, M., Sharma, S.D. (eds.) Pedestrian and Evacuation Dynamics, pp. 227–266. Springer, Heidelberg (2002)
6. Higashikawa, Y., Augustine, J., Cheng, S.W., Golin, M.J., Katoh, N., Ni, G., Su, B., Xu, Y.: Minimax regret 1-sink location problem in dynamic path networks. Theor. Comput. Sci. **31**(2) (2014)
7. Higashikawa, Y., Golin, M.J., Katoh, N.: Minimax regret sink location problem in dynamic tree networks with uniform capacity. In: Pal, S.P., Sadakane, K. (eds.) WALCOM 2014. LNCS, vol. 8344, pp. 125–137. Springer, Heidelberg (2014)
8. Kouvelis, P., Yu, G.: Robust Discrete Optimization and its Applications. Kluwer Academic Publishers, London (1997)
9. Li, H., Xu, Y., Ni, G.: Minimax regret 2-sink location problem in dynamic path networks. J. Comb. Optim. (2014)
10. Mamada, S., Makino, K., Fujishige, S.: Optimal sink location problem for dynamic flows in a tree network. IEICE Trans. Fundam. **E85–A**, 1020–1025 (2002)
11. Mamada, S., Uno, T., Makino, K., Fujishige, S.: An $O(n \log^2 n)$ algorithm for a sink location problem in dynamic tree networks. Discrete Appl. Math. **154**, 2387–2401 (2006)
12. Wang, H.: Minmax regret 1-facility location on uncertain path networks. Eur. J. Oper. Res. **239**(3), 636–643 (2014)

Approximate Aggregation for Tracking Quantiles in Wireless Sensor Networks

Zaobo He[1], Zhipeng Cai[1]([✉]), Siyao Cheng[1,2], and Xiaoming Wang[3]

[1] Department of Computing Science, Georgia State University, Atlanta, USA
zcai@gsu.edu
[2] School of Computer Science and Technology, Harbin Institute of Technology, Harbin, China
[3] School of Computer Science, Shaanxi Normal University, Xi'an, China

Abstract. We consider the problem of tracking quantiles in wireless sensor networks with efficient communication cost. Compared with the algebraic aggregations such as Sum, Count, or Average, holistic aggregations such as quantiles can better characterize data distribution. Let $S(t) = (d_1, \ldots, d_n)$ be the multi-set of sensory data that have arrived until time t in the entire network, which is a sequence of data orderly collected by nodes $s_1, s_2, \ldots, s_k$. The goal is to continuously track ϵ-approximate ϕ-quantiles ($0 \leq \phi \leq 1$) of $S(t)$ at the sink for all ϕ's with efficient total communication cost and balanced individual communication cost. In this paper, a deterministic tracking algorithm based on a dynamic binary tree is proposed to track ϵ-approximate ϕ-quantiles ($0 \leq \phi \leq 1$) in wireless sensor networks, whose total communication cost is $O(k/\epsilon \cdot \log n \cdot \log^2(1/\epsilon))$, where k is the number of the nodes in a network, n is the total number of the data items, and ϵ is the required approximation error.

1 Introduction

Wireless Sensor Networks (WSNs) consist of many nodes which interact with each other through wireless channel. They are now being widely deployed to monitor physical information, such as temperature, pressure, light intensity and so forth [1,2,6–8,10,18,21]. With the development of technologies, the scale of a WSN can be very large [19]. However, the most severe constraint imposed on the extensive applications is the limited power supply as the on-board power is still the main power source which is not rechargeable in most cases. Compared with data computation or storage control, communications among nodes consume more energy. According to [25], the energy consumption for sending one bit data is equal to that for executing 1000 instructions for one sensor. Thus, how to extract information from a huge amount of sensor data with efficient communication cost becomes a crucial problem.

Much effort has been spent on studying various aggregation operations (denoted by function f), including algebraic aggregations such as Sum, Count, or Average [5,16,17], holistic aggregations such as Heavy hitters [20], Quantiles

© Springer International Publishing Switzerland 2014
Z. Zhang et al. (Eds.): COCOA 2014, LNCS 8881, pp. 161–172, 2014.
DOI: 10.1007/978-3-319-12691-3_13

[9,13,15,24], and complex correlation queries in the database area such as Distributed joins [22]. Quantile allows one to extract the order statistics information from the dataset which is widely used in network monitoring [4,29] and database query optimization [11], so that data distribution can be much better characterized. In-network aggregation algorithms are efficient techniques to track algebraic aggregations through computing partial results at intermediate nodes during the process of routing data to the sink [5,13,15,17,24]. By preventing nodes from forwarding all the data to the sink, in-network aggregation algorithms significantly reduce energy consumption. In-network aggregation algorithms can be conducted efficiently for algebraic functions due to the decomposable property of these aggregations [3]. Unfortunately, unlike Sum, Count, or Average, quantiles are not decomposable so that the traditional in-network aggregation algorithms do not work well to track quantiles [9]. The ϕ-quantile ($0 \le \phi \le 1$) of an ordered dataset S is the data x such that $\phi|S|$ elements of S are less than or equal to x and no more than $(1 - \phi)|S|$ elements are larger than x, particularly, the $\frac{1}{2}$-quantile is the median of S.

Since exact results always require huge storage space and large communication cost in WSNs, approximate results are generally expected. The work in [27] shows that a random sample of size $\Theta(1/\epsilon^2)$ is needed to be drawn from a dataset to compute ϵ-approximate quantiles with a constant probability. Moreover, in many applications, the approximate results, rather than exact ones, are good enough for users to perform analysis and make decisions, such as trend analysis [9], anomaly detection [26], and so on. Based on these reasons, an ϵ-approximate ϕ-quantile is expected which can be formally defined as follows:

Definition 1. ϵ-approximate ϕ-quantiles: The ϵ-approximate ϕ-quantiles are those elements in dataset S such as element x that satisfies $(\phi - \epsilon)n \le r(x) \le (\phi + \epsilon)n$ where $r(x)$ is the rank of x in S and n is the total number of the data items.

For quantile-tracking objective, the data model can be divided into three classes: static model, single-stream model and multi-stream model. For the static model, data is predetermined and stored at nodes and f is computed over the union of these multiple datasets. For the single-stream model, there is only one node and data arrives at it in an online fashion. The goal is to track f over the items that have arrived with the minimum storage space or communication cost. Nowadays, the multi-distributed streaming model attracts a lot of attention since it is more general in the physical environment. In this model, data streams into each node in a distributed way and the tracking results are returned in a logical coordinator. If all the nodes are connected to one coordinator directly, it is called a flat model. The nodes in a WSN is organized into a spanning tree and the tracking results are returned at the sink. Moreover, for tracking results, the querying mode can be divided into two classes: single ϕ-quantile and all ϕ-quantile. For the single ϕ-quantile tracking mode, a certain summary always is maintained by a coordinator to compute a certain ϕ-quantile. Comparatively, the data structure or summary preserved by a coordinator for all the ϕ-quantiles can be used to compute any ϕ simultaneously. The bottleneck of single ϕ-quantile is that frequent tracking operations, such as multiple sampling, are needed to satisfy different user-defined ϕ's.

The aforementioned reasons motivate us to track quantiles in WSNs in a general way, which can be described as follows. The sensor nodes are organized into a spanning tree and sensory data streams into each node in an online fashion. The intermediate nodes not only need to relay data of its descendants, but also hold a local dataset for itself. $S(t)$ is the multi-set of items of the entire network that have arrived until time t. $S(t) = (d_1, \ldots, d_n)$ is a sequence of data that is collected orderly by nodes $s_1, s_2, \ldots, s_k$. The goal is to continuously track ϵ-approximate ϕ-quantiles ($0 \leq \phi \leq 1$) of $S(t)$ at the sink for all ϕ's.

The main contribution of this work can be summarized as follows: First, quantiles can be tracked over the arrived data at any time t rather than through a one-time computation over a predetermined dataset. Second, quantiles are computed based on an arbitrary topological spanning tree rather than the centralized flat model. Third, a data structure can be maintained in the tree from which all the ϕ-quantiles can be tracked simultaneously rather than for just a specific ϕ.

Thus, our tracking operation is conducted on a platform that combines multi-steam and all ϕ-quantile computations, but it is also significantly complex either. Finally, our algorithm can continuously track the ϕ-quantiles over dataset $S(t)$ for all ϕ's and has a total communication cost of $O\left(\frac{k}{\epsilon} \log n \log^2 \frac{1}{\epsilon}\right)$, where k is the number of the sensor nodes in the network, n is the total number of the data items, and ϵ is the required approximation error.

2   Related Works

The previous quantile tracking techniques can be divided into three categories, which are the exact algorithms, deterministic algorithms and probabilistic algorithms. For a given ϕ, the exact algorithms are to return the exact ϕ-quantile result to users. According to [23], the space complexity for computing the exact median with p passes is $\Omega(n^{1/p})$. Clearly, the space complexity of the exact algorithms is high, especially when the number of the passes p is small.

To further reduce the time and space complexities during tracking quantiles, the deterministic algorithms are proposed, such as the recent works [12,15,24,28]. Unlike the exact algorithms, the deterministic algorithms return an ϵ-approximate ϕ-quantiles of a dataset. Since the deterministic algorithms just require approximate results, they have lower space and communication complexities. In 2005, Cormode et al. [9] proposed an all-tracking algorithm with the cost of $O\left(\frac{k}{\epsilon^2} \log n\right)$. The work in [28] improves this result by a $\Theta(\frac{1}{\epsilon})$ factor, whose result has an upper bound $O\left(\frac{k}{\epsilon} \log n\right)$. Note that the work in [28] discusses the all ϕ-quantiles tracking problem under the flat model, however, it is unclear how to track quantiles in the tree model.

Considering that the approximate quantile with a probability guarantee can be accepted by users in most cases, the complexity of tracking the quantile can be further reduced. Thus, a group of probabilistic algorithms [5,15,17] were proposed. Different from the above two types of the algorithms, the probabilistic algorithms require that the ϵ-approximate ϕ-quantile result is guaranteed with a probability. For example, the work in [15] proposes a quantile estimator and partitions the routing tree to compute ϵ-approximate quantiles within constant probability with the total communication cost of $O(\sqrt{kH}/\epsilon)$ where H is the hight of the routing tree. However, this work just carries out one-time computation over the predetermined dataset, so it is not clear whether it works well for the data stream model.

3   Problem Definition

Without loss of generality, we assume that there are k sensor nodes in a WSN, denoted by $\{s_1, s_2, ..., s_k\}$. Meanwhile, we assume that each node samples a sensory value from the monitored environment at each time slot and it holds a small dataset before the algorithm is initiated. Δt is used to denote the length of the interval between two adjacent time slots and $s_i(t)$ denotes the sensory dataset sampled by node i $(1 \le i \le k)$ until time t. Therefore, a sensory dataset can be obtained at node i for any given time $t \in [0, +\infty)$, and $S(t) = s_1(t) \cup s_2(t) \cup ... \cup s_i(t)$ denotes the entire sensory dataset in the network at time $t \in [0, +\infty)$. We assume at the initial time, each node s_i preserves an initial dataset $s_i(0)$.

For any given ϕ $(0 \le \phi \le 1)$ and integer n $(0 \le n \le 1)$, if there exists $t' \in [0, +\infty]$ satisfying $n = |S(t')|$, then the proposed algorithm is to return the ϵ-approximate ϕ-quantiles at t $(\forall t \in [0, t'])$. Note that the definition of ϵ-approximate ϕ-quantile is given in Definition 1. Specifically, the problem studied in this paper is defined as follows.

Input:
1) ϵ $(\epsilon \ge 0)$ and ϕ $(0 \le \phi \le 1)$.
2) n and Δt.
3) $\{s_i(0) \mid i = 1, 2, \ldots, k\}$.
Output:
ϵ-approximate ϕ-quantile for any $t \in [0, t']$, where t' satisfies that $n = |S(t')|$.

4   The Proposed Algorithm

In order to efficiently track quantiles in WSNs, the whole network is organized by a spanning tree rooted at the sink. The nodes in the spanning tree can be distinguished as the leaf nodes and the intermediate nodes, where the communication cost of the intermediate nodes is large since they not only need to maintain local datasets but also need to relay data of its descendants. Therefore, one key problem of tracking quantiles in WSNs is to reduce the communication cost of the intermediate nodes. To achieve this goal, we develop a global data structure over the routing tree and maintain it dynamically with a bounded communication cost.

We divide the entire tracking period into $O(\log n)$ rounds, denoted by m_i ($i = 1, 2, \ldots \log n$). Whenever $|S(t)|$ has increased by a constant factor, $e.g.$, $|S(t)|$ is doubled, a new round is started. Assuming at time t'', round m_i is launched. We use M_i to denote the set of data at the beginning of round m_i, $i.e.$, $M_i = S(t'')$. M_i is fixed throughout round m_i, $i.e.$, $|M_i| \leq |S(t)|$ and $t'' \leq t$. It is always true that $\epsilon|S(t)| = \Theta(\epsilon|C \cdot S(t)|)$ for constant C and $|M_i| = C \cdot |S(t)|$ is ensured in one round. Thus, in round m_i, we have $\epsilon|S(t)| = \Theta(\epsilon|M_i|)$.

Based on this reason, our goal can be described in another way: along with data streaming into network continuously, the goal is to maintain a data structure over $S(t)$, based on which the rank of any data item x ($x \in S(t)$) can be extracted with error $O(\epsilon M_i)$, where $M_i \subseteq S(t) \subseteq M_{i+1}$ and $1 \leq i \leq \log n$.

Since the operations of initialization, maintenance and tracking are similar in each round, we first focus on one round and then obtain the total cost for all rounds naturally. For simplicity, we assume that all the data values are distinct. In summary, the algorithm for tracking quantiles in WSNs is described as follows.

First, initialize a binary tree T based on $\{s_i(0) \mid 1 \leq i \leq k\}$. The detailed structure of T is provided in Sect. 4.1 and the specific steps of the initialization algorithm are presented in Sect. 4.2.

Second, an accumulatively updating algorithm given in Sect. 4.3, is carried out to reduce the transmission cost when new data arrives.

Third, we need to maintain the binary tree T so that the height and the leaf nodes of T satisfy some requirements. The detail of the binary tree maintenance algorithm is given in Sect. 4.4.

Fourth, the sink computes the rank of x ($x \in S(t)$), $i.e.$, $r(x)$, based on the binary tree T.

Finally, the above four steps are executed iteratively until the number of the execution times reaches $\log n$, where n is the total number of the sensory values in the network as given in Sect. 3.

4.1   The Data Structure

The data structure for querying is a binary tree T that is initialized at the beginning of each round and maintained throughout one round. We take a specific round m_i as an example to describe the data structure. T is constructed in the following way. The root of T is the approximate median of M_i, which divides M_i into two subsets. Each subset is recursively split by selecting their approximate median as the root of the subtree. The splitting process is iteratively executed until the number of the items in each subset is no more than $\epsilon|M_i|/\beta$, where β is a constant satisfying $\beta > 1$ and ϵ is a user-defined error parameter as shown in Sect. 3. Thus, the data structure is a binary tree with ϵ/β as the error parameter.

Obviously, T has $\Theta(\beta/\epsilon)$ nodes in total and the height of T is $h = \Theta(\log \beta/\epsilon)$. Meanwhile, each node in T, denoted by b, corresponds to an interval $I_b = [l_b, u_b]$, where l_b and u_b are the smallest and largest values in the subtree rooted at b respectively.

The exact results can be obtained if we update I_b whenever a new sensory value arrives. However, it incurs a huge communication cost if both the size of the network and sampling frequency of each sensor are large. In practice, each node b just needs to correspond to an approximate interval A_b so that the corresponding interval does not need to be updated every time, where A_b satisfies $|I_b| - \mu \leq |A_b| \leq |I_b|$, μ satisfies $h\mu + \epsilon|M_i|/\beta = \epsilon|M_i|$, and h is the height of binary tree T.

The tracking process for $r(x)$ is a traversal process over T from the root to leaf node v such that $x \in A_v$. For each root-to-leaf path, whenever following a right child, the approximate interval size of its left sibling is summed up. Since there are at most h such intervals, the total error introduced by the traversal process is at most $h\mu$. Finally, since the interval size corresponded by a leaf node is less than $\epsilon|M_i|/\beta$, one can query $r(x)$ in $S(t)$ with absolute error of $O(h\mu + \epsilon|M_i|/\beta)$. If let $h\mu + \epsilon|M_i|/\beta = \epsilon|M_i|$ with corresponding parameters μ and β, the error of $r(x)$ is $O(\epsilon|M_i|)$.

4.2   Initialization of the Binary Tree

Algorithm Description. The initialization algorithm is initiated at the beginning of each round to build a global binary tree T. For any node a in the spanning tree, tr_a denotes the tree rooted at a and c_a denotes the number of children for node a. k_{tr_a} is used to denote the number of the nodes in tr_a. $|tr_a(0)|$ denotes the total number of the data items preserved by the nodes within tr_a at the initial time. r is the sink of the network. p and q denote the number of the leaf nodes and the intermediate nodes in the spanning tree respectively, where the leaf node set is denoted by $\{v_i \mid i = 1, 2, \ldots, p\}$ and the intermediate node set is denoted by $\{u_i \mid i = 1, 2, \ldots, q\}$. The binary tree built by node a is denoted by T_a and the binary tree built for spanning tree tr_a is denoted by T_{tr_a}.

The initialization algorithm in one round has the following 5 steps.

Step 1. Based on its initial dataset, each node s_i $(1 \leq i \leq k)$ builds its own approximate balanced binary tree T_{s_i} with ϵ/β as the error parameter. Now querying any $r(x)$ in $s_i(0)$ has an error of $\epsilon|s_i(0)|/\beta$.

Step 2. Each leaf node v_i $(1 \leq i \leq p)$ transmits its binary tree T_{v_i} to its parent node in the spanning tree.

Step 3. Assume v_i $(1 \leq i \leq c_{u_j})$ is the child node of node u_j $(1 \leq j \leq q)$. Based on T_{v_i} $(1 \leq i \leq c_{u_j})$ and T_{u_j}, u_j can compute any $r(x)$ within tr_{u_j} with an error of $\sum_{i=1}^{c_{u_j}} \epsilon|s_i(0)|/\beta = \epsilon|tr_{u_j}(0)|/\beta$, which is enough for u_j $(1 \leq j \leq q)$ to build a binary tree $T_{tr_{u_j}}$ with ϵ/β as the error parameter. Finally, T_{u_j} is transmitted to the parent node of u_j in the spanning tree.

Step 4. Step 3 is iteratively executed until sink r is reached. Then r broadcasts T to the network through the spanning tree.

Step 5. After receiving T, each node s_i $(1 \leq i \leq k)$ computes the exact number of items in each interval and transmits to its parent node, where the total number of intervals corresponded by the global binary tree T is $\Theta(\beta/\epsilon)$.

It is clear that each interval size of T is exact at the initial time.

Communication Cost. The communication cost of the initialization algorithm in one round is analyzed as follows. In Step 2, the communication cost of the network is $O(p\beta/\epsilon)$ since each leaf node v_i $(1 \le i \le p)$ needs to transmit the binary tree T_{v_i} to its parent node and the size of T_{v_i} is $\Theta(\beta/\epsilon)$. Similarly, the communication cost generated in Step 3 is $O(q\beta/\epsilon)$ since the intermediated nodes also need to report the binary tree to their parents. In Step 4, the communication cost is $O(k\beta/\epsilon)$ since the global binary tree T needs to be broadcasted to k nodes and the size of T is $\Theta(\beta/\epsilon)$. Finally, each node needs to transmit the number of the items in each interval to its parent along the spanning tree, and the communication cost is $O(k\beta/\epsilon)$. In summary, the communication cost of the initialization algorithm in one round is $O(k\beta/\epsilon)$.

4.3   Updating the Interval Size Accumulatively

Algorithm Description. After initialization, each node s_i $(1 \le i \le k)$ preserves a global binary tree T. The naive method of updating T is to report all sensory values sampled by the nodes to the sink leading to the communication cost of $O(nH)$, where H is the height of the spanning tree. Obviously, the cost is very huge since n is generally far larger than k and $1/\epsilon$, otherwise, we just need to send each sensory data to the sink. Thus, an accumulatively updating algorithm is proposed to reduce the communication cost in the updating phase with an accumulative report strategy. Although the error is generated during the quantile tracking process, the proposed algorithm dramatically reduces the communication cost for updating the global binary tree T.

Int is used to denote an arbitrary interval of T. As described in Sect. 4.1, each Int has an exact interval size and an approximated interval size denoted by $|I|$ and $|A|$ respectively. Meanwhile, each node keeps a set of counters for counting the size of each Int.

The accumulatively updating algorithm includes the following two steps:

Step 1. With new sensory data continuously streaming into s_i $(1 \le i \le k)$, s_i monitors Int continuously.

Step 2. If the local count of Int at s_i $(1 \le i \le k)$ has increased by a threshold since its last communication to its parent about the local count of Int, s_i must report an updated local count for Int to its parent. Then, each s_i $(1 \le i \le k)$ resets the counter to 0 and continuously monitors Int.

Among the steps of the accumulatively updating algorithm, determining the threshold is very important since it affects the communication cost and accuracy of the algorithm. Fortunately, this problem can be solved by Theorems 1 and 2.

Theorem 1. *To satisfy that any $r(x)$ $(x \in S(t))$ can be extracted with error $O(\epsilon M_j)$, where $M_j \subseteq S(t) \subseteq M_{j+1}$ and $1 \le j \le \log n$, the condition $|I| - \mu \le |A| \le |I|$ should be ensured, where $\mu = (1 - 1/\beta) \cdot \epsilon/h \cdot |M_j|$.*

The proof of Theorem 1 is given in our technique report [14].

Theorem 2. *Assuming s_i $(1 \leq i \leq k)$ is a node located at layer l_i $(0 \leq l_i \leq H)$ in the spanning tree, its ancestor node set is $\{P_f \mid 0 \leq f \leq l_i - 1, P_f$ is the parent of P_{f+1} and P_{l_i-1} is the parent of $s_i\}$. c_{P_f} is the number of children of P_f. If s_i $(1 \leq i \leq k)$ reports an updated local count for Int to its parent when the local count of sensory data in Int at s_i has increased by a threshold δ_i, where $\delta_i = (\prod_{f=1}^{l_i} c_{P_f})^{-1} \cdot \mu$, querying any $r(x)$ in T has error $O(\epsilon|M_j|)$.*

The proof of Theorem 2 is given in our technique report [14].

Communication Cost. Now we analyze the communication cost for the accumulative updating process in one round. Note that the cost for one time communication is regarded as one unit. When s_i $(1 \leq i \leq k)$ sends an updated count message for $|A|$ to its parent, the communication cost incurred by the accumulated δ_i sensory data is viewed as one unit so that the average cost of one item is $O(1/\delta_i)$. Since each item may incur h times of such a message shipping process, the average cost incurred by an item is $O(h/\delta_i)$. After substituting the expression of δ_i, one can obtain $O(h/\delta_i) = O((1 - 1/\beta)^{-1} \cdot h^2/\epsilon \cdot |M_j|^{-1} \cdot \prod_{f=1}^{l_i} c_{P_f})$. Since in one round, the number of the data items streaming into s_i $(1 \leq i \leq k)$ is $\Theta(|M_j(i)|)$, the cost of updating the local count of $|A|$ at s_i $(1 \leq i \leq k)$ is $O(h|M_j(i)|/\delta_i) = O((1 - 1/\beta)^{-1} \cdot h^2/\epsilon \cdot |M_j(i)|/|M_j| \cdot \prod_{f=1}^{l_i} c_{P_f})$. Thus, combing the condition $h = \Theta(\log(\beta/\epsilon))$, the total cost of updating the binary tree T in one round is $O((1 - 1/\beta)^{-1} \cdot 1/\epsilon \cdot \log^2(\beta/\epsilon) \cdot \sum_{i=1}^{k}(|M_j(i)|/|M_j| \prod_{f=1}^{l_i} c_{P_f}))$. Assuming data streams into each node with a similar speed i.e., $|M_j(i)|/|M_j| = 1/k$, the above expression can be rewritten as $O((1 - 1/\beta)^{-1} \cdot 1/\epsilon \cdot \log^2(\beta/\epsilon) \cdot 1/k \cdot \sum_{i=1}^{k}(\prod_{f=1}^{l_i} c_{P_f}))$. It is easy to derive that $\sum_{i=1}^{k}(\prod_{f=1}^{l_i} c_{P_f}) \leq k^2$, then the communication cost for the accumulative updating process in one round is $O((1 - 1/\beta)^{-1} \cdot k/\epsilon \cdot \log^2(\beta/\epsilon))$.

4.4 Maintaining the Binary Tree

Algorithm Description. The global binary tree T may become unbalanced with new items arriving in the data stream, which leads to a high communication cost for tracking quantiles. Thus, the approximate median as a splitting element should not deviate from the exact median too much. Meanwhile, the data structure requires that the interval size corresponded by the leaf nodes of T should not be beyond $\epsilon|M_i|/\beta$ in round m_i $(1 \leq i \leq \log n)$. Thus, our goal is to maintain the height of the binary tree T as $h = \Theta(\log(\beta/\epsilon))$ and the interval size corresponded by the leaf nodes of T. Before presenting the specific maintenance algorithm, we first introduce Lemma 1 and Theorem 3.

For any intermediate node u in binary tree T, let v and w be the left and right child of u respectively. $|A_u|$, $|A_v|$ and $|A_w|$ denote the approximate interval size of T corresponded by node u, v and w respectively. Meanwhile, $|I_u|$, $|I_v|$ and $|I_w|$ denote the exact interval size of T corresponded by node u, v and w respectively.

Lemma 1. *For any intermediate node u with left child v and right child w in T and parameter λ $(0 < \lambda < 1/2)$, if $\lambda|I_u| \leq |I_v| \leq (1-\lambda)|I_u|$ is ensured, it is always true that $h = \Theta(\log(\beta/\epsilon))$.*

The proof of Lemma 1 is given in our technique report [14].

Theorem 3. *For each $|A_u|$, $|A_v|$ and $|A_w|$, if condition*

$$\eta|A_u| \leq |A_v| \leq (1-\eta)|A_u| \tag{1}$$

is always satisfied, the height of T is bounded by $h = \Theta(\log(\beta/\epsilon))$, where $\eta = (\lambda h_i + 1)/(h_i - 1)$, $0 < \lambda < 1/2$ and h_i is the height of T at the beginning of a round.

The proof of Theorem 3 is given in our technique report [14].

Thus, Theorem 3 provides the critical condition whether binary tree T is unbalanced or not. Next, we will introduce the maintenance algorithm to let the height of T always satisfy $h = \Theta(\log(\beta/\epsilon))$.

When the critical condition (1) in Theorem 3 is violated, *i.e.*, one of the two conditions $|A_v| < \eta|A_u|$ or $|A_v| > (1-\eta)|A_u|$ is satisfied for the binary tree rooted at u, a partial rebuilding is needed to restore condition (1) for the partial binary tree rooted at u. If one of these two conditions is satisfied at several nodes in T simultaneously, we rebuild the highest tree rooted at one of these nodes. Operation of rebuilding the binary tree rooted at u needs to initialize the binary tree rooted at u. The initialization algorithm is shown in Sect. 4.2.

Meanwhile, as data items arrive, we need to make sure that the interval size of T corresponded by each leaf node of T is not larger than $\epsilon|M_i|/\beta$, *i.e.*, $|A_v| \leq \epsilon|M_i|/\beta$, where v is a leaf node of T. In each round, each approximate interval $|A_v|$ will be monitored and v will be split by adding two children for v as new leaves whenever $|A_v| > \epsilon|M_i|/\beta - \mu$. Because A_v has error of at most μ, $|A_v| > \epsilon|M_i|/\beta - \mu$ will ensure that $|A_v| \leq \epsilon|M_i|/\beta$. The splitting process is also to initialize the interval of T corresponded by v.

Communication Cost. If the sink detects that the binary tree rooted at u is unbalanced, it just needs to initialize the partial binary tree rooted at u, so that the cost of this partial initialization operation is $O\left(\frac{k\beta}{\epsilon} \cdot \frac{|I_u|}{|M_i|}\right)$. Since $I_v \subseteq I_u$, it means that a new rebuilding operation for the binary tree rooted at u is needed iff $|I_u|$ has increased by a constant factor. Thus, the average cost for an item to rebuild the tree rooted at u once is $O(\frac{k\beta}{\epsilon} \cdot \frac{1}{|M_i|})$. Since each item is contained in $O(h)$ intervals, *i.e.*, one item may incur $O(h)$ times of rebuilding, the average cost for an item to rebuild the tree rooted at u is $O(\frac{k\beta}{\epsilon} \cdot \frac{h}{|M_i|})$. Thus, the communication cost for maintaining the balance of T in one round is $O(\frac{k\beta}{\epsilon} \cdot \frac{h}{|M_i|} \cdot |M_i|) = O(\frac{k\beta}{\epsilon} \cdot h) = O(\frac{k\beta}{\epsilon} \cdot \log(\beta/\epsilon))$.

To split the leaf node v of T, the sink launches the initialization algorithm for the interval corresponded by v. Since the initialization operation is conducted on the interval corresponded by leaf v, it incurs a cost of $O\left(\frac{k\beta}{\epsilon} \cdot \frac{|I_v|}{|M_i|}\right)$.

Since the interval size corresponded by v is less than $\epsilon|M_i|/\beta$, i.e., $|I_v| \le \epsilon|M_i|/\beta$, one can derive $O\left(\frac{k\beta}{\epsilon} \cdot \frac{|I_v|}{|M_i|}\right) = O(k)$. Since T has at most β/ϵ leaf nodes, the cost for the splitting operation in one round is $O(k\beta/\epsilon)$.

In summary, the communication cost of the maintenance algorithm in one round is $O(\frac{k\beta}{\epsilon} \cdot \log(\beta/\epsilon))$.

4.5   Total Communication Cost

The above analysis shows that the communication cost for accumulative updating algorithm is dominant. Since β is a constant factor, we can obtain the following conclusion:

Proposition. There is a deterministic algorithm that can continuously track ϵ-approximate ϕ-quantiles in WSNs for all ϕ $(0 \le \phi \le 1)$ with a communication cost of $O\left(k/\epsilon \cdot \log n \cdot \log^2 \frac{1}{\epsilon}\right)$.

5   Conclusion

This paper studies the problem of tracking quantiles in WSNs. A binary tree based data structure is proposed to achieve continuous tracking of ϵ-approximate ϕ-quantiles $(0 \le \phi \le 1)$ over the arrived sensory data for all ϕ's. The communication cost of the proposed algorithm is $O\left(k/\epsilon \cdot \log n \cdot \log^2 \frac{1}{\epsilon}\right)$. Compared with the previous works, the proposed algorithm can (1) track quantiles over distributed data stream; (2) obtain quantiles over an arbitrary topological spanning tree; and (3) track all ϕ-quantiles simultaneously. Therefore, the proposed algorithm can better satisfy the requirements of quantile computation for the distributed stream data model with efficient a communication cost.

References

1. Cai, Z., Chen, Z.-Z., Lin, G.: A 3.4713-approximation algorithm for the capacitated multicast tree routing problem. Theoret. Comput. Sci. **410**(52), 5415–5424 (2008)
2. Cai, Z., Lin, G., Xue, G.: Improved approximation algorithms for the capacitated multicast routing problem. In: Wang, L. (ed.) COCOON 2005. LNCS, vol. 3595, pp. 136–145. Springer, Heidelberg (2005)
3. Calvo, T., Mayor, G., Mesiar, R. (eds.): Aggregation Operators: New Trends and Applications. Physica-Verlag GmbH, Heidelberg (2002)
4. Cao, J., Li, L.E., Chen, A., Bu, T.: Incremental tracking of multiple quantiles for network monitoring in cellular networks
5. Cheng, S., Li, J., Cai, J.: O(ϵ)-approximation to physical world by sensor networks. In: INFOCOM, pp. 3084–3092 (2013)
6. Cheng, X., Du, D., Baogang, X.: Relay sensor placement in wireless sensor networks. Wireless Netw. **14**(3), 347–355 (2008)
7. Cheng, X., Huang, X., Li, D., Weili, W., Du, D.: A polynomial-time approximation scheme for the minimum-connected dominating set in ad hoc wireless networks. Networks **42**(4), 202–208 (2003)

8. Cheng, X., Thaeler, A., Xue, G., Chen, D.: Tps: A time-based positioning scheme for outdoor wireless sensor networks. In: IEEE INFOCOM 2004, pp. 2685–2696, Hong Kong, China, 7–11 March 2004

9. Cormode, G., Garofalakis, M.: Holistic aggregates in a networked world: Distributed tracking of approximate quantiles. In: SIGMOD, pp. 25–36 (2005)

10. Ding, M., Chen, D., Xing, K., Cheng, X.: Localized fault-tolerant event boundary detection in sensor networks. In: IEEE INFOCOM 2005, pp. 902–913, Miami, USA, 13–17 March 2005

11. Gilbert, A.C., Kotidis, Y., Muthukrishnan, S., Strauss, M.J.: Domain-driven data synopses for dynamic quantiles. IEEE Trans. Knowl. Data Eng. **17**(7), 927–938 (2005)

12. Greenwald, M., Khanna, S.: Space-efficient online computation of quantile summaries. In: SIGMOD '01, pp. 58–66. ACM, New York (2001)

13. Greenwald, M.B., Khanna, S.: Power-conserving computation of order-statistics over sensor networks. In: PODS '04, pp. 275–285. ACM, New York (2004)

14. He, Z., Cai, Z., Cheng, S., Wang, X.: Appendix: Approximate aggregation for tracking quantiles in wireless sensor networks. http://www.cs.gsu.edu/zcai/reports/2014/COCOAAppendix.pdf

15. Huang, Z., Wang, L., Yi, K., Liu, Y.: Sampling based algorithms for quantile computation in sensor networks. In: SIGMOD '11, pp. 745–756. ACM, New York (2011)

16. Keralapura, R., Cormode, G., Ramamirtham, J.: Communication-efficient distributed monitoring of thresholded counts. In: SIGMOD '06, pp. 289–300. ACM, New York (2006)

17. Li, J., Cheng, S.: (ϵ, δ)-approximate aggregation algorithms in dynamic sensor networks. IEEE Trans. Parallel Distrib. Syst. **23**(3), 385–396 (2012)

18. Li, J., Cheng, S., Gao, H., Cai, Z.: Approximate physical world reconstruction algorithms in sensor networks. IEEE Trans. Parallel Distrib. Syst. (2014)

19. Liu, Y., He, Y., Li, M., Wang, J., Liu, K., Mo, L., Dong, W., Yang, Z., Xi, M., Zhao, J., Li, X.-Y.: Does wireless sensor network scale? a measurement study on greenorbs. In: 2011 Proceedings IEEE INFOCOM, pp. 873–881, April 2011

20. Metwally, A., Agrawal, D., El Abbadi, A.: An integrated efficient solution for computing frequent and top-k elements in data streams. ACM Trans. Database Syst. **31**(3), 1095–1133 (2006)

21. Mo, L., He, Y., Liu, Y., Zhao, J., Tang, S.-J., Li, X.-Y., Dai, G.: Canopy closure estimates with greenorbs: Sustainable sensing in the forest. In: SenSys '09, pp. 99–112. ACM, New York (2009)

22. Moon, B., Fernando Vega Lopez, I., Immanuel, V.: Efficient algorithms for large-scale temporal aggregation. IEEE Trans. Knowl. Data Eng. **15**(3), 744 (2003)

23. Munro, J.I., Paterson, M.S.: Selection and sorting with limited storage. In: SFCS '78, pp. 253–258. IEEE Computer Society, Washington, DC (1978)

24. Shrivastava, N., Buragohain, C., Agrawal, D., Suri, S.: Medians and beyond: New aggregation techniques for sensor networks. In: SenSys '04, pp. 239–249. ACM, New York (2004)

25. Siew, Z.W., Wong, C.H., Kiring, A., Chin, R.K.Y., Teo, K.T.K.: Fuzzy logic based energy efficient protocol in wireless sensor networks. ICTACT J. Commun. Technol. (IJCT) **3**(4), 639–645 (2012)

26. Thatte, G., Mitra, U., Heidemann, J.: Parametric methods for anomaly detection in aggregate traffic. IEEE/ACM Trans. Networking **19**(2), 512–525 (2011)

27. Vapnik, V., Chervonenkis, A.: On the uniform convergence of relative frequencies of events to their probabilities. Theory Probab. Its Appl. **16**(2), 264–280 (1971)

28. Yi, K., Zhang, Q.: Optimal tracking of distributed heavy hitters and quantiles. Algorithmica **65**(1), 206–223 (2013)
29. Yu, B.: Comment: Monitoring networked applications with incremental quantile estimation. Stat. Sci. **21**(4), 483–484 (2006)

Interference-Free k-barrier Coverage in Wireless Sensor Networks

Hongwei Du[1], Haiming Luo[1(✉)], Jing Zhang[1], Rongrong Zhu[1], and Qiang Ye[2]

[1] Department of Computer Science and Technology, Harbin Institute of Technology
Shenzhen Graduate School, Shenzhen, China
hwdu@hitsz.edu.cn, {cshmluo,cshitzhj,hitzrr13}@gmail.com
[2] Department of Computer Science and Information Technology, University of Prince
Edward Island, Charlottetown, Canada
qye@upei.ca

Abstract. Barrier coverage is a hot issue in wireless sensor networks. Most literatures study barrier coverage under one-hop wireless sensor networks. In this paper, we consider more practical environment: multi-hop wireless sensor networks. We study the problem of how to achieve interference-free k-barrier coverage which has not been discussed before. Firstly, we analyze how k-barrier coverage suffers from interference in multi-hop networks. Then we propose an effective algorithm to solve the problem which is called computing maximal k-connected barriers without interference. In this algorithm, we construct none-crossing k-barrier coverage firstly and then we construct a communication interference graph to compute a maximal k-connected barriers without interference. Simulations show our algorithm can prolong the network lifetime of k-barrier coverage and is more effective than the existing method.

Keywords: k-barrier coverage · Interference-free · Multi-hop wireless sensor networks

1 Introduction

Wireless sensor networks(WSNs) are multi-hop and self-organized networks that are composed of distributed wireless sensor nodes deployed in a monitoring region. Barrier coverage in wireless sensor networks is inspired by the moats which is a famous defense mechanism to detect intruders [1]. The barriers are formed by the wireless sensor nodes whose sensing ranges overlap in the horizontal direction instead of the moats as shown in Fig. 1. Any crossing path of an intruder can be detected by at least one sensor node of the barriers. Maximizing the network lifetime to detect intruders trespassing a monitoring region has become a challenging problem.

Most literatures mainly consider sensing in barrier coverage and focus on devising sleep-wakeup scheduling schemes. They assume that ordinary nodes can communicate with the sink node directly hence communication interference

© Springer International Publishing Switzerland 2014
Z. Zhang et al. (Eds.): COCOA 2014, LNCS 8881, pp. 173–183, 2014.
DOI: 10.1007/978-3-319-12691-3_14

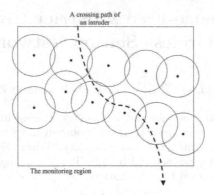

Fig. 1. An example of barrier coverage.

has a weak impact on the network lifetime of barrier coverage. But in reality, nodes are deployed randomly and the communication radius is not large enough. Thus, compared with one-hop communication environment, multi-hop communication environment is more practical. Under multi-hop communication environment of k-barrier coverage, communication links of neighboring barriers usually conflict with each other when they are transmitting simultaneously. It cost more energy of sensor nodes such that the performance of the network can be degraded. Therefore, communication interference is a critical factor in maximizing the network lifetime of barrier coverage.

In this paper, we study the problem of interference-free k-barrier coverage under multi-hop communication environment. We adopt the protocol interference model [2] to analyze communication interference between neighboring barriers. To solve the problem, we propose an effective algorithm to achieve interference-free k-barrier coverage. We compare the network lifetime of k-barrier coverage computed by our algorithm with that in [7] under the multi-hop network environment. We name these two kinds of the environment in the following:

(1) Conflict barriers in multi-hop networks which are original. (CBMN)
(2) Interference-free barriers in multi-hop networks. (IFBMN)

Simulations show that, even IFBMN discards some barriers, it has better performance than CBMN in medium-size networks. The rest of the paper is organized as follows: Sect. 2 presents the related work. Section 3 presents preliminaries. Section 4 presents an effective algorithm to achieve interference-free k-barrier coverage. Section 5 presents our simulation results. Finally, Sect. 6 concludes the paper.

2   Related Work

D.W. Gage [3] firstly proposed the concept of barrier coverage. Kumar et al. [1] defined the strong barrier coverage and the weak barrier coverage and demonstrated the critical condition of constructing weak barrier coverage after randomly deploying sensor nodes over a monitoring region. Liu et al. [4] demonstrated the critical condition of strong barrier coverage and they also present an efficient distributed algorithm to construct strong barrier coverage on long strip areas of irregular shape without any constraint on crossing paths via using the critical condition.

Network lifetime and energy efficiency are two hot issues in barrier coverage. Du et al. [5] study the problem of maximum lifetime connected coverage with two active-phase sensors and compute the polynomial-time of the problem. Wu and Du [6] develop two polynomial-time approximation algorithms for minimum connected sensor cover problem. Kumar et al. [7] proposed a sleep and wake-up scheduling algorithm for strong barrier coverage. They used max-flow algorithm to construct strong barrier coverage and obtained the number of the maximum node-disjoint paths m. Given the strength of barrier coverage k, they demonstrated the optimal network lifetime is m/k. Ban et al. [8] proposed an efficient distributed algorithm for scheduling weak barrier coverage and prolonged the network lifetime. Yang et al. [9] proposed two heuristic algorithms to solve the problem of minimum energy cost k-barrier coverage. Du et al. [10] focused on maximizing the network lifetime under a novel k-discrete barrier coverage model, whose goal is to cover some specific discrete points of interest by deploying sensors in k lines to form barriers. For more information about barrier coverage or other kinds of coverage, the reader can read the literature [11] and the literature [12].

3   Preliminaries

In this part, we give some important definitions and assumptions firstly and then explain why the network lifetime of k-barrier coverage is degraded suffering from communication interference.

A. Definitions and Assumptions

Definition 3.1 (Multi-hop networks). A multi-hop network is formed by the wireless sensor nodes which are randomly deployed in the monitoring region. There are two kinds of nodes which are the sink node and ordinary nodes. The sink node is placed somewhere in the monitoring region. Ordinary nodes communicate with the sink node directly or utilize other nodes as relay nodes and communicate with the sink node in multi-hop.

Assumption 3.1. We adopt disk model to study a sensor's sensing, communication and interference. We assume a sensor has sensing radius R_s, communication radius R_c and interference radius R_i. Practically the relationship among them is $R_s < R_c \leq R_i$. For simplicity, we assume R_c is twice as R_s and R_c is equal to R_i in this paper. Sensing(communication, interference) range is a disk area whose radius is sensing(communication, interference) radius of an individual sensor.

Definition 3.2 (Connected Coverage Graph CCG(V,E)). A connected Coverage Graph is generated from the sensor network N. We denote V as the set of

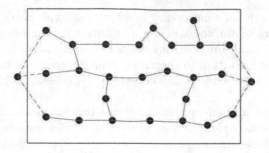

Fig. 2. An example of CCG(V,E)

sensor nodes and E as the set of edges. Ordinary nodes are randomly deployed in the monitoring region while the sink node is on the right border of the area. There is an edge between two nodes if their sensing ranges overlap. As the communication radius is assumed twice the sensing radius, sensing ranges overlap represent these two nodes are connected. At last, we add a virtual node s to V on the left border of the monitoring region. There is an edge between s and ordinary sensors whose sensing range covers the left border. Figure 2 shows an example of CCG(V,E).

Assumption 3.2. Ordinary nodes have two states: sleep and wake-up states. When an ordinary node is in the state of wake-up, it transmits a polling message to the sink node periodically in order to report the environment of the monitoring region. When an ordinary node detects an intruder, it transmits a warning message to the sink node. Ordinary nodes communicate with the sink node in multi-hops.

Definition 3.3 (Interference model). An interference model is used to determine whether two communication links are interfering with each other. We adopt protocol interference model [2] in this paper. For any two pairs of communication links (u_1,v_1) and (u_2,v_2), u_1 and u_2 are the senders while v_1 and v_2 are the receivers. If Euclidian distance $\parallel u_1,v_2 \parallel < R_i$, link($u_1,v_1$) causes interference to link(u_2,v_2), on the contrary, if Euclidian distance $\parallel u_2,v_1 \parallel < R_i$, link($u_2,v_2$) causes interference to link(u_1,v_1). Figure 3 illustrates an example of communication interference.

Assumption 3.3. When a link is interfered by others, the receiver of the link can not receive any message from the sender and the sender adopts retransmission mechanism.

Definition 3.4 (Communication Link(i,j)). A communication link(i,j) means sensor nodes i and j is connected.

Definition 3.5 (Interference range of a link). The union interference range of this link's two endpoints.

B. k-barrier coverage suffers from interference

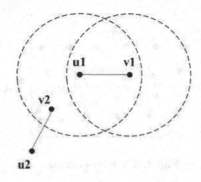

Fig. 3. The dashed circle on the left is interference range of sensor u_1 and the dashed circle on the right is interference range of sensor v_1. The straight line is communication links. If u_1 is sending data package to v_1 meanwhile u_2 is sending data package to v_2, then link(u_2,v_2) is interfered by link(u_1,v_1) because Euclidian distance $\|u_1,v_2\|<R_i$.

Lemma 3.1. For a sensor node u and its two-hop neighbor n_2. The Euclidian distance $\|u,n_2\| \geq R_c$.

Proof: Suppose $\|u,n_2\|<R_c$, then there exists an edge between u and n_2 because n_2 is in the communication range of u. n_2 is one-hop neighbor of u which is contradict with the condition. So our lemma holds.

Theorem 3.1. For any node u in a barrier B_1, if it needs two or more hops routing from u to any node v in another barrier B_2, then barrier B_1 and barrier B_2 are interference-free.

Proof: Suppose node u belongs to a barrier B_1 while v belongs to a barrier B_2. According to the condition it needs two or more hops routing from u to v. Consider the case of lower bound: routing from u to v needs two hops. Then we draw a conclusion that the Euclidian distance $\|u,v\|\geq R_i$ by **Assumption 3.1** and **Lemma 3.1**. Therefore, for a link L_1 in barrier B_1 and a link L_2 in barrier B_2, there is not any endpoint of link L_1 in the interference of link L_2 (vice versa). So our theorem holds.

Theorem 3.2. For any node u in a barrier B_1, if there is a one-hop neighbor of u belonging to another barrier B_2, then there exists interference between B_1 and B_2.

Proof: Suppose barrier B_1 and B_2 are interference-free and node u in B_1 has a one-hop neighbor n_1 which belongs to B_2. Then $\| u,n_1 \|\leq R_i$ which means node u is in the interference range of a link including endpoint n_1 or n_1 is in the interference range of a link including endpoint u. That causes a contradiction. So our theorem holds.

The following is an example of how k-barrier coverage suffers from interference:

In Fig. 4, it is 3-barrier coverage. From top to down we name the barriers B_1, B_2 and B_3. All nodes in the barriers are in the working state. They transmit a polling

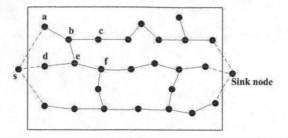

Fig. 4. 3 barrier-coverage

message to the sink node in multi-hop periodically. According to **Theorem 3.2**, there exists interference between barrier B_1 and barrier B_2 while barrier B_2 and barrier B_3 are interference-free. Here we give two cases that how barrier B_1 and barrier B_2 suffer from interference.

Case1: (One intruder)

When node a in barrier B_1 detects an intruder, it begins to send a warning message to its neighbor node b. But node b is in the interference range of link *(e,f)* according to **Definition 3.5**, if node e in barrier B_2 is sending a polling message to node f at the same time then node b can not receive any message from node a correctly such that node a has to adopt retransmission mechanism.

Case2: (Multiple intruders)

In k-barrier coverage, the detecting mission includes not only detecting the intruder, but also tracing the crossing path of the intruder. Suppose an intruder is detected by node d after it passes through barrier B_1. At this time, another intruder is trespassing the monitoring region and is detected by node b in barrier B_1. Link*(d,e)* and link*(b,c)* are transmitting messages simultaneously. Link*(d,e)* is interfered by link*(b,c)* according to **Definition 3.3**. Node d has to retransmit a warning message to node e.

Without loss of generality, there are also some other links that generate interference in barrier B_1 and barrier B_2 which means it costs nodes more energy to retransmit messages to the sink node. Thus, the performance of k-barrier coverage is degraded by the interference of neighboring barriers. Our goal is to solve the problem of the interference in neighboring barriers.

4 Achieve Interference-Free k-barrier Coverage

In this section, we propose an effective algorithm to achieve interference-free k-barrier coverage. Firstly, we construct none-crossing k-barrier coverage and then we construct a communication interference graph to compute a maximal k-connected barriers without interference. Our algorithm is called computing maximal k-connected barriers without interference.

Algorithm 1. Compute maximal k-connected barriers without interference

---

1: **Step1:** Input a sensor network and construct a connected coverage graph $CCG(V,E)$.

2: **Step2:** Use the method proposed in [13] to compute none-crossing node-disjoint paths P_m between the virtual node s and the sink node.

3: **Step3:** Construct a communication interference graph $CIG(V',E')$ of barriers, then compute maximal independent set MIS of $CIG(V',E')$ and output maximal k-connected barriers without interference.

4: **Return** The maximal k-connected barriers without interference.

The first step of **Algorithm 1** is to compute the connected coverage graph $CCG(V,E)$ from a sensor network. V contains all the sensor nodes deployed in the monitoring region. For any two nodes u and v, if their Euclidian distance $\|u,v\| \leq 2R_s$, then add an edge between u and v to E which means u and v can construct a local barrier segment and they can communicate with each other according to **Assumption 3.1**. In [4,7], a flow network is generated from a coverage graph easily by making the flow capacity of each edge 1. The standard max-flow algorithm *Edmond-Karp* is called to compute a maximum flow which is equivalent to the maximum node-disjoint paths P_m between two virtual nodes. However, in randomly deployed network, *Edmond-Karp* algorithm does not guarantee neighboring barriers are not crossing. Apparently, crossing barriers have strong interference which should be avoided. Intruders can utilize the interval of switching between barriers to trespass the monitoring region without being detected. Thus, we use a method proposed in [13] to avoid crossing barriers instead of using *Edmond-Karp* algorithm to construct barriers.

In the second step, the method in [13] which is used to compute none-crossing node-disjoint paths P_m without is based on divide-and-conquer to reduce the complexity and avoid crossing node-disjoint paths. We develop this method in the following: firstly, we adopt *Dijkstra* algorithm to compute a shortest path between the source node s and the sink node such that we can maintain consistency with *Edmond-Karp* because *Edmond-Karp* uses *BFS* to search augmenting-path and the path is also shortest. Then based on the shortest path we divide the remaining nodes into two groups: one group whose nodes are above the shortest path and another whose noses are below the shortest path. Recursively, we adopt *Dijkstra* algorithm in each group until there are no paths from the source node to the sink node.

The third step of **Algorithm 1** mainly determines a group of barriers without interference whose number is as large as possible. This problem can be solved by constructing a communication interference graph $CIG(V',E')$ and find a maximal independent set of $CIG(V',E')$.

The algorithm to construct $CIG(V',E')$ is a distributed algorithm. Though each node broadcasts a message to its neighbor, it stops forwarding messages after messages's field hop number reaches one. Therefore, when all nodes broadcast messages at the same time, it cost less time to find out which barriers are interfering with each other and construct $CIG(V',E')$. Then based on

---

Algorithm 2. Constructing CIG(V',E')

1: **Input:** The set of node-disjoint paths P_m and CCG(V,E)
2: **Initially** All nodes in the same path are assigned with a path ID which is a positive number; nodes in different path have different path IDs. Nodes that do not belong to any node-disjoint path are assigned an Path ID which is ∞. Let V' be the set of vertexes which represent the node-disjoint paths and set $E'=\varnothing$;
3: **Step1** Each node broadcasts a message to its neighbors except for the virtual node and the sink node. Virtual node s does not send and receive a message. This message contains two fields which are sender's path ID and the hop number which is initialized as 0.
4: **Step2** Each node which receives a message from its neighbor obtains the path ID and the hop number. Then it increases the hop number by one. After that, it takes measures in two cases: (Suppose the path ID of its own is id_i and the path ID obtained is id_j.)
5:     **Case 1** ($id_j=id_i$) If both id_j and id_i are not equal to ∞, then it discards this message. If id_j and id_i are equal to ∞, it forwards this message.
6:     **Case 2** ($id_j\neq id_i$) If both id_j and id_i are not equal to ∞ and the hop number is equal to one, then add an edge between vertexes which represent path i and path j respectively to the set E' ($1\leq i\leq m$, $1\leq j\leq m$). And it does not forwards this message any more. If id_j or id_i is equal to ∞, it just forwards this message but not adding any edge to E'.
7: **Return** Communication interference graph CIG(V',E').

CIG(V',E'), finding interference-free barriers can be transformed to maximum independent set problem. We utilize Du et al.'s [14] algorithm to construct a maximum independent set I to finish the remaining work.

5   Simulation

In this section, we do extensive simulations by Java and Matlab to evaluate the performance of our algorithm. We suppose the strength of barrier coverage k is the number of barriers in a maximum independent set and compare the network lifetime of k-barrier coverage under two different environment: conflict barriers in multi-hop networks(CBMN), interference-free barriers in multi-hop networks(IFBMN). If two nodes u and v are connected, they satisfy two conditions simultaneously which are $\|u,v\|<2R_s$ and $\|u,v\|<R_c$. Without loss of generality, we set sensing radius is 10 m and communication radius is 20 m initially. The size of monitoring region is 200*100 m. The initial energy of each sensor is 360 J. Every second, it needs 0.1 J for sensing. When a sensor transmits a polling message, it consumes 0.3 J. When a sensor transmits a warning message, it consumes 0.5 J.

Figure 5 presents the performance of CBMN and IFBMN in terms of the network lifetime versus deployed nodes. With the condition k is equal to the number of barriers in a maximum independent set, the network lifetime of IFBMN is constant when all barriers are active simultaneously in each case in spite of

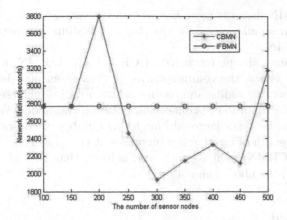

Fig. 5. Network lifetime versus number of sensor nodes

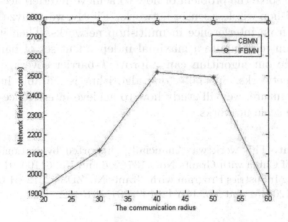

Fig. 6. Network lifetime versus communication radius

increasing nodes. At the beginning, the performance of CBMN and IFBMN is the same. This is reasonable because the number of the constructed barriers is little with a few deployed nodes. There is no interference between neighboring barriers or there is only 1-barrier coverage. When we increase the number of deployed nodes, the network lifetime of CBMN decreases on the whole. It is better than IFBMN at first. However, when the number of deployed nodes reaches a certain value, IFBMN outperforms CBMN and lasts for long. At last, when the number of deployed nodes is large, IFBMN is surpassed by CBMN. The results conform to our theory: IFBMN discards some barriers and the strength of interference is weak in the early stage, CBMN is better than IFBMN. With the number of deployed nodes increasing, the density of nodes becomes large such that the strength of interference increases. The performance of CBMN degrades and is lower than IFBMN. But, excessive deployed nodes makes the density too large, as a result, IFBMN discards more barriers and is surpassed by CBMN. Generally speaking, our simulations show that for the medium-size network, IFBMN outperforms CBMN.

Figure 6 presents the performance of CBMN and IFBMN in terms of the network lifetime versus the communication radius given 300 deployed nodes. We suppose the sensing radius and communication radius increase in equal proportion due to the condition of connection between two nodes. When the communication radius becomes large, although the number of constructed barriers increases, the strength of interference increases at the same time. Thus the network lifetime of CBMN changes gently and is lower than that of IFBMN which is consistent with the above simulations.

6 Conclusion

In this paper, we solve the problem of how to achieve interference-free k-barrier coverage in multi-hop wireless sensor networks. Firstly, we analyze how k-barrier coverage suffers from interference in multi-hop networks. Then we propose an effective algorithm to compute a maximal independent set of barriers. Simulations demonstrate our algorithm can improve k-barrier coverage in multi-hop wireless sensor networks. Specially, our algorithm is suitable in medium-size networks. In the future, we will study how to achieve interference-free k-barrier coverage in large scale networks.

Acknowledgment. This work was financially supported by National Natural Science Foundation of China with Grants No. 61370216 and No. 61100191, and Shenzhen Strategic Emerging Industries Program with Grants No. ZDSY20120613125016389, No. JCYJ20120613151201451 and No. JCYJ20130329153215152.

References

1. Kumar, S., Lai, T.H., Arora, A.: Barrier coverage with wireless sensors. In: Proceedings of the 11th Annual International Conference on Mobile Computing and Networking (MobiCom), August 2005
2. Wan, P.J., Ma, C., Wang, Z., et al.: Weighted wireless link scheduling without information of positions and interference. In: Proceedings of 2011 INFOCOM, April 2011
3. Gage, D.W.: Command control for many-robot systems. In: Proceedings of the Nineteenth Annual AUVS Technology Symposium (AUVS-92) (1992)
4. Liu, B., Dousse, O., Wang, J., Saipulla, A.: Strong barrier coverage of wireless sensor networks. In: Proceedings of the 9th ACM International Symposium on Mobile Ad Hoc Networking and Computing (MobiHoc) (2008)
5. Du, H., Pardalos, P.M., Wu, W., Wu, L.: Maximum lifetime connected coverage with two active-phase sensors. J. Glob. Optim. 56(2), 559–568 (2013)
6. Wu, L., Du, H., Wu, W., Li, D., Lv, J., Lee, W.: Approximations for minimum connected sensor cover. In: Proceedings of the 32nd IEEE International Conference on Computer Communications (INFOCOM) (2013)

7. Kumar, S., Lai, T.H., Posner, M.E., Sinha, P.: Maximizing the lifetime of a barrier of wireless sensors. IEEE Trans. Mob. Comput. (TMC) **9**(8), 1161–1172 (2010)
8. Ban, D., Feng, Q., Han, G., Yang, W., Jiang, J., Dou, W.: Distributed scheduling algorithm for barrier coverage in wireless sensor networks. In: Proceedings of the 2011 Third International Conference on Communications and Mobile Computing (CMC), April 2011
9. Yang, H., Li, D., Zhu, Q., Chen, W., Hong, Y.: Minimum energy cost k-barrier coverage in wireless sensor networks. In: Pandurangan, G., Anil Kumar, V.S., Ming, G., Liu, Y., Li, Y. (eds.) WASA 2010. LNCS, vol. 6221, pp. 80–89. Springer, Heidelberg (2010)
10. Du, J., Wang, K., Liu, H., et al.: Maximizing the lifetime of k-discrete barrier coverage using mobile sensors. Int'l J. IEEE Sens. (2013)
11. Felemban, E.: Advanced border intrusion detection and surveillance using wireless sensor network technology. Int'l J. Commun. Netw. Syst. Sci. **6**(5) (2013)
12. Thai, M.T., Wang, F., Jia, X.: Coverage problems in wireless sensor networks: designs and analysis. Int. J. Sens. Netw. **3**(3), 191–200 (2008)
13. Luo, H., Du, H., Huang, H., Ye, Q., Zhang, J.: Barrier coverage in wireless sensor networks with discrete level of sensing and transmission power. In: The 8th China Conference on China Wireless Sensor Networks CWSN 2014 (submitted)
14. Du, H., Wu, W., Ye, Q., Li, D., Lee, W., Xu, X.: CDS-based virtual backbone construction with guaranteed routing cost in wireless sensor networks. IEEE Trans. Parallel Distrib. Syst. **24**(4), 652–661 (2013)

Performance Analysis and Improvement for the Construction of MCDS Problem in 3D Space

Jun Li[1], Xiaofeng Gao[1(✉)], Guihai Chen[1], Fengwei Gao[1], and Ling Ding[2]

[1] Department of Computer Science and Technology, Shanghai Jiao Tong University,
Shanghai 200240, China
lijun2009@sjtu.edu.cn, {gao-xf,gchen}@cs.sjtu.edu.cn, gfwsteven@gmail.com
[2] Institute of Technology, University of Washington Tacoma, Tacoma, USA
lingding@u.washington.edu

Abstract. In this paper, we discuss the problem of finding a minimum connected dominating set (MCDS) in 3-dimensional space, where the communication model is a unit ball graph (UBG). MCDS in UBG is proved to be an NP-complete problem, and currently the best approximation is 14.937 in [1]. However, their projection method during the approximation deduction process is incorrect, which overthrows its final bound completely. As a consequence, in this paper we will first propose a new projection method to overcome their problem, illustrate the cardinality upper bound of independent points in a graph (which will be used to analyze the approximation ratio), and then optimize the algorithms to select MCDS with prune techniques. The major technique we use is an adaptive jitter scheme, which solves the open question in this area.

1 Introduction

A connected dominating set (CDS) is widely used for many network applications. For instance, it can form a virtual backbone to take charge of routing and message transmission process. It can also be denoted as sink tunnels for data gathering or sensor detection. Hence, there are lots of researches which related to it in the literature [2,3]. A CDS is defined to be a subset of V in a given graph $G = (V, E)$, such that every vertex of V is either in this subset or adjacent to a vertex in this subset and this subset can induce a connected subgraph.

Most literature discussed CDS in two-dimensional space, and use a unit disk graph (UDG) to model the network. However, such model cannot precisely describe the non-flat area such as mountainous region or underwater environment. Correspondingly, we can use a unit ball graph (UBG) to model such a

---

This work has been supported in part by the National Natural Science Foundation of China (Grant No.61202024), Shanghai Pujiang Program (No.13PJ1403900), Shanghai Educational Development Foundation (Chenguang Grant No.12CG09), and the Natural Science Foundation of Shanghai (Grant No.12ZR1445000).

© Springer International Publishing Switzerland 2014
Z. Zhang et al. (Eds.): COCOA 2014, LNCS 8881, pp. 184–199, 2014.
DOI: 10.1007/978-3-319-12691-3_15

network in 3-dimensional space. In a UBG $G = (V, E)$, any two vertices are adjacent (or connected) if and only if the Euclidean distance between them is at most 1.

Since UBG can formulate a network environment more precisely than UDG, CDS in UBG can represent more applications than that in UDG. For instance, Wang and Li [4] constructed 3D landmark maps with vision data extracted from camera images, and then used 3D-CDS to improve data association in application of simultaneous localization and mapping (SLAM). Yang [5] implemented 3D-CDS as clusters to find an optimal topology control strategy in 3D wireless sensor networks. In all, it is significant to design fast algorithms for selecting an appropriate CDS set from a given network and analyze their performance. Typically, a CDS with minimum cardinality is the most efficient choice for practical use, and we refer it as MCDS.

Computing minimum CDS (MCDS) is a well-known NP-complete problem, and lots of approximation algorithms were proposed during last decade. Those algorithms often include two phases. Firstly, they choose an maximal independent set (MIS) from G. Second, they add some extra nodes from G to connect this MIS, usually by Steiner trees. An MIS in a graph $G = (V, E)$ is a subset $M \subseteq V$ such that any two vertices from M are not connected and we cannot add another vertex from $M \backslash V$ to form a bigger MIS. Easy to see, in UDG or UBG, the distance between any two vertices in M should be more than 1. To analyze the performance of those approximations, the ratio $mis(G)/mcds(G)$ plays an important role, where $mis(G)$ is the size of MIS the algorithm selected and $mcds(G)$ is the size of an optimal MCDS. In 2-dimensional situation, this approximation ratio has been widely studied. Based on the fact that the neighborhood area of any node can contain at most five independent points, Wan et al. [6] proposed that $mis(G) \leq 4mcds(G) + 1$. Later, Wu et al. [7] improved this ratio to 3.8 by proving that the neighborhood of any two adjacent nodes can contain at most 8 nodes. In [8], Gao et al. showed the bound can be at most 3.453 and Li et al. improved the ratio into 3.4305 in [9]. Recently, Du and Du [10] showed that $mis(G) \leq 3.399mcds(G) + 4.874$, which is the best result up to now.

Although finding minimum CDS in UBG is very similar as in UDG, the analysis of those approximation ratios in UBG are much harder. Because, instead of disk packing, sphere packing has more complicated properties. To the best of our knowledge, few papers studied the approximation ratio for MCDS problem in UBG. In the earlier stage, Hansen and Schmutz [11] discussed the expected size of a CDS in a random UBG and compared the performance of existing algorithms. Later, Butenko and Ursulenko [12] proved that the ratio of $mis(G)/mcds(G)$ in UBG is at most 11 by using the well-known fact that a sphere can touch at most twelve spheres of the same size, which induced an approximation ratio of 22 for MCDS in UBG. Zhong et al. [13] claimed that such ratio could be reduced to 16. Zou et al. [14] further reduced this ratio to $13 + \ln 10$. Recently, Kim et al. [1] referred the idea in [7] and tried to answer how many independent points can be contained in two adjacent unit balls. Finally, they improved the ratio of $mis(G)/mcds(G)$ into 10.917 by showing that there are at most 22 independent points in two adjacent unit balls, and finally got an approximation ratio of MCDS as 14.937, which is the best result up to now.

However, after careful investigation, we find that during the deduction process in [1], one of the intermediate assertion is incorrect, which overthrows the final result completely. The main technique they implemented in their proof is a projection method to the ball surface and then applying some famous graph theories, and the problem comes under some scenarios when the projection result cannot guarantee the distance lower bound of two independent points. Researchers later found that designing a projection method to guarantee the distance lower bound is not an easy step, and it remains an open question in recent years [15]. As a consequence, in this paper we will first introduce a new projection method to guarantee the distance bound, and then illustrate the bound of $mis(G)/mcds(G)$ with some new analyzing techniques. Since the mistake in [1] only influences the selection of MIS, we will only focus on the first phase of MCDS construction. Next, we will further optimize the algorithms for MCDS selection in [1] with prune process and validate the efficiency of our design by numerical experiments.

The rest of the paper is organized as follows: Sect. 2 illustrates the problem in [1] with counter examples. Section 3 introduces our new projection method to analyze the ratio of $mis(G)/mcds(G)$. Section 4 discusses how to improve MCDS algorithm while Sect. 5 exhibits the simulation results with different parameter settings. Finally, Sect. 6 summarizes this paper.

2   Independent Points in Two Adjacent Unit Balls

Similar with the analysis in [7], once we have the answer of "*two-ball problem*", we can deduce a better upper bound for the ratio $mis(G)/mcds(G)$ and reduce the overall approximation ratio.. *two-ball problem* means the problem of "how many independent points can be contained in two adjacent unit balls". Here two adjacent unit ball means the Euclidean distance between two balls with unit radius is at most 1, while any two points are called independent points if and only if their Euclidean distance is at least 1.

Actually, what Kim et al. did in [1] follows this idea. However, their method have an unavoidable error. In Subsect. 2.1 we will review their method to prove two-ball problem, and then in Subsect. 2.2 we will precisely point out where their problem lies and provide an counter example to validate our claim.

2.1   Review Kim's Method in [1]

In [1], Kim et al. referred the idea in [7] and improved the ratio of $mis(G)/mcds(G)$ into 10.917 by showing that there are at most 22 independent points in two adjacent unit balls. Their answer to the two-ball problem is the most important contribution in their paper. In order to solve the two-ball problem, Kim et al. extended the approach for solving the famous Gregory-Newton problem [16]. They considered two adjacent unit balls, say, B_1 and B_2 with centers u_1 and u_2. To get an upper bound of MIS in these two adjacent balls, they assumed that the Euclidean distance between u_1 and u_2 is equal to 1, since the total volume of $B_1 \cap B_2$ is larger when the distance between these two adjacent nodes increases, and consequently more independent nodes can be contained in $B_1 \cap B_2$. They then divided all the independent nodes into two categories: the nodes located in $(B_1 \cup B_2) \backslash (B_1 \cap B_2)$ and the nodes in $B_1 \cap B_2$. They mainly focused on the former part and claimed that the size of MIS in this region is at most 20 with a special "projection" method.

Their projection method is a mapping rule to project all the independent nodes in $(B_1 \cup B_2) \backslash (B_1 \cap B_2)$ to the surface of $B_1 \cup B_2$. The detailed description can be shown as follows: For each independent node v, if $v \in B_1 \backslash B_2$ (respectively, $B_2 \backslash B_1$), then draw a radial from u_1 (respectively, u_2) going through v, and intersect the outer surface of B_1 (denoted by $Sur(B_1)$, and respectively $Sur(B_2)$) at point P. By this mapping rule, they got the projection points set $\{P_1, P_2, \cdots, P_t\}$, where t is the size of MIS in $(B_1 \cup B_2) \backslash (B_1 \cap B_2)$.

Next, for any two points P_i and P_j, if their Euclidean distance (denoted by $d(P_i, P_j)$) is between 1 and $3 \arccos(1/7)\pi$, they made a curve from P_i to P_j on $Sur(B_1 \cup B_2)$ in the specified way as shown in Sect. 3.2.1 in [1]. These curves partition $Sur(B_1 \cup B_2)$ into some tiny faces. By analyzing the lower bounds of those faces' areas and using Euler's formula, they proved that $t \leq 20$. Combined with the fact that a unit ball can pack at most 12 independent nodes [16], they finally concluded that the number of MIS in the union of two adjacent unit balls is at most 22.

2.2 The Problem of Kim's Method with Counter Examples

After careful investigation, we find that in Sect. 3.2 of [1], one of the intermediate assertion is incorrect, which overthrows the final result completely. This assertion says: **"According to their mapping rule, on $\mathbf{Sur(B_1 \cup B_2)}$, for any $\mathbf{P_i}$ and $\mathbf{P_j}$, $\mathbf{d(P_i, P_j) > 1}$."** This assertion is a foundation of their work. With this assertion, they could conclude that no two curves on $Sur(B_1 \cup B_2)$ can intersect, which is a declaration to guarantee the correctness of the lower bound for the tiny faces' areas on $Sur(B_1 \cup B_2)$.

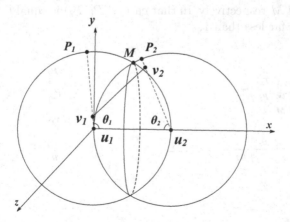

Fig. 1. An Example to show that under some cases $d(P_1, P_2) < 1$

However, although in many cases this assertion seems correct, it is not valid for every possible scenario. Let us provide an example to show why it is incorrect. In this example, v_1 and v_2 are two independent nodes located in two different balls and their projection points are P_1 and P_2 respectively, as Fig. 1 shows. Specially, we can let v_1, v_2, u_1 and u_2 locate in a same plane. Moreover, we set $\theta_1 + \theta_2 \leq \pi$, where θ_1, θ_2 denote $\angle v_1 u_1 u_2$ and $\angle v_2 u_2 u_1$ respectively.

Now we prove that $d(P_1, P_2) \leq 1$ in this situation. Since all points in this case are on a same plane, we can convert this case as a 2D plane in Fig. 2. Consider $\triangle P_1 M P_2$ in Fig. 2, by the Law of Cosines, we have

$$d^2(P_1, P_2) = d^2(P_1, M) + d^2(M, P_2) - 2d(P_1, M)d(M, P_2) \cos \angle P_1 M P_2.$$

From Fig. 2, we find that $\angle P_1 M P_2 = \angle T_1 M T_2 + \angle T_1 M P_1 + \angle T_2 M P_2$, where $\angle u_1 M T_1 = \angle u_2 M T_2 = \pi/2$. Here $T_1 M$ and $T_2 M$ are tangent lines to $disk(u_1)$ and $disk(u_2)$ respectively ($disk(u)$ is the cycle centered at u with radius 1). According to Alternate Segment Theorem,

$$\angle T_1 M P_1 = \angle P_1 u_1 M/2, \ \angle T_2 M P_2 = \angle P_2 u_2 M/2.$$

Therefore, $\angle P_1 M P_2 = (\theta_1 + \theta_2)/2 + \pi/3$. Also, it is easy to get $\angle P_1 u_1 M = \theta_1 - \pi/3$. Then,

$$d(P_1, M) = 2\sin(\theta_1/2 - \pi/6), \ d(P_2, M) = 2\sin(\theta_2/2 - \pi/6).$$

Hence, the Euclidean distance between P_1 and P_2 is:

$$d^2(P_1, P_2) = 4\cos^2(\frac{\theta_1 + \theta_2}{2}) - 4\cos(\frac{\theta_1 + \theta_2}{2})\cos(\frac{\theta_1 - \theta_2}{2}) + 1.$$

Since $\theta_1 + \theta_2 \leq \pi$ in this case, $0 \leq \cos(\frac{\theta_1 + \theta_2}{2}) \leq \cos(\frac{\theta_1 - \theta_2}{2})$. Therefore, $d^2(P_1, P_2) \leq 1$, which is a counter example for Kim's assertion.

Actually, we can also get a lower bound for $d(P_1, P_2)$ when v_1 and v_2 move to points u_1 and M respectively. In that case, $d(P_1, P_2)$ is equal to $2\sin(\pi/12) \approx 0.5176$, which is far less than 1.

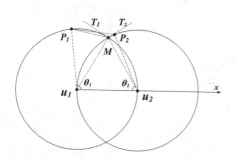

Fig. 2. An counter example in 2D

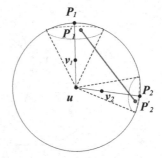

Fig. 3. The new projection of P_1' and P_2'

3   Our New Projection Method

To achieve an approximation ratio of 10.917, we have to guarantee $d(P_1', P_2') \geq 1$ for any pair of independent points on the surface of two adjacent unit balls. In this section we introduce a new projection method to overcome Kim's flaw. The main idea of our method is an adaptive jitter scheme in the projection process. We will first introduce the motivation of our new projection rule. Then, we will give the definition of the new projection and prove $d(P_1', P_2') \geq 1$ holds under this projection.

Before introducing our new method, we need to introduce some notations and definitions, which are frequently used in the rest of this section. For any two nodes P_1 and P_2 in a three dimensional space, $d(P_1, P_2)$ denotes the Euclidean distance between them, while $|P_1 P_2|$ denotes the length of geodesic arc between P_1 and P_2. $disk(v)$ denotes the unit disk with center 1. Let $Sur(B)$ denote the outer surface of a geometric object B. In order to apply geometric principle to solve our problem, we allow the distance between a pair of independent nodes to be equal to one. (Actually, we will often use this critical distance in the coming part of analysis.) Let B_1 and B_2 be two adjacent unit balls with centers u_1 and u_2, and $d(u_1, u_2) = 1$. For any vertex v, the plane going through point v, u_1 and u_2 is called the principal plane of v. In addition to "principal plane", we also define a plane called "normal plane", which is the perpendicular bisector of segment $u_1 u_2$, and denote by L the intersection of the normal plane with $Sur(B_1 \cup B_2)$. Besides, we call the projection in Kim's method "direct projection".

According to our observation, when $d(P_1, P_2) < 1$ occurs, at least one of v_1 and v_2 is much closer to its dominator (u_1 or u_2), which can be found from Fig. 1. Without loss of generality, we say the closer node is v_1. In the unit ball B_1, we know the size of its MIS is at most 12. But if we want to put all the 12 independent nodes in B_1, the efficient way is to put all of them on $Sur(B_1)$. Consequently, if v_1 is much closer to u_1, it will greatly affect the total number of MIS in B_1, and the size of MIS in $B_1 \cup B_2$ will also be affected. We consider this MIS number decrease as the sacrifice to shorten $d(P_1, P_2)$. In order to quantitatively describe and use this property, we provide Lemma 1 as follows.

Lemma 1. *A unit ball B with center u contains two independent points v_1, v_2 and $d(u, v_1) = r_1$, $d(u, v_2) = r_2$. Their direct projection points are P_1 and P_2. P_1' (respectively, P_2') is an arbitrary point in the cycle region on $Sur(B_1)$ (respectively, $Sur(B_2)$) with center P_1 (respectively, P_2) and spherical radius $\arccos(r_1/2) - \pi/3$ (respectively, $\arccos(r_2/2) - \pi/3$) as Fig. 3 shows. Then $d(P_1', P_2') \geq 1$.*

Proof. First, we consider the 2D situation as Figs. 4 and 5 show. In Fig. 4, point v_1 is in $disk(u)$. M_1, M_2 are intersection points of uv_1' perpendicular bisector with $disk(u)$. Since all independent nodes with v_1 are outside $disk(v_1)$, nodes in $disk(u)$ which are independent with v_1 cannot locate above line $M_1 M_2$. From Lemma 1, it is easy to know the available region of P_1' on $disk(u)$ is from P_1^l to P_1^r, where $|P_1^l P_1| = |P_1 P_1^r| = \arccos(r_1/2) - \pi/3$ (shown in Fig. 4). Further,

$\angle M_2 u v_1 = \arccos(r_1/2)$. Hence, $|P_1^r M_2| = \pi/3$ and $d(P_1^r, M_2) = 1$. Similarly, $d(P_1^l, M_1) = 1$. Based on the location of v_2, there are two situations to discuss.

Case 1: If v_2 is on the circle of $disk(u)$, then P_2, P_2' and v_2 are the same point. Since v_2 must locate below $M_1 M_2$ and P_1' is on the arc between P_1^l and P_1^r, it is obvious to conclude that $d(P_1' P_2') \geq 1$.

Case 2: When v_2 is not on the circle of $disk(u)$ (as Fig. 5 shows), P_1' (P_2', respectively) locates between P_1^l and P_1^r (P_2^l and P_2^r, respectively). Next, we will prove that $d(P_1', P_2')$ is minimum when P_1' is at point P_1^r and P_2' is at point P_2^l.

As Fig. 5 shows, segment uM is the midperpendicular of $P_1^r P_2^l$. And lines $P_1^r T_1$ and $P_2^l T_2$ are parallel to uM. Then, all points in the arc from P_1^l to P_1^r and the arc from P_2^l to P_2^r are outside the parallel lines $P_1^r T_1$ and $P_2^l T_2$. Therefore, $d(P_1', P_2') \geq d(P_1^r P_2^l)$. On the other side, $d(P_1^r P_2^l) \geq 1$ is equivalent to $|P_1^r P_2^l| \geq \pi/3$. Combining with Law of Cosines, we have

$$|P_1^r P_2^l| = \angle P_1^r P_2^l = \angle P_1 u P_2 - \angle P_1 u P_1^r - \angle P_2 u P_2^l$$
$$= \arccos\left(\frac{r_1^2 + r_2^2 - 1}{2 r_1 r_2}\right) - \left[\arccos\left(\frac{r_1}{2}\right) - \pi/3\right] - \left[\arccos\left(\frac{r_2}{2}\right) - \pi/3\right].$$

When r_1 is given, it can be proved that the value of $|P_1^r P_2^l|$ decreases with the value of r_2 increases. Thus, when r_2 equals one, $|P_1^r P_2^l|$ will be minimized, which is exactly **Case 1** where v_2 is on the circle of $disk_1(u)$. Therefore, we have $d(P_1', P_2') \geq 1$.

Similarly, it is easy to extend the 2-dimensional situation to 3-dimensional situation. □

According to Lemma 1, we come up with a new region called "Effective Projection Region" to describe the extra feasible moving space of P_1' or P_2'.

Definition 1 (Effective Projection Region). *For node v in a unit ball B, its direct projection point is P. The region on $Sur(B)$ whose center is point P and spherical radius is $\arccos(r/2) - \pi/3$ is called v's effective projection region, where r is the Euclidean distance between v and B's center.*

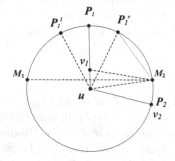 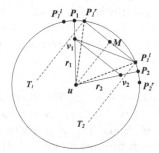

Fig. 4. Projection region with $r_2 = 1$ **Fig. 5.** Projection region with $r_2 < 1$

Obviously, in 3D space, for any two independent nodes v_1 and v_2 in a unit ball B, their direct projection points are P_1 and P_2. According to Lemma 1, if we arbitrarily move P_1 and P_2 along $Sur(B)$ inside their effective projection regions, $d(P_1, P_2)$ will always ≥ 1. Next, we will discuss the situation where v_1 and v_2 are in two different balls. Before that, we first define our new projection rule.

Definition 2 (Region Projection). *For any independent node v, its direct projection point is P, then any point which locates in v's effective projection region is a Region Projection point of v.*

Definition 3 (Final Projection). *For any point v in $(B_1 \cup B_2) \backslash (B_1 \cap B_2)$, the v's principal plane intersects with L and we assume M is the closer intersection point to v. P' is v's final projection if and only if it satisfies conditions as follows:*

(1) P' is a region projection point of v.
(2) P' is on v's principal plane.
(3) Among all the points satisfying (1) and (2), P' is the farthest from M.

Theorem 1. *For any two independent nodes v_1 and v_2 in $(B_1 \cup B_2) \backslash (B_1 \cap B_2)$, their final projection points are P_1' and P_2'. Then, $d(P_1', P_2') \geq 1$.*

Next, we will prove the correctness of Theorem 1. To make it simple, we first discuss the two-dimensional situation as a special case in Sect. 3.1. Afterwards, we generalize our conclusion for three-dimensional situation in Sect. 3.2.

By Lemma 1, if v_1 and v_2 are in the same unit ball, $d(P_1', P_2') \geq 1$. Thus, we only need to consider the situation when v_1, v_2 are in different balls. Without loss of generality, let v_1 in B_1 and v_2 in B_2.

3.1 Proof of Theorem 1 in 2-Dimensional Space

When $u_1 v_1$ and $u_2 v_2$ are in the same principal plane, our problem turns into a 2D problem as shown in Fig. 6.

In Fig. 6, P_1 and P_2 are the direct projection points; θ_1 and θ_2 denote $\angle P_1 u_1 u_2$ and $\angle P_2 u_2 u_1$; r_1 and r_2 denote $d(v_1, u_1)$ and $d(v_2, u_2)$ respectively. In addition, let α, β and γ denote $|P_1'M|$, $|P_2'M|$ and $\angle P_1'MP_2'$. To simplify our

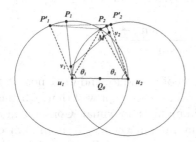

Fig. 6. Discussion in 2-dimension

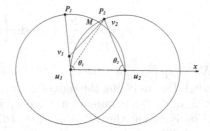

Fig. 7. Value of r_1 in 2D space

problem, we assume $d(v_1, v_2) = d(u_1, u_2) = 1$. In $\triangle P_1'MP_2'$, it is easy to figure out that $P_1'M = 2sin\frac{\alpha}{2}$, $P_2'M = 2sin\frac{\beta}{2}$, and $\gamma = \frac{2\pi}{3} + \frac{\alpha+\beta}{2}$. By cosine law, we have

$$d^2(P_1', P_2') = \left(2sin\frac{\alpha}{2}\right)^2 + \left(2sin\frac{\beta}{2}\right)^2 - 2\left(2sin\frac{\alpha}{2}\right)\left(2sin\frac{\beta}{2}\right)cos\angle P_1'MP_2'$$

$$= 4cos^2(\frac{\alpha+\beta}{2} + \frac{\pi}{3}) - 4cos(\frac{\alpha+\beta}{2} + \frac{\pi}{3})cos(\frac{\alpha-\beta}{2}) + 1. \qquad (1)$$

Let $x = cos(\frac{\alpha+\beta}{2} + \frac{\pi}{3})$ and $y = cos(\frac{\alpha-\beta}{2})$. Construct function $f(x, y) = 4x^2 - 4yx + 1$, where $y \geq 0$. Easy to see that $x \leq 0 \Rightarrow f(x, y) \geq 1$. That means $\alpha + \beta \geq \frac{\pi}{3}$. Moreover, $f(x, y) \geq 1 \Leftrightarrow x \leq 0 \quad or \quad x \geq y$. If $x \geq y$, then $cos(\frac{\alpha+\beta}{2} + \frac{\pi}{3}) \geq cos(\frac{\alpha-\beta}{2}) \geq 0$. Hence, $\alpha + \beta \leq \frac{\pi}{3}$ and $\frac{\alpha+\beta}{2} + \frac{\pi}{3} \leq \frac{\alpha-\beta}{2} \Rightarrow \beta + \frac{\pi}{3} \leq 0$, which is impossible. Therefore, $d(P_1', P_2') \geq 1$ is equivalent to $\alpha + \beta \geq \frac{\pi}{3}$.

According to Definition 3, we have $\alpha = \theta_1 + arccos(r_1/2) - 2\pi/3$, and $\beta = \theta_1 + arccos(r_1/2) - 2\pi/3$. Thus, our goal is to prove

$$\theta_1 + \theta_2 + arccos(\frac{r_1}{2}) + arccos(\frac{r_2}{2}) \geq \frac{5\pi}{3}. \qquad (2)$$

As $arccos(r_1/2) \geq \pi/3$ and $arccos(r_2/2) \geq \pi/3$, we only need to consider the case where $\theta_1 + \theta_2 < \pi$.

Note that $d(v_1, v_2) = 1$ and $\theta_1 + \theta_2 < \pi$, v_1v_2 can not be parallel to u_1u_2. Without loss of generality, assume v_1 is more closer to u_1u_2 as Fig. 6 shows. Because $\theta_1 + \theta_2 < \pi$, it can be proved that $\angle v_1P_2u_2 < \pi/2$. Thus, $d(v_1, P_2) > 1$. If we move v_2 to P_2 and keep anything else unchanged, the new state will produce shorter $P_1'P_2'$. In that state, $r_2 = 1$.

Hence, it is sufficient to prove Eq. (2) under condition $r_2 = 1$. Since $d(v_1, v_2) = 1$, we can figure out the relation between θ_1 and r_1, θ_2. As Fig. 7 shows, we build a polar coordinates on u_1. In $\triangle v_2u_1u_2$, it is easy to figure out that $d(v_2, u_1) = 2sin(\theta_2/2)$. Then, the coordinates of v_1 and v_2 are (θ_1, r_1) and $(\pi/2 - \theta_2/2, 2sin(\theta_2/2))$. By mathematic knowledge, we have

$$\theta_1 = \pi - \frac{\theta_2}{2} - arcsin\left(\frac{r_1^2 - 1 + 4sin^2\frac{\theta_2}{2}}{4r_1 sin\frac{\theta_2}{2}}\right). \qquad (3)$$

With Eq. (3), the value of $|P_1'M|$ is a function of r_1 and θ_2, and we denote it as $\alpha(r_1, \theta_2)$. By analyzing the sign of $\partial\alpha(r_1, \theta_2)/\partial r_1$, it can be proved that, when θ_2 is fixed, $\alpha(r_1, \theta_2)$ is minimum when r_1 is minimum or maximal. Consequently, when θ_2 is given, the value of $d(P_1', P_2')$ is minimum when r_1 is minimum or maximal. Therefore, what we need to do next is to verify $d(P_1', P_2') \geq 1$ for the two situations where r_1 is minimum and maximum.

Case 1 (r_1 is minimal): For this case, we first need to figure out the minimal value of r_1 when θ_2 is fixed. Considering v_1 locates in $disk(u_1)\backslash disk(u_2)$, it is obvious that, when v_1 locates on the intersection of $disk(v_2)$ with $disk(u_2)$, r_1 reaches minimal. Then, $d(v_1, u_2) = 1$, $\angle v_1 u_2 v_2 = \pi/3$, $\theta_1 = 2\pi/3 - \theta_2/2$ and $r_1 = 2\cos\theta_1$. Consequently, $\theta_1 + \theta_2 + \arccos(r_1/2) + \arccos(r_2/2) = 5\pi/3$ which meets inequality (2). Hence, in this case, $d(P_1', P_2') \geq 1$.

Case 2 (r_1 is maximal): It is easy to figure out that the maximal value of r_1 is 1 when v_1 locates on the boundary of $disk(u_1)$. In this case, P_1' and v_1 are the same. $d(P_1', P_2') = d(v_1, v_2) = 1$ which also meets the requirement in Theorem 1.

3.2   Proof of Theorem 1 in 3-Dimensional Space

When $u_1 v_1$ and $u_2 v_2$ are in different principal planes, we can follow the ideas in 2-dimension situation and give the same result.

First we explore the equivalent condition for $d(P_1', P_2') \geq 1$. In Fig. 8, δ denotes the dihedral angle between those two principal planes, Γ_1 and Γ_2; α and β denote $|P_1' M_1|$ and $|P_2' M_2|$ respectively. Similar as analysis in Sect. 3.1, we have $d(P_1', M) = 2\sin\frac{\alpha}{2}$, and $d(P_2', M) = 2\sin\frac{\beta}{2}$. Using method of analytical geometry, we have

$$d^2(P_1', P_2') = 4\cos^2\left(\frac{\alpha+\beta}{2} + \frac{\pi}{3}\right) - 4\cos\left(\frac{\alpha+\beta}{2} + \frac{\pi}{3}\right)\cos\left(\frac{\alpha-\beta}{2}\right) + 1$$

$$+ \left[\cos(\alpha-\beta) - \cos\left(\alpha+\beta+\frac{2\pi}{3}\right)\right](1 - \cos\delta). \tag{4}$$

Note that, when $\delta = 0$, Eq. (4) turns to Eq. (1). Since positive δ also contributes to $d(P_1', P_2')$ from Eq. (4), then $d(P_1', P_2')$ will be definitely larger than 1 when $\theta_1 + \theta_2 > \pi$. Therefore, we still just need to consider the situation where $\theta_1 + \theta_2 < \pi$. Let $x = \cos((\alpha+\beta)/2 + \pi/3)$, $y = \cos(\alpha-\beta)/2$, and $z = \cos\delta$. Thus,

$$d^2(P_1', P_2') = 2(1+z)x^2 - 4yx + 2(1-z)y^2 + 1.$$

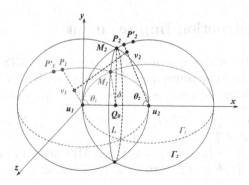

Fig. 8. Discussion in 3-dimensional space

Construct function $g(x, y) = 2(1 + z)x^2 - 4yx + 2(1 - z)y^2 + 1$. It is easy to find the solution of inequality $g(x, y) \geq 1$ and its effective solution is

$$\frac{x}{y} \leq \frac{1 - z}{1 + z}. \tag{5}$$

Hence, $d(P_1', P_2') \geq 1$ is equivalent to Eq. (5).

Similar with 2D situation we discussed in Sect. 3.1, we can also conclude that the value of $d(P_1', P_2')$ is smaller under condition $r_2 = 1$ and it is sufficient to prove Theorem 1 under this condition. Besides, when δ and θ_2 are fixed, the value of $d(P_1', P_2')$ is minimum when r_1 is minimum or maximum. Thus, we just need to verify Eq. (5) for these two situations.

Case 1 (r_1 is minimal): For this case, we first need to figure out the minimal value of r_1 when θ_2 and δ are fixed. Considering v_1 locates in $B_1 \backslash B_2$, it is obvious that, when v_1 locates on the intersection of $disk(u_2)$ on plane Γ_1 with the unit ball whose center locates on v_2, r_1 reaches minimal. In this state, we use θ_1^{min} to denote the current θ_1. Then, we have $r_1^{min} = 2\cos\theta_1^{min}$ and $2\sin 2\theta_1^{min} \sin\theta_2 \cos\delta - 2\cos 2\theta_1^{min} \cos\theta_2 = 1$. Hence, when θ_2 and δ are given, we can figure out r_1^{min} and θ_1^{min}.

Besides, in this situation the ranges of θ_2 and δ we need consider are as follows:

$$\begin{cases} \delta \in [0, \arccos\frac{1}{3}] \\ \theta_2 \in [\frac{\pi}{3}, 2\arctan(\frac{1 - 2\tan^2(\delta/2)}{\sqrt{3}}) + \frac{\pi}{3}]. \end{cases} \tag{6}$$

By numerical method, we can verify Eq. (5) under conditions (6).

Case 2 (r_1 is maximal): It is easy to figure out that the minimal value of r_1 is 1 when v_1 locates on $Sur(B_1)$. In this case, P_1' and v_1 are the same. $d(P_1', P_2') = d(v_1, v_2) = 1$ which meets the requirement in Theorem 1.

In conclusion, according to the analysis in Sect. 3.1 and Sect. 3.2, Theorem 1 always follows. Therefore, we can use our new projection rule to fix the incorrectness in [1].

4   MCDS Construction Improvement

So far, we have analyzed the approximation ratio of the MIS in UBG. In this section, we will introduce two prune methods to improve Kim's CDS construction algorithm. The following are some notations used in this section:

(1) For node u, $N(u) = \{v | v \in V(G) \backslash u$ and $d(u, v) \leq 1\}$, $N[u] = N(u) \cup \{x\}$.
(2) For node set C, $N(C) = (\cup_{v \in C} N(v)) \backslash C$.
(3) For u and C, $M_{u,C} = \{v | d(u, v) \leq 1, v$ is a MIS node and $v \notin C\}$.
(4) For u, $dom(u) = \{v | d(u, v) \leq 1$ and v is a CDS node$\}$.

4.1   Algorithm for Computing CDS

The algorithm introduced by Kim is formally described in Algorithm 1, which has two steps. It firstly generates an MIS M such that every node in M is two hops away from its nearest node in M. We can use Butenko and Ursulenko's algorithm to compute such MIS. The second step is to connect this MIS. Kim used a greedy strategy, which starts with the original node and repeats round by round. In each round, it picks a node v adjacent to the connected component C computed in the previous rounds that makes $|M_{v,C}|$ maximal, and add it to C, It terminates when all the points in MIS are connected.

4.2   Improve the Generated CDS

In this 2-step algorithm, there is some redundancy in the given CDS C. Firstly, through the computing of MIS, some nodes which have only one neighbor may be added to the MIS in order to maintain the properties of MIS. But when the connectors are added, those points would be useless for the whole CDS, and it is better to adjust it to non-CDS nodes. Also, the redundancy may occur in the inner side of the CDS due to the increased density of CDS nodes. Since the redundancy occurs after the algorithm terminates, we can add two more steps afterward to reduce CDS size with the help of prune techniques.

Algorithm 1. C-CDS-UBG($G(V, E)$)

1: Set $M = \Phi$, $B = \Phi$, $V' = V$.
2: Pick a root $r \in V'$ that r has the biggest degree in V'.
3: Set $M = \{r\}$, $B = N(r)$, $V' = V' \backslash N[r]$.
4: **while** $V' \neq \Phi$ **do**
5:     Pick a node $u \in N(B)$ such that $|N(u) \cap V'|$ is maximized.
6:     Set $M = M \cup \{u\}$, $B = B \cup (M \cap V')$, and $V' = V' \backslash N(u)$.
7: **end while**
8: Set $C = \{r\}$ and $M' = M = \{r\}$.
9: **while** $M' \neq \Phi$ **do**
10:     Pick a node $v \in N(C)$ such that $|M_{v,C}| = max\{|M_{u,c}|u \in N(C)\}$.
11:     Set $C = C \cup \{v\} \cup M_{v,C}$ and $M' = M' \backslash M_{v,C}$.
12: **end while**
13: Return C.

Notice that once remove a CDS node, the remaining CDS must maintain all its original properties. Thus for a CDS node $u \in G$, it could be removed iff:

(1) Every point that u dominates must have at least one alternative dominator.
(2) $G(C \backslash \{u\})$ is connected.

Correspondingly, we design two prune methods to reduce CDS size. The first method is to reduce some leaf nodes instantly. We use postorder traversal to traverse the CDS tree and reduce redundant points in it, as shown in Algorithm 2.

Lemma 2. *For any CDS C, After Algorithm 2 is executed, C is also a CDS.*

Algorithm 2. CUTLEAF(u)

1: Set $P = dom(u)$.
2: **while** $P \neq \Phi$ **do**
3:     Pick a node $x \in P$.
4:     **if** x has not been visited **then**
5:         $CUTLEAF(x)$.
6:     **end if**
7:     Set $P = P\backslash\{x\}$.
8: **end while**
9: **if** ($|dom(v)| \geq 2$ for all $v \in N(u)$ in graph G) and $|dom(u)| = 1$ **then**
10:     $C = C\backslash\{u\}$.
11: **end if**
12: Return C.

Proof. Let C' be the CDS after Algorithm 2 is executed. If $C' = C$, then Lemma 2 holds. Otherwise, let C_0 be the initial CDS, C_i be the CDS after the i-th reduction and u_i be the node reduced in this iteration. We show that if C_i is a CDS, then C_{i+1} is a CDS, for $i \geq 0$. According to Line 9, all $N(u_{i+1})$ has at least 2 dominators. Also, $|dom(u_{i+1}) = 1|$ ensures that only one CDS node is adjacent to u_{i+1}. So $C_{i+1} = C_i\backslash\{u_{i+1}\}$ is also connected. Hence, C_{i+1} is a CDS. Since C_0 is a CDS, recursively after Algorithm 2 is executed, C' is also a CDS.

Algorithm 3. CUTINSIDE($G(V,E)$,C)

1: Let G' be the subgraph generated by C, compute all the congestion nodes $C' \in C$.
2: Pick up a node $u \in V$ s.t. $u \notin C'$ and ($|dom(v)| \geq 2$ for all $v \in N(u)$ in G).
3: $C = C\backslash\{u\}$. Return.

Lemma 3. *For any CDS C, After Algorithm 3 is executed, C is also a CDS.*

Proof. Similar to Lemma 2, Line 2 ensures that all $N(u)$ has at least 2 dominators. Moreover, u is not a congestion node ensures that $C\backslash\{u\}$ is connected.

Algorithm 4. R-C-CDS-UBG($G(V,E)$)

1: $P =$C-CDS-UBG($G(V,E)$), Pick a root $r \in V'$ that r has the biggest degree in V'.
2: Run CUTLEAF(r).
3: Run CUTINSIDE($G(V,E)$,P) for several(30) times.
4: Return P.

With two algorithms above, Algorithm 4 is an improvement for computing CDS.

Theorem 2. *The time and space complexity of Algorithm 4 R-C-CDS-UBG is $O(n^2)$, where n is the number of nodes in a given input UBG.*

Proof. Since it is necessary to store the graph, the space complexity of Algorithm 4 is $O(n^2)$. Then we show that the time complexity is $O(n^2)$.

Firstly, the input time complexity for Algorithm 4 is $O(n)$. For the first step of Algorithm 1 (Lines 1–7), each round of the while loop add one point to the MIS, so the while loop ends in $O(n)$ rounds. In each round, a node x should be picked. Since we can store and update it instantly, the time complexity of node selection is $O(n)$ each round. Thus the time complexity of the whole loop is $O(n^2)$.

For the second step of Algorithm 1 (Lines 8–13), since each round of the while loop joint at least one MIS point to the CDS, and the MIS has $O(n)$ nodes, the loop ends in $O(n)$ rounds. During each round, we use an array to store $|M_{v,C}|$, and the maintenance time complexity is $O(n)$, since only points in $N(N(v))$ would change its $|M_{v,C}|$. Also, the time complexity to select a v is $O(n)$ each round. Hence, the time complexity of Algorithm 1 is $O(n^2)$.

For Algorithm 2, we can store $|dom(u)|$ for each $u \in G$. Once a CDS node v is reduced, only $dom(N(v))$ nodes need to change, so the maintenance complexity is $O(n)$. Next, each edge in G will be visited for a constant times, so the overall time complexity if $O(n^2)$. For Algorithm 3, we can use the Tarjan's strongly connected components algorithm for computing congestion set C', whose time complexity is $O(n^2)$. Hence, the time complexity of Algorithm 3 is $O(n^2)$.

For Algorithm 4, it runs Algorithm 1 once, Algorithm 2 once, and Algorithm 3 for a constant time, each with time $O(n^2)$. Therefore, the time complexity of Algorithm 4 is $O(n^2)$.

5   Simulation Results

In this section, we compare Algorithm 4 with Kim's Algorithm 1 to solve MCDS in UBGs. For the simulations, we deploy wireless nodes in a $20 \times 20 \times 20$ three-dimensional virtual space. We also ensure that the graph induced by all nodes is connected. The number of nodes varies from 100 to 1000 by increasing 100. We use 1 as the maximum transmission range of the nodes. Through the random graph generation process, we control the lower bound of distance between two nodes at 0.25, 0.5, 0.75. Thus, we can have graphs with different node density. In Fig. 9, we use R-C-CDS-UBG to identify our algorithm, and C-CDS-UBG to identify Kim's.

Figure 9 shows the comparison of the performance between two algorithms when the lower bound of distance between two points is 0.25, 0.5, and 0.75 respectively. Through the figure, we can see that whatever the graph is, our algorithm can give a better answer than Kim's averagely. the ratio between our answer and Kim's is nearly 0.78. Also, through the comparison, we can figure out that in most situations, our improvement is steady.

Figure 10 shows a sample of UBG which has 500 points. The first picture is the result by Kim's algorithm with 208 CDS points, while the second is the

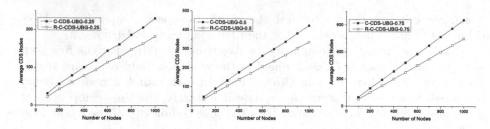

Fig. 9. Comparison with different parameter settings

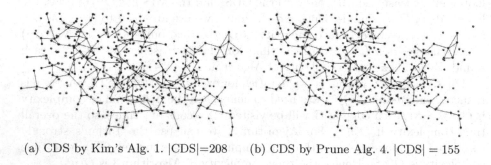

(a) CDS by Kim's Alg. 1. |CDS|=208   (b) CDS by Prune Alg. 4. |CDS| = 155

Fig. 10. An example solution with $n = 500$ points.

result by prune algorithm with 155 CDS points. In this example our algorithm reduces CDS size by 25 %, and it can be found in the graph that lots of the boundary nodes are dropped to make the CDS much smaller.

6   Conclusion

In this paper, we first pointed out the problem in Kim's method [1]. Then, we proposed a new projection method for solving the *two-ball problem*. With this new projection method, we successfully improved the ratio of $mis(G)/mcds(G)$ in UBG into 10.917. Moreover, we also optimized the algorithm for minimum connected dominating set selection in [1] with prune process and validate the efficiency of our design by numerical experiments.

References

1. Kim, D., Zhang, Z., Li, X., Wang, W., Wu, W., Du, D.Z.: A better approximation algorithm for computing connected dominating sets in unit ball graphs. IEEE Trans. Mob. Comput. **9**(8), 1108–1118 (2010)
2. Kouider, M., Vestergaard, P.D., et al.: Generalized connected domination in graphs. Theoret. Comput. Sci. **8**(1), 57–64 (2006)
3. Gaspers, S., Liedloff, M., et al.: A branch-and-reduce algorithm for finding a minimum independent dominating set. Theoret. Comput. Sci. **14**(1), 29–42 (2012)

4. Wang, X., Li, P.: Improved data association method in binocular vision-SLAM. In: ICICTA, vol. 2, pp. 502–505. IEEE (2010)
5. Yang, D.: An immunity-based ant colony optimization topology control algorithm for 3D wireless sensor networks. Sens. Transducers J. **150**(3), 125–129 (2013)
6. Wan, P.J., Alzoubi, K.M., Frieder, O.: Distributed construction of connected dominating set in wireless ad hoc networks. In: INFOCOM, vol. 3, pp. 1597–1604. IEEE (2002)
7. Wu, W., Du, H., Jia, X., Li, Y., Huang, S.C.H.: Minimum connected dominating sets and maximal independent sets in unit disk graphs. Theoret. Comput. Sci. **352**(1), 1–7 (2006)
8. Gao, X., Wang, Y., Li, X., Wu, W.: Analysis on theoretical bounds for approximating dominating set problems. Discret. Math. Algorithms Appl. **1**(01), 71–84 (2009)
9. Li, M., Wan, P.J., Yao, F.: Tighter approximation bounds for minimum CDS in unit disk graphs. Algorithmica **61**(4), 1000–1021 (2011)
10. Du, Y.L., Du, H.W.: A new bound on maximum independent set and minimum connected dominating set in unit disk graphs. J. Combin. Optim. 1–7 (2013). doi:10.1007/s10878-013-9690-0
11. Hansen, J.C., Schmutz, E.: Comparison of two CDS algorithms on random unit ball graphs. In: ALENEX/ANALCO, pp. 206–211 (2005)
12. Butenko, S., Ursulenko, O.: On minimum connected dominating set problem in unit-ball graphs. Preprint Submitted to Elsevier Science (2007)
13. Zhong, X., Wang, J., Hu, N.: Connected dominating set in 3-dimensional space for ad hoc network. In: WCNC, pp. 3609–3612. IEEE (2007)
14. Zou, F., Li, X., Kim, D., Wu, W.: Construction of minimum connected dominating set in 3-dimensional wireless network. In: Li, Y., Huynh, D.T., Das, S.K., Du, D.-Z. (eds.) WASA 2008. LNCS, vol. 5258, pp. 134–140. Springer, Heidelberg (2008)
15. Du, D.-Z., Wan, P.-J.: Connected Dominating Set: Theory And Applications. Springer Optimization and Its Applications, vol. 77. Springer, Heidelberg (2012)
16. Zong, C.: Sphere Packings. Springer, New York (1999)

A Practical Greedy Approximation
for the Directed Steiner Tree Problem

Dimitri Watel[1,2]($\boxtimes$) and Marc-Antoine Weisser[1]

[1] Computer Science Department, SUPELEC System Sciences,
91192 Gif Sur Yvette, France
{dimitri.watel,marc-antoine.weisser}@supelec.fr
[2] University of Versailles, 45 avenue des Etats-Unis, 78035 Versailles, France

Abstract. The Directed Steiner Tree (DST) NP-hard problem asks, considering a directed weighted graph with n nodes and m arcs, a node r called *root* and a set of k nodes X called *terminals*, for a minimum cost directed tree rooted at r spanning X. The best known polynomial approximation ratio for DST is a $O(k^\varepsilon)$-approximation greedy algorithm. However, a much faster k-approximation, returning the shortest paths from r to X, is generally used in practice. We give in this paper a new $O(\sqrt{k})$-approximation greedy algorithm called Greedy$_{\text{FLAC}}^{\triangleright}$, derived from a new fast k-approximation algorithm called Greedy$_{\text{FLAC}}$ running in time at most $O(nmk^2)$.

We provide computational results to show that, Greedy$_{\text{FLAC}}$ rivals the running time of the fast k-approximation and returns solution with smaller cost in practice.

Keywords: Directed steiner tree · Approximation algorithm · Greedy algorithm

1 Introduction

The *Undirected Steiner Tree* problem asks, in an undirected weighted graph, for a minimum cost tree spanning a set of specific nodes called *terminals*. It is a classical NP-hard problem [1] and can be polynomially approximated with constant ratio [2–4]. We focus here on its directed extension, the *Directed Steiner Tree* problem (DST).

Problem 1. Given a *directed* graph $G = (V, A)$ with n nodes and m arcs, a node r called *root*, a set X of k nodes called *terminals* and non negative weights ω on the arcs, the Directed Steiner Tree problem (DST) consists in finding a minimum cost directed tree rooted at r spanning every terminal.

These two problems are known to have applications essentially in multicast routing where one wants to minimize bandwidth consumption [5–7]. DST is used when the model of a symmetric network is not sufficient.

Contrary to the undirected version, there is no polynomial approximation algorithm achieving a constant ratio for DST. Indeed, as a generalization of

© Springer International Publishing Switzerland 2014
Z. Zhang et al. (Eds.): COCOA 2014, LNCS 8881, pp. 200–215, 2014.
DOI: 10.1007/978-3-319-12691-3_16

the NP-hard Set Cover problem [1], it is known to be inapproximable within a $O(\log(k))$ ratio unless $NP \subseteq DTIME[n^{O(\log \log n)}]$, where k is the number of terminals [8]. It was later proved under the same assumption that for $\varepsilon > 0$ there is no $O(\log^{2-\varepsilon}(k))$ approximation algorithm [9].

Charikar *et al.* introduced a greedy algorithm [10,11] which extends a $\log(k)$-approximation algorithm for Set Cover [12,13]. At each step, it searches for a tree T of cost $\omega(T)$ covering only a part $X(T)$ of the terminals and minimizing the *density* $d(T) = \frac{\omega(T)}{|X(T)|}$. A small density is characteristic of a tree which makes a compromise between covering many terminals and reducing its cost. However, finding a minimum density Directed Steiner Tree is an NP-hard problem and has to be polynomially approximated too. Nonetheless, an efficient approximation for this problem implies a greedy algorithm with efficient ratio for DST [10]. This method gives the current best known approximation, CH_l, with ratio $2l^2(l - 1)\sqrt[l]{k/2} = O(k^{\frac{1}{l}})$ for any $l > 1$ [10,11].

In order to achieve the best approximation ratio with its algorithm, Charikar *et al.* do not work with the instance itself but with the shortest paths instance.

Definition 1. *Let $\mathcal{I} = (G, r, X, \omega)$ be a directed Steiner tree instance. We define as $\omega^\triangleright(u, v)$ the cost of a shortest path linking u to v, or $+\infty$ if such a path does not exist. The shortest paths instance $\mathcal{I}^\triangleright = (G^\triangleright = (V, A^\triangleright), r, X, \omega^\triangleright)$ is a complete directed graph where each arc (u, v) is weighted by $\omega^\triangleright(u, v)$.*

From a feasible solution $T^\triangleright$ of $\mathcal{I}^\triangleright$, one can build a feasible solution T of $\mathcal{I}$ by selecting for each arc (u, v) in $T^\triangleright$ any shortest path linking u and v. The cost of T is smaller than the cost of $T^\triangleright$.

Considering an approximation algorithm $\mathcal{B}$ for DST, one can build an approximation algorithm $\mathcal{B}^\triangleright$ for DST using Algorithm 1.

Algorithm 1. An approximation algorithm $\mathcal{B}^\triangleright$ for DST from an approximation algorithm $\mathcal{B}$ for DST

Require: A DST instance $\mathcal{I} = (G, r, X, \omega)$, an algorithm $\mathcal{B}$ for DST.
Ensure: A feasible solution for $\mathcal{I}$.
1: Build $\mathcal{I}^\triangleright$
2: $T^\triangleright \leftarrow \mathcal{B}(\mathcal{I}^\triangleright)$
3: $T \leftarrow \bigcup_{(u,v) \in T^\triangleright}$ a shortest path from u to v in $\mathcal{I}$
4: **return** T

Generally, $\mathcal{B}^\triangleright$ achieves a smaller approximation ratio than $\mathcal{B}$. However, $\mathcal{B}^\triangleright$ is necessarily longer than $\mathcal{B}$, especially when the original graph is sparse.

If the shortest path instance is precomputed, the running time of CH_l is $O(n^l k^{2l})$ with n as the number of nodes. As l increases, this grows exponentially. Consequently, a much faster k-approximation is frequently used, specifically when the application needs an answer fast and is satisfied by any correct solution [14–16]. It returns the shortest paths from r to all terminals. In a network, it would be equivalent to replacing the multicast tree by k unicast paths.

In order to compare algorithms, a test set named *SteinLib* aggregates most of the generated Undirected Steiner Problem instances that can be found in literature [17], and is used to compare exact or approximation algorithms for this problem [18–20]. However, the DST instances are still under construction. As a consequence, comparing DST algorithms requires transforming undirected instances into directed ones. This is usually done by replacing each edge by two opposites arcs [21–23]. Nonetheless, this method seems too restrictive as this kind of graph does not reflect the variety of possible instances and applications.

Our results. We provide an heuristic to build solutions covering only a part of the terminals with small density, following the ideas given in [10] and use it to build a greedy algorithm called Greedy$_{\text{FLAC}}$ for DST.

Additionally, we build a second approximation Greedy$_{\text{FLAC}}$$^{\triangleright}$ by applying Greedy$_{\text{FLAC}}$ to $\mathcal{I}^{\triangleright}$ as explained with Algorithm 1. Table 1 summarizes the two algorithms properties.

Table 1. Approximation algorithms provided in this paper, with their approximation ratios and Time complexities.

Algorithm	Approximation ratio	Time complexity
Greedy$_{\text{FLAC}}$$^{\triangleright}$	$4\sqrt{2k}$	$O(n^3 k^2)$
Greedy$_{\text{FLAC}}$	k	$O(nmk^2)$

We have built three test sets to compare our algorithms, partially derived from the SteinLib test set. We have compared Greedy$_{\text{FLAC}}$ and Greedy$_{\text{FLAC}}$$^{\triangleright}$ with four other algorithms including two versions of the shortest path algorithms. It is highlighted that Greedy$_{\text{FLAC}}$ returns solutions with nearly same cost than Greedy$_{\text{FLAC}}$$^{\triangleright}$ in almost all the cases, that Greedy$_{\text{FLAC}}$ returns solutions with smaller cost than the four other tested algorithms, and finally that the running time of Greedy$_{\text{FLAC}}$ rivals that of the fastest shortest path algorithm, surpassing every other algorithms including Greedy$_{\text{FLAC}}$$^{\triangleright}$.

This paper is organized in this way. In the next section, we introduce our algorithms and prove the two approximation ratios. Section 3 is dedicated firstly to improving the implementation of Greedy$_{\text{FLAC}}$. Finally, in Sect. 4, we present two new methods to obtain a directed graph from an undirected instance and an evaluation of performances of our algorithms.

2 The Algorithm

2.1 The Density

The *density* of a tree describes its ability to span the most terminals by paying the least.

Definition 2. *The* density $d(T)$ *of a tree* T *of cost* ω *spanning* $X' \subset X$ *is* $\frac{\omega}{|X'|}$.

From any algorithm $\mathcal{A}$ returning a directed tree spanning any subset X' of X, one can build a greedy algorithm $\text{Greedy}_{\mathcal{A}}$ for DST by applying $\mathcal{A}$ to cover some terminals, removing them from X and repeating until X is empty. Algorithm 2 describes this technique.

Algorithm 2. Greedy Algorithm $\text{Greedy}_{\mathcal{A}}$ from an algorithm $\mathcal{A}$

Require: A DST instance $\mathcal{I} = (G, r, X, \omega)$, an algorithm $\mathcal{A}$ returning a directed tree rooted in r covering a subset X' of X
Ensure: A feasible solution for $\mathcal{I}$.
 1: $T = \emptyset$
 2: **while** $X \neq \emptyset$ **do**
 3:     $T_0 \leftarrow \mathcal{A}(G, r, X, \omega)$
 4:     $T = T \cup T_0$
 5:     $X = X \backslash (X \cap T_0)$
    return T

The key idea, developed in [10], is that the best is the density of the trees returned by $\mathcal{A}$, the best is the approximation ratio of $\text{Greedy}_{\mathcal{A}}$.

In the next subsection, we provide an algorithm to find partial solutions with small density. The next subsection does not describe an algorithm but a thought experiment as it is easier this way to understand why and how our algorithm works to find a low density tree. We will then describe an implementation.

2.2 Finding Partial Solution with a Low Density

We define the thought experiment as if each arc were a pipe able to carry water. Each terminal is a source of water. It feeds each of its entering arcs with one litre of water per second. The volume of an arc, in litres, is equal to its cost. When an arc is completely full, it is *saturated*. Each terminal sends one litre of flow per second to each arc it can reach with a path of saturated arcs. Conversely, the flow rate, in litres per second, inside an arc, is the number of terminals that arc is linked to with paths of saturated arcs. More precisely, if a terminal reaches 3 arcs with paths of saturated arcs, each arc is filled with 1 litre per second, and not $\frac{1}{3}$ litre per second.

Definition 3. *If a terminal reaches an arc a with two or more distinct paths of saturated arcs or with a path containing a cycle, we say the flow is* degenerate.

At time 0, no flow has come out from the terminals. We increase the flow until the root is reached, we then focus on the subgraph made of saturated arcs linking the root to terminals. If multiple arcs saturates at the same time, we arbitrarily order them. Each time the flow is degenerate, we remove the last saturated arc to cancel the degeneration, so that we assure the flow is never degenerate when the root is reached.

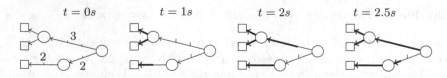

Fig. 1. This example illustrates the flow growing in a tree. Unspecified arc costs are 1. Water is represented by bold arc parts. Note that, at the last step, because the arc is fed with 2 terminals, it takes $0.5s$ to saturate it, and the arc below is only $25\,\%$ full.

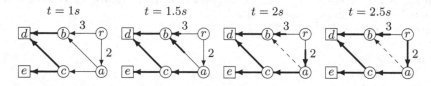

Fig. 2. This example illustrates the flow increasing in a graph. Unspecified arc costs are 1. Note that when (a, b) is saturated at $t = 2s$, the flow is degenerate. We remove (a, b) from the graph and the (r, a) flow rate remains 2. Note also that, at $t = 0s$, the two entering arcs of the terminal d are fed with $1\ L/s$ each and not $\frac{1}{2}$.

An example in a tree is shown in Fig. 1, and in a directed graph in Fig. 2.

A subgraph with low density emerges, because the time that the root is reached depends on the number of terminals joined to accelerate the flow and on the existence of paths of low cost. Those two properties are strongly associated with a low density. As shown in Fig. 1, the flow does not immediately describe a directed Steiner tree but a partial solution spanning only the first terminals reaching the root: we will repeat our algorithm to build a feasible solution.

Remark 1. This flow is not a conservative flow as in the maximum flow problem. Actually, we describe a *dual ascent* heuristic like those described in [21,24,25] where primal variables x_a for $a \in A$, between 0 and 1, include a in the feasible solution if $x_a = 1$. This flow describes how the primal variables evolve: x_a is the ratio of flow inside a divided by $\omega(a)$.

2.3   Naïve Implementation

We provide here an algorithm implementing the experiment. This implementation does not guarantee the running time given in Sect. 1 but guarantees the approximation ratio. A faster implementation is given in Sect. 3.

Let $\mathcal{I} = (G = (V, A), r, X, \omega)$ be an instance of DST. To describe the implementation, we will use the following useful notations and definitions.

The quantity of flow inside the arc a is $f(a)$. We define an auxiliary graph $G_{SAT} = (V, A_{SAT} \subset A)$. An arc is said to be *saturated* if G_{SAT} contains it. The number of terminals that a node v or an arc $a = (u, v)$ is connected to in G_{SAT} is $k(v) = k(a)$. We define the current elapsed time as t and the remaining time before a is saturated by $t(a) = \frac{\omega(a) - f(a)}{k(a)}$. If $k(a) = 0$, $t(a)$ is infinite.

Finally, the flow is said to be *degenerate* if G_{SAT} contains a cycle or two distinct paths from any node to any terminal. To ensure the algorithm returns a tree, any time the flow is degenerate, we remove the last saturated arc from G_{SAT} and add it to the set $\mathcal{M}$ of *marked*. No flow goes through a marked arc.

We use Algorithm 3, called FLAC, to increase the flow in a graph. As explained previously, FLAC does not return a feasible solution for DST but only a partial solution. An example is given in Table 2.

Algorithm 3. FLow Algorithm Computation (FLAC)

Require: A DST instance $\mathcal{I} = (G, r, X, \omega)$.
Ensure: A directed tree rooted in r covering some nodes in X.
1: $G_{SAT} \leftarrow (V, \emptyset)$, $t \leftarrow 0$, $\mathcal{M} \leftarrow \emptyset$ # Marked arcs.
2: **for** $a \in A$ **do** $f(a) = 0$
3: **while** True **do**
4:     **for** $(a \in A \backslash (\mathcal{M} \cup G_{SAT}))$ **do** $t(a) \leftarrow (\omega(a) - f(a))/k(a)$
5:     $(u, v) \leftarrow \arg\min(t(a), a \in A \backslash (\mathcal{M} \cup G_{SAT}))$
6:     **for** $(a \in A \backslash (\mathcal{M} \cup G_{SAT}))$ **do** $f(a) \leftarrow f(a) + k(a) \cdot t(u, v)$
7:     $t \leftarrow t + t(u, v)$
8:     $G_{SAT} \leftarrow G_{SAT} \cup (u, v)$
9:     **if** $(u = r)$ **return** the tree T_0 in G_{SAT} linking r to the terminals
10:    **if** the flow is degenerate **then**
11:        $\mathcal{M} \leftarrow \mathcal{M} \cup (u, v)$
12:        Remove (u, v) from G_{SAT}

At Line 9, there is only one tree T_0 in G_{SAT} linking the root to the terminals, because of the arcs added to $\mathcal{M}$ each time the flow is degenerated. One can build it with breadth-first search from the root to the terminals.

Table 2. This example illustrates how Algorithm 3 runs on the graph shown in Fig. 2, and how it differs from the experiment. It details the arc saturated at the beginning of each iteration, with the content of $\mathcal{M}$, the values of t and $t(a)$ for each arc a.

Iteration	(u,v)	t	$\mathcal{M}$	$t(b,d)$	$t(c,d)$	$t(c,e)$	$t(a,c)$	$t(a,b)$	$t(r,a)$	$t(r,b)$
-	-	0	$\emptyset$	1	1	1	$+\infty$	$+\infty$	$+\infty$	$+\infty$
1	(b,d)	1	$\emptyset$	0	0	0	$+\infty$	1	$+\infty$	3
2	(c,d)	1	$\emptyset$	-	0	0	1	1	$+\infty$	3
3	(c,e)	1	$\emptyset$	-	-	0	0.5	1	$+\infty$	3
4	(a,c)	1.5	$\emptyset$	-	-	-	0	0.5	1	2.5
5	(a,b)	2	$\{(a,b)\}$	-	-	-	-	0	0.5	2
6	(r,a)	2.5	$\{(a,b)\}$	-	-	-	-	-	0	1.5

We can now build the two approximation algorithms for DST from FLAC. The Greedy$_{\mathrm{FLAC}}$ algorithm is build using Algorithm 2 with FLAC. We build the Greedy$_{\mathrm{FLAC}}{}^{\triangleright}$ algorithm using Algorithm 1 with Greedy$_{\mathrm{FLAC}}$. We now prove Greedy$_{\mathrm{FLAC}}$ is a k-approximation and Greedy$_{\mathrm{FLAC}}{}^{\triangleright}$ is an $4\sqrt{2k}$-approximation.

2.4   Approximation Ratio of Greedy$_{\text{FLAC}}$

Assuming we start the j-th iteration of the loop, let $a = (w, x)$ be an arc, we define $f_j(a)$ as the flow inside a, $k_j(x)$ as the value of $k(x)$, $t_j(a) = \frac{w(a) - f_j(a)}{k_j(x)}$ as the value of $t(a)$, and finally t_j as the value of t.

Firstly, because $k_i(a)$ can only increase as i grows, with Lines 6 and 7, we can prove the following lemma:

Lemma 1. *If during the $(l-1)$-th iteration, $a \notin G_{SAT} \cup \mathcal{M}$, then for each $j \leq l - 1$, $\omega(a) \geq f_l(a) \geq f_j(a) + (t_l - t_j) \cdot k_j(a)$.*

Lemma 2. *If $t_l > t_j + t_j(a)$ for some $j \leq l - 1$, then a was saturated or marked during or before the $(l-1)$-th iteration.*

Proof. We assume a was neither saturated nor marked during or before the $(l-1)$-th iteration. By Lemma 1, $\omega(a) \geq f_l(a) \geq f_j(a) + (t_l - t_j)k_j(a) > f_j(a) + t_j(a)k_j(a) = \omega(a)$, which is a contradiction. □

Let $f - 1$ be the final iteration when T_0 is returned at Line 9 (t has then value t_f). Let x be the closest terminal to the root r, and P a shortest path linking r to x in $\mathcal{I}$.

Lemma 3. *t_f is no more than the cost of P.*

Proof. Let $v_l = r$, $v_{l-1}, \ldots, v_1$, $v_0 = x$ be the successive nodes of P and ω_i the cost of the subpath of P from v_i to v_0. By induction on i we prove that either $t_f \leq \omega_i$ or there is an iteration j_i with $k_{j_i}(v_{i+1}, v_i) \leq 1$, and $t_{j_i} \leq \omega_i$.

If $i = 0$, with $j_0 = 1$, the property is true.

If the property is true for some $i - 1$, and if $t_f > \omega_i \geq \omega_{i-1}$, by induction, there is an iteration j_{i-1} with $t_{j_{i-1}} \leq \omega_{i-1}$ and $k_{j_{i-1}}(v_i, v_{i-1}) \leq 1$. Consequently, $t_{j_{i-1}}(v_i, v_{i-1}) = (\omega_i - \omega_{i-1} - f_{j_{i-1}}(a))/k_{j_{i-1}}(v_i, v_{i-1}) \leq \omega_i - \omega_{i-1}$.

Then, by Lemma 2 and because $t_f > \omega_i$, there is an iteration $j_i < f$ where (v_i, v_{i-1}) is saturated or marked, then $k_{j_i}(v_{i+1}, v_i) \leq 1$, and $t_{j_i} \leq t_{j_{i-1}} + t_{j_{i-1}}(v_i, v_{i-1}) = \omega_i$. By induction on i, $t_f \leq \omega_l$ and the lemma is proved. □

Lemma 4. *t_f is the density of the tree T_0 returned at line 9.*

Proof. We first recall that T_0 is fully saturated by definition, no arc of T_0 is marked, and every arc entering T_0 is either marked in G_{SAT} or not, then there is no iteration during which an arc not included in T_0 sends flow into T_0.

For a node $v \in T_0$, let T_v, k_v, ω_v be the subtree rooted in v, the number of terminals and the cost of T_v. Let t_v be the value of t when the last arc of T_v is saturated. Finally, let ε_v be the flow that went through v at that iteration. One can see ε_v as the quantity of flow inside a fictive arc entering v with infinite cost when the last arc of T_v is saturated.

We prove hereinafter by induction on their height that for every node v, $t_v \cdot k_v = \omega_v + \varepsilon_v$. As no flow came through r before the f-th iteration, $t_r \cdot k_r = \omega_r$ and the lemma is proved.

If v is a leaf, T_v is empty and is then fully saturated at the first iteration when $t = 0$, then the property is true. If now the property is true for every node of height h, we prove it for a node v of height $h + 1$.

Let v_1, v_2, ... v_s be the children of v with $\omega(v, v_i) = \alpha_i$. The inductive hypothesis claims that for each $i \in [\![1; s]\!]$, $t_{v_i} \cdot k_{v_i} = \omega_{v_i} + \varepsilon_{v_i}$.

Let t'_{v_i} be the time when every arc of $(v, v_i) \cup T_{v_i}$ is saturated. Note that (v, v_i) can be saturated before the last saturated arc of T_{v_i}, depending on the value of ε_{v_i}. Firstly, if $\alpha_i \leq \varepsilon_{v_i}$, $t'_{v_i} = t_{v_i}$. In that case, (v, v_i) is not the last saturated arc of $(v, v_i) \cup T_{v_i}$ and $\varepsilon_{v_i} - \alpha_i$ is the volume of flow sent through v before the time t'_{v_i}. Secondly, if $\alpha_i > \varepsilon_{v_i}$, $t'_{v_i} = t_{v_i} + \frac{\alpha_i - \varepsilon_{v_i}}{k_{v_i}}$ by Lemma 1. Indeed, after the time t_{v_i}, every terminal of T_{v_i} feeds (v, v_i). No flow is sent through v before t'_{v_i} as (v, v_i) is in that case the last saturated arc of $(v, v_i) \cup T_{v_i}$.

After each time t'_{v_i}, each arc of $(v, v_i) \cup T_{v_i}$ is saturated and the terminals of T_{v_i} sends, by Lemma 1, $k_{v_i} \cdot (t_v - t'_{v_i})$ units of flow through v. Thus, $t_v = \max\limits_{i \in [\![1;s]\!]} t'_{v_i}$.

$$\varepsilon_v = \sum_{i=1}^{s} (k_{v_i} \cdot (t_v - t'_{v_i})) + \sum_{i \setminus \alpha_i \leq \varepsilon_{v_i}} (\varepsilon_{v_i} - \alpha_i)$$

$$\varepsilon_v = t_v \cdot k_v - \sum_{i \setminus \alpha_i \leq \varepsilon_{v_i}} (k_{v_i} \cdot t'_{v_i}) - \sum_{i \setminus \alpha_i > \varepsilon_{v_i}} (k_{v_i} \cdot t'_{v_i}) + \sum_{i \setminus \alpha_i \leq \varepsilon_{v_i}} (\varepsilon_{v_i} - \alpha_i)$$

$$\varepsilon_v = t_v \cdot k_v - \sum_{i \setminus \alpha_i \leq \varepsilon_{v_i}} (\omega_{v_i} + \alpha_i) - \sum_{i \setminus \alpha_i > \varepsilon_{v_i}} (\omega_{v_i} + \alpha_i)$$

Thus $\varepsilon_v = t_v \cdot k_v - \omega_v$. □

We now recall the lemma (adapted with our terminology) from [10].

Lemma 5 ([10], Lemma 1). *Let f be a function such that $\frac{f(k)}{k}$ is non strictly decreasing. Suppose for any DST instance with k terminals and ω^* as the minimum solution cost, that FLAC returns a tree T_0 with $d(T_0) \leq f(k) \cdot \frac{\omega^*}{k}$. Then $Greedy_{FLAC}$ is a $\int_0^k \frac{f(u)}{u} du$-approximation for DST.*

Theorem 1. *$Greedy_{FLAC}$ is a k-approximation for the DST problem.*

Proof. By Lemmas 3 and 4, $d(T_0) \leq \omega(P)$. Because any optimal solution contains a path from r to y, $d(T_0) \leq \omega^*$ where ω^* is the optimal cost of the instance. And then $d(T_0) \leq k\frac{\omega^*}{k}$. By Lemma 5, $Greedy_{FLAC}$ ratio is $\int_0^k du = k$. □

2.5 Approximation Ratio of Greedy$_{FLAC}^{\triangleright}$

We recall Greedy$_{FLAC}^{\triangleright}$ applies Greedy$_{FLAC}$ in the shortest path instance $\mathcal{I}^{\triangleright}$ in order to build a feasible solution $T^{\triangleright}$ of $\mathcal{I}^{\triangleright}$. From that solution, a feasible solution of $\mathcal{I}$ is build by selecting for each arc (u, v) in $T^{\triangleright}$ a shortest path linking u to v. Note that the cost of $T^{\triangleright}$ is greater than the cost of T.

Let T_2 be in $\mathcal{I}^{\triangleright}$ a minimum density tree of height at most 2. Note that if T_2 is a disjoint union of trees rooted at r, they all have the same density $d(T_2)$. We keep one such tree, so that r has only one child v. Let $x_1, x_2, \cdots, x_l$ be the children terminals of v. Let $\alpha = \omega(r, v)$ and $\beta_i = \omega(v, x_i)$ such that $\beta_1 \leq \beta_2, \leq \cdots \leq \beta_l$.

Lemma 6. t_f *is no more than the density of* T_2.

Proof. As T_2 has minimum density among all the trees of height 2, for all $i \leq l$, $\beta_i \leq d(T_2)$. If $t_f \leq \beta_i \leq d(T_2)$ for some i, the lemma is proved. If not, by Lemma 2, each arc (v, x_i) is saturated at iteration $j_i < f$ and $t_{j_i} \leq \beta_i$. Consequently, at the first iteration j when $t_j \geq \beta_l$, by Lemma 1, v sent at least $\sum_{i \leq l}(t_j - \beta_i)$ units of flow inside (r, v). Thus $t_j + t_j(r, v) \leq t_j + \frac{\alpha - \sum_{i \leq l}(t_j - \beta_i)}{l} = d(T_2)$. By Lemma 2, $t_f \leq d(T_2)$. $\square$

Theorem 2. *Greedy*$_{FLAC^{\triangleright}}$ *is a* $4\sqrt{2k}$-*approximation for DST.*

Proof. We follow the proof given in [10]. A tree T_2 of height 2 with best density has a lower density that any tree T_2^* of height 2 spanning X with best cost. The cost of the tree T_2^* is a $4\sqrt{k/2}$-approximation of any minimum solution T^* of DST [11,26]. By Lemmas 4 and 6, $d(T_0) \leq d(T_2) \leq 4\sqrt{k/2}\frac{\omega(T^*)}{k} = 2\sqrt{2k}\frac{\omega(T^*)}{k}$. By Lemma 5, Greedy$_{FLAC}$, applied in $G^{\triangleright}$ by Greedy$_{FLAC^{\triangleright}}$, returns a solution $T^{\triangleright}$ of cost at most $\int_0^k \frac{2\sqrt{2u}}{u} du \cdot \omega(T^*) = 4\sqrt{2k} \cdot \omega(T^*)$.

Finally, note that the cost of T is lower than the cost of $T^{\triangleright}$, thus is also a $4\sqrt{2k}$-approximation. $\square$

Next Section is dedicated to describing a faster implementation of Greedy$_{FLAC}$ and, consequently, of Greedy$_{FLAC^{\triangleright}}$.

3   A Faster Implementation

This section provides a faster implementation of the algorithm described in Sect. 2.2 and proves that, with this faster version of FLAC, Greedy$_{FLAC}$ runs in time $O(k^2 mn)$. Three steps need to be detailed: saturating an arc, detecting a degenerate flow and building the returned tree T_0.

Note that, at each iteration, all the unsaturated and unmarked entering the arcs of a node v have the same flow $f(u, v)$ and flow rate $k(v)$ and are saturated or marked by order of weight. Thus, instead of storing flow with the arcs, we store them with v. We store the nodes in a Fibonacci heap $\mathcal{F}$, where each node v is associated with the next time t_v one of its entering arcs is saturated.

For each node u, we define T_u as the set of nodes w from which there is a path to u in G_{SAT}. When the flow is degenerate, we mark the arc and remove it from G_{SAT}, thus T_u is a tree. We store the list $\Gamma^-(v)$ of arcs entering v, sorted by weights, the sets $\Gamma_{SAT}^-(v)$ and $\Gamma_{SAT}^+(v)$ of saturated arcs entering v and outgoing from v and the next saturated arc a_v entering v. Note that, for each node u, we can iterate over T_u in linear time by returning u and then recursively iterating over T_w for each node w such that (w, u) is in $\Gamma_{SAT}^-(u)$. We also store the terminals X_v feeding v in a boolean array of size k: $X_v[y]$ is true if G_{SAT} contains a path from v to y. We finally keep the current time t.

Saturating arcs. We have to find, at each step, the next saturated arc. At each step, we remove the first node v from the heap $\mathcal{F}$ in time $O(\log(n))$. The next saturated arc is a_v. We update the value of t to the key t_v of v in $\mathcal{F}$.

Detecting a degenerate flow. A flow is degenerate if there are two distinct paths in G_{SAT} or a path with a cycle from a node to a terminal. When $a_v = (u, v)$ becomes saturated, there is a new path in G_{SAT} from each node w in T_u to each terminal v is linked to. As a consequence, to detect a degenerate flow, we have to check for each node w in T_u whether X_w and X_v are disjoint or not: is there a terminal y with $X_v[y] = X_w[y] = true$? We check the sets intersection in time $O(k)$. The complete detection takes at most $O(nk)$ steps.

Updating the flow rates. If the flow is not degenerate, we have to update all the nodes affected by the saturation of (u, v). We first add a_v to $\Gamma_{SAT}^-(v)$ and to $\Gamma_{SAT}^+(u)$. If the arc entering v are saturated, we do nothing more. Otherwise, we get the minimum cost unsaturated arc a_v' entering v in $O(1)$ as $\Gamma^-(v)$ is sorted by weight. And v can be reinjected inside $\mathcal{F}$ with $t_v' = t + \frac{\omega(a_v') - \omega(a_v)}{|X_v|}$.

Then we update all the nodes w in T_u. If w was not in $\mathcal{F}$, if a_w is its minimum cost unsaturated entering arc, $X_w = X_v$ and $t_w = t + \frac{\omega(a_w)}{|X_v|}$. If it was, the new set X_w' is $X_w \uplus X_v$, computed in time $O(k)$. Next we decrease the key t_w in $\mathcal{F}$ to $t + \frac{(t_w - t) \cdot |X_w|}{|X_w'|}$. This takes $O(nk)$ operations.

Building the returned tree. All the saturated arcs to which the root is linked constitute a tree spanning terminals. There are at most $n - 1$ such arcs. Thus this is done once in time $O(n)$ by returning r and recursively iterating over the saturated arcs w is linked to for each node w such that (r, w) is in $\Gamma_{SAT}^+(r)$.

The running time of Algorithm Greedy$_{\text{FLAC}}$. To return a DST feasible solution, the FLAC algorithm is called at most k times as each call spans at least one terminal. For each of them, m arcs at most are saturated, leading to a conflict detection, and updates in time $O(\log(n) + 2nk)$. Finally the solution is built in linear time. Consequently, we return the solution in time $O(k^2mn)$.

Remark 2. If a $O(\sqrt{k})$-approximation is required, we use Greedy$_{\text{FLAC}}^{\triangleright}$ and have to compute FLAC in $\mathcal{I}^{\triangleright}$. The Dijkstra algorithm can be used to build $\mathcal{I}^{\triangleright}$ in time $O(nm + n^2 \log(n))$, and then the algorithm Greedy$_{\text{FLAC}}^{\triangleright}$ runs in time $O(k^2n^3)$.

4 Evaluations of Performance

4.1 Ratio Study

We firstly compare Greedy$_{\text{FLAC}}$ with Greedy$_{\text{FLAC}}^{\triangleright}$ over practical cases. Surprisingly, Greedy$_{\text{FLAC}}$ returns solutions with smaller or same costs than Greedy$_{\text{FLAC}}^{\triangleright}$ in nearly all cases. Considering that the running time of Greedy$_{\text{FLAC}}$ is always smaller than that of Greedy$_{\text{FLAC}}^{\triangleright}$, one should use Greedy$_{\text{FLAC}}$ in practice.

We then compared four algorithms with Greedy$_{\text{FLAC}}$: two versions of the mostly used in practice the Shortest Paths algorithm (ShP$_1$ and ShP$_2$), the first and fastest Dual Ascent algorithm (DuAs) [24] and the CH$_2$ algorithm [10]. ShP$_1$ was implemented using Dijkstra's algorithm to compute all shortest paths from the root to all terminals. ShP$_2$ selects a shortest path from the root to its closest terminal, sets the weights of the arcs of that path to 0, and restarts until all the terminals are reached. CH$_2$ was implemented with the Roos modified algorithm described in [22][1]. We now detail the test set used during the evaluation.

Description of the test set. To run the experiments, we transformed the SteinLib instances [17] into three directed instances test sets. Each undirected graph was transformed into one or more directed graphs with the same optimal cost. We first created the bidirected test set (BTS) containing 999 instances, where each edge was replaced by two opposite arcs with the same cost and chose any terminal as the root. As those instances seem too specific to reflect the variety of all instances and applications, we also created an acyclic instances test set (ATS) and a strongly connected instances test set (SCTS), containing 353 instance each. In order to build those test sets, we first computed an optimal tree T^* of each instance (currently, we have processed one third of all the instances). Then we transformed T^* into a directed tree rooted at any of its nodes. In (ATS), we directed all other edges in order to get an acyclic digraph with breadth-first search from the root. In (SCTS), every other edge was directed uniformly at random. Then, until we got a strongly connected graph, an arc was added between a random couple of unconnected nodes with cost equal to that of the shortest path cost in the original instance.

The test sets were generated from the following SteinLib groups: WRP3, WRP4, ALUE, ALUT, DIW, DXMA, GAP, MSM, TAQ, LIN, SP, X, MC, I080 to I640, ES10FST to ES10000FST, B,C,D,E, P6E and P6Z. Other groups, for instance P4E and P4Z, were not considered because they contain incorrect files.

Comparison of Greedy$_{\text{FLAC}}$ with Greedy$_{\text{FLAC}}^{\triangleright}$. Considering the cost ω of a solution returned by Greedy$_{\text{FLAC}}$, the cost $\omega^{\triangleright}$ of a solution returned by Greedy$_{\text{FLAC}}^{\triangleright}$, and the optimal cost ω^*, we are interested in the relative distance $\delta = \frac{\omega - \omega^{\triangleright}}{\omega^{\triangleright}}$ and the cases when $\omega = \omega^*$ or $\omega^{\triangleright} = \omega^*$.

We removed from the test set (BTS) the biggest instances where Greedy$_{\text{FLAC}}^{\triangleright}$ was not able to return a result due to required memory. As a consequence, it contains 471 instances for this evaluation. Table 3 summarises the results.

From this evaluation, we conclude that:

- Greedy$_{\text{FLAC}}^{\triangleright}$ returns strictly smaller costs for less than 10 % of the instances;
- according to the columns $\delta = 0$ and $|\delta| < 5\%, 10\%, 15\%$, the costs ω and $\omega^{\triangleright}$ are frequently equal or close;

---

[1] The four algorithms were run with Java 1.7.0_025 on Ubuntu 12.10 with Intel Core 3.10 GHz processors. The code source can be found at https://github.com/mouton5000/DSTAlgoEvaluation.

Table 3. Comparison of the costs ω and $\omega^\triangleright$ over each test set. The left part details the number of instances for which ω is lower, equal or greater than $\omega^\triangleright$. The middle part explains how close are those costs. The last part gives the number of instances for which the algorithms return an optimal solution.

	$\delta < 0$	$\delta = 0$	$\delta > 0$	$|\delta| < 5\%$	$|\delta| < 10\%$	$|\delta| < 15\%$	$\omega = \omega^*$	$\omega^\triangleright = \omega^*$
BTS	54 %	34 %	10 %	83 %	98 %	99 %	20 %	15 %
ATS	17 %	71 %	10 %	94 %	98 %	99 %	48 %	46 %
SCTS	16 %	76 %	6 %	95 %	98 %	99 %	63 %	57 %

– Greedy$_{\mathrm{FLAC}}$ returns an optimal solution in more cases than Greedy$_{\mathrm{FLAC}}{}^\triangleright$ (on the contrary, the number of instances optimally solved by Greedy$_{\mathrm{FLAC}}{}^\triangleright$ but not by Greedy$_{\mathrm{FLAC}}$ is less than 3 %).

Due to the computation of $\mathcal{I}^\triangleright$ and to the number of degenerate flows that Greedy$_{\mathrm{FLAC}}{}^\triangleright$ encounters in a directed complete graph, its running time is much greater than that of Greedy$_{\mathrm{FLAC}}$. Thus it is preferable to use Greedy$_{\mathrm{FLAC}}$ in practice than Greedy$_{\mathrm{FLAC}}{}^\triangleright$. This happens, contrary to what their approximation ratios suggest, because the shortest path instance, as a complete directed graph, contains shortcuts between the terminals and the root that cannot be employed by Greedy$_{\mathrm{FLAC}}$, this favours solutions where terminals share their flows.

Comparison of Greedy$_{\mathrm{FLAC}}$ with other approximation algorithms. Considering the cost ω of the returned solution compared to the optimal cost ω^*, we are interested in the relative error $e = \frac{\omega - \omega^*}{\omega^*}$. We represent in Fig. 3 the error cumulative distributions: for each e and each algorithm $\mathcal{A}$, we plots the number of instances for which the relative error of $\mathcal{A}$ is lower than e.

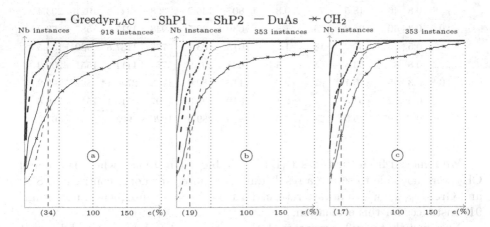

Fig. 3. Cumulative distribution function of the relative error e in percent, for each algorithm. The tree test sets are (a) BTS, (b) ATS, and (c) SCTS. The maximum error of solutions obtained with Greedy$_{\mathrm{FLAC}}$ is indicated in parentheses.

Table 4. Average running times in milliseconds and standard deviations for each class of graphs, depending on their number of nodes, their expected number of arcs and their number of terminals.

n	p	$\frac{k}{n}$	Greedy$_{FLAC}$		ShP$_1$		ShP$_2$		DuAs		CH$_2$	
			mean	s. dev.	mean	s. dev.	mean	s. dev.	mean	s. dev.	mean	s. dev.
50	0.3	0.3	<1		<1		1	0.5	2	1.9	2	0.8
		0.5	<1		<1		1	0.6	2	2.2	3	1.0
		0.8	1	0.5	<1		2	0.6	2	2.5	5	1.3
	0.5	0.3	1	0.7	<1		1	0.6	3	3.4	2	0.7
		0.5	1	0.6	<1		2	0.7	3	3.1	3	1.0
		0.8	1	0.7	<1		3	1.0	3	2.9	6	1.6
	0.8	0.3	1	0.7	<1		2	0.6	4	4.5	2	0.8
		0.5	1	0.8	<1		3	0.8	4	4.9	4	1.5
		0.8	1	0.8	<1		5	0.7	4	4.2	7	1.8
100	0.3	0.3	2	1.0	<1		8	1.1	15	20.9	11	2.3
		0.5	3	1.2	<1		13	1.8	14	14.0	19	3.1
		0.8	4	1.5	<1		21	3.5	15	14.9	31	4.0
	0.5	0.3	4	1.5	1	0.5	13	2.1	20	20.9	16	2.7
		0.5	5	1.5	1	0.4	21	3.0	17	22.8	26	3.7
		0.8	6	1.6	1	0.5	34	5.2	19	24.2	44	5.4
	0.8	0.3	6	1.7	2	0.5	20	2.1	48	56.4	25	2.8
		0.5	7	2.0	2	0.6	34	3.5	33	39.3	41	3.3
		0.8	8	2.2	2	0.6	53	6.2	27	35.4	66	6.7
250	0.3	0.3	31	14.7	7	0.8	343	39.1	515	921.3	354	39.3
		0.5	37	14.4	7	0.7	513	74.9	542	728.8	615	74.4
		0.8	46	18.5	7	0.8	807	136.2	398	441.0	1021	203.4
	0.5	0.3	60	28.3	13	1.3	639	46.5	948	1217.1	587	73.2
		0.5	75	36.7	13	1.2	1073	103.7	777	1137.8	980	133.5
		0.8	82	33.6	13	1.5	1638	201.2	781	1111.2	1551	299.2
	0.8	0.3	95	46.1	22	2.0	1116	80.6	1763	2789.1	867	166.2
		0.5	128	68.3	22	2.1	1819	193.3	1697	2510.1	1511	216.0
		0.8	127	65.5	23	4.0	2859	289.0	1226	1903.6	2331	463.3

We removed from the test set (BTS) the biggest instances where DuAs and CH$_2$ were not able to return a result due to required memory (contrary to ShP and Greedy$_{FLAC}$ which always returned a result). As a consequence, it contains 918 instances for this evaluation.

Note first that Fig. 3 aggregates all the instances of each test set and does not take into account the number of nodes or terminals distribution, or the different SteinLib categories. Considering this, Greedy$_{FLAC}$ gives better practical results, followed by ShP$_2$, in all test sets, as most of the instances are solved with less than 5 % error. ShP_1 returns the worst results for the precision inferior to 30 % and then overtakes CH_2. (ATS) and (SCTS) are better solved by DuAs than the shortest paths algorithms. This is because (ATS) and (SCTS) instances contain fewer feasible solutions than (BTS), but contain almost the same shortest paths from the root to terminals. We conclude that $Greedy_{FLAC}$ outperforms the shortest path algorithms, DuAs and CH_2. We study in the next subsection the execution time results to determine the practical interest of $Greedy_{FLAC}$.

4.2   Time Study

We used a different test set for this evaluation because we want to study the impact on the running time of 3 parameters: the number of nodes n, of arcs m, and of terminals k. As we do not control those parameters with the SteinLib data set, we generated a set of random instances and ran all algorithms, except $Greedy_{FLAC}^{\triangleright}$ which running time is very high compared to others.

The instances were generated with 3 parameters: n, the probability to link two nodes with an arc p, and the ratio $\frac{k}{n}$. We first add n nodes to an empty instance, then select the first one as root and the k following nodes as terminals. Finally, for each couple (u, v) of nodes, we add the arc (u, v) to the graph with probability p. The weight of the arc is chosen uniformly between 0 and a maximum chosen between 1 and 100. The probability p is the expected value of $\frac{m}{n^2}$. We built 10500 instances of 50 nodes, 5250 instances of 100 nodes and 1050 instances of 250 nodes. Table 4 summarises the results.

It is highlighted that ShP_1 is the fastest algorithm, followed by $Greedy_{FLAC}$ which is competitive, according to its mean and its low standard deviation.

The algorithms ShP_2 and CH_2 are slower than other algorithms, due to k distinct applications of the Dijkstra algorithm. Finally, note the high standard deviation of DuAs, due to the random part of that algorithm. Indeed, its execution depends on the order in which the terminals are given to the algorithm.

5   Conclusion and Perspectives

We introduced a fast greedy algorithm for the Directed Steiner Tree problem. When it is opposed to the fastest Shortest Paths algorithm mostly used in today's applications, it shows lower error rates while maintaining a fast running time.

Even so, this algorithm was implemented assuming a central processing unit exists. As the main application of the Directed Steiner Tree is the multicast tree in networks, it would be interesting to work on a distributed implementation, minimizing the number of exchanged messages, following the technique of [21].

Moreover, we built two new test sets of directed instances, in addition to the bidirected instances, in order to create hard instances for which some of the approximation algorithms would return high relative errors. However, the three test sets seems equally hard to solve. Further works are needed to build instances for which approximation algorithms could be efficiently compared and clustered.

Finally, in Sect. 2, to be competitive and simple to implement, FLAC avoids degenerate flows by removing the last saturated arc. This solution is not sufficient to reduce the approximation ratio of Greedy$_{\text{FLAC}}{}^{\triangleright}$ under $O(\sqrt{k})$, but other ways could be explored to achieve better theoretical performances.

References

1. Karp, R.M.: Reducibility Among Combinatorial Problems. Springer, New York (1972)
2. Kou, L., Markowsky, G., Berman, L.: A fast algorithm for steiner trees. Acta Inf. **15**(2), 141–145 (1981)
3. Zelikovsky, A.Z.: An 11/6-approximation algorithm for the network steiner problem. Algorithmica **9**(5), 463–470 (1993)
4. Byrka, J., Grandoni, F., Rothvoss, T., Sanità, L.: Steiner tree approximation via iterative randomized rounding. J. ACM (JACM) **60**(1), 6:1–6:33 (2013)
5. Cheng, X., Du, D.Z.: Steiner Trees in Industry, vol. 11. Springer, New York (2001)
6. Voß, S.: Steiner tree problems in telecommunications. In: Resende, M.G.C., Pardalos, P.M. (eds.) Handbook of Optimization in Telecommunications, pp. 459–492. Springer, New York (2006)
7. Novak, R., Rugelj, J., Kandus, G.: A note on distributed multicast routing in point-to-point networks. Comput. Oper. Res. **28**(12), 1149–1164 (2001)
8. Feige, U.: A threshold of ln n for approximating set cover. J. ACM (JACM) **45**(4), 634–652 (1998)
9. Halperin, E., Krauthgamer, R.: Polylogarithmic inapproximability. In: Proceedings of the Thirty-Fifth Annual ACM Symposium on Theory of Computing, pp. 585–594 (2003)
10. Charikar, M., Chekuri, C., Cheung, T.Y., Dai, Z., Goel, A., Guha, S., Li, M.: Approximation algorithms for directed steiner problems. J. Algorithms **33**(1), 73–91 (1999)
11. Helvig, C.S., Robins, G., Zelikovsky, A.: An improved approximation scheme for the group steiner problem. Networks **37**(1), 8–20 (2001)
12. Johnson, D.S.: Approximation algorithms for combinatorial problems. In: Proceedings of the Fifth Annual ACM Symposium on Theory of Computing, pp. 38–49 (1973)
13. Chvatal, V.: A greedy heuristic for the set-covering problem. Math. Oper. Res. **4**(3), 233–235 (1979)
14. Olsson, P.M., Kvarnstrom, J., Doherty, P., Burdakov, O., Holmberg, K.: Generating uav communication networks for monitoring and surveillance. In: 2010 11th International Conference on Control Automation Robotics & Vision (ICARCV), pp. 1070–1077. IEEE (2010)
15. Gundecha, P., Feng, Z., Liu, H.: Seeking provenance of information using social media. In: Proceedings of the 22nd ACM International Conference on Information & Knowledge Management, pp. 1691–1696. ACM (2013)
16. Lappas, T., Terzi, E., Gunopulos, D., Mannila, H.: Finding effectors in social networks. In: Proceedings of the 16th ACM SIGKDD International Conference on Knowledge Discovery and Data Mining, pp. 1059–1068. ACM (2010)
17. Koch, T., Martin, A., Voß, S.: SteinLib: an updated library on Steiner tree problems in graphs. In: Cheng, X.Z., Du, D.-Z. (eds.) Steiner Trees in Industry, pp. 285–325. Springer, New York (2001)
18. Chimani, M., Woste, M.: Contraction-based steiner tree approximations in practice. In: Asano, T., Nakano, S., Okamoto, Y., Watanabe, O. (eds.) ISAAC 2011. LNCS, vol. 7074, pp. 40–49. Springer, Heidelberg (2011)

19. Stanojevic, M., Vujosevic, M.: An exact algorithm for steiner tree problem on graphs. Int. J. Comput. Commun. Control **1**(1), 41–46 (2006)
20. Uchoa, E., Werneck, R.F.F.: Fast local search for steiner trees in graphs. In: ALENEX, vol. 10, pp. 1–10. SIAM (2010)
21. Drummond, L., Santos, M., Uchoa, E.: A distributed dual ascent algorithm for steiner problems in multicast routing. Networks **53**(2), 170–183 (2009)
22. Hsieh, M.I., Wu, E.H.K., Tsai, M.F.: Fasterdsp: a faster approximation algorithm for directed steiner tree problem. J. Inf. Sci. Eng. **22**, 1409–1425 (2006)
23. de Aragão, M.P., Uchoa, E., Werneck, R.F.: Dual heuristics on the exact solution of large steiner problems. Electron. Notes Discrete Math. **7**, 150–153 (2001)
24. Wong, R.T.: A dual ascent approach for steiner tree problems on a directed graph. Math. Program. **28**(3), 271–287 (1984)
25. Melkonian, V.: New primal-dual algorithms for steiner tree problems. Comput. Oper. Res. **34**(7), 2147–2167 (2007)
26. Zelikovsky, A.: A series of approximation algorithms for the acyclic directed steiner tree problem. Algorithmica **18**(1), 99–110 (1997)

Spanning Properties of Theta-Theta Graphs

Mirela Damian$^{(\boxtimes)}$ and Dumitru V. Voicu

Department of Computer Science, Villanova University, Villanova, PA 19085, USA
{mirela.damian,dvoicu}@villanova.edu

Abstract. We study the spanning properties of Theta-Theta graphs. Similar in spirit with the Yao-Yao graphs, Theta-Theta graphs partition the space around each vertex into a set of k cones, for some fixed integer $k > 1$, and select at most one edge per cone. The difference is in the way edges are selected. Yao-Yao graphs select an edge of minimum length, whereas Theta-Theta graphs select an edge of minimum orthogonal projection onto the cone bisector. It has been established that the Yao-Yao graphs with parameter $k = 6k'$ have spanning ratio 11.67, for $k' \geq 6$. In this paper we establish a first spanning ratio of 7.82 for Theta-Theta graphs, for the same values of k. We also extend the class of Theta-Theta spanners with parameter $6k'$, and establish a spanning ratio of 16.76 for $k' \geq 5$. We surmise that these stronger results are mainly due to a tighter analysis in this paper, rather than Theta-Theta being superior to Yao-Yao as a spanner. We also show that the spanning ratio of Theta-Theta graphs decreases to 4.64 as k' increases to 8. These are the first results on the spanning properties of Theta-Theta graphs.

Keywords: Yao graph · Theta graph · Yao-Yao · Theta-Theta · Spanner

1 Introduction

Let S be a set of n points in the plane, and let G be an undirected plane graph with vertex set S. The *length* of a path in G is the sum of the Euclidean lengths of its constituent edges. The distance in G between any two points $a, b \in S$ is the length of a shortest path between a and b. We say that G is a *spanner* if it preserves distances between each pair of points in S, up to a given factor. Specifically, for a fixed integer $t \geq 1$, we say that G is a *t-spanner* if any two points $a, b \in S$ at distance $|ab|$ in the plane are at distance at most $t \cdot |ab|$ in G. The smallest integer t for which this property holds is called the *spanning ratio* of G. Clearly there is a tradeoff between the spanning ratio and the sparsity of G: the smaller the spanning ratio, the denser the spanner and the better the approximation of the original distances.

One way to control the tradeoff between the spanning ratio and the sparsity of the spanner is to partition the space around each point into equiangular cones

This work was supported by NSF grant CCF-1218814.

Z. Zhang et al. (Eds.): COCOA 2014, LNCS 8881, pp. 216–230, 2014.
DOI: 10.1007/978-3-319-12691-3_17

of angle $\theta = 2\pi/k$, for some integer $k \geq 1$, and connect each point to a "nearest" point in each cone. Intuitively, this construction promises a short detour between any two points $a, b \in S$, by following the edge from a aiming in the direction of b (the one lying in the cone with apex a containing b). The definition of a "nearest" point comes in two flavors, in the context of Yao graphs [31] and $Theta$-graphs (or Θ-graphs) [14,22]. For Yao graphs, the "nearest" point is simply the point that minimizes the L_2-distance, whereas for Theta graphs, the "nearest" point in a cone C is the point whose orthogonal projection onto the bisector of C minimizes the L_2-distance. Both Yao and Theta graphs are parameterized by a positive integer $k \geq 1$, which controls the cone angle $\theta = 2\pi/k$. In the following we will refer to the Yao graph as Y_k and Theta graphs as Θ_k, for a fixed $k \geq 1$. Both Y_k and Θ_k are known to be efficient spanners, for $k \geq 6$. The spanning ratios of these graphs are summarized in Table 1.

Table 1. Spanning ratios of Yao and Theta graphs for various $\theta = 2\pi/k$ values.

Parameter k	Spanning Ratio			
	Y_k	Θ_k	YY_k	$\Theta\Theta_k$
< 4	∞ [25]			
4	696.1 [9]	237 [3]	∞ [17]	
5	3.74 [2]	9.96 [11]	OPEN	∞ [23]
6	5.8 [2]	2 [4]	∞ [25]	OPEN
$k > 6$		$\frac{1}{1-\sin(\theta/2)}$ [27]	11.67 for	16.76 for
$4k + 2$	$\frac{1}{1-2\sin(\theta/2)}$ [8]	$1 + 2\sin(\theta/2)$	$k = 6k'$ and	$k = 6k'$ and
$4k + 4$		$1 + \frac{2\sin(\theta/2)}{\cos(\theta/2)-\sin(\theta/2)}$ [12,7]	$k' \geq 6$	$k' \geq 5$
$4k + 3, 4k + 5$	$\frac{1}{1-2\sin(3\theta/8)}$ [2]	$\frac{\cos(\theta/4)}{\cos(\theta/2)-\sin(3\theta/4)}$	[16]	[HERE]

Interest in Yao and Theta graphs has increased with the advancement of wireless ad hoc networks and the need for efficient communication (see [15, 21, 26, 28] and the references therein). Designing routing algorithms for wireless ad hoc networks is an extremely difficult task and research in this area is still in progress. The overlay communication graph formed by the wireless links should be a spanner to ensure fast delivery of information, and should also have low degree to ensure a low maintenance cost and reduced MAC-level contention and interference [20]. We observe that both Yao and Theta graphs obey the first requirement (as detailed in Table 1), but fail to satisfy the second requirement. One simple example consists of $n - 1$ points equally distributed around a circle centered at an nth point p. Then, for $k \geq 6$, both Θ_k and Y_k will have an edge directed from each of the $n-1$ points towards p, because p is "nearest" in one of their cones. So each of Θ_k and Y_k has out-degree k, but in-degree $n-1$. To reduce the in-degree, alternate spanner structures based on Yao and Theta graphs have been proposed, such as Yao-Yao [30], Sink [1,24], Stable Roommates [6], and Ordered-Yao [29].

The *Yao-Yao* graph with integer parameter $k \geq 1$, denoted YY_k, is a subgraph of Y_k obtained by applying a second Yao step to the set of incoming edges in each cone. More precisely, for each point p and each cone with apex a containing two or more incoming edges, YY_k retains only a shortest incoming edge and discards the rest. Ties are broken arbitrarily. This construction guarantees a degree of at most $2k$ at each node in YY_k (one incoming and one outgoing edge per cone), however the spanning property of YY_k is still under investigation. The only existing result shows that $YY_{6k'}$, for $k' \geq 6$, is a spanner with spanning ratio 11.67. For $k' \geq 8$, the spanning ratio of $YY_{6k'}$ drops to 4.75 [16].

Sink spanners [1,24] transform bounded outdegree spanners, such as Y_k and Θ_k, into bounded degree spanners, by replacing each directed star consisting of all links directed into a point p and lying in a cone with apex p, by a tree of bounded degree with "sink" p. The result is a spanner with degree at most $k(k+2)$ and spanning ratio $1/(1 - 2\sin(\theta/2))^2$.

The *Stable Roommates* spanner introduced in [6] has degree at most k and spanning ratio matching the spanning ratio of Y_k, so this spanner combines both qualities – low spanning ratio and low degree – of the Yao and Yao-Yao graphs, respectively. The only drawback of this approach is that it processes pairs of points in non-decreasing order by their distances, making it unsuitable for a fast local implementation. (The authors present a distributed implementation that requires $O(n)$ rounds of communication.)

The *ordered* Theta approach [10] reduces the potentially linear degree of the Theta graph to a logarithmic degree. Similar to the stable roommates approach, the ordered Theta approach imposes a particular ordering on the input points. The authors show that careful orderings can produce graphs with spanning ratio $1/(\cos\theta - \sin\theta)$ and degree $O(k \log n)$.

Similar in spirit with the Yao-Yao graph, in this paper we introduce the *Theta-Theta* graph $\Theta\Theta_k$, parameterized by integer $k \geq 1$, and study the spanning properties of this graph. The graph $\Theta\Theta_k$ is obtained by applying a filtering step to the edges of Θ_k as follows. For each point p and each cone C with apex p, we consider all edges in C directed into p, and maintain only a "shortest" edge while discarding the rest. Recall that in the context of Theta graphs, a "shortest" edge minimizes the length of its projection on the cone bisector. Ties are arbitrarily broken.

Our main result shows that $\Theta\Theta_{6k'}$ is a spanner, for any $k' \geq 5$. This result relies on a result by Bonichon et al. [4], who prove that Θ_6 is a 2-spanner. Our main contribution is showing that $\Theta\Theta_{6k'}$ contains a short path between the endpoints of each edge in Θ_6. More precisely, we show that for each edge $ab \in \Theta_6$, there is a path between a and b in $\Theta\Theta_{6k'}$ no longer than $8.38|ab|$, for $k' \geq 5$. This, combined with the fact that Θ_6 is a 2-spanner, yields an upper bound of 16.76 on the spanning ratio of $\Theta\Theta_{6k'}$. A similar approach has been used in [16] to establish that $YY_{6k'}$ has spanning ratio 11.67, for $k' \geq 6$. We observe that the spanning ratio of $\Theta\Theta_{6k'}$ decreases to 7.82, 5.63 and 4.64 as k' increases to 6, 7, and above 8, respectively. The spanning ratios established in this paper for $\Theta\Theta_{6k'}$ are stronger than the ones obtained in [16] for $YY_{6k'}$, for the same parameter values $k' \geq 6$. We surmise that this is mainly due to the tighter analysis in this paper, rather than $\Theta\Theta_{6k'}$ being superior to $YY_{6k'}$ as a spanner.

1.1 Definitions

Throughout the paper, S will refer to a fixed set of n points in the plane. The directed Yao graph Y_k with integer parameter $k \geq 1$ on S is constructed as follows. For each point $a \in S$, starting with the direction of the positive x-axis, extend k equally spaced rays $r_1, r_2, \ldots, r_k$ originating at a, in counterclockwise order (see Fig. 1a for $k = 6$). These rays divide the plane into k cones, denoted by $C_{\{k,1\}}(a), C_{\{k,2\}}(a), \ldots, C_{\{k,k\}}(a)$, each of angle $\theta = 2\pi/k$. To avoid overlapping boundaries, we assume that each cone is half-open and half-closed, meaning that $C_{\{k,i\}}(a)$ includes r_i but excludes r_{i+1} (here $r_{k+1} \equiv r_1$ wraps around). In each cone of a, draw a directed edge from a to its "closest" point b in that cone (the one that minimizes the L_2-distance $|ab|$). Ties are broken arbitrarily. These directed edges collectively form the edge set for the directed Yao graph. The undirected Yao graph (or simply Yao graph) on S is obtained by simply ignoring the directions of these edges. The Theta graph Θ_k is defined in a similar way, with the only difference being in the definition of "closest": in each cone C with apex a, draw a directed edge from a to the point b that minimizes the distance between a and the orthogonal projection of b on the bisector of the cone. For example, looking at the cone $C_{\{6,1\}}(a)$ in Fig. 1b, notice that b_1 minimizes the L_2-distance to a, whereas b_2 minimizes the L_2-distance between its projection onto the cone bisector and a. Consequently, $\overrightarrow{ab_1}$ will be added to Y_6, and $\overrightarrow{ab_2}$ to Θ_6. Similarly, $\overrightarrow{ad_1} \in C_{\{6,3\}}(a)$ will be added to Y_6, and $\overrightarrow{ad_2}$ to Θ_6. Figure 2a shows the Yao graph Y_6 for the point set depicted in Fig. 1b, and Fig. 2c shows the Theta graph Θ_6 for the same point set.

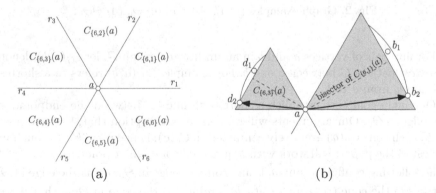

(a)   (b)

Fig. 1. Definitions. (a) Rays defining the cones at point a. (b) Theta edges ab_2, ad_2.

The Yao-Yao graph $YY_k \subseteq Y_k$ is obtained from Y_k by applying a reverse Yao step to the set of incoming Yao edges in Y_k. That is, for each node a and each cone with apex a containing two or more incoming edges, YY_k retains a shortest incoming edge and discards the rest. Ties are broken arbitrarily. The Theta-Theta graph $\Theta\Theta_k \subseteq \Theta_k$ is obtained from Θ_k in a similar way, with the only difference being in the requirement that a "shortest" incoming edge in a cone minimizes the length of its projection onto the cone bisector. Figure 2b shows the graph YY_6 derived from the graph Y_6 depicted in Fig. 2a, and Fig. 2d shows the graph $\Theta\Theta_6$ derived from the graph Θ_6 depicted in Fig. 2c.

When the choice of a particular cone is either irrelevant or is clear from the context, we ignore the cone subscript and use $C_k(a)$ to denote any of the cones $C_{\{k,1\}}(a), C_{\{k,2\}}(a), \ldots C_{\{k,k\}}(a)$. For any two points $a, b \in S$, let $C_k(a, b)$ denote the cone with apex a that contains b. Let $\triangle_k(a, b)$ be the canonical triangle with two of its sides along the rays bounding $C_k(a, b)$, and the third side orthogonal to the bisector of $C_k(a, b)$ and passing through b. For example, shaded in Fig. 1b are the canonical triangles $\triangle_6(a, b_2)$ and $\triangle_6(a, d_2)$.

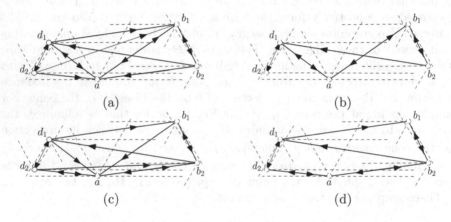

Fig. 2. Graph examples (a) Y_6 (b) YY_6 (c) Θ_6 (d) $\Theta\Theta_6$.

For any pair of vertices a and b in an undirected graph G, let $\xi_G(a, b)$ denote a *shortest* path in G between a and b. For example, $\xi_{\Theta_6}(a, b)$ refers to a shortest path in Θ_6 from a to b.

Our main goal is to establish a short path in $\Theta\Theta_k$ between the endpoints of each edge in Θ_6. Our arguments will rely on the assumption that, for each point $a \in S$, each cone $C_k(a)$ is entirely contained in $C_6(a)$, hence $k = 6k'$. Throughout the rest of the paper, will work with a quadruple of distinct points $a, b, b', a' \in S$ in the following configuration: $\overrightarrow{ab}$ is an arbitrary edge in Θ_6; $\overrightarrow{ab'}$ is the edge in Θ_k that lies in the cone $C_k(a, b) \subset C_6(a, b)$; and $\overrightarrow{a'b'}$ is the edge in $\Theta\Theta_k$ that lies in the cone $C_k(b', a) \subset C_6(b', a)$. We will refer to this configuration as a *canonical Θ-configuration*, to avoid repeating these definitions in different contexts. For a snapshot of a canonical Θ-configuration, see ahead to Fig. 4a. We will further assume, without loss of generality, that in a canonical Θ-configuration $\overrightarrow{ab}$ lies in $C_{\{6,1\}}(a)$, and the bisector of $C_k(a, b)$ lies below, or aligns with, the bisector of

$\triangle_6(a, b)$. Any other configuration is equivalent to this canonical Θ-configuration under rotational and/or reflectional symmetry.

2   Preliminaries

In this section we present a few isolated lemmas that will be used in our main proof from Sect. 3. Due to space constraints, most of these lemmas are stated without proof. Please refer to [19] for complete proofs of these lemmas. We encourage the reader to skip ahead to Sect. 3, and refer back to these lemmas from the context of Theorem 2, where their role will become evident. We begin this section with the statement of an existing result.

Theorem 1. [4] *For any pair of points $a, b \in S$, there is a path in Θ_6 whose total length is bounded above by $2|ab|$.*

The key ingredient in the result of Theorem 1 is a specific subgraph of Θ_6, called *half-Θ_6*. This graph preserves half of the edges in Θ_6, those belonging to non-consecutive cones. Bonichon et al. [4] show that half-Θ_6 is a *triangular-distance*[1] Delaunay triangulation, computed as the dual of the Voronoi diagram based on the triangular distance function. Combined with Chew's proof that any triangular-distance Delaunay triangulation is a 2-spanner [13], this result settles Theorem 1. The structure of Θ_6, viewed as the union of two planar 2-spanners, has been used in establishing spanning properties of other graphs as well [5, 16, 18].

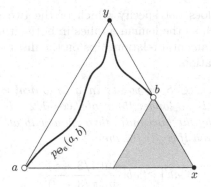

Fig. 3. Lemma 1: $|\xi_{\Theta_6}(a, b)| \leq |ay| + |by|$.

Before stating our preliminary results, we define the term $T(\alpha)$ parameterized by angle $\alpha \in [0, \pi/3]$ as

$$T(\alpha) = \frac{\sin(\pi/3 - \alpha) - \sin\alpha}{\sin(\pi/3)} \leq 1 \qquad (1)$$

---

[1] The *triangular distance* from a point a to a point b is the side length of the smallest equilateral triangle centered at a that touches b and has one horizontal side.

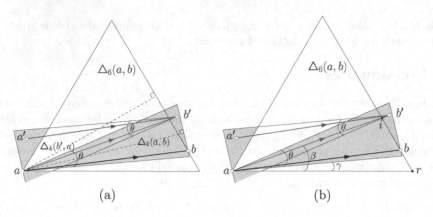

Fig. 4. (a) Canonical Θ-configuration: $ab \in \Theta_6$, $ab' \in \Theta_k$ and $a'b' \in \Theta\Theta_k$ (b) bounding $|ab'|$, $|a'b'|$ and $|bb'|$.

This term will occur frequently in our analysis, and this definition will come in handy. The upper bound of 1 follows from the fact that $T(\alpha)$ decreases as α increases, therefore $T(\alpha) \leq T(0) = 1$. The following lemma plays a central role in the proofs of Lemmas 3 and 4.

Lemma 1. [16] *Let $a, b \in S$ and let x and y be the other two vertices of $\triangle_6(a, b)$. If $\triangle_6(b, x)$ is empty of points in S, then $|\xi_{\Theta_6}(a, b)| \leq |ay| + |by|$. Moreover, each edge of $\xi_{\Theta_6}(a, b)$ is no longer than $|ay|$.* [Refer to Fig. 3.]

Note that Lemma 1 does not specify which of the two sides ax and ay lies clockwise from $\triangle_6(a, b)$, so the lemma applies in both situations. The following lemma establishes fundamental relationships on the distances between points in a canonical Θ-configuration.

Lemma 2. *Let $a, b, b', a' \in S$ be points in a canonical Θ-configuration. Then each of $|ab'|$ and $|a'b'|$ is no longer than $|ab|/\cos(\theta/2)$. In addition, if β and γ are the angles formed by the horizontal through a with ab' and the lower ray of $C_k(a, b)$, respectively, and if $\beta \leq \pi/6$, then*

$$|ab'| \geq |ab| \frac{\sin(\pi/3 + \gamma)}{\sin(\pi/3 + \beta)}$$

[Refer to Fig. 4b.]

Lemmas 3 through 5 isolate specific situations that will arise in the analysis of our main result. We state them independently in this section.

Lemma 3. *Let $a, b, b', a' \in S$ be points in a canonical Θ-configuration, with the additional constraint that $a' \in C_{\{6,2\}}(a)$. Let β and γ be the angles formed by the horizontal through a with ab' and the lower ray of $C_k(a, b)$, respectively. Then*

$$|\xi_{\Theta_6}(a, a')| + |\xi_{\Theta_6}(b, b')| \leq (|ab| + |a'b'|) \cdot T(\gamma) - 2|ab'| \cdot T(\beta)$$

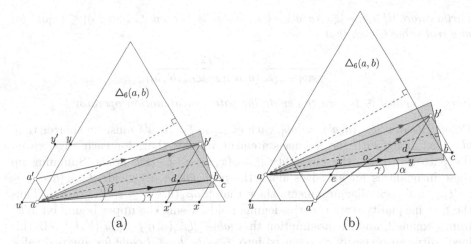

Fig. 5. Bounding $|\xi_{\Theta_6}(a, a')| + |\xi_{\Theta_6}(b, b')|$ (a) Lemma 3: a' above a (b) Lemma 4: a' below a.

Here the term T is as defined in (1). Furthermore, each edge of $\xi_{\Theta_6}(a, a')$ and $\xi_{\Theta_6}(b, b')$ is strictly smaller than ab, for $\theta \leq \pi/6$. [Refer to Fig. 5a.]

Lemma 4. *Let $a, b, b', a' \in S$ be points in a canonical Θ-configuration, with the additional constraints that $a' \in C_{\{6,6\}}(a)$, and the angle α formed by ab with the horizontal through a is at most $\pi/6$. Then*

$$|\xi_{\Theta_6}(a, a')| + |\xi_{\Theta_6}(b, b')| \leq |ab| - |a'b'| \cdot \frac{\sin(\pi/3 - \alpha - \theta) - \sin\theta}{\sin(\pi/3 - \alpha)}$$

Furthermore, each edge of $\xi_{\Theta_6}(a, a')$ and $\xi_{\Theta_6}(b, b')$ is strictly shorter than ab, for $\theta \leq \pi/12$. [Refer to Fig. 5b.]

Lemma 5. *Let $a, b, b', a' \in S$ be points in a canonical Θ-configuration, with the additional constraint that either $a' \in C_{\{6,5\}}(a)$, or $a' \in C_{\{6,6\}}(a)$ and the angle formed by ab with the horizontal through a is above $\pi/6$. Then*

$$|\xi_{\Theta_6}(a, a')| + |\xi_{\Theta_6}(b, b')| \leq 8|ab| \sin(\theta/2)$$

Furthermore, each edge of $\xi_{\Theta_6}(a, a')$ and $\xi_{\Theta_6}(b, b')$ is strictly shorter than ab, for $\theta \leq \pi/15$.

Our approach to finding a short path in $\Theta\Theta_k$ between the endpoints of each edge in Θ_6 uses induction on the Euclidean lengths of the edges in Θ_6. The following lemma will be useful in proving the inductive step in various situations.

Lemma 6. *Let $a, b, b', a' \in S$ be points in a canonical Θ-configuration, and let $t \geq 1$ be a fixed real value. Assume that, for each edge $xy \in \Theta_6$ no longer than ab, the inequality $|\xi_{\Theta\Theta_k}(x, y)| \leq t \cdot |xy|$ holds. Let $\xi_{\Theta\Theta_k}(a, b) = \xi_{\Theta\Theta_k}(a, a') \oplus a'b' \oplus \xi_{\Theta\Theta_k}(b', b)$. If $|\xi_{\Theta_6}(a, a')| < |ab|$ and $|\xi_{\Theta_6}(b, b')| < |ab|$, then*

$$|\xi_{\Theta\Theta_k}(a, b)| \leq t \cdot |\xi_{\Theta_6}(a, a')| + t \cdot |\xi_{\Theta_6}(b', b)| + |a'b'|$$

Furthermore, if $|ab| - |\xi_{\Theta_6}(a, a')| - |\xi_{\Theta_6}(b', b)| > 0$, *then* $|\xi_{\Theta\Theta_k}(a, b)| \leq t \cdot |ab|$ *for any real value t such that*

$$t \geq \frac{|ab|/\cos(\theta/2)}{|ab| - |\xi_{\Theta_6}(a, a')| - |\xi_{\Theta_6}(b', b)|}. \tag{2}$$

Here the symbol $\oplus$ *is used to denote the path concatenation operator.*

Proof. Because $|\xi_{\Theta_6}(a, a')| < |ab|$, each edge on $\xi_{\Theta_6}(a, a')$ must be shorter than ab. This along with the lemma statement implies that, for each edge xy on the path $\xi_{\Theta_6}(a, a')$, the inequality $|\xi_{\Theta\Theta_k}(x, y)| \leq t \cdot |xy|$ holds. Summing up these inequalities for all edges along the path $\xi_{\Theta_6}(a, a')$ yields $|\xi_{\Theta\Theta_k}(a, a')| \leq t \cdot |\xi_{\Theta_6}(a, a')|$. Similar arguments show that $|\xi_{\Theta\Theta_k}(b, b')| \leq t \cdot |\xi_{\Theta_6}(b, b')|$. Thus the first inequality stated by this lemma holds. Using the upper bound on $|a'b'|$ from Lemma 2, and the assumption that $|ab| - |\xi_{\Theta_6}(a, a')| - |\xi_{\Theta_6}(b', b)| > 0$, this inequality can be easily reorganized into $|\xi_{\Theta\Theta_k}(a, b)| \leq t \cdot |ab|$ for any real value t that satisfies (2). $\square$

3 $\Theta\Theta_{6k'}$ is a Spanner, for $k' \geq 4$

This section presents our main result, which shows that $\Theta\Theta_k$ is a spanner, provided that $k = 6k'$ and $k' \geq 5$ (and so $\theta \leq \pi/15$). In particular, we show that for each edge $ab \in \Theta_6$, there is a path in $\Theta\Theta_{6k'}$ no longer than $8.38|ab|$. This, combined with the result of Theorem 1, yields our main result that $\Theta\Theta_{6k'}$ is a 16.76-spanner, for $k' \geq 5$. The spanning ratio decreases to 7.82 for $k' \geq 6$, which is superior to the spanning ratio of 11.67 established in [18] for $YY_{6k'}$, with $k' \geq 6$. We also show that the spanning ratio of $\Theta\Theta_{6k'}$ drops to 4.64 for $k' \geq 8$.

   Our approach takes advantage of the fact that each edge $ab \in \Theta_6$ is embedded in an equilateral triangle $\triangle_6(a, b)$ empty of points in S. The restriction $k = 6k'$ is necessary in our analysis to guarantee that each cone used in constructing Θ_k and $\Theta\Theta_k$ is a subset of a cone used in constructing Θ_6, therefore it inherits a large area empty of points in S. This property is crucial in establishing a "short" path in $\Theta\Theta_k$ between the endpoints of each edge in Θ_6. Although we search for *undirected* paths in the undirected version of $\Theta\Theta_k$, we sometimes point out the direction of an edge if significant in the context.

Theorem 2. *Let* $k = 6k'$ *be a positive integer, with* $k' \geq 5$. *For each edge* $\overrightarrow{ab} \in \Theta_6$, *a shortest path in* $\Theta\Theta_k$ *between* a *and* b *satisfies* $|\xi_{\Theta\Theta_k}(a, b)| \leq t \cdot |ab|$, *where* t *is a positive real with values* 8.38, 3.91, 2.811 *and* 2.32 *corresponding to* k' *values* 5, 6, 7, *and above* 8, *respectively.*

Proof. Recall that $\theta = 2\pi/k$, so in the context of this theorem $\theta \leq \pi/15$. Throughout this proof will refer to the value t from the theorem statement as the *stretch* factor, with the understanding that it measures the "stretch" in $\Theta\Theta_k$ of an edge $ab \in \Theta_6$, and to be distinguished from the spanning ratio of $\Theta\Theta_k$ (which by Theorem 1 is at most $2t$).

The proof is by induction on the Euclidean length of the edges in Θ_6. The base case corresponds to a shortest edge $\overrightarrow{ab} \in \Theta_6$. In this case we show that $\overrightarrow{ab} \in \Theta_k$ and $\overrightarrow{ab} \in \Theta\Theta_k$. Assume to the contrary that $\overrightarrow{ab} \notin \Theta_k$ and let $\overrightarrow{ab'} \in \Theta_k$ be the edge that lies in $C_k(a,b)$. Lemma 3 does not impose any restrictions on the relative position of the b and b', therefore the result that each edge on $\xi_{\Theta_6}(b,b')$ is strictly shorter than ab applies in this context. This contradicts our assumption that ab is a shortest edge in Θ_6. This shows that $\overrightarrow{ab} \in \Theta_k$. Similar arguments, used in conjunction with Lemmas 3, 4 and 5 (which distinguish between different locations of a' relative to a), show that $\overrightarrow{ab} \in \Theta\Theta_k$.

Our inductive hypothesis states that the theorem holds for all edges in Θ_6 of length strictly lower than some fixed value $\delta > 0$. To prove the inductive step, pick a shortest edge $\overrightarrow{ab} \in \Theta_6$ of length δ or higher, and find a "short" path $\xi_{\Theta\Theta_k}(a,b)$ that satisfies the conditions of the theorem. Let a' and b' be the other two points in S which, along with a and b, complete a canonical Θ-configuration: $\overrightarrow{ab'} \in \Theta_k$ lies in $C_k(a,b)$, and $\overrightarrow{a'b'} \in \Theta\Theta_k$ lies in $C_k(b',a)$. Refer to Fig. 4a. Also recall our general assumptions that in a canonical Θ-configuration $ab \in C_{\{6,1\}}(a)$, and the bisector of $C_k(a,b)$ aligns with, or lies below, the bisector of $\triangle_6(a,b)$. The locus of b' is $\triangle_k(a,b) \setminus \triangle_6(a,b)$, which is an area completely inside $C_{\{6,1\}}(a)$. The locus of a' is $\triangle_k(b',a) \setminus \triangle_6(a,b)$, which is an area that may overlap two or three of the cones $C_{\{6,2\}}(a)$, $C_{\{6,5\}}(a)$ and $C_{\{6,6\}}(a)$. Note that a' may not lie in $C_{\{6,3\}}(a)$, due to our assumption that the bisector of $\triangle_k(a,b)$ is no higher than the bisector of $\triangle_6(a,b)$.

Our intent is to use the result of Lemma 6 to establish the existence of a path between a and b of length at most $t \cdot |ab|$, for some fixed real constant $t > 1$. The two key ingredients needed by Lemma 6 are "short" paths in Θ_6 between a and a', and between b and b'. We discuss three cases, depending on whether a' lies in $C_{\{6,2\}}(a)$, $C_{\{6,5\}}(a)$ or $C_{\{6,6\}}(a)$. The case $a' \in C_{\{6,5\}}(a)$ is the simplest, so we will save it for last. Let α, β and γ be the angles formed by the horizontal through a with ab, ab', and the lower ray of $C_k(a,b)$, respectively.

Case $a' \in C_{\{6,2\}}(a)$. This case is depicted in Fig. 5a. By Lemma 3, we have

$$|\xi_{\Theta_6}(a,a')| + |\xi_{\Theta_6}(b,b')| \leq (|ab| + |a'b'|) \cdot T(\gamma) - 2|ab'| \cdot T(\beta) \qquad (3)$$

where T is as defined in (1). Notice the restrictions on the angles β and γ:

$$0 \leq \gamma \leq \pi/6 - \theta/2$$
$$\gamma \leq \beta \leq \gamma + \theta \qquad (4)$$

The upper bound on γ is due to our assumption that the bisector of $\triangle_k(a,b)$ is no higher than the bisector of $\triangle_6(a,b)$. The bounds on β follow immediately from the definitions of γ and β. Next we determine a maximum for the quantity on the right hand side of (3). We consider two situations, depending on ranges of β, which affect the sign of $T(\beta)$. Observe that $T(\gamma)$ is always positive, since $\pi/3 - \gamma > \gamma$ for any $\gamma < \pi/6$.

Assume first that $\beta \leq \pi/6$, so ab' is no higher than the bisector of $\triangle_6(a, b)$. In this case $\beta \leq \pi/3 - \beta$ and $\sin \beta \leq \sin(\pi/3 - \beta)$, therefore $T(\beta)$ is positive. Substituting in (3) the upper bound on $|a'b'|$ and the lower bound on $|ab'|$ from Lemma 2 yields

$$\frac{|\xi_{\Theta_6}(a, a')| + |\xi_{\Theta_6}(b, b')|}{|ab|} \leq T(\gamma) + \frac{T(\gamma)}{\cos(\theta/2)} - 2T(\beta) \cdot \frac{\sin(\pi/3 + \gamma)}{\sin(\pi/3 + \beta)}$$

Let $X(\theta, \gamma, \beta)$ denote the quantity on the right hand side of the inequality above. Note that $X(\theta, \gamma, \beta)$ increases as θ increases, therefore $X(\theta, \gamma, \beta) \leq X(\pi/15, \gamma, \beta)$ for $\theta \leq \pi/15$. Figure 6a shows how $X(\theta, \gamma, \beta)$ varies with $\gamma \in [0, \pi/6 - \theta/2]$ and $\beta \in (\gamma, \min\{\pi/6, \gamma + \theta\}]$, for fixed $\theta = \pi/15$. It can be verified that $|\xi_{\Theta_6}(a, a')| + |\xi_{\Theta_6}(b, b')| < 0.88|ab|$, for any $0 < \theta \leq \pi/15$. This along with Lemma 6 yields a stretch factor $t = 8.3760$ for the path in $\Theta\Theta_k$ between a and b. The stretch factor t decreases with θ as shown in the second column of Table 2.

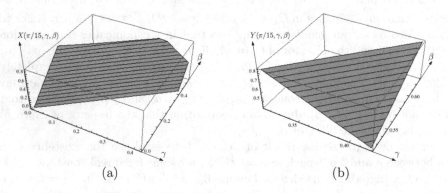

Fig. 6. Case $a' \in C_{\{6,2\}}(a)$: upper bound on $|\xi_{\Theta_6}(a, a')| + |\xi_{\Theta_6}(b, b')|$ for $\theta = \pi/15$ and $\gamma \in [0, \pi/6 - \theta/2]$ (a) $\alpha \in (\gamma, \min\{\pi/6, \gamma + \theta\}]$ (b) $\alpha \in (\pi/6, \min\{\pi/6 + \theta/2, \gamma + \theta\}]$

Table 2. Case $a' \in C_{\{6,2\}}(a)$: real constant t from Lemma 6 for various θ values.

θ	Case $a' \in C_{\{6,2\}}(a)$: stretch factor t from Lemma 6	
	$0 < \beta \leq \pi/6$	$\pi/6 < \beta \leq \pi/6 + \theta/2$
$\pi/15$	8.3760	6.2720
$\pi/18$	3.9058	3.3377
$\pi/21$	2.8109	2.5014
$\pi/24$	2.3159	2.1057

Assume now that $\pi/6 < \beta \leq \pi/6 + \theta/2$, so ab' lies above the bisector of $\triangle_k(a, b)$. In this case $T(\beta)$ is negative, and by (4) we have $\gamma \geq \pi/6 - \theta$. Substituting in (3) the upper bound on $|ab'|$ and $|a'b'|$ from Lemma 2 yields

$$\frac{|\xi_{\Theta_6}(a, a')| + |\xi_{\Theta_6}(b, b')|}{|ab|} \leq T(\gamma) + \frac{T(\gamma) - 2T(\beta)}{\cos(\theta/2)} \tag{5}$$

Let $Y(\theta, \gamma, \beta)$ denote the quantity on the right hand side of the inequality above. Because $T(\gamma)$ is positive and $T(\beta)$ is negative, $T(\gamma) - 2T(\beta)$ is positive and therefore $Y(\theta, \gamma, \beta)$ increases as θ increases. It follows that $Y(\theta, \gamma, \beta) \leq Y(\pi/15, \gamma, \beta)$ for $\theta \leq \pi/15$. Figure 6b shows how $Y(\theta, \gamma, \beta)$ varies with $\gamma \in [\pi/6 - \theta, \pi/6 - \theta/2]$ and $\beta \in (\pi/6, \gamma + \theta]$, for $\theta = \pi/15$. It can be verified that $|\xi_{\Theta_6}(a, a')| + |\xi_{\Theta_6}(b, b')| < 0.8397|ab|$, for any $0 < \theta \leq \pi/15$. This along with Lemma 6 yields a stretch factor $t = 6.2720$ for the path in $\Theta\Theta_k$ between a and b. The stretch factor t decreases with θ as shown in the third column of Table 2.

Case $a' \in C_{\{6,6\}}$. This case is depicted in Fig. 5b. We discuss two situations, depending on whether ab lies above or below the bisector of $\triangle_6(a, b)$. Assume first that ab is no higher than the bisector of $\triangle_6(a, b)$, so $\alpha \leq \pi/6$. Thus we are in the context of Lemma 4, which gives us an upper bound $|\xi_{\Theta_6}(a, a')| + |\xi_{\Theta_6}(b, b')| \leq |ab| - |a'b'| \cdot Z(\theta, \alpha)$, where

$$Z(\theta, \alpha) = \frac{\sin(\pi/3 - \theta - \alpha) - \sin\theta}{\sin(\pi/3 - \alpha)}$$

Note that $Z(\theta, \alpha)$ decreases as θ increases, therefore $Z(\theta, \alpha) \geq Z(\pi/15, \alpha)$ for any $\theta \leq \pi/15$. It can be verified that $Z(\theta, \alpha) \geq m = 0.2022$, for any $\theta \leq \pi/15$. By Lemma 6, we have

$$\xi_{\Theta\Theta_k}(a, b) \leq t|ab| - t|a'b'| \cdot Z(\theta, \alpha) + |a'b'|$$

Simple calculations show that the right hand side of the inequality above does not exceed $t|ab|$ for any $t \geq 4.945 \geq 1/m$. This bound decreases with θ as shown in the second column of Table 3.

Assume now that ab lies above the bisector of $\triangle_6(a, b)$, so $\alpha > \pi/6$. Intuitively, this forces a and a' to lie close to each other (for sufficiently small θ values), and similarly for b and b', so we can work with somewhat looser upper bounds without exceeding the spanning ratio established so far. Our context matches the context of Lemma 5, which tells us that $|\xi_{\Theta_6}(a, a')| + |\xi_{\Theta_6}(b, b')| \leq X(\theta) = 8|ab|\sin(\theta/2)$. The bound $X(\theta)$ increases with θ, therefore $X(\theta) \leq X(\pi/15) \leq 0.8363$. This together with Lemma 6 yields $|\xi_{\Theta\Theta_k}(a, b)| < t \cdot |ab|$ for any $t \geq 6.1397$. This bound decreases with θ as shown in the third column of Table 3.

Case $a' \in C_{\{6,5\}}$. The bound on $|\xi_{\Theta_6}(a, a')| + |\xi_{\Theta_6}(b, b')|$ provided by Lemma 5 applies here as well, therefore the analysis for this case is identical to the one for the previous case (with $b' \in C_{\{6,6\}}$ and ab above the bisector of $\triangle_6(a, b)$), yielding the spanning ratios listed in the third column of Table 3.

To derive the results listed in Tables 2 and 3, we worked with a quadruplet of *distinct* points a, b, a', b' in a Θ-configuration. The cases where a and a' coincide, or b and b' coincide, are special instances of this general case and yield lower stretch factors. The results listed in Tables 2 and 3 indicate that the stretch factor is highest when a' lies above a and ab' is below the bisector of $\triangle_6(a, b)$. The largest stretch factor value is $t = 8.376$ for $\theta = \pi/15$, and it drops to 3.91,

Table 3. Case $a' \in C_{\{6,6\}}(a)$: real constant t from Lemma 6 for various θ values.

θ	Case $a' \in C_{\{6,6\}}(a)$: stretch factor t from Lemma 6	
	$0 \le \alpha \le \pi/6$	$\pi/6 < \alpha \le \pi/6 + \theta/2$
$\pi/15$	4.9454	6.1397
$\pi/18$	2.9697	3.3157
$\pi/21$	2.3117	2.4936
$\pi/24$	1.9829	2.1020

2.82 and 2.32 for θ values $\pi/18$, $\pi/21$ and $\pi/24$, respectively. This completes the proof.   □

Combined with the result of Theorem 1, the result of Theorem 2 yields the main result of this paper, stated by Theorem 3 below.

Theorem 3. *The graph $\Theta\Theta_k$, with $k = 6k'$ and $k' \ge 5$, is a 16.76-spanner. The spanning ratio decreases to 7.82, 5.63 and 4.64 as k' increases to 6, 7, and above 8, respectively.*

4   Conclusions

In this paper we present the first results on the spanning property of $\Theta\Theta_k$-graphs. We show that, for any integer $k' \ge 5$, the graph $\Theta\Theta_{6k'}$ is a spanner with spanning ratio 16.76. The spanning ratio drops to 7.82 for $k' \ge 6$, which is superior to the spanning ratio of 11.67 established in [18] for $YY_{6k'}$, with $k' \ge 6$. The framework of our analysis seems inadequate to handle all graphs $\Theta\Theta_k$, for all $k > 6$, because it relies on the fact that each cone used in constructing $\Theta\Theta_k$ is a subset of a cone used in constructing Θ_6. It is unclear whether a fundamentally new technique is required to handle all $\Theta\Theta_k$ graphs, for $k \ge 6$. Proving or disproving that these graphs are spanners remains the main open problem in this area.

References

1. Arya, S., Das, G., Mount, D.M., Salowe, J.S., Smid, M.: Euclidean spanners: short, thin, and lanky. In: STOC '95: Proceedings of the 27th Annual ACM Symposium on Theory of Computing, pp. 489–498. ACM, New York (1995)
2. Barba, L., Bose, P., Damian, M., Fagerberg, R., Keng, W.L., O'Rourke, J., van Renssen, A., Taslakian, P., Verdonschot, S., Xia, G.: New and improved spanning ratios for Yao graphs. In: Proceedings of the 30th Annual Symposium on Computational Geometry, SOCG'14, pp. 30–39. ACM, New York (2014)
3. Barba, L., Bose, P., De Carufel, J.-L., van Renssen, A., Verdonschot, S.: On the stretch factor of the Theta-4 graph. In: Dehne, F., Solis-Oba, R., Sack, J.-R. (eds.) WADS 2013. LNCS, vol. 8037, pp. 109–120. Springer, Heidelberg (2013)

4. Bonichon, N., Gavoille, C., Hanusse, N., Ilcinkas, D.: Connections between Theta-graphs, Delaunay triangulations, and orthogonal surfaces. In: Thilikos, D.M. (ed.) WG 2010. LNCS, vol. 6410, pp. 266–278. Springer, Heidelberg (2010)

5. Bonichon, N., Gavoille, C., Hanusse, N., Perković, L.: Plane spanners of maximum degree six. In: Abramsky, S., Gavoille, C., Kirchner, C., Meyer auf der Heide, F., Spirakis, P.G. (eds.) ICALP 2010. LNCS, vol. 6198, pp. 19–30. Springer, Heidelberg (2010)

6. Bose, P., Carmi, P., Chaitman, L., Collette, S., Katz, M.J., Langerman, S.: Stable roommates spanner. Comput. Geom. Theory Appl. **46**(2), 120–130 (2013). (Special issue of selected papers from the 22nd Canadian Conference on Computational Geometry, CCCG'10)

7. Bose, P., De Carufel, J.-L., Morin, P., van Renssen, A., Verdonschot, S.: Towards tight bounds on Theta-graphs. CoRR abs/1404.6233 (2014)

8. Bose, P., Damian, M., Douïeb, K., O'Rourke, J., Seamone, B., Smid, M.H.M., Wuhrer, S.: Pi/2-angle Yao graphs are spanners. CoRR, abs/1001.2913 (2010)

9. Bose, P., Damian, M., Douïeb, K., O'Rourke, J., Seamone, B., Smid, M.H.M., Wuhrer, S.: Pi/2-angle Yao graphs are spanners. Int. J. Comput. Geom. Appl. **22**(1), 61–82 (2012)

10. Bose, P., Gudmundsson, J., Morin, P.: Ordered Theta graphs. Comput. Geom. Theory Appl. **28**(1), 11–18 (2004)

11. Bose, P., Morin, P., van Renssen, A., Verdonschot, S.: The Theta-5 graph is a spanner. CoRR abs/1212.0570 (2014) (To appear in Computational Geometry: Theory and Applications)

12. Bose, P., van Renssen, A., Verdonschot, S.: On the spanning ratio of Theta-graphs. In: Dehne, F., Solis-Oba, R., Sack, J.-R. (eds.) WADS 2013. LNCS, vol. 8037, pp. 182–194. Springer, Heidelberg (2013)

13. Chew, L.P.: There are planar graphs almost as good as the complete graph. J. Comput. Syst. Sci. **39**(2), 205–219 (1989)

14. Clarkson, K.L.: Approximation algorithms for shortest path motion planning. In: Proceedings of the 19th Annual ACM Conference on Theory of Computing, STOC'87, pp. 56–65 (1987)

15. Cowen, L.J., Wagner, C.G.: Compact roundtrip routing in directed networks (extended abstract). In: Proceedings of the 19th Annual ACM Symposium on Principles of Distributed Computing, PODC '00, pp. 51–59. ACM, New York (2000)

16. Damian, M., Bauer, M.: An infinite class of sparse-Yao spanners. In: Proceedings of the 24th ACM-SIAM Symposium on Discrete Algorithms, SODA'13, pp. 184–196, 6–8 January 2013

17. Damian, M., Molla, N., Pinciu, V.: Spanner properties of $\pi/2$-angle Yao graphs. In: Proceedings of the 25th European Workshop on Computational Geometry, pp. 21–24, March 2009

18. Damian, M., Raudonis, K.: Yao graphs span Theta graphs. Discret. Math. Algorithms Appl. **4**(2), 181–194 (2012)

19. Damian, M., Voicu, D.M.: Spanning properties of theta-theta graphs. CoRR abs/1407.3507 (2014)

20. Hamdaoui, B., Ramanathan, P.: Energy efficient and MAC-aware routing for data aggregation in sensor networks. Sensor Network Operations, pp. 291–308 (2006)

21. Kanj, I.A., Perkovic, L., Xia, G.: Local construction of near-optimal power spanners for wireless ad hoc networks. IEEE Trans. Mob. Comput. **8**(4), 460–474 (2009)

22. Keil, J.M.: Approximating the complete Euclidean graph. In: Karlsson, R., Lingas, A. (eds.) SWAT'88. LNCS, vol. 318, pp. 208–213. Springer, Heidelberg (1988)

23. Keng, W.L., Xia, G.: The Yao graph y_5 is a spanner. CoRR abs/1307.5030 (2013)
24. Li, M., Wan, P.-J., Wang, Y.: Power efficient and sparse spanner for wireless ad hoc networks. In: Proceedings of the 10th International Conference on Computer Communications and Networks, pp. 564–567 (2001)
25. Molla, N.: Yao spanners for wireless ad hoc networks. Technical report, M.S. Thesis, Department of Computer Science, Villanova University, December 2009
26. Roditty, I., Thorup, M., Zwick, U.: Roundtrip spanners and roundtrip routing in directed graphs. ACM Trans. Algorithms 4(3), 1–17 (2008)
27. Ruppert, J., Seidel, R.: Approximating the d-dimensional complete Euclidean graph. In: Proceedings of the 3rd Canadian Conference on Computational Geometry, CCCG'91, pp. 207–210 (1991)
28. Scheideler, C.: Overlay networks for wireless systems. New Topics in Theoretical Computer Science, pp. 213–251 (2008)
29. Song, W.-Z., Wang, Y., Li, X.-Y., Frieder, O.: Localized algorithms for energy efficient topology in wireless ad hoc networks. In: Proceedings of the 5th ACM International Symposium on Mobile Ad Hoc Networking and Computing, MobiHoc'04, pp. 98–108. ACM, New York (2004)
30. Wang, Y., Li, X.-Y., Frieder, O.: Distributed spanners with bounded degree for wireless ad hoc networks. Int. J. Found. Comput. Sci. 14(2), 183–200 (2003)
31. Yao, A.C.-C.: On constructing minimum spanning trees in k-dimensional spaces and related problems. SIAM J. Comput. 11(4), 721–736 (1982)

A Bicriteria Approximation Algorithm
for DVRP with Time Windows

Hao Gu[1,2], Liang Song[1,2], Hejiao Huang[1,2(✉)], and Hongwei Du[1,2]

[1] Harbin Institute of Technology Shenzhen Graduate School, Shenzhen, China
[2] Shenzhen Key Laboratory of Internet Information Collaboration, Shenzhen, China
guhaozc@gmail.com,
{songliang,hjhuang,hwdu}@hitsz.edu.cn
http://carc.hitsz.edu.cn/MICCRC/index.html

Abstract. In this paper, we study a distance constrained vehicle rout-ing problem with time windows (DVRPTW). DVRPTW is defined as follows: given a metric space on a set of vertices, a release time and a deadline for each vertex, a length bound D, find a minimum cardinality set of tours originating at the depot that covers all vertices, such that each tour has length at most D, and visit as many vertices as possible within their time windows. We give a bicriteria approximation algorithm for DVRPTW on the metric plane, and all the distances satisfy the tri-angle inequality.

Keywords: Metric space · DVRPTW · Approximation algorithm

1 Introduction

The vehicle routing problem was firstly introduced by Dantzig and Ramser [2] in 1959 with the name of 'The Trunk Dispatching Problem'. In the following fifty years, the VRP has been widely studied by the researchers in Operational Research and Computer Science. Many variants, such as capacitated VRP, dis-tance constrained VRP, have been derived from the most basic form of VRP, which are used to reduce transportation costs via computerized models. In this paper, we consider two main constraints of VRP, the distance constraint and time windows. The length of tours must be controlled, because of the fuel tank capacity, and the working time (assume the speed is constant, time can be con-verted into distance). On the other side, all the customers want to receive their goods in their favorite time, so vehicles need to visit customers within their time windows as far as possible. However, finding a set of tours that all customers are serviced within time windows is difficult, even impossible sometimes. So, one goal

Heijiao Huang: This work was financially supported by National Natural Sci-ence Foundation of China with Grants No.11071271, No. 11371004, No. 61100191 and No. 61370216, and Shenzhen Strategic Emerging Industries Program with Grants No. ZDSY20120613125016389, No. JCYJ20120613151201451 and No. JCYJ20130329153215152.

© Springer International Publishing Switzerland 2014
Z. Zhang et al. (Eds.): COCOA 2014, LNCS 8881, pp. 231–238, 2014.
DOI: 10.1007/978-3-319-12691-3_18

in this paper is to maximize the number of customers which are visited within their time windows. The other one is the same as classic DVRP: minimize the number of vehicle used for transporting goods to all the customers. We provide an objective function considering the two goals in Sect. 2, which is different from general objective functions of VRP.

VRP with time windows (VRPTW) is one variant of VRP. So far, a lot of papers have been published to solve the VRPTW. Various heuristics, such as genetic algorithm [5], column generation [9], and local search [10] are used to solve the VRPTW. In addition, approximation algorithms [6–8] are also considered. Bansal et al. [11] firstly gave approximation guarantees for the general case with arbitrary deadlines or time-windows. Meanwhile, distance constrained VRP is also widely studied. Laporte et al. [1] gave two exact algorithms by means of integer program formulation: one based on Gomory cutting planes and one on branch and bound. Li et al. [3] studied DVRP under considering two objective functions: minimize the total distance traveled by vehicles and minimize the number of vehicles. They demonstrated a close relationship between the optimal solutions for the two objective functions and perform a worst case analysis for a class of heuristics. Then, they presented a heuristic that provides a good worst case result when the number of vehicles is relatively small. V. Nagarajan et al. [4] put forward approximation algorithms for DVRP on both tree metrics and general metrics. One is a 2-approximation algorithm for DVRP on tree metrics, and the other is a $(O(\log 1/\epsilon), 1+\epsilon)$ bicriteria approximation algorithm on general metrics. However, they didn't take time windows into consideration. Our work builds on the approach that Nagarajan [4] used to solve DVRP on general metrics. In our paper, time windows are also taken into consideration, and a new objective function and a bicriteria approximation algorithm for DVRPTW are presented.

The remaining part of this paper is organized as follows. In Sect. 2, the DVRPTW and the formulation are introduced. Section 3 shows our algorithm and proof. We list our results of computational experiments in Sect. 4, and draw some conclusions in Sect. 5.

2   Problem Description

We formulate this problem on a complete graph $G = (V, A)$, with $|V|$ nodes and $|A|$ arcs. The node set is partitioned as $V = \{0\} \cup I$, where node 0 is the depot and I is the set of customers. Table 1 lists symbols and their definitions as used in this paper. Our work takes two objectives into consideration. One is to minimize the number of vehicles, and the other is to maximize the number of customers that are visited within their time windows. Then we present the new formulation (1) for DVRPTW as follows:

$$\min \sum_{k \in K} y_k - \eta \sum_{i \in V, i \neq 0} z_i \qquad (1)$$

Table 1. Symbols and definitions

G	A completed graph on which DVRP is defined
A	A set of arcs in G
D	The length of each tour is at most D
K	A set of available vehicles
d_{ij}	The distance required to traverse from node i to node j
t_{ij}	Travel time from node i to node j
s_i	Service time at node i
a_i	Release time at node i
b_i	Deadline at node i
τ_i	Arrival time at node i. So $\tau_i = \tau_{i-1} + s_i + t_{ij}$, if $\tau_0 = 0$.
x_{ijk}	Binary variables equal to 1 if vehicle k travels from node i to j and 0 otherwise
y_k	Binary variables equal to 1 if vehicle k is used and 0 otherwise
z_i	Binary variables equal to 1 if $a_i \leq \tau_i \leq b_i$ and 0 otherwise
η	A parameter denotes which object is more decisive

Subject to:

$$\sum_{j \in V} \sum_{k \in K} x_{ijk} = 1 \qquad \forall i \in V \tag{2}$$

$$\sum_{j \in V} x_{ijk} = \sum_{j \in V} x_{jik} \qquad \forall i \in V, k \in K \tag{3}$$

$$\sum_{i \in V} \sum_{j \in V} d_{ij} x_{xjk} \leq D \qquad \forall k \in K \tag{4}$$

Constraint (2) indicates that every vertex is visited exactly once by exactly one vehicle. Constraint (3) ensures conservation at both customer nodes as well as depot nodes, thereby ensuring all vehicle routes start and end at the same depot. Next, constraint (4) represents that the driving distance of each vehicle is at most D.

3 A Bicriteria Approximation Algorithm for DVRPTW

In this part, we present a bicriteria approximation for DVRPTW. Before the introduction of this algorithm, a definition of unrooted DVRP is given below: find a minimum cardinality set of tours, which are allowed to start and end at any two vertices, that covers all customers.

The basic idea of our algorithm is as follows: if we partition all the vertices in G into parts according to their distance from the depot, we can solve each part as the unrooted DVRP. Of course the length bound of each part is different

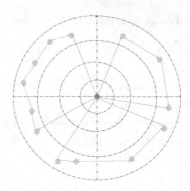

Fig. 1. The vertices are paritioned into several parts according their distances from depot. Each vehicle may visit the costomers only in one part.

(see Fig. 1). In Algorithm 1, vertices in G are partitioned into several parts according to their distances from depot. Formula (5) is used to partition the vertices. Algorithm 2 is a subroutine of Algorithm 1. In Algorithm 2, minimum spanning tree algorithm and greedy policy are used to find a set of tours that minimize objective function value. In this paper, We prove the numbers of vehicles in our solution is at most $O(\log 1/\epsilon)$ times the optimal numbers. However, we haven't given a performance ratio to the second objective.

$$
V_j = \begin{cases} \{v : (1-\epsilon) \cdot \frac{D}{2} < d(r,v) \le \frac{D}{2}\} & if\ j = 0 \\ \{v : (1-2^j\epsilon) \cdot \frac{D}{2} < d(r,v) \le (1-2^{j-1}\epsilon) \cdot \frac{D}{2}\} & if\ 1 \le j \le t-1 \\ \{v : 0 < d(r,v) \le (1-2^{t-1}\epsilon) \cdot \frac{D}{2}\} & if\ j = t \end{cases} \quad (5)
$$

Algorithm 1. Main Algorithm

Input: Graph, Time Windows, Length bound D, Parameter $\epsilon \in (0,1)$
Output: a feasible solution S

1 Define vertex sets $V_0, V_1, \cdots, V_t$ by formula (5), where t=$\lceil \log_2 1/\epsilon \rceil$.
2 For $j = 0, 1, \cdots, t$, run the Algorithm 2 for unrooted DVRPTW, for each vertex set V_j,with the distance constraint $2^{j-1}\epsilon \cdot D$. Let Π_j denote the set of paths obtained.
3 For every path in Π_j, append both its end points with edges from the depot 0, to obtain the set of r-tours $\{0 \cdot \pi \cdot 0 | \pi \in \Pi_j\}$. Let S denote the set of r-tours obtained.
4 Return S.

Theorem 1. *Algorithm 2 is a polynomial-time 3-approximation algorithm.*

Proof. The greedy minimum spanning tree algorithm runs in polynomial time, and other steps in Algorithm 2 are clearly polynomial. The optimal solution is

---

Algorithm 2. Algorithm For Unrooted DVRPTW

---

1 Find a minimum spanning tree T of G.
2 Change each edge e in T to two (parallel) edges between the same pair of
   vertices. Call the resulting graph H .
3 **for** $i = 0, 1, \cdots, |v_j|$ **do**
4   choose v_i as the start node;
5      $k = 0$;
6   **while** *unvisited node!=0* **do**
7   find an unvisited adjacent node along the edges in H, that minimize the
             objective formulation value with visiting each edge in H once;
8   if add the node to route, we have $Length_{tour} > D$, k=k+1, and choose
             this node as the origin of a new tour;
9   **end**
10 **end**
11 Return the set of tours that minimize the objective function value.

---

composed of k paths with the length bound D_j. Therefore, $l(OPT) \leq kD_j$, and
imply that $\sum_{i=1}^{k} l(TOUR_i) \leq 2kD_j$. Let N(i) denote the number of vehicles in
route i, we have that:

$$N(i) \leq \lceil \frac{l(TOUR_i)}{D_j} \rceil \leq \frac{l(TOUR_i)}{D_j} + 1 \tag{6}$$

Therefore,

$$|SOL| = \sum_{i=1}^{k} N(i) \leq \frac{1}{D_j} \sum_{i=1}^{k} l(TOUR_i) + k \leq \frac{2kD_j}{D_j} + k = 3k \tag{7}$$

From (6) and (7) above, we prove that this algorithm can return a solution with
at most $3 \cdot k^*$ paths.

Theorem 2. *For every $0 < \epsilon < 1$, Algorithm 1 is an $O((\log 1/\epsilon), 1+\epsilon)$ bicriteria
approximation algorithm for DVRPTW.*

Proof. Firstly, we show that each r-tour getting by Algorithm 1 has length at
most $(1 + \epsilon)D$. For $j = 0$, each r-tour added two edges from the depot 0, with
length at most $(\epsilon/2)D$. So such a tour has length at most $2 \cdot (D/2) + \epsilon \cdot (D/2) \leq
(1+\epsilon)D$. For $1 \leq j \leq t$, every vertex of v_j has the length at most$(1-2^{j-1}\epsilon) \cdot (D/2)$
from the depot. So each r-tour has the length at most $2^{j-1}\epsilon D + (1-2^{j-1}\epsilon) \cdot D = D$.

We now prove the performance guarantee of this algorithm. Below OPT denotes
the optimal number of r-tours (each of length at most D). Let Γ_{opt} denote an
optimal solution for DVRPTW. Consider one r-tour $\delta \in \Gamma_{opt}$ and δ_j denotes
the part of δ in V_j. The length of δ_j is at most $D - 2 \cdot (D/2)(1 - 2^j\epsilon) = 2^j\epsilon D$,
because every vertex in V_j is located at distance at least $(1 - 2^j\epsilon)(D/2)$ from
depot. So we can visit all the vertics in δ_j by two vehicles with distance bound
$2^{j-1}\epsilon D$. Split other r-tours in Γ_{opt} in this manner, and we can get a feasible

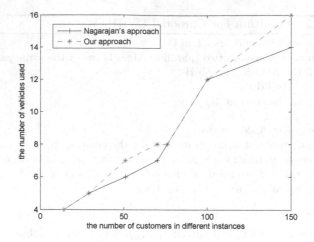

Fig. 2. Compare with Nagarajan's approach

solution to unrooted DVRPTW, which cover all the vertices in V_j. Let SOL_j denote the optimal solution for unrooted DVRPTW on V_j, and K denote the feasible solution, then:

$$|SOL_j| \leq K \leq 2 \cdot OPT \tag{8}$$

Therefore, using **Theorem 1**, we have that:

$$|SOL| \leq 3 \sum_{j=1}^{t} SOL_j \leq 6(t+1) \cdot OPT \tag{9}$$

So the total number of vehicles in our algorithm is at most $6(t+1) \cdot OPT$.

4   Computational Results

This part shows our computational results. Algorithm 1 is just used for partitioning the vertices into several parts, such that each part can be solved as an unrooted DVRPTW. So our experiment is for unrooted DVRPTW. We give several instances with different number of customers, and assume that all the vehicles deliver goods at same speed. So the time windows can be denoted by distance. The time windows are properly relaxed, because if time windows are too small, few vertices can obey these constraints. All experiments were run on Intel(R) Core(TM) i3-3220 @ 3.30 GHz with 8 GB of RAM. Operating System is Windows 8. Algorithms are programmed by Matlab.

The experiments show that the number of vehicles in our solution is almost same as Nagarajan's in [4] (Fig. 2), even time windows are taken into consideration in our approach. In our experiments, we let $\eta = 1$.

In Table 2, C_NUM denotes the number of customers visited within time windows when the objective function value is minimum. The time windows in

Table 2. Computational results

Instances	C_NUM
n_14	8
n_29	16
n_51	28
n_70	32
n_76	41
n_100	48
n_150	71

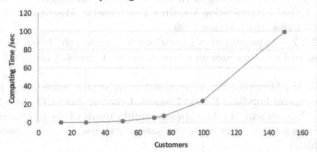

Fig. 3. Computing time

our experiments are arbitrary. So the number of customers serviced on time is just nearly half. In practical, the number of these customers can increase, if logistics companies designate several time windows during which the customers would like to receive goods. Then, we will show the running time (see Fig. 3).

From Fig. 3, it is obvious that the running time increases a lot with the increase of the number of customers. Theorem 2 show that our algorithm gives a solution in polynomial time. However, our algorithm still take a long time if the number of customers is too large. Therefore, some measures should be taken to reduce the running time in practical applications.

5   Conclusions

We present a bicriteria approximation algorithm for DVRPTW on the Euclidean plane. Theorems 1 and 2 prove the number of vehicles given by our algorithm is at most $O(\log 1/\epsilon)$ times the optimal number of vehicles. Furthermore, time windows are also taken into consideration in our algorithm. To logistics company, visiting more customers within their time windows means providing more exceptional customer service. Therefore, our algorithm is very pratical, which find a minimum cardinality set of routes, as well as more customers are serviced within time windows in each tour. Finally, our experiments indicate that more customers are visited on time if the number of vehicles are relaxed.

References

1. Laporte, G., Desrochers, M., Nobert, Y.: Two exact algorithms for the distance constrained vehicle routing problem. Networks **14**, 47–61 (1984)
2. Dantzig, G.B., Ramser, J.H.: The truck dispatching problem. Manage. Sci. **6**, 80 (1959)
3. Li, C., Simchi-Levi, D., Desrochers, M.: On the distance constrained vehicle routing problem. Oper. Res. **40**, 790–799 (1992)
4. Nagarajan, V., Ravi, R.: Approximation algorithms for distance constrained vehicle routing problems. Networks **3**, 209–214 (2012)
5. Thangiah, S.: Vehicle routing with time windows using genetic algorithms. In: Chambers, L. (ed.) Application Handbook of Genetic Algorithms: New Frontiers, vol. II. CRC Press, Boca Raton (1995)
6. Bao, X., Liu, Z.: Approximation algorithms for single vehicle scheduling problems with release and service times on a tree or cycle. Theoret. Comput. Sci. **434**, 1–10 (2012)
7. Nagamochi, H., Ohnishi, T.: Approximating a vehicle scheduling problem with time windows and handling times. Theoret. Comput. Sci. **393**, 133–146 (2008)
8. Karuno, Y., Nagamochi, H.: An approximability result of the multi-vehicle scheduling problem on a path with release and handling times. Theoret. Comput. Sci. **312**, 267–280 (2004)
9. Desrochers, M., Desrosiers, J., Solomon, M.: A new optimization algorithm for the vehicle routing problem with time windows. Oper. Res. **40**, 342–354 (1992)
10. Savelsbergh, M.: Local search for routing problems with time windows. Ann. Oper. Res. **4**, 285–305 (1985)
11. Bansal, N., Blum, A., Chawla, S., Meyerson, A.: Approximation algorithms for deadline-TSP and vehicle routing with time-windows. In: STOC '04 Proceedings of the Thirty-Sixth Annual ACM Symposium on Theory of Computing, pp. 166–174 (2004)

Optimization in Graphs

Data-Oblivious Graph Algorithms
in Outsourced External Memory

Michael T. Goodrich$^{(\boxtimes)}$ and Joseph A. Simons

Department of Computer Science, University of California, Irvine, USA
{goodrich,jsimons}@uci.edu

Abstract. Motivated by privacy preservation for outsourced data, *data-oblivious external memory* is a computational framework where a client performs computations on data stored at a semi-trusted server in a way that does not reveal her data to the server. We give new efficient data-oblivious algorithms in the outsourced external memory model for a number of fundamental graph problems.

1 Introduction

In this paper, we work within the *data-oblivious outsourced external memory* (DO-OEM) model, which is our name for the model used in recent papers on algorithms and systems for data-oblivious outsourced storage solutions (e.g., see [4,5,9,17,18]). We assume that a large data set of size N is stored on a server, who we will call "Bob," and that a client, "Alice," has access to this data through an I/O interface that allows her to make read and write requests of Bob using messages of size B as atomic actions. We also assume Alice has a small amount of secure, private working memory, of size $M = \Omega(\log N)$.

The server, Bob, is "honest-but-curious," which means that he will correctly perform every task requested, but he will also try to learn as much as possible about Alice's data. This, of course, introduces privacy constraints for the DO-OEM model not found in the traditional I/O model (such as in [2]). Thus, Alice must encrypt her data and then decrypt it and re-encrypt it with each read and write request, using a semantically-secure encryption scheme. Alice can safely perform any computation in her private memory, but her sequence of data accesses on the server must also not leak information about her data. That is, it must be *data oblivious*. The access sequence may depend on the function being computed, but it should be independent of the input data values.

Formally, we suppose Alice wants to perform an algorithm, A, which computes some function, f, on her data stored with Bob. In the context of graph algorithms, the input to f is a graph, usually formatted as an array of edges, with V and E being the number of the graph's vertices and edges, respectively. The output of f may either be a property of the graph, such as whether or not the graph is biconnected, or another graph, such as a spanning tree, which will also be stored with Bob. Alice performs the algorithm A by issuing read and write requests to Bob.

© Springer International Publishing Switzerland 2014
Z. Zhang et al. (Eds.): COCOA 2014, LNCS 8881, pp. 241–257, 2014.
DOI: 10.1007/978-3-319-12691-3_19

We say that A is *data-oblivious* and can compute f in the DO-OEM model if every probabilistic polynomial time adversary has only a negligable advantage over random guessing in a *input-indistinguishability* game. In this game, Alice flips a fair coin and based on its outcome either uses A to compute f on her input or on a random syntactically-correct input. Bob observes her memory accesses and then must decide if she was computing f for her input or not. Given a function f and public input parameters (e.g. an upper bound on the size of the graph), γ, the probability that an algorithm to compute f in the DO-OEM model executes a particular access sequence S must be equally likely for any two inputs X, Y satisfying parameters γ. That is, $P(S|f, \gamma, X) = P(S|f, \gamma, Y)$, or, from the Bob's perspective, $P(X|f, \gamma, S) = P(Y|f, \gamma, S)$. We can achieve data-obliviousness if, knowing the size of the input, the function being computed, and the access sequence, all inputs are equally likely.

Previous Related Results. Goldreich and Ostrovsky [6] introduce the oblivious RAM model and show that an arbitrary RAM algorithm can be simulated (in internal memory) with an overhead of $O(\log^3 N)$ through the use of constant-time random oracles, and this has subsequently been improved (e.g., see [7–10]), albeit while still using constant-time random oracles.

Chiang *et al.* [2] study (non-oblivious) external-memory graph algorithms and Blanton *et al.* [1] give data-oblivious algorithms for breadth-first search, single-source-single-target shortest paths and minimum spanning tree with running time $O(v^2)$, and maximum flow with running time $O(v^3 E \log V)$. However, their approach is based on computations on the adjacency matrix of the input graph, and thus only optimal on very dense graphs, whereas our approach is based on reductions to sorting the edge list, and is efficient on graphs of all densities.

Our Results. Let $\mathrm{DO} - \mathrm{Sort}\,(N)$ denote the number of I/Os required to sort an input of size N in the DO-OEM model. For instance, Goodrich and Mitzen-macher [7] show $\mathrm{DO} - \mathrm{Sort}\,(N) = O((N/B) \log^2_{M/B}(N/B))$ I/Os, assuming $M > 3B^4$. We develop efficient algorithms in the DO-OEM model for a number of fundamental graph problems:

- We show how to construct a minimum spanning tree of a graph G in time depending on the input parameters V, E, density, and class of G, in the DE-OEM model:

Density	Class	Running Time	Constants
$E = O(V \log^\gamma V)$	Any	$O(\mathrm{DO} - \mathrm{Sort}\,(E) \log V / \log \log V)$	$\gamma \geq 0$
$E = \Theta(V 2^{\log^\delta V})$	Any	$O(\mathrm{DO} - \mathrm{Sort}\,(E) \log^{1-\delta} V)$	$0 < \delta < 1$
$E = \Omega(V^{1+\epsilon})$	Any	$O(\mathrm{DO} - \mathrm{Sort}\,(E))$	$0 < \epsilon \leq 1$
Any	Minor Closed	$O(\mathrm{DO} - \mathrm{Sort}\,(E))$	—

- Given a tree T, we can perform any associative traversal computation over T in $O(\mathrm{DO} - \mathrm{Sort}\,(V))$ time in the DO-OEM model.

- Given a tree $T = (V, E)$ and a set, S, of pairs of vertices, we can compute the LCA for each pair in S in $O(\text{DO} - \text{Sort}(|S| + V))$ time in the DO-OEM model.
- Given a graph G and a spanning tree of G, we can compute the biconnected components of G in $O(\text{DO} - \text{Sort}(E))$ time in the DO-OEM model.
- Given a biconnected graph G, and a spanning tree of G, we can construct an open ear decomposition in $O(\text{DO} - \text{Sort}(E))$ time in the DO-OEM model.
- Given a biconnected graph G and its open ear decomposition, we can find an st-numbering of G in $O(\text{DO} - \text{Sort}(E))$ time in the DO-OEM model.

None of our algorithms use constant-time random oracles. Instead, they are based on a number of new algorithmic techniques and non-trivial adaptations of existing techniques. This paper outlines the main results; details can be found in the full version [12].

2 Data-Oblivious Algorithm Design

Compressed-Scanning. Goodrich *et al.* [11] introduced *compressed-scanning* as a data-oblivious algorithm design technique for internal memory. We extend compressed-scanning to the DO-OEM model, and we use an algorithm design technique, where an algorithm is formulated so that it processes the input in a series of t rounds as follows:

1. Scan each item of input exactly once; a random permutation hides the access pattern.
    - Read a block of B items from input, possibly including some dummy items.
    - Perform some computation in private memory.
    - Write a block of B items to output, possibly including some dummy items.
2. Sort the output data-obliviously.
3. Truncate the output, ignoring a portion, $L(i, \gamma)$, of the items, which may depend on the index of the round i and input parameters γ (e.g. input size), but not on any data values. For example, L could be 0 (no items are discarded) or $N/2^i$ (the last half of the output is discarded in each round).
4. Use the output as input for the next round.

Theorem 1. *Let A be any compressed-scanning algorithm for which t and B depend only on N. Let N_i denote the size of the input passed to round i. Then, A can be simulated in the DO-OEM model in $O(\sum_{i=1}^{t} \text{DO} - \text{Sort}(N_i))$ time[1] without the use of constant-time random oracles.*

Proof. By definition, we can run A in the DO-OEM model if and only if A satisfies the input-indistinguishability game. The algorithm runs in t rounds, and each round has three phases: scan, sort, and truncate. The adversary can win the game if in any round, in any phase, the sequence of memory accesses to the

---

[1] Thoughout this paper we measure time in terms of I/Os with the server.

shared encrypted array is different with non-negligible probability between the adversary's input and input generated uniformly at random. However, by construction, the distribution of memory access by A is the same at every phase for all inputs with the same input parameters. In the scan phase, each item of input is accessed once and in random order, regardless of the actual input values. Thus the scan phase conveys no advantage to the adversary. In the sorting phase, the sequence of memory access is likewise independent of the data values by definition, since we use a data-oblivious sorting algorithm. Hence, the sorting phase also conveys no advantage to the adversary. Finally, in the truncate phase, the portion of memory the algorithm chooses to ignore depends on the input parameters, but will be identical, regardless of whether the input is the one chosen by the adversary or the one generated uniformly at random. So the truncate phase also conveys no advantage to the adversary. In every phase of every round, the adversary gains no information as to which input was chosen by the coin toss. Therefore, every probabilistic polynomial time adversary has a negligible advantage over random guessing. The running time is a straightforward sum of the cost to data-obliviously sort the input in each round.   □

3   Tree-Traversal Computations

Many traditional graph algorithms are based on a traversal of a spanning tree of the graph, for example, using depth first search. However, the data access pattern of depth first search fundamentally depends on the structure of the graph, and it is not clear how to perform DFS efficiently in the DO-OEM model. Instead, we use *Euler Tours* [19], adapted for data-oblivious tree-traversal computation [11]. Given an undirected rooted tree T, we imagine that each edge $\{p(v), v\}$ is composed of two directed edges $(p(v), v)$ and $(v, p(v))$, called an *advance* edge and *retreat* edge respectively. An *Euler tour* of T visits these directed edges in the same order as they would be visited in a depth first search of T. However, an Euler tour implemented with compressed-scanning does not reveal information to the adversary because each data item is accessed once and in random order.

Some tree statistics are straightforward to compute using Euler Tours. For example, Goodrich *et al.* [11] show how to compute the size of the subtree for each node $v \in T$ using an Euler Tour and compressed-scanning pass over the edges of T. The calculation is straightforward once we observe that $\texttt{size}(v) = (\texttt{E-order}(v, p(v)) - \texttt{E-order}(p(v), v))/2 + 1$, since for each proper descendant of v, we will traverse one advance edge and one retreat edge. Thus, the number of edges traversed between $(p(v), v)$ and $(v, p(v))$ is twice the number of proper descendants of v, and we add one to also include v in $\texttt{size}(v)$. Moreover, Euler tour construction can be done data-obliviously in external memory in $O(\text{DO} - \text{Sort}(|T|))$ I/Os by a data-oblivious compressed-scanning implementation of the algorithm by Chiang *et al.* [2].

However, Euler-Tours are insufficient to compute most functions where the value at a vertex is dependent on its parent or children. Therefore, in the following, we describe more sophisticated techniques for tree *traversal computations*, suitable for computing functions in which the value at a vertex depends on its parent or children.

Bottom-Up Computation. Let T be a tree rooted at r. First, we show how to compute recursive functions on the vertices bottom up using a novel data-oblivious algorithm inspired by the classic parallel tree contraction of Miller and Reif [15]. Like Miller and Reif's algorithm, our algorithm compresses a tree down to a single node in $O(\log V)$ rounds. However, unlike the original algorithm, we are able to compress long paths of degree two nodes into a single edge in a single iteration and guarantee the size of the graph decreases by half in each round.

Each round of the tree contraction algorithm is divided into two operations: rake, which removes all the leaves from T, and compress, which compresses long paths by contracting edges for which the parent node only has a single child.

First, we label each vertex with its degree by scanning the edge list in adjacency list order. Then, for each vertex $v \neq r$, if v has degree 1, it is a leaf, and it is marked for removal by the rake operation. Otherwise, if it has degree 2, then it is marked for contraction by the compress operation. The marks are stored with the endpoints of each edge.

Now, we perform the rake operation via an Euler tour of T. For each unmarked edge we read, we write back its value unchanged. If an advance edge is marked as a leaf, then we mark the edge for removal. The next edge we read is the corresponding retreat edge from the leaf. We evaluate the leaf and output the computed value together with the label of the parent vertex instead of the original retreat edge. Next we distribute this information to the other incident edges; we sort the edge list so that for each vertex we first see all the evaluated leaves and then see the remaining outgoing edges. In a compressed scanning pass we are able to store the function evaluation from each leaf in its parent. Thus we complete the rake operation.

Next, we perform the compress operation via another Euler tour of T. We remove each marked advance edge by writing dummy values in its place. For each retreat edge to a degree two node, we contract the edge by composing the functions at the parent and child and storing this value in private memory. For all but the last edge in a path of degree two nodes, we mark the edge for removal. For the last edge in the path, we output the label of the parent vertex together with the composition of all functions along the path. Although the path may not have constant size, for the functions considered in this paper (such as min), the composition across values of nodes along the path can be expressed in $O(1)$ space by partially evaluating the function as we go. We pass this information to other edges incident to the last vertex in the path, for all the compressed paths, using a single compressed scanning round. Thus, we complete the compress operation. (See Fig. 1.)

Finally, we perform one last compressed scanning pass to set aside all edges marked for removal. These edges are placed at the end of the list by the sort and are not required for subsequent processing. Thus we complete one round of the algorithm. However, we may continue to access some dummy edges in subsequent rounds so that in round $i+1$ we always access a constant fraction of

Round

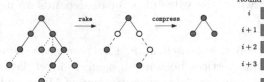

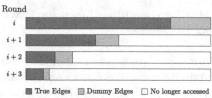

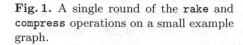

■ True Edges ■ Dummy Edges □ No longer accessed

Fig. 1. A single round of the `rake` and `compress` operations on a small example graph.

Fig. 2. The portion of memory accessed by the algorithm decreases by a constant fraction in each round.

the memory accessed in round i, thus maintaining the data-oblivious property of our algorithm (see Fig. 2).

We now analyze the total time it takes to contract a tree down to the root node. Each round of `rake` and `compress` on a tree T_i of size V_i takes $O(\text{DO} - \text{Sort}\,(V_i))$ time to perform $O(1)$ Euler tours and compressed scanning rounds. We begin with an initial tree $T_0 = T$ of size $V_0 = V$. Without loss of generality, we can partition the nodes of any tree T_i into three sets: B_i, the branch nodes with at least 2 children; P_i, the path nodes with 1 child; and L_i, the leaf nodes. Clearly $|B_i| + |P_i| + |L_i| = V_i$, and $|L_i| \geq 2|B_i|$. The rake operation removes all of L_i, and all but at most one node in each path in P_i. Thus, $V_{i+1} \leq |B_i| + \frac{1}{2}|P_i| < \frac{1}{2}(3|B_i| + |P_i|) \leq \frac{1}{2}V_i$. Hence, $\sum_i V_i$ is a geometric sum, and the total running time of all rounds is $O(\text{DO} - \text{Sort}\,(V))$.

Throughout the algorithm the children of each branch and leaf node are finalized before we process the node. However, some path nodes may have been compressed and set aside before all of their descendants were finalized. Thus, in a final post-processing step, we perform one more round of compressed scanning and Euler tour over the full edge list to finalize the value of the internal path nodes.

Computation of `low`. Suppose we are given a spanning tree T of a graph G. We illustrate rake and compress by computing the following simple recursive function.

$$
\begin{aligned}
\texttt{low}(v) = \min(&\{\texttt{preorder}(v))\} \\
&\cup \{\texttt{low}(w) \mid w \text{ is a child of } v \text{ in } T\} \\
&\cup \{\texttt{preorder}(w) \mid (v, w) \in G - T\})
\end{aligned}
$$

That is, for each vertex $v \in T$, `low`(v) is the lowest preorder number of a vertex that is a descendant of v in T, or adjacent to a descendant via a non-tree edge. This function is a key part of the biconnected components algorithm, and key functions in our other algorithms are computed similarly.

First, we compute the preorder numbers of each vertex by an Euler tour of T. Next, we preprocess the edge list. In $O(1)$ compressed-scanning rounds, we compute for each vertex the minimum preorder number between that vertex and all its neighbors in $G - T$. We store this data in the endpoints of each edge (u, v) in the edge list as the initial values for $\texttt{low}(u)$ and $\texttt{low}(v)$. This preprocessing requires $O(\text{DO} - \text{Sort}\,(E))$ time.

Now, we use rake and compress to compute the recursive portion of $\texttt{low}$. Each iteration of rake and compress proceeds as follows. The low value of each leaf is already finalized. We store $\texttt{low}(v) = \min\left(\{\texttt{low}(v)\} \cup \{\texttt{low}(w) \mid w \text{ is a child of } v\}\right)$ as the function for each internal node v. When we rake a leaf ℓ, we update its parent p, $\texttt{low}(p) = \min(\texttt{low}(p), \texttt{low}(\ell))$. During the compress step, when we contract an edge $(p(v), v)$ we update the function stored at in $p(v)$ as follows: $\texttt{low}(p(v)) = \min(\texttt{low}(p(v)), \texttt{low}(v))$, and $\texttt{children}(p(v)) = \texttt{children}(v)$. Note that vertex may have many children, but always a single parent. Thus, the easiest way to change the assignment of children is to set $\texttt{label}(p(v)) = \texttt{label}(v)$ and then relabel the edge $(p(p(v)), p(v)) = (p(p(v)), v)$. We may need to perform this relabeling and calculation of $\texttt{low}$ over a long path, but we always process nodes bottom up, and we maintain the values of the previous edge processed in private memory. We output dummy values for all but the final edge in the path, which stores the minimum $\texttt{low}$ of the whole path, together with the labels of the first and last vertex on the path. Finally, we synchronize each edge with the new values of its endpoints via compressed-scanning, which completes the iteration of rake and compress. After at most $O(\log V)$ iterations and $O(\text{DO} - \text{Sort}\,(V))$ I/Os, we complete the rake and compress algorithm. Thus, the preprocessing time dominates, and the total time required to compute $\texttt{low}$ for all vertices is $O(\text{DO} - \text{Sort}\,(E))$.

Top-Down Computation. We now show how to run our compression algorithm "in reverse" in order to efficiently compute top-down functions where each vertex depends on the value of its parent. First, we simulate the compression algorithm described above, and label each edge e with $\texttt{contract}(e)$, the order in which it would have been removed from the graph. Thus, all edges incident to leaves in the initial graph are given a label smaller than any interior nodes, and all edges incident to the root are given larger labels than edges incident to nodes of depth > 1.

Next, we sort the edges in reverse order according to their $\texttt{contract}$ labels. We process the edges in this order in $\lceil \log V \rceil$ stages. We mark the root as *finished*. Then, in each stage i, we perform the following on the first 2^i edges in the sorted order: For each advance edge $e = (p(v), v)$, if $p(v)$ is marked as *finished*, we evaluate the function at v, augment e with its value, and mark v as *reached*. Between stages i and $i + 1$, we process the first 2^{i+1} edges, and distribute the new values at reached vertices from the previous stage to any incident edges belonging to the next stage. Finally, we mark each reached vertex as finished. Thus, the function at a parent is always evaluated before the function at its children, and each child edge has been augmented with the value from the parent before the edge is processed.

Each stage requires $O(1)$ rounds of compressed scanning. Since the number of edges processed in each stage is 2^i, the running time of the final stage dominates all other stages, and thus the total time is $O(\text{DO} - \text{Sort}\,(V))$. Note that since our algorithm essentially reduces to data-oblivious sorting and compressed scanning, the sequence of data accesses made by the algorithm are independent of the input values. Thus, no probabilistic polynomial adversary has more than negligible advantage in the input-indistinguishability game. We summarize our results in the following theorem:

Theorem 2. *Given a tree T, we can perform any top-down or bottom-up tree-traversal computation over T in $O(\mathrm{DO} - \mathrm{Sort}\,(V))$ time in the DO-OEM model.*

LCA computation. Suppose we are given a connected graph $G = (V, E)$ and a spanning tree $T = (V, E_T)$ of G rooted at t. We can preprocess G and augment each edge $(u, v) \in E$ with additional information such that we can find the least common ancestor $\mathrm{LCA}(u, v)$ with respect to T in constant time. Given two integers x, y, let $\mathtt{rzb}(x)$ denote the number of rightmost zero bits in the binary representation of x, and let $x\&y$ denote the bitwise logical AND of x and y. The following preprocessing algorithm is adapted from the parallel algorithm of Schieber and Vishkin [16].

For each node $v \in T$, we compute $\mathtt{preorder}(v)$, and $\mathtt{size}(v)$. We also set $\mathtt{inlabel}(v)$ to $\max_{w \in T_v} \mathtt{rzb}(\mathtt{preorder}(w))$, that is, the maximal number of rightmost zero bits of any of the preorder numbers of the vertices in the subtree rooted at v. As in our computation of $\mathtt{low}$ in Sect. 3, we can compute these functions via rake and compress together with a $O(1)$ Euler tour and compressed-scanning steps so that each edge $(u, v) \in G$ stores the augmented information associated with its endpoints. We also initialize $\mathtt{ascendent}(t) = 2^{\lfloor \log V \rfloor}$.

We compute $\mathtt{ascendant}(v)$ for each vertex $v \in V$ as follows: If $\mathtt{inlabel}(v)$ is equal to $\mathtt{inlabel}(p(v))$, then set $\mathtt{ascendent}(v)$ to $\mathtt{ascendent}(p(v))$. Otherwise, set $\mathtt{ascendent}(v)$ to $\mathtt{ascendent}(p(v)) + 2^i$, where $i = \log(\mathtt{inlabel}(v) - [\mathtt{inlabel}(v)\&(\mathtt{inlabel}(v) - 1)])$ is the index of the rightmost non-zero bit in $\mathtt{inlabel}(v)$. Thus, $\mathtt{ascendent}$ is a top-down function, and we can evaluate it at all nodes in the tree in $O(\mathrm{DO} - \mathrm{Sort}\,(V))$ time using the method of Sect. 3.

Finally, we perform an Euler-Tour traversal to compute a table $\mathtt{head}$. We set $\mathtt{head}(\mathtt{inlabel}(v))$ to be the vertex of minimum depth $d(u)$ among all the vertices u such that $\mathtt{inlabel}(u) = \mathtt{inlabel}(v)$ on the path from the t to v in T. Note that since there are at most $\log V$ distinct $\mathtt{inlabel}$ numbers, the size of $\mathtt{head}$ is at most $O(\log V)$ and can fit in private memory. By a constant number of compressed-scanning steps, we store $\mathtt{head}(\mathtt{inlabel}(u))$ and $\mathtt{head}(\mathtt{inlabel}(v))$ with each edge $(u, v) \in G$.

Given the additional information now stored in each edge, we can compute the $\mathrm{LCA}(u, v)$ for any edge $(u, v) \in G$ in constant time by a few simple algebraic computations as shown by Schieber and Vishkin [16].

We summarize this result in the following theorem:

Theorem 3. *Given a tree T and a set of pairs of vertices $S \subset V \times V$, we can compute $\mathrm{LCA}(u, v)$ for all $u, v \in S$ in $O(\mathrm{DO} - \mathrm{Sort}\,(|S| + V))$ time in the DO-OEM model.*

4 Algorithms

In this section we present a novel algorithm to compute the minimum spanning tree of a general graph in the DO-OEM model. Our algorithm has additional input parameters of the density and class of the graph, and its running time depends on these parameters. Thus, our algorithm necessarily reveals asymptotically the vertex and edge counts, and whether the input graph is minor-closed. However, revealing this information does not convey any advantage to the adversary in the input-indistinguishability game. In fact, these input parameters can be freely chosen by the adversary. Of course, if we don't want to allow these input parameters and the corresponding gains in efficiency, we can avoid revealing this information by working with an adjacency matrix instead of an edge list, or we can also achieve a tradeoff between privacy and efficiency by padding the input with dummy edges.

In the case of somewhat dense graphs, or graphs from a minor closed family, our runtime is $O(DO - \text{Sort}(E))$. We conjecture that this is optimal since we require this time to perform even a single round of compressed scanning in our model. For graphs of other classes and densities, our algorithm still beats the previously best known method for computing *any* spanning tree in the DO-OEM model by logarithmic factors.

We also leverage our minimum spanning tree and tree-traversal computation algorithms to design new data-oblivious algorithms to compute biconnected components of a graph, and an open ear decomposition and st-numbering of a biconnected graph.

Minimum Spanning Tree. Our MST algorithm requires the following subroutines: `trim`, `select`, `contract` and `cleanup`.

`trim`(G, α):
  We scan the edge list of G and trim the outgoing edges from each node, depending on the value of an input parameter α. If the degree of a node is at most α, then we leave its outgoing edges unchanged. If the degree of a node is less than α, then we implicitly pad its outgoing edge list up to size α with additional dummy edges of weight ∞. However, if the degree of a node is greater than α, then the node keeps its α smallest outgoing edges and discards the rest. For a given edge each endpoint independently chooses to keep or discard it as an outgoing edge. Note that we can trim the edge list in $O(1)$ rounds of compressed scanning. Afterwards, the graph may no longer be connected (see Fig. 3).
`select`(G):
  Each node selects the minimum outgoing edge from its adjacency list. If two edges have the same weight, we break ties lexicographically. We mark the edges as selected as follows: First, sort the edges lexicographically by source vertex, weight. Then, in a single compressed scanning round we mark the minimum weight edge from each source vertex as selected. The set of selected edges partition G into connected components, and in each connected

component, one edge has been selected twice (see Fig. 3). For each double-selected edge, we arbitrarily choose to keep one copy and mark the other as dummy. We can gather all the dummy edges to the end of the list in $O(1)$ compressed scanning rounds. The output is a spanning forest of G such that in each tree all the edges are oriented from the root to the leaves.

contract(G, E_s):

The input is a graph G and a set of selected edges E_s which induce a spanning forest $F \subseteq G$. For each connected component in F, we merge the nodes in the component into a single pseudo-node by contracting all the selected edges in that component (Fig. 3). Using our top-down tree-traversal algorithm, we re-label each node with the label of the root of its component. Then, in two compressed scanning rounds we relabel the endpoints of the all edges in G to reflect the relabeled nodes, possibly creating loops and parallel edges. In the first round we relabel the source for each outgoing edge of each node. In the second round we sort the edges to group them by incoming edges with each node, and relabel the target for each edge. In a final oblivious sort, we restore the edge list of G to adjacency list (lexicographical) order.

cleanup(G):

We detect and remove duplicate, parallel and loop edges in a single compressed-scanning round. When we encounter parallel edges, we remove all but the minimum weight edge between two nodes. As we scan the edge list, we remove an edge by writing a dummy value in its place, and then we perform a final oblivious sort to place all the dummy values at the end of the list.

Minimum Spanning Forest (MSF). Let $G_0 = \text{trim}(G, \alpha)$ for an appropriate choice of parameter α to be discussed later. Then, the core of our algorithm is

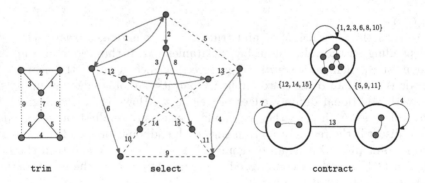

Fig. 3. Left: the result of trim on a small example graph with $\alpha = 2$. Removed edges are denoted by dotted lines. Center and Right: the result of select and contract on a small example graph. In select, unselected edges are dashed, and selected edges are solid and oriented towards the vertex which selected them. The selected edges form a spanning forest of G; each connected component is a tree in which one edge has been selected twice. In contract, each connected component is contracted to a single pseudo-node.

as follows: for $i \in [1, \alpha]$, let

$$G_i = \texttt{trim}(\texttt{cleanup}(\texttt{contract}(G_{i-1}, \texttt{select}(G_{i-1}))), \alpha).$$

That is, we perform α iterations in which we select the minimum edge out of each node, contract the connected components, "remove" unwanted edges from the resulting graph by labeling them as dummies, and pass the cleaned and trimmed graph to the next iteration. Each subsequent iteration accesses a constant fraction of the memory accessed in the previous iteration, possibly including some dummy edges (see Fig. 2). In a final pass, we $\texttt{contract}$ and $\texttt{cleanup}$ all the edges of G with respect to the connected components represented by the nodes of G_α.

Throughout our algorithm, the set of selected edges induce a spanning forest of G. Each pseudo-node represents an entire tree in this forest. We define a weight function $w(v)$ for each pseudo-node, which corresponds to the number of true nodes contained in the tree represented by the pseudo-node. Initially each node has weight 1.

At each $\texttt{trim}$ step, we remove a subset of edges. The remaining edges induce a set of *potential* connected components $C_1, \ldots, C_t$. For each C_i, let $w(C_i) = \sum_{v \in C_i} w(v)$. We maintain the following invariant throughout all iterations of our algorithm: $w(C_i) \geq \alpha$ for all C_i.

The invariant remains true after the initial $\texttt{trim}$ step; we know that each C_i contains at least $\alpha + 1$ nodes, since G was connected, and $\texttt{trim}$ only removes edges from nodes of degree more than α. Subsequent $\texttt{trim}$ steps also maintain the invariant since nodes of degree $\leq \alpha$ are not effected and nodes of degree $> \alpha$ will still be connected to at least α other nodes after the trim.

Next, in the $\texttt{select}$ step, each node selects one outgoing edge for contraction. For each edge (u, v) that we contract in the $\texttt{contract}$ step, we create a new pseudo-node x of weight $w(x) = w(u) + w(v)$. Thus, the total weight of each C_i does not change over the contract step. However, we select at least $|C_i|/2$ edges in each component. Thus, the number of nodes within each component C_i is reduced by half.

The $\texttt{cleanup}$ step only removes redundant edges, and does not effect the number of nodes or the weight in any components.

Hence, after $O(\log \alpha)$ iterations, the size of each C_i is reduced by a factor of α, and the total weight of each C_i remains at least α. Therefore, the resulting graph has at most $O(V/\alpha)$ pseudo-nodes.

We now consider the running time of the MSF algorithm. The first run of $\texttt{trim}(G)$ and the final pass both take $O(\text{DO} - \text{Sort}(E))$ time. Each sub-routine used in each iteration takes $O(\text{DO} - \text{Sort}(|G_i|))$ time. Moreover, for each i, $|G_{i+1}| \leq |G_i|/2$. Therefore, the time used in all iterations is a geometric sum, and the time required by all iterations is $O(\text{DO} - \text{Sort}(|G_0|)) = O(\text{DO} - \text{Sort}(\alpha \cdot V))$. Hence, the total running time of one repetition of MSF is $O(\text{DO} - \text{Sort}(E) + \text{DO} - \text{Sort}(\alpha \cdot V))$. Moreover, we can repeat our algorithm k times to achieve the following result. Let $G^j(\alpha)$ denote the graph for the jth repetition of the MSF algorithm with parameter α, that is, the graph consisting of the pseudo-nodes (subsets of vertices of G) from the output of the previous iteration, and the α smallest edges out of each pseudo-node. The running time of k repetitions is

$$\sum_{j=0}^{k-1} O(\text{DO} - \text{Sort}\,(E) + \text{DO} - \text{Sort}\,(\alpha|G^j(\alpha)|)) = O(k \cdot \text{DO} - \text{Sort}\,(E) + \text{DO} - \text{Sort}\,(\alpha V))$$

since $\sum |G^j(\alpha)|$ is a geometric sum. We require that the final output $|G^k(\alpha)| = 1$ is a single pseudo-node representing a spanning tree of G, which implies that $k = \frac{\log V}{\log \alpha}$. Thus, we choose parameter $\alpha = \frac{\beta}{\log \beta}$, where $\beta = \frac{E \log V}{V}$ to minimize the total running time, depending on the density of the graph. Hence, the running time is

$$O\left(\text{DO} - \text{Sort}\,(E)\,\frac{\log V}{\log \frac{\beta}{\log \beta}} + \text{DO} - \text{Sort}\,\left(E\frac{\log V}{\log \beta}\right)\right)$$

If the graph G is somewhat dense with $E = \Omega(V^{1+\epsilon})$ for any arbitrarily small constant $\epsilon > 0$, then this implies a running time of $O(\text{DO} - \text{Sort}\,(E))$. If G has density $E = \theta(V 2^{\log^\delta V})$, for any constant $0 < \delta < 1$ then this implies a running time of $O(\text{DO} - \text{Sort}\,(E) \cdot \log^{1-\delta} V)$. If G is sparse with $E = O(V \log^\gamma V)$, for any constant $\gamma \geq 0$, then we achieve a running time of $O(\text{DO} - \text{Sort}\,(E) \cdot \log V / \log \log V)$. However, if G is from a minor closed family, e.g. if it is a planar graph, we can be even more efficient and simplify our algorithm.

Mareš [14] gave a related algorithm in the standard RAM model (not data-oblivious). He showed that for any non-trivial minor closed families of graphs, a Borůvka-style round of edge contractions will always decrease the size of the graph by a constant factor. Therefore, we will have a similar geometric sum in the running time of our algorithm for any input graph drawn from a non-trivial minor closed family of graphs, including any graph with bounded genus.

Thus, in this case we do not need the `trim` sub-routine at all, since the size of the graph will be decreasing geometrically by the minor closed property. Then we run our algorithm with parameters $\alpha = E$ and $k = 1$. Therefore, the total time required for all iterations is $O(\text{DO} - \text{Sort}\,(E))$. Furthermore, since we always selected the minimum edge out of each component, by the cut property of minimum spanning trees we are guaranteed that the algorithm produces a minimum spanning tree of G. Moreover, The algorithm makes choices depending on the input parameters but the fundamental components of our algorithm are data-oblivious sorting and compressed scanning and the memory access pattern never depends on the data values. Thus, no probabilistic polynomial time adversary has more than negligible advantage in the input-indistinguishability game.

We summarize our result in the following theorem:

Theorem 4. *In the DO-OEM model, we can construct a minimum spanning tree of a graph G in time depending on the input parameters V, E, density, and class of G. If G belongs to a minor-closed family of graphs, such as any graph with bounded genus, then the running time is $O(\text{DO} - \text{Sort}\,(E))$. We achieve the following run-times:*

Density	Class	Running Time	Constants
$E = O(V \log^\gamma V)$	Any	$O(\text{DO} - \text{Sort}(E) \log V / \log \log V)$	$\gamma \geq 0$
$E = \Theta(V 2^{\log^\delta V})$	Any	$O(\text{DO} - \text{Sort}(E) \log^{1-\delta} V)$	$0 < \delta < 1$
$E = \Omega(V^{1+\epsilon})$	Any	$O(\text{DO} - \text{Sort}(E))$	$0 < \epsilon \leq 1$
Any	Minor Closed	$O(\text{DO} - \text{Sort}(E))$	—

Given the minimum spanning tree algorithm and tree-traversal computation technique outlined above, we can also achieve the following results for a biconnected graph. Details can be found in the full version [12].

Theorem 5 (Biconnected Components). *Given a graph G and a spanning tree, we can compute the biconnected components of G in $O(\text{DO} - \text{Sort}(E))$ time in the DO-OEM model.*

Open Ear Decomposition. Let $T = (V, E_T)$ be a spanning tree of G rooted at t such that there is only a single edge $(s, t) \in E_T$ incident to t. The edges of $E - E_T$ are denoted *non-tree edges*, and the edges $E_T - (s, t)$ are denoted *tree edges*. The edge (s, t) is treated separately. Let G be a biconnected graph.

We structure our algorithm around the same steps used by Maon et al. [13] in their parallel ear-decomposition algorithm. However, the details of each step are necessarily different in important ways as we show how to efficiently implement each step in the DO-OEM model.

1. Find a spanning T of G rooted at t such that (s, t) is the only edge incident to t. We remove the vertex t from G and run the above MST algorithm on $G[V - \{t\}]$ to get a set of edges E' which form a spanning tree of $V - \{t\}$. Let $E_T = E' \cup (s, t)$. Then the desired spanning tree of G is given by $T = (V, E_T)$.
2. (a) For each tree edge $(u, v) \in E_T$, we compute $d(u)$, $d(v)$, $p(u)$ and $p(v)$ via an Euler-Tour traversal of T. We also preprocess G using the LCA algorithm given in Corollary 3.
   (b) Number the edges of G, assigning each $e \in E$ an arbitrary integer serial number $\mathtt{serial}(e) \in [E]$. Define a lexicographic order on the non-tree edges f according to $\mathtt{number}(f) = (d(\text{LCA}(f)), \mathtt{serial}(f))$
3. Each non-tree edge f induces a simple cycle in $(V, E_T \cup f)$. We will adapt an algorithm of Vishkin [20] to compute a function $\mathtt{master}(e)$ for each tree edge e.

Fact 6 *[13]. Each non-tree edge f, together with the set of edges e_i such that $\mathtt{master}(e_i) = f$ form a simple path or cycle, called the* ear *of f.*

Fact 7 *[13]. The lexicographical order on $\mathtt{number}(f)$ over the non-tree edges induces an order on the ears, which yields an ear decomposition (which is not necessarily open).*

Fact 8 *[13]. Let (u, v) be a non-tree edge. Let $x = \text{LCA}(u, v)$. Let e_u and e_v be the first edges on the path from x to u and v in T respectively. Then, the* *ear induced by* (u, v) *is closed if and only if* e_u *and* e_v *both choose* (u, v) *as their* `master`.

As shown by Maon *et al.* [13], the resulting ear-decomposition is not necessarily open, and we must first refine the order defined on the edges. Therefore, we refine the ordering on non-tree edges sharing a common LCA by updating the assignment of `serial` with the following additional steps.

4. Construct a bipartite graph $H_x = (V_x, E_x)$ for each vertex $x \in V - \{t\}$. Each vertex in V_x corresponds to an edge in G. Specifically, V_x is the set of edges (u, v) such that $x = \mathrm{LCA}(u, v)$. Note that this includes tree edges (x, w) for which $x = p(w)$. Thus, the graphs H_x partition the edges of G.

   There is an edge in E_x between a tree edge e and a non-tree edge (y, z) if and only if e is the first edge on the path from x to y or z in T. There are no other edges in E_x. We can test all such edges by creating a list of each endpoint in V_x sorted by preorder number. That is, all tree edges (x, w) will appear once ordered by `preorder`(w) and each non-tree edge (y, z) will appear twice, once ordered by `preorder`(y) and once ordered by `preorder`(z). Then, each non-tree edge is incident to the first tree edge that comes before and after it according to the above ordering.

   We compute the set E_x for all the graphs H_x "in parallel"; that is, we do not require separate passes over the edge list of G to compute each H_x. First, in $O(1)$ rounds of compressed scanning, we label each edge e with $\mathrm{LCA}(e)$. Then, we sort the edge list by $\mathrm{LCA}(e)$, `E-order`(e). In an additional $O(1)$ rounds of compressed scanning, we compute V_x and E_x for each H_x.

5. Construct a spanning forest for each H_x. Note that each edge in E appears in exactly one H_x, and that the degree of each non-tree node is at most 2. Hence, the total size of all such graphs H_x is $\theta(E)$. We construct all the spanning forests in two Borůvka-style rounds as follows. In the first round, each non-tree node in V_x selects one of its incident edges. This results in a spanning forest. Then in a cleanup phase, we contract each connected component into a pseudo-node, and remove loops and duplicate edges between pseudo-nodes. In the second round, each non-tree node in V_x selects its second edge (if present). We perform a second cleanup phase. The remaining selected edges form a forest consisting of a spanning tree over each connected component of each H_x. Clearly we can implement the selection and cleanup phases in a constant number of compressed-scanning rounds.

6. **Fact 9** *[13]. For each connected component of each* H_x, *there exists at least one tree edge* $e \in V_x$ *such that* $d(\mathrm{LCA}(\texttt{master}(e))) < d(x)$.

   In light of this fact, we do the following:
   (a) For each connected component C of each H_x, find such an edge e guaranteed by the above fact and construct an Euler Tree T_C over C rooted at e.
   (b) Compute the pre-order numbers `preorder`(e) of each non-tree edge e with respect to T_C by performing an Euler tour traversal of T_C.
   (c) Recall that the edges were ordered according to the number assigned above: `number`$(f) = (d(\mathrm{LCA}(f)), \texttt{serial}(f))$. We reorder the non-tree

edges by replacing their old serial numbers with the new preorder numbers as follows: `newnumber`$(f) = (d(\text{LCA}(f)), \texttt{preorder}(f))$.

7. Now we are ready to compute the assignment of `master` based on the new ordering of the non-tree edges. We proceed as follows:

   (a) Compute `preorder`(v) and `postorder`(v) for each vertex $v \in V$ by an Euler Tour traversal of T.

   (b) For each vertex v, let E_v denote the set of non-tree edges incident to v. If $\text{LCA}(E_v) \neq v$, then let $(u, v) \in E_v$ be an edge such that $\text{LCA}(u, v) = \text{LCA}(E_v)$. That is, the edge (u, v) such that $d(\text{LCA}(u, v))$ is minimized. If there is more than one such edge for a given vertex, choose a single edge arbitrarily.

   Assign a tuple `serial`$(u, v) = (d(\text{LCA}(u, v)), \texttt{serial}(u, v))$ to each chosen edge (u, v) combining the depth of the least common ancestor and its old serial number. Note that the edge (u, v) can be found by scanning the adjacency list of v in $E - E_T$ after the above LCA preprocessing. Thus, we can relabel the serial numbers of all chosen edges in a single compressed-scanning pass.

   As shown by Vishkin [20], all the non-tree edges that were not chosen in the previous step can be discarded for the remaining computation of `master`.

   (c) For each tree edge $e = (p(v), v)$, initialize `master`$(e) = f$, where f is the non-tree edge incident to v with minimum serial number among the edges chosen in the previous step. Note that the initial assignment of `master` can be computed for all tree edges in a single scan after sorting the chosen edges and tree edges in adjacency list order.

   (d) Perform a bottom-up traversal computation on T. For each tree edge $e = (p(v), v)$, we assign a new value for `master`$(e) = g$, where g is the edge with minimum `master`(g) among all the edges between $(p(v), v)$ and $(v, p(v))$ in the Euler Tour traversal of T.

Fact 10 *[13]. The set of non-tree edges chosen as `master` partition the tree edges into the subsets that chose them. Each `master` edges induces an ear, and the ordering of the corresponding non-tree edges results in an open ear decomposition of G.*

This fact, together with the above algorithm, give us the following theorem.

Theorem 11. *Given a biconnected graph G, and a spanning tree of G, we can construct an open ear decomposition in $O(\text{DO} - \text{Sort}(E))$ time in the DO-OEM model.*

Building on the previous theorem, we can also achieve the following result; details can be found in the full version [12].

Theorem 12. *Given a biconnected graph G and a spanning tree for G, we can find an st-numbering of G in $O(\text{DO} - \text{Sort}(E))$ time in the DO-OEM model.*

5   Conclusion

We provided several I/O-efficient algorithm for fundamental graph problems in the DO-OEM model, which are more efficient than simulations of known graph algorithms using existing ORAM simulation methods (e.g., see [4,5,9,17,18]). Moreover, our methods are based on new techniques and novel adaptations of existing paradigms to the DO-OEM model (such as our bottom-up and top-down tree computations).

References

1. Blanton, M., Steele, A., Aliasgari, M.: Data-oblivious graph algorithms for secure computation and outsourcing. In: ASIACCS, pp. 207–218 (2013)
2. Chiang, Y.J., Goodrich, M.T., Grove, E.F., Tamassia, R., Vengroff, D.E., Vitter, J.S.: External-memory graph algorithms. In: Symposium on Discrete Algorithms (SODA), pp. 139–149 (1995)
3. Damgård, I., Meldgaard, S., Nielsen, J.B.: Perfectly secure oblivious ram without random oracles. In: Ishai, Y. (ed.) TCC 2011. LNCS, vol. 6597, pp. 144–163. Springer, Heidelberg (2011)
4. Emil Stefanov, E.S., Song, D.: Towards practical oblivious RAM. In: Sion, R. (ed.) NDSS 2012 (2012)
5. Gentry, C., Goldman, K.A., Halevi, S., Julta, C., Raykova, M., Wichs, D.: Optimizing ORAM and using it efficiently for secure computation. In: De Cristofaro, E., Wright, M. (eds.) PETS 2013. LNCS, vol. 7981, pp. 1–18. Springer, Heidelberg (2013)
6. Goldreich, O., Ostrovsky, R.: Software protection and simulation on oblivious RAMs. J. ACM **43**(3), 431–473 (1996)
7. Goodrich, M.T., Mitzenmacher, M.: Privacy-preserving access of outsourced data via oblivious RAM simulation. In: Aceto, L., Henzinger, M., Sgall, J. (eds.) ICALP 2011, Part II. LNCS, vol. 6756, pp. 576–587. Springer, Heidelberg (2011)
8. Goodrich, M.T., Mitzenmacher, M., Ohrimenko, O., Tamassia, R.: Oblivious RAM simulation with efficient worst-case access overhead. In: CCSW, pp. 95–100 (2011)
9. Goodrich, M.T., Mitzenmacher, M., Ohrimenko, O., Tamassia, R.: Practical oblivious storage. In: CODASPY, pp. 13–24 (2012)
10. Goodrich, M.T., Mitzenmacher, M., Ohrimenko, O., Tamassia, R.: Privacy-preserving group data access via stateless oblivious RAM simulation. In: Symposium on Discrete Algorithms (SODA), pp. 157–167 (2012)
11. Goodrich, M.T., Ohrimenko, O., Tamassia, R.: Graph drawing in the cloud: privately visualizing relational data using small working storage. In: Didimo, W., Patrignani, M. (eds.) GD 2012. LNCS, vol. 7704, pp. 43–54. Springer, Heidelberg (2013)
12. Goodrich, M.T., Simons, J.: Data-oblivious graph algorithms in outsourced external memory. ArXiv 1409.0597 (2014). http://arxiv.org/abs/1409.0597
13. Maon, Y., Schieber, B., Vishkin, U.: Parallel ear decomposition search (EDS) and st-numbering in graphs. Theor. Comput. Sci. **47**(3), 277–298 (1986)
14. Mareš, M.: Two linear time algorithms for MST on minor closed graph classes. Arch. Math. (Brno) **40**(3), 315–320 (2004)
15. Miller, G.L., Reif, J.H.: Parallel tree contraction and its application. In: FOCS, pp. 478–489 (1985)

16. Schieber, B., Vishkin, U.: On finding lowest common ancestors: Simplification and parallelization. SIAM J. Comput. **17**(6), 1253–1262 (1988)
17. Stefanov, E., van Dijk, M., Shi, E., Fletcher, C.W., Ren, L., Yu, X., Devadas, S.: Path ORAM: an extremely simple oblivious RAM protocol. IACR Cryptology ePrint Archive 2013, p. 280 (2013)
18. Stefanov, E., Shi, E.: Oblivistore: high performance oblivious cloud storage. In: IEEE Security and Privacy, pp. 253–267 (2013)
19. Tarjan, R.E., Vishkin, U.: An efficient parallel biconnectivity algorithm. SIAM J. Comput. **14**(4), 862–874 (1985)
20. Vishkin, U.: On efficient parallel strong orientation. Inf. Process. Lett. **20**(5), 235–240 (1985)

A Dichotomy for Upper Domination in Monogenic Classes

Hassan AbouEisha[1], Shahid Hussain[1], Vadim Lozin[2(✉)],
Jérôme Monnot[3,4], and Bernard Ries[3,4]

[1] King Abdullah University of Science and Technology, Thuwal, Saudia Arabia
{hassan.aboueisha,shahid.hussain}@kaust.edu.sa
[2] DIMAP and Mathematics Institute,
University of Warwick, Coventry CV4 7AL, UK
V.Lozin@warwick.ac.uk
[3] PSL, Université Paris-Dauphine, 75775 Paris Cedex 16, France
[4] CNRS, LAMSADE UMR 7243, Paris, France
{jerome.monnot,bernard.ries}@dauphine.fr

Abstract. An upper dominating set in a graph is a minimal (with respect to set inclusion) dominating set of maximum cardinality. The problem of finding an upper dominating set is NP-hard for general graphs and in many restricted graph families. In the present paper, we study the computational complexity of this problem in monogenic classes of graphs (i.e. classes defined by a single forbidden induced subgraph) and show that the problem admits a dichotomy in this family. In particular, we prove that if the only forbidden induced subgraph is a P_4 or a $2K_2$ (or any induced subgraph of these graphs), then the problem can be solved in polynomial time. Otherwise, it is NP-hard.

1 Introduction

In a graph $G = (V, E)$, a *dominating set* is a subset of vertices $D \subseteq V$ such that any vertex outside of D has a neighbour in D. A dominating set D is *minimal* if no proper subset of D is dominating. An *upper dominating set* is a minimal dominating set of maximum cardinality. The UPPER DOMINATING SET problem (i.e. the problem of finding an upper dominating set in a graph) is known to be NP-hard [2]. On the other hand, in some restricted graph families, the problem can be solved in polynomial time, which is the case for bipartite graphs [3], chordal graphs [8], generalized series-parallel graphs [7] and graphs of bounded clique-width [4].

In the present paper, we study the complexity of the problem in monogenic classes of graphs, i.e. classes defined by a single forbidden induced subgraph. Our main result is that the problem admits a dichotomy in this family: for each

Vadim Lozin: The author gratefully acknowledges support from DIMAP - the Center for Discrete Mathematics and its Applications at the University of Warwick, and from EPSRC, grant EP/L020408/1.

© Springer International Publishing Switzerland 2014
Z. Zhang et al. (Eds.): COCOA 2014, LNCS 8881, pp. 258–267, 2014.
DOI: 10.1007/978-3-319-12691-3_20

class in the family the problem is either NP-hard or can be solved in polynomial time. Up to date, a complete dichotomy in monogenic classes was available only for VERTEX COLORING [11], MINIMUM DOMINATING SET [10] and MAXIMUM CUT [9].

The organization of the paper is as follows. In Sect. 2, we introduce basic definitions and notations related to the topic of the paper and prove some preliminary results about minimal dominating sets. In Sects. 3 and 4, we prove some NP-hardness and polynomial-time results, respectively. In Sect. 5, we summarize our arguments in a final statement.

2 Preliminaries

All graphs in this paper are simple, i.e. undirected, without loops and multiple edges. The *girth* of a graph G is the length of a shortest cycle in G. As usual, we denote by K_n, P_n and C_n a complete graph, a chordless path and a chordless cycle with n vertices, respectively. Also, $2K_2$ is the disjoint union of two copies of K_2 and a *star* is a connected graph in which all edges are incident to the same vertex, called the *center* of the star.

Let $G = (V, E)$ be a graph with vertex set V and edge set E, and let u and v be two vertices of G. If u is adjacent to v, we write $uv \in E$ and say that u and v are neighbours. The neighbourhood of a vertex $v \in V$ is the set of its neighbours; it is denoted by $N(v)$. The degree of v is the size of its neighbourhood. If the degree of each vertex of G equals 3, then G is called *cubic*.

The *complement* of a graph G, denoted $\overline{G}$, is the graph with the same vertex set in which two vertices are adjacent if and only if they are not adjacent in G. A subgraph of G is *induced* if two vertices of the subgraph are adjacent if and only if they are adjacent in G. If a graph H is isomorphic to an induced subgraph of a graph G, we say that G contains H. Otherwise we say that G is H-free.

In a graph, a *clique* is a subset of pairwise adjacent vertices, and an *independent set* is a subset of vertices no two of which are adjacent. A graph is *bipartite* if its vertices can be partitioned into two independent sets. It is well-known that a graph is bipartite if and only if it is free of odd cycles.

We say that an independent set I is *maximal* if no other independent set properly contains I. The following simple lemma connects the notion of a maximal independent set and that of a minimal dominating set.

Lemma 1. *Every maximal independent set is a minimal dominating set.*

Proof. Let $G = (V, E)$ be a graph and let I be a maximal independent set in G. Then every vertex $u \notin I$ has a neighbour in I (else I is not maximal) and hence I is dominating.

The removal of any vertex $u \in I$ from I leaves u undominated. Therefore, I is a minimal dominating set.   □

Definition 1. *Given a dominating set D and a vertex $x \in D$, we say that a vertex $y \notin D$ is a private neighbour of x if x is the only neighbour of y in D.*

Lemma 2. *Let D be a minimal dominating set in a graph G. If a vertex $x \in D$ has a neighbour in D, then it also has a private neighbour outside of D.*

Proof. If a vertex $x \in D$ is adjacent to a vertex in D and has no private neighbour outside of D, then D is not minimal, because the set $D - \{x\}$ is also dominating. □

Lemma 3. *Let G be a connected graph and D a minimal dominating set in G. If there are vertices in D that have no private neighbour outside of D, then D can be transformed in polynomial time into a minimal dominating set D' with $|D'| \leq |D|$ in which every vertex has a private neighbour outside of D'.*

Proof. Assume D contains a vertex x which has no private neighbours outside of D. Then x is isolated in D (i.e. it has no neighbours in D) by Lemma 2. On the other hand, since G is connected, x must have a neighbour y outside of D. As y is not a private neighbour of x, it is adjacent to a vertex z in D. Consider now the set $D_0 = (D - \{x\}) \cup \{y\}$. Clearly, it is a dominating set. If it is a minimal dominating set in which every vertex has a private neighbour outside of the set, then we are done. Otherwise, it is either not minimal, in which case we can reduce its size by deleting some vertices, or it has strictly fewer isolated vertices than D. Therefore, by iterating the procedure, in at most $|V(G)|$ steps we can transform D into a minimal dominating set D' with $|D'| \leq |D|$ in which every vertex has a private neighbour outside of the set. □

3 NP-hardness Results

Theorem 1. *The UPPER DOMINATING SET problem restricted to the class of planar graphs with maximum vertex degree 6 and girth at least 6 is NP-hard.*

Proof. We use a reduction from the MAXIMUM INDEPENDENT SET problem (IS for short) in planar cubic graphs, where IS is NP-hard [6]. The input of the decision version of IS consists of a simple graph $G = (V, E)$ and an integer k and asks to decide if G contains an independent set of size at least k.

Let $G = (V, E)$ and an integer k be an instance of IS, where G is a planar cubic graph. We denote the number of vertices and edges of G by n and m, respectively. We build an instance $G' = (V', E')$ of the UPPER DOMINATING SET problem by replacing each edge $e = uv \in E$ with two induced paths $u - v_e - u_e - v$ and $u - v'_e - u'_e - v$, as shown in Fig. 1.

Clearly, G' can be constructed in time polynomial in n. Moreover, it is not difficult to see that G' is a planar graph with maximum vertex degree 6 and girth at least 6.

We claim that G contains an independent set of size at least k if and only if G' contains a minimal dominating set of size at least $k + 2m$.

Suppose G contains an independent set S with $|S| \geq k$ and without loss of generality assume that S is maximal with respect to set-inclusion (otherwise, we greedily add vertices to S until it becomes a maximal independent set). Now we consider a set $D \subset V'$ containing

Fig. 1. Replacement of an edge by two paths

- all vertices of S,
- vertices v_e and v'_e for each edge $e = uv \in E$ with $v \in S$,
- exactly one vertex in $\{u_e, v_e\}$ (chosen arbitrarily) and exactly one vertex in $\{u'_e, v'_e\}$ (chosen arbitrarily) for each edge $e = uv \in E$ with $u, v \notin S$.

It is not difficult to see that D is a maximal independent, and hence, by Lemma 2, a minimal dominating, set in G'. Moreover, $|D| = |S| + 2m \geq k + 2m$.

To prove the inverse implication, we first observe the following:

- *Every minimal dominating set in G' contains either exactly two vertices or no vertex in the set $\{u_e, v_e, u'_e, v'_e\}$ for every edge $e = uv \in E$.* Indeed, assume a minimal dominating set D in G' contains at least three vertices in $\{u_e, v_e, u'_e, v'_e\}$, say u_e, v_e, u'_e. But then D is not minimal, since u_e can be removed from the set. If D contains one vertex in $\{u_e, v_e, u'_e, v'_e\}$, say u_e, then both u and v must belong to D (otherwise it is not dominating), in which case it is not minimal (u_e can be removed).
- *If a minimal dominating set D in G' contains exactly two vertices in the set $\{u_e, v_e, u'_e, v'_e\}$, then*
  - *one of them belongs to $\{u_e, v_e\}$ and the other to $\{u'_e, v'_e\}$.* Indeed, if both vertices belong to $\{u_e, v_e\}$, then both u and v must also belong to D (to dominate u'_e, v'_e), in which case D is not minimal (u_e and v_e can be removed).
  - *at most one of u and v belongs to D.* Indeed, if both of them belong to D, then D is not minimal dominating, because u and v dominate the set $\{u_e, v_e, u'_e, v'_e\}$ and any vertex of this set can be removed from D.

Now let $D \subseteq V'$ be a minimal dominating set in G' with $|D| \geq k + 2m$. If D contains exactly two vertices in the set $\{u_e, v_e, u'_e, v'_e\}$ for every edge $e = uv \in E$, then, according to the discussion above, the set $D \cap V$ is independent in G and contains at least k vertices, as required.

Assume now that there are edges $e = uv \in E$ for which the set $\{u_e, v_e, u'_e, v'_e\}$ contains no vertex of D. We call such edges D-*clean*. Obviously, both endpoints of a D-clean edge belong to D, since otherwise this set is not dominating. To prove the theorem in the situation when D-clean edges are present, we transform D into another minimal dominating set D' with no D'-clean edges and with $|D'| \geq |D|$. To this end, we do the following. For each vertex $u \in V$ incident to at least one D-clean edge, we first remove u from D, and then for each D-clean edge $e = uv \in E$ incident to u, we introduce vertices v_e, v'_e to D. Under this transformation vertex v may become redundant (i.e. its removal may result in a dominating set), in which case we remove it. It is not difficult to see that the set D' obtained in this way is a minimal dominating set with no D'-clean edges and with $|D'| \geq |D|$. Therefore, $D' \cap V$ is an independent set in G of cardinality at least k. □

Theorem 2. *The* UPPER DOMINATING SET *problem restricted to the class of complements of bipartite graphs is NP-hard.*

Proof. We use a reduction from the MINIMUM DOMINATING SET problem, which is known to be NP-hard [5]. The input of the decision version of this problem consists of a simple graph $G = (V, E)$ and an integer k. The problem asks to determine if G contains a dominating set of size at most k.

Assume an instance of the MINIMUM DOMINATING SET problem is given by a graph $G = (V, E)$ with n vertices and m edges and an integer $k \leq n - 3$. Without loss of generality, we may further assume that G is connected. We build an instance $G' = (V', E')$ of the UPPER DOMINATING SET problem where G' is the complement of a bipartite graph as follows.

- $V' = V \cup V_E \cup \{a, b\}$, where $V_E = \{v_e : e \in E\}$;
- $V \cup \{a\}$ and $V_E \cup \{b\}$ are cliques. Also, a vertex $v \in V$ is connected to a vertex $v_e \in V_E$ if and only if v is incident to $e \in E$ in G. Finally, a is connected to every vertex of $V_E \cup \{b\}$.

Clearly, this construction can be done in time polynomial in n. We claim that there is a dominant set in G of size at most k if and only if there is a minimal dominating set in G' of size at least $n - k$.

Suppose G contains a dominating set D with $|D| \leq k$. Without loss of generality, we assume that D is a minimal dominating set (otherwise we can remove some vertices from D to make it minimal). Moreover, we will assume that D satisfies Lemma 3, i.e. every vertex of D has a private neighbour outside of the set. Since D is a dominating set, for every vertex u outside of D, there is an edge e_u connecting it to a vertex in D. We claim that the set $D' = \{v_{e_u} : u \notin D\}$ is a minimal dominating set in G'. By construction, D' dominates $V_E \cup \{a, b\} \cup (V - D)$. To show that it also dominates D, assume by contradiction that a vertex $w \in D$ is not dominated by D' in G'. By Lemma 3 we know that w has a private neighbour u outside of D. But then the edge $e = uw$ is the only edge connecting u to a vertex in D. Therefore, v_e belongs to D' and hence it dominates w, contradicting our assumption. In order to show that D' is a minimal dominating set, we observe that if we remove from D' a vertex v_{e_u} with $e_u = uv$, $u \notin D$, $v \in D$, then u becomes undominated in G'. Finally, since $|D'| = n - |D|$, we conclude that $|D'| \geq n - k$.

Conversely, let $D' \subseteq V'$ be a minimal dominating set in G' with $|D'| \geq n - k$ and $n - k \geq 3$ (by assumption $k \leq n - 3$). Then D' cannot intersect both $V \cup \{a\}$ and $V_E \cup \{b\}$, since otherwise it contains exactly one vertex in each of these sets (else it is not minimal, because each of them is a clique), in which case $|D'| = 2$. Also, D' cannot be a subset of $V \cup \{a\}$, since otherwise it contains a (because a

is the only vertex of $V \cup \{a\}$ dominating b) and hence it coincides with $\{a\}$ (else it is not minimal, because a dominates the graph), in which case $|D'| = 1$. Therefore, $D' \subseteq V_E \cup \{b\}$. Also, $b \notin D'$, since otherwise D' is not minimal (i.e. b can be removed from D'). Therefore, there exists a subset of edges $F \subseteq E$ such that $D' = \{v_e \; : \; e \in F\}$. Let us denote the subgraph of G formed by the edges of F (and all their endpoints) by G_F and prove the following:

- G_F *is a spanning forest of* G, because F covers V (else D' is not dominating) and G_F is acyclic (else D' is not minimal).
- G_F *is* P_4-*free, i.e. each connected component of* G_F *is a star*, since otherwise D' is not minimal, because any vertex of D' corresponding to the middle edge of a P_4 in G_F can be removed from D'.

Let D be the set of the centers of the stars of G_F. Then D is dominating in G (since F covers V) and $|D| = n - |F| = n - |D'| \leq k$, as required. $\square$

4   Polynomial-Time Results

As we have mentioned in the introduction, the UPPER DOMINATING SET problem can be solved in polynomial time for bipartite graphs [3], chordal graphs [8] and generalized series-parallel graphs [7]. It also admits a polynomial-time solution in any class of graphs of bounded clique-width [4]. Since P_4-free graphs have clique-width at most 2 (see e.g. [1]), we make the following conclusion.

Proposition 1. *The* UPPER DOMINATING SET *problem can be solved for* P_4-*free graphs in polynomial time.*

In what follows, we develop a polynomial-time algorithm to solve the problem in the class of $2K_2$-free graphs.

We start by observing that the class of $2K_2$-free graphs admits a polynomial-time solution to the MAXIMUM INDEPENDENT SET problem (see e.g. [12]). By Lemma 2 every maximal (and hence maximum) independent set is a minimal dominating set. These observations allow us to restrict ourselves to the analysis of minimal dominating sets X such that

- X contains at least one edge,
- $|X| > \alpha(G)$,

where $\alpha(G)$ is the independence number, i.e. the size of a maximum independent set in G.

Let G be a $2K_2$-free graph and let ab an edge in G. Assuming that G contains a minimal dominating set X containing both a and b, we first explore some properties of X. In our analysis we use the following notation. We denote by

- N the neighbourhood of $\{a, b\}$, i.e. the set of vertices outside of $\{a, b\}$ each of which is adjacent to at least one vertex of $\{a, b\}$,
- A the anti-neighbourhood of $\{a, b\}$, i.e. the set of vertices adjacent neither to a nor to b,

- $Y := X \cap N$,
- $Z := N(Y) \cap A$, i.e. the set of vertices of A each of which is adjacent to at least one vertex of Y.

Since a and b are adjacent, by Lemma 2 each of them has a private neighbour outside of X. We denote by

- a^* a private neighbour of a,
- b^* a private neighbour of b.

By definition, a^* and b^* belong to $N - Y$ and have no neighbours in Y. Since G is $2K_2$-free, we conclude that

Claim 1. A is an independent set.

We also derive a number of other helpful claims.

Claim 2. $Z \cap X = \emptyset$ and $A - Z \subseteq X$.

Proof. Assume a vertex $z \in Z$ belongs to X. Then $X - \{z\}$ is a dominating set, because z does not dominate any vertex of A (since A is independent) and it is dominated by its neighbor in Y. This contradicts the minimality of X and proves that $Z \cap X = \emptyset$. Also, by definition, no vertex of $A - Z$ has a neighbour in $Y \cup \{a, b\}$. Therefore, to be dominated $A - Z$ must be included in X.   □

Claim 3. If $|X| > \alpha(G)$, then $|Y| = |Z|$ and every vertex of Z is a private neighbor of a vertex in Y.

Proof. Since every vertex y in Y belongs to X and has a neighbour in X (a or b), by Lemma 2 y must have a private neighbor in Z. Therefore, $|Z| \geq |Y|$. If $|Z|$ is strictly greater than $|Y|$, then $|X| \leq |A \cup \{a\}| \leq \alpha(G)$ (since A is independent), which contradicts the assumption $|X| > \alpha(G)$. Therefore, $|Y| = |Z|$ and every vertex of Z is a private neighbor of a vertex in Y.   □

Claim 4. If $|Y| > 1$ and $|X| > \alpha(G)$, then $Y \subseteq N(a) \cap N(b)$.

Proof. Let y_1, y_2 be two vertices in Y and let z_1, z_2 be two vertices in Z which are private neighbours of y_1 and y_2, respectively.

Assume a is not adjacent to y_1, then b is adjacent to y_1 (by definition of Y) and a^* is adjacent to z_1, since otherwise the vertices a, a^*, y_1, z_1 induce a $2K_2$ in G. Also, a^* is adjacent to z_2, since otherwise a $2K_2$ is induced by a^*, z_1, y_2, z_2. But now the vertices a^*, z_2, b, y_1 induce a $2K_2$. This contradiction shows that a is adjacent to y_1. Since y_1 has been chosen arbitrarily, a is adjacent to every vertex of Y, and by symmetry, b is adjacent to every vertex of Y.   □

Claim 5. If $|Y| > 1$ and $|X| > \alpha(G)$, then a^* and b^* have no neighbours in Z.

Proof. Assume by contradiction that a^* is adjacent to a vertex $z_1 \in Z$. By Claim 3, z_1 is a private neighbour of a vertex $y_1 \in Y$. Since $|Y| > 1$, there exists another vertex $y_2 \in Y$ with a private neighbor $z_2 \in Z$. From Claim 4, we know that b is adjacent to y_2. But then the set $\{b, y_2, a^*, z_1\}$ induces a $2K_2$. This contradiction shows that a^* has no neighbours in Z. By symmetry, b^* has no neighbours in in Z. $\qquad\Box$

The above series of claims leads to the following conclusion, which plays a key role for the development of a polynomial-time algorithm.

Lemma 4. *If $|X| > \alpha(G)$, then $|Y| = 1$ and $Y \subseteq N(a) \cap N(b)$.*

Proof. First, we show that $|Y| \le 1$. Assume to the contrary that $|Y| > 1$. By definition of a^* and Claim 2, vertex a^* has no neighbours in $A - Z$, and by Claim 5, a^* has no neighbours in Z. Therefore, $A \cup \{a^*, b\}$ is an independent set of size $|X| = |Y| + |A - Z| + 2$. This contradicts the assumption that $|X| > \alpha(G)$ and proves that $|Y| \le 1$.

Suppose now that $|Y| = 0$. Then, by Claim 3, $|Z| = 0$ and hence, by Claim 2, $X = A \cup \{a, b\}$. Also, by definition of a^*, vertex a^* has no neighbours in A. But then $A \cup \{a^*, b\}$ is an independent set of size $|X|$, contradicting that $|X| > \alpha(G)$.

From the above discussion we know that Y consists of a single vertex, say y. It remains to show that y is adjacent to both a and b. By definition, y must be adjacent to at least one of them, say to a. Assume that y is not adjacent to b. By definition of a^*, vertex a^* has no neighbours in $\{y\} \cup (A - Z)$, and by definition of Z, vertex y has no neighbours in $A - Z$. But then $(A - Z) \cup \{a^*, b, y\}$ is an independent set of size $|X| = |Y| + |A - Z| + 2$. This contradicts the assumption that $|X| > \alpha(G)$ and shows that y is adjacent to both a and b. $\qquad\Box$

Corollary 1. *If a minimal dominating set in a $2K_2$-free graph G is larger than $\alpha(G)$, then it consists of a triangle and all the vertices not dominated by the triangle.*

In what follows, we describe an algorithm $\mathcal{A}$ to find a minimal dominating set M with maximum cardinality in a $2K_2$-free graph G in polynomial time. In the description of the algorithm, given a graph $G = (V, E)$ and a subset $U \subseteq V$, we denote by $A(U)$ the anti-neighbourhood of U, i.e. the subset of vertices of G outside of U none of which has a neighbour in U.

Algorithm $\mathcal{A}$

Input: A $2K_2$-free graph $G = (V, E)$.
Output: A minimal dominating set M in G with maximum cardinality.

1. Find a maximum independent set M in G.
2. For each triangle T in G:
   - Let $M' := T \cup A(T)$.
   - If M' is a minimal dominating set and $|M'| > |M|$, then $M := M'$.
3. Return M.

Theorem 3. *Algorithm $\mathcal{A}$ correctly solves the* UPPER DOMINATING SET *problem for $2K_2$-free graphs in polynomial time.*

Proof. Let G be a $2K_2$-free graph with n vertices. In $O(n^2)$ time, one can find a maximum independent set M in G (see e.g. [12]). Since M is also a minimal dominating set (see Lemma 1), any solution of size at most $\alpha(G)$ can be ignored.

If X is a solution of size more than $\alpha(G)$, then, by Corollary 1, it consists of a triangle T and its anti-neighbourhood $A(T)$. For each triangle T, verifying whether $T \cup A(T)$ is a minimal dominating set can be done in $O(n^2)$ time. Therefore, the overall time complexity of the algorithm can be estimated as $O(n^5)$. □

5   Main Result

Theorem 4. *Let H be a graph. If H is a $2K_2$ or P_4 (or any induced subgraph of $2K_2$ or P_4), then the* UPPER DOMINATING SET *problem can be solved for H-free graphs in polynomial time. Otherwise the problem is NP-hard for H-free graphs.*

Proof. Assume H contains a cycle C_k, then the problem is NP-hard for H-free graphs

- either by Theorem 1 if $k \leq 5$, because in this case the class of H-free graphs contains all graphs of girth at least 6,
- or by Theorem 2 if $k \geq 6$, because in this case the class of H-free graphs contains the class of $\overline{K}_3$-free graphs and hence all complements of bipartite graphs.

Assume now that H is acyclic, i.e. a forest. If it contains a claw (a star whose center has degree 3), then the problem is NP-hard for H-free graphs by Theorem 2, because in this case the class of H-free graphs contains all $\overline{K}_3$-free graphs and hence all complements of bipartite graphs.

If H is a claw-free forest, then every connected component of H is a path. If H contains at least three connected components, then the class of H-free graphs contains all $\overline{K}_3$-free graphs, in which case the problem is NP-hard by Theorem 2. Assume H consists of two connected components P_k and P_t.

- If $k + t \geq 5$, then the class of H-free graphs contains all $\overline{K}_3$-free graphs and hence the problem is NP-hard by Theorem 2.
- If $k + t \leq 3$, then the class of H-free graphs is a subclass of P_4-free graphs and hence the problem can be solved in polynomial time in this class by Proposition 1.
- If $k + t = 4$, then
  - either $k = t = 2$, in which case $H = 2K_2$ and hence the problem can be solved in polynomial time by Theorem 3,
  - or $k = 4$ and $t = 0$, in which case $H = P_4$ and hence the problem can be solved in polynomial time by Proposition 1,
  - or $k = 3$ and $t = 1$, in which case the class of H-free graphs contains all $\overline{K}_3$-free graphs and hence the problem is NP-hard by Theorem 2. □

References

1. Brandstädt, A., Engelfriet, J., Le, H.-O., Lozin, V.V.: Clique-width for 4-vertex forbidden subgraphs. Theory Comput. Syst. **39**(4), 561–590 (2006)
2. Cheston, G.A., Fricke, G., Hedetniemi, S.T., Jacobs, D.P.: On the computational complexity of upper fractional domination. Discrete Appl. Math. **27**(3), 195–207 (1990)
3. Cockayne, E.J., Favaron, O., Payan, C., Thomason, A.G.: Contributions to the theory of domination, independence and irredundance in graphs. Discrete Math. **33**(3), 249–258 (1981)
4. Courcelle, B., Makowsky, J.A., Rotics, U.: Linear time solvable optimization problems on graphs of bounded clique-width. Theory Comput. Syst. **33**(2), 125–150 (2000)
5. Garey, M.R., Johnson, D.S.: Computers and Intractability: A Guide to the Theory of NP-Completeness. W. H Freeman, New York (1979)
6. Garey, M.R., Johnson, D.S., Stockmeyer, L.J.: Some simplified NP-Complete graph problems. Theor. Comput. Sci. **1**(3), 237–267 (1976)
7. Hare, E.O., Hedetniemi, S.T., Laskar, R.C., Peters, K., Wimer, T.: Linear-time computability of combinatorial problems on generalized-series-parallel graphs. In: Johnson, D.S., et al. (eds.) Discrete Algorithms and Complexity, pp. 437–457. Academic Press, New York (1987)
8. Jacobson, M.S., Peters, K.: Chordal graphs and upper irredundance, upper domination and independence. Discrete Math. **86**(1–3), 59–69 (1990)
9. Kamiński, M.: MAX-CUT and containment relations in graphs. Theor. Comput. Sci. **438**, 89–95 (2012)
10. Korobitsyn, D.V.: On the complexity of determining the domination number in monogenic classes of graphs. Diskretnaya Matematika **2**(3), 90–96 (1990)
11. Král', D., Kratochvíl, J., Tuza, Z., Woeginger, G.J.: Complexity of Coloring Graphs without Forbidden Induced Subgraphs. In: Brandstädt, A., Le, V.B. (eds.) WG 2001. LNCS, vol. 2204, pp. 254–262. Springer, Heidelberg (2001)
12. Lozin, V.V., Mosca, R.: Independent sets in extensions of $2K_2$-free graphs. Discrete Appl. Math. **146**(1), 74–80 (2005)

Algorithms for the Maximum Weight Connected k-Induced Subgraph Problem

Ernst Althaus, Markus Blumenstock[✉], Alexej Disterhoft,
Andreas Hildebrandt, and Markus Krupp

Institut für Informatik, Johannes Gutenberg-Universität, Mainz, Germany
markusblumenstock@hotmail.com,
{ernst.althaus,andreas.hildebrandt,kruppm}@uni-mainz.de,
alexej@disterhoft.de

Abstract. Finding differentially regulated subgraphs in a biochemical network is an important problem in bioinformatics. We present a new model for finding such subgraphs which takes the polarity of the edges (activating or inhibiting) into account, leading to the problem of finding a connected subgraph induced by k vertices with maximum weight. We present several algorithms for this problem, including dynamic programming on tree decompositions and integer linear programming. We compare the strength of our integer linear program to previous formulations of the k-cardinality tree problem. Finally, we compare the performance of the algorithms and the quality of the results to a previous approach for finding differentially regulated subgraphs.

Keywords: Linear programming · k-cardinality tree · Tree decomposition · Heuristics · Bioinformatics · Gene regulation

1 Introduction

1.1 Problem Definition

We are considering the following problem: given a simple graph $G = (V, E)$, edge weights $w : E \mapsto \mathbb{R}$, and an integer $k \in \{1, \ldots, |V|\}$, find a subset $V' \subseteq V$ of k vertices (i.e. $|V'| = k$) such that the subgraph induced by V' is connected and has maximum total edge weight (i.e. $\sum_{e \in E \cap (V' \times V')} w(e)$ is maximized). We call this the maximum weight connected k-induced subgraph (MWCIS) problem. If connectivity is not required, we refer to it as the MWIS problem. Both problems are easily seen to be NP-complete by a reduction from the CLIQUE problem.

There are several variants of this problem. We can have vertex scores only or additionally (i.e. the weight-function is $\sum_{e \in E \cap (V' \times V')} w(e) + \sum_{v \in V'} s(v)$), or we can sum the weights of all edges with at least one endpoint in V' (i.e. the weight-function is $\sum_{e \in E | e \cap V' \neq \emptyset} w(e)$).

Notice that the latter can be solved with the induced edge-weight objective and additional vertex scores by setting $s(v) = \sum_{uv \in E} w(u, v)$ and flipping the sign of all edge weights. As all our algorithms are capable of handling vertex scores, we restrict to the induced edge-weight objective unless stated otherwise.

© Springer International Publishing Switzerland 2014
Z. Zhang et al. (Eds.): COCOA 2014, LNCS 8881, pp. 268–282, 2014.
DOI: 10.1007/978-3-319-12691-3_21

1.2   Application to Bioinformatics

The motivation for our work comes from bioinformatics, where interactions between biochemical entities (proteins, metabolites, DNA, ...) are often represented as graphs named biochemical networks. An important application of such networks is the detection of differentially regulated (or *deregulated*) pathways or subnetworks, where we are asked to determine which parts of the network react most drastically upon environmental changes, or changes of the phenotype. To this end, we are given a biochemical network and a set of quantitative measurements for each vertex as a function of the environmental change or the phenotype. Typically, the vertices represent proteins, and the quantitative data comes in the form of expression values. An exemplary study might, for instance, want to determine which subnetworks are significantly activated or deactivated as a result of a certain type of cancer. Hence, the input would consist of a biochemical network representing the current knowledge on protein interactions in humans, as well as of measured expression values for a number of patients and a healthy control group.

Detecting deregulated subgraphs requires a measure of deregulation for a subnetwork. While simple measures, such as the addition of vertex-based expression values, can be easily established, statistically significant results require much more elaborate procedures, such as the popular Gene Set Enrichment Analysis (GSEA) [STM+05] and its variants [DPM+09], which uses careful sampling and a Kolmogorov-Smirnov test to establish whether the set of expression changes on the vertices of a given subgraph is of statistical significance. To detect subgraphs of interest, one then tries to find connected sets of vertices with maximal total vertex score, often under additional assumptions, such as the existence of regulatory cascades, where a single so-called *key player* controls the regulation of several downstream genes. The resulting subgraphs are then scored using the full GSEA procedure to establish their statistical significance.

It has recently been argued that this approach often overestimates the significance, as it ignores inconsistencies in the data: interactions can have both a *direction* (which can be integrated easily into the procedures described above) and a *polarity*, i.e., one vertex can activate or inhibit expression of the other. As a result of noise both in the network models as well as in the expression data, expression differences are often inconsistent with the polarity of the interaction. We often find, e.g., cases where two proteins A and B both show increased expression levels in, say, the diseased sample, even though expression of A is supposed to inhibit expression of B. GSEA analysis would reward the inclusion of A and B in a subset of differentially regulated genes, as both are connected and show a differential regulation. On the other hand, the inconsistency should instead make us suspicious of their relevance.

To include consistency into the scoring of subnetworks, Geistlinger et al. have recently proposed the so-called Gene Graph Enrichment Analysis (GGEA) approach [GCK+11], which replaces the vertex-based scores of GSEA with edge weights which are computed from the expression changes of both vertices incident to the edge, as well as from its polarity. While this change has been shown to lead to improved scoring of subnetworks [GCK+11], previous optimization approaches to detect the most strongly differentially regulated parts do not directly apply, as they are based on vertex scores instead of edge weights.

2 Related Work

2.1 Other Approaches for Finding Deregulated Networks

Backes et al. [BRK+11] modeled the problem of finding a deregulated subgraph differently in two ways. On the one hand, they use a vertex-score function, in which a vertex has a high score if the corresponding gene is deregulated. They do not take into account whether the sign of the deregulation is consistent with its predecessor gene. On the other hand, they consider a directed network, where the direction indicates the causing and the affected gene. Hence they require a designated root vertex that corresponds to a key player gene responsible for deregulation, from which all other vertices are reachable. We will review the integer linear programming approach by Backes et al. in Sect. 3.1.

ILP approaches for undirected, edge-weighted deregulated networks include finding paths [ZWCA08] and finding maximum connected subgraphs with vertex scores (often also called weights). The latter is called the MWCS problem, which was solved by transformation into the prize-collecting Steiner tree problem by Dittrich et al. [DKR+08].

2.2 Similar Problems

Álvarez-Miranda et al. [AMLM13] compared an ILP formulation for the prize-collecting Steiner tree problem to an ILP formulation of MWCS and the Backes approach by polyhedral comparison.

In the related k-cardinality tree problem, one searches for a tree with k edges that minimizes the sum of all edge weights (sometimes nonnegative) or vertex scores. ILP Formulations for the k-cardinality tree (and closely related problems) were given by Fischetti et al. [FHJM94], Garg [Gar96], and Ljubić [Lju04]. Recent works are by Quintão et al. [QdCM08, QadCML10], which use the Miller-Tucker-Zemlin constraints [MTZ60], and Chimani et al. [CKLM10], which compares the approaches by Fischetti et al., Garg, and Ljubić polyhedra-wise and gives the best separation routine in practice. Approximation algorithms were given by Blum et al. [BRV99] and Arora and Karakostas [AK00], and for metaheuristics, we refer to Blum and Blesa [BB05] for a comparison.

Another similar problem is the densest k-subgraph problem (DKS), which is defined by the average vertex degree of graphs. The average vertex degree of a simple graph $G = (V, E)$ is defined as $ad(G) := \frac{2|E|}{|V|}$. The maximum average degree of G is defined as the maximum of the average degrees of all subgraphs, $mad(G) := \max_{H \subseteq G} ad(H)$. This will appear again later in Sect. 3.3.

Finding the subgraph with maximum average vertex degree, i.e. the densest subgraph, is computable in polynomial time using flow techniques [Law76]. If the subgraph size is restricted to k vertices, the problem is NP-hard by a reduction from CLIQUE and can be seen as a special case of MWIS with unit weights.

Feige et al. [FKP01] give a polynomial-time algorithm that solves DKS with an approximation ratio of $\mathcal{O}(|V|^{1/3-\epsilon})$. They show that such an algorithm can be used to approximate the MWIS problem for nonnegative edge weights with a loss in the approximation ratio of $\mathcal{O}(\log |V|)$. Bhaskara et al. [BCC+10] give an algorithm using linear and semidefinite programming relaxations that achieves an $\mathcal{O}(|V|^{1/4+\epsilon})$-approximation in polynomial time and an $\mathcal{O}(|V|^{1/4})$-approximation in $\mathcal{O}(|V|^{\log |V|})$ time. Moreover, it was shown by Khot [Kho06] that under the assumption that there are no subexponential algorithms for NP-complete problems, there is no polynomial-time approximation scheme for the DKS problem.

3   Integer Linear Programming Formulations

In all following ILP approaches, there are binary variables y_v indicating whether a vertex v is selected for the subgraph, and binary variables z_{uv} indicating selected edges for the objective function $\max \sum_{e \in E} z_e w(e)$. The induced edges can be modeled by the constraints $z_{uv} \geq y_u + y_v - 1$ and $z_{uv} \leq y_u, y_v$. The objective where an edge contributes its weight when at least one end vertex is selected is modeled by $z_{uv} \leq y_u + y_v$ and $z_{uv} \geq y_u, y_v$.

3.1   Adapting the Approach by Backes et al.

In the following, we adapt the approach by Backes et al. for the MWCIS problem. The constraints for the variables y_v are adopted from Backes et al. with removal of the root vertex constraints. The most interesting set of constraints consists of those enforcing connectivity which is done as follows. We require that for every set $C \subseteq V$ with $|C| < k$, at least one adjacent vertex is also selected, i.e. $\sum_{w \in In(C)} y_w \geq y_v \; \forall v \in C \; \forall C \subseteq V$ with $|C| < k$, where $In(C)$ is the set of all vertices in $V \setminus C$ with at least one edge incident to a vertex in C.

Instead of generating this exponentially large set of constraints beforehand, a branch-and-cut procedure is used. The relaxation of the ILP is solved with the basic constraints, and we search for connected components in the subgraph induced by the (rounded) fractional solution. If there is more than one connected component, the corresponding connectivity constraints for every set of vertices that forms a connected component are added and the solving continues. If there is one connected component, it constitutes the solution.

3.2   Formulations for the k-Cardinality Tree Problem

The ILP formulation by Fischetti et al. uses binary variables y_v for vertices and b_{uv} for spanning edges in the undirected sense. Apart from the straightforward constraints that ensure k vertices and $k-1$ spanning edges (see (2) and (3) in the next section), there are an exponential number of constraints, the generalized subtour elimination constraints (GSEC). For every set $S \subseteq V$ with $|S| \geq 2$ and for $t \in S$, the constraint

$$\sum_{uv \in E(S)} b_{uv} \leq \sum_{v \in S} y_v - y_t \tag{1}$$

is added to the model, where $E(S)$ denotes the edges induced by S.

The directed cut formulation (DCUT) by Chimani et al. transforms the problem into the k-arborescence problem with binary variables $x_{u,v}$ for the directed edges and an additional root vertex with directed edges to all vertices.

Both approaches use maximum-flow problems for separation on fractional solutions. One or more minimum cuts are extracted from the solution and their corresponding constraints are added to the model. We will introduce a novel formulation in the next subsection and compare it to the existing approaches.

3.3   A Novel k-Cardinality Tree Formulation Based on the Maximum Average Degree Problem

We propose a novel formulation with binary variables for vertices and undirected spanning edges and $\mathcal{O}(|V| + |E|)$ constraints. The idea to enforce acyclicity is that the maximum average degree of a tree of k vertices is $2(k-1)/k = 2 - 2/k$, while a cyclic graph has a maximum average degree of at least two.

The following has been proven by Cohen [Coh10]: For a graph of maximum average degree z, we can distribute a value of 2 for each edge (the degree generated by it) to its endpoints, i.e. define continuous edge flow values $f_{uv,u}$ and $f_{uv,v}$ with $f_{uv,u} + f_{uv,v} = 2$, such that the total amount assigned to a vertex is at most z, i.e. $\sum_{uv \in E} f_{uv,v} \leq z$ for all $v \in V$. Furthermore, this is not possible for any value z smaller than the maximum average degree. This leads to the following mixed integer linear programming formulation, where we use an edge flow of one instead of two since the model is linear:

Variables

$$y_v \in \{0, 1\} \quad \forall\, v \in V$$
$$b_e \in \{0, 1\} \quad \forall\, e \in E$$
$$f_{uv,u}, f_{uv,v} \in \mathbb{R}_0^+ \quad \forall\, uv \in E$$

Constraints

$$\sum_{v \in V} y_v = k \tag{2}$$

$$\sum_{e \in E} b_e = k - 1 \tag{3}$$

$$b_{uv} \leq y_u, y_v \quad \forall\, uv \in E \tag{4}$$

$$f_{uv,u} + f_{uv,v} = b_{uv} \quad \forall\, uv \in E \tag{5}$$

$$\sum_{uv \in E} f_{uv,v} \leq 1 - \frac{1}{k} \quad \forall\, v \in V \tag{6}$$

The integrality constraints ensure that a non-selected vertex v does not receive any flow because the incident spanning edge variables $\{b_{uv}\}_{uv\in E}$ must be zero as well and therefore do not generate any flow to v.

However, we can use the stronger constraint $\sum_{uv\in E} f_{uv,v} \leq \left(1 - \frac{1}{k}\right) y_v$ instead of (6) to forbid some solutions of the relaxed model that would otherwise be feasible. If a vertex then received less than $\left(1 - \frac{1}{k}\right) y_v$, the remaining flow could not be absorbed because of (2) and (3), so we can even write

$$\sum_{uv\in E} f_{uv,v} = \left(1 - \frac{1}{k}\right) y_v \quad \forall\, v \in V. \tag{6a}$$

In reverse, either (2) or (3) can be dropped if (6a) is added. Furthermore, it implies the inequality $f_{uv,v} \leq \left(1 - \frac{1}{k}\right) y_v$ for all edges and since we solve b_{uv} to integrality, we can add

$$f_{uv,u}, f_{uv,v} \geq \frac{1}{k} b_{uv} \quad \forall\, uv \in E, \tag{7}$$

which reduces the amount of possible relaxed solutions even more. We call the above formulation with the improved constraints the strong Cohen formulation (Cs).

3.4   Polyhedral Comparison to Existing k-Cardinality Approaches

Chimani et al. have shown that GSEC, DCUT and the multi-commodity flow formulation by Ljubić are equivalent and strictly stronger than Garg's formulation. To compare two formulations, one compares the polyhedra defined by the constraints without integrality requirements. A formulation is said to be stronger than another if its relaxation gives a tighter upper bound than the other. The polyhedron of the strong Cohen formulation is

$$\mathcal{P}_{\mathrm{Cs}} := \{(y, b, f_u, f_v) \in [0,1]^{|V|+3|E|} \mid (y, b, f_u, f_v) \text{ satisfies } (3)-(5), (6a), (7)\}$$

and we denote the identical projection on the (y, b) subspace as $\mathrm{proj}(\mathcal{P}_{\mathrm{Cs}})$.

Likewise, we have the GSEC polyhedron

$$\mathcal{P}_{\mathrm{GSEC}} = \{(y, b) \in [0,1]^{|V|+|E|} \mid (y, b) \text{ satisfies } (1)-(3)\}.$$

Lemma 3.1. GSEC *and* Cs *are not comparable.*

Proof. Consider the graph with two connected components consisting of three interconnected vertices each and $k = 6$. Cs has an LP-solution by setting $f_{uv,u} = f_{uv,v} = 5/12$ for every edge uv (Fig. 1), but GSEC is clearly infeasible since for the set of three vertices in each component, the total value of the induced edges can be two at most, but all edge variables must sum to five.

On the other hand, consider the graph consisting of a path of three vertices and an isolated vertex, and $k = 2$. GSEC allows an LP solution with $y = (\frac{1}{2}, \frac{1}{2}, \frac{1}{2}, \frac{1}{2})$ by setting the two edge variables to $\frac{1}{2}$ (Fig. 2).

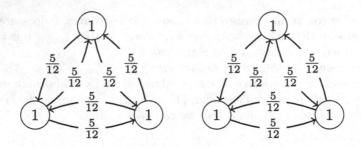

Fig. 1. A CS LP solution for $k = 6$ on a graph with six vertices and two connected components. There are no GSEC and DCUT LP solutions for this instance.

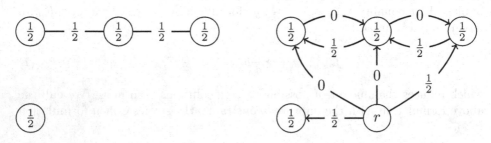

Fig. 2. A graph with four vertices with GSEC (left) and DCUT (right) LP solutions for $k = 2$. There is no CS LP solution for this choice of $y = (\frac{1}{2}, \frac{1}{2}, \frac{1}{2}, \frac{1}{2})$.

CS has no LP-solution with $y = (\frac{1}{2}, \frac{1}{2}, \frac{1}{2}, \frac{1}{2})$ because then, the isolated vertex would have to absorb an edge flow of exactly $\frac{1}{4}$ by (6a), but it has no incident edges to produce it. Note that constraint (7) is not required for this part of the proof.   □

Theorem 3.2. CS *becomes strictly stronger than* GSEC *and its equivalent formulations by adding the generalized subtour elimination constraints.*

Proof. Add the generalized subtour elimination constraints (1) to the CS formulation (note that (4) is implied by (1)). We obtain $\text{proj}(\mathcal{P}_{CS}) \cap \mathcal{P}_{GSEC}$ on the variable space (y, b). By Lemma 3.1, it is a proper subset of $\mathcal{P}_{GSEC}$.   □

4   Dynamic Programming on a Tree Decomposition

Trees have many desirable properties: they are sparse, connected acyclic graphs that decompose into smaller trees when an edge or a vertex is removed. This often allows us to consider separated subproblems in a dynamic programming algorithm.

The tree decomposition and treewidth of a graph generalize this concept. A good introduction can be found in Kleinberg and Tardos [KT05]. Graphs similar to a tree have small treewidth and many NP-complete become tractable for graphs with bounded treewidth.

Definition 4.1. *Let $G = (V, E)$ be an undirected graph. A tree decomposition $(T, \{V_t\}_{t \in T})$ consists of a tree T on a set of vertices different from V, which we will call nodes to avoid confusion, and a subset $V_t \subseteq V$ for every node $t \in T$, which is called bag (or piece). The following properties must be satisfied:*

1. (Node coverage) Every vertex $v \in V$ belongs to at least one bag V_t.
2. (Edge coverage) For every edge $\{u, v\} \in E$, at least one bag V_t contains u and v.
3. (Coherence) Let $t_1, t_2, t_3 \in T$ where t_2 lies on a path from t_1 to t_3. Then, if a vertex $v \in V$ belongs to both V_{t_1} and V_{t_3}, it also belongs to V_{t_2}.

Definition 4.2. *The width of a tree decomposition $(T, \{V_t\}_{t \in T})$ for a graph G is defined as $tw(T, \{V_t\}) = \max_t |V_t| - 1$.*

The treewidth of G, denoted by $tw(G)$, is the minimum width of all tree decompositions for G.

Deciding whether a graph has treewidth k is NP-complete, but solvable in linear time if k is constant [Bod96].

It is useful to introduce a special type of tree decomposition which is the analogue of a rooted binary tree:

Definition 4.3. *A tree decomposition $(T, \{V_t\}_{t \in T})$ for a graph G is nice if*

1. There is a root $r \in T$.
2. Every node t has at most 2 children.
3. If a node t is a leaf, then $|V_t| = 1$.
4. If a node t has exactly one child t', then V_t and $V_{t'}$ differ by exactly one vertex. If $|V_t| = |V_{t'}| + 1$, then t is called an *introduce* node, and if $|V_t| + 1 = |V_{t'}|$, then t is called a *forget* node.
5. If a node t has exactly two children s and x, then $V_x = V_t = V_s$, and t is called a *join* node.

Any tree decomposition can be refined into a nice decomposition with the same treewidth with a linear addition of nodes.

We now describe a dynamic programming algorithm on nice tree decompositions for the MWCIS problem. The algorithm is similar to those for the (prize-collecting) Steiner tree problem and the k-cardinality tree problem as proposed by Chimani et al. [CMZ12]. The key idea is to build the solution bottom-up (leaves to root) on tables tab_i for every bag i of the (nice) tree decomposition. Each table holds all possible sub-solutions for the vertex set of its bag. Since the treewidth is bounded, so is the size of every bag and exponentials become constants. An efficient encoding scheme for the solutions is used to save runtime and memory. Given a tree decomposition of width tw, the algorithm for the k-cardinality tree problem runs in $\mathcal{O}(B_{tw+2}^2(tw + k^2)|V|)$ time where B_i denotes the i-th Bell number.

For our problem, consider an undirected graph $G = (V, E)$ and a nice tree decomposition $(T, \{V_t\}_{t \in T})$ rooted at $r \in T$. The dynamic programming algorithm computes bottom-up a table $W(t, \mathcal{P}, a)$ of values for each node $t \in T$ and a configuration $(\mathcal{P}, a)$: Let $\mathcal{P}$ be an arbitrary partition of a subset of V_t, $\mathcal{P} = \{P_1, ..., P_l\}$, then $W(t, \mathcal{P}, a)$ is the maximum weight of a subgraph induced by a vertices in connected components $V_1, ..., V_l \subset V_t$ such that $P_i = V_i \cap (\bigcup_{j=1}^{l} P_j)$.

The maximum of $W(t, \mathcal{P}, k)$ for $t \in T$ and $\mathcal{P}$ containing a single set is the value of the maximum connected induced subgraph. An entry $W(t, \mathcal{P}, a)$ is computed bottom-up from the children of t in the following way.

Leaf node. Let t be a leaf node with $V_t = \{v\}$. We create entries $W(t, \{\}, 0) = 0$ and $W(t, \{\{v\}\}, 1) = 0$.

Introduce node. Let t be an introduce node with child t' and $\{v\} = V_t \setminus V_{t'}$ the vertex which is added. For every $W(t', \mathcal{P}, a)$, create an entry $W(t, \mathcal{P}, a) = W(t', \mathcal{P}, a)$ and $W(t, \mathcal{P}^v, a + 1) = W(t, \mathcal{P}, a) + \sum_{uv \in E: u \in \bigcup_{P \in \mathcal{P}} P} w(u, v)$, where $\mathcal{P}^v$ is defined to be the partition $\mathcal{P}$ where $\{v\}$ is added and all sets adjacent to v are united with $\{v\}$.

Forget node. Let t be a forget node with child t' and $\{v\} = V_{t'} \setminus V_t$ the vertex which is deleted. For every configuration $(\mathcal{P}', a)$ of t', we set $W(t, \mathcal{P}, a)$ to $W(t', \mathcal{P}', a)$ where $\mathcal{P}$ is the partition $\mathcal{P}'$ with v removed from the containing set, if present at all.

Join node. Let t be a join node with children t_1, t_2. For every $(t_1, \mathcal{P}_1, a_1)$ and $(t_2, \mathcal{P}_2, a_2)$ where $\mathcal{P}_1$ and $\mathcal{P}_2$ are partitions of the same subset, set $W(t, \mathcal{P}, a_1 + a_2 - \#\mathcal{P}_1)$ to $W(t_1, \mathcal{P}_1, a_1) + W(t_2, \mathcal{P}_2, a_2)$ where $\mathcal{P}$ is the partition obtained by uniting all subsets U_1, U_2 of $\mathcal{P}_1$ such that there exist $u_1 \in U_1$ and $u_2 \in U_2$ that are in a common set in $\mathcal{P}_2$, and $\#\mathcal{P}_1$ denotes the total number of vertices contained in the partition $\mathcal{P}_1$.

5   Heuristics

5.1   Vertex Score Heuristic

Edge-weights can be approximated by vertex weights. This needs less variables in the ILP formulations and often speeds up the computation significantly. Setting $s(v) = \frac{1}{2} \sum_{uv \in E} w(uv)$ for a vertex v is a compromise between the induced weight objective and the objective which counts every edge where at least one incident vertex is selected.

5.2   A Greedy Heuristic

Our simple greedy heuristic grows the subgraph as one connected component, starting with a source vertex $s \in V$. The next vertex to be added is chosen in a greedy manner until k vertices have been found. This procedure is started once in every vertex to ensure the whole graph is explored. This does not yield the optimal result in general. A similar algorithm was given by Fischetti et al. [FHJM94].

5.3   Simulated Annealing

Simulated annealing is a local search procedure capable of escaping from local optima. In every iteration, the current solution is changed a little. This change is accepted with some probability depending on the difference in the objective values and a gradually decreasing virtual temperature.

Simulated annealing has been used in similar problems for finding subnetworks in regulatory networks [IOSS02]. For our problem, we start the simulated annealing algorithm with an arbitrary connected subgraph with k vertices. In every iteration, one randomly chosen vertex is removed. Then, a vertex adjacent to any of the remaining vertices is chosen randomly and added to the subgraph. We evaluate this new subgraph and accept the change depending on the evaluation difference to the previous subgraph and the current temperature. If the change is not accepted, it is reverted. It is easy to see that full search space is maintained (assuming the graph is connected) because there is always a sequence of changes that leads us to any given connected k-induced subgraph.

5.4   Heuristic Based on the Tree Decomposition Approach

Even though the treewidth of our input graphs is too large to use our algorithm based on the tree decomposition, we can use it in a heuristic: we create a large subgraph of the input with small treewidth and solve the problem on this graph. We compute a subgraph of small treewidth in a Kruskal-fashion: we add the edges in the order of decreasing weight as long as the treewidth stays small. As we have to compute the treewidth $|E|$ times, we use a heuristic to compute it.

We note that with this heuristic, we find much better subgraphs of small treewidth as in the approach by Fix et al. [FCBZ12].

6   Experiments

6.1   Benchmark Setting and Input

We performed tests on a real-world graph arising in the study of regulatory networks (6270 vertices, 8650 edges). This graph was formulated by the integration of the Reactome pathway knowledgebase [CMH+14]. This open-data resource for human biochemical interactions is a widespread analysis tool in which each vertex represents an encoded gene, protein or chemical compound, and each edge represents a directed biochemical interaction. Furthermore, we conducted tests on randomly generated Erdős-Rényi graphs with $|V| = 3,000$ and $|E| = 10,000$ and edge weights from a normal distribution.

The tests were carried out on an Intel Core 2 P8600 CPU with 8 GB DDR3-RAM. The tree decomposition algorithm was implemented in C++ using the TreeD library [Sub07]. All other algorithms were programmed in Java 7 [Ora12]. The integer linear programs were solved using Gurobi 5.6 [Gur14], which is free for academic purposes. The solution of the greedy heuristic was given as an initial feasible solution for the ILP solvers, which provides them with a lower bound to the optimum value.

The strong Cohen formulation (Cs) for which the results are shown did not include GSEC or equivalent constraints. A naïve implementation of the DCUT separation did not improve performance for our instances. However, according to Chimani et al. [CKLM10], advanced techniques to extract more minimum cuts from the maximum flow problem can significantly speed up the computation.

The simulated annealing algorithm was run for 100, 000 steps per instance.

6.2   Quality of Solutions

We compared the objective values obtained by the different algorithms for our real-world graph. Here, we also included edges where one endpoint is chosen for the subgraph. The results are shown in Fig. 3. The tests run on randomly generated graphs showed similar results.

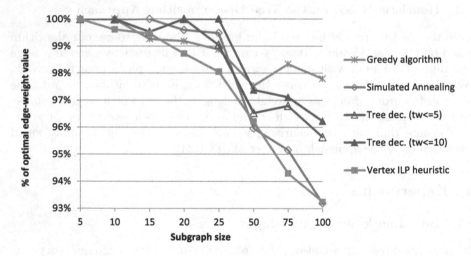

Fig. 3. Quality results of heuristics for an edge-weighted real-world graph.

To assess the biological reliability of the algorithms, we modeled an optimal biological subgraph with 15 vertices into our real-world graph. In this subgraph, the polarity (activation and inhibition) of each edge as well as its corresponding vertices were designed such that they represent a consistent biological process.

Additional instances of this modified graph were generated which exhibit several levels of additive Gaussian noise. We analyzed how well the optimal biological subgraph could be recovered from these instances. As a result, our model outreaches Backes et al. and identified more vertices in the graph for all noise levels with the induced edge weight objective (Fig. 4). Objectives that

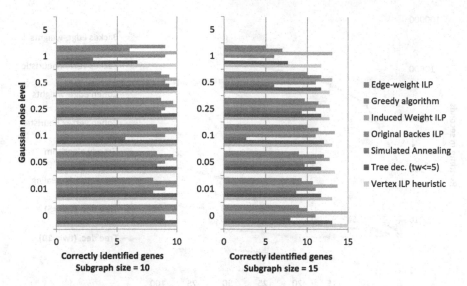

Fig. 4. Number of correctly identified genes (of 15) for $k = 10$ (left) and $k = 15$ (right) for different levels of additive Gaussian noise, averaged over three runs.

include not only induced, but all incident edges exhibit the problem that the algorithm can collect weight from crucial vertices without selecting them.

6.3   Performance Results

The benchmark results show that the integer linear programming based algorithms can compete with the greedy heuristic and simulated annealing for small values of k, but their runtimes increase quickly with the subgraph size (Fig. 5) and for worse data. The tree decomposition algorithm was generally fast for a treewidth bound of five, and its runtime is more predictable. Note that the time to build the tree decomposition with a heuristic is negligible for this bound.

Our new integer linear program performed better than the one adopted from Backes et al. for $k \geq 25$ with the edge-weight objectives, which was confirmed on the randomly generated instances. Both ILP approaches needed up to several gigabytes of RAM to solve with Gurobi, where the Backes approach usually needs more space. The space and time needed for the dynamic programming algorithm heavily depends on the treewidth bound. For a treewidth bound of five, the algorithm produced good results in minutes while needing less than 200MB of memory. For a treewidth bound of ten, it needed 5 GB, several hours, and the results improved just slightly.

The vertex-score integer linear programs were often faster than their edge-weight counterparts because they need less variables and constraints.

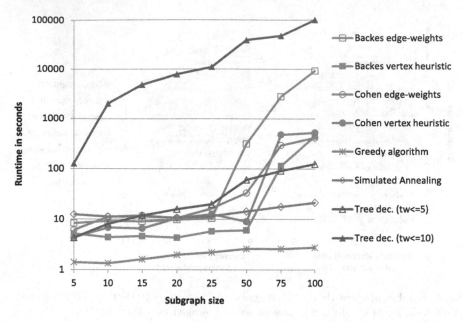

Fig. 5. Performance results of different approaches for an edge-weighted real-world graph.

7   Conclusion and Outlook

We defined a new model to compute deregulated subgraphs in a regulatory network, based on the GGEA concept introduced by Geistlinger et al. [GCK+11], that results in finding a connected induced subgraph of maximal weight.

Several algorithms were tested on the model, including a novel mixed integer linear programming method and an algorithm based on tree decompositions. The linear program can be combined with previous ILP approaches to get a stronger formulation, which will be addressed in future implementations. Due to immense treewidths of regulatory networks, the tree decomposition algorithm uses a heuristic, which initially identifies a subgraph with large weight and small treewidth. The new model yields better results than the previous approach by Backes et al. [BRK+11], as the polarity of the edges is taken into account. Our algorithms showed better performance than the adaption of the Backes approach for undirected graphs for large subgraph sizes.

References

[AK00] Arora, S., Karakostas, G.: A $2 + \epsilon$ approximation algorithm for the k-MST problem. In: Proceedings of the Eleventh Annual ACM-SIAM Symposium on Discrete Algorithms, SODA '00, pp. 754–759. Society for Industrial and Applied Mathematics, Philadelphia (2000)

[AMLM13] Álvarez Miranda, E., Ljubić, I., Mutzel, P.: The maximum weight connected subgraph problem. In: Jünger, M., Reinelt, G. (eds.) Facets of Combinatorial Optimization, pp. 245–270. Springer, Heidelberg (2013)

[BB05] Blum, C., Blesa, M.J.: New metaheuristic approaches for the edge-weighted K-cardinality tree problem. Comput. Oper. Res. **32**(6), 1355–1377 (2005)

[BCC+10] Bhaskara, A., Charikar, M., Chlamtac, E., Feige, U., Vijayaraghavan, A.: Detecting high log-densities - an $O(n^{1/4})$ approximation for densest k-subgraph. In: Proceedings of the Forty-second ACM Symposium on Theory of Computing, STOC '10, pp. 201–210. ACM, New York (2010)

[Bod96] Bodlaender, H.L.: A linear-time algorithm for finding tree-decompositions of small treewidth. SIAM J. Comput. **25**(6), 1305–1317 (1996)

[BRK+11] Backes, C., Rurainski, A., Klau, G.W., Müller, O., Stöckel, D., Gerasch, A., Küntzer, J., Maisel, D., Ludwig, N., Hein, M., Keller, A., Burtscher, H., Kaufmann, M., Meese, E., Lenhof, H.-P.: An integer linear programming approach for finding deregulated subgraphs in regulatory networks. Nucleic Acids Res. (2011)

[BRV99] Blum, A., Ravi, R., Vempala, S.: A constant-factor approximation algorithm for the k-MST problem. J. Comput. Syst. Sci. **58**(1), 101–108 (1999)

[CKLM10] Chimani, M., Kandyba, M., Ljubić, I., Mutzel, P.: Obtaining optimal K-cardinality trees fast. J. Exp. Algorithmics **14**, 5:2.5–5:2.23 (2010)

[CMH+14] Croft, D., Mundo, A.F., Haw, R., Milacic, M., Weiser, J., Wu, G., Caudy, M., Garapati, P., Gillespie, M., Kamdar, M.R., Jassal, B., Jupe, S., Matthews, L., May, B., Palatnik, S., Rothfels, K., Shamovsky, V., Song, H., Williams, M., Birney, E., Hermjakob, H., Stein, L., D'Eustachio, P.: The reactome pathway knowledgebase. Nucleic Acids Res. **42**(D1), D472–D477 (2014)

[CMZ12] Chimani, M., Mutzel, P., Zey, B.: Improved Steiner tree algorithms for bounded treewidth. J. Discrete Algorithms **16**, 67–78 (2012)

[Coh10] Cohen, N.: Several graph problems and their LP formulation (explanatory supplement for the Sage graph library), July 2010. http://hal.inria.fr/inria-00504914

[DKR+08] Dittrich, M.T., Klau, G.W., Rosenwald, A., Dandekar, T., Müller, T.: Identifying functional modules in protein–protein interaction networks: an integrated exact approach. Bioinformatics **24**(13), 223–231 (2008)

[DPM+09] Dinu, I., Potter, J.D., Mueller, T., Liu, Q., Adewale, A.J., Jhangri, G.S., Einecke, G., Famulski, K.S., Halloran, P., Yasui, Y.: Gene-set analysis and reduction. Briefings Bioinf. **10**, 24–34 (2009)

[FCBZ12] Fix, A., Chen, J., Boros, E., Zabih, R.: Approximate MRF inference using bounded treewidth subgraphs. In: Fitzgibbon, A., Lazebnik, S., Perona, P., Sato, Y., Schmid, C. (eds.) ECCV 2012, Part I. LNCS, vol. 7572, pp. 385–398. Springer, Heidelberg (2012)

[FHJM94] Fischetti, M., Hamacher, H.W., Jørnsten, K., Maffioli, F.: Weighted k-cardinality trees: complexity and polyhedral structure. Networks **24**(1), 11–21 (1994)

[FKP01] Feige, U., Kortsarz, G., Peleg, D.: The dense k-subgraph problem. Algorithmica **29**(3), 410–421 (2001)

[Gar96]   Garg, N.: A 3-approximation for the minimum tree spanning k vertices. In: Proceedings of the 37th Annual Symposium on Foundations of Computer Science, pp. 302–309, October 1996

[GCK+11]   Geistlinger, L., Csaba, G., Küffner, R., Mulder, N., Zimmer, R.: From sets to graphs: towards a realistic enrichment analysis of transcriptomic systems. Bioinformatics 27(13), i366–i373 (2011)

[Gur14]   Gurobi Optimization, Inc., Gurobi Optimizer Reference Manual (2014). http://www.gurobi.com/documentation/5.0/reference-manual/

[IOSS02]   Ideker, T., Ozier, O., Schwikowski, B., Siegel, A.F.: Discovering regulatory and signalling circuits in molecular interaction networks. Bioinformatics 18(Suppl. 1), S233–S240 (2002)

[Kho06]   Khot, S.: Ruling out PTAS for graph min-bisection, dense k-subgraph, and bipartite clique. SIAM J. Comput. 36, 1025–1071 (2006)

[KT05]   Kleinberg, J., Tardos, E.: Algorithm Design. Addison-Wesley Longman Publishing Co. Inc., Boston (2005)

[Law76]   Lawler, E.: Combinatorial Optimization: Networks and Matroids. Holt, Rinehart and Winston, New York (1976)

[Lju04]   Ljubić, I.: Exact and memetic algorithms for two network design problems. Ph.D. thesis, Technische Universität Wien (2004). https://www.ads.tuwien.ac.at/publications/bib/pdf/ljubicPhD.pdf

[MTZ60]   Miller, C.E., Tucker, A.W., Zemlin, R.A.: Integer programming formulation of traveling salesman problems. J. ACM 7(4), 326–329 (1960)

[Ora12]   Oracle corporation. Java Platform, Standard Edition 7 (2012). http://docs.oracle.com/javase/7/docs/api/

[QadCML10] Quintão, F.P., da Cunha, A.S., Mateus, G.R., Lucena, A.: The k-cardinality tree problem: reformulations and lagrangian relaxation. Discrete Appl. Math. 158(12), 1305–1314 (2010)

[QdCM08]   Quintão, F.P., da Cunha, A.S., Mateus, G.R.: Integer programming formulations for the k-cardinality tree problem. Electron. Notes Discrete Math. 30, 225–230 (2008)

[STM+05]   Subramanian, A., Tamayo, P., Mootha, V.K., Mukherjee, S., Ebert, B.L., Gillette, M.A., Paulovich, A., Pomeroy, S.L., Golub, T.R., Lander, E.S., Mesirov, J.P.: Gene set enrichment analysis: a knowledge-based approach for interpreting genome-wide expression profiles. Proc. Natl Acad. Sci. U.S.A. 102(43), 15545–15550 (2005)

[Sub07]   Sathiamoorthy Subbarayan. TreeD: A Library for Tree Decomposition, July 2007. http://itu.dk/people/sathi/treed/

[ZWCA08]   Zhao, X.-M.M., Wang, R.-S.S., Chen, L., Aihara, K.: Uncovering signal transduction networks from high-throughput data by integer linear programming. Nucleic Acids Res. 36(9), e48 (2008)

Algorithms for Cut Problems on Trees

Iyad Kanj[1], Guohui Lin[2], Tian Liu[3], Weitian Tong[2], Ge Xia[4(✉)], Jinhui Xu[5], Boting Yang[6], Fenghui Zhang[7], Peng Zhang[8], and Binhai Zhu[9]

[1] School of Computing, DePaul University, 243 S. Wabash Avenue,
Chicago, IL 60604, USA
ikanj@cs.depaul.edu
[2] Department of Computing Science, University of Alberta,
Edmonton, Alberta T6G 2E8, Canada
{guohui,weitiang}@ualberta.ca
[3] School of Electronic Engineering and Computer Science, Peking University,
Beijing 100871, China
lt@pku.edu.cn
[4] Department of Computing Science, Lafayette college, Easton, PA 18042, USA
xiag@lafayette.edu
[5] Department of Computer Science and Engineering, State University of New York
at Buffalo (SUNY Buffalo), 338 Davis Hall, Buffalo, NY 14260, USA
jinhui@buffalo.edu
[6] Department of Computer Science, University of Regina,
Regina, Saskatchewan S4S 0A2, Canada
boting@cs.uregina.ca
[7] Google Kirkland, 747 6th Street South, Kirkland, WA 98033, USA
fhzhang@gmail.com
[8] School of Computer Science and Technology, Shandong University,
Jinan 250101, China
algzhang@sdu.edu.cn
[9] Department of Computer Science, Montana State University,
Bozeman, MT 59717, USA
bhz@cs.montana.edu

Abstract. We study the MULTICUT ON TREES and the GENERALIZED MULTIWAY CUT ON TREES problems. For the MULTICUT ON TREES problem, we present a parameterized algorithm that runs in time $O^*(\rho^k)$, where $\rho = \sqrt{\sqrt{2}+1} \approx 1.555$ is the positive root of the polynomial $x^4 - 2x^2 - 1$. This improves the current-best algorithm of Chen *et al.* that runs in time $O^*(1.619^k)$. By reducing GENERALIZED MULTIWAY CUT ON TREES to MULTICUT ON TREES, our results give a parameterized algorithm that solves the GENERALIZED MULTIWAY CUT ON TREES problem in time $O^*(\rho^k)$. We also show that the GENERALIZED MULTIWAY CUT ON TREES problem is solvable in polynomial time if the number of terminal sets is a fixed constant.

© Springer International Publishing Switzerland 2014
Z. Zhang et al. (Eds.): COCOA 2014, LNCS 8881, pp. 283–298, 2014.
DOI: 10.1007/978-3-319-12691-3_22

1   Introduction

We consider the following problems on trees:

MULTICUT ON TREES (MCT)
Given: A tree T and a set R of pairs of vertices of T called *terminals*:
$R = \{(u_1, v_1), \ldots, (u_q, v_q)\}$
Parameter: k
Question: Is there a set of at most k edges in T whose removal disconnects each u_i from v_i, for $i = 1, \ldots, q$?

For convenience, we will refer to a pair of terminals $(u_i, v_i) \in R$ as a *request*, and we will also say that u_i *has a request to* v_i, and vice versa.

GENERALIZED MULTIWAY CUT ON TREES (GMWCT)
Given: A tree T and a collection of terminal sets $S_1, \ldots S_q$
Parameter: k
Question: Is there a set of at most k edges in T whose removal disconnects each pair of terminals in the same terminal set S_i, for $i = 1, \ldots, q$?

The GMWCT problem generalizes the well-known MULTIWAY CUT ON TREES (MWCT) problem in which there is only one terminal set (i.e., $q = 1$).

The MCT problem has applications in networking [7], and weighted versions of the problem were studied before (e.g., see [17]). The MCT problem is known to be NP-complete, and its optimization version is APX-complete and admits a 2-approximation [12]. Under the Unique Games Conjecture, the MCT problem cannot be approximated within $2 - \epsilon$ for any positive ϵ [15]. From the parameterized complexity perspective, Guo and Niedermeier [13] showed that MCT is fixed-parameter tractable by giving an $O^*(2^k)$ time algorithm. (The asymptotic notation $O^*(f(k))$ suppresses any polynomial factor in the input length.) They also showed that MCT has an exponential-size kernel. Bousquet *et al.* improved the upper bound on the kernel size for MCT to $O(k^6)$ [3], which was improved very recently by Chen *et al.* to $O(k^3)$ [4]. Chen *et al.* also gave a parameterized algorithm for the problem running in time $O^*(1.619^k)$ [4].

On the other hand, the MWCT problem was proven to be solvable in polynomial time in [6,8]: Chopra and Rao [6] first gave a polynomial-time greedy algorithm; later, Costa and Billionnet [8] presented a linear time dynamic programming algorithm. Very recently, Liu and Zhang [18] generalized the MWCT problem from one terminal set to allowing multiple terminal sets — the GMWCT defined above. They showed that the GMWCT problem is fixed-parameter tractable by reducing it to the MCT problem [18]. Clearly, GMWCT is NP-complete by a trivial reduction from the MCT problem. Liu and Zhang asked about the complexity of the GMWCT problem if the number of terminal sets is a constant [18]. We note that generalizations of the GMWCT problem to graphs were also studied (e.g., see [1]).

The MULTICUT and MULTIWAY CUT problems on general graphs are very important problems that have been studied extensively. Marx [19] studied the parameterized complexity of several graph separation problems, including MULTICUT and MULTIWAY CUT on general graphs. Recently, the MULTICUT problem on general graphs was shown to be fixed-parameter tractable independently by Bousquet et al. [2] and by Marx and Razgon [21], answering an outstanding open problem in parameterized complexity theory. Very recently, Chitnis et al. [5] proved that the MULTIWAY CUT problem on directed graphs is fixed-parameter tractable when parameterized by the size of the solution. Also very recently, Klein and Marx [16] and Marx [20] gave upper bounds and lower bounds, respectively, on the parameterized complexity of the PLANAR MULTIWAY CUT problem parameterized by the number of terminals.

In this paper[1] we present a parameterized algorithm for the MCT problem that runs in time $O^*(\rho^k)$, where $\rho = \sqrt{\sqrt{2}+1} \approx 1.555$ is the positive root of the polynomial $x^4 - 2x^2 - 1$. This improves the current-best algorithm of Chen et al. [4] that runs in time $O^*(1.619^k)$. This improvement is obtained by a deeper analysis of the connection between the MCT problem and the VERTEX COVER problem, first used in the paper of Chen et al. [4]. For the GMWCT problem, we show that it is solvable in polynomial time when the number of terminal sets is a constant; this answers the open question posed by Liu and Zhang. By reducing the (weighted) GMWCT problem to the MCT problem, our result implies that the GMWCT problem is also solvable in $O^*(\rho^k)$ time. We note that, very recently and independently, Deng et al. [9] gave a polynomial time algorithm for the GENERALIZED MULTIWAY CUT problem on graphs of bounded treewidth, which implies a polynomial time algorithm for the GMWCT problem. However, the running time of their algorithm is worse than that presented in the current paper by an exponential factor of q.

2   Preliminaries

We assume familiarity with basic graph theory and parameterized complexity. For more information, we refer the reader to [10,11,22,23].

For a graph H we denote by $V(H)$ and $E(H)$ the sets of vertices and edges of H, respectively. For a vertex $v \in H$, $H - v$ denotes $H[V(H) \setminus \{v\}]$, and for a subset of vertices $S \subseteq V(H)$, $H - S$ denotes $H[V(H) \setminus S]$. By *removing* a subgraph H' of H we mean removing $V(H')$ from H to obtain $H - V(H')$. Two vertices u and v in H are said to be *adjacent* or *neighbors* if $uv \in E(H)$. For two vertices $u, v \in H$, we denote by $H - uv$ the graph $(V(H), E(H) \setminus \{uv\})$. By *removing* an edge uv from H we mean setting $H = H - uv$. For a subset of edges $E' \subseteq E(H)$, we denote by $H - E'$ the graph $(V(H), E(H) \setminus E')$. For a vertex $v \in H$, $N(v)$ denotes the set of neighbors of v in H; the *degree* of a vertex v in H, denoted $\deg_H(v)$, is $|N(v)|$. The *degree* of H, denoted $\Delta(H)$, is $\Delta(H) = \max\{\deg_H(v) : v \in H\}$. A *vertex cover* for a graph H is a set of vertices such that each edge in H is incident to at least one vertex in this set. We denote by $\tau(H)$ the size/cardinality of a *minimum* vertex cover of H.

---

[1] See [14] for a complete version of this paper with full proofs.

Given a *tree* T, a non-leaf vertex is called *internal*. For two vertices $u, v \in T$, the *distance* between u and v in T, denoted $dist_T(u, v)$, is the length of the (unique) path connecting them (defined as the number of edges on the path). A leaf $x \in T$ is said to be *attached* to vertex u if u is the (unique) neighbor of x.

Let T be a tree with root r. For a vertex $u \neq r$ in $V(T)$, we denote by $\pi(u)$ the parent of u in T. A *sibling* of u is a child $v \neq u$ of $\pi(u)$ (if exists); an *uncle* u of v is a sibling of $\pi(v)$, and in this case v is said to be a *nephew* of u. For a vertex $u \in V(T)$, T_u denotes the subtree of T rooted at u. The *children* of a vertex u in $V(T)$ are the vertices in $N(u)$ if $u = r$, and in $N(u) - \pi(u)$ if $u \neq r$. A vertex u is a *grandparent* of a vertex v if $\pi(v)$ is a child of u, and v is a *grandchild* of u.

A *parameterized problem* is a set of instances of the form (x, k), where $x \in \Sigma^*$ for a finite alphabet set Σ, and k is a non-negative integer called the *parameter*. A parameterized problem Q is *fixed parameter tractable* (FPT), if there exists an algorithm that on input (x, k) decides if (x, k) is a yes-instance in time $f(k)|x|^{O(1)}$, where f is a computable function independent of $|x|$.

Let T be a tree and R a set of requests. A subset of edges $E' \subseteq E(T)$ is said to be an (edge) *cut* for R if for every request $(u, v) \in R$, there is no path between u and v in $T - E'$. A cut E' is *minimum* if its size/cardinality is the minimum among all cuts. Let (T, R, k) be an instance of the (parameterized, in the sequel dropped whenever there is no confusion) MCT. For an edge $uv \in E(T)$, if we know that it can be included in the target solution, we can remove uv from T and decrement the parameter k by 1; we say in this case that we *cut* edge uv. By *cutting* a leaf we mean cutting the unique edge incident to it. If T is a rooted tree and $u \in T$ is not the root, we say that we *cut* u to mean that we cut the edge $u\pi(u)$. On the other hand, if we know that edge uv can be excluded from the target solution, we say in this case that edge uv is *kept*, and we can *contract* it to create a new vertex w with neighbors $(N(u) \cup N(v)) \setminus \{u, v\}$; subsequently, any request in R of the form (u, x) or (v, x) is replaced by a request (w, x).

For a vertex $u \in T$, we define an auxiliary graph G_u as follows. The vertices of G_u are the leaves in T attached to u (if any). Two vertices x and y in G_u are adjacent in G_u if and only if there is a request between x and y in R. It is not difficult to see that if C is a vertex cover for G_u then the edge-set $E_C = \{uw \in E(T) \mid w \in C\}$ cuts every request between a pair of leaves attached to u. On the other hand, for any cut K for R, the leaves in T that are incident to the edges in K form a vertex cover for G_u. It follows that the number of edges in K that are incident to the leaves corresponding to the vertices in G_u is $\geq \tau(G_u)$.

3   Known Results

Most terminologies, observations, reductions, and branching in this section were introduced in [4].

Let (T, R, k) be an instance of MCT. We assume that T has more than two vertices and is rooted at some internal vertex r in the tree (chosen arbitrarily). A non-leaf vertex $u \in V(T)$ is *important* if all its children are leaves.

Lemma 1 [4]. Let (T, R, k) be an instance of MULTICUT ON TREES. Suppose that T is rooted at r. There exists a minimum cut E_{min} for the requests of R in T such that, for every important vertex $u \in V(T)$, the subset of edges in E_{min} that are incident to the children of u corresponds to a minimum vertex cover of G_u.

Definition 1 [4]. Let $w \neq r$ be an important vertex in T. A request between a vertex in $V(T_w)$ and a vertex in $V(T_{\pi(w)}) \setminus V(T_w)$ is called a *cross request*.

Let $w \neq r$ be an important vertex in T, and u be a child of w such that u is contained in some minimum vertex cover of G_w. If edge $w\pi(w)$ is in some minimum cut of T, then the edges incident to the leaves of any minimum vertex cover of G_w are contained in some minimum cut: simply replace all the edges that are incident to the children of w in a minimum cut that contains $w\pi(w)$ with the edges incident to the leaves corresponding to the desired minimum vertex cover of G_w. Since u is contained in some minimum vertex cover of G_w, there is a minimum cut that contains wu. Therefore, if we choose edge $w\pi(w)$ to be in the solution, then we can choose the edge wu to be in the solution as well.

Definition 2 [4]. Let w be an important vertex and u be a child of w. We say that we *favor* vertex u if whenever we cut $w\pi(w)$ during branching we cut uw. By contrapositivity, if we decide not to cut u in a branch, then we can assume that w will not be cut as well in the same branch.

Observation 1 [4]. Let w be an important vertex in T, and $v \in G_w$. We can assume that the set of edges in T_w contained in the target solution corresponds to a minimum vertex cover of G_w. Since any minimum vertex cover of G_w either contains v, or excludes v but contains its neighbors, we can branch by cutting v in the first side of the branch, and by cutting the neighbors of v in G_w in the second side of the branch. Note that there must be at least one request between v and another child of w (otherwise, the edge vw can be contracted), and hence, $\deg_{G_w}(v) \geq 1$.

Observation 1 leads to the following branching rule:

BranchRule 1 [4]. *Let w be an important vertex in T. If there exists a vertex $v \in G_w$ such that $\deg_{G_w}(v) \geq 3$, then branch by cutting v in the first side of the branch, and by cutting the neighbors of v in G_w in the second side of the branch. Cutting v reduces the parameter k by 1, and cutting the neighbors of v in G_w reduces k by at least 3. Therefore, the number of leaves in the search tree of the algorithm, $L(k)$, satisfies the recurrence relation $L(k) \leq L(k-1) + L(k-3)$.*

We can now assume that for any important vertex w, $\Delta(G_w) \leq 2$. Hence, G_w consists of a collection of disjoint paths and cycles. Moreover, we can assume that no child of w would have a cross request to $\pi(w)$. We have the following proposition.

Proposition 1 [4]. *The instance (T, R, k), where T is rooted at a vertex r, can be reduced in polynomial time to satisfy the following:*

(i) For any vertex $u \in V(T)$, there exists no request between u and $\pi(u)$.

(ii) For any vertex $u \neq r$ in $V(T)$, there exists a request between some vertex in $V(T_u)$ and some vertex in $V(T_{\pi(u)}) \setminus V(T_u)$.

(iii) For any internal vertex $u \in V(T)$, there exists at least one request between two vertices in $V(T_u) - u$.

(iv) For any important vertex $w \in V(T)$ and any child u of w, there exists a request between u and a sibling of u.

(v) For any important vertex $w \in V(T)$, G_w contains no connected component that is a path of even length.

(vi) For every leaf $\ell \in V(T)$, ℓ is in a minimum vertex cover of $G_{\pi(\ell)}$.

(vii) For any important vertex $w \neq r$ in $V(T)$, there is no minimum vertex cover of G_w such that cutting the leaves in this minimum vertex cover cuts all the cross requests from the vertices in $V(T_w)$.

4   The Algorithm

The algorithm is a branch-and-bound algorithm, and its execution can be depicted by a search tree. The algorithm distinguishes 16 cases and performs the branching according to each case. The running time of the algorithm is upper bounded by the number of root-leaf paths, or equivalently, by the number of leaves in the search tree, multiplied by the time spent along each such path, which will be polynomial in k. Therefore, the main step in the analysis of the algorithm is to derive an upper bound on the number of leaves $L(k)$ in the search tree corresponding to the algorithm.

The algorithm exploits the connection between the MCT problem and the VERTEX COVER problem. This connection was first used in the paper of Chen et al. [4]. The idea is to introduce two auxiliary graphs, defined based on the neighborhood of a suitably chosen important vertex, and then to utilize properties of the minimum vertex covers of the two auxiliary graphs to achieve efficient branching. One of these two auxiliary graphs was first introduced in [4], and the main contribution of the current paper over [4] is in performing a deeper and a refined analysis based on these two auxiliary graphs. The algorithm also relies heavily on the notion of favoring vertices for branching (see Definition 2).

The instance (T, R, k) is said to be *reduced* if all the statements in Proposition 1 hold. We shall assume that instance (T, R, k) is reduced before every branch of the algorithm; we shall also assume that the branches are considered in the listed order. In particular, when a branch is considered, (T, R, k) is reduced and none of the preceding branches applies. From the results in the last section, we branch with a recurrence relation that is not worse than $L(k) \leq L(k-1) + L(k-3)$ to ensure $\Delta(G_w) \leq 2$ for every important vertex w in the resulting forest. This section contains the branching cases of the algorithm.

Observation 2. Let w be an important vertex in T. If G_w contains a path P of odd length greater than 3, let u be an endpoint of P. Observe that there exists exactly one minimum vertex cover C_u for P, which contains u. By Lemma 1, if we decide to cut u, then we can cut $|C_u| = (|P| + 1)/2 \geq 3$ edges between w and the vertices in C_u. On the other hand, if wu is kept then the neighbor of u in G_w is cut.

Observation 2 leads to the following branching rule:

BranchRule 2. *Let w be an important vertex in T such that $\Delta(G_w) \leq 2$. If G_w contains a path P of odd length greater than 3, let u be an endpoint of P and C_u be the (unique) minimum vertex cover of P containing u. Branch by cutting the vertices in C_u in the first side of the branch, and by cutting the neighbor of u in P in the second side of the branch. Since $|C_u| = (|P| + 1)/2 \geq 3$, $L(k)$ satisfies the recurrence relation $L(k) \leq L(k-3) + L(k-1)$.*

Now for any important vertex w in T, G_w consists of a collection of disjoint cycles and paths of lengths 3 or 1. Note that every vertex in G_w is contained in some minimum vertex cover of G_w. Assume T is rooted at r, and let $w \in T$ be a farthest important vertex. The algorithm distinguishes the following 16 cases when branching. We shall assume that none of BranchRule 1 and BranchRule 2 is applicable before any of the cases.

Case 1. Vertex w has a cross request to a non-leaf sibling w'. In this case at least one of w, w' must be cut. We branch by cutting w in the first side of the branch, and cutting w' in the second side of the branch. Note that by part (iii) of Proposition 1, the size of a minimum vertex cover in G_w is at least 1, and similarly for $G_{w'}$ because w' is a non-leaf vertex. Moreover, a minimum vertex cover for each of G_w and $G_{w'}$ can be computed in polynomial time since both graphs have maximum degree at most 2 (note that by the choice of w, w' is an important vertex as well). Therefore, in the first side of the branch we end up cutting the edges corresponding to a minimum vertex cover of G_w, which reduces the parameter further by at least 1. Similarly, we end up reducing the parameter further by at least 1 in the second side of the branch. Therefore, we have $L(k) \leq 2L(k-2)$ in this case.

Case 2. There exists a child u of w such that $\deg_{G_w}(u) = 2$ and u has a cross request. We favor u. Note that since we can assume that the solution contains a minimum vertex cover of G_w, we can branch by cutting u in the first side of the branch, and by keeping u and cutting the two neighbors of u in G_w in the second side of the branch. If the cross request is between u and an uncle w' of u, then we branch as follows. In the first side of the branch we cut u. In the second side of the branch we keep edge uw, and cut the two neighbors of u in G_w. Since u is not cut and u is favored, w is not cut as well, and hence w' must be cut. Therefore, $L(k)$ in this case satisfies the recurrence relation $L(k) \leq L(k-3) + L(k-1)$. If the cross request is between u and a cousin u' of u, let $w' = \pi(u')$ and note that $\pi(w) = \pi(w')$. We favor u'; thus if u' is not cut then w' is not cut as well. In this case we branch as follows. In the first side of the branch we cut u. In the second side of the branch uw is kept and we cut the two neighbors of u in G_w. Since in the second side of the branch uw is kept, $w\pi(w)$

is kept as well, and u' must be cut (otherwise, w' is not cut as well because u' is favored) since $(u, u') \in R$. Therefore, $L(k)$ in this case satisfies the recurrence relation $L(k) \leq L(k-1) + L(k-3)$.

Case 3. There exists a child u of w such that u is an endpoint of a path of length 3 in G_w and u has a cross request. Let the path containing u in G_w be $P = (u, x, y, z)$. We favor u. Note that since we can assume that the solution contains a minimum vertex of G_w, we can branch by cutting u in the first side of the branch, and in this case y can be cut as well, and by cutting x in the second side of the branch. If the cross request is between u and an uncle w' of u, then we branch as follows. In the first side of the branch we cut u and y. In the second side of the branch we keep uw and cut x. Since u is not cut in the second side of the branch and u is favored, w is not cut as well, and hence w' must be cut. Therefore, $L(k)$ in this case satisfies the recurrence relation $L(k) \leq 2L(k-2)$. If the cross request is between u and a cousin u' of u, let $w' = \pi(u')$ and note that $\pi(w) = \pi(w')$. We favor u'; thus if u' is not cut then w' is not cut as well. In this case we branch as follows. In the first side of the branch we cut u and y, and in the second side of the branch uw is kept and we cut x. Since uw is kept in the second side of the branch, $w\pi(w)$ is kept as well, and u' must be cut (otherwise w' is not cut) since $(u, u') \in R$. Therefore, $L(k)$ in this case satisfies the recurrence relation $L(k) \leq 2L(k-2)$.

Case 4. There exists a child u of w such that u has a cross request to a non-leaf uncle w'. Let v be the neighbor of u in G_w, and note that uv must be an isolated edge in G_w, and hence, exactly one of u, v is in any minimum vertex cover of G_w. We favor u. We branch by cutting u in the first side of the branch, and cutting v in the second side of the branch. In the second side of the branch wu is kept, and so is $w\pi(w)$. Since $(u, w') \in R$, w' must be cut. By part (iii) of Proposition 1, the size of a minimum vertex cover of $G_{w'}$ is at least 1, and by the choice of w, w' is a farthest vertex from the r, and hence $\Delta(G_{w'}) \leq 2$. Therefore, a minimum vertex cover for $G_{w'}$ has size at least 1 and can be computed in polynomial time. It follows that the parameter is reduced by at least 3 in the second side of the branch. We have $L(k) \leq L(k-1) + L(k-3)$ in this case.

Let us summarize what we have at this point. If all the previous cases do not apply, then we can assume that, for any farthest important vertex w, no child of w that has a cross request is of degree 2 in G_w, and no endpoint of a path of length 3 in G_w has a cross request. Therefore, no child of w that belongs to a cycle or a path of odd length ≥ 3 in G_w has a cross request. The only children of w that may have cross requests are the endpoints of the single edges in G_w. Moreover, if w has a cross request then it must be to a leaf-sibling; if a child of w has a cross request to an uncle, then it must be to a leaf-uncle.

Case 5. There exists a child u of w such that u has at least 2 cross requests. By the above discussion we have $deg_{G_w}(u) = 1$. Let v be the neighbor of u in G_w, and note that exactly one of u, v is in any minimum vertex cover of

G_w. Let u' and u'' be two vertices that u has cross requests to. We distinguish the following subcases:

SubCase 1. $\pi(u') \neq \pi(u'')$ or $\pi(u') = \pi(u'') = \pi(w)$.

We favor vertex u and the vertices in $\{u', u''\}$ that are not children of $\pi(w)$, and branch as follows. In the first side of the branch we cut v and keep edge wu. Since edge uw is kept and u is favored, edge $w\pi(w)$ is kept as well. Since the vertices in $\{u', u''\}$ that are not children of $\pi(w)$ are favored, u' and u'' are cut. In the second side of the branch we cut u. This gives $L(k) \leq L(k-1) + L(k-3)$.

SubCase 2. $\pi(u') = \pi(u'') = w'$.

If there exists a minimum vertex cover of $G_{w'}$ containing both u' and u'', then we favor $\{u', u''\}$ and branch as follows. In the first side of the branch we cut v. In this case wu is kept, and so is $w\pi(w)$. Moreover, u' and u'' are cut. In the second side of the branch u is cut. This gives $L(k) \leq L(k-1) + L(k-3)$.

If there does not exist a minimum vertex cover of $G_{w'}$ containing both u' and u'', then since T is reduced and w' is an important vertex, by part (v) of Proposition 1, u' and u'' must be neighbors in $G_{w'}$. We favor u and branch as follows. In the first side of the branch we cut v and keep wu, and in the second side of the branch we cut u. When we keep wu in the first side of the branch $w\pi(w)$ is kept as well. Since at least two edges in $\{\pi(w')w', w'u', w'u''\}$ must be cut (since $(u, u'), (u, u''), (u', u'') \in R$ and $uw, w\pi(w)$ are kept), it is safe to cut edges $w'\pi(w')$ and any of the two edges $w'u', w'u''$. This gives $L(k) \leq L(k-1) + L(k-3)$.

We can assume henceforth that every child of w has at most 1 cross request.

Case 6. Vertex w has a cross request to a leaf-sibling w', and the size of a minimum vertex cover of G_w is at least 2. In this case at least one of the edges $w\pi(w), w'\pi(w)$ must be cut. We branch by cutting w in the first side of the branch, and cutting w' in the second side of the branch. Since the size of a minimum vertex cover of G_w is at least 2, in the first side of the branch we can cut the edges corresponding to a minimum vertex cover of G_w, which reduces the parameter further by at least 2. Therefore, we have $L(k) \leq L(k-3) + L(k-1)$ in this case.

Case 7. Vertex w has a cross request to a leaf-sibling w', and either w has a request to a sibling $w'' \neq w'$ or a child u of w has a cross request to a vertex other than w'. Suppose that w has a cross request to a sibling $w'' \neq w'$. Then branch by cutting w in the first side of the branch, and cutting both w' and w'' in the second side of the branch. Observing that when w is cut the parameter is reduced further by 1 due to cutting one of the two children of w (arbitrarily chosen), we obtain $L(k) \leq 2L(k-2) \leq L(k-3) + L(k-1)$ in this case. If u has a cross request to a sibling $w'' \neq w'$ of w, then we favor u and branch by cutting u in the first side of the branch, and keeping uw and cutting v in the second side of the branch. In the second side of the branch, w is kept (since u is kept and is favored), and hence both w' and w'' must be cut. We

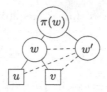

Fig. 1. A special quadruple $\{w, w', u, v\}$.

obtain $L(k) \leq L(k-3) + L(k-1)$ in this case. Similarly, if u has a cross request to a cousin x, then we favor both u and x. In the first side of the branch u is cut, and in the second side of the branch v (v is a neighbor of u in G_w), w', x are all cut. We obtain $L(k) \leq L(k-3) + L(k-1)$.

Case 8. Vertex w has a cross request to a leaf-sibling w' and w' has a request to a vertex in $V(T_{\pi(w')})$ that is not a child of w. If w' has a request to a sibling $w'' \neq w'$, then we branch by cutting w' in the first side of the branch and cutting both w and w'' in the second side of the branch. Observing that when w is cut the parameter is further reduced by 1 due to cutting one of the two children of w (arbitrarily chosen), we obtain $L(k) \leq L(k-1) + L(k-3)$ in this case. If w' has a request to a vertex x that is a nephew of w', then we favor x. We branch by cutting w' in the first side of the branch, and cutting both w and x in the second side of the branch. Observing that when w is cut the parameter is further reduced by 1, we obtain $L(k) \leq L(k-1) + L(k-3)$ in this case.

From the above discussion, if a farthest important vertex w has a cross request to a vertex w' in $T_{\pi(w)}$, then w' must be a leaf-sibling of w (note that by Case 4 and by symmetry, w does not have a cross request to a nephew) and the following must hold: (1) w has exactly two children, (2) w has no request to any vertex in $V(T_{\pi(w)})$ except to w', (3) both children of w have cross requests only to w' (note that by part (vii) of Proposition 1 both children of w must have cross requests in this case), and (4) w' has no request to any vertex in $V(T_{\pi(w')}) \setminus V(T_w)$. In this case we call such a set of vertices $\{w, w', u, v\}$ a *special quadruple*. The structure of a special quadruple is depicted in Fig. 1.

Case 9. A leaf-sibling w' of w that is not contained in a special quadruple has at least three requests to leaf-siblings or nephews. If w' has at least three requests to leaf siblings w_1, w_2, w_3, then we can branch by cutting w' in the first side of the branch, and cutting all of w_1, w_2, w_3 in the second side of the branch. This gives $L(k) \leq L(k-1) + L(k-3)$. If w' has two requests to nephews x and y such that x and y have the same parent w'' and $(x, y) \in R$, let z be a vertex other than x and y that w' has a request to; if z is a nephew of w' then favor it. Branch by cutting w' in the first side of the branch, and cutting w'', one of x, y, and z in the second side of the branch (note that since Case 4 does not apply, $d_G(z) = 1$). The reason why we can cut w'' in the second side of the branch follows from the fact that we would need to cut both x and y

otherwise. This gives $L(k) \leq L(k-1) + L(k-3)$. Finally, if the above does not apply, then we can favor all nephews of w' that w' has requests to, and branch by cutting w' in the first side of the branch, and by cutting the siblings and nephews that w' has requests to in the second side of the branch. This gives $L(k) \leq L(k-1) + L(k-3)$ as well.

Case 10. There exist two edges uv and xy in G_w such that all vertices u, v, x, y have cross requests. We can show that in this case we can branch with $L(k) \leq 3L(k-4) + L(k-2)$.

Before proceeding to the next case, we need the following results.

Proposition 2. *Let T be a reduced tree with root r, and let $w \neq r \in T$ be an important vertex that is farthest from r. If none of BranchRule 1, BranchRule 2 and the above cases applies, then the following hold true:*

  (i) *For every child w' of $\pi(w)$ (sibling of w or w) that is an important vertex (non-leaf), $G_{w'}$ consists of disjoint edges, length-3 paths, and cycles. No vertex that is contained in a cycle or a length-3 path in $G_{w'}$ has any cross requests, and every endpoint of an edge in $G_{w'}$ has at most one cross request.*

  (ii) *For every child w' of $\pi(w)$ that is an important vertex, there exist exactly two children u, v of w' such that $(u, v) \in R$ and both u and v have cross requests.*

 (iii) *Every leaf child w' of $\pi(w)$ that is not contained in a special quadruple has at least one request, and at most two requests, to vertices in $V(T_{\pi(w)})$ that are either leaf siblings or nephews of w'.*

  (iv) *Every non leaf child of $\pi(w)$ that is not contained in a special quadruple has no cross requests.*

Definition 3. *Let T be a reduced tree with root r, and let $w \neq r \in T$ be an important vertex that is farthest from r. Suppose that none of BranchRule 1, BranchRule 2, or the above cases applies. We define the auxiliary graph $G^*_{\pi(w)}$ as follows. The vertices of $G^*_{\pi(w)}$ are the leaf children and the grandchildren of $\pi(w)$ that are not contained in any special quadruple. Two vertices x and y in $G^*_{\pi(w)}$ are adjacent if and only if $(x, y) \in R$. Note that the edges in $G^*_{\pi(w)}$ correspond to either a request between two grandchildren of $\pi(w)$ that have the same parent, a request between two leaf children of $\pi(w)$, or a request between a leaf-child and a grandchild of $\pi(w)$.*

The following proposition is the dual of Proposition 1:

Proposition 3. *Let T be a reduced tree with root r, and let $w \in T$, where $\pi(w) \neq r$, be an important vertex that is farthest from r. Suppose that none of BranchRule 1, BranchRule 2, or the above cases applies. Consider the graph $G^*_{\pi(w)}$. Then the following are true:*

  (i) $\Delta(G^*_{\pi(w)}) \leq 2$, *and hence $G^*_{\pi(w)}$ consists of disjoint paths and cycles.*

(ii) *For every path P in $G^*_{\pi(w)}$ such that at least one endpoint of P is a grand-child of $\pi(w)$, there exists a minimum cut of T that cuts the vertices in some minimum vertex cover of P.*

(iii) *For every path P and every cycle C in $G^*_{\pi(w)}$, there exists a minimum cut C_{min} such that the number of edges in C_{min} that are incident to the vertices in P or their parents in case these vertices are grandchildren of $\pi(w)$, is equal to the size of a minimum vertex cover of P, and the number of edges in C_{min} that are incident to the vertices in C or their parents in case these vertices are grandchildren of $\pi(w)$, is equal to the size of a minimum vertex cover of C.*

The following reduction rule follows from parts (ii) and (iii) of Proposition 3:

Reduction Rule 1. *Let T be a reduced tree with root r, and let $w \in T$ be an important vertex that is farthest from r, and such that $\pi(w) \neq r$. Suppose that none of BranchRule 1, BranchRule 2, or the above cases applies. If there exists a path P in $G^*_{\pi(w)}$ of even length then cut the vertices in P that correspond to the unique vertex cover of P.*

The following branching rule follows from parts (ii) and (iii) of Proposition 3, after noticing that for a path of odd length in $G^*_{\pi(w)}$, there is a unique set of edges of cardinality $\tau(P)$ in $E(T_{\pi(w)})$ that cuts all requests corresponding to the edges of P in addition to cutting an endpoint of P:

BranchRule 3. *Let T be a reduced tree with root r, and let $w \in T$ be an important vertex that is farthest from r, and such that $\pi(w) \neq r$. Suppose that none of BranchRule 1, BranchRule 2, or the above cases applies. If there exists a path P in $G^*_{\pi(w)}$ of odd length such that $|P| > 3$, let u be an endpoint of P and let C_u be the (unique) minimum vertex cover of P containing u. Branch by cutting the vertices in C_u and contracting the edges between w and vertices in $V(P) - C_u$ in the first side of the branch, and by cutting the neighbor of u in P in the second side of the branch. Since $|C_u| = (|P| + 1)/2 \geq 3$, $L(k)$ satisfies the recurrence relation: $L(k) \leq L(k - 3) + L(k - 1)$.*

The following branching rule follows from parts (ii) and (iii) of Proposition 3 after noticing that, for a cycle of even length in $G^*_{\pi(w)}$ there are exactly two sets of edges in $E(T_{\pi(w)})$, each of cardinality $\tau(P)$, such that each cuts all requests corresponding to the edges of C:

BranchRule 4. *Let T be a reduced tree with root r, and let $w \in T$ be an important vertex that is farthest from r, and such that $\pi(w) \neq r$. Suppose that none of BranchRule 1, BranchRule 2, or the above cases applies. If there exists a cycle C in $G^*_{\pi(w)}$ of even length, branch into a two-sided branch: in the first side of the branch cut the vertices corresponding to one of the minimum vertex covers of C, and in the second side of the branch cut the vertices corresponding to the other minimum vertex cover of C. Since $|C| \geq 4$, and hence $\tau(C) \geq 2$, we get $L(k) \leq 2L(k - 2)$.*

BranchRule 5. *Let T be a reduced tree with root r, and let $w \in T$ be an important vertex that is farthest from r, and such that $\pi(w) \neq r$. Suppose that none of BranchRule 1, BranchRule 2, or the above cases applies. If there exists a cycle of odd length $C = (u_1, u_2, \ldots, u_{2\ell+1})$ in $G^*_{\pi(w)}$ such that $\ell \geq 3$ and C is not a cycle in $G_{w'}$ for some child w' of $\pi(w)$, then branch as follows. First observe that since C is not a cycle in $G_{w'}$ for some child w' of $\pi(w)$, $|C|$ is odd, and T is reduced, at least one vertex, say u_1 on C must be a leaf child of $\pi(w)$. We favor the vertices in $\{u_2, u_{2\ell+1}\}$ that are grandchildren of $\pi(w)$ (if any). In the first branch u_1 is kept and $u_2, u_{2\ell+1}$ are cut. In the second side of the branch u_1 is cut, and the cycle becomes a path of odd length at least 5; therefore, we can further branch according to BranchRule 3. This yields $L(k) \leq 2L(k-2) + L(k-4)$.*

Now let T be a reduced tree with root r, and let $w \in T$ be an important vertex that is farthest from r, and such that $\pi(w) \neq r$. Suppose that none of the branching rules of the above cases applies. Then each vertex in $V(T_{\pi(w)}) - \pi(w)$ is contained in one of the following structures/groups in $G^*_{\pi(w)}$. Group I, abbreviated GP_1, are paths of length 1 (edge) in $G^*_{\pi(w)}$ between two children of an important child of $\pi(w)$, Group II, abbreviated GP_2, are paths of length 1 in $G^*_{\pi(w)}$ between two leaf children of $\pi(w)$, Group III, abbreviated GP_3, are paths of length 3 in $G^*_{\pi(w)}$ but not in G'_w for any important child of $\pi(w)$, Group IV, abbreviated GP_4 are cycles of length 3 in $G^*_{\pi(w)}$ but not in G'_w for any important child of $\pi(w)$, Group V, abbreviated GP_5, are cycles of lengths 5 in $G^*_{\pi(w)}$ but not in G'_w for any important child of $\pi(w)$, Group VI, abbreviated GP_6 are special quadruples, and Group VII, abbreviated GP_7, are paths of length 3 or cycles in $G_{w'}$, for some important child w' of $\pi(w)$. Note that no vertex in GP_7 can have a cross request.

Let $w_1 = \pi(w)$, and let $w_2, \ldots, w_l$ be the siblings of w_1. Each sibling w_i of w_1 is either a leaf, an important vertex, or T_{w_i} has a similar structure to T_{w_1}.

Case 11. For some w_i, there is no request from a vertex in T_{w_i} to a vertex in some tree T_{w_j}, for any $j \neq i$. In this case, contract $w_i\pi(w_i)$.

This can be seen as follows. If w_i is cut by some minimum cut C_{min}, then since there are no requests between vertices in T_{w_i} and a vertex in T_{w_j}, for any $j \neq i$ in $\{1, \ldots, l\}$, edge $w_i\pi(w_i)$ can be replaced by the edge between $\pi(w_i)$ and its parent to yield a minimum cut that excludes the edge between w_i and its parent. Therefore, we can assume that, for any $i \in \{1, \ldots, l\}$, there exists a request between some vertex in T_{w_i} and a vertex in some T_{w_j}. Moreover, for any vertex u in T_{w_i}, there exists a minimum cut that cuts u. Further, if any edge e on the path between $\pi(w_i)$ and u is part of a minimum cut, then there is a minimum cut that includes e and cuts u as well. Therefore, u can be favored, and if u is kept in a certain branch then all edges on the path between u and w_i are kept.

Consider now w_i for some fixed i. Let u be a vertex in T_{w_i} that has a request to a vertex x in T_{w_j}, for some $j \neq i$. We favor both u and x. We distinguish the following cases.

Case 12. u is important. Since u is important, u must have two children y, z such that $(y, z) \in R$ and both z and y have cross requests. Suppose that z has a cross request to z' in T_{w_i}. Favor z' (if z' is a grandchild of w_i) and z. Branch by cutting y in the first side of the branch and keeping z, and by cutting z in the second side of the branch. In the first side of the branch z is kept and so is u. Therefore, z' and x must be cut. This gives $L(k) \leq L(k-3) + L(k-1)$.

We can assume now that u is not an important vertex in T_{w_i}. Therefore, u must be a vertex in $G^*_{w_i}$; let d_u be the degree of u in $G^*_{w_i}$.

Case 13. $d_u = 2$. If in a certain branch u is kept, then two edges can be cut. This can be seen as follows. If u is a GP_3 vertex, then both neighbors of u in $G^*_{w_i}$ can be cut (favor the neighbors that are not leaf children of w_i). If u is a GP_4 vertex, let (u, u_1, u_2) be the length-3 cycle containing u. If u is a leaf child of w_i, then the parent of u_1, u_2 can be cut, in addition to one of u_1, u_2 (chosen arbitrarily). If u is not an leaf child of w_i, then u_1 and u_2 can be cut (since u is favored). If u is a GP_5 vertex, then by favoring any neighbor of u in $G^*_{w_i}$ that is not a leaf child of w_i (if the neighbor is a leaf child of w_i then there is no need to favor it), it can be easily seen that when u is kept then its two neighbors can be cut. If u is a GP_6 vertex, then the same analysis carries as when u is GP_4 vertex. Finally, if u is a GP_7 vertex, then it can be easily seen that both neighbors of u can be cut. Therefore, if $d_u = 2$, then we can branch by cutting u in the first side of the branch, and keeping u and cutting its two neighbors in $G^*_{w_i}$, in addition to x in the second side of the branch. This gives $L(k) \leq L(k-1) + L(k-3)$.

Case 14. $d_u = 1$. In this case either u is a GP_3 or a GP_7 vertex that is an endpoint of a length-3 path, or u is a GP_1 or a GP_2 vertex. If u is an endpoint of a length-3 path (u, u_1, u_2, u_3), then by part (b) of Proposition 3, if u is cut, then u_2 must be cut as well. On the other hand, if u is kept then u_1 and x must be cut. This gives $L(k) \leq 2L(k-2)$.

We can now assume that all requests between the T_{w_i}'s go from GP_1 or GP_2 vertices to GP_1 or GP_2 vertices.

Case 15. u is an endpoint of a GP_1 group. If u is an endpoint of a GP_1 group, let $w = \pi(u)$, and let v be the child of w such that $(u, v) \in R$. Note that w is an important vertex, and hence, there exists z, y, children of w, such that $(z, y) \in R$ and both z and y have cross requests. We favor u and branch as follows. In the first side of the branch we cut u and keep v, and in the second side of the branch we keep u and cut v. Let us analyze the second side of the branch when u is kept. In this case x must be cut. Since u is kept and is favored, w is kept as well. Since both z and y have cross requests, z and y are either part of a GP_3, GP_4, GP_5, or GP_6 group. If z and y are contained in a GP_6 or a GP_4 group, then their uncle must be cut leading to a further reduction of the parameter by at least 1. If z and y are contained in a GP_3 group (s, z, y, t), then by part (b) of Proposition 3 either s, y or z, t must be cut, so we can branch further into these two branches. If z and y are part of a GP_5 group (s, z, y, p, q), where p and q are children of an important child of w_i, then since w is kept we branch on z: if z is cut, then y is kept and p is cut (since w is kept), and if z is kept then y and s are cut. In the worst case, we get $L(k) \leq L(k-1) + 2L(k-4)$.

Case 16. All requests between the T_{w_i}'s go between GP_2 vertices. If for every GP_2 group in T_{w_i} at most one vertex has a request to some T_{w_j}, where $j \neq i$, then there exists a cut of T_{w_i} that cuts all requests to T_{w_j}, and whose cardinality is equal to the set of edges in a minimum cut that are contained in T_{w_i}; therefore, edge $w_i \pi(w_i)$ can be contracted. Hence, we can assume that for every T_{w_i}, there exists a GP_2 in T_{w_i} whose both vertices u, v have requests to vertices in other trees; suppose that u has a request to u' and v to v', where u' and v' are not in T_{w_i}. Moreover, we can assume that T_{w_i} contains an important vertex (choose a tree among the T_{w_j}'s that contains an important vertex, and by the above argument, there exists a GP_2 in T_{w_j} whose both vertices u, v have requests to vertices in other trees). Since each important vertex must have two children with cross requests, and since there is a GP_2 in T_{w_i}, any minimum cut must cut at least three vertices in $V(T_{w_i}) - w_i$. We branch as follows. Either w_i is cut or is kept. When w_i is cut, at least 3 edges in $E(T_{w_i})$, corresponding to any minimum cut of T_{w_i} can be cut. When w_i is kept, we branch by cutting u and favoring v' in the first side of the branch, and cutting v and favoring u' in the second side of the branch. When u is cut, v is kept, and hence v' must be cut (since w_i is kept). When v is cut, u is kept and hence u' is cut. This gives $L(k) \leq 2L(k-2) + L(k-4)$.

Theorem 3. *The* MCT *problem is solvable in time* $O^*(1.555^k)$.

Proof. The above cases exhaust all possibilities. The worst case corresponds to the branch $L(k) \leq 2L(k-2) + L(k-4)$, which yields the desired time bound. □

There is a simple reduction from GMWCT to MCT. For an instance $(T, \{S_1, \ldots, S_q\}, k)$ of GMWCT, construct the instance (T, R, k), where the set of requests R is given as follows. For each terminal set S_i, $i = 1, \ldots, q$, and for every pair of distinct terminals $u, v \in S_i$, add the request (u, v) to R. Combining this reduction with Theorem 3 we obtain:

Corollary 1. *The* GMWCT *problem is solvable in time* $O^*(1.555^k)$.

5 A Linear Time Algorithm for GMWCT

We show that the GMWCT problem is solvable in linear time when the number of terminal sets is a constant. (The problem is NP-complete by a simple reduction from the MCT problem.)

Theorem 4. *The* GMWCT *problem with q terminal sets can be solved in time* $O(3^q |V(T)|)$, *which is linear time when q is a constant.*

References

1. Avidor, A., Langberg, M.: The multi-multiway cut problem. Theor. Comput. Sci. **377**(1–3), 35–42 (2007)
2. Bousquet, N., Daligault, J., Thomassé, S.: Multicut is FPT. In: STOC, pp. 459–468 (2011)
3. Bousquet, N., Daligault, J., Thomassé, S., Yeo, A.: A polynomial kernel for multicut in trees. In: STACS, pp. 183–194 (2009)
4. Chen, J., Fan, J., Kanj, I., Liu, Y., Zhang, F.: Multicut in trees viewed through the eyes of vertex cover. J. Comput. Syst. Sci. **78**, 1637–1650 (2012)
5. Chitnis, R., Hajiaghayi, M., Marx, D.: Fixed-parameter tractability of directed multiway cut parameterized by the size of the cutset. In: SODA, pp. 1713–1725 (2012)
6. Chopra, S., Rao, M.: On the multiway cut polyhedron. Networks **21**, 51–89 (1991)
7. Costa, M., Letocart, L., Roupin, F.: Minimal multicut and maximal integer multiflow: a survey. Eur. J. Oper. Res. **162**, 55–69 (2005)
8. Costa, M.-C., Billionnet, A.: Multiway cut and integer flow problems in trees. Electron. Notes Discrete Math. **17**, 105–109 (2004)
9. Deng, X., Lin, B., Zhang, C.: Multi-multiway cut problem on graphs of bounded branch width. In: Fellows, M., Tan, X., Zhu, B. (eds.) FAW-AAIM 2013. LNCS, vol. 7924, pp. 315–324. Springer, Heidelberg (2013)
10. Downey, R., Fellows, M.: Parameterized Complexity. Springer, New York (1999)
11. Flüm, J., Grohe, M.: Parameterized Complexity Theory. Springer, Heidelberg (2010)
12. Garg, N., Vazirani, V.V., Yannakakis, M.: Primal-dual approximation algorithms for integral flow and multicut in trees. Algorithmica **18**, 3–20 (1997)
13. Guo, J., Niedermeier, R.: Fixed-parameter tractability and data reduction for multicut in trees. Networks **46**, 124–135 (2005)
14. Kanj, I.A., Lin, G., Liu, T., Tong, W., Xia, G., Xu, J., Yang, B., Zhang, F., Zhang, P., Zhu, B.: Algorithms for cut problems on trees. CoRR, abs/1304.3653 (2013)
15. Khot, S., Regev, O.: Vertex cover might be hard to approximate to within $2 - \epsilon$. J. Comput. Syst. Sci. **74**, 335–349 (2008)
16. Klein, P.N., Marx, D.: Solving PLANAR k-TERMINAL CUT in $O(n^{c\sqrt{k}})$ time. In: Czumaj, A., Mehlhorn, K., Pitts, A., Wattenhofer, R. (eds.) ICALP 2012, Part I. LNCS, vol. 7391, pp. 569–580. Springer, Heidelberg (2012)
17. Levin, A., Segev, D.: Partial multicuts in trees. Theor. Comput. Sci. **369**(1–3), 384–395 (2006)
18. Liu, H., Zhang, P.: On the generalized multiway cut in trees problem. J. Comb. Optim. **27**(1), 65–77 (2014)
19. Marx, D.: Parameterized graph separation problems. Theor. Comput. Sci. **351**, 394–406 (2006)
20. Marx, D.: A tight lower bound for planar multiway cut with fixed number of terminals. In: Czumaj, A., Mehlhorn, K., Pitts, A., Wattenhofer, R. (eds.) ICALP 2012, Part I. LNCS, vol. 7391, pp. 677–688. Springer, Heidelberg (2012)
21. Marx, D., Razgon, I.: Fixed-parameter tractability of multicut parameterized by the size of the cutset. In: STOC, pp. 469–478 (2011)
22. Niedermeier, R.: Invitation to Fixed-Parameter Algorithms. Oxford University Press, Oxford (2006)
23. West, D.B.: Introduction to Graph Theory. Prentice Hall Inc., Upper Saddle River (1996)

The Minimum Vulnerability Problem on Graphs

Yusuke Aoki[1][(✉)], Bjarni V. Halldórsson[2], Magnús M. Halldórsson[2],
Takehiro Ito[1], Christian Konrad[2], and Xiao Zhou[1]

[1] Graduate School of Information Sciences, Tohoku University, Sendai, Japan
{y.aoki,takehiro,zhou}@ecei.tohoku.ac.jp
[2] School of Computer Science, Reykjavík University, Reykjavik, Iceland
{mmh,bjarnivh,christiank}@ru.is

Abstract. Suppose that each edge e of an undirected graph G is associated with three nonnegative integers $\mathsf{cost}(e)$, $\mathsf{vul}(e)$ and $\mathsf{cap}(e)$, called the cost, vulnerability and capacity of e, respectively. Then, we consider the problem of finding k paths in G between two prescribed vertices with the minimum total cost; each edge e can be shared without cost by at most $\mathsf{vul}(e)$ paths, and can be shared by more than $\mathsf{vul}(e)$ paths if we pay $\mathsf{cost}(e)$, but cannot be shared by more than $\mathsf{cap}(e)$ paths even if we pay the cost of e. This problem generalizes the disjoint path problem, the minimum shared edges problem and the minimum edge cost flow problem for undirected graphs, and it is known to be NP-hard. In this paper, we study the problem from the viewpoint of specific graph classes, and give three results. We first show that the problem remains NP-hard even for bipartite series-parallel graphs and for threshold graphs. We then give a pseudo-polynomial-time algorithm for bounded treewidth graphs. Finally, we give a fixed-parameter algorithm for chordal graphs when parameterized by the number k of required paths.

1 Introduction

In this paper, we study the minimum vulnerability problem on undirected graphs, originally introduced by Assadi et al. [1]. This problem has strong relationships to several well-known problems such as the disjoint path problem [6], the minimum shared edges problem [1,9,10] and the minimum edge cost flow problem [6,8]. It is defined as follows.

Let $G = (V, E)$ be an undirected and connected graph; we sometimes denote by $V(G)$ and $E(G)$ the vertex set and edge set of G, respectively. Suppose that each edge $e \in E(G)$ is associated with three nonnegative integers $\mathsf{cost}(e)$, $\mathsf{vul}(e)$ and $\mathsf{cap}(e)$, called the *cost*, *vulnerability* and *capacity* of e, respectively. (See Fig. 1(a) as an example.) Let $\mathcal{P}$ be a multi-set of paths in G. Then, for each

Magnús M. Halldórsson and Christian Konrad are supported by Icelandic Research Fund grant-of-excellence no. 120032011.
Takehiro Ito: This work is partially supported by JSPS KAKENHI 25106504 and 25330003.

Z. Zhang et al. (Eds.): COCOA 2014, LNCS 8881, pp. 299–313, 2014.
DOI: 10.1007/978-3-319-12691-3_23

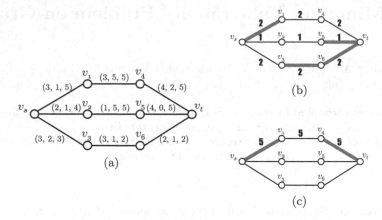

Fig. 1. (a) An instance for the minimum vulnerability problem, where the triple attached to each edge e represents the three weights $(\mathsf{cost}(e), \mathsf{vul}(e), \mathsf{cap}(e))$. (b) A feasible solution for $k = 5$ such that $\lambda(\mathcal{P}) = 12$, where the bold number attached to each edge e represents $\mu(e, \mathcal{P})$ and we have to pay the cost for the (red) thick edges. (c) An optimal solution $\mathcal{P}^*$ for $k = 5$, where $\lambda(\mathcal{P}^*) = \mathsf{OPT}_5(G) = 7$ (Color figure online).

edge $e \in E(G)$, we define $\mu(e, \mathcal{P})$ to be the number of paths in $\mathcal{P}$ that contain e, and define the *penalty* $\lambda(e, \mathcal{P})$ of e on $\mathcal{P}$ as follows:

$$\lambda(e, \mathcal{P}) = \begin{cases} 0 & \text{if } 0 \leq \mu(e, \mathcal{P}) \leq \mathsf{vul}(e); \\ \mathsf{cost}(e) & \text{if } \mathsf{vul}(e) < \mu(e, \mathcal{P}) \leq \mathsf{cap}(e); \\ +\infty & \text{otherwise}, \end{cases}$$

that is, each edge e can be shared without cost by at most $\mathsf{vul}(e)$ paths in $\mathcal{P}$, and cannot be shared by more than $\mathsf{cap}(e)$ paths even if we pay the cost of e. Then, the *penalty* of $\mathcal{P}$, denoted by $\lambda(\mathcal{P})$, is defined as $\lambda(\mathcal{P}) = \sum_{e \in E(G)} \lambda(e, \mathcal{P})$. See Fig. 1(b, c) for an example.

Given an edge-weighted graph G with two specified vertices $v_s, v_t \in V(G)$ and a nonnegative integer k, the *minimum vulnerability problem* is to find a set $\mathcal{P}$ consisting of exactly k (not necessarily distinct) paths in G between v_s and v_t, called $v_s v_t$-*paths* in short, such that the penalty $\lambda(\mathcal{P})$ of $\mathcal{P}$ is minimized. We denote by $\mathsf{OPT}_k(G, v_s, v_t)$, in short by $\mathsf{OPT}_k(G)$, the optimal value of the minimum vulnerability problem for a given graph G.

This problem arises in the context of communication network design, reliable multicast communications, and distributed communication protocols [1,9]. For example, consider the data transfer of some important data from a server v_s to another one v_t via a computer network formulated as a graph G. To avoid the attack of hackers, we first divide the data into k smaller chunks of data, and we wish to send them along disjoint paths in G. If G does not have k disjoint $v_s v_t$-paths, then some data must share edges (i.e., links in the network); assume that links use different security protocols, and hence they have different capacities, vulnerabilities and costs to be shared. Thus, the minimum vulnerability problem formulates this situation, and an optimal solution provides a way to send the data at minimum cost while satisfying given security requirements.

1.1   Known and Related Results

The minimum vulnerability problem was originally defined on digraphs (i.e., directed graphs) [1], in which we wish to find k directed paths from v_s to v_t. The problem on digraphs generalizes the minimum shared edges problem [1,9,10] and the minimum edge cost flow problem [6,8] as follows: (In the following, we denote by n the number of vertices in a graph G, and by m the number of edges or arcs in G.)

- The minimum shared edges problem:
  This problem corresponds to the minimum vulnerability problem on digraphs restricted to the case where $\mathsf{cost}(e) = 1$, $\mathsf{vul}(e) = 1$ and $\mathsf{cap}(e) = k$ for all arcs e in a given digraph. The problem is known to be NP-hard [9], and even hard to approximate within a factor of $2^{\log^{1-\varepsilon} n}$ for any constant $\varepsilon > 0$ [9]. On the other hand, Assadi et al. [1] showed that the minimum shared edges problem can be approximated in polynomial time within a factor of $\min\{n^{\frac{3}{4}}, m^{\frac{1}{2}}\}$, or a factor of $\lfloor k/2 \rfloor$.
  Furthermore, there exists a pseudo-polynomial-time algorithm for the minimum shared edges problem when restricted to undirected graphs G with bounded treewidth [10]. Its running time is $O\left(n(k + 1)^{2^{(t(t+1)/2)}} + n(k + 1)^{(t+4)(2t+8)}\right)$, where t is the treewidth of G. (The definition of treewidth will be given in Sect. 3.).
- The minimum edge cost flow problem:
  This problem corresponds to the minimum vulnerability problem on digraphs restricted to the case where $\mathsf{vul}(e) = 0$ for all arcs e in a given digraph. This problem is known to be strongly NP-hard even for bipartite digraphs [8], and hard even to approximate within a factor of $2^{\log^{1-\varepsilon} n}$ for any constant $\varepsilon > 0$ [5].
  The minimum edge cost flow problem remains NP-hard even for series-parallel digraphs [8], but it admits a fully polynomial-time approximation scheme (FPTAS) for series-parallel digraphs [8].

The Minimum Vulnerability Problem. We now explain known results for our problem. Since the minimum vulnerability problem on digraphs is a generalization of the minimum shared edges problem and the minimum edge cost flow problem, all hardness results obtained for the latter two problems hold for the problem on digraphs, too. Furthermore, the strong NP-hardness proof given in [8] can easily be extended to bipartite undirected graphs, and hence the minimum vulnerability problem remains strongly NP-hard even for bipartite undirected graphs.

However, this relation does not hold in the other direction, that is, algorithms obtained for the two problems above do not always work for the minimum vulnerability problem even on undirected graphs. Thus, Assadi et al. [1] developed an $O(n^3 m^{2(k-1)})$-time algorithm which exactly solves the minimum vulnerability problem on any digraph for the case where all arcs e have identical positive vulnerability $\mathsf{vul}(e) = r \geq 1$. They also gave an approximation result for the case where $r \geq 0$: The best known approximation ratio is $\lfloor \frac{k}{r+1} \rfloor$ for general digraphs [1]. As far as we know, these are the only positive results known for the minimum vulnerability problem on digraphs.

1.2 Our Contributions

In this paper, we study the minimum vulnerability problem on undirected graphs from the viewpoint of specific graph classes, and mainly give the following three results. (We will later define the graph classes mentioned below.)

First, we show that the problem remains NP-hard for undirected graphs, more specifically, for bipartite series-parallel graphs and for threshold graphs, even if $\mathsf{cap}(e) \geq 1$ and $\mathsf{vul}(e) \geq 1$ holds for all edges e in a graph G. Therefore, it is very unlikely that the problem can be solved in polynomial time even for these very restricted graph classes. It is important that the result holds under the condition that $\mathsf{cap}(e) \geq 1$ and $\mathsf{vul}(e) \geq 1$ holds for all edges $e \in E(G)$, because otherwise any graph can be represented as a complete graph (which is a threshold graph) by appropriately choosing edge-costs.

Second, we give a pseudo-polynomial-time algorithm for bounded treewidth graphs, which form a super-class of series-parallel graphs; note that our algorithm works also for the case where $\mathsf{cap}(e) = 0$ or $\mathsf{vul}(e) = 0$ holds for some edges e. Thus, this algorithm solves the minimum shared edges problem and the minimum edge cost flow problem for undirected graphs, too. Furthermore, our algorithm improves the best running time known for the minimum shared edges problem on undirected graphs with bounded treewidth [10].

Third, by taking the number k of required $v_s v_t$-paths as a parameter, we give a fixed-parameter algorithm for chordal graphs G such that $\mathsf{vul}(e) \geq 1$ holds for all edges $e \in E(G)$. Note that the problem is NP-hard for chordal graphs, because chordal graphs form a super-class of threshold graphs.

2 Computational Hardness

In this section, we clarify the complexity status of the minimum vulnerability problem. First, we give a reduction showing that the problem is NP-hard for bipartite series-parallel graphs. We then show that this reduction can be extended to an NP-hardness proof for threshold graphs.

2.1 Bipartite Series-Parallel Graphs

Subdividing an edge (u, v) of a graph is the operation of deleting the edge (u, v) and adding a path between u and v through several newly added vertices of degree two. A graph G is said to be a *subdivision* of a graph G' if G is obtained from G' by subdividing some of the edges of G'. A graph is *series-parallel* if

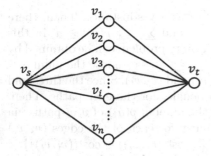

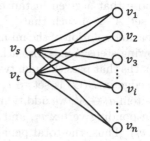

Fig. 2. A series-parallel graph used in the reduction.

Fig. 3. A threshold graph used in the reduction.

it does not contain a subdivision of a complete graph K_4 on four vertices as a subgraph [4].

Theorem 1. *The minimum vulnerability problem is NP-hard for bipartite series-parallel graphs, even if $\mathsf{cap}(e) \geq 1$ and $\mathsf{vul}(e) = r$ hold for all edges $e \in E(G)$, where $r \geq 1$ is any fixed constant.*

Proof. We give a polynomial-time reduction from KNAPSACK [6]. In an instance of KNAPSACK, we are given a set A of n items $a_1, a_2, \ldots, a_n$, a positive integer weight $w(a_i)$ and a positive integer profit $p(a_i)$ for each item $a_i \in A$, and two positive integers c and d. Then, the KNAPSACK problem is to determine whether there exists a subset $A' \subseteq A$ such that the total weight of A' is at most c and the total profit of A' is at least d. This problem is known to be NP-complete [6].

We indeed prove that the following decision version of the minimum vulnerability problem is NP-hard: Given a graph G with two specified vertices $v_s, v_t \in V(G)$ associated with three nonnegative integers $\mathsf{cost}(e)$, $\mathsf{vul}(e)$ and $\mathsf{cap}(e)$, and two nonnegative integers k and c, is there a set $\mathcal{P}$ consisting of exactly k $v_s v_t$-paths on G such that the penalty $\lambda(\mathcal{P})$ of $\mathcal{P}$ is at most c?

We first construct the corresponding graph G associated with three integers $\mathsf{cost}(e)$, $\mathsf{vul}(e)$ and $\mathsf{cap}(e)$. For each item a_i, $1 \leq i \leq n$, we add a vertex v_i to $V(G)$ corresponding to a_i. Then, we add two vertices v_s and v_t to $V(G)$, and for each i, $1 \leq i \leq n$, we add two edges (v_s, v_i) and (v_i, v_t) to $E(G)$. We set three integers $\mathsf{cost}(e)$, $\mathsf{vul}(e)$ and $\mathsf{cap}(e)$ for each edge $e \in E(G)$ as follows: Let $r \geq 1$ be any fixed constant. For each i, $1 \leq i \leq n$, we set $\mathsf{cost}((v_s, v_i)) = 0$, $\mathsf{cost}((v_i, v_t)) = w(a_i)$, $\mathsf{vul}((v_s, v_i)) = \mathsf{vul}((v_i, v_t)) = r$, and $\mathsf{cap}((v_s, v_i)) = \mathsf{cap}((v_i, v_t)) = r + p(a_i)$. Clearly, G is a bipartite series-parallel graph with $\mathsf{cap}(e) \geq 1$ and $\mathsf{vul}(e) = r$ for all edges $e \in E(G)$, as shown in Fig. 2. Finally, we set the number k of required $v_s v_t$-paths as $k = nr + d$, and the upper bound c on the penalty as the same upper bound on the total weight (i.e., capacity) of KNAPSACK. This corresponding instance can be constructed in polynomial time.

We show that a given instance of KNAPSACK is a yes-instance if and only if the corresponding instance of the minimum vulnerability problem is a yes-instance.

Suppose that a given instance of KNAPSACK is a yes-instance. Then, there exists a set $A' \subseteq A$ such that $\sum_{a_i \in A'} w(a_i) \le c$ and $\sum_{a_i \in A'} p(a_i) \ge d$. In this case, a feasible solution of the minimum vulnerability problem can be obtained by the following steps. First, for each vertex v_i, $1 \le i \le n$, we choose the number r of $v_s v_t$-paths that pass through two edges (v_s, v_i) and (v_i, v_t). Since the threshold for each edge in $E(G)$ is set to r, there is no penalty for these nr paths. Then, for each item $a_i \in A'$, we additionally choose the number $p(a_i)$ of $v_s v_t$-paths via the corresponding vertex v_i, and pay the penalty for each of the edges (v_s, v_i) and (v_i, v_t). Thus, the total penalty is $\sum_{a_i \in A'} \{\mathsf{cost}((v_s, v_i)) + \mathsf{cost}((v_i, v_t))\} = \sum_{a_i \in A'} w(a_i) \le c$. Since $\sum_{a_i \in A'} p(a_i) \ge d$, the total number of chosen $v_s v_t$-paths is at least $nr + d = k$. Therefore, the chosen $v_s v_t$-paths form a feasible solution, and hence the corresponding instance of the minimum vulnerability problem is a yes-instance.

Conversely, suppose that the corresponding instance of the minimum vulnerability problem is a yes-instance. Then, there exists a set $\mathcal{P}$ consisting of $k = nr + d$ of $v_s v_t$-paths on G such that the penalty $\lambda(\mathcal{P})$ of $\mathcal{P}$ is at most c. Let $B \subseteq V(G)$ be the set of all vertices v_i in G such that the edges (v_s, v_i) and (v_i, v_t) incident to v_i are passed through by more than r paths in $\mathcal{P}$. Namely, we have to pay the penalties for the edges (v_s, v_i) and (v_i, v_t) if and only if $v_i \in B$. Let A' be the set of all items in A that correspond to the vertices in B. Then, we clearly have $\sum_{a_i \in A'} w(a_i) = \sum_{v_i \in B} \{\mathsf{cost}((v_s, v_i)) + \mathsf{cost}((v_i, v_t))\}$, and hence $\sum_{a_i \in A'} w(a_i) \le c$. Since we pay the penalties for the edges (v_s, v_i) and (v_i, v_t) such that $v_i \in B$, the total number of $v_s v_t$-paths passing through these edges is more than $\sum_{v_i \in B} \mathsf{vul}((v_s, v_i))$ and at most $\sum_{v_i \in B} \mathsf{cap}((v_s, v_i))$. On the other hand, for the edges (v_s, v_i) and (v_i, v_t) such that $v_i \notin B$, we do not pay the penalties for them, and hence the total number of $v_s v_t$-paths passing through these edges is at most $\sum_{v_i \in V(G) \setminus B} \mathsf{vul}((v_s, v_i))$. Therefore, we have

$$\sum_{v_i \in B} \mathsf{cap}((v_s, v_i)) + \sum_{v_i \in V(G) \setminus B} \mathsf{vul}((v_s, v_i)) \ge |\mathcal{P}| = nr + d.$$

Since $\mathsf{cap}((v_s, v_i)) = r + p(a_i)$ and $\mathsf{vul}((v_s, v_i)) = r$,

$$\sum_{v_i \in B} \mathsf{cap}((v_s, v_i)) + \sum_{v_i \in V(G) \setminus B} \mathsf{vul}((v_s, v_i)) = \sum_{a_i \in A'} p(a_i) + \sum_{a_i \in A} r$$

$$= \sum_{a_i \in A'} p(a_i) + nr.$$

Therefore, $\sum_{a_i \in A'} p(a_i) \ge d$ holds for the subset A' of items. Thus, the set A' is a feasible solution for the given instance of KNAPSACK, and hence it is a yes-instance. $\qquad\square$

2.2 Threshold Graphs

A graph G is a *threshold graph* if there exists a real number α and a mapping $w : V(G) \to \mathbb{R}$ such that $(x, y) \in E(G)$ if and only if $w(x) + w(y) \ge \alpha$, where $\mathbb{R}$ is the set of all real numbers [4].

Theorem 2. *The minimum vulnerability problem is NP-hard for threshold graphs, even if* $\mathsf{cap}(e) \geq 1$ *and* $\mathsf{vul}(e) = r$ *hold for all edges* $e \in E(G)$, *where* $r \geq 1$ *is any fixed constant.*

Proof. We modify the instance constructed in the proof of Theorem 1, as follows: Add an edge $e = (v_s, v_t)$ to the graph and set $\mathsf{cost}(e) = 1$, $\mathsf{vul}(e) = r$ and $\mathsf{cap}(e) = r$. (See Fig. 3.) Then, reset the number k of required $v_s v_t$-paths to $k = (n+1)r + d$. Clearly, the graph is a threshold graph such that $\mathsf{cap}(e) \geq 1$ and $\mathsf{vul}(e) = r \geq 1$ hold for all edges $e \in E(G)$.

Note that the edge (v_s, v_t) can be passed through by at most r paths, and these r paths do not cause any extra penalty. Therefore, the same arguments in the proof of Theorem 1 establish the theorem.   □

3   Algorithm for Bounded Treewidth Graphs

In this section, we give an algorithm for bounded treewidth graphs.

A *tree-decomposition* of a graph G is a pair $\langle \{X_i : i \in V_T\}, T \rangle$, where $T = (V_T, E_T)$ is a rooted tree, such that the following four conditions (1)–(4) hold [2]:

(1) Each X_i is a subset of $V(G)$, and is called a *bag*;
(2) $\bigcup_{i \in V_T} X_i = V(G)$;
(3) for each edge $(u, v) \in E(G)$, there is at least one node $i \in V_T$ such that $u, v \in X_i$; and
(4) for each vertex $v \in V(G)$, the set $\{i \in V_T : v \in X_i\}$ induces a connected subgraph in T.

For example, Fig. 4(b) illustrates a tree-decomposition of the graph G in Fig. 4(a). We will refer to a *node* in V_T in order to distinguish it from a vertex in $V(G)$. The *width* of a tree-decomposition $\langle \{X_i : i \in V_T\}, T \rangle$ is defined as $\max\{|X_i| - 1 : i \in V_T\}$, and the *treewidth* of G is the minimum t such that G has a tree-decomposition of width t. We denote by $\mathsf{tw}(G)$ the treewidth of G.

Recall that the minimum vulnerability problem is NP-hard even for series-parallel graphs (Theorem 1), which are of treewidth at most two. In this section, we thus give a pseudo-polynomial-time algorithm for bounded treewidth graphs.

Theorem 3. *Let G be a graph whose treewidth is bounded by a fixed constant t. Then, $\mathsf{OPT}_k(G)$ can be computed in time $(k+1)^{O(t^{t+1})} n$, where $n = |V(G)|$.*

As a proof of Theorem 3, we give such an algorithm in the remainder of this section. Our algorithm can easily be modified to actually find k $v_s v_t$-paths on G with the minimum penalty $\mathsf{OPT}_k(G)$.

3.1   Nice Tree-Decomposition

A tree-decomposition $\langle \{X_i : i \in V_T\}, T \rangle$ of G is called a *nice tree-decomposition* if the following conditions (5)–(10) hold [2]:

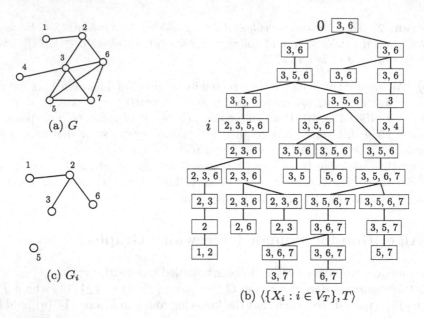

Fig. 4. (a) Graph G, (b) a nice tree-decomposition $\langle\{X_i : i \in V_T\}, T\rangle$ of G, and (c) the subgraph G_i of G for the node $i \in V_T$.

(5) $|V_T| = O(t^2 n)$, where $n = |V(G)|$ and $t = \max\{|X_i| - 1 : i \in V_T\}$;

(6) every node in V_T has at most two children in T;

(7) if a node $i \in V_T$ has two children l and r, then $X_i = X_l = X_r$;

(8) if a node $i \in V_T$ has only one child j, then
- $|X_i| = |X_j| - 1$ and $X_i \subset X_j$ (such a node i is called a *forget* node); or
- $|X_i| = |X_j| + 1$ and $X_i \supset X_j$ (such a node i is called an *introduce* node);

(9) for each edge $(u, v) \in E(G)$, there is a leaf node $i \in V_T$ such that $u, v \in X_i$; and

(10) the bag of every leaf node in V_T contains exactly two vertices.

Figure 4(b) illustrates a nice tree-decomposition $\langle\{X_i : i \in V_T\}, T\rangle$ of the graph G in Fig. 4(a) whose treewidth is three. Let t be a fixed constant. Then, for a given graph G, there is a linear-time algorithm which either outputs $\text{tw}(G) > t$ or gives a nice tree-decomposition of G whose width is at most t [2].

Given a graph G, we assume without loss of generality that $(v_s, v_t) \in E(G)$. If $(v_s, v_t) \notin E(G)$, then add (v_s, v_t) to $E(G)$ and set $\text{cap}((v_s, v_t)) = 0$. The treewidth of this graph is bounded by $\text{tw}(G) + 2$.

Let $\langle\{X_i : i \in V_T\}, T\rangle$ be a nice tree-decomposition with width at most t of a graph G. We regard the node i with $X_i = \{v_s, v_t\}$ as the root 0 of T. For each edge $e = (u, v) \in E(G)$, since there are some leaves whose bags contain both u and v, we arbitrarily choose one of such bags, say X_i, which is called a *representation* of e and denoted by $\text{rep}(e) = i$.

We recursively define a vertex set $V_i \subseteq V(G)$ and an edge set $E_i \subseteq E(G)$ for each node i of T, similar to the way used in [7], as follows:

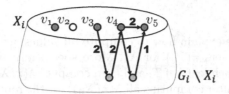

Fig. 5. An example of an (s, a)-path set.

(a) If i is a leaf, then let $V_i = X_i$ and $E_i = \{e \in E(G) : \mathsf{rep}(e) = i\}$; and
(b) if i is an internal node having only one child j, then let $V_i = V_j \cup X_i$ and $E_i = E_j$, where V_j and E_j are the vertex and edge sets for j, respectively; and
(c) if i is an internal node having two children l and r, then let $V_i = V_l \cup V_r$ and $E_i = E_l \cup E_r$, where V_l and E_l are the vertex and edge sets for l, respectively, and V_r and E_r are the vertex and edge sets for r, respectively.

Then, for each node i of T, we denote by G_i the graph with vertex set V_i and edge set E_i and hence $G_i = (V_i, E_i)$. In this way, for any node i with the children l and r, there exists no edge that is contained in both E_r and E_l.

3.2 Definitions

Let $\mathcal{S}(X_i)$ be the set of all permutations of the vertices in X_i, and let $\mathcal{A}(X_i)$ be the set of all $|X_i|$-tuples of nonnegative integers at most k. A path P in G_i joining two vertices $v, v' \in X_i$ is *inner* if $V(P)\backslash\{v, v'\} \subseteq V_i\backslash X_i$. For a pair of $s = (v_1, v_2, \ldots, v_{|X_i|}) \in \mathcal{S}(X_i)$ and $a = (a_1, a_2, \ldots, a_{|X_i|-1}) \in \mathcal{A}(X_i)$, a set $\mathcal{P}$ of paths in G_i is called an (s, a)-*path set* if $|\mathcal{P}| = \sum_{j=1}^{|X_i|-1} a_j$ and $\mathcal{P}$ has exactly a_j inner $v_j v_{j+1}$-paths for each j, $1 \leq j \leq |X_i| - 1$. Figure 5 illustrates an (s, a)-path set, where $s = (v_1, v_3, v_4, v_5, v_2)$ and $a = (0, 2, 3, 0)$. Let $\pi(X_i)$ be a set of pairs (s, a) such that $s \in \mathcal{S}(X_i)$ and $a \in \mathcal{A}(X_i)$. Then the set $\pi(X_i) = \{(s_1, a_1), (s_2, a_2), \ldots, (s_{|\pi(X_i)|}, a_{|\pi(X_i)|})\}$ is *active* if there exists a set $\mathcal{P}$ of inner paths in G_i and a partition $\mathcal{P}_1, \mathcal{P}_2, \ldots, \mathcal{P}_{|\pi(X_i)|}$ of $\mathcal{P}$ such that each $\mathcal{P}_j$, $1 \leq j \leq |\pi(X_i)|$, forms an (s_j, a_j)-path set. Such a set $\mathcal{P}$ is called a $\pi(X_i)$-*path set*. Finally, for a set $\pi(X_i)$, we define

$$\lambda(\pi(X_i)) = \min\{\lambda(\mathcal{P}) : \mathcal{P} \text{ is a } \pi(X_i)\text{-path set}\}$$

if $\pi(X_i)$ is active and hence there exists a $\pi(X_i)$-path set, otherwise we let $\lambda(\pi(X_i)) = +\infty$.

Our algorithm computes $\lambda(\pi(X_i))$ for all sets $\pi(X_i)$ for each bag X_i of T, from the leaves of T to the root 0 of T, by means of dynamic programming. Then, $\mathsf{OPT}_k(G)$ can be computed at the root 0 from the values $\lambda(\pi(X_0))$, as described in Sect. 3.3.

3.3   Algorithm

In this subsection, we explain how to compute all values $\lambda(\pi(X_i))$ for each node $i \in V_T$ of T and all sets $\pi(X_i)$ for X_i. More specifically, we first compute all values $\lambda(\pi(X_i))$ for each leaf i of T, and then compute $\lambda(\pi(X_i))$ for each internal node i in T. Finally, after computing all $\lambda(\pi(X_0))$ for the root 0 of T, we compute $\mathsf{OPT}_G(k)$.

[1. The node i is a leaf of T.]
    In this case, by the Definition (10) of a nice tree-decomposition, there are exactly two vertices v_1 and v_2 in X_i, and $e = (v_1, v_2) \in E(G)$. Then, a set $\pi(X_i) \in 2^{\mathcal{S}(X_i) \times \mathcal{A}(X_i)}$ is active if and only if the following two conditions hold:

(i) $|\pi(X_i)| = 1$; and
(ii) $a_1 \le \mathsf{cap}(e)$ for the pair $(s, a) \in \pi(X_i)$ with $a = (a_1)$.

For an active set $\pi(X_i)$, we let

$$\lambda(\pi(X_i)) = \begin{cases} 0 & \text{if } 0 \le a_1 \le \mathsf{vul}(e); \\ \mathsf{cost}(e) & \text{if } \mathsf{vul}(e) < a_1 \le \mathsf{cap}(e). \end{cases}$$

For the other sets $\pi(X_i)$, we let $\lambda(\pi(X_i)) = +\infty$.

[2. The node i is an internal node.]
    Since $\langle \{X_i : i \in V_T\}, T \rangle$ is a nice tree-decomposition of G and the node i is an internal node, either i has two children, is a forget node, or is an introduce node. Therefore we have the following three cases to consider.

Case 1: The node i has two children l and r.

In this case, each set of paths in G_i can be obtained by merging two sets of paths in G_l and G_r. Therefore, a set $\pi(X_i) \in 2^{\mathcal{S}(X_i) \times \mathcal{A}(X_i)}$ is active if and only if there exist two active sets $\pi(X_l) \in 2^{\mathcal{S}(X_l) \times \mathcal{A}(X_l)}$ and $\pi(X_r) \in 2^{\mathcal{S}(X_r) \times \mathcal{A}(X_r)}$ such that for each $(s, a) \in \pi(X_i)$, there exist $(s_l, a_l) \in \pi(X_l)$ and $(s_r, a_r) \in \pi(X_r)$ satisfying $s = s_l = s_r$ and $a = a_l + a_r$, where $a_l + a_r$ is defined as the addition of each element of a_l and a_r. Then, $\lambda(\pi(X_i)) = \min\{\lambda(\pi(X_l)) + \lambda(\pi(X_r))\}$, where the minimum is taken over all pairs of such active $\pi(X_l)$ and $\pi(X_r)$.

Case 2: The node i is a forget node.

Let j be the child of the node i, and let v be the vertex such that $X_j \backslash X_i = \{v\}$. Then, a set $\pi(X_i) \in 2^{\mathcal{S}(X_i) \times \mathcal{A}(X_i)}$ is active if and only if there exists an active set $\pi(X_j) \in 2^{\mathcal{S}(X_j) \times \mathcal{A}(X_j)}$ such that for each $(s, a) \in \pi(X_i)$, there exists some pair $(s', a') \in \pi(X_j)$ with $s' = (v_1', v_2', \ldots, v_{|X_j|}')$ and $a' = (a_1', a_2', \ldots, a_{|X_j|-1}')$ satisfying the following two conditions:

(i) $s = s' \backslash v$, where $s' \backslash v$ is the sequence obtained from s' by deleting v; and
(ii) $a_{l-1}' = a_l'$ and $a = (a_1', a_2', \ldots, a_{l-1}', a_{l+1}', \ldots, a_{|X_j|-1}')$ for the index l such that $v_l' = v$ in s'.

Then, $\lambda(\pi(X_i))$ is the minimum value of $\lambda(\pi(X_j))$, taken over all such active sets $\pi(X_j)$.

Case 3: The node i is an introduce node.

Let j be the child of the node i. In this case, since $|X_i \setminus X_j| = 1$, let v be the vertex in $X_i \setminus X_j$. Then, a set $\pi(X_i) \in 2^{\mathcal{S}(X_i) \times \mathcal{A}(X_i)}$ is active if and only if the following two conditions hold:

(i) $a_{l-1} = a_l = 0$ for each pair $(s, a) \in \pi(X_i)$ such that $s = (v_1, v_2, \ldots, v_{|X_i|})$ with $v_l = v$ and $a = (a_1, a_2, \ldots, a_{|X_i|-1})$; and

(ii) there exists a set $\pi(X_j) \in 2^{\mathcal{S}(X_j) \times \mathcal{A}(X_j)}$ which is active such that $\pi(X_j) = \{(s \setminus v, a') \mid (s, a) \in \pi(X_i)\}$, where $v_l = v$ in $s = (v_1, v_2, \ldots, v_{|X_i|})$, $a = (a_1, a_2, \ldots, a_{|X_i|-1})$ and $a' = (a_1, a_2, \ldots, a_{l-1}, a_{l+1}, \ldots, a_{|X_i|-1})$.

Then, $\lambda(\pi(X_i))$ is the minimum value of $\lambda(\pi(X_j))$, taken over all such active sets $\pi(X_j)$.

[3. The node i is the root of T.]
In this case, $i = 0$. We first compute the values $\lambda(\pi(X_0))$ for all sets $\pi(X_0)$ for X_0, according to one of the three cases 2–4 above. Then, our algorithm computes $\mathrm{OPT}_k(G)$ from the values $\lambda(\pi(X_0))$ for all active sets $\pi(X_0) \in 2^{\mathcal{S}(X_0) \times \mathcal{A}(X_0)}$. Since $X_0 = \{v_s, v_t\}$, we only need to count the number of inner paths connecting v_s and v_t. Therefore, $\mathrm{OPT}_k(G) = \min \lambda(\pi(X_0))$, where the minimum is taken over all active sets $\pi(X_0) = \{(s, a)\} \in 2^{\mathcal{S}(X_0) \times \mathcal{A}(X_0)}$ such that $a = (k)$.

3.4   Running Time

Recall that $|X_i| \le t + 1$ for each node $i \in V_T$, where t is an upper bound on the treewidth of G. Then, $|\mathcal{S}(X_i)| = (t + 1)! = O(t^{(t+1)})$. Furthermore, since $\mathcal{A}(X_i)$ is the set of all $|X_i|$-tuples of nonnegative integers at most k, we have $|\mathcal{A}(X_i)| \le (k + 1)^{(t+1)}$. Thus, the number of all sets $\pi(X_i) \in 2^{\mathcal{S}(X_i) \times \mathcal{A}(X_i)}$ for each node $i \in V_T$ can be bounded by

$$\left((k + 1)^{(t+1)}\right)^{(t+1)!} \le (k + 1)^{O\left(t^{(t+1)}\right)}.$$

Recall that $|X_i| = 2$ for each leaf $i \in V_T$. Then, according to the case 1 above, one can compute the value $\lambda(\pi(X_i))$ in $O(1)$ time for each set $\pi(X_i)$. Therefore, the values $\lambda(\pi(X_i))$ for all sets $\pi(X_i) \in 2^{\mathcal{S}(X_i) \times \mathcal{A}(X_i)}$ can be computed in $(k + 1)^{O\left(t^{(t+1)}\right)}$ time. By the Definition (5) of a nice tree-decomposition, T has at most $O(t^2 n)$ leaves; one can thus compute $\lambda(\pi(X_i))$ for all leaves of T in $n(k + 1)^{O\left(t^{(t+1)}\right)}$ time.

Similarly, for each internal node i, each of the update formulas above can be computed in $(k + 1)^{O\left(t^{(t+1)}\right)}$ time for each set $\pi(X_i)$. Therefore, the values $\lambda(\pi(X_i))$ for all sets $\pi(X_i) \in 2^{\mathcal{S}(X_i) \times \mathcal{A}(X_i)}$ can be computed in $(k + 1)^{O\left(t^{(t+1)}\right)}$ time for each internal node i. By the Definition (5) of a nice tree-decomposition, T has at most $O(t^2 n)$ internal nodes, and hence one can compute the values $\lambda(\pi(X_i))$ for all internal nodes of T in $n(k + 1)^{O\left(t^{(t+1)}\right)}$ time.

Finally, for the root 0 of T, we can compute $\mathsf{OPT}_k(G)$ in $(k+1)^{O(t^{(t+1)})}$ time from the values $\lambda(\pi(X_0))$.

In this way, our algorithm runs in $n(k+1)^{O(t^{(t+1)})}$ time in total. This completes the proof of Theorem 3.   □

4   Parameterized Algorithm for Chordal Graphs

A graph G is *chordal* if every cycle in G of length at least four has a chord, which is an edge joining non-consecutive vertices in the cycle [4].

Recall that the minimum vulnerability problem is NP-hard for threshold graphs, which form a subclass of chordal graphs, even when $\mathsf{vul}(e) \geq 1$ and $\mathsf{cap}(e) \geq 1$ hold for all edges $e \in E(G)$ (Theorem 2). In this section, we thus give an FPT algorithm for chordal graphs when parameterized by the number k of required $v_s v_t$-paths.

Theorem 4. *Let G be a chordal graph with n vertices and m edges such that $\mathsf{vul}(e) \geq 1$ and $\mathsf{cap}(e) \geq 1$ hold for all edges $e \in E(G)$. Then, $\mathsf{OPT}_k(G)$ can be computed in time $m + (k+1)^{O(k^{(3k+2)})} n$.*

As a proof of Theorem 4, we give such an algorithm in the remainder of this section. If a graph G is a chordal graph, then there exists a tree-decomposition $\langle \{X_i : i \in V_T\}, T \rangle$ such that each bag X_i forms a clique, and such a tree-decomposition can be found in linear time [3].

For a vertex subset V' of a graph G, let $G[V']$ be the subgraph of G induced by V'. First, we prove the following two lemmas.

Lemma 1. *Let G be a graph with a cut-set X such that $\mathsf{vul}(e) \geq 1$ and $\mathsf{cap}(e) \geq 1$ hold for all edges $e \in E(G)$. Suppose that X is a clique and $|X| \geq 3k$. If there is a connected component C in $G \backslash X$ such that $v_s, v_t \in V(C) \cup X$, then $\mathsf{OPT}_k(G, v_s, v_t) = \mathsf{OPT}_k(G', v_s, v_t)$, where $G' = G[V(C) \cup X]$.*

Proof. Since G' is a subgraph of G, we have $\mathsf{OPT}_k(G, v_s, v_t) \leq \mathsf{OPT}_k(G', v_s, v_t)$. Therefore, it suffices to prove that $\mathsf{OPT}_k(G, v_s, v_t) \geq \mathsf{OPT}_k(G', v_s, v_t)$.

Let G_X be the graph obtained from G by contracting all vertices in X into one vertex v_X. Each resulting edge (v, v_X) has the same values of vulnerability, capacity and cost as its original edge. Note that if $v_s \in X$ then let $v_s = v_X$, and if $v_t \in X$ then let $v_t = v_X$. Clearly $\mathsf{OPT}_k(G, v_s, v_t) \geq \mathsf{OPT}_k(G_X, v_s, v_t)$, and hence it suffices to prove that $\mathsf{OPT}_k(G_X, v_s, v_t) \geq \mathsf{OPT}_k(G', v_s, v_t)$.

Let $\mathcal{P}_X$ be a set of k $v_s v_t$-paths as an optimal solution in G_X. Then we will construct a set $\mathcal{P}'$ of k $v_s v_t$-paths in G' from $\mathcal{P}_X$ such that $\lambda(\mathcal{P}') = \lambda(\mathcal{P}_X)$ as follows. For each path $P \in \mathcal{P}_X$, there are the following two cases to consider.

Case 1: The edges in G' corresponding to the edges on P in G_X form a $v_s v_t$-path P' in G'.

In this case we add the path P' to $\mathcal{P}'$.

Case 2: The edges in G' corresponding to the edges on P in G_X form exactly two paths in G'; one is a $v_s v_1$-path P_1 and the other is a $v_2 v_t$-path P_2, where $v_1, v_2 \in X$.

In this case, we choose an arbitrary vertex $u \in X$ which is not on any path in $\mathcal{P}'$ so far and is not an end of some edges corresponding to the edges on some paths in $\mathcal{P}_X$. Since $|\mathcal{P}_X| = k$, there are at most $2k$ edges on paths in $\mathcal{P}_X$ adjacent to vertices in X. Furthermore $|X| \geq 3k$. Therefore, there are at least $|X| - 2k \geq k$ vertices in X which can be chosen in total. After chosen, by adding two edges $e_1 = (v_1, u)$ and $e_2 = (u, v_2)$ to join P_1 to P_2, the resulting path P' is a $v_s v_t$-path in G', and add it to $\mathcal{P}'$. Since $\mathsf{vul}(e_1) \geq 1$, $\mathsf{cap}(e_1) \geq 1$, $\mathsf{vul}(e_2) \geq 1$, $\mathsf{cap}(e_2) \geq 1$, and these two edges e_1 and e_2 are not used by another path in $\mathcal{P}'$, we have $\lambda(\mathcal{P}') = \lambda(\mathcal{P}_X)$.

We have completed to construct a set $\mathcal{P}'$ of k $v_s v_t$-paths in G' from $\mathcal{P}_X$ such that $\lambda(\mathcal{P}') = \lambda(\mathcal{P}_X)$, and hence $\mathsf{OPT}_k(G', v_s, v_t) \leq \lambda(\mathcal{P}') = \lambda(\mathcal{P}_X) = \mathsf{OPT}_k(G_X, v_s, v_t)$.   □

Lemma 2. *Let G be a graph with a cut-set X such that $\mathsf{vul}(e) \geq 1$ and $\mathsf{cap}(e) \geq 1$ hold for all edges $e \in E(G)$. Suppose that X is a clique and $|X| \geq 3k$. If there are two connected components C_1 and C_2 in $G \backslash X$, then $\mathsf{OPT}_k(G, v_s, v_t) = \mathsf{OPT}_k(G_1, v_s, v) + \mathsf{OPT}_k(G_2, v, v_t)$, where v is an arbitrary vertex in $X \backslash \{v_s, v_t\}$, $G_1 = G[V(C_1) \cup X]$ and $G_2 = G[V(C_2) \cup X]$.*

Proof. Since G_1 and G_2 are subgraphs of G, it is trivial that $\mathsf{OPT}_k(G, v_s, v_t) \leq \mathsf{OPT}_k(G_1, v_s, v) + \mathsf{OPT}_k(G_2, v, v_t)$. Therefore, it suffices to prove that

$$\mathsf{OPT}_k(G, v_s, v_t) \geq \mathsf{OPT}_k(G_1, v_s, v) + \mathsf{OPT}_k(G_2, v, v_t).$$

Let G_X be the graph obtained from G by contracting all vertices in X as one vertex v_X. Each resulting edge (v, v_X), $v \in V(G_X) \backslash \{v_X\}$, has the same values of vulnerability, capacity and cost of its corresponding edge. Note that if $v_s \in X$ then let $v_s = v_X$, and if $v_t \in X$ then let $v_t = v_X$. Clearly $\mathsf{OPT}_k(G, v_s, v_t) \geq \mathsf{OPT}_k(G_X, v_s, v_t)$, and hence it suffices to prove that

$$\mathsf{OPT}_k(G_X, v_s, v_t) \geq \mathsf{OPT}_k(G_1, v_s, v) + \mathsf{OPT}_k(G_2, v, v_t).$$

Let $\mathcal{P}_X$ be a set of k $v_s v_t$-paths as an optimal solution in G_X. Then we will construct a set $\mathcal{P}_1$ of k $v_s v$-paths in G_1, and a set $\mathcal{P}_2$ of k vv_t-paths in G_2, such that $\lambda(\mathcal{P}_X) = \lambda(\mathcal{P}_1) + \lambda(\mathcal{P}_2)$ as follows.

For each path $P_X \in \mathcal{P}_X$, the edges in $E(G_1) \cup E(G_2)$ corresponding to the edges on P_X in G_X form exactly two paths: One is a $v_s v_1$-path P_1 in G_1, and the other is a $v_2 v_t$-path P_2 in G_2, where $v_1, v_2 \in X$.

Then, we choose an arbitrary vertex $u \in X$ which is not on any path in $\mathcal{P}_1$ so far and is not an end of some edges corresponding to the edges on some paths in $\mathcal{P}_X$. Since $|\mathcal{P}_X| = k$, there are at most $2k$ edges on paths in $\mathcal{P}_X$ adjacent to vertices in X. Furthermore $|X| \geq 3k$. Therefore there are at least $|X| - 2k \geq k$ vertices in X which can be chosen in total. After chosen, by adding two edges $e_1 = (v_1, u)$ and $e_2 = (u, v)$ to join P_1 to v, the resulting path P_1 is a $v_s v$-path in G_1, and add it to $\mathcal{P}_1$. Since $\mathsf{vul}(e_1) \geq 1$, $\mathsf{cap}(e_1) \geq 1$, $\mathsf{vul}(e_2) \geq 1$, $\mathsf{cap}(e_2) \geq 1$, and these two edges e_1 and e_2 are not used by another path in $\mathcal{P}_1$, we do not pay any costs on e_1 and e_2.

Similarly, we choose an arbitrary vertex $u \in X$ which is not on any path in $\mathcal{P}_2$ so far and is not an end of some edges corresponding to the edges on some paths in $\mathcal{P}_X$. After chosen, by adding two edges $e_1 = (v, u)$ and $e_2 = (u, v_2)$ to join P_2 to v, the resulting path P_2 is a vv_t-path in G_2, and add it to $\mathcal{P}_2$. Since these two edges e_1 and e_2 are not used by another path in $\mathcal{P}_2$, we do not pay any costs on e_1 and e_2.

We have completed to construct two sets, $\mathcal{P}_1$ of k $v_s v$-paths in G_1 and $\mathcal{P}_2$ of k vv_t-paths in G_2 from $\mathcal{P}_X$ such that $\lambda(\mathcal{P}_1) + \lambda(\mathcal{P}_2) = \lambda(\mathcal{P}_X)$, and hence

$$\mathsf{OPT}_k(G_1, v_s, v) + \mathsf{OPT}_k(G_2, v, v_t) \leq \lambda(\mathcal{P}_1) + \lambda(\mathcal{P}_2)$$
$$= \lambda(\mathcal{P}_X)$$
$$= \mathsf{OPT}_k(G, v_s, v_t).$$

This completes the proof of Lemma 2. □

By using Lemmas 1 and 2, we thus have the following FPT algorithm $\mathsf{Alg}(G, v_s, v_t, k)$ that returns $\mathsf{OPT}_k(G, v_s, v_t)$ for a chordal graph G.

Algorithm 1. $\mathsf{Alg}(G, v_s, v_t, k)$

1: let $\langle \{X_i : i \in V_T\}, T \rangle$ be a tree-decomposition of the chordal graph G with treewidth t.
2: **if** there are all $i \in V_T$ such that $|X_i| \leq 3k - 1$ **then**
3:     compute $\mathsf{OPT}_k(G, v_s, v_t)$ by Theorem 3 and return it;
4: **else**
5:     /* in this case, there is a node $i \in V_T$ such that $|X_i| \geq 3k$ */
6:     Let i be a node in V_T such that $|X_i| \geq 3k$;
7:     **if** $v_s, v_t \in V(G_i)$ **then**
8:         **return** $\mathsf{Alg}(G_i, v_s, v_t, k)$;
9:     **else if** $v_s, v_t \notin V(G_i)$ **then**
10:        **return** $\mathsf{Alg}(\overline{G_i}, v_s, v_t, k)$, where $\overline{G_i} = G[(V(G)\backslash V(G_i)) \cup X_i]$;
11:     **else**
12:        suppose without loss of generality that $v_s \in V(G_i)$ and $v_t \notin V(G_i)$;
13:        let v be an arbitrary vertex in $X_i \backslash \{v_s, v_t\}$;
14:        **return** $\mathsf{Alg}(G_i, v_s, v, k) + \mathsf{Alg}(\overline{G_i}, v, v_t, k)$;
15:     **end if**
16: **end if**

We are now ready to prove Theorem 4.

Proof of Theorem 4. By Lemmas 1 and 2 and Theorem 3, $\mathsf{Alg}(G, v_s, v_t, k)$ above correctly computes $\mathsf{OPT}_k(G)$. Therefore, it suffices to prove the time-complexities. Lines 4–15 can be performed in combined $O(tn)$ time. By Theorem 3, Line 2–3 can be done in $(k + 1)^{O(k^{(3k+2)})} n$ time. This completes the proof of Theorem 4. □

References

1. Assadi, S., Emamjomeh-Zadeh, E., Norouzi-Fard, A., Yazdanbod, S., Zarrabi-Zadeh, H.: The minimum vulnerability problem. In: Chao, K.-M., Hsu, T., Lee, D.-T. (eds.) ISAAC 2012. LNCS, vol. 7676, pp. 382–391. Springer, Heidelberg (2012)
2. Bodlaender, H.L.: A linear-time algorithm for finding tree-decompositions of small treewidth. SIAM J. Comput. **25**, 1305–1317 (1996)
3. Boutiche, M.A., Ait Haddadène, H., Le Thi, H.A.: Maintaining graph properties of weakly chordal graphs. Appl. Math. Sci. **6**, 765–778 (2012)
4. Brandstädt, A., Le, V.B., Spinrad, J.P.: Graph Classes: A Survey. SIAM, Philadelphia (1999)
5. Even, G., Kortsarz, G., Slany, W.: On network design problems: fixed cost flows and the covering steiner problem. ACM Trans. Algorithms **1**, 74–101 (2005)
6. Garey, M.R., Johnson, D.S.: Computers and Intractability: A Guide to the Theory of NP-Completeness. Freeman, San Francisco (1979)
7. Isobe, S., Zhou, X., Nishizeki, T.: A polynomial-time algorithm for finding total colorings of partial k-trees. Int. J. Found. Comput. Sci. **10**, 171–194 (1999)
8. Krumke, S.O., Noltemeier, H., Schwarz, S., Wirth, H.-C., Ravi, R.: Flow improvement and network flows with fixed costs. In: Proceedings of OR, vol. 1998, pp. 158–167 (1998)
9. Omran, M.T., Sack, J.-R., Zarrabi-Zadeh, H.: Finding paths with minimum shared edges. J. Comb. Optim. **26**, 709–722 (2011)
10. Ye, Z.Q., Li, Y.M., Lu, H.Q., Zhou, X.: Finding paths with minimum shared edges in graphs with bounded treewidth. In: Proceedings of FCS vol. 2013, pp. 40–46 (2013)

The List Coloring Reconfiguration Problem
for Bounded Pathwidth Graphs

Tatsuhiko Hatanaka[✉], Takehiro Ito, and Xiao Zhou

Graduate School of Information Sciences, Tohoku University,
Aoba-yama 6-6-05, Sendai 980-8579, Japan
{hatanaka,takehiro,zhou}@ecei.tohoku.ac.jp

Abstract. We study the problem of transforming one list (vertex) coloring of a graph into another list coloring by changing only one vertex color assignment at a time, while at all times maintaining a list coloring, given a list of allowed colors for each vertex. This problem is known to be PSPACE-complete for bipartite planar graphs. In this paper, we first show that the problem remains PSPACE-complete even for bipartite series-parallel graphs, which form a proper subclass of bipartite planar graphs. We note that our reduction indeed shows the PSPACE-completeness for graphs with pathwidth two, and it can be extended for threshold graphs. In contrast, we give a polynomial-time algorithm to solve the problem for graphs with pathwidth one. Thus, this paper gives precise analyses of the problem with respect to pathwidth.

1 Introduction

Graph coloring is one of the most fundamental research topics in the field of theoretical computer science. Let $C = \{1, 2, \ldots, k\}$ be the set of k colors. A (proper) k-coloring of a graph $G = (V, E)$ is a mapping $f : V \to C$ such that $f(v) \neq f(w)$ for every edge $vw \in E$. In *list coloring*, each vertex $v \in V$ has a set $L(v) \subseteq C$ of colors, called the *list* of v. Then, a k-coloring f of G is called an *L-coloring* of G if $f(v) \in L(v)$ holds for every vertex $v \in V$. Figure 1(b) illustrates four *L*-colorings of the same graph G with the same list L depicted in Fig. 1(a); the color assigned to each vertex is attached to the vertex. Clearly, a k-coloring of G is an *L*-coloring of G for which $L(v) = C$ holds for every vertex v of G, and hence *L*-coloring is a generalization of k-coloring.

Graph coloring has several practical applications, such as in scheduling and frequency assignments. For example, in the frequency assignment problem, each vertex corresponds to a base station and each edge represents the physical proximity and hence the two corresponding base stations have the high potential of interference. Each color represents a channel of a particular frequency, and we wish to find an assignment of channels to the base stations without any interference. Furthermore, in list coloring, each base station can have a list of channels that can be assigned to it.

© Springer International Publishing Switzerland 2014
Z. Zhang et al. (Eds.): COCOA 2014, LNCS 8881, pp. 314–328, 2014.
DOI: 10.1007/978-3-319-12691-3_24

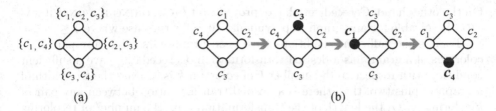

Fig. 1. (a) Graph G and its list L, and (b) a sequence of L-colorings of G.

1.1 Our Problem

However, a practical issue in channel assignments requires that the formulation should be considered in more dynamic situations. One can imagine a variety of practical scenarios where an L-coloring (e.g., representing a feasible channel assignment) needs to be transformed (to use a newly found better assignment or to satisfy new side constraints) by individual color changes (keeping the network functionality and preventing the need for any coordination) while maintaining feasibility (so that the users receive service even during the reassignment).

In this paper, we thus study the following problem: Suppose that we are given two L-colorings of a graph G (e.g., the leftmost and rightmost ones in Fig. 1(b)), and we are asked whether we can transform one into the other via L-colorings of G such that each differs from the previous one in only one vertex color assignment. We call this decision problem the LIST COLORING RECONFIGURATION problem. For the particular instance of Fig. 1(b), the answer is "yes," as illustrated in Fig. 1(b), where the vertex whose color assignment was changed from the previous one is depicted by a black circle.

1.2 Known and Related Results

Recently, similar settings of problems have been extensively studied in the framework of reconfiguration problems [13], which arise when we wish to find a step-by-step transformation between two feasible solutions of a problem such that all intermediate solutions are also feasible. This reconfiguration framework has been applied to several well-studied combinatorial problems, including satisfiability [11,21], independent set [6,12,13,20,22], set cover, matching [13], shortest path [3,4,19], list edge-coloring [14,16], list $L(2,1)$-labeling [15], and so on.

In particular, the reconfiguration problem for (non-list) k-colorings is one of the most well-studied reconfiguration problems. The k-COLORING RECONFIGURATION problem is LIST COLORING RECONFIGURATION such that $L(v) = \{1, 2, \ldots, k\}$ holds for every vertex v. This problem has been studied intensively from various viewpoints [1,2,5,7,9,10,18,25], as follows.

Bonsma and Cereceda [5] proved that k-COLORING RECONFIGURATION is PSPACE-complete for $k \geq 4$; they also proved that LIST COLORING RECONFIGURATION is PSPACE-complete, even for bipartite planar graphs and $k = 4$.

On the other hand, Cereceda et al. [10] proved that k-COLORING RECONFIGURA-
TION is solvable for any graph in polynomial time for the case where $1 \le k \le 3$.

Then, some sufficient conditions have been proposed so that any pair of k-colorings of a graph has a desired transformation. Cereceda [9] gave a sufficient condition with respect to the number k of colors: if k is at least the treewidth of a graph G plus two, then there is a desired transformation between any pair of k-colorings of G; the length of the transformation (i.e., the number of recoloring steps) is estimated by Bonamy and Bousquet [2]. Bonamy et al. [1] gave some sufficient condition with respect to graph structures: for example, chordal graphs and chordal bipartite graphs satisfy their sufficient condition.

Recently, Bonsma et al. [7] and Johnson et al. [18] independently developed a fixed-parameter algorithm to solve k-COLORING RECONFIGURATION when parameterized by $k + \ell$, where k is the number of colors and ℓ is the number of recoloring steps. In contrast, if the problem is parameterized only by ℓ, then it is W[1]-hard [7] and does not admit a polynomial kernelization unless the polynomial hierarchy collapses [18].

In this way, even for the non-list version, only a few results are known from the viewpoint of polynomial-time solvability. Furthermore, as far as we know, no algorithmic result has been obtained for the list version.

1.3 Our Contribution

In this paper, we study the LIST COLORING RECONFIGURATION problem from the viewpoint of graph classes, especially pathwidth of graphs. (The definition of pathwidth will be given in Sect. 2.)

We prove that the problem remains PSPACE-complete even for graphs with pathwidth two. In contrast, we give a polynomial-time algorithm to solve the problem for graphs with pathwidth one. Thus, this paper gives precise analyses of the problem with respect to pathwidth.

Indeed, our reduction for the PSPACE-completeness proof constructs a bipartite series-parallel graph (whose treewidth is two), which is a bipartite planar graph. We note that the problem of finding one L-coloring of a given graph can be solved in polynomial time for bounded treewidth graphs (and hence for bounded pathwidth graphs) [17]. However, our proof shows that the reconfiguration variant is PSPACE-complete even if treewidth and pathwidth are two. Furthermore, as a byproduct, our reduction can be extended for threshold graphs.

Due to the page limitation, we omit some proofs from this extended abstract.

2 Preliminaries

We assume without loss of generality that graphs are simple and connected. Let $G = (V, E)$ be a graph with vertex set V and edge set E; we sometimes denote by $V(G)$ and $E(G)$ the vertex set and edge set of G, respectively. For a vertex v in G, we denote by $d(v)$ the degree of v in G. For a vertex subset $V' \subseteq V$, we denote by $G[V']$ the subgraph of G induced by V'.

We now define the notion of pathwidth [24]. A *path-decomposition* of a graph G is a sequence of subsets X_i of vertices in G such that

(1) $\bigcup_i X_i = V(G)$;
(2) for each $vw \in E(G)$, there is at least one subset X_i with $v, w \in X_i$; and
(3) for any three indices p, q, r such that $p \leq q \leq r$, $X_p \cap X_r \subseteq X_q$.

The *width* of a path-decomposition is defined as $\max_i |X_i| - 1$, and the *pathwidth* of G is the minimum t such that G has a path-decomposition of width t.

To develop our algorithm in Sect. 4, it is important to notice that every connected graph of pathwidth one is a caterpillar [23]. A caterpillar will be defined in Sect. 4, but an example can be found in Fig. 4.

For a graph G with a list L, we define the *reconfiguration graph* R_G^L as follows: each node of R_G^L corresponds to an L-coloring of G, and two nodes of R_G^L are joined by an edge if their corresponding L-colorings f and f' satisfy $|\{v \in V : f(v) \neq f'(v)\}| = 1$, that is, f' can be obtained from f by changing the color assignment of a single vertex v. We will refer to a *node* of R_G^L in order to distinguish it from a vertex of G. Since we have defined L-colorings as the nodes of R_G^L, we use graph terms such as adjacency and path for L-colorings. For notational convenience, we sometimes identify a node of R_G^L with its corresponding L-coloring of G if it is clear from the context.

Given a graph G with a list L and two L-colorings f_0 and f_r of G, the LIST COLORING RECONFIGURATION problem asks whether the reconfiguration graph R_G^L has a path between the two nodes f_0 and f_r.

3   PSPACE-Completeness

In this section, we first prove that the problem is PSPACE-complete even for graphs with pathwidth two. Then, we show in Sect. 3.3 that the reduction can be extended to proving the PSPACE-completeness of threshold graphs.

A graph is *series-parallel* if it does not contain a subdivision of a complete graph K_4 on four vertices [8]. Note that series-parallel graphs may have super-constant pathwidth, although their treewidth can be bounded by two. We give the following theorem.

Theorem 1. *The* LIST COLORING RECONFIGURATION *problem is* PSPACE-*complete even for bipartite series-parallel graphs of pathwidth two.*

It is known that LIST COLORING RECONFIGURATION is in PSPACE [5]. Therefore, as a proof of Theorem 1, we give a polynomial-time reduction from the SHORTEST PATH REROUTING problem [4] (defined in Sect. 3.1) to our problem for bipartite series-parallel graphs of pathwidth two.

3.1   Reconfiguration Problem for Shortest Path

Let H be an unweighted graph, and let s and t be two vertices in H. We call a shortest path in H between s and t simply an *S-path* in H. Note that, since H

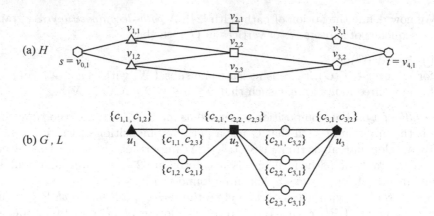

Fig. 2. (a) Graph H for SHORTEST PATH REROUTING, and (b) the corresponding graph G and its list L for LIST COLORING RECONFIGURATION.

is unweighted, an S-path indicates a path between s and t having the minimum number of edges. We say that two S-paths P and P' in H are *adjacent* if they differ in exactly one vertex, that is, $|V(P') \setminus V(P)| = 1$ and $|V(P) \setminus V(P')| = 1$ hold. Given two S-paths P_0 and P_r in H, the SHORTEST PATH REROUTING problem asks whether there exists a sequence $\mathcal{P} = \langle P_0, P_1, \ldots, P_\ell \rangle$ of S-paths such that $P_\ell = P_r$, and P_{i-1} and P_i are adjacent for each $i = 1, 2, \ldots, \ell$. This problem is known to be PSPACE-complete [4].

To construct our reduction, we introduce some terms. Let d be the number of edges of an S-path in a graph H. For two vertices v and w in H, we denote by $\mathsf{dist}(v, w)$ the number of edges of a shortest path in H between v and w; then $\mathsf{dist}(s, t) = d$. For each $i \in \{0, 1, \ldots, d\}$, the *layer $D_{s,i}$ from s* is defined to be the set of all vertices in H that are placed at distance i from s, that is, $D_{s,i} = \{v \in V(H) : \mathsf{dist}(s, v) = i\}$. Similarly, let $D_{t,j} = \{v \in V(H) : \mathsf{dist}(t, v) = j\}$ for each $j \in \{0, 1, \ldots, d\}$. Then, for each $i \in \{0, 1, \ldots, d\}$, the *layer D_i of H* is defined as follows: $D_i = D_{s,i} \cap D_{t,d-i}$. Notice that any S-path in H contains exactly one vertex from each layer D_i, $0 \le i \le d$. Then, observe that two S-paths P and P' in H are adjacent if and only if there exists exactly one index $j \in \{1, 2, \ldots, d-1\}$ such that $V(P) \cap D_j \ne V(P') \cap D_j$. Therefore, we may assume without loss of generality that H consists of only vertices in $\bigcup_{0 \le i \le d} D_i$, as illustrated in Fig. 2(a). In the example of Fig. 2(a), all vertices in the same layer D_i are depicted by the same shape, that is, $D_0 = \{s\}$, $D_1 = \{v_{1,1}, v_{1,2}\}$, $D_2 = \{v_{2,1}, v_{2,2}, v_{2,3}\}$, $D_3 = \{v_{3,1}, v_{3,2}\}$ and $D_4 = \{t\}$. Note that both $D_0 = \{s\}$ and $D_d = \{t\}$ always hold.

3.2   Reduction

Given an instance (H, P_0, P_r) of SHORTEST PATH REROUTING, we construct the corresponding instance (G, L, f_0, f_r) of LIST COLORING RECONFIGURATION.

Construction of G and L. We first construct the corresponding graph G with a list L. For each $i \in \{1, 2, \ldots, d-1\}$, let $D_i = \{v_{i,1}, v_{i,2}, \ldots, v_{i,q}\}$ be the layer of H; then we introduce a vertex u_i, called a *layer vertex*, to G. The list of each layer vertex u_i is defined as $L(u_i) = \{c_{i,1}, c_{i,2}, \ldots, c_{i,q}\}$, where each color $c_{i,j}$ in $L(u_i)$ corresponds to the vertex $v_{i,j}$ in D_i; assigning color $c_{i,j}$ represents selecting the vertex $v_{i,j}$ as the i-th vertex of an S-path in H. We denote by U_{layer} the set of all layer vertices $u_1, u_2, \ldots, u_{d-1}$ in G. In Fig. 2(b), each layer vertex u_i is illustrated as a black vertex which is the same shape as the vertices in D_i.

We then connect layer vertices in G by *forbidden paths* of length two, as follows. Let $v_{i,x} \in D_i$ and $v_{i+1,y} \in D_{i+1}$ be an arbitrary pair of vertices in H such that $v_{i,x}v_{i+1,y} \notin E(H)$. We introduce a vertex w to G, and join w and each of the two layer vertices u_i and u_{i+1} by an edge. The list $L(w)$ of w consists of two colors $c_{i,x}$ and $c_{i+1,y}$ which correspond to the vertices $v_{i,x}$ and $v_{i+1,y}$ in H, respectively. (See the vertices depicted by white circles in Fig. 2(b).) We call such a vertex w in G a $(v_{i,x}, v_{i+1,y})$-*forbidden vertex* or simply a *forbidden vertex*. Notice that there is no proper L-coloring f such that $f(u_i) = c_{i,x}$ and $f(u_{i+1}) = c_{i+1,y}$; otherwise there is no color in $L(w)$ that can be assigned to w. This property ensures that every L-coloring of G corresponds to an S-path in H. Let U_{forbid} be the set of all forbidden vertices, then $|U_{\text{forbid}}| = O(|E(\bar{H})|)$ where $\bar{H}$ is the complement graph of H.

This completes the construction of G and L. Since $|V(G)| = O(d + |E(\bar{H})|) = O(|V(H)|^2)$ and $|E(G)| = O(|E(\bar{H})|) = O(|V(H)|^2)$, we can construct G in polynomial time. Clearly, G is a bipartite series-parallel graph of pathwidth two, whose bipartition consists of U_{layer} and U_{forbid}.

Construction of f_0 and f_r. We now construct two L-colorings f_0 and f_r of G which correspond to S-paths P_0 and P_r in H, respectively. For each $i \in \{1, 2, \ldots, d-1\}$, let $v_{i,0}$ be the vertex in the layer D_i passed through by the S-path P_0; then we let $f_0(u_i) = c_{i,0}$ for each layer vertex $u_i \in U_{\text{layer}}$. For each $(v_{i,x}, v_{i+1,y})$-forbidden vertex $w \in U_{\text{forbid}}$, we choose an arbitrary color from $c_{i,x}$ and $c_{i+1,y}$ which is assigned to neither u_i nor u_{i+1}. We note that such an available color always exists, because P_0 has an edge between the two vertices corresponding to the colors $f_0(u_i)$ and $f_0(u_{i+1})$, and hence at least one of $f_0(u_i) \neq c_{i,x}$ and $f_0(u_{i+1}) \neq c_{i+1,y}$ holds for each $(v_{i,x}, v_{i+1,y})$-forbidden vertex. Similarly, we construct f_r. This completes the construction of the corresponding instance (G, L, f_0, f_r).

Correctness of the reduction. To show the correctness of this reduction, we give the following lemma.

Lemma 1. (H, P_0, P_r) *is a yes-instance if and only if* (G, L, f_0, f_r) *is a yes-instance.*

Proof. We first prove the if-part. Suppose that the reconfiguration graph R_G^L has a path between the two nodes f_0 and f_r. We can classify the recoloring steps into the following two types: (1) recoloring a layer vertex in U_{layer}, and (2) recoloring a forbidden vertex in U_{forbid}. Therefore, the path in R_G^L can be divided into sub-paths, intermittently at each edge corresponding to a recoloring step of

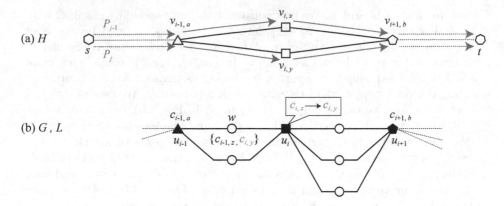

Fig. 3. (a) Two adjacent S-paths P_{j-1} (blue) and P_j (red), and (b) the corresponding recoloring steps (Color figure online).

type (1) above; all edges in a sub-path correspond to recoloring steps of type (2) above. Therefore, all nodes in each sub-path correspond to the same S-path in H. Furthermore, any two consecutive sub-paths correspond to two adjacent S-paths (that differ in only one vertex), because the sub-paths are divided by a recoloring step of type (1). Thus, we can construct a sequence of adjacent S-paths that transforms P_0 into P_r.

We then prove the only-if-part. Suppose that there exists a sequence $\mathcal{P} = \langle P_0, P_1, \ldots, P_\ell \rangle$ of S-paths such that $P_\ell = P_r$, and P_{j-1} and P_j are adjacent for each $j = 1, 2, \ldots, \ell$. For two adjacent S-paths P_{j-1} and P_j, $j \in \{1, 2, \ldots, \ell\}$, assume that P_j is obtained from P_{j-1} by replacing $v_{i,x} \in V(P_{j-1}) \cap D_i$ with $v_{i,y} \in V(P_j) \cap D_i$. (See Fig. 3(a).) This rerouting step from $v_{i,x}$ to $v_{i,y}$ corresponds to recoloring the layer vertex $u_i \in U_{\mathsf{layer}}$ from the color $c_{i,x}$ to $c_{i,y}$. (See Fig. 3(b).) To do so, if there is a forbidden vertex $w \in U_{\mathsf{forbid}}$ which is adjacent with u_i and receives the color $c_{i,y}$, we first need to recolor w from $c_{i,y}$ to another color $c_{q,z} \in L(w) \setminus \{c_{i,y}\}$, where $q \in \{i-1, i+1\}$. (Note that, since P_{j-1} corresponds to a feasible L-coloring of G, no forbidden vertex adjacent with u_i receives the color $c_{i,x}$.) Since w is a forbidden vertex, we know that $|L(w)| = 2$. Let $v_{q,a}$ be the vertex in the layer D_q which is adjacent with $v_{i,x}$ and $v_{i,y}$ in P_{j-1} and P_j, respectively; and hence the layer vertex u_q receives the color $c_{q,a}$. (Figure 3 illustrates the case where $q = i - 1$.) Then, since each forbidden vertex is placed only for a pair of vertices in H which is *not* joined by an edge, we observe that the color $c_{q,z} \in L(w) \setminus \{c_{i,y}\}$ is different from $c_{q,a}$. Therefore, we can recolor w from $c_{i,y}$ to another color $c_{q,z}$ without recoloring any other vertices. Since U_{forbid} forms an independent set of G, we can apply this recoloring step to all forbidden vertices independently. Now, any neighbor of u_i is colored with neither $c_{i,x}$ nor $c_{i,y}$, and hence we can recolor u_i from $c_{i,x}$ to $c_{i,y}$. Thus, R_G^L has a path between f_0 and f_r which corresponds to $\mathcal{P}$. $\qquad\square$

3.3 Threshold Graphs

In this subsection, we extend our reduction in Sect. 3.2 to threshold graphs. A graph G is *threshold* if there exist a real number s and a mapping $\omega : V(G) \to \mathbb{R}$ such that $xy \in E(G)$ if and only if $\omega(x) + \omega(y) \geq s$ [8], where $\mathbb{R}$ is the set of all real numbers.

Theorem 2. *The* LIST COLORING RECONFIGURATION *problem is* PSPACE-*complete even for threshold graphs.*

Proof. We modify the graph G constructed in Sect. 3.2, as follows: join all pairs of vertices in U_{layer}, and join each vertex in U_{forbid} with all vertices in U_{layer}. Let G' be the resulting graph, then $G'[U_{\mathsf{layer}}]$ forms a clique and $G'[U_{\mathsf{forbid}}]$ forms an independent set. Notice that G' is a threshold graph; set the threshold $s = 1$, and the mapping $\omega(v) = 1$ for each layer vertex $v \in U_{\mathsf{layer}}$ and $\omega(u) = 0$ for each forbidden vertex $u \in U_{\mathsf{forbid}}$.

Consider any pair of vertices u and v such that $uv \in E(G') \setminus E(G)$, that is, they are joined by a new edge for constructing G' from G. Then, by the construction in Sect. 3.2, the two lists $L(u)$ and $L(v)$ contain no color in common. Therefore, adding new edges to G does not affect the existence of L-coloring: more formally, any L-coloring of G is an L-coloring of G', and vice versa. Thus, by Lemma 1 an instance (H, P_0, P_r) of SHORTEST PATH REROUTING is a yes-instance if and only if (G', L, f_0, f_r) is a yes-instance.   □

4 Algorithm for Graphs with Pathwidth One

In contrast to Theorem 1, we give the following theorem in this section.

Theorem 3. *The* LIST COLORING RECONFIGURATION *problem can be solved in polynomial time for graphs with pathwidth one.*

As a proof of Theorem 3, we give such an algorithm. However, since every connected graph of pathwidth one is a caterpillar [23], it suffices to develop a polynomial-time algorithm for caterpillars.

A *caterpillar* G is a tree whose vertex set $V(G)$ can be partitioned into two subset V_S and V_L such that $G[V_S]$ forms a path and each vertex in V_L is incident to exactly one vertex in V_S. We may assume without loss of generality that the two endpoints of the path $G[V_S]$ are of degree one in the whole graph G.

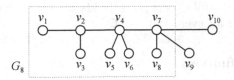

Fig. 4. A caterpillar G and its vertex ordering, where the subgraph surrounded by a dotted rectangle corresponds to G_8.

(See v_1 and v_{10} in Fig. 4.) We call each vertex in V_S a *spine vertex* of G, and each vertex in V_L a *leaf* of G.

We assume that all vertices in G are ordered as $v_1, v_2, \ldots, v_n$ by the breadth-first search starting with the endpoint (degree-1 vertex) of the path $G[V_S]$ with the priority to leaves; that is, when we visit a spine vertex v, we first visit all leaves of v and then visit the unvisited spine vertex. (See Fig. 4 for example.) For each index $i \in \{1, 2, \ldots, n\}$, we let $V_i = \{v_1, v_2, \ldots, v_i\}$ and $G_i = G[V_i]$. Then, clearly $G_n = G$. For each index $i \in \{1, 2, \ldots, n\}$, let $\mathsf{sp}(i)$ be the latest spine vertex in V_i, that is, $\mathsf{sp}(i) = v_i$ if v_i is a spine vertex, otherwise $\mathsf{sp}(i)$ is the unique neighbor of v_i. Then, v_i is adjacent with only the spine vertex $\mathsf{sp}(i-1)$ in G_i for each $i \in \{2, 3, \ldots, n\}$.

We first prove that the length of each list can be restricted without loss of generality. Note that the following lemma holds for any graph.

Lemma 2. *For an instance (G', L', f_0', f_r'), one can obtain another instance (G, L, f_0, f_r) in polynomial time such that $2 \leq |L(v)| \leq d(v) + 1$ for each vertex $v \in V(G)$, and (G', L', f_0', f_r') is a yes-instance if and only if (G, L, f_0, f_r) is a yes-instance.*

Proof. If $|L(v)| = 1$ for a vertex $v \in V(G')$, then any L-coloring of G' assigns the same color $c \in L(v)$ to v. Therefore, c is never assigned to any neighbor u of v. We can thus delete v from G' and set $L(u) := L(u) \setminus \{c\}$ for all neighbors u of v in G'. Clearly, this modification does not affect the reconfigurability (i.e., the existence or non-existence of a path in the reconfiguration graph).

If $|L(v)| \geq d(v) + 2$ for a vertex $v \in V(G')$, we simply delete v from G' without any modification of lists; let G be the resulting graph. Let f be any L-coloring of G, and consider a recoloring step for a neighbor u of v from the current color $c = f(u)$ to another color c'. We claim that this recoloring step can be simulated in G', as follows. If c' is not assigned to v in G', we can directly recolor u from c to c'. Thus, suppose that c' is assigned to v in G'. Then, since $|L(v)| \geq d(v) + 2$, there is at least one color $c^* \in L(v)$ which is not c' and is not assigned to any of $d(v)$ neighbors of v. Therefore, we first recolor v from c' to c^*, and then recolor u from c to c'. In this way, any recoloring step in G can be simulated in G', and hence the modification does not affect the reconfigurability.

Thus, we can obtain an instance such that $2 \leq |L(v)| \leq d(v) + 1$ holds for each vertex v without affecting the reconfigurability. Clearly, the modified instance can be constructed in polynomial time. $\qquad\square$

Therefore, in the remainder of this section, we assume that G is a (connected) caterpillar and $2 \leq |L(v)| \leq d(v) + 1$ holds for every vertex $v \in V(G)$. In particular, $|L(v)| = 2$ for every leaf v of G.

4.1   Idea and Definitions

The main idea of our algorithm is to extend techniques developed for SHORTEST PATH REROUTING [3], and apply them to LIST COLORING RECONFIGURATION for caterpillars. Our algorithm employs a dynamic programming method based on the vertex ordering $v_1, v_2, \ldots, v_n$ of G.

For each $i \in \{1, 2, \ldots, n\}$, let $\mathsf{R}^L_{G_i}$ be the reconfiguration graph for the subgraph G_i and the list L. Then, $\mathsf{R}^L_{G_i}$ contains all L-colorings of G_i as its nodes. Our algorithm efficiently constructs $\mathsf{R}^L_{G_i}$ for each $i = 1, 2, \ldots, n$, in this order. However, of course, the number of nodes in $\mathsf{R}^L_{G_i}$ cannot be bounded by a polynomial size in general. We thus use the property that the vertex v_{i+1} (will be added to G_i) is adjacent with only the spine vertex $\mathsf{sp}(i)$ in G_{i+1}; and we "encode" the reconfiguration graph $\mathsf{R}^L_{G_i}$ into a polynomial size with keeping the information of (1) the color assigned to $\mathsf{sp}(i)$ and (2) the connectivity of nodes in $\mathsf{R}^L_{G_i}$.

Before explaining the encoding methods, we first note that it suffices to focus on only one connected component in $\mathsf{R}^L_{G_i}$ which contains the restriction of f_0, where the *restriction* of an L-coloring f of a graph G to a subgraph G' is an L-coloring g of G' such that $g(v) = f(v)$ hold for all vertices $v \in V(G')$. For notational convenience, we denote by $f[V_i]$ the restriction of an L-coloring f of a caterpillar G to its subgraph G_i. Then, we have the following lemma.

Lemma 3. *Let g be an L-coloring of G_i such that $f_0[V_i]$ and g are contained in the same connected component in $\mathsf{R}^L_{G_i}$. Then, for each $j \in \{1, 2, \ldots, i - 1\}$, $f_0[V_j]$ and $g[V_j]$ are contained in the same connected component in $\mathsf{R}^L_{G_j}$.*

From now on, we thus focus on only the connected component of $\mathsf{R}^L_{G_i}$ which contains $f_0[V_i]$. Since the list is fixed to be L in the remainder of this section, we simply denote by R_i the reconfiguration graph $\mathsf{R}^L_{G_i}$, and by R^0_i the connected component of $\mathsf{R}_i = \mathsf{R}^L_{G_i}$ containing $f_0[V_i]$.

Encoding graph. We now partition the nodes of R^0_i into several subsets with respect to (1) the color assigned to $\mathsf{sp}(i)$, and (2) the connectivity of nodes in R^0_i. For two nodes g and g' of R^0_i with $g(\mathsf{sp}(i)) = g'(\mathsf{sp}(i))$, we write $g \sim_{\mathsf{sp}(i)} g'$ if R^0_i has a path $\langle g_1, g_2, \ldots, g_\ell \rangle$ such that $g_1 = g$, $g_\ell = g'$, and $g_j(\mathsf{sp}(i)) = g(\mathsf{sp}(i)) = g'(\mathsf{sp}(i))$ holds for every $j \in \{1, 2, \ldots, \ell\}$, that is, g can be reconfigured into g' without recoloring the color assigned to the vertex $\mathsf{sp}(i)$. Since the adjacency relation on L-colorings is symmetric (i.e., R_i is an undirected graph), it is easy to see that $\sim_{\mathsf{sp}(i)}$ is an equivalence relation. Thus, the node set of R^0_i can be uniquely partitioned by the relation $\sim_{\mathsf{sp}(i)}$. We denote by $\mathcal{G}^0_i$ the partition of the node set of R^0_i into equivalence classes with respect to $\sim_{\mathsf{sp}(i)}$.

We finally define our dynamic programming table. For each subgraph G_i, $i \in \{1, 2, \ldots, n\}$, our algorithm keeps track of four information $(H_i, \mathsf{col}_i, \mathsf{ini}_i, \mathsf{tar}_i)$, defined as follows.

- The *encoding graph* H_i of R^0_i which can be obtained from R^0_i by contracting each node set in $\mathcal{G}^0_i$ into a single node. (See Fig. 5 as an example.) We will refer to an *e-node* of H_i in order to distinguish it from a node of R^0_i. (Thus, each node refers to an L-coloring of G_i, and each e-node refers to a set of L-colorings of G_i.) For each e-node $x \in V(H_i)$, we denote by $\Phi_i(x)$ the set of all nodes in R^0_i that were contracted into x. (Note that we do not compute $\Phi_i(x)$, but use it only for definitions and proofs.)

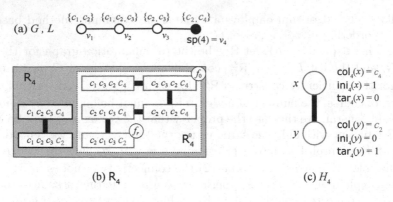

Fig. 5. (a) A caterpillar $G = G_4$, (b) the reconfiguration graph R_4 consisting of all L-colorings of G, and (c) the encoding graph H_4 of R_4^0 consisting of two e-nodes x and y, where each L-coloring in (b) is represented as the sequence of colors assigned to the vertices in G from left to right.

- The color $col_i(x) \in L(sp(i))$ for each e-node $x \in V(H_i)$, which is assigned to $sp(i)$ in common by the nodes (i.e., L-colorings of G_i) in $\Phi_i(x)$.
- The label $ini_i(x) \in \{0, 1\}$ for each e-node $x \in V(H_i)$, such that $ini_i(x) = 1$ if $f_0[V_i] \in \Phi_i(x)$, otherwise $ini_i(x) = 0$.
- The label $tar_i(x) \in \{0, 1\}$ for each e-node $x \in V(H_i)$, such that $tar_i(x) = 1$ if $f_r[V_i] \in \Phi_i(x)$, otherwise $tar_i(x) = 0$.

To prove Theorem 3, we give a polynomial-time algorithm which computes $(H_i, col_i, ini_i, tar_i)$ for each subgraph G_i, $i \in \{1, 2, \ldots, n\}$, by means of dynamic programming. Then, the problem can be solved as in the following lemma.

Lemma 4. (G, L, f_0, f_r) *is a yes-instance if and only if the encoding graph* H_n *contains a node* x *such that* $tar_n(x) = 1$.

Proof. Since H_n contains a node x such that $tar_n(x) = 1$, we have $f_r[V_n] = f_r \in \Phi_n(x)$. Recall that H_n is the encoding graph of R_n^0 which contains the L-coloring f_0 of G as a node. Since $\Phi_n(x) \subseteq V(R_n^0)$, the lemma clearly follows. □

4.2 Algorithm

As the initialization, we first consider the case where $i = 1$, that is, we compute $(H_1, col_1, ini_1, tar_1)$. (See Fig. 6(d) as an example.) Note that G_1 consists of a single vertex v_1, and recall that v_1 is a spine vertex of degree one. By Lemma 2 we then have $|L(v_1)| = 2$. Therefore, the reconfiguration graph R_1 is a complete graph on $|L(v_1)| = 2$ nodes such that each node corresponds to an L-coloring of G_1 which assigns a distinct color to the vertex $sp(1) = v_1$. Since R_1 is complete and contains the node $f_0[V_1]$, we have $R_1^0 = R_1$. Furthermore, $H_1 = R_1^0$ since all nodes in R_1^0 assign distinct colors in $L(v_1)$ to $sp(1) = v_1$. Then, for each e-node x of H_1 corresponding to the set consisting of a single L-coloring g of G_1, we set

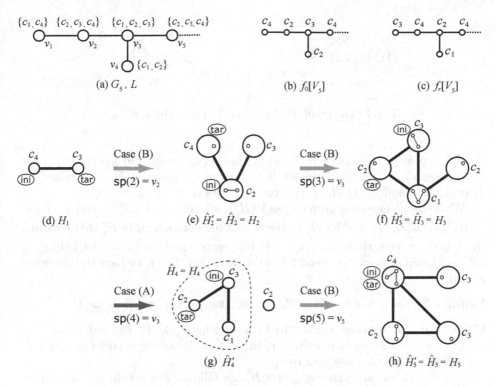

Fig. 6. Application of our algorithm to the instance depicted in (a)–(c). In (d)–(h), $col_i(x) \in L(sp(i))$ is attached to each e-node x, and the e-nodes x with $ini_i(x) = 1$ and $tar_i(x) = 1$ have the labels "ini" and "tar," respectively. Furthermore, in (e), (f) and (h), the small graph contained in each e-node x of H_i represents the subgraph of H_{i-1} induced by $EN(x)$.

$$col_1(x) = g(v_1);$$

$$ini_1(x) = \begin{cases} 1 & \text{if } g(v_1) = f_0(v_1), \\ 0 & \text{otherwise;} \end{cases}$$

$$tar_1(x) = \begin{cases} 1 & \text{if } g(v_1) = f_r(v_1), \\ 0 & \text{otherwise.} \end{cases}$$

For $i \geq 2$, suppose that we have already computed $(H_{i-1}, col_{i-1}, ini_{i-1}, tar_{i-1})$. Then, we compute $(H_i, col_i, ini_i, tar_i)$, as follows.

Case (A): v_i **is a leaf in** V_L. (See Figs. 6(g) and 7(a).)

By Lemma 2 we have $|L(v_i)| = 2$ in this case; let $L(v_i) = \{c_1, c_2\}$. Recall that v_i is adjacent with only the spine vertex $sp(i-1)$ in G_i. Furthermore, $sp(i) = sp(i-1)$ in this case.

Let $H_{i-1}^{c_1}$ be the subgraph of H_{i-1} obtained by deleting all e-nodes y in H_{i-1} with $col_{i-1}(y) = c_1$. Then, $H_{i-1}^{c_1}$ encodes all nodes of R_{i-1}^0 that do not assign the color c_1 to $sp(i-1)$. Thus, we can extend each L-coloring h of G_{i-1} encoded

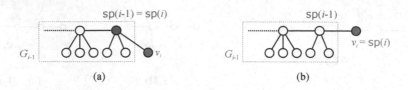

Fig. 7. The graph G_i for (a) $v_i \in V_L$ and (b) $v_i \in V_S$.

in $H_{i-1}^{c_1}$ to an L-coloring g of G_i such that $g(v_i) = c_1$ and $g(v) = h(v)$ for all vertices $v \in V_{i-1}$. Similarly, let $H_{i-1}^{c_2}$ be the subgraph of H_{i-1} obtained by deleting all e-nodes z in H_{i-1} with $\mathsf{col}_{i-1}(z) = c_2$.

We define an encoding graph $\hat{H}_i'$ as $V(\hat{H}_i') = V(H_{i-1}^{c_1}) \cup V(H_{i-1}^{c_2})$ and $E(\hat{H}_i') = E(H_{i-1}^{c_1}) \cup E(H_{i-1}^{c_2})$; and let $\hat{H}_i$ be the connected component of $\hat{H}_i'$ that contains the e-node x such that $\mathsf{ini}_{i-1}(x) = 1$. For each e-node x in $\hat{H}_i$, let $\hat{\mathsf{col}}_i(x) = \mathsf{col}_{i-1}(x)$, $\hat{\mathsf{ini}}_i(x) = \mathsf{ini}_{i-1}(x)$ and $\hat{\mathsf{tar}}_i(x) = \mathsf{tar}_{i-1}(x)$. Then, we have the following lemma.

Lemma 5. *For a leaf $v_i \in V_L$, $(H_i, \mathsf{col}_i, \mathsf{ini}_i, \mathsf{tar}_i) = (\hat{H}_i, \hat{\mathsf{col}}_i, \hat{\mathsf{ini}}_i, \hat{\mathsf{tar}}_i)$.*

Case (B): v_i is a spine vertex in V_S. (See Figs. 6(e), (f), (h) and 7(b).)

In this case, notice that $\mathsf{sp}(i) = v_i$ in G_i, and hence we need to update col_i according to the color assigned to v_i.

We first define an encoding graph $\hat{H}_i'$, as follows. For a color $c \in L(v_i)$, let H_{i-1}^c be the subgraph of H_{i-1} obtained by deleting all e-nodes y in H_{i-1} with $\mathsf{col}_{i-1}(y) = c$. For each connected component in H_{i-1}^c, we add a new e-node x to $\hat{H}_i'$ such that $\hat{\mathsf{col}}_i(x) = c$; we denote by $\mathsf{EN}(x)$ the set of all e-nodes in H_{i-1}^c that correspond to x. We apply this operation to all colors in $L(v_i)$. We then add edges to $\hat{H}_i'$: two e-nodes x and y in $\hat{H}_i'$ are joined by an edge if and only if $\mathsf{EN}(x) \cap \mathsf{EN}(y) \neq \emptyset$.

We now define $\hat{\mathsf{ini}}_i(x)$ and $\hat{\mathsf{tar}}_i(x)$ for each e-node x in $\hat{H}_i'$, as follows:

$$\hat{\mathsf{ini}}_i(x) = \begin{cases} 1 & \text{if } \hat{\mathsf{col}}_i(x) = f_0(v_i) \text{ and} \\ & \mathsf{EN}(x) \text{ contains an e-node } y \text{ with } \mathsf{ini}_{i-1}(y) = 1; \\ 0 & \text{otherwise,} \end{cases}$$

and

$$\hat{\mathsf{tar}}_i(x) = \begin{cases} 1 & \text{if } \hat{\mathsf{col}}_i(x) = f_r(v_i) \text{ and} \\ & \mathsf{EN}(x) \text{ contains an e-node } y \text{ with } \mathsf{tar}_{i-1}(y) = 1; \\ 0 & \text{otherwise.} \end{cases}$$

Let $\hat{H}_i$ be the connected component of $\hat{H}_i'$ that contains the e-node x such that $\hat{\mathsf{ini}}_i(x) = 1$. Then, we have the following lemma.

Lemma 6. *For a spine vertex $v_i \in V_S$, $(H_i, \mathsf{col}_i, \mathsf{ini}_i, \mathsf{tar}_i) = (\hat{H}_i, \hat{\mathsf{col}}_i, \hat{\mathsf{ini}}_i, \hat{\mathsf{tar}}_i)$.*

4.3 Running Time

We now estimate the running time of our algorithm in Sect. 4.2. The following is the key lemma for the estimation.

Lemma 7. *For each index* $i \in \{1, 2, \ldots, n\}$,

$$|V(H_i)| \leq \begin{cases} 2 & \textit{if } i = 1; \\ |V(H_{i-1})| + d(v_i) & \textit{otherwise.} \end{cases}$$

In particular, $|V(H_n)| = O(n)$, *where* n *is the number of vertices in* G.

By Lemma 7 each encoding graph H_i is of size $O(n)$ for each $i \in \{1, 2, \ldots, n\}$. Therefore, our algorithm runs in polynomial time.

This completes the proof of Theorem 3.   □

5 Concluding Remarks

In this paper, we gave precise analyses of the LIST COLORING RECONFIGURATION problem with respect to pathwidth: the problem is solvable in polynomial time for graphs with pathwidth one, while it is PSPACE-complete for graphs with pathwidth two.

Very recently, Wrochna [25] gave another proof for the PSPACE-completeness of LIST COLORING RECONFIGURATION for graphs with pathwidth two. His reduction is constructed from a PSPACE-complete problem, called H-WORD REACHABILITY.

Acknowledgments. We are grateful to Daichi Fukase and Yuma Tamura for fruitful discussions with them. This work is partially supported by JSPS KAKENHI Grant Numbers 25106504 and 25330003.

References

1. Bonamy, M., Johnson, M., Lignos, I., Patel, V., Paulusma, D.: Reconfiguration graphs for vertex colourings of chordal and chordal bipartite graphs. J. Comb. Optim. **27**, 132–143 (2014)
2. Bonamy, M., Bousquet, N.: Recoloring bounded treewidth graphs. Electron. Notes Discrete Math. **44**, 257–262 (2013)
3. Bonsma, P.: Rerouting shortest paths in planar graphs. In: Proceedings of FSTTCS 2012, LIPIcs 18, pp. 337–349 (2012)
4. Bonsma, P.: The complexity of rerouting shortest paths. Theoret. Comput. Sci. **510**, 1–12 (2013)
5. Bonsma, P., Cereceda, L.: Finding paths between graph colourings: PSPACE-completeness and superpolynomial distances. Theoret. Comput. Sci. **410**, 5215–5226 (2009)
6. Bonsma, P., Kamiński, M., Wrochna, M.: Reconfiguring independent sets in claw-free graphs. In: Ravi, R., Gørtz, I.L. (eds.) SWAT 2014. LNCS, vol. 8503, pp. 86–97. Springer, Heidelberg (2014)

7. Bonsma, P., Mouawad, A.E.: The complexity of bounded length graph recoloring (2014). arXiv:1404.0337
8. Brandstädt, A., Le, V.B., Spinrad, J.P.: Graph Classes: A Survey. SIAM, Philadelphia (1999)
9. Cereceda, L.: Mixing graph colourings. Ph.D. thesis, London School of Economics and Political Science (2007)
10. Cereceda, L., van den Heuvel, J., Johnson, M.: Finding paths between 3-colorings. J. Graph Theor. **67**, 69–82 (2011)
11. Gopalan, P., Kolaitis, P.G., Maneva, E.N., Papadimitriou, C.H.: The connectivity of Boolean satisfiability: computational and structural dichotomies. SIAM J. Comput. **38**, 2330–2355 (2009)
12. Hearn, R.A., Demaine, E.D.: PSPACE-completeness of sliding-block puzzles and other problems through the nondeterministic constraint logic model of computation. Theoret. Comput. Sci. **343**, 72–96 (2005)
13. Ito, T., Demaine, E.D., Harvey, N.J.A., Papadimitriou, C.H., Sideri, M., Uehara, R., Uno, Y.: On the complexity of reconfiguration problems. Theoret. Comput. Sci. **412**, 1054–1065 (2011)
14. Ito, T., Kamiński, M., Demaine, E.D.: Reconfiguration of list edge-colorings in a graph. Discrete Appl. Math. **160**, 2199–2207 (2012)
15. Ito, T., Kawamura, K., Ono, H., Zhou, X.: Reconfiguration of list $L(2,1)$-labelings in a graph. Theoret. Comput. Sci. **544**, 84–97 (2014)
16. Ito, T., Kawamura, K., Zhou, X.: An improved sufficient condition for reconfiguration of list edge-colorings in a tree. IEICE Trans. Inf. Syst. **E95–D**, 737–745 (2012)
17. Jansen, K., Scheffler, P.: Generalized coloring for tree-like graphs. Discrete Appl. Math. **75**, 135–155 (1997)
18. Johnson, M., Kratsch, D., Kratsch, S., Patel, V., Paulusma, D.: Colouring reconfiguration is fixed-parameter tractable (2014). arXiv:1403.6347
19. Kamiński, M., Medvedev, P., Milanič, M.: Shortest paths between shortest paths. Theoret. Comput. Sci. **412**, 5205–5210 (2011)
20. Kamiński, M., Medvedev, P., Milanič, M.: Complexity of independent set reconfigurability problems. Theoret. Comput. Sci. **439**, 9–15 (2012)
21. Makino, K., Tamaki, S., Yamamoto, M.: An exact algorithm for the Boolean connectivity problem for k-CNF. Theoret. Comput. Sci. **412**, 4613–4618 (2011)
22. Mouawad, A.E., Nishimura, N., Raman, V., Simjour, N., Suzuki, A.: On the parameterized complexity of reconfiguration problems. In: Gutin, G., Szeider, S. (eds.) IPEC 2013. LNCS, vol. 8246, pp. 281–294. Springer, Heidelberg (2013)
23. Proskurowski, A., Telle, J.A.: Classes of graphs with restricted interval models. Discrete Math. Theor. Comput. Sci. **3**, 167–176 (1999)
24. Robertson, N., Seymour, P.: Graph minors. I Excluding a forest. J. Comb. Theor. Ser. B **35**, 39–61 (1983)
25. Wrochna, M.: Reconfiguration in bounded bandwidth and treedepth (2014). arXiv:1405.0847

Two Paths Location of a Tree with Positive or Negative Weights

Jianjie Zhou[1], Liying Kang[1], and Erfang Shan[2(✉)]

[1] Department of Mathematics, Shanghai University,
Shanghai 200444, People's Republic of China
[2] School of Management, Shanghai University,
Shanghai 200444, People's Republic of China
efshan@shu.edu.cn

Abstract. This paper studies two problems of locating two paths in a tree with positive and negative weights. The first problem has objective to minimize the sum of minimum weighted distance from every vertex of the tree to the two paths, while the second is to minimize the sum of the weighted minimum distance from each vertex to the two paths. We develop an $O(n^2)$ algorithm based on the optimal properties for the first problem, and also an $O(n^3)$ algorithm for the second problem.

Keywords: Facility location · Path · Tree · Semi-obnoxious

1 Introduction

Network location problem is to find the optimal locations of service facilities in a network. The location problem usually has important applications in transportation and communication and thus has received much attention from researchers. According to the realistic demands, the shapes of the facilities can be points, paths, or trees. In the past, a variety of network location problems have been defined and studied in the literature [1,7–11,13,18].

The core of tree is defined as a path in the graph with minimum sum of the distance of all vertices of the graph from the path (see [13] for this definition). Morgan and Slater [13] studied a linear time algorithm for finding the core of a tree. Becker and Perl [2] presented two algorithms for the problem of finding the two-core of a tree. The first requires $O(n^2)$ time and the second requires $O(dn)$ time, where d is the maximum number of edges of any simple path in the tree. Wang [20] reduced the computational complexity of this problem to linear time.

Sometimes facilities maybe temporality unavailable to provide service to customers due to unknown or unpredictable situations. Perto et al. [15] considered the reliability problems in multiple path-shaped facility location on networks. They gave an $O(n^2)$ time complexity algorithm when the graph is a tree and

This research was partially supported by the National Nature Science Foundation of China (Nos. 11471210, 11171207).

Z. Zhang et al. (Eds.): COCOA 2014, LNCS 8881, pp. 329–342, 2014.
DOI: 10.1007/978-3-319-12691-3_25

also proved that the problem is NP-hard even when the graph is a cactus. In some cases, there is a bound on the total length of the located paths. Becker et al. [4] considered the problem of locating central-median paths with bounded length on trees, they presented two problems and gave an $O(nlog^2n)$ divide-and-conquer algorithm [4]. It was shown that the problem of locating a median path of length at most l is NP-complete on the simple classes of cactus and grid graphs [3,12,17].

In reality, due to the complexity of the world, some vertices are desirable and others are undesirable, the problem is referred to as the semi-obnoxious location problem. To model this problem, Burkard et al. [6] considered 2-medians problems in trees with positive or negative (for simplicity we write pos/neg) weights, they formulated two objective functions and gave the corresponding algorithms. Zaferanieh and Fathali [21] presented that the core of a tree with pos/neg weights can be found in linear time.

In this paper, we consider the problem of locating two paths of a tree with pos/neg weights. We formulate two different objective functions: (1) the sum of the minimum weighted distances of the type vertex-path over all vertices; (2) the sum of the weighted minimum distances of the type vertex-path over all vertices. In Sect. 2, we give a formal formulation of two problems. Section 3 analyses the optimal properties of the first problem and gives an $O(n^2)$ algorithm. In the last section, we show that the second problem has some optimal properties and present an $O(n^3)$ algorithm for the problem.

2   Problem Formulation

Let $T = (V, E)$ be a tree consisting of a set V of vertices and a set E of edges. Each edge $e = (v_i, v_j)$ of T has a positive length $l(e) = l(v_i, v_j)$ and each vertex $v_i \in V$ has a real wight $w(v_i)$. When the tree is rooted at r it is denoted by T_r. For any vertex v_i, let T_{v_i} be the subtree of T_r rooted at vertex v_i, $S(v_i)$ the set of the children of v_i in T_r and $p(v_i)$ the parent of v_i in T_r.

Let $d(v, u)$ be the shortest distance between points v and u, where "points" can be vertices or points on an edge. The length of the shortest path between vertex v and path P is $d(v, P)$ defined as follows $d(v, P) = \min_{u \in P} d(u, v)$. A path is discrete if both its endpoints are vertices of T, otherwise it is continuous.

We consider the location of two path shaped facilities P_1, P_2. For a given location $P = \{P_1, P_2\}$, the two meaningful objective functions mentioned in the introduction are given as follows:

$$F_1(P) = \sum_{i=1}^{n} \min_{j=1,2} w(v_i)d(v_i, P_j), \tag{1}$$

$$F_2(P) = \sum_{i=1}^{n} w(v_i) \min_{j=1,2} d(v_i, P_j). \tag{2}$$

The corresponding two optimization problems are: find in T a location of two path P_1 and P_2 such that (1) is minimized, this problem is called *problem L_1*;

find in T a location of two path P_1 and P_2 such that (2) is minimized, this problem is called *problem L_2*.

A solution $P^* = \{P_1^*, P_2^*\}$ is called an *optimal solution* of problem L_1 (resp. L_2) if P^* minimize (1) (resp. (2)) and there is no sub-path $P_1' \subset P_1^*$ and $P_2' \subset P_2^*$ such that $P' = \{P_1', P_2'\}$ has the same objective function as P^*.

There is a tie when the two paths P_1, P_2 are equal distance from vertex v_i. In this case we have the following assumption: in problem L_1, if $w(v_i) \geq 0$, it is served by P_1, otherwise it is served by P_2; in problem L_2, it is served by P_1. For any $v \in V(T)$, if v is served by P_i ($i = 1$ or $i = 2$) and there exists a point x such that $d(v, P_i) = d(v, x)$, we call v is *served by x*.

3   Properties and Algorithm for the Problem L_1

3.1   Optimal Properties of Problem L_1

Assume $P^* = \{P_1, P_2\}$ is an optimal solution of problem L_1. It is easily seen that if the weight of vertex v is nonnegative, it is served by the closer path from vertex v; otherwise it is served by the farther path for vertex v. 🔴

Proposition 1. *(1) If $P_1 \cap P_2 = \emptyset$ and $x \in P_1, y \in P_2$ are the points such that $d(x, y) = d(P_1, P_2)$, then the vertices of T with negative weights are served by x or y.*
(2) If $P_1 \cap P_2 \neq \emptyset$ and $P_1 \cap P_2$ is the path from point x to point y, then the vertices of T with negative weights are served by x, y or vertices in $V(P_1 \cap P_2)$.

Proof. (1). For any $v \in V(T)$ with negative weight, if $d(v, P_1) = \min\{d(v, P_1), d(v, P_2)\}$, then v is served by P_2. Since $d(x, y) = d(P_1, P_2)$, $d(v, P_2) = d(v, y)$. Then v is served by y. If $d(v, P_2) = \min\{d(v, P_1), d(v, P_2)\}$, then v is served by P_1. Since $d(x, y) = d(P_1, P_2)$, $d(v, P_2) = d(v, y)$. Then v is served by x.

(2). The proof is similar to the proof of (1), we omit it.   □

Proposition 2. *In an optimal solution of problem L_1 the two optimal paths are always discrete.*

Proof. Suppose that $P^* = \{P_1 = P(v_i, x), P_2 = P(v_s, v_t)\}$ is an optimum solution for problem L_1, $v_i, v_s, v_t \in V(T)$ and x is an inner point of edge (v_r, v_q), where $v_r \in P_1$. Let V_x be set of vertices served by x. Then $d(v, P_1) = d(v, x)$ for $v \in V_x$. This implies that $d(v, x) = d(v, v_q) + d(v_q, x)$ for $v \in V_x$. Then the total cost of all vertices in V_x is

$$f(x)[V_x] = \sum_{v \in V_x} w(v)d(v, x)$$

$$= \sum_{v \in V_x} w(v)d(v, v_q) + \sum_{v \in V_x} w(v) \cdot d(x, v_q)$$

$$= \sum_{v \in V_x} w(v)d(v, v_q) + \left(\sum_{v \in V_x} w(v)\right) \cdot d(x, v_q). \tag{3}$$

We first give the following claims.

Claim A. $P_1 \cap P_2 = \emptyset$.

Proof. Suppose to the contrary that $P_1 \cap P_2 \neq \emptyset$. Then we move the endpoint x of P_1 in the direction of v_q with a small ϵ if $\sum_{v \in V_x} w(v) > 0$, and we move the endpoint x of P_1 in the direction of v_r with a small ϵ if $\sum_{v \in V_x} w(v) < 0$. We assume that ϵ is small enough so that x remains in (v_r, v_q). In both cases, by (3) we can obtain another solution with smaller objection function value than the original one, a contradiction. If $\sum_{v \in V_x} w(v) = 0$, we move the endpoint x of P_1 in the direction of v_r with a small ϵ such that x remains in (v_r, v_q). We can obtain a shorter path P_1' than P_1 such that $\{P_1', P_2\}$ have the same objection function value as P^*, a contradiction. The claim follows. □

Claim B. Endpoint x is on the path from P_1 to P_2.

Proof. By Claim A, $P_1 \cap P_2 = \emptyset$. If x is not on the path from P_1 to P_2, Proposition 1 implies that no vertex with negative weight is served by x, then $\sum_{v \in V_x} w(v) \geq 0$. If $\sum_{v \in V_x} w(v) > 0$, we move the endpoint x of P_1 in the direction of v_q with a small ϵ, by (3) we can obtain another solution with smaller objection function value than the original one, a contradiction. If $\sum_{v \in V_x} w(v) = 0$, we move the endpoint x of P_1 in the direction of v_r with a small ϵ such that x remains in (v_r, v_q). We can obtain a shorter path P_1' than P_1 such that $\{P_1', P_2\}$ have the same objection function value as P^*, a contradiction. The claim follows. □

By Claim B, we have $d(x, P_2) = d(P_1, P_2)$. Let m be the midpoint of the path between x and P_2. We claim that m is a vertex of T. Otherwise, we assume that m is an inner point of an edge. Using the similar discussion, we get a contradiction.

In the following, let V_m^+ and V_m^- be the set of vertices with nonnegative and negative weights that are equidistance from the two paths, respectively. We assume the vertices in V_m^+ are served by P_1 and the vertices in V_m^- are served by P_2.

If $\sum_{v \in V_x} w(v) > 0$, we move the endpoint x of P_1 in the direction of v_q with a small ϵ. According to the problem formulation, this extension does not change the service partition of vertices. By (3) we get another solution for problem L_1 with smaller objection function value than that of P^*, a contradiction.

If $\sum_{v \in V_x} w(v) \leq 0$, we move the endpoint x of P_1 in the direction of v_r with a small ϵ to x'. Because the midpoint m of the path between x and P_2 is a vertex, the above operation will change the service partition of vertices. So vertices in V_m^+ will be served by P_2, vertices in V_m^- are served by P_1 and other vertices service partition remains the same.

For the optimal solution $P^* = \{P_1, P_2\}$, let V_x^+ be the vertices of V_x with nonnegative weight, V_x^- be the vertices of V_x with negative weight. Similarly, for the new solution $\{P_1' = P(v_i, x'), P_2 = (v_s, v_t)\}$, let $V_{x'}^+$ and $V_{x'}^-$ be the vertices of $V_{x'}$ with nonnegative weight and negative weight, respectively. We have the following relation:

$$V_{x'} = V_x - V_m^+ + V_m^-.$$

Then

$$\sum_{v \in V_x - V_m^+} w(v) \cdot d(x', v_q) - \sum_{v \in V_x - V_m^+} w(v) \cdot d(x, v_q)$$

$$= (\sum_{v \in V_x - V_m^+} w(v)) \cdot d(x', x) \le 0$$

So

$$\sum_{v \in V_x - V_m^+} w(v) \cdot d(x', v_q) \le \sum_{v \in V_x - V_m^+} w(v) \cdot d(x, v_q). \qquad (4)$$

Similarly,

$$\sum_{v \in V_m^-} w(v) \cdot d(x', v_q) \le \sum_{v \in V_m^-} w(v) \cdot d(x, v_q). \qquad (5)$$

Then

$$\sum_{v \in V_{x'}} w(v) \cdot d(x', v_q) = \sum_{v \in V_x - V_m^+} w(v) d(x', v_q) + \sum_{v \in V_m^-} w(v) \cdot d(x', v_q)$$

$$\le \sum_{v \in V_x - V_m^+} w(v) d(x, v_q) + \sum_{v \in V_m^-} w(v) \cdot d(x, v_q) \quad \text{By (4) and (5)}$$

$$\le \sum_{v \in V_x} w(v) \cdot d(x, v_q).$$

The cost of any vertex $v \in V - V_x + V_m^+ - V_m^-$ remains the same. We can also get a better paths location than P^*, which is a contradiction to the optimal of P^*. So x is a vertex of T.

When more than one endpoint of the two paths are not vertices of T, we can use the similar arguments as the above to show that one can always extend or reduce the two paths to find two better discrete paths. The result follows.   □

3.2   Algorithm for Problem L_1

The algorithm based on searching intersecting and disjoint paths location separately. We use the similar approach as in [16]. We begin with the two optimal paths that are intersecting. We distinguish two cases: (i) they intersect in at least one edge (r_1, r_2); (ii) they intersect in exactly one vertex r. We analyse these two cases separately, since they rely on two different solution strategies.

Case (i). For each edge (r_1, r_2), we need to find the pair of paths intersecting at least in edge (r_1, r_2) that minimizing the total cost. Let $P^1_{(r_1, r_2)}$ and $P^2_{(r_1, r_2)}$

be the two optimal paths. Consider an edge (r_1, r_2), T_{r_1} and T_{r_2} are subtrees of $T - (r_1, r_2)$ containing r_1 and r_2, respectively. The two subtrees T_{r_1} and T_{r_2} are trees rooted at vertices r_1 and r_2, respectively. For a given root $r_i (i = 1, 2)$, we denote by $P^1_{r_i}$ and $P^2_{r_i}$ the two branches $P^1_{(r_1, r_2)}$ and $P^2_{(r_1, r_2)}$ in T_{r_i}, respectively. The idea is to construct the optimal paths starting form r_i $(i = 1, 2)$ to find the two branches $P^1_{r_i}$ and $P^2_{r_i}$ in T_{r_i}. This can be done by independently visiting top-down the two subtrees T_{r_1} and T_{r_2}. We focus our analysis only on T_{r_1}, the one for T_{r_2} being the same.

Consider the binary rooted tree T_{r_1}, if it is not binary, we can transform it into a binary tree by applying the linear time provided in [19]. Let v_1 and v_2 be the left and right children of a given vertex v of T_{r_1}, when vertex v has only one child, we assume it is the left child of v. In the top-down visit of T_{r_1} three different situations may arise at a given vertex v:

1. The two paths follows the same track up to vertex v, but they separate after v.
2. The two paths follows the same track up to vertex v and after v, they proceed together towards either v_1 or v_2.
3. The two paths follows the same track up to some ancestor of v, but just one of them passes through v into T_v.

To cope with the three above cases, we associate to each vertex v of T_{r_1}, three quantities labeled as follows:

1. $S^2_\wedge(v)$ is the maximum saving in the objective function when the two paths follows the same track up to vertex v, but they separate after v.
2. $S^2_\parallel(v)$ is the maximum saving in the objective function when the two paths follows the same track up to vertex v and after v, they proceed together towards either v_1 or v_2.
3. $S^1(v)$ is the maximum saving in the objective function when the two paths follows the same track up to some ancestor of v, but just one of them passes through v into T_v.

The above quantities can be computed recursively during a bottom-up visit of T_{r_1}. Let $W(T_v) = \sum_{v_i \in V_v} w(v_i)$ be the sum of the vertex weight in T_v and $W^+(T_v) = \sum_{v_i \in V_v, w(v_i) > 0} w(v_i)$ be the sum of the positive vertex weight in T_v. The bottom-up computation is well-known and straightforward [14,15].

The quantities $S^1(v)$ can be computed as follows:

$$S^1(v) = \begin{cases} 0 & \text{if } v \text{ is a leaf,} \\ \max\{W^+(T_{v_1})d(v, v_1) + S^1(v_1); \\ \qquad W^+(T_{v_2})d(v, v_2) + S^1(v_2)\} & \text{otherwise.} \end{cases} \qquad (6)$$

When there is no vertex in T_v has positive weight, i.e. $W^+(T_v) = 0$, it is no need to compute $S^1(v)$ and we set $S^1(v) = 0$.

The quantities $S^2_\wedge(v)$ can be computed applying the following recursive formulas:

$$S^2_\wedge(v) = \begin{cases} 0 & \text{if } v \text{ is a leaf,} \\ W^+(T_{v_1})d(v,v_1) + S^1(v_1) \\ \quad + W^+(T_{v_2})d(v,v_2) + S^1(v_2) & \text{otherwise.} \end{cases} \tag{7}$$

We note that when a vertex v has only one child the above quantities cannot be computed and we set $S^2_\wedge(v) = 0$.

The quantities $S^2_\parallel(v)$ are determined as follows:

$$S^2_\parallel(v) = \begin{cases} 0 & \text{if } v \text{ is a leaf,} \\ \max\{W(T_{v_1})d(v,v_1) + MS(v_1); \\ \quad W(T_{v_2})d(v,v_2) + MS(v_2)\} & \text{otherwise.} \end{cases} \tag{8}$$

where

$$MS(v_i) = max\{S^2_\wedge(v_i); S^2_\parallel(v_i)\}, i = 1, 2.$$

The objective function value associated to the pair of best paths $P^1_{(r_1,r_2)}$ and $P^2_{(r_1,r_2)}$ is then given by:

$$F_1(\{P^1_{(r_1,r_2)}, P^2_{(r_1,r_2)}\}) = F_1(P^1_{r_1}, P^2_{r_1}) + F_1(P^1_{r_2}, P^2_{r_2}), \tag{9}$$

where, for $i = 1, 2$, one has:

$$F_1(P^1_{r_i}, P^2_{r_i}) = \sum_{v \in V_{r_i}} w(v)d(v, r_i) - max\{S^2_\wedge(r_i); S^2_\parallel(r_i)\}.$$

Among all possible pairs of best paths $P^1_{(r_1,r_2)}$ and $P^2_{(r_1,r_2)}$ that can be obtained by considering all the different edges (r_1, r_2) of T, the best location is given the pair with the minimum service cost (9). If one wants to provide the structure of the two paths, it suffices to trace back the sequence of choices made in the bottom-up computation of the above recursive formulas.

Case (ii). Consider the tree rooted at a vertex r which is not a leaf, obviously, there is no need to consider the leaf. Let $T_{r_1}, \cdots, T_{r_m}$, with $m \geq 2$, be the subtree rooted at each child of r. Note that we can locate only one branch of the two paths in each subtree $T_{r_i}, i = 1, \cdots, m$.

For each $T_{r_i}, i = 1, \cdots, m$, we assume it is binary and find the above branches by computing the recursive quantities $S^1(v)$ bottom-up from the leaves of T_{r_i} to the root r. After the evaluation of all $T_{r_i} \cup (r_i, r), i = 1, \cdots, m$, for the root r we have m different quantities that we denote by $S^1_{r_i}(r), i = 1, \cdots, m$.

Then in order to avoid branches of the two paths intersecting in an edge (r_i, r) for some i, we evaluate at most the 4 largest values for $S^1_{r_i}(v)$ and select the two paths P^1_r, P^2_r intersecting only in vertex r. Suppose that the maximum total saving is $Sav(r)$. The objective function value associated to this solution is given by

$$F_1[T_r] = \sum_{v \in V_r} w(v)d(v, r) - Sav(r).$$

Now we analysis disjoint case. Let P_1, P_2 be two disjoint paths and $m \in (v_i, v_j)$ (if m is a vertex, suppose it is v_i) be the midpoint in the shortest path between P_1 and P_2. Let T_{v_i} and T_{v_j} be the subtrees of $T - (v_i, v_j)$ containing v_i and v_j, respectively. Obviously, $P_1 \subseteq T_{v_i}, P_2 \subseteq T_{v_j}$ and all vertices with nonnegative weights in T_i and all vertices with negative weights in T_j are served by P_1. The remaining vertices are served by P_2.

The idea of the approach is to find the optimal pair P_1, P_2 by searching for the two paths separately in subtrees T_{v_i} and T_{v_j} such that the following objective function is minimized:

$$F_1(P_1, P_2)[e = (v_i, v_j)] = \sum_{v \in T_{v_i} \wedge w(v) \geq 0} w(v)d(v, P_1) + \sum_{v \in T_{v_j} \wedge w(v) < 0} w(v_i)d(v, P_1)$$

$$+ \sum_{v \in T_{v_j} \wedge w(v) \geq 0} w(v)d(v, P_2) + \sum_{v \in T_{v_i} \wedge w(v) < 0} w(v)d(v, P_2)$$

Based on this observation, for every edge $e = (v_i, v_j)$, construct two trees T_e^1, T_e^2, which differ from the original tree only in the weights of their vertices. For the tree T_e^1 the weights of its vertices are defined as

$$w^2(v) = \begin{cases} w(v) & \text{if } v \in T_{v_i} \wedge w(v) \geq 0 \text{ or } v \in T_{v_j} \wedge w(v) < 0 \\ 0 & \text{if } v \in T_{v_i} \wedge w(v) < 0 \text{ or } v \in T_{v_j} \wedge w(v) \geq 0 \end{cases}$$

For the tree T_e^2 the weights of its vertices are defined analogously:

$$w^2(v) = \begin{cases} 0 & \text{if } v \in T_{v_i} \wedge w(v) \geq 0 \text{ or } v \in T_{v_j} \wedge w(v) < 0, \\ w(v) & \text{if } v \in T_{v_i} \wedge w(v) < 0 \text{ or } v \in T_{v_j} \wedge w(v) \geq 0. \end{cases}$$

Then we find the core for each of the trees T_e^1 and T_e^2 with pos/neg weight by applying the linear time algorithm of Zaferanieh, and Fathali [21]. An optimal disjoint path solution of problem L_1 can be found among the pairs of core of the two adjusted trees obtained by the deletion of some edge $e = (v_i, v_j)$ as described above. This property is direct generalization of Theorem 5 in [5], so we do not present the proof which is basically the same as the one of Theorem 5 in [5], which can be illustrate in the following:

Proposition 3. *Let P_e^1, P_e^2 be a pair of disjoint paths such that*

$$F_1(P_e^1, P_e^2)[e = (v_i, v_j)] = \min_{P_1 \in T_{v_i}, P_2 \in T_{v_j}} F_1(P_1, P_2)[e = (v_i, v_j)].$$

Let P_1'', P_2'' be an optimal disjoint paths solution to problem L_1. Then

$$F_1(P_1'', P_2'') = \min_{e \in T}\{F_1(P_e^1, P_e^2)[e = (v_i, v_j)]\}.$$

To conclude this section, we summarize the whole procedure to solve problem L_1.

Algorithm 1

1. For each edge (r_1, r_2) of T, transform the two subtrees T_{r_i} into binary trees rooted at r_i, compute $F_1(\{P^1_{(r_1,r_2)}, P^2_{(r_1,r_2)}\})$ and find the corresponding pair of paths $P^1_{(r_1,r_2)}$ and $P^2_{(r_1,r_2)}$.
2. For each vertex r of T that is not a leaf, transform the subtrees T_{r_i} into binary trees rooted at each children of r, compute $F_1[T_r]$ and find the corresponding pair of paths P^1_r and P^2_r.
3. Among all the pair of paths obtained above, choose the best pair, say $\{P'_1, P'_2\}$.
4. For each edge (v_i, v_j) of T, compute the adjusted weights of the vertices in T^1_e and T^2_e, find the core in T^1_e, T^2_e and denote them by P^1_e, P^2_e.
5. Among all the pairs P^1_e, P^2_e, for $e \in T$, choose the best pair and denote it by $\{P''_1, P''_2\}$.
6. Choose the best paths location between $\{P'_1, P'_2\}$ and $\{P''_1, P''_2\}$.

Theorem 1. *For a given weighted tree T with n vertices. The problem L_1 can be solved in $O(n^2)$ time.*

Proof. The validity of the Algorithm 1 can be proved by the above propositions. Transforming a tree T_{r_i} into a binary tree requires $O(|T_{r_i}|)$ time by applying the procedure in [19]. The quantities (6)–(8) can be computed in $O(n)$ time by visiting a rooted tree bottom-up, so that finding the best pair $\{P'_1, P'_2\}$ along with its objective function value takes $O(n^2)$ time.

For solving disjoint paths location, after the removal of an edge $e = (v_i, v_j)$, the adjusted vertex weights and the cores in T^1_e and T^2_e can be found in linear time. Thus, also finding the optimal pair of vertex disjoint paths $\{P''_1, P''_2\}$ requires $O(n^2)$ time.

4   Properties and Algorithm for the Problem L_2

4.1   Optimal Properties of Problem L_2

Proposition 4. *If the two paths of an optimal solution of problem L_2 are intersecting, then the two paths are discrete.*

Proof. Suppose that $P^* = \{P_1 = P(v_i, x), P_2 = P(v_s, v_t)\}$ is an optimal solution for the problem L_2, and P_1, P_2 are intersecting, $v_i, v_r, v_q \in V(T)$, $x \in (v_r, v_q)$ and $v_r \in P_1$. Let V_x be set of vertices served by x. Then for all $v \in V_x$, we have $d(v, x) = d(v, v_q) + d(v_q, x)$. Thus the total cost of all vertices in V_x is

$$f(x)[V_x] = \sum_{v \in V_x} w(v) d(v, x)$$

$$= \sum_{v \in V_x} w(v) d(v, v_q) + \sum_{v \in V_x} w(v) \cdot d(x, v_q)$$

$$= \sum_{v \in V_x} w(v) d(v, v_q) + \left(\sum_{v \in V_x} w(v) \right) \cdot d(x, v_q).$$

If $\sum\limits_{v \in V_x} w(v) > 0$, we extend the path $P_1 = P(v_i, x)$ to $P_1' = P(v_i, v_q)$. The total cost of all vertices in V_x is reduced to $F_1(x)[V_x] = \sum\limits_{v \in V_x} w(v)d(v, v_q)$. If $\sum\limits_{v \in V_x} w(v) < 0$, we reduce the path $P_1 = P(v_i, x)$ to $P_1' = P(v_i, v_r)$. The total cost of all vertices in V_x is reduced to $F_1(x)[V_x] = \sum\limits_{v \in V_x} w(v)d(v, v_r)$. If $\sum\limits_{v \in V_x} w(v) = 0$, we reduce the path $P_1 = P(v_i, x)$ to $P_1' = P(v_i, v_r)$ that does not change the total cost of all vertices in V_x. In all above cases, the operations do not change the vertices service partition and the cost of every vertex $v \in V - V_x$ remains the same. We can get a better solution than P^*, which is a contradiction to the optimal of P^*. Therefore, x is a vertex of the tree.

When more than one endpoint of the two paths are not vertices of T, we can use the similar arguments as the above to show that one can always extend or reduce the two paths to find a better paths location. The result follows.   □

Proposition 5. *If the two paths of an optimal solution of L_2 are disjoint, then at least one of the two paths is discrete.*

Proof. Suppose that $P^* = \{P_1 = P(v_i, x), P_2 = P(v_j, y)\}$ is an optimum solution for problem L_2 and P_1 and P_2 are disjoint, $v_i, v_j \in V(T)$, x, y are two inner points of edge $(v_r, v_q), (v_s, v_t)$ respectively, where $v_r \in P_1, v_t \in P_2$. If one of x, y is not on the shortest path between P_1 and P_2, without loss of generality, we assume it is x. Let V_x be the set of vertices served by point x of P_1. The total cost of all vertices in V_x is

$$f(x)[V_x] = \sum_{v \in V_x} w(v)d(v, v_q) + \Big(\sum_{v \in V_x} w(v)\Big) \cdot d(x, v_q).$$

We can get a contradiction by the similar arguments as in Proposition 4. So we suppose that both x and y are on the shortest path between P_1 and P_2, i.e. $d(x, y) = d(P_1, P_2)$. Let V_y be set of vertices served by point y of P_2. The total cost of all vertices in V_x and V_y is

$$
\begin{aligned}
&f(y)[V_x] + f(y)[V_y] \\
&= \sum_{v \in V_x} w(v)d(v, v_q) + \Big(\sum_{v \in V_x} w(v)\Big) \cdot d(x, v_q) + \sum_{v \in V_y} w(v)d(v, v_s) \\
&\quad + \Big(\sum_{v \in V_y} w(v)\Big) \cdot d(y, v_s) \\
&= \sum_{v \in V_x} w(v)d(v, v_q) + \sum_{v \in V_y} w(v)d(v, v_s) + \Big(\sum_{v \in V_x} w(v)\Big) \cdot d(x, v_q) \\
&\quad + \Big(\sum_{v \in V_y} w(v)\Big) \cdot d(y, v_s).
\end{aligned}
$$

If $\sum\limits_{v \in V_x} w(v) + \sum\limits_{v \in V_y} w(v) > 0$, we extend P_1 and P_2 with the same length until x reach v_q or y reach v_s. If $\sum\limits_{v \in V_x} w(v) + \sum\limits_{v \in V_y} w(v) < 0$, we reduce P_1 and P_2 with the same length until x reach v_r or y reach v_t. If $\sum_{v \in V_x} w(v) + \sum_{v \in V_y} w(v) = 0$, we reduce P_1 and P_2 with the same length until x reach v_r or y reach v_t, we get a shorter paths location. The midpoint of the two paths P_1 and P_2 does not change after all the above operations. This means we do not change the vertices service partition. But we can get a better solution than P^*, which is a contradiction to the optimal of P^*. Thus, at least one of x, y is a vertex of the tree.

When more than two endpoints of the two paths are not vertices of T, we can use the similar arguments as the above to show that one can always extend or reduce the two paths to find a better paths location. The result follows. □

Proposition 6. *If the two paths P_1, P_2 of an optimal solution of problem L_2 is disjoint and one of them is continuous, then the midpoint of the path between P_1 and P_2 is a vertex of the tree.*

Proof. According to Proposition 5, we suppose that $P^* = \{P_1 = P(v_i, x), P_2 = P(v_s, v_t)\}$ is an optimum solution for problem L_2 and $P_1 \cap P_2 = \emptyset$, x is an inner point of edge (v_r, v_q), and $v_i, v_s, v_t \in V(T)$, $v_r \in P_1$. Let V_x be set of vertices served by point x. The total cost of all vertices in V_x is

$$f(x)[V_x] = \sum_{v \in V_x} w(v)d(v, v_q) + (\sum_{v \in V_x} w(v)) \cdot d(x, v_q).$$

We can show that x is on the shortest path between P_1 and P_2 by the similar method as in Proposition 4. Let $m \in (v_a, v_b)$ be the midpoint of the path between x and P_2. If m is not a vertex of the tree T. As in the proof of Proposition 4, we can get a contradiction by moving the endpoint x of the path P_1 in either direction with some length until one of x reaches a vertex. By assumption, P_1 is not discrete, so m is a vertex of the tree. □

4.2 Algorithm for Problem L_2

Similar as problem L_1, we search the optimal paths of problem L_2 which may intersecting or disjoint separately. From Proposition 4 we know that the two optimal intersecting paths for problem L_2 are discrete. Hence, we can use the similar method as finding the optimal intersecting paths for problem L_1 to find the optimal intersecting paths for problem L_2 after the following revisions: replace

$$S^1(v) = \begin{cases} 0 & \text{if } v \text{ is a leaf,} \\ \max\{W^+(T_{v_1})d(v, v_1) + S^1(v_1); & \\ \quad W^+(T_{v_2})d(v, v_2) + S^1(v_2)\} & \text{otherwise.} \end{cases}$$

with

$$S^1(v) = \begin{cases} 0 & \text{if } v \text{ is a leaf,} \\ \max\{W(T_{v_1})d(v, v_1) + S^1(v_1); & \\ \quad W(T_{v_2})d(v, v_2) + S^1(v_2)\} & \text{otherwise.} \end{cases}$$

Replace

$$S^2_\wedge(v) = \begin{cases} 0 & \text{if } v \text{ is a leaf,} \\ W^+(T_{v_1})d(v,v_1) + S^1(v_1) & \\ +W^+(T_{v_2})d(v,v_2) + S^1(v_2) & \text{otherwise} \end{cases}$$

with

$$S^2_\wedge(v) = \begin{cases} 0 & \text{if } v \text{ is a leaf,} \\ W(T_{v_1})d(v,v_1) + S^1(v_1) & \\ +W(T_{v_2})d(v,v_2) + S^1(v_2) & \text{otherwise.} \end{cases}$$

Now we search optimal disjoint paths. First, we search optimal disjoint discrete paths. Let P_1, P_2 be two disjoint discrete paths and $m \in (v_i, v_j)$ (if m is a vertex, suppose it is v_i) be the midpoint in the shortest path between P_1 and P_2. Let T_{v_i} and T_{v_j} be the subtrees of T containing v_i and v_j, respectively, which are obtained when the edge (v_i, v_j) is deleted. Obviously, all vertices in T_{v_i} are served by P_1 and all vertices in T_{v_j} are served by P_2.

Let $r_1 \in P_1$ and $r_2 \in P_2$ such that $d(r_1, r_2) = d(P_1, P_2)$. Let T'_{r_1} and T'_{r_2} be the subtree obtained from T by deleting all edges on $P(r_1, r_2)$ which containing r_1 and r_2, respectively. Then P_1 is contained in T'_{r_1} and P_2 is contained in T'_{r_2}. Let $Sav(r_i) = \sum_{v \in T'_{r_i}} w(v)d(v, r_i) - \sum_{v \in T'_{r_i}} w(v)d(v, P_i) (i = 1, 2)$. Then the objective function value of P_1, P_2 is given by:

$$F_2(P_1, P_2) = \sum_{v \in T_{v_i}} w(v)d(v, r_1) - Sav(r_1) + \sum_{v \in T_{v_j}} w(v)d(v, r_2) - Sav(r_2) \quad (10)$$

Obviously, the path P_i $(i = 1, 2)$ represents the maximum saving path in T'_{r_i} that pass through r_i $(i = 1, 2)$. We can obtain the saving functions $Sav(r_i)(i = 1, 2)$ and P_i by the same method as in Sect. 3.2 case (ii). Thus, in order to find the optimal disjoint discrete paths, it suffice to consider all pairs of $\{r_1, r_2\}$ in T.

From Proposition 6, we know that if one of the two optimal disjoint paths P_1, P_2 is continuous, then the two paths have the following structure: one path is discrete, another path is continuous and one endpoint which is not a vertex is on the shortest path between P_1 and P_2 and the midpoint between P_1 and P_2 is a vertex.

Suppose that $P_1 = P(v_i, x), P_2 = P(v_j, v_s)$ is an optimum disjoint solution for problem L_2 such that x is an inner point of edge (v_r, v_q) and $v_r \in P_1$. Let the midpoint between these two paths be v_m and $v_j \in P_2$ such that $d(x, v_j) = d(P_1, P_2)$. For a vertex v_j there are at most $O(n)$ vertices v_m in T such that $d(v_j, v_m) = d(v_m, x)$. Thus, for a fixed vertex v_j there are at most $O(n^2)$ pairs $\{x, v_j\}$ such that $d(v_j, x) = d(P_1, P_2)$ and $x \in P_1, v_j \in P_2$.

For a given vertex v_j, consider the rooted tree T_{v_j}. Suppose that $Sav(v_j)$ is the maximum saving of one path which pass through v_j among the branches (except the branch that v_m lies) of $T - \{v_j\}$. Then the objective value associated with the pair of disjoint paths P_1, P_2 is given by:

$$F_2(\{P_1, P_2\}) = \sum_{v \in T_{v_j} - T_{v_m}} w(v)d(v, v_j) - Sav(v_j) + \sum_{v \in T_{v_m}} w(v)d(v, v_r)$$

$$- (W(T_{v_m}) - W(T_{v_r})) \cdot d(x, v_r) - S^1(v_r). \tag{11}$$

To conclude this section, we summarize the whole procedure to solve problem L_2.

Algorithm 2:

1. For each edge (r_1, r_2) of T, transform the two subtrees T_{r_i} into binary trees rooted at r_i, compute $F_1(\{P^1_{(r_1, r_2)}, P^2_{(r_1, r_2)}\})$ and find the corresponding pair of paths $P^1_{(r_1, r_2)}$ and $P^2_{(r_1, r_2)}$.
2. For each vertex r of T that is not a leaf, transform the subtrees T_{r_i} into binary trees rooted at each children of r, compute $F_1[T_r]$ and find the corresponding pair of paths P^1_r and P^2_r.
3. Among all the pair of paths obtained above, choose the best pair, say $\{P'_1, P'_2\}$.
4. For each pair vertices $\{r_1, r_2\}$ of T, compute function (10), find the pair disjoint discrete paths and denote them by $P^1_{r_1, r_2}, P^2_{r_1, r_2}$.
5. For each vertex v_j, and all points x such that the midpoint is a vertex v_m, compute function (11) and find the best pair $P^1_{v_j, x}, P^2_{v_j, x}$.
6. Among all the pairs $P^1_{r_1, r_2}, P^2_{r_1, r_2}$ and $P^1_{v_j, x}, P^2_{v_j, x}$, choose the best pair and denote it by $\{P''_1, P''_2\}$.
7. Choose the best paths location between $\{P'_1, P'_2\}$ and $\{P''_1, P''_2\}$.

Theorem 2. *For a given weighted tree T with n vertices. The problem L_2 can be solved in $O(n^3)$ time.*

Proof. The validity of the Algorithm 2 can be proved by the above propositions. Similar as Algorithm 1 finding the best pair $\{P'_1, P'_2\}$ along with its objective function value takes $O(n^2)$ time. For each vertex pair $\{r_1, r_2\}$, we can find $P^1_{r_1, r_2}, P^2_{r_1, r_2}$ in $O(n)$ time. Since there are $O(n^2)$ pairs $\{r_1, r_2\}$, we need $O(n^3)$ time to find optimal disjoint discrete paths.

For a given vertex v_j and all $v \in T_{v_j}$, $S^1(v)$ and the maximum saving path starts v in T_v can be obtained in linear time. Thus, for a given pair $\{v_j, x\}$, we can find the pair paths $P^1_{v_j, x}, P^2_{v_j, x}$ in constant time. There are at most $O(n^3)$ candidate pair points $\{v_j, x\}$. Thus, finding the optimal disjoint paths requires $O(n^3)$ time. ∎

5   Conclusions

In this paper we investigated the problems of locating two path shaped facilities in trees with pos/neg weights. According to the feature of the problems, we formulate two objective functions. Both of the two problems L_1 and L_2 have good combinatorial properties and can be solved in polynomial time. Finally, we conclude this paper by giving directions for further studies. One direction is to solve the k ($k \geq 3$) paths location problems on trees. Another direction is to design polynomial algorithms for two paths location problems on other kind of graphs.

References

1. Armon, A., Gamzu, I., Segev, D.: Mobile facility location: combinatorial filtering via weighted occupancy. J. Comb. Optim. **28**, 358–375 (2014)
2. Becker, R.I., Perl, Y.: Finding the two-core of a tree. Discrete Appl. Math. **11**, 103–113 (1985)
3. Becker, R.I., Lari, I., Scozzari, A., Storchi, G.: The location of median paths on grid graphs. Ann. Oper. Res. **150**, 65–78 (2001)
4. Becker, R.I., Lari, I., Scozzari, A.: Algorithms for central-median paths with bounded length on trees. Eur. J. Oper. Res. **179**, 1208–1220 (2007)
5. Berman, O., Krass, D., Menezes, M.: Facility reliability issues in network p-median problems: strategic centralization and co-location effects. Oper. Res. **55**, 332–350 (2007)
6. Burkard, R.E., Cela, E., Dollani, H.: 2-medians in trees with pos/neg-weights. Discrete Appl. Math. **60**, 51–71 (2000)
7. Burkard, R.E., Hatzl, J.: Median problems with positive and negative weights on cycles and cacti. J. Comb. Optim. **20**, 27–46 (2010)
8. Elloumi, S.: A tighter formulation of the p-median problem. J. Comb. Optim. **19**, 69–83 (2010)
9. Goldman, A.J.: Optimal center location in simple networks. Trans. Sci. **5**, 212–221 (1971)
10. Kariv, O., Hakimi, S.L.: An algorithmic approach to network location problems, Part I: The p-centers. SIAM J. Appl. Math. **37**, 513–518 (1979)
11. Kariv, O., Hakimi, S.L.: An algorithmic approach to network location problems, Part II: The p-medians. SIAM J. Appl. Math. **37**, 539–560 (1979)
12. Lari, I., Ricca, F., Scozzari, A.: Comparing different metaheuristic approaches for the median path problem with bounded length. Eur. J. Oper. Res. **190**, 587–597 (2008)
13. Morgan, C.A., Slater, P.J.: A linear algorithm for the core of a tree. J. Algorithms **1**, 247–258 (1980)
14. Peurto, J., Ricca, F., Scozzari, A.: The continuous and discrete path-variance problems on trees. Networks **53**, 221–228 (2009)
15. Peurto, J., Ricca, F., Scozzari, A.: Extensive facility location problems on networks with equity measures. Discrete Appl. Math. **157**, 1069–1085 (2009)
16. Peurto, J., Ricca, F., Scozzari, A.: Reliability problems in multiple path-shaped facility location. Discrete Optim. **12**, 61–72 (2014)
17. Richey, M.B.: Optimal location of a path or tree on a network with cycles. Networks **20**, 391–407 (1990)
18. Slater, P.J.: Locating central paths in a graph. Transp. Sci. **16**, 1–18 (1982)
19. Tamir, A.: An $O(pn^2)$ algorithm for the p-median and related problems on tree graphs. Oper. Res. Lett. **19**, 59–64 (1996)
20. Wang, F.: Finding a two-core of a tree in linear time. SIAM J. Discrete Math. **15**, 193–210 (2002)
21. Zaferanieh, M., Fathali, J.: Finding a core of a tree with pos/neg weight. Math. Meth. Oper. Res. **76**, 147–160 (2012)

Approximation Algorithms for Optimization Problems in Random Power-Law Graphs

Yilin Shen[1], Xiang Li[2(✉)], and My T. Thai[2(✉)]

[1] Samsung Research America, San Jose, CA 95134, USA
yilin.shen@samsung.com
[2] CISE Department, University of Florida, Gainesville, FL 32611, USA
{xixiang,mythai}@cise.ufl.edu

Abstract. Many large-scale real-world networks are well-known to have the power law distribution in their degree sequences: the number of vertices with degree i is proportional to $i^{-\beta}$ for some constant β. It is a common belief that solving optimization problems in power-law graphs is easier. Unfortunately, many problems have been proven NP-hard, along with their inapproximability factors in power-law graphs. Therefore, it is of great importance to develop an algorithm framework such that these optimization problems can be approximated in power-law graphs, with provable theoretical approximation ratios.

In this paper, we propose an algorithmic framework, called *Low-Degree Percolation* (LDP) Framework, for solving Minimum Dominating Set, Minimum Vertex Cover and Maximum Independent Set problems in power-law graphs. Using this framework, we further show a theoretical framework to derive the approximation ratios for these optimization problems in two well-known random power-law graphs. Our numerical analysis shows that, these optimization problems can be approximated into near 1 factor with high probability, using our proposed LDP algorithms, in power-law graphs with exponential factor $\beta \geq 1.5$, which belongs to the range of most real-world networks.

Keywords: Power-law graphs · Random graphs · Approximation algorithms · Probabilistic analysis

1 Introduction

A great number of large-scale networks in real life are discovered to follow a power-law distribution in their degree sequences, ranging from the Internet [8], the World-Wide Web (WWW) [3] to social networks [15]. That is, the number of vertices with degree i is proportional to $i^{-\beta}$ for some constant β in these graphs, which is called power-law graphs. The observations show that the exponential factor β ranges between 1.5 and 3 for most real-world networks [4].

This work was finished when Yilin Shen was with CISE Department, University of Florida.

Z. Zhang et al. (Eds.): COCOA 2014, LNCS 8881, pp. 343–355, 2014.
DOI: 10.1007/978-3-319-12691-3_26

There are many papers studying the computational complexity of optimization problems in power-law networks. Ferrante *et al.* [9] had an initial attempt on power-law graphs to show the NP-hardness of Maximum Clique and Minimum Graph Coloring by constructing a bipartite graph to embed a general graph into a power-law graph and NP-hardness of Minimum Dominating Set (MDS), Minimum Vertex Cover (MVC) and Maximum Independent Set (MIS) problems based on their optimal substructure properties. Recently, Shen *et al.* [16] proposed two new techniques on optimal substructure problems to further show the APX-hardness and the inapproximability result of MIS, MDS, and MVC problems on general power-law graphs and simple power-law graphs.

From algorithmic perceptive, some experiments had been developed to evaluate the simple algorithms for optimization algorithm in power-law graphs [7,14]. Later on, the upper bound of flow problem was provided in random power-law graphs in [10] and an $1 - o(1)$ approximation algorithm for maximum clique problem was proposed in [12]. However, these works did not provide a algorithm framework for solving a set of problems in power-law graphs with degree distribution property, let alone a theoretical analysis framework for analyzing approximation ratios.

In this paper, we focus on addressing the following questions: Can the property of power-law degree distribution help us to design an effective algorithm framework for NP-hard optimization problems? How can we provide a theoretical framework for analyzing approximation ratios of these problems using this power-law degree property? Will these approximation ratios change dramatically for different exponential factors β, i.e. in power-law graphs with different densities?

Our Contributions: We propose an algorithm framework, called Low-Degree Percolation (LDP), to approximate the optimization problems in power-law networks, including MIS, MDS, and MVC problems. The idea of LDP framework is to percolate the graph starting from a large number of low-degree nodes in a power-law graph, which allows us to develop a theoretical framework, which can be used to analysis the approximation ratios via probabilistic analysis. In particular, we apply this theoretical framework to show the approximation ratios for these problems on two well-known random power-law models in [2,5]. At last, numerical analysis of our proposed approaches not only validates our theoretical analysis but also illustrates the effectiveness of our approaches empirically.

Organization: In Sect. 2, we present the two well-known random power-law graph models, and the definitions of classic optimization problems. The Low-Degree Percolation (LDP) algorithm framework is proposed in Sect. 3, along with two different versions for MDS/MVC and MIS problems respectively. In Sect. 4, using LDP framework, we provide a theoretical framework for analyzing approximation ratios, and further extend to achieve the corresponding approximation factors for these two random models. Numerical analysis of our proposed approaches are illustrated in Sect. 5. Related work is presented in Sect. 6 and Sect. 7 concludes the whole paper and provides further discussions.

2 Random Power-Law Models and Problem Definitions

In the section, we first present the two well-known random power-law models, Expected Random Power-Law (ERPL) Graph and Structural Random Power-Law (SRPL) Graph. Then, we recall the definitions of several classical optimization problems.

2.1 Random Power-Law Graph Models

First, in order to represent the degree sequence property of power-law graphs, we consider the following random graph, (α, β) graph $G_{(\alpha,\beta)}$, with its power-law degree distribution depending on two given values α and β.

Definition 2.1 ((α, β) Graph $G_{(\alpha,\beta)}$). *Given an undirected graph $G = (V, E)$ having $|V| = n$ nodes and $|E| = m$ edges, it is called a (α, β) power-law graph if its maximum degree is $\Delta = \lfloor e^{\alpha/\beta} \rfloor$ and the number of nodes with degree i is*

$$y_i = \begin{cases} \lfloor \frac{e^\alpha}{i^\beta} \rfloor, & \text{if } i > 1 \text{ or } \sum_{i=1}^{\Delta} \lfloor \frac{e^\alpha}{i^\beta} \rfloor \text{ is even} \\ \lfloor e^\alpha \rfloor + 1, & \text{otherwise} \end{cases} \quad (2.1)$$

Note that the number of nodes $n = e^\alpha \zeta(\beta) + O(e^{\frac{\alpha}{\beta}} - 1)$ and the number of edges $m = e^\alpha \zeta(\beta - 1) + O(e^{\frac{\alpha}{2\beta}} - 1)$, where $\zeta(\beta) = \sum_{i=1}^{\infty} \frac{1}{i^\beta}$ is the Riemann Zeta function. For simplicity, since there is only a very small error $o(1)$ w.r.t. n and m when $\beta > 2$ when counting the number of both nodes and edges, we denote them as $n \doteq e^\alpha \zeta(\beta)$ and edges $m \doteq e^\alpha \zeta(\beta - 1)$.

The following two well-known models [2,5] were proposed in the literature to construct the (α, β) graphs.

Expected Random Power-Law (ERPL) Graph. Given the parameters α and β, the ERPL model is to construct an expected (α, β) power-law graph according to its degree sequence $\boldsymbol{d}$, which consists of a sequence of integers $(1, \ldots, 1, 2, \ldots, 2, \ldots, \Delta)$ where the number of i is equal to y_i defined in the above Definition 2.1.

Definition 2.2 (ERPL Graph). *Given $\boldsymbol{d} = (d_1, d_2, \ldots, d_n)$ be a sequence of integers $(1, \ldots, 1, 2, \ldots, 2, \ldots, \Delta)$ where the number of i is equal to y_i, the ERPL model generates a random graph in which edges are independently assigned to each pair of vertices (i, j) with probability $d_i d_j \rho$ where $\rho = \frac{1}{\sum_{i=1}^{n} d_i} \doteq \frac{1}{e^\alpha \zeta(\beta-1)}$.*

Structural Random Power-Law (SRPL) Graph. Given the parameters α and β, the SRPL model is proposed as an structural approach to construct a (α, β) power-law graph according to its degree sequence $\boldsymbol{d}$, which consists of a sequence of integers $(1, \ldots, 1, 2, \ldots, 2, \ldots, \Delta)$ where the number of i is equal to y_i defined in the above Definition 2.1.

Definition 2.3 (SRPL Graph). *Given* $d = (d_1, d_2, \ldots, d_n)$ *be a sequence of integers* $(1, \ldots, 1, 2, \ldots, 2, \ldots, \Delta)$ *where the number of i is equal to y_i, the SRPL model generates a random graph as follows. Consider $D = \sum_{i=1}^{n} d_i$ mini-nodes lying in n clusters of each size d_i where $1 \leq i \leq n$, we construct a random perfect matching among the mini-nodes and generate a graph on the n original nodes as suggested by this perfect matching in the natural way: two original nodes are connected by an edge if and only if at least one edge in the random perfect matching connects the mini-nodes of their corresponding clusters.*

2.2   Problem Definitions

Definition 2.4 (Minimum Dominating Set (MDS)). *Given an undirected graph $G = (V, E)$, find a subset $S \subseteq V$ with the minimum size such that for each vertex $v_i \in V \setminus S$, at least one neighbor of v_i belongs to S.*

Definition 2.5 (Minimum Vertex Cover (MVC)). *Given an undirected graph $G = (V, E)$, find a subset $S \subseteq V$ with the minimum size such that for each edge E at least one endpoint belongs to S.*

Definition 2.6 (Maximum Independent Set (MIS)). *Given an undirected graph $G = (V, E)$, find a subset $S \subseteq V$ with the maximum size such that no two vertices in S are adjacent.*

3   Low-Degree Percolation (LDP) Algorithm Framework

In this section, we propose an algorithm framework to solve optimization problems based on the degree sequence property in power-law graphs. As one can see, the most fundamental property of power-law graphs is that they contain a large number of low-degree nodes, while only a small number of high-degree nodes. Therefore, the idea of our proposed Low-Degree Percolation (LDP) algorithm framework is to sort the nodes by their degrees and percolate the graph from the nodes of lowest degree. The process continues in residual graph iteratively until no more nodes, which are surely in optimal solution, can be detected. At last, we apply existing approximation approaches in the remaining graph.

For MDS and MVC problems, as shown in Algorithm 1, since the node incident to a node of degree 1 certainly belongs to an optimal solution, we percolate the graph by adding all the neighbors of nodes with degree 1 in each iteration. Until no more nodes of degree 1 exists in residual graph, we apply existing approximation algorithm in [17] for MDS (or [13] for MVC) to obtain the solution in this residual graph.

On the other hand, Algorithm 2 shows the algorithm for MIS. In this case, the nodes of degree 1 will belong to the optimal solution, and in the meanwhile, it is certain that their neighbors cannot be in optimal solution any more. Therefore, in order to obtain MIS, we select all nodes of degree 1 into the solution in each iteration. At last, we apply the approximation algorithm in [11] to obtain the MIS in the remaining graph.

Here, we note that in a special case that two nodes of degree 1 are connected, the optimal solution of MDS (or MVC, MIS) contains either one of them.

Algorithm 1. LDP Algorithm for MDS/MVC Problems

    Input   : Power-law graph G
    Output: MDS (or MVC) S
1 **while** $\exists$ *Nodes of degree 1* **do**
2   **foreach** *Node v of degree 1* **do**
3   Add its neighbor $N(v)$ into S;
4   Remove v from G;
5   **end**
6   Remove all nodes incident to S from graph G;
7 **end**
8 Determine the leftover MDS (or MVC) in G using existing approximation algorithm in [17] (or [13]) and add them into S;
9 **return** S;

Algorithm 2. LDP Algorithm for MIS Problem

    Input   : Power-law graph G
    Output: MIS S
1 **while** $\exists$ *Nodes of degree 1* **do**
2   **foreach** *Node v of degree 1* **do**
3   Add v into S;
4   Remove v and all its neighbors $N(v)$ from G;
5   **end**
6 **end**
7 Determine the leftover MIS in G using existing approximation algorithm in [11] and add them into S;
8 **return** S;

---

4   Approximation Ratio Analysis

In this section, we show the approximation ratio analysis of LDP Algorithms in both structural and expected random power-law networks. To do this, we first provide a theoretical framework, using LDP algorithm, to analyze the approximation ratio based on the probability that a node does not connect to any node of degree 1. Then, this framework is applied to show the ratio of optimization problems in two different models.

4.1   Theoretical Framework

In this theoretical framework, as the connected component of size 2 is trivial, we mainly focus on the ratio analysis in the rest part of power-law graphs. To begin with, we first provide a formal proof of the following Lemma 4.1 (Similar argument for Corollary 4.1), which has been briefly discussed the LDP algorithms.

Lemma 4.1. *If we do not consider the case of connected components with size 2, the optimal solutions to MDS and MVC contain all nodes incident to at least one node of degree 1, and no nodes of degree 1.*

Proof. In the proof, let u be a node of degree 1 incident to another v of arbitrary degree larger than 1, we consider several cases: (1) If neither u and v is selected in optimal solution, no neighbor is select for u and u is not selected as well, this leads to an infeasible solution; (2) If both u and v are selected, it is easy to see that the solution is no longer optimal; (3) If u is selected instead of v, we have to select a set of nodes to satisfy v if v has degree no less than 2; (4) If v is selected instead of u, both u and v are already satisfied, which means the size of the solution less than the size in a solution containing u. Thus, the proof is complete.

Corollary 4.1. *If we do not consider the case of connected components with size 2, the optimal solution to MIS contains all nodes of degree 1, and no nodes incident to at least one node of degree 1.*

Next, we define $\mu(\alpha, \beta, i)$ to be the probability that a node v of degree i not incident to any nodes of degree 1 in a power-law graph $G_{(\alpha,\beta)}$. Our purpose is to analyze the approximation ratio based on $\mu(\alpha, \beta, i)$ in this graph $G_{(\alpha,\beta)}$.

Let X_i^u be a random variable that a node u of degree i does not connect to any nodes of degree 1. Then, we have

$$X_i^u = \begin{cases} 1, u \in D_1 \\ 0, u \notin D_1 \end{cases}$$

where D_1 is a set of nodes incident to at least one node of degree 1. Note that for all nodes of the same degree, they have the same random variables. For simplicity, we define X_i to be a random variable that some node of degree i. Therefore, we have the expected value of node u not incident to any nodes of degree 1 as

$$E(X_i) = \mu(\alpha, \beta, i)$$

Since the number of nodes of degree i is equal to e^α / i^β, by letting $\Delta = e^{\alpha/\beta}$ and $X = \sum_{i=2}^{\Delta} \frac{e^\alpha}{i^\beta} X_i$, we have the following lemma:

Lemma 4.2. *The expected number of nodes, which have degree no less than 2, and are not incident to any nodes of degree 1, is*

$$\sum_{i=2}^{\Delta} \frac{e^\alpha}{i^\beta} \mu(\alpha, \beta, i)$$

Proof. The expected number of nodes not incident to any nodes of degree 1 is the sum of all nodes of degree no less than 2, i.e. $X = \sum_{i=2}^{\Delta} \frac{e^\alpha}{i^\beta} X_i$. Then we have

$$E(X) = \sum_{i=2}^{\Delta} \frac{e^\alpha}{i^\beta} E(X_i) = \sum_{i=2}^{\Delta} \frac{e^\alpha}{i^\beta} \mu(\alpha, \beta, i)$$

Lemma 4.3. *The variance of X is upper bounded by*

$$e^{2\alpha} \sum_{i=2}^{\Delta} \sum_{j=2}^{\Delta} \frac{\sqrt{\chi(\alpha, \beta, i)\chi(\alpha, \beta, j)}}{(ij)^\beta}$$

where $\chi(\alpha, \beta, i) = \mu(\alpha, \beta, i)(1 - \mu(\alpha, \beta, i))$.

Proof. For a random variable corresponds to a node of degree i not incident to any nodes of degree 1, the variance is

$$Var(X_i) = \left(1 - \frac{i}{\zeta(\beta-1)}\right)\left(1 - \left(1 - \frac{i}{\zeta(\beta-1)}\right)\right) = \mu(\alpha,\beta,i)(1 - \mu(\alpha,\beta,i))$$

For any two variables correspond to two nodes of degree i and j not incident to any nodes of degree 1, according to *Cauchy-Schwarz Inequality*, we have

$$|Cov(X_i, X_j)| \leq \sqrt{Var(X_i)Var(X_j)} = \sqrt{\chi(\alpha,\beta,i)\chi(\alpha,\beta,j)}$$

Then, we sum them up and obtain

$$\sum_{X_i,X_j} |Cov(X_i, X_j)| \leq \sum_{X_i,X_j} \sqrt{Var(X_i)Var(X_j)}$$

$$\leq \sum_{i=2}^{\Delta} \frac{e^\alpha}{i^\beta} \left(\sum_{j=2}^{\Delta} \frac{e^\alpha}{j^\beta} \sqrt{\chi(\alpha,\beta,i)\chi(\alpha,\beta,j)}\right) = e^{2\alpha} \sum_{i=2}^{\Delta} \sum_{j=2}^{\Delta} \frac{\sqrt{\chi(\alpha,\beta,i)\chi(\alpha,\beta,j)}}{(ij)^\beta}$$

Therefore, we have the variance of X to be

$$Var(X) = \sum_{X_i,X_j} |Cov(X_i, X_j)| \leq e^{2\alpha} \sum_{i=2}^{\Delta} \sum_{j=2}^{\Delta} \frac{\sqrt{\chi(\alpha,\beta,i)\chi(\alpha,\beta,j)}}{(ij)^\beta}$$

Lemma 4.4. *The number of nodes, which have degree no less than 2 and are not incident to any nodes of degree 1, is larger than $\lambda \sum_{i=2}^{\Delta} \frac{e^\alpha}{i^\beta}$ with probability at most*

$$\frac{1}{\dfrac{\left(\sum_{i=2}^{\Delta} \frac{1}{i^\beta}\left(\lambda - \mu(\alpha,\beta,i)\right)\right)^2}{\sum_{i=2}^{\Delta} \sum_{j=2}^{\Delta} \frac{\sqrt{\chi(\alpha,\beta,i)\chi(\alpha,\beta,j)}}{(ij)^\beta}} + 1}$$

Proof. Let $\phi = \lambda \sum_{i=2}^{\Delta} \frac{e^\alpha}{i^\beta}$, according to *One-Sided Chebyshev Inequality*,

$$Pr[X \geq \phi] = Pr\left[X - E(X) \geq \frac{\phi - E(X)}{\sqrt{Var(X)}} \sqrt{Var(X)}\right]$$

$$\leq \frac{1}{\frac{(\phi-E(X))^2}{Var(X)} + 1} \leq \frac{1}{\dfrac{\left(\sum_{i=2}^{\Delta} \frac{1}{i^\beta}\left(\lambda - \mu(\alpha,\beta,i)\right)\right)^2}{\sum_{i=2}^{\Delta} \sum_{j=2}^{\Delta} \frac{\sqrt{\chi(\alpha,\beta,i)\chi(\alpha,\beta,j)}}{(ij)^\beta}} + 1}$$

For simplicity, we define the following p_λ and obtain the Corollary 4.2.

$$p_\lambda = \frac{1}{\dfrac{\left(\sum_{i=2}^{\Delta} \frac{1}{i^\beta}\left(\lambda - \mu(\alpha,\beta,i)\right)\right)^2}{\sum_{i=2}^{\Delta} \sum_{j=2}^{\Delta} \frac{\sqrt{\chi(\alpha,\beta,i)\chi(\alpha,\beta,j)}}{(ij)^\beta}} + 1} \tag{4.1}$$

As shown in Eq. (4.1), it is easy to see that $p_\lambda \in (0, 1)$. Particularly, when λ becomes larger and close to 1, p_λ becomes close to 0 as $\mu(\alpha, \beta, i)$ is very small in power-law graphs due to the fact that there are a great number of nodes with degree 1.

Corollary 4.2. *The number of nodes, which have degree no less than 2, and are not incident to any nodes of degree 1, is at least $(1 - \lambda) \sum_{i=2}^{\Delta} \frac{e^\alpha}{i^\beta}$ with probability at least $1 - p_\lambda$.*

Then, based on Lemma 4.1, we derive the following approximation ratios of MDS and MVC in a power-law graph $G_{(\alpha,\beta)}$:

Theorem 4.1 (Main Theorem (MDS&MVC)). *In a power-law graph $G_{(\alpha,\beta)}$, by using Algorithm 1, MDS and MVC can be approximated into*

$$1 + (\Psi - 1)\lambda$$

with probability at least $1 - p_\lambda$, where Ψ is the approximation ratio of MDS (or MVC) in Algorithm [17] (or [13]) in general graphs w.r.t. a graph of size at most $e^\alpha \sum_{i=2}^{\Delta} \frac{1}{i^\beta}$.

Proof. Let ℓ be the number of nodes incident to degree 1 in some power-law graph $G_{(\alpha,\beta)}$. We have the approximation ratio as

$$\frac{\ell + \Psi OPT}{\ell + OPT} \le \frac{\ell + \Psi(\sum_{i=2}^{\Delta} \frac{1}{i^\beta} - \ell)}{\ell + \sum_{i=2}^{\Delta} \frac{1}{i^\beta} - \ell}$$

According to Corollary 4.2, we have $\ell \ge \sum_{i=2}^{\Delta} \frac{1}{i^\beta} - \lambda \sum_{i=2}^{\Delta} \frac{1}{i^\beta}$ with probability at least $1 - p_\lambda$. The proof is complete.

In terms of MIS, we have the approximation ratio as follows:

Theorem 4.2 (Main Theorem (MIS)). *In a power-law graph $G_{(\alpha,\beta)}$, by using Algorithm 2, MIS can be approximated into*

$$\frac{N + e^\alpha \left(\lambda \sum_{i=2}^{\Delta} \frac{1}{i^\beta} \right)}{N + \frac{1}{\Psi} e^\alpha \left(\lambda \sum_{i=2}^{\Delta} \frac{1}{i^\beta} \right)}$$

with probability at least $1 - p_\lambda$, where N is the number of nodes with degree 1, Ψ is the approximation ratio of MIS in Algorithm [11] in general graphs w.r.t. a graph of size at most $e^\alpha \sum_{i=2}^{\Delta} \frac{1}{i^\beta}$.

The proof is omitted due to its similarity of the proof in Theorem 4.1.

4.2   Expected Random Power-Law (ERPL) Graph

To achieve the approximation ratios of these problems in ERPL graph, we only need to calculate the following μ_{ERPL} so as to substitute it into Theorems 4.1 and 4.2.

Lemma 4.5. *In graph G, the probability that a node v of degree i not incident to any nodes of degree 1 is*

$$\mu_{ERPL}(\alpha, \beta, i) = 1 - \prod_{j=1}^{n}\left(1 - \frac{d_i d_j}{\rho}\prod_{x \neq i}\left(1 - \frac{d_i d_x}{\rho}\right)\right)$$

Proof. Consider a node v of degree i in G of degree i. According to the definition of ERPL graph, it has probability $d_i d_j\ \rho$ to connect to a node of degree expected j. Therefore, v will connect to at least one node of degree 1 if and only if there exists a node u of expected degree d_j only incident to v, in which the probability can be calculated as

$$1 - \frac{d_i d_j}{\rho}\prod_{x \neq i}\left(1 - \frac{d_i d_x}{\rho}\right)$$

Then, the probability that v is not incident to any nodes of degree 1 can be calculated by eliminating all these cases and the proof follows.

4.3   Structural Random Power-Law (SRPL) Graph

In SRPL graph, the straightforward computation of $\mu_{\mathrm{SRPL}}(\alpha, \beta, i)$ is intractable due to the difficulty to calculate all possible combinations. To this end, we consider each case that there are particular number of connected components of size 2 in SRPL. At last, the approximation factors can be derived from the law of total probability. In the rest of this subsection, we show the probability to have τ connected component of size 2 in a SRPL graph and each $\mu_{\mathrm{SRPL}}^{\tau}(\alpha, \beta, i)$ respectively, and apply them to obtain the approximation ratios.

Lemma 4.6. *The probability $Pr[C_2 = \tau]$ that there are τ connected components of size 2 in a SRPL graph is*

$$\frac{\binom{w}{2\tau}(2\tau)!!\binom{N-w}{w-2\tau}(w-2\tau)!}{N!!/(N-2w-1+2\tau)!!}$$

where $N = e^{\alpha}\zeta(\beta - 1)$, $w = e^{\alpha}$ is the size of nodes of degree 1.

Proof. In order to have τ connected component of size 2, 2τ mini-nodes are selected first from all w nodes of degree 1. Moreover, there are $(2\tau - 1)!!$ possibilities to match these 2τ mini-nodes. Since the number of perfect matching $f(n)$ for n mini-nodes is $(n-1)!!$, the probability can be calculated by simplifying the following equation.

$$Pr[C_2 = \tau] = \frac{\binom{w}{2\tau}(2\tau)!!\binom{N-w}{w-2\tau}(w-2\tau)!f(N-2w+2\tau)}{f(N)}$$

Lemma 4.7. *In a SRPL graph G, if there are τ connected component of size 2, the probability that a node v of degree i not incident to any nodes of degree 1 is*

$$\mu_{SRPL}^{\tau}(\alpha, \beta, i) = \begin{cases} \prod_{k=0}^{w-1} \frac{N^{\tau}-i-w^{\tau}-k}{N^{\tau}-w^{\tau}-k}, & \text{If } N^{\tau} - i - w^{\tau} > w^{\tau}; \\ 0, & \text{otherwise.} \end{cases}$$

where $N^{\tau} = e^{\alpha}\zeta(\beta - 1) - 2\tau$, $w^{\tau} = e^{\alpha} - 2\tau$ is the size of nodes of degree 1.

Proof. Let D_1 be a set of nodes incident to at least one node of degree 1. Consider that the whole mini-nodes are composed of three subsets, i.e., i nodes correspondent to v, w^{τ} nodes correspondent to all nodes of degree 1 and all left-over nodes, which is referred to as N_i and N_w and $N^{\tau} \setminus \{N_i \cup N_w\}$ respectively. When $N^{\tau} - i - w^{\tau} < w^{\tau}$, there are not enough mini-nodes to match all nodes of degree 1, the probability that $v \notin D_1$ is 0. Otherwise, in order for $v \notin D_1$, we have to select the nodes incident to all nodes in N_w from $N \setminus \{N_i \cup N_w\}$.

$$Pr[v \notin D_1] = \frac{\binom{N^{\tau}-i-w^{\tau}}{w^{\tau}} w^{\tau}! f(N^{\tau} - 2w^{\tau})}{\binom{N^{\tau}-w^{\tau}}{w^{\tau}} w^{\tau}! f(N^{\tau} - 2w^{\tau})} = \prod_{k=0}^{w^{\tau}-1} \frac{N^{\tau} - i - w^{\tau} - k}{N^{\tau} - w^{\tau} - k}$$

where $f(n) = (n-1)!!$, representing the number of perfect matching for n nodes.

Theorem 4.3. *In a SRPL graph G, by using Algorithm 1, MDS and MVC can be approximated into*

$$1 + (\Psi - 1)\lambda$$

with probability at least

$$\sum_{\tau=0}^{\lfloor e^{\alpha}/2 \rfloor} Pr[C_2 = \tau](1 - p_{\lambda}^{\tau})$$

where $p_{\lambda}^{\tau} = \frac{1}{\frac{(\lambda^{\tau} - \mu(\alpha,\beta,2))^2}{\chi(\alpha,\beta,2)} + 1}$ in which $\lambda^{\tau} = \lambda + \frac{\tau}{\sum_{i=2}^{\Delta} \frac{1}{i^{\beta}}}$.

Proof. Consider one case that there are τ connected components in the power-law graph. Thus, according to Theorem 4.1, the probability that the approximation ratio is smaller than $1 + (\Psi - 1)\lambda$ is $1 - p_{\lambda}^{\tau}$ for $p_{\lambda}^{\tau} = \frac{1}{\frac{(\lambda^{\tau} - \mu(\alpha,\beta,2))^2}{\chi(\alpha,\beta,2)} + 1}$ where $\lambda^{\tau} = \lambda + \frac{\tau}{\sum_{i=2}^{\Delta} \frac{1}{i^{\beta}}}$. Therefore, according to the law of total probability, the theorem follows by taking into account all τ, which ranges from 0 up to $\lfloor e^{\alpha}/2 \rfloor$.

For MIS problem, the approximation ratio can be obtained as

$$\frac{N + e^{\alpha}\left(\lambda \sum_{i=2}^{\Delta} \frac{1}{i^{\beta}}\right)}{N + \frac{1}{\Psi} e^{\alpha}\left(\lambda \sum_{i=2}^{\Delta} \frac{1}{i^{\beta}}\right)}$$

with probability at least $\prod_{\tau=0}^{\lfloor e^{\alpha}/2 \rfloor} Pr[C_2 = \tau](1 - p_{\lambda}^{\tau})$, where $p_{\lambda}^{\tau'} = \frac{1}{\frac{(\lambda^{\tau'} - \mu(\alpha,\beta,2))^2}{\chi(\alpha,\beta,2)} + 1}$ in which $\lambda^{\tau'} = \lambda + \frac{\tau}{\sum_{i=2}^{\Delta} \frac{1}{i^{\beta}}}$.

5   Numerical Analysis

Figure 1 illustrates the performance of our LDP algorithms in random power-law graphs, the relation between different β and the corresponding approximation ratios, along with the better experimental results compared with theoretical bounds.

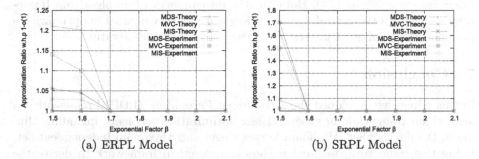

(a) ERPL Model   (b) SRPL Model

Fig. 1. Numerical results of our LDP algorithms on different β ($\alpha = 5$): (1) Theoretical results shows the approximation ratios with probability at least $1 - o(1)$. As one can see, our LDP algorithms can obtain the optimal solution for all these problems after β gets larger than 1.6 and 1.7 in ERPL and SRPL respectively, which covers the range of β in most real-world networks [4]. For the other smaller exponential factors β, we can see that the approximation ratios are a little bit higher, especially up to 5 for MDS and MIS problems for SRPL model. However, the probabilities that these two problems can obtain the approximation ratios less than 1.5 using LDP algorithms are at least 0.95 (only a little bit lower than $1 - o(1)$). (2) Experimental results further reveals that our LDP algorithms can achieve even better solutions compared with the above theoretical bounds. (We tests on 100 cases and choose the average.) As illustrated in Fig. 1, the approximation ratios of all MDS,MVC,MIS problems is no larger than 1.2 and 2.5 even when $\beta = 1.3$ in ERPL and SRPL models respectively.

6   Related Work

Many experimental results were proposed in random power-law graphs for solving optimization problems. Eubank *et al.* [7] empirically showed that a simple greedy algorithm leads to a $1 + o(1)$ approximation factor on MDS and MVC on power-law graphs (without any formal proof) although MDS and MVC has been proved NP-hard to be approximated within $(1 - \epsilon) \log n$ and 1.366 on general graphs respectively [6]. In [14], Gopal also claimed that there exists a polynomial time algorithm that guarantees a $1 + o(1)$ approximation of the MVC problem with probability at least $1 - o(1)$. Unfortunately, there is no such formal proof for this claim either.

Several papers also have some theoretical guarantees for some specific problems on power-law graphs. Gkantsidis *et al.* [10] proved the flow through each link is at most $O(n \log^2 n)$ on power-law random graphs where the routing of

$O(d_u d_v)$ units of flow between each pair of vertices u and v with degrees d_u and d_v. In [10], the authors take advantage of the property of power-law distribution by using the structural random model [1,2] and show the theoretical upper bound with high probability $1 - o(1)$ and the corresponding experimental results. Likewise, Janson *et al.* [12] gave an algorithm that approximated Maximum Clique within $1 - o(1)$ on power-law graphs with high probability on the random poisson model $G(n, \alpha)$ (i.e. the number of vertices with degree at least i decreases roughly as n^{-i}). However, the development of an algorithm framework as well as a theoretical provable framework for solving a set of optimization problems is still of great interest, yet remains open in the literature.

7   Conclusion

In this paper, we developed the *Low-Degree Percolation* (LDP) Framework, an algorithmic framework for solving classic optimization problems, including Minimum Dominating Set, Minimum Vertex Cover and Maximum Independent Set. Using this framework, we further show a theoretical framework to derive the approximation ratios for these optimization problems in two well-known random power-law graphs, Expected Random Power-Law Graph and Structural Random Power-Law Graph. Our numerical analysis further validated that our LDP algorithm enables the very good near-optimal solutions for all these optimization problems with high probability.

Acknowledgment. This work is partially supported by the NSF CCF-1422116 and DTRA YIP HDTRA-1-09-1-0061 grants.

References

1. Aiello, W., Chung, F., Lu, L.: A random graph model for massive graphs. In: STOC '00, pp. 171–180. ACM, New York (2000)
2. Aiello, W., Chung, F., Lu, L.: A random graph model for power law graphs. Exp. Math. **10**, 53–66 (2000)
3. Albert, R., Jeong, H., Barabasi, A.L.: The diameter of the world wide web. Nature **401**, 130–131 (1999)
4. Bornholdt, S., Schuster, H.G. (eds.): Handbook of Graphs and Networks: From the Genome to the Internet. Wiley, New York (2003)
5. Chung, F., Lu, L.: Connected components in random graphs with given expected degree sequences. Ann. Comb. **6**(2), 125–145 (2002)
6. Dinur, I., Safra, S.: On the hardness of approximating minimum vertex cover. Ann. Math. **162**, 439–485 (2005)
7. Eubank, S., Kumar, V.S.A., Marathe, M.V., Srinivasan, A., Wang, N.: Structural and algorithmic aspects of massive social networks. In: SODA '04, pp. 718–727. Society for Industrial and Applied Mathematics, Philadelphia (2004)
8. Faloutsos, M., Faloutsos, P., Faloutsos, C.: On power-law relationships of the internet topology. In: Proceedings of the Conference on Applications, Technologies, Architectures, and Protocols for Computer Communication, SIGCOMM '99, pp. 251–262. ACM, New York (1999)

9. Ferrante, A., Pandurangan, G., Park, K.: On the hardness of optimization in power-law graphs. Theoret. Comput. Sci. **393**(1–3), 220–230 (2008)
10. Gkantsidis, C., Mihail, M., Saberi, A.: Conductance and congestion in power law graphs. SIGMETRICS Perform. Eval. Rev. **31**(1), 148–159 (2003)
11. Halldórsson, M., Radhakrishnan, J.: Greed is good: approximating independent sets in sparse and bounded-degree graphs. In: Proceedings of the Twenty-sixth Annual ACM Symposium on Theory of Computing, STOC '94, pp. 439–448. ACM, New York (1994)
12. Janson, S., Luczak, T., Norros, I.: Large cliques in a power-law random graph(2009)
13. Karakostas, G.: A better approximation ratio for the vertex cover problem. ACM Trans. Algorithms **5**(4), 41:1–41:8 (2009)
14. Pandurangan, G.: (2006). https://sites.google.com/site/gopalpandurangan/papers-by-date
15. Redner, S.: How popular is your paper? an empirical study of the citation distribution. Eur. Phys. J. B - Condens. Matter Complex Syst. **4**(2), 131–134 (1998)
16. Shen, Y., Nguyen, D.T., Xuan, Y., Thai, M.T.: New techniques for approximating optimal substructure problems in power-law graphs. Theoret. Comput. Sci. **447**, 107–119 (2012)
17. Vazirani, V.V.: Approximation Algorithms. Springer-Verlag New York Inc., New York (2001)

A Comparison Between the Zero Forcing Number and the Strong Metric Dimension of Graphs

Cong X. Kang$^{(\boxtimes)}$ and Eunjeong Yi$^{(\boxtimes)}$

Texas A&M University at Galveston, Galveston, TX 77553, USA
{kangc,yie}@tamug.edu

Abstract. The *zero forcing number*, $Z(G)$, of a graph G is the minimum cardinality of a set S of black vertices (whereas vertices in $V(G) - S$ are colored white) such that $V(G)$ is turned black after finitely many applications of "the color-change rule": a white vertex is converted black if it is the only white neighbor of a black vertex. The *strong metric dimension*, $sdim(G)$, of a graph G is the minimum among cardinalities of all strong resolving sets: $W \subseteq V(G)$ is a *strong resolving set* of G if for any $u, v \in V(G)$, there exists an $x \in W$ such that either u lies on an $x - v$ geodesic or v lies on an $x - u$ geodesic. In this paper, we prove that $Z(G) \leq sdim(G) + 3r(G)$ for a connected graph G, where $r(G)$ is the cycle rank of G. Further, we prove the sharp bound $Z(G) \leq sdim(G)$ when G is a tree or a unicyclic graph, and we characterize trees T attaining $Z(T) = sdim(T)$. It is easy to see that $sdim(T + e) - sdim(T)$ can be arbitrarily large for a tree T; we prove that $sdim(T + e) \geq sdim(T) - 2$ and show that the bound is sharp.

Keywords: Zero forcing number · Strong metric dimension · Cycle rank · Tree · Unicyclic graph

1 Introduction

Let $G = (V(G), E(G))$ be a finite, simple, undirected, and connected graph of order $|V(G)| \geq 2$. The *path cover number*, $P(G)$, of G is the minimum number of vertex disjoint paths, occurring as induced subgraphs of G, that cover all the vertices of G. The *degree* $\deg_G(v)$ of a vertex $v \in V(G)$ is the number of edges incident to the vertex v in G; a *leaf* (or *pendant*) is a vertex of degree one. We denote the number of leaves of G by $\sigma(G)$. For $S \subseteq V(G)$, we denote by $\langle S \rangle$ the subgraph induced by S. The *distance* between two vertices $u, v \in V(G)$, denoted by $d_G(u, v)$, is the length of a shortest path in G between u and v. We omit G when ambiguity is not a concern.

The notion of a zero forcing set, as well as the associated zero forcing number, of a simple graph was introduced by the aforementioned "AIM group" in [1] to bound the minimum rank of graphs. Let each vertex of a graph G be given one of two colors, dubbed "black" and "white" by convention. Let S denote the

© Springer International Publishing Switzerland 2014
Z. Zhang et al. (Eds.): COCOA 2014, LNCS 8881, pp. 356–365, 2014.
DOI: 10.1007/978-3-319-12691-3_27

(initial) set of black vertices of G. The *color-change rule* converts the color of a vertex from white to black if the white vertex u_2 is the only white neighbor of a black vertex u_1; we say "u_1 forces u_2" in this case. The set S is said to be *a zero forcing set* of G if all vertices of G will be turned black after finitely many applications of the color-change rule. The *zero forcing number*, $Z(G)$, of G is the minimum of $|S|$, as S varies over all zero forcing sets of G.

Since its introduction by the "AIM group", zero forcing number has become a graph parameter studied for its own sake, as an interesting invariant of a graph. For example, for discussions on the number of steps it takes for a zero forcing set to turn the entire graph black (the graph parameter has been named the *iteration index* or the *propagation time* of a graph), see [6,12]. In [13], a probabilistic interpretation of zero forcing in graphs is introduced. It's also noteworthy that physicists have independently studied the zero forcing parameter, referring to it as the *graph infection number*, in conjunction with the control of quantum systems (see [3,4,19]).

A vertex $x \in V(G)$ *resolves* a pair of vertices $u, v \in V(G)$ if $d(u, x) \neq d(v, x)$. A vertex $x \in V(G)$ *strongly resolves* a pair of vertices $u, v \in V(G)$ if u lies on an $x - v$ geodesic or v lies on an $x - u$ geodesic. A set of vertices $W \subseteq V(G)$ *(strongly) resolves* G if every pair of distinct vertices of G is (strongly) resolved by some vertex in W; then W is called a *(strong) resolving set* of G. For an ordered set $W = \{w_1, w_2, \ldots, w_k\} \subseteq V(G)$ of distinct vertices, the *metric representation* of $v \in V(G)$ with respect to W is the k-vector $D_G(v|W) = (d(v, w_1), d(v, w_2), \ldots, d(v, w_k))$. The *metric dimension* of G, denoted by $dim(G)$, is the minimum among cardinalities of all resolving sets of G. The *strong metric dimension* of G, denoted by $sdim(G)$, is the minimum among cardinalities of all *strong* resolving sets of G.

Metric dimension was introduced by Slater [20] and, independently, by Harary and Melter [11]. Applications of metric dimension can be found in robot navigation [15], sonar [20], combinatorial optimization [18], and pharmaceutical chemistry [5]. Strong metric dimension was introduced by Sebö and Tannier [18]; they observed that if W is a strong resolving set, then the vectors $\{D_G(v|W) \mid v \in V(G)\}$ uniquely determine the graph G (also see [14] for more detail); whereas for a resolving set U of G, the vectors $\{D_G(v|U) \mid v \in V(G)\}$ may not uniquely determine G. It is noted that determining the (strong) metric dimension of a graph is an NP-hard problem (see [10,16]).

In this paper, we initiate a comparative study between the zero forcing number and the strong metric dimension of graphs. The zero forcing number and the strong metric dimension coincide for paths P_n, complete graphs K_n, complete bi-partite graphs $K_{s,t}$ $(s + t \geq 3)$, for examples; they are 1, $n - 1$, and $s + t - 2$, respectively. The Cartesian product of two paths shows that zero forcing number can be arbitrarily larger than strong metric dimension; cycles C_n show that strong metric dimension can be arbitrarily larger than zero forcing number. We prove the sharp bound that $Z(G) \leq sdim(G)$ when G is a tree or a unicyclic graph, and we characterize trees T attaining $Z(T) = sdim(T)$. It is easy to see that $sdim(T + e) - sdim(T)$ can be arbitrarily large for a tree T; we prove that

$sdim(T+e) \geq sdim(T)-2$ and show that the bound is sharp. In the final section, we show, for any graph G with cycle rank $r(G)$, that $Z(G) \leq sdim(G) + 3r(G)$ and pose an open problem pertaining to its refinement.

2 The Zero Forcing Number and the Strong Metric Dimension of Trees

In this section, we show that $Z(T) \leq sdim(T)$ for a tree T, and we characterize trees T satisfying $Z(T) = sdim(T)$. We first recall some results that will be used here.

Theorem 1. *Let T be a tree. Then*

(a) [1] $Z(T) = P(T)$,
(b) [18] $sdim(T) = \sigma(T) - 1$.

Theorem 2. [17] *Let G be a graph with cut-vertex $v \in V(G)$. Let $V_1, V_2, \ldots, V_k$ be the vertex sets for the connected components of $\langle V(G) - \{v\} \rangle$, and for $1 \leq i \leq k$, let $G_i = \langle V_i \cup \{v\} \rangle$. Then $Z(G) \geq [\sum_{i=1}^{k} Z(G_i)] - k + 1$.*

The following terminology are defined for a graph G. A vertex of degree at least three is called a *major vertex*. A leaf u is called *a terminal vertex of a major vertex v* if $d(u,v) < d(u,w)$ for every other major vertex w. The *terminal degree*, $ter(v)$, of a major vertex v is the number of terminal vertices of v. A major vertex v is an *exterior major vertex* if it has positive terminal degree. An *exterior degree two vertex* is a vertex of degree 2 that lies on a shortest path from a terminal vertex to its major vertex, and an *interior degree two vertex z* is a vertex of degree 2 such that a shortest path from z to any terminal vertex includes a major vertex.

Theorem 3. [8] *Let T be a tree. Then*

(a) $dim(T) \leq Z(T)$,
(b) $dim(T) = Z(T)$ if and only if T has no interior degree two vertex and each major vertex v of T satisfies $ter(v) \geq 2$.

It is shown in [9] that $P(T) \leq \sigma(T) - 1$; this and Theorem 1 imply the following:

Theorem 4. *For any tree T, $Z(T) \leq sdim(T)$.*

Next, we characterize trees T satisfying $Z(T) = sdim(T)$.

Theorem 5. *For any tree T, we have $Z(T) = sdim(T)$ if and only if T has an interior degree two vertex on every $v_i - v_j$ path, where v_i and v_j are major vertices of T.*

Proof. ($\Longrightarrow$) Suppose that there exist a pair of major vertices, say v_1 and v_2, in T such that no interior degree two vertex lies in the $v_1 - v_2$ path. We may assume $v_1 v_2 \in E(T)$. If not, replace v_2 with the vertex adjacent to v_1 on the $v_1 - v_2$ path. We consider two disjoint subtrees $T_1, T_2 \subset T$ such that $v_1 \in V(T_1)$, $v_2 \in V(T_2)$, $V(T) = V(T_1) \cup V(T_2)$ and $E(T) = E(T_1) \cup E(T_2) \cup \{v_1 v_2\}$. By Theorem 4, $P(T_1) \leq \sigma(T_1) - 1$ and $P(T_2) \leq \sigma(T_2) - 1$. So, $P(T) \leq P(T_1) + P(T_2) \leq \sigma(T_1) + \sigma(T_2) - 2 = \sigma(T) - 2$, i.e., $Z(T) \leq sdim(T) - 1$.

($\Longleftarrow$) We will induct on $m(T)$, the number of major vertices of the tree T. If $m(T) = 0$, then $Z(T) = 1 = sdim(T)$; if $m(T) = 1$, then $Z(T) = P(T) = \sigma(T) - 1 = sdim(T)$. Suppose the statement holds for all trees T with $2 \leq m(T) \leq k$. Let x be a degree 2 vertex lying between two major vertices u and v of a tree T with $m(T) = k + 1$. Let ℓ and r be the two edges of T incident with x, and denote by T_ℓ (T_r, resp.) the subtree of $T - r$ ($T - \ell$, resp.) containing x. Clearly, T is the vertex sum of T_ℓ and T_r at the vertices being labeled x. The induction hypothesis applies to T_ℓ and T_r, since each has at most k major vertices; thus, $Z(T_\ell) = \sigma(T_\ell) - 1$ and $Z(T_r) = \sigma(T_r) - 1$. Now by Theorem 2, $Z(T) \geq (Z(T_\ell) + Z(T_r)) - 1 = (\sigma(T_\ell) - 1 + \sigma(T_r) - 1) - 1 = \sigma(T) - 1 = sdim(T)$; thus, by Theorem 4, $Z(T) = sdim(T)$. $\qquad\square$

Remark 1. Notice $dim(T) \leq Z(T) \leq sdim(T)$ by Theorems 3(a) and 4, where the equalities are characterized by Theorems 3(b) and 5.

3 The Zero Forcing Number and the Strong Metric Dimension of Unicyclic Graphs

A graph is *unicyclic* if it contains exactly one cycle. Notice that a connected graph G is unicyclic if and only if $|E(G)| = |V(G)|$. By $T + e$, we shall mean a unicyclic graph obtained from a tree T by attaching the edge e joining two non-adjacent vertices of T. In this section, we show that $Z(G) \leq sdim(G)$ for a unicyclic graph G and the bound is sharp. We first recall some results that will be used here.

We say that $x \in V(G)$ is *maximally distant* from $y \in V(G)$ if $d_G(x, y) \geq d_G(z, y)$, for every $z \in N_G(x) = \{v \in V(G) \mid xv \in E(G)\}$. If x is maximally distant from y and y is maximally distant from x, then we say that x and y are *mutually maximally distant* and denote this by x MMD y. It is pointed out in [16] that if x MMD y in G, then any strong resolving set of G must contain either x or y. Noting that any two distinct leaves of a graph G are MMD, we have the following:

Observation 6. *For any connected graph G, all but one of the $\sigma(G)$ leaves must belong to any strong resolving set of G.*

Theorem 7. *Let G be a connected graph of order $n \geq 2$. Then*

(a) *[7] $Z(G) - 1 \leq Z(G + e) \leq Z(G) + 1$ for $e \in E(\overline{G})$, where $\overline{G}$ denotes the complement of G,*

(b) [21] sdim(G) = 1 if and only if G = P_n.

Proposition 1. *Let T be a tree of order at least three. Then $sdim(T + e) \geq sdim(T) - 2$ for $e \in E(\overline{T})$, and the bound is sharp.*

Proof. Since $\sigma(T) - 2 \leq \sigma(T + e) \leq \sigma(T)$, the desired inequality follows from Theorem 1(b) and Observation 6. For the sharpness of the bound, let T be the "comb" with $k \geq 4$ exterior major vertices (see Fig. 1). Then $sdim(T) = \sigma(T) - 1 = k + 1$. Since $\{\ell_i \mid 1 \leq i \leq k - 1\}$ forms a strong resolving set for $T + e$, $sdim(T + e) \leq k - 1 = sdim(T) - 2$; thus $sdim(T + e) = sdim(T) - 2$. $\square$

Remark 2. We note that $sdim(T + e) - sdim(T)$ can be arbitrarily large. For example, suppose that $T = P_n$ and $T + e = C_n$; then $sdim(T) = 1$ and, as noted in [16], $sdim(C_n) = \lceil \frac{n}{2} \rceil$.

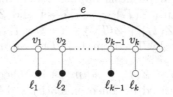

Fig. 1. Unicyclic graph $T + e$ satisfying $sdim(T + e) = sdim(T) - 2$

Theorems 4, and 7(a), and Proposition 1 imply that $Z(T+e) \leq sdim(T+e)+3$. We will show that, in fact, $Z(T + e) \leq sdim(T + e)$.

As defined in [2], a *partial n-sun* is the graph H_n obtained from C_n by appending a leaf to each vertex in some $U \subseteq V(C_n)$, and a *segment* of H_n refers to any maximal subset of consecutive vertices in U. By a *generalized partial n-sun*, we shall mean a graph obtained from C_n by attaching a finite, and not necessarily equal, number of leaves to each vertex $v \in V(C_n)$. See Fig. 2.

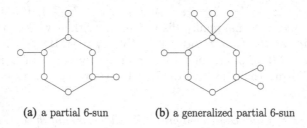

(a) a partial 6-sun   (b) a generalized partial 6-sun

Fig. 2. A partial 6-sun and a generalized partial 6-sun

Theorem 8. *[17] Let H_n be a partial n-sun with segments $U_1, U_2, \ldots, U_t$. Then*

$$Z(H_n) = \max\left\{2, \sum_{i=1}^{t}\left\lceil\frac{|U_i|}{2}\right\rceil\right\}.$$

Corollary 1. *Let H_n be a partial n-sun. Then $Z(H_n) \leq sdim(H_n)$.*

Proof. The formula in Theorem 8 implies that $Z(H_n) \leq \lceil\frac{n}{2}\rceil$. Considering MMD vertices, it's clear that $sdim(H_n) \geq sdim(C_n)$ and, as noted in [16], $sdim(C_n) = \lceil\frac{n}{2}\rceil$. □

Following [2], for a given unicyclic graph G, a vertex $v \in V(G)$ is called an *appropriate vertex* if at least two components of $G - v$ are paths; a vertex $\ell \in V(G)$ is called a *peripheral leaf* if $\deg_G(\ell) = 1$, $\ell u \in E(G)$, and $\deg_G(u) = 2$ (whereas $\deg_G(u) \leq 2$ in [2]). The *trimmed form* of G is an induced subgraph obtained by a sequence of deletions of appropriate vertices, isolated paths, and peripheral leaves until no more such deletions are possible. Further, define $sdim(G) = sdim(G_1) + sdim(G_2)$ (*additivity of sdim over disjoint components*), when G is the disjoint union of G_1 and G_2. This is a natural extension of the (original) definition of $sdim$ for a connected graph; it is needed for the inductive arguments to come.

Remark 3. [17] Let G be a unicyclic graph. Then

(a) for an appropriate vertex v in G, $Z(G - v) - 1 = Z(G)$;
(b) for an isolated path P in G, $Z(G - V(P)) + 1 = Z(G)$;
(c) for a peripheral leaf ℓ in G, $Z(G - \ell) = Z(G)$.

Lemma 1. *Let G be a unicyclic graph, and let C be the unique cycle in G.*

(a) If v is an appropriate vertex in G such that $v \notin V(C)$, then $sdim(G-v)-1 \leq sdim(G)$.
(b) If P is an isolated path in G, then $sdim(G - V(P)) + 1 = sdim(G)$.
(c) If ℓ is a peripheral leaf in G, then $sdim(G - \ell) = sdim(G)$.

Proof. Let $\mathcal{M}_H(x) = \{y \in V(H) : y \text{ MMD } x\}$.

(a) Denote the connected components of $G-v$ by G_1 (with $C \subseteq G_1$) and $T_1, \ldots T_k$ ($k \geq 2$), of which T_1 and T_2 (and possibly more trees) are isolated paths; let u denote the sole neighbor of v in $V(G_1)$. Let S be a minimum strong resolving set of G. Let L denote the set of leaves in $G - G_1$. By Observation 6, $0 \leq |L - S| \leq 1$. If $|L - S| = 0$, then $S \cup \{u\}$ forms a strong resolving set for $G - v$, since a geodesic between any $\ell \in L$ and any $x \in V(G_1)$ necessarily passes through u; thus we have $sdim(G - v) - 1 \leq sdim(G)$. So, suppose $|L - S| = 1$. Since L strongly resolves the complement of G_1 in $G-v$, it suffices to prove the following:

Claim. $S \cap V(G_1)$ strongly resolves G_1.

Proof of Claim. Let $\ell_0 \in L - S$. Let $x, y \in V(G_1)$ be strongly resolved by $\ell \in L \cap S$; we will show that x and y are strongly resolved by some $z \in S \cap V(G_1)$. If x or

y, say x, does not lie on $\mathcal{C}$, then there must exist a leaf $\ell' \in V(G_1) \cap S$ which strongly resolves x and y, and we are done. So, suppose both x and y lie on $\mathcal{C}$. Let u' denote the vertex on $\mathcal{C}$ which is closest to u. There must exist a $w \in V(G_1)$ satisfying w MMD ℓ_0 and such that $d(u', w')$ equals the diameter of $\mathcal{C}$; here w' denotes the vertex on $\mathcal{C}$ which is closest to w. This w lies in S, since $\ell_0 \notin S$. Notice that x and y together lie on the same one of the two semi-circles defined by u' and w'; otherwise, $u' - x$ geodesic does not contain y and $u' - y$ geodesic does not contain x; the relevance here being that a geodesic from $\ell \in L$ to either x or y must pass through u'. Thus, without loss of generality, we may assume a $u' - y$ geodesic contains x. Then, a $w' - x$ geodesic, hence also a $w - x$ geodesic, contains y. It follows that $w \in S \cap V(G_1)$ strongly resolves x and y. □

(b) This follows from the fact $sdim(P) = 1$ and the additivity of $sdim$ over disjoint components.

(c) Since ℓ is a peripheral leaf in G, there exists a vertex $u \in V(G)$ such that $\ell u \in E(G)$ with $\deg_G(u) = 2$. Let $G' = G - \ell$. Since $\mathcal{M}_G(u) = \emptyset$ and $\mathcal{M}_{G'}(u) = \mathcal{M}_G(\ell)$, $sdim(G - \ell) = sdim(G)$. □

Remark 4. Let G be a unicyclic graph, and let $\mathcal{C}$ be the unique cycle of G.

(a) For an appropriate vertex $v \in V(G)$, $sdim(G) - sdim(G - v)$ can be arbitrarily large. If G is a unicyclic graph as in (a) of Fig. 3, then $sdim(G) = \lceil \frac{n}{2} \rceil + k - 1$ and $sdim(G - v) = k + 1$.

(b) There exists G such that, for an appropriate vertex $v \in V(\mathcal{C})$, $sdim(G-v) = sdim(G) + 2$. If G is a unicyclic graph as in (b) of Fig. 3, then $sdim(G) = 6$ (the solid vertices form a minimum strong resolving set of G) and $sdim(G - v) = 8$.

Lemma 2. *Let H be a **generalized** partial n-sun. Then $Z(H) \le sdim(H)$.*

Proof. It's clear that our claim holds for a H which has only one major vertex. Thus, we may assume that H contains at least two major vertices. Let H^0 be a maximal partial n-sun contained in H; then $Z(H^0) \le sdim(H^0)$ by Corollary 1. For $i \ge 0$, let H^{i+1} denote the graph obtained as the vertex sum of a P_2 with H^i at a major vertex of H^i, so that $H = H^k$ for some $k \ge 0$. By the choice of H^0, we have $sdim(H^{i+1}) = sdim(H^i) + 1 \ge Z(H^i) + 1 \ge Z(H^{i+1})$ for each $0 \le i \le k - 1$, where the left inequality is given by the induction hypothesis. □

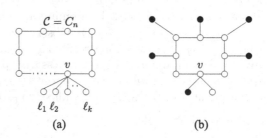

Fig. 3. Unicyclic graph G and an appropriate vertex $v \in V(G)$

Now, we arrive at our main result.

Theorem 9. *If G is a unicyclic graph, then $Z(G) \leq sdim(G)$.*

Proof. Assume $Z(G) > sdim(G)$ for some unicyclic graph G. By trimming as much as possible, but NOT trimming at any vertex lying on the unique cycle C of G, we arrive at a generalized partial n-sun $H \subseteq G$. We descend from the given G to H by, for each trim at an allowed vertex x of G', discarding all components of $G' - x$ except the connected component G'' containing C. Let $G' - x = G'' + T_1 + \ldots + T_m$, where $+$ denotes disjoint union. Remark 3 and Lemma 1 imply $Z(G'' + T_1 + \ldots + T_m) > sdim(G'' + T_1 + \ldots + T_m)$ which, by the additivity of both Z and $sdim$, is equivalent to

$$Z(G'') + \sum_{i=1}^{m} Z(T_i) > sdim(G'') + \sum_{i=1}^{m} sdim(T_i). \qquad (1)$$

Since $Z(T_i) \leq sdim(T_i)$ for each tree T_i by Theorem 4, inequality (1) implies $Z(G'') > sdim(G'')$. Through this process of "descent", we eventually reach $Z(H) > sdim(H)$, which is the desired contradiction to Lemma 2.   □

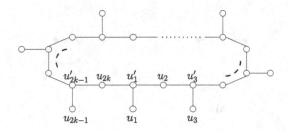

Fig. 4. Unicyclic graphs G with $Z(G) = sdim(G)$

Remark 5. There exists a unicyclic graph G satisfying $Z(G) = sdim(G)$. For an odd integer $k \geq 3$, let G be a partial $2k$-sun with the unique cycle C given by $u_1' u_2 u_3' u_4 \ldots u_{2k-1}' u_{2k}$ such that $ter(u_{2j}) = 0$ and $ter(u_{2j-1}') = 1$, where $1 \leq j \leq k$ (see Fig. 4). Then $Z(G) = k$ by Theorem 8, and $sdim(G) = k$: (i) $sdim(G) \geq k$ since u_j MMD u_{j+k} for each $j \in \{1, 2, \ldots k\}$; (ii) $sdim(G) \leq k$ since $\{u_{2j-1} \mid 1 \leq j \leq k\}$ forms a strong resolving set for G.

4   A Concluding Thought

The cycle rank $r(G)$ of a connected graph G is defined as $|E(G)| - |V(G)| + 1$. In the preceding sections, we have provided *sharp* bounds (relating $Z(G)$ and $sdim(G)$) when $r(G)$ equals 0 or 1; now, we offer a rough bound which, notably, places no restriction on $r(G)$.

Proposition 2. *Let G be a connected graph with cycle rank $r(G)$. Then $Z(G) \leq sdim(G) + 3 \cdot r(G)$.*

Proof. Let T be a spanning tree of G obtained through the deletion of $r = r(G)$ edges of G. We have $Z(G) \leq Z(T) + r \leq sdim(T) + r$, where the left and right inequalities are respectively given by Theorems 7(a) and 4. Since the removal of an edge e from G results in at most two more leaves in $G - e$, we have $\sigma(T) \leq 2r + \sigma(G)$. Since $sdim(T) = \sigma(T) - 1$ by Theorem 1(b), we have $Z(G) \leq 2r + \sigma(G) - 1 + r$. Since $\sigma(G) - 1 \leq sdim(G)$ by Observation 6, we obtain $Z(G) \leq sdim(G) + 3r$. $\qquad\square$

Question. What is the best k such that $Z(G) \leq sdim(G) + k \cdot r(G)$ for any connected graph G?

We conjecture $0 < k < 1$, as suggested by the following example.

Example 1. Let $G = P_s \square P_s$ be the Cartesian product of P_s with itself, where $s \geq 2$. Then $Z(G) = s$ (see [1]) and $sdim(G) = 2$. Notice that $r(G) = (s-1)^2$. So, $Z(G) = sdim(G) + \frac{s-2}{(s-1)^2} r(G)$. See Fig. 5 when $s = 3$, where the solid vertices in Fig. 5(a) form a minimum zero forcing set for G and the solid vertices in Fig. 5(b) form a minimum strong resolving set for G.

(a)   (b)

Fig. 5. $Z(P_3 \square P_3) = 3$ and $sdim(P_3 \square P_3) = 2$

References

1. Barioli, F., Barrett, W., Butler, S., Cioabă, S.M., Cvetković, D., Fallat, S.M., Godsil, C., Haemers, W., Hogben, L., Mikkelson, R., Narayan, S., Pryporova, O., Sciriha, I., So, W., Stevanović, D., Van der Holst, H., Wehe, A.W.: (AIM minimum rank-special graphs work group): zero forcing sets and the minimum rank of graphs. Linear Algebra Appl. **428**, 1628–1648 (2008)
2. Barioli, F., Fallat, S., Hogben, L.: On the difference between the maximum multiplicity and path cover number for tree-like graphs. Linear Algebra Appl. **409**, 13–31 (2005)
3. Burgarth, D., Giovannetti, V.: Full control by locally induced relaxation. Phys. Rev. Lett. **99**, 100501 (2007)
4. Burgarth, D., Maruyama, K.: Indirect Hamiltonian identification through a small gateway. New J. Phys. **11**, 103019 (2009)
5. Chartrand, G., Eroh, L., Johnson, M.A., Oellermann, O.R.: Resolvability in graphs and the metric dimension of a graph. Discrete Appl. Math. **105**, 99–113 (2000)

6. Chilakamarri, K., Dean, N., Kang, C.X., Yi, E.: Iteration index of a zero forcing set in a graph. Bull. Inst. Combin. Appl. **64**, 57–72 (2012)
7. Edholm, C.J., Hogben, L., Huynh, M., LaGrange, J., Row, D.D.: Vertex and edge spread of zero forcing number, maximum nullity, and minimum rank of a graph. Linear Algebra Appl. **436**, 4352–4372 (2012)
8. Eroh, L., Kang, C.X., Yi, E.: A comparison between the metric dimension and zero forcing number of trees and unicyclic graphs. arXiv:1408.5943
9. Eroh, L., Kang, C.X., Yi, E.: Metric dimension and zero forcing number of two families of line graphs. Math. Bohem. **139**, 467–483 (2014)
10. Garey, M.R., Johnson, D.S.: Computers and Intractability: A Guide to the Theory of NP-completeness. Freeman, New York (1979)
11. Harary, F., Melter, R.A.: On the metric dimension of a graph. Ars Combin. **2**, 191–195 (1976)
12. Hogben, L., Huynh, M., Kingsley, N., Meyer, S., Walker, S., Young, M.: Propagation time for zero forcing on a graph. Discrete Appl. Math. **160**, 1994–2005 (2012)
13. Kang, C.X., Yi, E.: Probabilistic zero forcing in graphs. Bull. Inst. Combin. Appl. **67**, 9–16 (2013)
14. Kang, C.X., Yi, E.: The fractional strong metric dimension of graphs. In: Widmayer, P., Xu, Y., Zhu, B. (eds.) COCOA 2013. LNCS, vol. 8287, pp. 84–95. Springer, Heidelberg (2013)
15. Khuller, S., Raghavachari, B., Rosenfeld, A.: Landmarks in graphs. Discrete Appl. Math. **70**, 217–229 (1996)
16. Oellermann, O.R., Peters-Fransen, J.: The strong metric dimension of graphs and digraphs. Discrete Appl. Math. **155**, 356–364 (2007)
17. Row, D.D.: A technique for computing the zero forcing number of a graph with a cut-vertex. Linear Algebra Appl. **436**, 4423–4432 (2012)
18. Sebö, A., Tannier, E.: On metric generators of graphs. Math. Oper. Res. **29**, 383–393 (2004)
19. Severini, S.: Nondiscriminatory propagation on trees. J. Phys. A: Math. Theor. **41**, 482002 (2008)
20. Slater, P.J.: Leaves of trees. Congr. Numer. **14**, 549–559 (1975)
21. Yi, E.: On strong metric dimension of graphs and their complements. Acta Math. Sin. (Engl. Ser.) **29**, 1479–1492 (2013)

Optimal Trees for Minimizing Average Individual Updating Cost

Sicen Guo[1], Minming Li[1]([⊠]), and Yingchao Zhao[2]

[1] City University of Hong Kong, Kowloon Tong, Hong Kong
scguo2@student.cityu.edu.hk, minming.li@cityu.edu.hk
[2] Caritas Institute of Higher Education, Kowloon Tong, Hong Kong
zhaoyingchao@gmail.com

Abstract. Key tree is a popular model to maintain the security of group information sharing by using a tree structure to maintain the keys held by different users. Previously, researchers proved that to minimize the worst case updating cost in case of single user deletion, one needs to use a special 2-3 tree. In this paper, we study the average case for user update. We prove that in the optimal tree, the branching degree of every node can be bounded by 3 and furthermore the structure of the optimal tree can be pretty balanced. We also show the way to construct the optimal tree when there are loyal users in the group.

Keywords: Key tree · Optimality · Individual re-keying

1 Introduction

Broadcasting is widely used in our daily lives, such as paid television channel and group-chat. For either privacy or profit reason, security (message integrity, confidentiality and authority) has become an important design issue in broadcasting systems. To satisfy the security requirement, key encryption must be adopted in broadcasting/multicasting communication. Only authorized users hold a shared group key, which is used for decrypting broadcasting messages. When a user leaves/joins the group, broadcasting a new shared group key is compulsory to achieve both Backward Access Control (the new users cannot access previous messages) and Forward Access Control (the leaving users cannot decrypt further messages) [11].

There are two re-keying strategies: individual re-keying, and batch re-keying. Individual re-keying means that the group needs to update keys for every user join/leave request, while batch re-keying updates keys only after a certain period instead of immediately. Wong et al. [7] proposed a key tree model to specify secure groups and discussed the performance of individual re-keying. They proved that the complexity of individual re-keying is $\mathcal{O}(log n)$, where n is the

This work was fully supported by a grant from the Research Grants Council of the Hong Kong Special Administrative Region, China [Project No. CityU 122512].

© Springer International Publishing Switzerland 2014
Z. Zhang et al. (Eds.): COCOA 2014, LNCS 8881, pp. 366–378, 2014.
DOI: 10.1007/978-3-319-12691-3_28

group size. Individual re-keying can achieve both Backward Access Control and Forward Access Control perfectly, but synchronization problem will occur when there is a batch of updates [10]. Meanwhile, batch re-keying can alleviate out-of-sync problems and also improve scalability. Since batch re-keying updates keys in certain period, it cannot entirely fulfill the Forward Access Control. In most situations, batch re-keying is acceptable if the update period is not too long. However, individual re-keying is more suitable than batch re-keying in certain situations.

In the key tree model proposed by Wong et al. [7], the group membership is maintained by a Group Controller (GC). There are three kinds of keys: the group key, individual keys and auxiliary keys, all of which are maintained through a key tree by the GC. The group key, a unique symmetric key (traffic encryption key, shorted as TEK), is used to encrypt the content and the GC should notify all the group users about this key. The individual key is the key known by each individual user and the GC only. In the key tree, the root stores the TEK and each leaf stores an individual key, while the rest (non-root internal node) stores auxiliary keys (key encryption key, shorted as KEK). Whenever a user joins or leaves, the TEK should be updated and notified to the remaining users to guarantee the content security. Since all the users (leaves) know about the keys in the path from itself to the root, the GC should rekey all these keys when certain user change happens. The updating procedure is in bottom-up fashion. For batch re-keying, Graham et al. [2] studied the optimal key tree structure with the assumption that all the users have the same probability p of being replaced by a new user in the batch period, which is extended from [11]. They proved that when $p > 1 - 3^{\frac{1}{3}} \approx 0.307$, the star is optimal, and when $p \leq 1 - 3^{\frac{1}{3}}$, the branching degree of each non-root internal node has an upper bound four. Based on these findings, they provided an $\mathcal{O}(n)$ algorithm for constructing the optimal tree for n users with a fixed probability p. Based on the work of [2], Chan et al. [1] extended the model by introducing loyal users to batch re-keying. They showed that when $p \geq 0.43$, the star is optimal and similar structural results are obtained which enable a dynamic programming algorithm.

For individual re-keying, Snoeyink et al. [6] investigated the optimal tree structure with n leaves where the worst case single deletion cost is minimum. Their result shows that the optimal tree is a special kind of 2-3 tree defined as follows.

(1) When $n \geq 5$, the root degree is 3 and the number of leaves in three subtrees of the root differs by at most 1. When $n = 4$, the tree is a complete binary tree. When $n = 2$ or $n = 3$, the tree has root degree 2 and 3 respectively.
(2) Each subtree of the root is a 2-3 tree defined recursively.

Optimal trees with other user join/leave behaviors are studied in [4,5,8,9].

In this paper we study the average case and aim to find the optimal tree structures to minimize the average cost for a single deletion.

The remaining part of this paper is organized as follows. In Sect. 2, we describe our model and problem in detail. In Sect. 3, we prove that a ternary balanced tree is optimal in the normal case. Later in Sect. 4, we show that the same result still holds for one loyal user case. Finally, we conclude our work in Sect. 5.

2 Preliminaries

In the key tree model, each leaf represents a user. Every user has an individual key which is only known by the user itself and the group controller (GC). The root stores a group key called traffic encryption key (TEK), which is used for transmitting encrypted contents. Except for the root, other internal nodes store a key encryption key (KEK), which is used for updating TEK and shared by all its leaf descendants.

In individual re-keying scenario for a popular broadcasting service, a user will be replaced by a new user upon leaving. The GC will assign a new individual key to the new user. At the same time, the GC needs to update all the keys which belong to the ancestors of the updating spot to ensure both Backward Access Control and Forward Access Control. The key updating process is done in bottom-up fashion. Suppose v is the parent node of the updated user and its original group key is k_v. At first, GC will assign the new user an individual key. After that, GC needs to update the key of node v. Assume that the new KEK for v is k_v^{new}. To inform all the children of v of k_v^{new}, the GC uses the individual key of every child of v to encrypt k_v^{new} and broadcasts the encrypted message. In this way, the old user is unable to get the new KEK k_v^{new}. Also, the new user is not aware of the old KEK k_v. In total, the GC needs to broadcast d_v messages, where d_v stands for the number of children of v. The updating process of $v's$ key is done. After that, the parent of v needs to update its key. The GC repeats the updating process until it comes to the root, where the stored TEK is updated.

If there is only one user leaving the system, this user can be any leaf in the tree. Notice that different user's leaving may bring different update costs. We are interested in the tree with minimum average updating cost. When the number of users in the key tree is n, then minimizing average updating cost is equivalent to minimizing total cost of updating any leaf.

Denote the set of leaves in key tree $T = (V, E)$ as $L(T)$ and denote $v's$ ancestor set to be $anc(v)$. Let d_v be the number of children of v. Throughout the paper, we also use degree to denote d_v. In other words, we use "degree" to mean "branching degree". For example, a node with degree 3 means that the node has three children. Our goal is to find a tree with n leaves to minimize the total updating cost $\sum_{v \in L(T)} \sum_{u:u \in anc(v)} d_u$. We use $OPT(n)$ to denote the total updating cost of such a tree with n leaves. Alternatively, this total updating cost can also be expressed as $\sum_{u \in V - L(T)} d_u N_u$ where N_u is the number of leaf descendants of u. When we use the alternative expression of the total cost, we say the cost is contributed by internal nodes.

3   Optimal Tree Structures in the Normal Case

In this section, we will discuss the structures of the optimal tree. First of all, to minimize the cost, it is easy to see that degree 1 internal nodes cannot exist in the optimal tree. Then we start bounding the degree from above. In all the following figures, a small circle always represents a leaf.

Lemma 1. *There exists an optimal tree where every node has degree at most 3.*

Proof. Suppose there is an internal node u with degree ≥ 4 in the optimal tree. We can do the following structure change as shown in Fig. 1, where we add two internal nodes u_1 and u_2 as children of u and split the original children of u into two groups with one group being the children of u_1 and the other group being the children of u_2. Since the original degree of u is at least 4, we can guarantee that each group has at least 2 members.

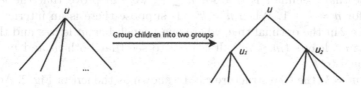

Fig. 1. Group branches

Before the change, the total cost is $C' + d_u N_u$, where $d_u N_u$ is the cost contributed by node u and C' represents the cost contributed by the other internal nodes. After the change, the total cost is $C' + 2N_u + d_{u_1} N_{u_1} + d_{u_2} N_{u_2}$. The change of cost ΔC can be calculated as

$$\begin{aligned}
\Delta C &= C' + 2N_u + d_{u_1} N_{u_1} + d_{u_2} N_{u_2} - (C' + d_u N_u) \\
&= d_{u_1} N_{u_1} + d_{u_2} N_{u_2} - (d_u - 2)N_u \\
&= d_{u_1} N_{u_1} + d_{u_2} N_{u_2} - (d_{u_1} + d_{u_2} - 2)(N_{u_1} + N_{u_2}) \\
&= (2 - d_{u_1})N_{u_2} + (2 - d_{u_2})N_{u_1} \\
&\leq 0
\end{aligned}$$

The last inequality holds because $d_{u_1} \geq 2$ and $d_{u_2} \geq 2$. Therefore, the above change does not increase the total updating cost of the tree. We can do this change as long as the tree still has some node with degree at least 4. Because every time we make the change, we increase the number of nodes in the tree by one, this process will finally stop because the total number of leaves in the tree is limited by n. The tree we obtain at that time will have no nodes with degree at least 4, which then proves the lemma.

Lemma 2. *There is an optimal tree where children of degree-2 internal nodes are all leaves.*

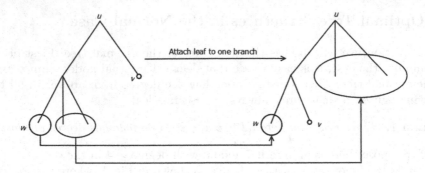

Fig. 2. Remove degree 2(1)

Proof. First of all, there is always an optimal tree satisfying Lemma 2 when $n \leq 6$ by enumeration.

Assume that Lemma 2 is true for $n \leq k$, we will prove that the lemma is still true for $n = k + 1$. When $n = k + 1$, suppose there is an internal node u with degree 2 in the optimal tree, where one branch has m leaves and the other branch has m' leaves ($m \leq m'$). It is easy to see that both m and m' are less than or equal to k.

When $m = 1$, the tree structure can be shown as the left of Fig. 2. According to our induction hypothesis, the branch with m' leaves can satisfy Lemma 2. The cost contributed from the subtree rooted at u is $C' + 3m' + 2(m' + 1)$, where C' is the cost contributed by the remaining parts of the subtree rooted at u. We can change the tree by attaching the one single leaf to be the sibling of one of the children w of the other branch and moving the original siblings of w to be the children of u as shown in Fig. 2. After the change, the cost becomes $C' + 3(m' + 1) + 2(N_w + 1)$. The change of cost ΔC can be calculated as

$$\Delta C = C' + 3(m' + 1) + 2(N_w + 1) - (C' + 2(m' + 1) + 3m')$$
$$= 2(N_w - m') + 3 < 0$$

The last inequality holds since each of other two branches have at least one leaf, which implies that $m' - N_w \geq 2$. Hence, this change can decrease the cost. We can keep pushing down the single leaf to the bottom.

When $m = 2$, the branch with m' leaves can satisfy Lemma 2, which implies that root degree of that subtree is 3. We change the tree as shown in Fig. 3. Before the change, the cost is $C' + 2(m' + 2) + 3m' + 4$. After rearranging the children of two branches, the cost become $C' + 3(m' + 2) + 2(m' + 1)$. The change of cost ΔC can be calculated as

$$\Delta C = C' + 3(m' + 2) + 2(m' + 1) - (C' + 2(m' + 2) + 3m' + 4)$$
$$= 0$$

In this case, degree 2 is pushed down to the next level without increasing the tree cost. We can keep doing this until all the degree 2 nodes only appear at the bottom without increasing any cost.

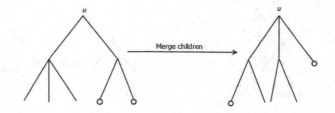

Fig. 3. Remove degree 2(2)

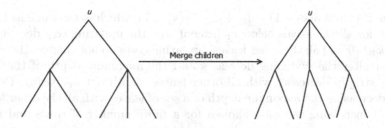

Fig. 4. Remove degree 2(3)

When $m \geq 3$, both branches satisfy Lemma 2. We change the tree as shown in Fig. 4. The cost before the change is $C' + 2(m + m') + 3m + 3m'$. After rearranging the children, the cost becomes $C' + 3(m + m') + 2(m + m')$. The change of cost ΔC can be calculated as

$$\Delta C = (C' + 3(m' + m) + 2(m' + m)) - (C' + 2(m' + m) + 3m' + 3m)$$
$$= 0$$

In this case, we can also push degree 2 to the next level if at least one of the children is not a leaf without increasing the cost.

Therefore, Lemma 2 still holds for $n = k + 1$ and this finishes the proof of the lemma using mathematical induction.

With the degree bounds obtained, we will continue to show a balanced property for the optimal tree in the following two lemmas.

Lemma 3. *The difference in depth of any two leaves in the optimal tree is less than or equal to 1.*

Proof. Assume that there are two leaves in the optimal tree with depth k_1 and k_2 ($k_1 > k_2 + 1$). According to Lemma 2, degree 2 nodes only have leaves as their children. Hence, updating a leaf in such a tree has cost $3(k - 1) + d$, where k is the depth of this leaf and d is the degree of the parent of this leaf. It is easy to see that a deeper leaf has a higher cost.

We can switch the leaf with depth k_2 (u_1) with the parent (v_1) of the leaf with depth k_1 (w) as shown in Fig. 5. Before the switching, the cost for updating u_1 and w is $3(k_2 - 1) + d_u + 3(k_1 - 1) + d_{v_1}$. After switching, the cost for updating

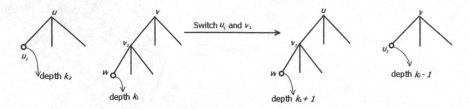

Fig. 5. Leaves on different levels

u_1 and w becomes $3(k_2 - 1) + d_{v_1} + d_u + 3(k_1 - 1)$, which is the same as before. However, for all the leaves below v_1 (except u_1), the updating cost decreases by $k_1 - 1 - k_2 > 0$ and all the other leaves' updating costs do not change. Hence, the total cost after the switching decreases and the maximum depth of the leaves below v_1 strictly decreases with all other leaves' depth not increasing. Because strictly decreasing the maximum depth of a set of leaves with all the other leaves' depth not increasing can only happen for a finite number of times and it will not cause loop in tree structure changes, the whole process can finally move all the leaves to have depth different at most 1, which proves the lemma.

Lemma 4. *There is an optimal tree where the number of leaves in different branches of root differs by at most 1.*

Proof. According to Lemma 3, all the leaves in the optimal tree are on two levels, k and $k + 1$. There could be several possibilities on a two level tree at bottom as shown in Fig. 6. The bottom level of structure c, d and g can have either 2 or 3 leaves.

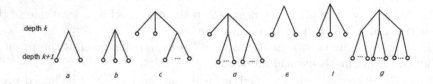

Fig. 6. Different structures at bottom

We first show that when b exists, c, d, e or f cannot exist in the optimal tree. Otherwise, we can remove one leaf from structure b and add it to be a sibling with some leaf on level k in e or f, or we can remove one leaf from structure b and change some c to d or change d to g. The decreased cost is $3k+3+2 = 3k+5$, while the increased cost is $3(k - 1) + 4 + 1 + 1 + 1 = 3k + 4$ when the receiving structure is f and $3(k - 1) + 3 + 1 + 1 = 3k + 2$ when the receiving structure is e and $3(k - 1) + 2 + 5 = 3k + 4$ when we are changing c to d or changing d to g. Thus, b cannot coexist with c, d, e or f in the optimal tree.

Hence, when structure b exists, no other structure except a (alternatively g) can exist. Therefore, all the leaves are on the same level in the optimal tree. We can exchange the structure a and b between branches without increasing the cost so that all the branches have nearly the same number of leaves and the tree is still optimal. Specifically, if in total we have m structure a and m' structure b in the tree, we must have $m + m' = 3^k$ because all the internal nodes on level $k - 1$ or above has degree 3. Let $m = 3p + q$ and $m' = 3p' + q'$, where $p, p' \in N$ and $q, q' \in \{0, 1, 2\}$. Since $m + m' = 3^k$, $(q + q') \bmod 3 = 0$. Thus, there are three possible value sets for $(q, q') : (0, 0), (1, 2)$ and $(2, 1)$. When (q, q') is equal to $(0, 0)$, we assign p structure a and p' structure b to each branch. Every branch has the same number of leaves. At the same time, $p + p' = \frac{m + m'}{3} = 3^{k-1}$. When (q, q') is equal to $(1, 2)$, we still assign p structure a and p' structure b to each branch. Then the one additional structure a goes to one branch, while the other two extra structure b go to the other two branches separately. Again each branch gets the same number of structures (3^{k-1} structures) and the difference in the number of leaves of three branches is less than or equal to 1. When (q, q') is equal to $(2, 1)$, we distribute p structure a and p' structure b to each branch. The extra two structure a and one structure b are allocated to each branch. All three branches have the same number of structures and the difference in the number of leaves is less than or equal to 1.

Without structure b, all the leaves in the deepest level are in structure a. We show that a and e cannot co-exist. Otherwise, we can remove a leaf from structure a and the cost is decreased by $3k + 4$. Then we add a leaf to e and the cost is increased by $3(k - 1) + 5 = 3k + 2$. This change decrease the total cost and therefore contradicts the optimality of the tree. This implies that all level $k + 1$ leaves have degree 2 parents while all level k leaves have degree 3 parents. If we switch structure a with any level k leaf, the total cost will stay the same. In this way, we can do the switch in the optimal tree until all the branches have nearly the same number of leaves. More specifically, suppose we have m structure a and m' single leaf which we want to distribute in the bottom level. Similarly, we must have $m + m' = 3^k$. Again we can set $m = 3p + q$ and $m' = 3p' + q'$, where $p, p' \in N$ and $q, q' \in \{0, 1, 2\}$. (q, q') has three possible value sets: $(0, 0), (1, 2)$ and $(2, 1)$. Like what we did with structure a and structure b, we firstly allocate p structure a and p' single leaf to each branch. When (q, q') is equal to $(0, 0)$, the distribution is done. All branches have the same number of leaves and structures. When (q, q') is equal to $(1, 2)$ or $(2, 1)$, we allocate three extra structures to each branch separately. The structures are distributed evenly and the difference in the number of leaves is less than or equal to 1.

Theorem 1. *For any n, a totally balanced ternary tree is an optimal tree.*

Proof. The theorem is implied by all the previous lemmas.

4 Optimal Tree Structures with Loyal Users

In this section, we consider the scenario with loyal users which was introduced by Chan et al. [1]. Loyal users refer to users who have no possibility of leaving the system and correspondingly normal users refer to users who will leave the system (in other words, normal users are the users we consider in the previous section). Thus, the updating cost of loyal user is 0. After introducing loyal users, we use n to represent the number of normal users and N_v to represent the number of normal leaves in the subtree rooted at v. We will explore the optimal tree structure in the presence of loyal users. In this section, we also use branching degree of v to represent the number of children of v. We also use normal user (loyal user) and normal leaf (loyal leaf) interchangeably.

Due to space limit and similarity of proofs, we move most of the proofs in this section to the Appendix.

4.1 Loyal User Node

We use the following definition of *Loyal User Node* (LUN) introduced in [1]. Loyal User Node is defined as an internal node whose children are all loyal users. We will prove that in optimal trees, there should be only one LUN and this LUN has all the loyal users as its children.

Lemma 5. *In the optimal tree, all the loyal users must be children of LUN unless there is only one loyal user.*

Proof. Assume that all the loyal users are distributed separately in the tree and there is no LUN. We can combine any two loyal users into one LUN. As showed in Fig. 7, we move the loyal user under v to be the sibling of the loyal user under u. After combining, the branching degree of u is still the same, so the updating cost of leaves under u stays the same. At the same time, the branching degree of v decreases by 1. Hence, the updating cost of each normal leaf under v decreases by 1.

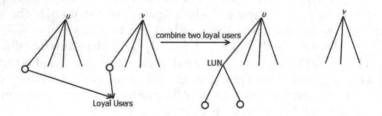

Fig. 7. Combine two loyal users

If there are already LUNs in the tree, we can attach loyal users to any LUN without increasing any cost. In Fig. 8, we move one separate loyal user w to be under one LUN. After moving, the whole LUN still has zero cost, while v, the original parent of w, has branching degree decreased by 1. Thus the updating cost of each normal leaf under v decreases by 1.

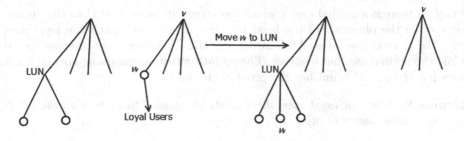

Fig. 8. Move one loyal user to LUN

Lemma 6. *If there are more than one loyal user, there is only one LUN in the optimal tree with all the loyal users as its children.*

Proof. According to Lemma 5, all the loyal users are under LUN. Assume that there are multiple LUNs in the tree. We can combine LUNs together and decrease the total cost. As showed in Fig. 9, the subtree rooted at v is not affected after combining, while w has less branching degree and each normal leaf descendent under w has its updating cost decreased by 1.

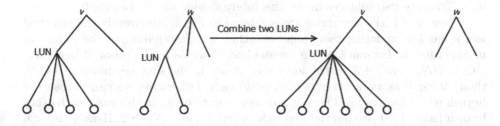

Fig. 9. Combine LUNs

We can keep combining LUNs until there is only one LUN in the tree.

If we replace one LUN with a loyal leaf (loyal user), the parent of LUN is not affected. Also, the remaining part of the tree has the same updating cost as well. Hence, we only need to consider the case of one loyal leaf when calculating the optimal tree.

4.2 Optimal Tree Structures

In this section, we will explore the optimal tree structure with loyal users.

Lemma 7. *In optimal trees, if the loyal user becomes a normal one, then its updating cost is no less than the updating cost of any other normal user.*

Proof. If there is a normal user s which has larger updating cost than the updating cost of the place where the loyal user lies at, we can switch the loyal user with the normal user at s. After switching, the loyal user has the same cost 0 while the normal user has less cost. The updating cost of remaining normal users does not change. This finishes the proof of the lemma.

Lemma 8. *With one loyal user, there exists an optimal tree where all the internal nodes have degree at most 3.*

Proof. Like the case without loyal users, we suppose there is an internal node with degree ≥ 4 in the optimal tree. Repeat what we have done in Lemma 1. We can bound branching degree to 3 no matter where the loyal user is.

Lemma 9. *With one loyal user, there is an optimal tree where internal nodes whose children are not all leaves have degree 3.*

Proof. For $1 \leq n \leq 5$, we enumerate all the possible trees and there is always an optimal tree satisfying Lemma 9.

Like what we did in the normal case, we assume that Lemma 9 is true for $n \leq k$ and prove that Lemma 9 is still true when $n \leq k+1$. When $n = k+1$, suppose in the optimal tree, there is an internal node with degree 2 whose branches have m and m' leaves ($m \leq m'$). Obviously both m and m' are less than or equal to k. Thus the two subtrees under this internal node satisfy Lemma 9.

When $m = 1$, the tree structure is showed in Fig. 2. Apparently the attached leaf v has less updating cost than any other leaf descendants of the parent of v. According to Lemma 7, v is a normal leaf. Same as in the proof of Lemma 2, $\Delta C = 2(N_w - m') + 3$. If the loyal user is not in the subtree shown in Fig. 2, then $\Delta C < 0$ as we already proved previously. Otherwise, we can choose the branch which has the loyal user as a descendant to be w. In this case, each other branch has at least one normal user, which implies $m' - N_w \geq 2$. Hence, $\Delta C < 0$ still holds. We can keep pushing down the single leaf to the bottom without increasing the total cost.

When $m = 2$, We use Fig. 3 to do the merge. Before merging, the two leaves shown in the figure have the least updating cost in this subtree (there are some leaves with updating cost larger than them), which by Lemma 7 shows that neither of these two leaves is a loyal leaf. The cost before change and the cost after change are the same as we analyzed in Lemma 2. Thus, the cost stays the same after the merging and we can push down degree 2 to the next level without increasing the total cost.

When $m \geq 3$, we do the merging as shown in Fig. 4. No matter where the loyal leaf is, the merge can push down degree 2 without increasing any cost.

Therefore, Lemma 9 is true for $n = k+1$ with one loyal user and the lemma is proved using mathematical induction.

Lemma 10. *With one loyal user, the difference in depth of any two leaves in the optimal tree is less than or equal to 1.*

Proof. Again, we pick two leaves in the optimal tree with depth k_1 and k_2 ($k_1 > k_2 + 1$). If both leaves are normal, the switch in Fig. 5 can decrease the cost. If one of the leaves is loyal, that must be w. Before switching, the cost of u_1 is $3(k_2 - 1) + d_u$. After switching, the cost of u_1 is $3(k_1 - 1)$. $\Delta C = 3(k_1 - k_2) - d_u$. Considering one other leaf descendant of v_1, the cost before change is $C' + 3(k_1 - 1)$ where C' indicates the cost under level k_1. After change, the cost becomes $C' + d_u + 3(k_2 - 1)$. $\Delta C' = 3(k_2 - k_1) + d_u$. Obviously $\Delta C + \Delta C' < 0$. The change will decrease the cost and using similar arguments as before, we can show the correctness of the lemma.

Lemma 11. *With one loyal user, there is an optimal tree where the number of leaves in different branches of root differs by at most 1.*

Proof. As we showed in Lemma 10, all the leaves, including the loyal leaf, are on two levels, k and $k + 1$. Figure 6 shows different structures at the bottom of the optimal tree. Clearly, the loyal leaf is in structure a or b by Lemma 7.

Firstly, we show that structure b cannot coexist with c, d, e or f. When b exists, if it does not contain the loyal leaf, then we already proved it in Lemma 4. Otherwise, b contains the loyal leaf, then we can remove one normal leaf from structure b and add it to c, d, e or f so that c becomes d, d becomes g, e becomes f and f becomes c. One can verify that the removal of the normal leaf decreases cost by $3k + 4$, while the addition of that leaf to other structures increases the cost by at most $3k + 4$. Thus, as long as b coexists with c, d, e or f in the optimal tree, we can always remove either all b structure or all the c, d, e and f structures without increasing the cost.

If we removed all the c, d, e and f in the above process, then only a and b can exist in the bottom level, which means that all the leaves are on the same level in the optimal tree. According to the proof of Lemma 4, we have a way to assign structure a and b so that the difference in the number of leaves of three branches is less than or equal to 1. The loyal leaf can lie in any structure b.

If we removed all the b in the above process, then only a can be on level $k + 1$, and the loyal leaf must be in a structure a by Lemma 7. If the a does not contain the loyal leaf, we already showed in the proof of Lemma 4 that a and e cannot coexist and the remaining proof just carries on. Otherwise, a contains the loyal leaf, in this case, we can remove one normal leaf in a (at the same time decreasing the level of the loyal leaf by one), which decreases the cost by $3k + 4$. If we add the leaf to e and turn e into f, the cost increases by $3k + 2$. This change still decreases the cost. Hence, a and e also cannot coexist in this case and the remaining proof of Lemma 4 is still valid.

By the lemmas that we have proved above, we are able to show the following theorem.

Theorem 2. *For any number of normal users and one loyal user, the totally balanced ternary tree is an optimal tree.*

If we have more than one loyal user, we can generate the optimal tree with one loyal user first. Then we replace the loyal leaf with an LUN.

5   Conclusion

In this paper, we study the optimal key tree structure for minimizing the average updating cost for individual re-keying. We show that there is an optimal tree where the degree of all nodes are at most 3 and the number of leaves in three branches differs by at most 1. This property shows that a totally balanced ternary tree is optimal in minimizing the average updating cost for individual re-keying. The same result still holds for the case where some users never leave the system.

References

1. Chan, Y.K., Li, M., Wu, W.: Optimal tree structure with loyal users and batch updates. J. Comb. Optim. **22**, 630–639 (2011)
2. Graham, R.L., Li, M., Yao, F.F.: Optimal tree structures for group key management with batch updates. SIAM J. Disc. Math. **21**(2), 532–547 (2007)
3. Li, X.Z., Yang, Y.R., Gouda, M.G., Lam, S.S.: Batch rekeying for secure group communications. In: WWW 2001 Proceedings of the 10th International Conference on World Wide Web, pp. 525–534. ACM, New York (2001)
4. Chen, Z.-Z., Feng, Z., Li, M., Yao, F.F.: Optimizing deletion cost for secure multicast key management. Theor. Comput. Sci. **401**(1–3), 52–61 (2008)
5. Li, M., Feng, Z., Zang, N., Graham, R.L., Yao, F.F.: Approximately optimal trees for group key management with batch updates. Theor. Comput. Sci. **410**, 1013–1021 (2009)
6. Snoeyink, J., Suri, S., Varghese, G.: A lower bound for multicast key distribution. In: Proceedings of the Twentieth Annual IEEE Conference on Computer Communications, pp. 422–431 (2001)
7. Wong, C.K., Gouda, M.G., Lam, S.S.: Secure group communications using key graphs. ACM SIGCOMM Comput. Commun. Rev. **28**(4), 68–79 (2000)
8. Wu, W., Li, M., Chen, E.: Optimal tree structures for group key tree management considering insertion and deletion cost. Theor. Comput. Sci. **410**, 2619–2631 (2009)
9. Wu, W., Li, M., Chen, E.: Optimal key tree structure for two-user replacement and deletion problems. J. Comb. Optim. **26**(1), 44–70 (2013)
10. Yang, R.Y., Li, X.S., Zhang, X.B., Lam, S.S.: Reliable group rekeying: a performance analysis. In: ACM SIGCOMM Computer Communication Review - Proceedings of the 2001 SIGCOMM Conferrence, vol. 31, no. 4, pp. 27–38 (2001)
11. Zhu, F., Chan, A., Noubir, G.: Optimal tree structure for key management of simultaneous join/leave in secure multicast. In: Proceedings of Military Communications Conference, pp. 773–778. IEEE Computer Society, Washington, DC (2003)

Cascading Critical Nodes Detection with Load Redistribution in Complex Systems

Subhankar Mishra[1]($\boxtimes$), Xiang Li[1], My T. Thai[1], and Jungtaek Seo[2]

[1] Department of Computer and Information Science and Engineering,
University of Florida, Gainesville, FL 32611, USA
{mishra,xixiang,mythai}@cise.ufl.edu
[2] The Attached Institute of ETRI, Seoul, Korea
seojt@ensec.re.kr

Abstract. In complex networked systems, the failures of a few critical components will cause a large cascade of component failures because of operational dependencies between components, resulting in the breakdown of the network. Therefore, it is crucial to identify these critical nodes in the study of complex network vulnerability under cascading failure. Unfortunately, we show that this problem is NP-hard to be approximated within a ratio of $O(n^{1-\epsilon})$. Accordingly, we design two approaches to solve this problem. The first one estimates the cascading potential of each node while the second one measures the cooperated impact of node failures under an ordered attack. Since smart-grids is an important complex networked infrastructure, we also demonstrate some safety setting for power grids using the designed algorithms.

Keywords: Complex network vulnerability · Cascading failure · Smart grids · Inapproximability

1 Introduction

The complex networked systems are now essential parts of the modern society. Unfortunately, they are extremely vulnerable to attacks. The failure of a part of the system propagates the local vulnerability to the rest, triggering the failure of the successive parts and hence leading to a much more devastating consequence. This is easily understood by the fact that failure of few nodes may change the balance of flows, which leads to avalanche also known as cascade of node failures because of the global redistribution of load over the whole network.

The cascading failure is observed in various social, economic and physical scenarios [1]. Let us consider a power grid, one of the critical infrastructure networks. Unfortunately it is exposed to the cascading failure, where one of the elements fails and distributes its load to the neighboring elements in the system. These receiving elements are pushed beyond their capacity, so they become overloaded and distribute their load to other elements. This iteration continues and leads to massive cascading failure of power grids. A single overloaded

© Springer International Publishing Switzerland 2014
Z. Zhang et al. (Eds.): COCOA 2014, LNCS 8881, pp. 379–394, 2014.
DOI: 10.1007/978-3-319-12691-3_29

line in a power transmission network can lead to a blackout spanning millions of homes [2]. Similar scenario is also observed in the information traffic on the Internet because of overload due to congestion or breakdown [3,4], in traffic networks, due to congestion by traffic accidents or road work, in protein interaction network because of the removal of protein or substrate [5].

The cascading failure has attracted a lot of attention and been studied in various perspective [6–9]. The structural vulnerability of power networks was studied in [7]. The authors showed that removing small fraction of highest degree nodes significantly reduces the connectivity of the network. Various models of cascading failures were later proposed to study the vulnerability of networks under the targeted attack [8,9].

However, these works mainly present different ranking methods for nodes and select the highest ranked nodes as the critical ones. These methods fail to address the effect of the cascading process and the ordered of nodes being attacked. Therefore, this paper introduces a new optimization problem, namely Cascading Critical Nodes Detection, which asks us to find an ordered subset of nodes in a given network, whose deletion (being attacked) in that order will cause the maximum damage to the whole network under cascading failures. This identification in turn helps us in designing better protection of the system. Unfortunately, it is computationally very challenging to identify such a subset due to the complicated interdependencies between the entities in the network.

In this paper, we show that this problem is NP-hard to be approximated within a ratio of $O(n^{1-\epsilon})$. We next design a novel metric, called cascading potential to rank the critical level of each node in the case of cascading. Based on this metric, we further develop an effective algorithm to identify this ordered subset. Even though the above idea worked very well under normal conditions, nodes with high load impact get selected without producing much damage to the network under the robust network condition. Hence, we develope another metric, called impact efficiency, which increases the number of failed nodes and avoids redistribution of load to impossible-to-fail nodes, and thus leading to cascade failure. The second metric is crucial in the case of cooperated attacks. In the evaluation section, we analyze the above algorithms in different safety settings and find the latter algorithm outperforming the former in the high failure tolerance network setting. We also take a step further and discuss the importance of safety setting and system tolerance with respect to network robustness.

The rest of the paper is structured as follows. Section 2 presents the network models and our problem formulation. The inapproximability is shown in Sect. 3. The two approaches are presented in Sects. 4 and 5 respectively. Section 6 discusses the safety issue of power networks and finally Sect. 7 concludes our paper.

2   Network Model and Problem Formulation

2.1   Graph Notations

The network is modeled by a weighted directed graph $G = (V, E)$ with the node set V of $|V| = n$ nodes and the edge set E of $|E| = m$ oriented connections between nodes. Each edge (u,v) is associated with a weight $w(u,v)$ representing the operating parameter of the network. The higher $w(u,v)$ is, the more load is distributed from u to v. In addition, each node u has a current load $L(u)$ and a capacity $C(u)$. The capacity $C(u)$ is the maximum load that node u can handle. Finally, we denote the set of incoming neighbors and outgoing neighbors of u by N_u^- and N_u^+, respectively.

2.2   Cascading Failure Model

In this paper, we adopt the Load Redistribution model (LR-model) which was widely used in the research community [8,10]. In this model, nodes are failed in the cascading manner due to the load redistribution of failed nodes. Initially, a set of nodes S are failed, then the failures are propagated to other nodes in time steps. When node u fails, its load is redistributed to its neighbors as illustrated in Fig. 1. Each alive neighbor v will receive an additional load which is proportional to weight $w(u,v)$ of edge from u to v. Precisely, v will receive additional load:

$$\Delta L(v) = L(u) \times \frac{w(u,v)}{\sum_{z \in N_u^+} w(u,z)}$$

Due to the load redistribution, the load of some nodes will exceed their capacities resulting in the failure of those nodes in the next time step. The process of load redistributing and node failing will stop when there are no more failed nodes. The set of failed nodes caused by the initial failure of S is denoted by $F(S)$.

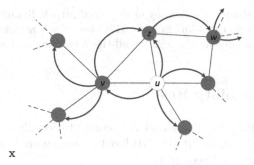

Fig. 1. When node u fails, its load is redistributed to the neighbor nodes. Among these nodes, v receives a high portion of load from u and becomes overloaded. The load of v is redistributed to its neighbors which makes z fail. Finally, the load from z continues to cause w fail and the cascading process stops

2.3   Attack Scenario

Due to the cascading failures, the failure of a small set of nodes S can result in a catastrophic number of failed nodes. These nodes become the targets for attackers, thus we need to identify them in order to protect the network. Additionally, given the same set of attacked nodes, different attacking orders lead to different outcomes. With the same attacking cost, the attacker can choose the best order with suitable time for each subset of attacked nodes. For simplicity, we consider each subset of size one in this paper, thus each node is attacked one after another. An example of the efficiency of this strategy is illustrated in Fig. 2. As can be inferred, not the simultaneous attack, but the order of attack determines the cascading effect. In particular, given an ordered set $S = \{s_1, s_2, \ldots, s_k\}$, a node s_i is attacked after the cascading process caused by attacking s_{i-1} stops. Let us denote $F(S)$ as the number of failed nodes when subsets in S are attacked serially and formally define the problem as follows: The next node is taken down when the cascading process stops. In particular, given an order set $S = \{s_1, s_2, \ldots, s_k\}$, node s_i is attacked after the cascading process caused by attacking s_{i-1} stops.

Definition 1 (Cascading Critical Nodes (CasCN) Identification Problem). *Given a network $G = (V, E)$ with their weighted, capacity, and load functions, and a positive integer k, the problem asks to find an **ordered** subset $S \subseteq V$ of size $|S| = k$ so as the serial failures of nodes in S maximizes the number of failed nodes $F(S)$ under the LR-model.*

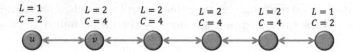

Fig. 2. An instance about the efficiency of the serial attack. Simultaneously attacking any two nodes causes no cascading failures. However, if u is attacked first and then v, the failure will be cascaded and as a result all the remaining nodes will also fail.

3   Inapproximability Results

Unfortunately, finding the ordered set S is computationally hard. Not only that the problem is NP-hard, but it also NP-hard to approximate it within the ratio of $O(n^{1-\epsilon})$, as shown in Theorem 1.

Theorem 1. *It is NP-hard to approximate the CasCN problem within a ratio of $O(n^{1-\epsilon})$ for any constant $1 > \epsilon > 0$ unless $P = NP$.*

Proof. We use the gap-introduction reduction [11] to prove the inapproximability of the CasCN problem. Using a polynomial time reduction from Set Cover, to the CasCN problem, we show that if there exists a polynomial time algorithm that approximates the later problem within $O(n^{1-\epsilon})$, then there exists a polynomial time algorithm to solve the former problem.

Definition 2 (Set Cover problem). *Given a universe $\mathcal{U} = \{e_1, e_2, \ldots, e_n\}$, a collection of subsets $\mathcal{S} = \{S_1, S_2, \ldots, S_m\} \subseteq 2^{\mathcal{U}}$, and an integer k, the Set Cover problem asks whether or not there are k subsets whose union is $\mathcal{U}$.*

Instead of using the hardness result of the general Set Cover problem, we use the result on a restricted variant MIN3SC2 of the Set Cover problem where the sizes of subsets are at most 3 and each element appears in exactly two subset [12].

Reduction. Given an instance of the Set Cover problem $\mathcal{I} = (\mathcal{U}, \mathcal{S}, k)$ where each element appears in exactly two subsets, $n_1 = |\mathcal{U}|$ and $m_1 = |\mathcal{S}|$, we construct an instance $\mathcal{I}'$ of the CasCN problem as illustrated in Fig. 3.

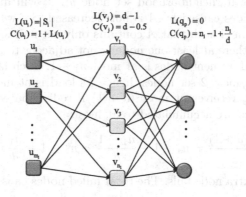

Fig. 3. Reduction form MIN3SC2 to CasCN

The node set V. Add a *set node* u_i for each set $S_i \in \mathcal{S}$, an *element node* v_j for each element $e_j \in \mathcal{U}$, and $d = (n_1 + m_1)^{\frac{2}{\epsilon}}$ *extra nodes* $q_1, q_2, \ldots, q_d$.

The edge set E. Add edge (u_i, v_j) if the element v_j in the set S_i. In addition, there is an edge from v_j to q_p $\forall 1 \le j \le n_1, 1 \le p \le d$. All edges have the weight of 1.

Node load and capacity. The load and capacity of the set node u_i are $L(u_i) = |S_i|$ and $C(u_i) = 1 + L(u_i)$. The load and capacity of the element node v_j are $L(v_j) = d - 1$ and $C(v_j) = d - 0.5$. All extra nodes have the load of 0 and capacity of $n_1 - 1 + \frac{n_1}{d}$.

Next, we prove that if $\mathcal{I}$ has a set cover of size k then there exist a seeding set $A \subset V$ such that $|F(A)| > d$. Otherwise, for all $A \subset V$ and $|A| \le k$, $|F(A)| < n_1 + m_1$.

Assume that $\mathcal{I}$ has a set cover $\mathcal{SC}$ of size k, then we will select the set $A = \{u_i | S_i \in \mathcal{SC}\}$ as the seeding set. Initially, each node $u_i \in S$ redistributes 1 unit of load to each of its $|S_i|$ neighbors. Since each element is covered by at least one set in $\mathcal{SC}$, each node v_j receives the additional load of at least 1. At the next round, v_j has the load at least $L(v_j) \ge ((d-1)+1)$, which is higher than its capacity, and is failed. When v_j fails, it equally redistributes $L(v_j)/d \ge 1$ load to its d extra neighbor nodes. The load of each extra node q_p after receiving load from all failed element nodes is $L(q_p) \geq n > C(q_p)$, hence all extra nodes are failed. The cascading process stops with $|F(A)| = d + n_1 + k$ failed nodes.

In the case $\mathcal{I}$ has no set cover of size k, we will show the optimal seeding set can cause at most $n_1 + k$ nodes to fail in $\mathcal{I}'$. Let A be an arbitrary optimal seeding set. We observe that a set node only fails when it is selected in the seeding set since it has no incoming edge. Thus, there are at least $m_1 - k$ set nodes which are not in A. We can replace any extra node $q_p \in A$, if there are any, by a unselected set node without decreasing the number of failed nodes. Next suppose, there exists an element node $v_j \in S$, we can also replace it by a set node. If v_j is adjacent to some node $u_i \in A$, we can remove v_j from A while maintaining the same number of failed nodes. If v_j is not adjacent to any node in A, we just replace v_j by one of its neighborhood set node u_i. u_i will make v_j fail, so the number of failed nodes caused by A is not decreased. So, we can replace extra and element nodes in A such that A contains only set nodes. Since, there is no set cover of size k, there at least one node is not adjacent to any node in A, i.e, the number of failed element nodes is at most $n_1 - 1$. Each failed element node v_j is adjacent to at most 2 set nodes, hence its load is at most $L(v_j) \leq d + 1$. Each extra node q_p receives at most $(d + 1)/d$ redistributed load from failed element nodes which are accumulated to at most:

$$\frac{(n_1 - 1)(d + 1)}{d} = n_1 - 1 + \frac{n_1 - 1}{d} < n_1 - 1 + \frac{n_1}{d} = C(q_p)$$

Thus, there is no extra node fails. The total failed nodes caused by A is at most $n_1 + k < n_1 + m_1$.

Now suppose that we have polynomial algorithm $\mathcal{A}$ which approximates CasCN problem within $n^{1-\epsilon}$, we can decide the set cover problem as follows. For any instance $\mathcal{I}$ of the Set Cover problem, we construct the instance $\mathcal{I}'$ as above in polynomial time as d is a polynomial function of n_1 and m_1. Now, if $\mathcal{I}$ has a set cover of size k, the optimal A_{opt} seeding set causes at least d nodes fail in $\mathcal{I}'$. The algorithm $\mathcal{A}$ approximate the optimal solution within $n^{1-\epsilon}$ ($n = n_1 + m_1 + d$, the number of nodes in $\mathcal{I}'$), so it finds a seeding set $\mathcal{A}(\mathcal{I}')$ whose causes at least $(m_1 + n_1)$ nodes fail:

$$|F(\mathcal{A}(\mathcal{I}'))| \geq \frac{|F(A_{opt})|}{n^{1-\epsilon}} > \frac{d}{(d+m+n)^{1-\epsilon}}$$
$$> \frac{d}{(2d)^{1-\epsilon}} > \frac{d^\epsilon}{2}$$
$$= \frac{(m_1 + n_2)^2}{2} > m_1 + n_1$$

On the other hand, if $\mathcal{I}$ has no set cover of size k, then the optimal seeding set A_{opt} of $\mathcal{I}'$ causes less than $(m_1 + n_1)$ nodes fail. We have:

$$|F(\mathcal{A}(\mathcal{I}'))| \geq |F(A_{opt})| < (m_1 + n_1)$$

It implies the $\mathcal{I}$ has a set cover of size k if and only if $|F(\mathcal{A}(\mathcal{I}'))| > (m_1 + n_1)$. Hence, we can use $\mathcal{A}$ to decide the Set Cover problem in polynomial time i.e. $P = NP$. $\qquad\square$

4   Cascading Potential and Derived Algorithms

In this section, we introduce a new metric to measure the node importance under the cascading failure, and then use it as a criterion to design efficient algorithms for CasCN. When multiple nodes are attacked in the network, the effect of the early nodes is strengthened by the later nodes. Hence, we introduce a new metric which considers both direct and mutual impact of the node.

4.1   Cascading Potential

The cascading potential of a node is defined as combination of all possible impacts a node causes in the network under the cascading effect. Let's consider the failure of node u. For any other node v, there are two possible impacts that u can induce on v:

- *Failure impact.* The failure of u leads to the failure of v.
- *Load impact.* The failure of u makes the load of v increase, but not enough to fail.

The overall *failure impact* and *load impact* of u in the network are defined as the number of failed nodes and the total increase in load of the unfailed nodes, respectively. The *cascading potential* of u is the linear combination of these factors:

$$C(u) = \frac{|F(\{u\})|}{n} + \frac{\sum_{v \in V - F(\{u\})} \Delta L_u(v)}{\sum_{v \in V - F(\{u\})} (C(v) - L(v))} \qquad (1)$$

where $F(\{u\})$ is the set of failed nodes when u fails and $\Delta L_u(v)$ is the additional load that v receives due to the failure of u.

In this formula, we normalize both the failure and load impacts to avoid the unit difference. The failure impact is divided by the number of nodes, hence is at most 1 when all other nodes fail. Similarly, the load impact is divided by the total of capacity-load difference of unfailed nodes and achieves the maximum value 1 when all remained nodes are at the edge of failure, i.e., the most vulnerable state of the network.

The role of the load impact. In the formulation of the cascading potential, the load impact plays an important role in providing a better assessment of the network vulnerability comparing to the metric in [9]. If only one node is attacked, it is obvious to choose the node which maximizes the number of failed nodes. However, when multiple nodes are attacked, we need to consider the co-impact of attacked nodes to trigger a large size cascading failure. The load impact is acting as a bridge connecting the impact of these nodes. If the total load impact of earlier attacked nodes is high, the load of remaining nodes is close to the capacity. As a result, later attacked nodes can make more nodes to fail easily. For example, if u has the maximum load impact of 1 and the network is strongly connected, then attacking any node after u can take down the whole network. Thus, the cascading potential evaluate the importance of nodes more comprehensively.

4.2   Adaptive Cascading Potential (ACP) Algorithm

Intuitively, we can use cascading potential directly to design an algorithm for CasCN. We first compute the cascading potential of all nodes, then select top k as attacked nodes. Although this method runs fast, but it neglects an important property. Let consider two nodes u and v which both have failure impact on node z. If u is selected before v, then v has no impact on z as z is already failed. As a consequence, some nodes that have a high impact initially may have a little impact at the later stage. To handle this situation, we propose to update the impact of remained nodes on the fly. More specifically, at the ith iteration, the impact (failure or load impact) of node u on failed nodes (due to the selection of first $i-1$ attacked nodes) will be subtracted from the initial impact of u. After that, the node with the highest remained impact will be selected.

The crucial problem is how to update the impact of nodes efficiently. A naive way of keeping the impacted nodes list at each node u will result in $\Omega(n^3)$ running time for each iteration, which is very time consuming. We reduce the time required to update impact over entire graph by reversing the process. Each node v will keep two lists of nodes: the list $FI[v]$ contains nodes which have failure impact on v and the list $LI(v)$ contains nodes which have load impact on v. Since the load impact of other nodes on v are different, we use $LI[v][u]$ to store the load impact of u on v after the normalization. We calculate $LI[v][u]$ as follows:

$$LI[v][u] \leftarrow \frac{\Delta L_u(v)}{\sum_{z \in V - F(\{u\})}(C(z) - L(z))}$$

When v is failed, the impact of nodes in its lists will be updated. The crucial point is that each node only fails once, thus the running time is reduced significantly. Since each node has impact on at most n nodes, the total size of all FI and LI lists are at most n^2. The number of updates is bounded by the total size of FI and LI lists. Therefore the total running time is $O(nm + n^2 + kn)$.

We can fully update the cascading potential of each node as follows. After selecting a new node, we simulate the cascading failure triggered by it and obtain a new graph of remaining nodes. In this graph, the load of a node is the load when the cascading process stops. We then evaluate the cascading potential of

---

Algorithm 1. ACP Algorithm

Data: A network $G = (V, E)$, an integer k.
Result: A set S of k attacked nodes.
1 Initialize $S \leftarrow \emptyset$
2 **for** $i = 1$ *to* k **do**
3   Compute the cascading potential of all nodes in G by [1]
4   Select u as the node with highest cascading potential
5      $S \leftarrow S \cup \{u\}$
6   Update node loads and remove all failed nodes in G with the failure of u
7 **Return** S

---

all nodes in the updated graph and select one with the highest value. We present the algorithm in Algorithm 1.

Time complexity. We need to compute the cascading potential of all nodes to select a new one with time $O(nm)$. Thus the total running time is $O(kmn)$. However, the algorithm may run much faster than the worst case time since the size of the updated graph decreases when a new node is selected.

5 Impact Efficient Algorithm

We now turn our attention to a case where the nodes have high failure tolerance, thus the network is more robust. In this case, the gap between the capacity and load is big and the failure impact of each node is small. Note that ACP tends to select nodes with high load impact. If the load of these nodes are scattered to many nodes, and the loads do not accumulate together in order to cause further failures, we just end up with a large number of nodes whose loads are increased and only a few failed nodes. Hence, we need a better strategy which builds a strong connection between selected nodes to increase the number of failed nodes, thereby maximizing the failure impact instead of the load impact. To fulfill this goal, the new strategy should satisfy following features:

- The redistributed load of selected nodes should be concentrated on certain nodes to fail them. If early selected nodes redistributed load to a set of nodes, then later selected nodes should also redistribute load to this set. It is said that selected nodes are cooperating in redistributing load to make more nodes to fail.
- Selected nodes should cooperate to make high load nodes fail. The failure of high load nodes can expand the cascading failures further. However, if high load node preference reduces the number of failed nodes, the new strategy should not blindly favoring to fail high load nodes.

We design a new evaluation function, called the impact efficiency, of nodes with properties that tailor the selection process to embrace both desired features. Firstly, we give higher evaluation to nodes which redistributes its load to load-increased nodes. If the failure of u pushes an additional load $\Delta L_u(v)$ on v, then the impact of u on v is defined by:

$$\gamma(u,v) = \frac{\Delta L_u(v)}{C(v) - L(v)}$$

when $\Delta L_u(v) + L(v) \leq C(v)$. Since it requires $C(v) - L(v)$ additional load to make v fail, we can interpret that u makes a fraction $\frac{\Delta L_u(v)}{C(v)-L(v)}$ of v fail.

The new impact function implies that if the more load a node has already received, the more impact it received under the same additional load. On the other hand, the evaluation of u is higher if the loads of its neighbors are increased. This implication is stated in Proposition 1.

Proposition 1. *For any node v at two points of time, if v receives more load at the second time point, i.e., $L_2(v) > L_1(v)$, then the impact of other node u with the same redistributed load ΔL is higher at the second time point: $\gamma_2(u,v) \geq \gamma_1(u,v)$.*

Proof. We consider following cases:

Case 1. $L_2(u) + \Delta L > L_1(u) + \Delta L > C(u)$, then:

$$\gamma_2(u, \Delta L) = \gamma_1(u, \Delta L) = 1$$

Case 2. $L_2(u) + \Delta L > C(u) \leq L_1(u) + \Delta L$, then:

$$\gamma_2(u, \Delta L) = 1 > \alpha \geq \gamma_1(u, \Delta L)$$

Case 3. $C(u) \geq L_2(u) + \Delta L > L_1(u) + \Delta L$, then

$$\gamma_2(u, \Delta L) = \frac{\alpha \Delta L}{C(u) - L_2(u)} > \frac{\alpha \Delta L}{C(u) - L_1(u)} = \gamma_1(u, \Delta L)$$

□

We assume that u redistributes the same load on v in the Proposition 1, i.e. the load of u is the same at two points of time. In fact, the load of u may increase due to the selection of previous nodes, thus the evaluation of u increases even more at the second point of time.

To fulfill the second feature, we assign higher values to high load nodes which are impacted. The value of a node with load L is:

$$\sigma(L) = \frac{e^L}{1 + e^L}$$

The function $\sigma(L)$ is monotone increasing and in the range $0.5 \leq \sigma(L) < 1$. The monotone increasing of the function shows the preference toward high load nodes. Recall that the main goal is to increase the number of failed nodes, so even nodes with the lowest load have the value at least half of the highest value nodes.

Next, we will define the efficiency of selecting u via the impact on v. Intuitively, u makes $\gamma(u,v)$ fraction of v fail and v has value of $\sigma(L(v))$, thus the efficiency of u represented on v is:

$$\lambda(u,v) = \gamma(u,v)\sigma(L(v))$$

Finally, we should take into account the number of failed nodes when evaluating node u. The overall efficiency of u is the total of the number of failed and the efficiency on unfailed nodes:

$$\lambda(u) = |F(\{u\})| + \sum_{v \in V \setminus F(\{u\})} \lambda(u,v) \tag{2}$$

The efficiency evaluation shows several notable properties which serves our design goal as followings:

Increase the number of failed nodes first. If u makes z fail and has efficiency $\lambda(u, v)$ on the unfailed node v, then the contribution of z to the overall efficiency of u is always higher than v since $1 \geq \gamma(u, v)\sigma(L(v))$.

Avoid redistributing load to impossible-to-fail nodes. If node v needs too much additional load before failing, it will be ignored in efficiency evaluation of nodes as stated in the Proposition 2.

Proposition 2. *Given two nodes u and v with fixed load $L(v)$, the efficiency of u on v is monotone decreasing and goes to 0 when the capacity $C(v)$ of v increases and goes to infinity.*

Proof. It is easy to see that $\gamma(u, v)$ is monotone decreasing and goes to 0 when $C(v)$ increases and goes to infinity. In addition, $\sigma(L(u))$ is a constant, so the efficiency $\lambda(u, v) = \gamma(u, v)\sigma(L(v))$ decreases and goes to 0. $\qquad\square$

Not favoring high load nodes with all cost. We consider the case the capacity is linear to the load, a common setting in the reality to guarantee the safety of nodes. In this case, even the load of node v is extremely large, it is still ignored as shown in Preposition 3.

Proposition 3. *Suppose that the capacity $C(v)$ is linear to the load $C(v) = T * L(v)$ with constant factor T. Then, the efficiency of any node u on v goes to 0 when the load $L(v)$ goes to infinity.*

Proof. We have:
$$\gamma(u, v)\sigma(L(u)) = \frac{\Delta L_u(v)}{C(v) - L(v)} \frac{e^{L(v)}}{1 + e^{L(v)}}$$
$$< \frac{\Delta L_u(v)}{(T-1)L(v)}$$

The function goes to 0 when $L(u)$ goes to infinity. $\qquad\square$

Based on the efficiency evaluation, we propose the Impact Efficient (IE) algorithm with the same manner as ACP. The algorithm also selects nodes one by one. After updating the state of the network, the node with the highest efficiency is selected. The whole algorithm is described in Algorithm 2.

Algorithm 2. Impact Efficient (IE) Algorithm

    Data: A network $G = (V, E)$, an integer k.
    Result: A set S of k attacked nodes.
1 Initialize $S \leftarrow \emptyset$ for $i = 1$ to k **do**
2   Evaluate the efficiency of all nodes in G based on [2]
3   Select u as the node with the highest efficiency
4      $S \leftarrow S \cup \{u\}$
5   Update node loads and remove all failed nodes in G with the failure of u
6 **Return** S

---

6   Experimental Evaluation

In this section, we demonstrate the experimental results of our proposed algorithms. For the experimental evaluation, we choose the scenario of complex network of power grid and experiment on both synthesized and real power networks. We first test the performance of the proposed algorithms in the comparison with current attacking strategies in the literature [8,9]. These strategies sort nodes based on some criterion and select top k nodes as attacked nodes. The sorting criteria are: (1) Highest load (HL); (2) Lowest load (LL); (3) Highest percentage of failure (POF). The percentage of failure of a node u is the fraction of nodes that fails when u fails; and (4) Highest risk if failure (RIF). RIF of a node u is the ration between its load and the total load of its neighbor nodes.

For the dataset, we tested both on real and synthesized networks.

Real networks. We use the Western North American (WNA) power grid network [13] with 4941 stations and 6594 transmission lines to run experiments. However, the dataset is lacking of load and capacity information of nodes, thus we use the similar method in [10] to assign the load and capacity for each node. The initial load of node u is given by $L(u) = d(u)^\beta$, where $d(u)$ is the total of incoming and outgoing degrees of u and β is a constant parameter. This assignment method is reasonable as the load of a node is shown to have high correlation with its degree [14]. In all experiments, the default value of β is 1, unless otherwise mentioned. Node capacities are assigned based on three different schemes:

Normal networks. In normal networks, the capacity $C(u)$ of each node u is proportional to its initial load $L(u)$:

$$C(u) = T * L(u)$$

where T is a constant representing the *system tolerance*. The larger T is, the more robust the network is under the cascading failure.

Safe networks. In safe networks, the node capacities are assigned in two phases. First, the capacity $C(u)$ of each node u is scaled as the normal network. Then capacities of all nodes are raised to satisfy the $N - 1$ failure tolerance criterion in which the failure of any node will cause no additional failed nodes. It means that any node u will not fail when it receives the redistributed load from any of its neighbor. The capacity of u will be:

$$C(u) = \max\{T * L(u), \max_{v \in N_u^-} \{L(u) + L(v)\frac{w(v,u)}{\sum_{z \in N_v^+} w(v,z)}\}\}$$

Scaled Safe networks. In contrast to safe networks, scaled safe networks are formed by raising the node capacities to satisfy $N - 1$ failure tolerance criterion first, then be scaled up later. In particular, the capacity of u is made safe by assigning as follows:

$$C(u) = T * (\max_{v \in N_u^-} \{L(u) + L(v)\frac{w(v,u)}{\sum_{z \in N_v^+} w(v,z)}\})$$

Synthesized Networks. We also run the experiments on synthesized networks generated by Erdos-Renyi random network model [15]. Each network has 5000 nodes with the average degree of 4. The other parameters of the network are generated similarly to the above schemes.

6.1 The Performance of Different Algorithms

We first evaluate the performance of all algorithms with various network settings and system tolerance values. We carry experiments with different system tolerance values: low value $T = 1.2$, medium value $T = 1.6$, and high value $T = 2.0$. The comparison is shown in Figs. 4, 5, and 6. Overall, ACP and IE are competitive with each other and outperform other algorithms. When the network is easy to attack, the performance of the FACP algorithm is better but is not too far away from that of the CA algorithm. On WNA (synthesized) network with normal and safe settings and $T \le 1.6$, the number of failed nodes provided by two algorithms is almost the same. On WNA network with normal and safe settings with $T = 2$, the ACP algorithm starts to surpass IE when $k = 5$, achieves peak when $k = 10$ with 18 % better. When $k > 10$, the number of failed nodes is close to the total number of nodes, thus the gap between two algorithms is reduced. Under the scaled safe setting, the setting with highest failure tolerance, IE takes advantage over ACP when nodes are attacked. In WNA network with $T = 2$, IE makes a considerable fraction of nodes fail when the number of attacked nodes exceeds 11 while ACP almost makes no effect until $k = 19$. However, the number of failed nodes is the same when $k > 30$. The reason is that when a large number of nodes are failed, the load of remaining nodes are much higher which is the favor scenario for ACP. In the synthesized network with the scaled safe setting,

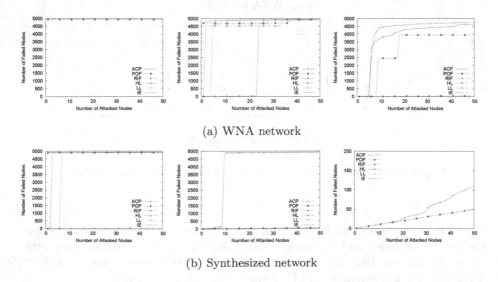

(a) WNA network

(b) Synthesized network

Fig. 4. Vulnerability of networks under the normal settings.

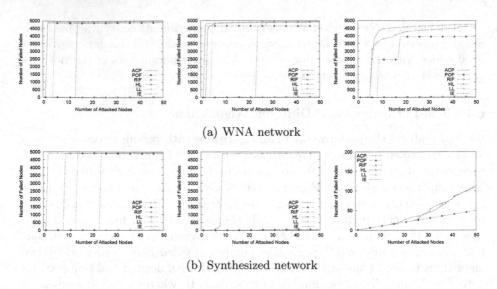

(a) WNA network

(b) Synthesized network

Fig. 5. Vulnerability of networks under the safety settings.

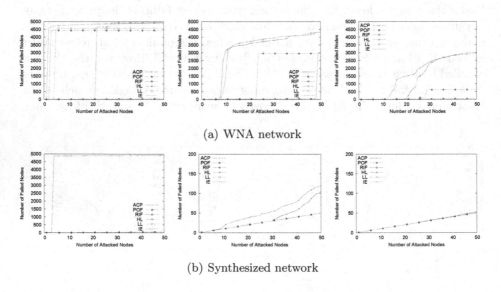

(a) WNA network

(b) Synthesized network

Fig. 6. Vulnerability of WSN network under the scaled safety settings.

$T = 1.6$, it is more difficult to attack the network. Since IE aims to increase the number of failed nodes as soon as possible, it produces better solution with about 20 % better than ACP. We can conclude that the better choice is the ACP if the network is vulnerable and IE if the network is more robust.

6.2 Network Robustness Under Different Settings

From the above figures, we observe that scaled safe networks are more robust than safe networks and safe networks are stronger than normal networks. However, there is a bias since with the same value of T, the same node in the scaled safe network has higher capacity than one in the safe and normal networks. Thus, we set up the experiment such that all types of networks have the same total capacity to verify the network robustness of each setting. We first generate the safe network, then we choose a suitable T value to generate normal and scaled safe networks such that the total capacity is the same. Note that the normal network has the larger value of T while the scaled safe network has the smaller value. Figure 7 shows that the scaled safe network is the most robust one. Normal and safe networks have the same robustness although the safe network can avoid the cascading failure after the first attack. Under the safe setting, a node can embrace the impact created by the failure of any neighbor (with the initial load) which is not true if its neighbors receive additional load. The failure of first nodes redistributed a certain load to other nodes which is enough to disqualify the $N - 1$ safety setting. Hence, the failure of later nodes triggers the same cascading failures in both normal and safe networks. Therefore, it recommends us to use the scaled safe setting to improve the robustness of the power grid instead of the $N - 1$ safety criterion.

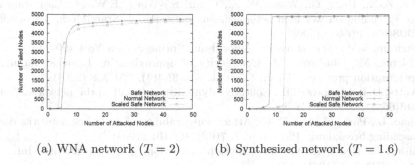

(a) WNA network $(T = 2)$ (b) Synthesized network $(T = 1.6)$

Fig. 7. Network robustness with different failure tolerance settings.

7   Conclusion

In summary, we study the critical node detection problem for cascading failures in complex networks. The identification of these nodes helps in locating the possible cooperated attacks, thereby providing a better protection scheme. We have shown the inapproximabilty and presented two algorithms. In addition, we use proposed algorithms to study the vulnerability of different safety settings. We discover that even with networks of the same topology, node load, and total capacity, the network safety depends a lot on the distribution of the protection cost (the gap between the capacity and the load).

Acknowledgment. This work is partially supported by NSF Career Award 0953284.

References

1. Watts, D.J.: A simple model of global cascades on random networks. Proc. Nat. Acad. Sci. (PNAS) **99**(9), 5766–5771 (2002)
2. Brummitt, C.D., D'Souza, R.M., Leicht, E.A.: Suppressing casades of load in interdependent networks. Proce. Nat. Acad. Sci. (PNAS) **109**(12), E680–E689 (2012)
3. Jacobson, V.: Congestion avoidance and control. SIGCOMM Comput. Commun. Rev. **18**(4), 314–329 (1988)
4. Petersen, I.: Fatal Defect Chasing Killer Computer Bugs. Vintage, New York (1995)
5. Sun, L., Wang, S., Li, K., Meng, D.: Analysis of cascading failure in gene networks. Front. Genet. **3**, 292 (2012)
6. Motter, A.E., Lai, Y.C.: Cascade-based attacks on complex networks. Phys. Rev. E **66**(6), 065102 (2002)
7. Albert, R., Jeong, H., Barabasi, A.L.: Error and attack tolerance of complex networks. Nature **406**(6794), 378–382 (2000)
8. Wang, J.-W., Rong, L.-L.: Cascade-based attack vulnerability on the us power grid. Saf. Sci. **47**(10), 1332–1336 (2009)
9. Wang, W., Cai, Q., Sun, Y., He, H.: Risk-aware attacks and catastrophic cascading failures in us power grid. In: 2011 IEEE Global Telecommunications Conference (GLOBECOM 2011), pp. 1–6. IEEE (2011)
10. Wu, Z.-X., Peng, G., Wang, W.-X., Chan, S., Wong, E.W.-M.: Cascading failure spreading on weighted heterogeneous networks. J. Stat. Mech. Theory Exp. **2008**(05), P05013 (2008)
11. Vazirani, V.V.: Approximation Algorithms. Springer, New York (2001)
12. Chlebik, M., Chlebikova, J.: Complexity of approximating bounded variants of optimization problems. Theor. Comput. Sci. **354**(3), 320–338 (2006)
13. Watts, D.J., Strogatz, S.H.: Collective dynamics of 'small-world' networks. Nature **393**(6684), 440–442 (1998)
14. Zhao, L., Park, K., Lai, Y.-C.: Attack vulnerability of scale-free networks due to cascading breakdown. Phys. Rev. E **70**(3), 035101 (2004)
15. Erdos, P., Rényi, A.: On the evolution of random graphs. Bull. Inst. Int. Stat. **38**(4), 343–347 (1961)

The Power of Rejection in Online Bottleneck Matching

Barbara M. Anthony[1] and Christine Chung[2]($\boxtimes$)

[1] Math and Computer Science Department, Southwestern University,
Georgetown, TX, USA
anthonyb@southwestern.edu

[2] Department of Computer Science, Connecticut College, New London, CT, USA
cchung@conncoll.edu

Abstract. We consider the online matching problem, where n server-vertices lie in a metric space and n request-vertices that arrive over time each must immediately be permanently assigned to a server-vertex. We focus on the egalitarian bottleneck objective, where the goal is to minimize the maximum distance between any request and its server. It has been demonstrated that while there are effective algorithms for the utilitarian objective (minimizing total cost) in the resource augmentation setting where the offline adversary has half the resources, these are not effective for the egalitarian objective. Thus, we propose a new Serve-or-Skip bicriteria analysis model, where the online algorithm may reject or skip up to a specified number of requests, and propose two greedy algorithms: GRiNN(t) and GRiN*(t). We show that the Serve-or-Skip model of resource augmentation analysis can essentially simulate the doubled-server-capacity model, and then characterize the performance of GRiNN(t) and GRiN*(t).

1 Introduction

We consider the well-studied problem of minimum-cost bipartite matching in a metric space. We are given n established points in the metric space, $s_1, s_2, \ldots, s_n$, referred to as *servers*, which form one side of the bipartition. Over time, the other n points, $r_1, r_2, \ldots, r_n$, appear in the metric space, and we refer to them as *requests*. We must permanently match or assign each request r_i, for $i = 1 \ldots n$, to a server upon its arrival, without any knowledge of future request locations $r_{i+1}, \ldots, r_n$.

The standard objective for this problem has been to minimize the *total cost* of the final matching. Formally, if $d(r, s)$ gives the *cost* or *distance* in the metric space from point r to point s, and we use $\mu(r_i)$ to denote the server matched to request r_i in the matching μ, the standard goal is to find a matching

$$\mu^* = \operatorname*{argmin}_{\mu} \sum_{i=1}^{n} d(r_i, \mu(r_i)).$$

© Springer International Publishing Switzerland 2014
Z. Zhang et al. (Eds.): COCOA 2014, LNCS 8881, pp. 395–411, 2014.
DOI: 10.1007/978-3-319-12691-3_30

In this work, however, we consider instead the objective of minimizing the maximum distance from any request to its assigned server. That is, we seek a matching

$$\mu^* = \operatorname*{argmin}_{\mu} \max_{i=1...n} d(r_i, \mu(r_i)).$$

This objective is also known as the bottleneck objective, and we refer to this problem as the *online minimum-bottleneck matching problem*.

From an economic perspective, the total-cost objective is a *utilitarian* objective that may be natural in many situations (for example, when trying to stay within an overall, centralized budget), while the bottleneck objective is *egalitarian* and seeks to ensure fairness. Issues such as fairness are a growing concern in algorithm design, as part of the wide-spread emergence of game theoretic settings in computer science, due to the proliferation of internet-based, distributed and ad-hoc computing. However, in spite of the egalitarian objective growing in importance and relevance, to the best of our knowledge, the bottleneck objective has remained largely unexplored.

There is growing evidence that the utilitarian total-cost objective for our online matching problem is easier than the egalitarian bottleneck objective. Indeed, the total-cost objective has been quite well-understood with respect to deterministic algorithms for over two decades. Kalyansunduram and Pruhs [14], and independently, Khuller et al. [18], first showed that the basic greedy algorithm that matches each arriving request to its nearest available server has an exponential competitive ratio, and that the best competitive ratio any deterministic algorithm can achieve is $2n - 1$. They also give an algorithm called PERMUTATION that achieves this competitive ratio. (Interestingly, when the underlying metric is restricted to the line, the worst-known lowerbound is 9.001 [5], yet no constant-competitive algorithm has been found.)

The fact that no sub-linear-competitive deterministic algorithm can be found for the utilitarian objective gave rise to a natural question: if the offline optimal solution is too lofty a benchmark, what natural adversary *can* an online algorithm hope to compete with? An answer was given in [15], in the context of the Online Transportation Problem, a generalization of the min-cost matching problem that additionally specifies the number of servers at each server location. They showed that when given twice the server capacity at each server location, the greedy algorithm BALANCE is constant-*halfOPT-competitive* with the offline optimal assignment. In other words, the online algorithm is constant-competitive with an offline adversary that has half the resources.

Unfortunately, such a "weakened adversary" or "resource augmentation" approach does not help when it comes to the bottleneck objective. It was demonstrated in [1] that PERMUTATION and BALANCE both fail to be better than $O(n)$-halfOPT-competitive for the bottleneck objective. Additionally, algorithms (such as BALANCE) that are designed based on the assumption that each server location has the capacity to serve at least two requests, do not apply to the classic matching setting of one server per request. It was also shown early on in an unpublished manuscript by Idury and Schäffer [13] that any deterministic algorithm must be at least $\approx 1.5n$-competitive (against the standard, unweakened, optimal, offline adversary).

These negative results motivate us to consider an alternate form of resource augmentation that gives a benchmark as simple and appealing as "half the server capacity," but perhaps powerful enough to compete with the offline optimal solution for the more difficult bottleneck objective. In the present work, we ask: what happens when the online algorithm is allowed to reject requests? In other words, what if the online algorithm is given a number of "free passes"?

Specifically, we propose the Serve-or-Skip (SoS) bicriteria analysis model: assume the online algorithm has an allowance of p *passes* or *skips*, which means the online algorithm may reject up to p of the requests without incurring any cost. This means that, in the case of the bottleneck (egalitarian) objective, a rejected request will not be a candidate for being the bottleneck match. We refer to an algorithm as $c\text{-}SoS(p)\text{-}competitive$ if, when rejecting no more than p of the requests, it is c-competitive with the offline optimal solution.

One might expect the resource augmentation model of doubling the capacity of each server to be just as powerful as having permission to ignore half the requests. However, we show that SoS can readily simulate the result of any algorithm under the doubled-capacity model of resource augmentation, suggesting that it may in fact be the more "powerful" resource augmentation model (i.e., the "weaker" weakened adversary model). We also propose two threshold-based algorithms and analyze their competitiveness against the offline optimal matching that is required to serve all requests.

We believe that SoS has, in some circumstances, more practical appeal as a benchmark than the previously proposed benchmark of doubling the resources, since in many real-world application areas, the question is one of "quality of service" rather than "service at all costs." We speculate that service providers in various real-world situations may not only be interested in quantifying the gains from rejecting a fraction of their incoming service requests, but also be more willing to entertain the notion of reducing their quality of service before they consider the idea of dramatically augmenting their resources.

1.1 Other Related Work

Incidentally, there has been an active line of recent work in online algorithms studying the power of allowing the online algorithm to have some recourse for its past actions. Megow et al. [19] study the online minimum spanning tree problem, where points arrive online and the online algorithm must connect the point to the existing tree as they arrive. They show that by allowing for a small number of re-arrangements on previously-placed edges, a nearly optimal tree can be maintained. They also apply their technique to online TSP. Gu et al. [10] then showed that with only a single retroactive edge-swap per step, a constant-competitive Steiner tree can be maintained online. Finally, Gupta et al. [11] show that in online b-matching (where a bipartite matching must be maintained online such that the degree of each node is at most b) allowing a constant number of total re-assignments of past matches suffices for a constant-competitiveness.

They also apply their technique to online scheduling with unrelated machines in the restricted assignment model, as well as to an online single-sink flow problem.

Resource augmentation analysis, or using a weakened adversary, is a technique that has been well-established and effectively and successfully used in many domains, including matching, scheduling, and algorithmic game theory. In the literature on matching, the resource augmentation has been in the form of doubling [1,15] or incrementing [2] server site capacities. In the online machine scheduling literature, it has taken the form of augmenting the processing speed of machines [16,21]. In auction theory, in an extensive line of work starting with [9], including [12], a weakened adversary is used as a trade-off for the lack information the auction mechanism has regarding the bidders true valuations for the items being sold. And in a seminal work of algorithmic game theory by [22], the optimal solution in a routing game is required to route twice as much traffic in exchange for centralized coordination among the players.

A very active and popular problem of study in the domain of online bipartite matching is known as the AdWords problem (see, e.g., [3,4,8,17,20]). There, however, there is no assumption of an underlying metric space (i.e., the edge weights need not satisfy the triangle inequality), each server may be matched to multiple requests as long as its given "budget" is not exceeded, and the objective is to maximize the weight of the final matching.

1.2 Our Results

We begin by showing that our proposed SoS model of resource augmentation analysis can simulate the doubled-server-capacity model, and hence any algorithm analyzed under that model immediately implies a corresponding algorithm with an SoS-competitive ratio at least as good as the original algorithm's halfOPT-competitive ratio.

We then propose a natural algorithm that assigns requests greedily, but skips any request that has no available servers within t times the distance to its nearest neighbor. We call the algorithm GRINN(t) for Greedy Inside Nearest Neighbor Threshold, and show that this algorithm has an SoS(p)-competitive ratio of $\Theta(2^{n-p})$ and that this ratio is tight. I.e., when allowed p free passes, the bottleneck cost of the GRINN(t) solution is guaranteed to be no more than $\Theta(2^{n-p})$ times that of the optimal offline assignment that matches all the requests. Hence each free pass improves the ratio by a factor of 2.

We then propose another greedy algorithm, which we refer to as GRIN*(t), which assigns requests greedily as long as there are available servers within t times the optimal offline bottleneck distance of the requests that have arrived so far. If there are no available servers within this threshold, the request is skipped. We then show that the SoS(p)-competitive ratio of GRIN*(1) is exponential in $n - p$, and that this is tight. So each free pass improves the competitive ratio by a factor of 2, as with GRINN(t). We also show that the SoS(p)-competitive ratio of GRIN*(2) is exponential in $n - 2p$, suggesting, interestingly, that each free pass improves the competitive ratio by a factor of 4. We conjecture that this ratio is tight for GRIN*(2). Finally, we also provide lowerbounds for the

SoS-competitiveness of $\mathrm{GRIN}^*(t)$ for any $t > 1$. Our results are summarized in the following table (Table 1).

Table 1. Summary of the results in this work. The bottom-most lowerbound on $\mathrm{GRIN}^*(t)$ is stronger (higher) than the one above it, but we include the weaker bound because we find its structure to be intuitive and instructive, and perhaps useful as a general technique for lower-bounding in the SoS model of weakened adversary analysis.

Algorithm	Applicable values of t	Applicable values of p	SoS(p)-competitive ratio c
$\mathrm{GRINN}(t)$	$1 < t < 2^{n-p} - 1$	$0 \leq p \leq n - 1$	$c = \Theta(2^{n-p})$
$\mathrm{GRIN}^*(t)$	$t = 1$	$0 \leq p \leq n - 1$	$c = 2^{n-p} - 1$
	$t = 2$	$0 \leq p \leq (n-1)/2$	$c = \Omega(p2^{n-2p})$
	$1 < t \leq 2^{(n-1)/p-1}$	$0 \leq p \leq \frac{n+1}{\lceil \log t \rceil + 1}$	$c \geq 2^{n-(\lceil \log t \rceil + 1)p - 1}$
	$1 < t \neq 2$	$0 \leq p \leq (n-1)/2$	$c \geq \frac{t^p - 2^p}{(t-2)t^p} + 2^{n-2p} - 1$

2   Preliminaries

We propose the Serve-or-Skip (SoS) model of resource augmentation analysis, where the online algorithm is endowed with an allotment of p "free *passes*" to be used as follows. Upon the arrival of each request, the online algorithm may choose to assign it irrevocably as usual, or, instead, use one of the p passes and reject the request, leaving it unassigned. After the online algorithm has used all p passes, it is required to assign all remaining incoming requests.

This is in contrast with the previously established resource augmentation model where the offline algorithm is assumed to have greater server capacity at each server location. Specifically, [15] defined a *c-halfOPT-competitive* algorithm to be one that is c-competitive with an optimal offline solution that uses half the capacity at each server location (so this characterization only applies to settings where there are at least two servers per server location). We call an algorithm *c-SoS(p)-competitive* if, when rejecting no more than p of the requests, it is c-competitive with the offline optimal solution.

2.1   SoS$\left(\frac{n}{2}\right)$-competitive vs halfOPT-competitive

We show that any online algorithm that is c-halfOPT-competitive can immediately be transformed into an algorithm that is c-SoS$\left(\frac{n}{2}\right)$-competitive, and hence that SoS$\left(\frac{n}{2}\right)$ is at least as easy a benchmark as halfOPT for online algorithms to compete with.

Theorem 1. *If an online algorithm X is c-halfOPT-competitive for minimum bottleneck matching, there is an online algorithm Y that is c-SoS$\left(\frac{n}{2}\right)$-competitive for minimum bottleneck matching.*

Proof. Given algorithm X that is c-halfOPT-competitive, transform it to a c-SoS$\left(\frac{n}{2}\right)$-competitive algorithm Y as follows.

1. Pretend there is a "secondary server" at the location of each server in the input.
2. Run X, and
   - whenever X wants to assign a request to an imaginary "secondary server," reject that request,
   - otherwise, choose the same server that X does.

We know that Y will not use more than $\frac{n}{2}$ free passes, because if it does, then X must have assigned more than n requests, but there are only n requests total.

Since all of the assignments made by Y are also made by X, the bottleneck assignment of Y will cost no more than the bottleneck assignment of X.   □

2.2 Greedy Threshold Algorithms

Greedy algorithms are a natural first choice to consider in the SoS(p) model. An algorithm that merely assigns each request to the closest available server, without skipping any requests, is the naive greedy algorithm in the classical model. Thus, we must specify how the algorithm determines which requests, if any, to skip. One logical choice is to use a *threshold*, where if the desired server is distance x away, the algorithm greedily picks the closest available server within distance $t \cdot x$ for some threshold $t \geq 1$; if no such server exists, it rejects the request. (Clearly it does not make sense to consider a threshold $t < 1$.)

It remains to define the 'desired' server. We first consider an algorithm we call GRINN(t), where the desired server is the *nearest neighbor* to the current request. Thus, when request r_i arrives, the algorithm with a threshold parameter $t \geq 1$ finds the minimum distance d_i to any server (available or not) from r_i. Formally, $d_i = \min_j d(r_i, s_j)$. It then greedily picks the closest *available* server within distance at most $t \cdot d_i$, assigning r_i to that server if it exists, and rejecting request r_i if there is no such server. If p rejections have already been made, the algorithm greedily assigns r_i to the nearest available server. While the simplicity of this algorithm is appealing, we show that its performance is exponential in the number of assigned requests for any threshold t.

We thus consider an improved version of the threshold algorithm which makes its decision about serving or skipping a request based on what the optimal solution would do with the set of requests that have arrived so far. We denote the algorithm GRIN*(t) with an associated parameter $t \geq 1$. Let OPT$_i$ refer to the optimal offline matching on the set of requests $\{r_1, r_2, \ldots r_i\}$, i.e., if M_i refers to the set of all partial matchings between the first i requests and any i of the n servers $\{s_1, \ldots, s_n\}$,

$$\mathrm{OPT}_i = \operatorname*{argmin}_{\mu_i \in M_i} \max_{j=1\ldots i} d(r_i, \mu_i(r_i)).$$

(We abuse notation to let OPT$_i$ represent either the set of assignments made, or the bottleneck cost of said set of assignments.) With the arrival of request r_i, if the nearest available server is within distance $t \cdot \mathrm{OPT}_i$ of r_i, then assign it to r_i, otherwise reject/skip r_i. Note that OPT_i can be computed efficiently (e.g., as described in [6,7]) from OPT_{i-1} and thus the assignments made by $\text{GRIN}^*(t)$ are polynomial-time computable.

2.3   Observations About $\text{GRIN}^*(t)$ and $\text{GRINN}(t)$

We now make some basic observations about $\text{GRIN}^*(t)$ and $\text{GRINN}(t)$ that will be useful in analyzing their performance in the $\text{SOS}(p)$ model. Let $\text{OPT} = \text{OPT}_n$.

Lemma 1. $\text{GRIN}^*(t)$ *(resp. $\text{GRINN}(t)$) always assigns the first request, with a cost of at most $\text{OPT}_1 \le \text{OPT}$.*

Proof. The first request can always be assigned to the server it was assigned to by OPT_1, which is its nearest neighbor. $\text{GRIN}^*(t)$ and $\text{GRINN}(t)$ will make exactly this assignment. Since the OPT_i values are non-decreasing, $\text{OPT}_1 \le \text{OPT}$.

We now provide an instance that requires $n - 1$ skips for $\text{GRIN}^*(1)$ to stay optimal, and then show that $\text{GRIN}^*(1)$ is in fact 1-SoS($n - 1$)-competitive (i.e., optimal when it can reject $n - 1$ requests). This instance is a *subdivided star* where all but one edge in the original star is subdivided into two edges.

Lemma 2. *If $\text{GRIN}^*(1)$ (resp. $\text{GRINN}(1)$) is 1-SoS(p)-competitive, then $p \ge n - 1$.*

Proof. Let there be a centrally-located server, called s_n, at the root of a subdivided star, with n requests each a distance of 1 from the center server. Servers $1 \dots n - 1$ are located at the leaves of the star, with server s_i emanating a distance of $1 + i\epsilon$ from r_i, for $i = 1 \dots n - 1$. Thus, the subdivided edges of the star each have a request at the division and a server at the leaf, while the one undivided edge has a request at the leaf. All distances $d(r, s)$ not given explicitly are equal to the cost of the path from r to s in the subdivided star.

Note that $\text{OPT}_i = 1 + (i - 1)\epsilon$ for $i = 1 \dots n$, as it always matches r_i to s_n and r_j to s_j for $j = 1 \dots i-1$. On the other hand, $\text{GRIN}^*(1)$ (or $\text{GRINN}(1)$) skips every request after the first request, which it assigns to the root. If $\text{GRIN}^*(1)$ (or $\text{GRINN}(1)$) is only allowed to skip $p < n - 1$ requests, it will skip requests $r_2 \dots r_{p+1}$, forcing it to assign all remaining requests, up to and including r_n. The bottleneck edge is the one associated with r_n, with a distance of $3 + \epsilon$ while $\text{OPT} = 1 + (n - 1)\epsilon$. $\quad\square$

Proposition 1. $\text{GRIN}^*(1)$ *(resp. $\text{GRINN}(1)$) is 1-SoS($n - 1$)-competitive for online min-bottleneck matching.*

Proof. Since Lemma 1 ensures that the first request is always assigned with a cost of at most OPT, $\text{GRIN}^*(1)$ (resp. $\text{GRINN}(1)$) can skip all of the remaining requests. With a threshold value of 1, $\text{GRIN}^*(1)$ will only assign later requests if the assignment cost is at most $\text{OPT}_i \le \text{OPT}$, guaranteeing that the bottleneck cost is at most OPT. Likewise, $\text{GRINN}(1)$ will only assign to the nearest neighbor; that distance is again at most OPT.

3 Upper Bounds for GRIN*(t) and GRINN(t)

In this section we prove upper bounds on the performance of GRIN*(t) and GRINN(t). For convenience, we define $k = n - p$ to be the minimum number of assignments that must be made by the online algorithm. We first consider $t = 1$ and then generalize to an arbitrary threshold $t \geq 1$. We show that the SoS(p)-competitive ratio of both GRIN*(t) and GRINN(t) is $O(2^{n-p})$. In the next section we show that this ratio is tight for GRIN*(1) as well as GRINN(t), for $t > 1$, so each free pass reduces the ratio by a factor of 2.

3.1 Upper Bound for $t = 1$

Theorem 2. *For $1 \leq p \leq n-1$, GRIN*(1) is $(2^{n-p} - 1)$-SoS(p)-competitive for online minimum bottleneck matching.*

Proof. By definition of GRIN*(1), and since the OPT_i are nondecreasing in i, if more than k requests are assigned, they must each have assignment cost at most $\mathrm{OPT} = \mathrm{OPT}_n$. Thus, we may assume that exactly k requests are assigned, and only consider these requests for the remainder of the proof.

We relabel these assigned requests to be $r_1, \ldots, r_k$, where the subscripts represent the relative order of arrival (and thus assignment) within these k requests. We will show inductively that the assignment cost of r_i is at most $(2^i - 1) \cdot \mathrm{OPT}$ for $i = 1 \ldots k$.

Base case: $i = 1$. By Lemma 1, the first request r_1 is always assigned with a cost of at most OPT, which satisfies the claim.

Inductive case: Assume the assignment cost of r_j is at most $(2^j - 1) \cdot \mathrm{OPT}$ for all $1 \leq j \leq i$. Consider the assignment cost of r_{i+1}. Let s_{i+1} be the server that OPT_{i+1} assigns to r_{i+1}. If s_{i+1} is available, the assignment cost is at most $\mathrm{OPT}_{i+1} \leq \mathrm{OPT}$.

Thus, we may assume that s_{i+1} is not available, and it is used by some r_j with $j \leq i$. We thus consider the graph consisting of edges in OPT_{i+1} and the i edges assigned thus far by GRIN*(1). Since r_{i+1} is not yet matched by GRIN*(1), there must be a path from r_{i+1} to some s_a that is used by OPT_{i+1} but not currently matched by GRIN*(1). Observe that said path must begin with an edge in OPT_{i+1} and alternate between edges in OPT_{i+1} and edges in GRIN*(1), terminating with an edge in OPT_{i+1}.

We can thus use triangle inequality to compute the distance from this available s_a to r_{i+1}, giving an upper bound on the assignment cost of r_{i+1}. Since all distances are nonnegative, additional edges either cause the total cost to increase or stay the same. Thus, in the worst case, the path includes all of the i assignments already made, as well as the $i + 1$ edges in OPT_i.

By strong induction, the cost of the i assignments already made is at most

$$\sum_{h=1}^{i}(2^h - 1) \cdot \mathrm{OPT} = \left(2^{i+1} - i - 2\right) \cdot \mathrm{OPT}.$$

Noting that the OPT_i are nondecreasing and adding in the cost of the $i + 1$ edges from OPT_{i+1} in the path bounds the total distance (and thus assignment cost) of r_{i+1} to some available server by $(2^{i+1}-1)\cdot\text{OPT}$, completing the inductive proof.

Corollary 1. *For $1 \leq p \leq n - 1$, $\text{GRINN}(1)$ is $(2^{n-p} - 1)\text{-}SoS(p)\text{-}competitive$ for online minimum bottleneck matching.*

Proof. The proof is identical to that of Theorem 2, with the observation that the distance from a request to its nearest neighbor is naturally at most OPT.

3.2 Upper Bound for $t > 1$

In generalizing the upper bound for $\text{GRIN}^*(t)$ to $t > 1$, any of the greedy assignments made by $\text{GRIN}^*(t)$ may in fact cost up to a factor of t more than the optimal solution on the set of requests thus far. As such, request r_{i+1} could in fact be assigned to a server at a distance of $t \cdot \text{OPT}$ when a more restrictive threshold may have in fact skipped this request. Thus, the proof parallels that of Theorem 2, but has some notable distinctions.

Theorem 3. *For $1 \leq p \leq n-1$, $\text{GRIN}^*(t)$ is $\max\{2^{n-p}-1, t\}\text{-}SoS(p)\text{-}competitive$ for online minimum bottleneck matching.*

Proof. Since $\text{GRIN}^*(t)$ only assigns a request if it has no skips left or the cost is below the threshold, if more than k requests are assigned, these additional requests (above k) must all have assignment costs of at most $t \cdot \text{OPT}$. Thus we may consider only the first k requests that are assigned for the remainder of the proof.

Relabeling the requests and using strong induction as in Theorem 2 then shows that the assignment cost of r_i is at most $\max\{2^i - 1, t\} \cdot \text{OPT}$. Specifically, the base case is unchanged, and the inductive case now relies on the observation that either the alternating path has the same bound as in Theorem 2 or that request r_{i+1} was assigned to some available server within distance $t \cdot \text{OPT}_{i+1}$. Thus, taking the larger of these two gives the upper bound for general t.

The result in Theorem 3 again extends naturally to the $\text{GRINN}(t)$ algorithm.

Corollary 2. *For $1 \leq p \leq n - 1$, $\text{GRINN}(t)$ is $\max\{2^{n-p} - 1, t\}\text{-}SoS(p)\text{-}competitive$ for online minimum bottleneck matching.*

In the next section we provide a matching lower bound for $\text{GRINN}(t)$ (and $\text{GRIN}^*(1)$) that, combined with the upper bound in this section, shows that each skip reduces the $SoS(p)$-competitiveness of $\text{GRINN}(t)$ (resp. $\text{GRIN}^*(1)$) by a factor of 2.

4   Lower Bounds

4.1   Lower Bound for GRINN(t) and GRIN*(1)

We provide a lowerbound on GRIN*(1), and use the same instance to provide a lower bound for the cost of GRINN(t) in the SoS(p) model for arbitrary $p = n - k$. We use an example whose structure is comprised of a broom with a subdivided star. While a standard broom graph consists of a path and a star, this specialized broom graph consists of a path, one of whose endpoints is the center of a subdivided star.

Theorem 4. *For $0 \leq p \leq n - 1$, if GRIN*(1) is c-SoS(p)-competitive for online minimum bottleneck matching, then $c \geq (2^{n-p} - 1 - \epsilon)$.*

Proof. The requests are numbered in order of arrival. (See Fig. 1.) The broom with a subdivided star consists of a star portion, centered at s_{p+2} with $p + 1$ leaves, and a longer handle. In particular, r_1 through r_{p+1} are each a distance 1 away from the center, s_{p+2}. For $i = 1 \ldots p + 1$, there is a server s_i that is at a distance of $1 + i\epsilon$ away from r_i, making these servers all a distance greater than 2 from the center. The remaining requests and servers lie along a line emanating from the center, terminating with request r_n. Specifically, until r_n is reached, r_{p+2} is a distance 1 from the center. Server s_{p+3} is 3 units further along the line, with request r_{p+3} 1 unit further. Server s_{p+4} is 7 units further, followed by request r_{p+4} 1 unit further. Server s_{p+5} is 15 units further, followed by request r_{p+5} 1 unit further. In general, s_{p+j} is $2^{j-1} - 1$ units past request r_{p+j-1}, and 1 unit before r_{p+j}.

The optimal solution assigns each request r_i in the star portion (i.e., $1 \leq i \leq p + 1$) to its corresponding leaf server s_i for a cost of $1 + i\epsilon$, and each request on the handle portion also to its corresponding server (r_i to s_i for $p + 2 \leq i \leq n$), for a cost of 1. Hence, OPT is $1 + (p + 1)\epsilon$.

We now consider the behavior of GRIN*(1). Again, the first request is always assigned, so r_1 is greedily assigned to the center of the star portion, s_{p+2}. With the arrival of r_2, OPT_2 would have assigned r_1 to s_1 and r_2 to s_{p+2}, so OPT_2 is $1 + \epsilon$, causing GRIN*(1) to skip r_2 (since the closest available server, s_2, is $1 + 2\epsilon$ away). Similarly, for the remaining requests on the star portion, that is, for $i = 2 \ldots p + 1$, when r_i arrives, OPT_i is $1 + (i - 1)\epsilon$ and the nearest available server, s_i is at a distance $1 + i\epsilon$ so the algorithm skips request r_i. Hence, it uses up all of the allowed p skips. Thus, all of the remaining requests must be assigned, and will be done so greedily. In particular, r_{p+2} greedily chooses the server s_{p+3} that is 3 to the right instead of paying $3 + \epsilon$ (or more) to use a server in the star portion. Similarly, each of the remaining requests will choose to go further along the handle when possible rather than paying slightly more to go back to the star portion. Thus, the final request r_n, the leaf of the handle, must traverse the entire length of the handle and then go to the closest leaf of the star (s_1) for a cost of $2^{n-p} - 1 + \epsilon$, growing exponentially in the number of requests assigned by the algorithm. $\square$

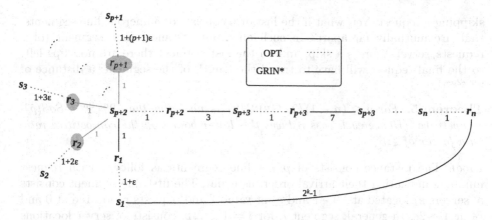

Fig. 1. The broom-with-subdivided-star instance. Requests skipped by GRIN*(1) are highlighted.

In contrast with GRIN*(1), GRIN*(t), $t \geq 2$, for example, does well on the broom with subdivided star. GRiNN(t), on the other hand, is not as effective. We can easily modify the distances in the graph to be more challenging for GRiNN(t).

Corollary 3. *For* $0 \leq p \leq n-1$, *if* GRiNN(t) *is* c-*SoS(p)-competitive for online minimum bottleneck matching, then* $c \geq \Omega(2^{n-p})$.

Proof. The instance here is similar to that of Theorem 4 (Fig. 1), however the distances $d(r_i, s_i)$, for $i = 1 \ldots p + 1$ are instead $t + \epsilon$. The edge from r_{p+2} to s_{p+3} now has cost $t + 2$, the edge from r_{p+3} to s_{p+4} has cost $2t + 5$, the edge from r_{p+4} to s_{p+5} has cost $4t + 11$, and in general the edge from r_{p+i} to s_{p+i+1} has cost $2^{i-2}t + 3 \cdot 2^{i-2} - 1$ for all $i = 2 \ldots k$.

Thus, the bottleneck edge, the edge out of r_n, has a cost of $2^{n-p-2}t + 3 \cdot 2^{n-p-2} - 1$. OPT assigns each r_i to the corresponding s_i, for a bottleneck cost of $t + \epsilon$.   □

We note that taken together with Corollary 2, we have now given a tight analysis of the SoS-competitiveness of GRiNN(t).

4.2 Lower Bounds for GRIN*(2)

We now return to GRIN*(2), having noted that it does well on some instances on which GRIN*(1) does poorly. We present two lowerbound instances in this section, the first weaker, but straightforward and instructive, and perhaps useful as a basic lower-bounding technique in the SoS analysis model.

It is well known (and we can see, i.e., from the "handle" portion of our broom instance) that a standard greedy algorithm that is not able to reject requests can do quite poorly, even when all the servers and requests lie on a single line. Intuitively, the SoS(p) model allows GRIN*(t) to recover from a poor decision by skipping a request. Yet, what if the instance consists of numerous line segments, that are mutually far apart? In such an input instance, each segment (of 2 requests/servers) forces a skip, and on the last segment there are no skips left, so the final request will have to travel the length of the segment: a distance of 2^{n-2p-1}.

Theorem 5. *For $p \leq (n-1)/2$, GRIN*$^*(2)$ *is no better than 2^{n-2p-1}-SoS(p)-competitive. (Hence each pass reduces this lower bound on the competitive ratio by a factor of 4.)*

Proof. The instance consists of $p + 1$ line segments as follows, with request numbers indicating their arrival order, as usual. The first line segment consists of servers s_1 located at $-1 - \epsilon$ and s_2 at value 1 and requests r_1 located at 0 and r_2 at $1 + 2\epsilon$. In general, segment i, for $i = 1, \ldots, p$, consists of server locations s_{2i-1} located at $-1 - \epsilon$ on line segment i, s_{2i} at value 1 on line segment i, and requests r_{2i-1} located at 0 on line segment i and r_{2i} at $1 + 2\epsilon$ on line segment i. Segment i, for $i = 1 \ldots p + 1$, is far enough (at least 2^n) away from all points on previous segments $j < i$. The final segment has the standard server locations of the known worst instance against the basic greedy algorithm: server s_j for $j = 2p + 1, \ldots, n$ are at locations $-1 - \epsilon, 1, 3, 7, \ldots, 2^{n-2p-1}$. Requests r_j for $j = 2p + 1, \ldots, n$ arrive at the following corresponding locations on the final segment: $0, 1, 3, 7, \ldots, 2^{n-2p-1}$.

OPT will assign each r_i to the corresponding s_i, for a bottleneck cost of $1 + \epsilon$ (with many assignments having cost 0). GRIN$^*(2)$ assigns r_1 to s_2 (the cheapest assignment) and then the cost of assigning r_2 to s_1 is $2 + 3\epsilon$, which exceeds the allowable cost of $2 \cdot \text{OPT} = 2 + 2\epsilon$, and all other servers are at least 2^n away, so r_2 is skipped. On the next segment, GRIN$^*(2)$ similarly assigns r_3 to s_4, and skips r_5, and so on for the first p segments. After p segments, the algorithm has used all of its skips, so its behavior on the remaining segment is to assign all requests to the servers to their immediate right, assigning the final request to the leftmost server of segment p for a cost of 2^{n-2p-1}. □

Note that when half the requests may be rejected ($p \approx n/2$), GRIN$^*(2)$ is constant-SoS-competitive on this instance. Unfortunately, our next theorem shows even when $p \approx n/2$, GRIN$^*(2)$ has an SoS-competitive ratio no better than linear.

While the previous instance required the algorithm to assign at least two assigned requests per skipped request, this next instance is able to force a skipped request for each assigned request. We conjecture that this is the worst-possible for GRIN$^*(2)$ and that there is a matching upperbound on the SoS-competitive ratio.

Theorem 6. *For $p \leq (n-1)/2$, GRIN*$^*(2)$ *has a SoS(p)-competitive ratio no better than $\Omega(p2^{n-2p})$.*

Proof. Consider the following input instance, pictured in Fig. 2. The structure of the instance is that requests arrive from left to right, with requests r_i, for $i = 1, 3, \ldots, n$ (n an odd integer) assigned by GRIN$^*(t)$ arriving on a line, and skipped requests r_i, $i = 2, 4, \ldots, n - 1$, arriving "above" the line after each

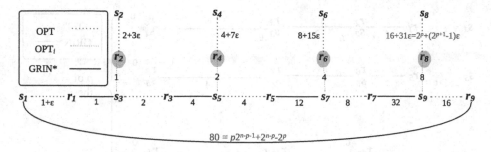

Fig. 2. Requests are numbered in order of arrival and requests skipped by the online algorithm GRIN* are highlighted. The special case of the instance where $p = k - 1 = (n - 1)/2$ is pictured here. Note that in this case GRIN* $= (n - p)2^p$.

assigned request. (Note that here we assume $p = k - 1 = (n - 1)/2$, but for any $p < k - 1$, we simply omit any excess requests arriving above the line, stopping after p have arrived: one "above" each assigned request until no more are needed. The remaining requests all arrive on the line following the same pattern as the other requests that have arrived on the line.) Each server s_i, for $i = 1, 3, \ldots n$, numbered according to the index of the request that is matched to it by OPT, we have $d(r_{i-1}, s_i) = 2^{(i-1)/2-1}$, and $d(r_i, s_i) = d(r_{i-1}, s_{i-1}) = 2^{(i-1)/2} + (2^{(i+1)/2} - 1)\epsilon$. Finally, each request r_i, for $i = 3, \ldots n$, has $d(r_i, s_{i+2}) = d(r_{i-2}, s_i) + d(r_i, s_i)$, with $d(r_1, s_3) = 1$. Each of the requests r_i, for $i = 1, 3, \ldots, n-2$ is initially assigned by both GRIN* and OPT$_i$ to the server to its right, s_{i+2}. But upon arrival of r_{i+1}, we note that OPT$_{i+1}$ will re-assign r_i to the server to its left, s_i, and that assignment remains through OPT$_n$. Each request r_j, for $j = 2, 4, \ldots, n-1$, OPT$_j$ assigns it to server s_{j+1} on the line, at a cost of 1, and therefore r_j is skipped by GRIN*. But upon arrival of request r_{j+1}, OPT$_{j+1}$ reassigns r_j to s_j, which doubles the bottleneck cost of OPT. Finally, when r_n arrives, GRIN* has no choice but to match it to s_1 for a final bottleneck cost of $(n - p)2^{n-p-1}$. For general $p \le k - 1$, the bottleneck cost comes to

$$p2^{k-1} + \sum_{j=1}^{k-p} 2^{k-j} = p2^{n-p-1} + 2^{n-p} - 2^p.$$

OPT, on the other hand, had a final bottleneck cost of $2^p + (2^{p+1} - 1)\epsilon$. Hence, GRIN*/OPT $\approx (p2^{n-p-1} + 2 \cdot 2^{n-p-1} - 2^p)/2^p = (p + 2)2^{n-2p-1} - 1$.

4.3   Lower Bounds for Grin*(t)

Extending the instance of Theorem 5 to larger thresholds simply entails lengthening the segments, as illustrated in Fig. 3. As the threshold increases we also get a lower ratio. Specifically, for a threshold of t, a segment of $\lceil \log t \rceil + 1$ requests is needed to force each free pass, yielding the following corollary to Theorem 5.

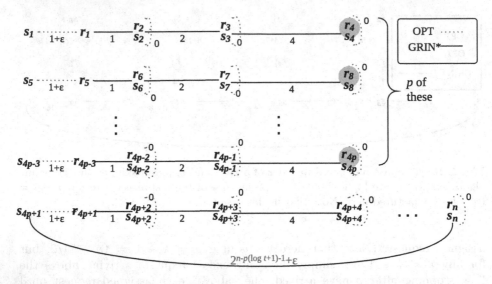

Fig. 3. An instance for $t = 6$ is illustrated here, yielding an SoS(p)-competitive ratio of $2^{n-p(\lceil \log t \rceil + 1) - 1} = 2^{n-4p-1} - 1$. Requests are numbered in order of arrival and requests skipped by the online algorithm GRIN* are highlighted.

Corollary 4. GRIN*(t) *has an* SoS(p)*-competitive ratio no better than* $2^{n-p(\lceil \log t \rceil + 1) - 1}$.

A somewhat tighter (worse) lowerbound on GRIN*(t), for all $t > 1$, is based on the instance of Theorem 6 in Fig. 2. In particular, we extend the distance from r_2 to s_2 in that instance to $t + (t + 1)\epsilon$, and use t for the distance from r_3 to s_3. We follow these initial distances and extend the remaining distances accordingly. (See Fig. 4, though note that epsilon terms on the distances have been omitted, for simplicity.) These adjustments yield the following corollary to Theorem 6.

Corollary 5. *Suppressing epsilon terms,* GRIN*(t) *has an* SoS(p)*-competitive ratio no better than*

$$\frac{t^p - 2^p}{(t - 2)t^p} + 2^{n-2p} - 1.$$

Proof. As in Fig. 2, Fig. 4 illustrates the special case of $p = k - 1 = \frac{n-1}{2}$, where

$$\text{GRIN}^*(t) = \sum_{j=0}^{k-1} t^j 2^{k-j-1} = \frac{t^k - 2^k}{t - 2}.$$

But for general $p \leq k - 1$, the skipped requests beyond p that are pictured above the main line segment would be incorporated into the remainder of the line segment, yielding a bottleneck cost of

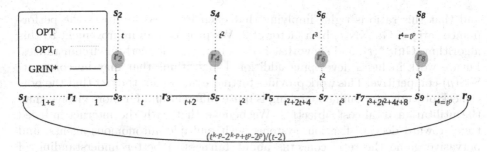

Fig. 4. Requests are numbered in order of arrival and requests skipped by GRIN* are highlighted. In the special case of the instance where $p = k - 1$, pictured here, GRIN* $= (t^k - 2^k)/(t - 2)$, but the general cost for any $p \le k - 1$ is GRIN* $= (t^p - 2^p + t^{k-p} - 2^{k-p})/(t - 2)$. Note that the epsilon terms of Fig. 2 have been removed from this illustration for readability.

$$\text{GRIN}^*(t) = \sum_{j=0}^{p-1} t^j 2^{k-j-1} + \sum_{j=p}^{k-1} t^p 2^{k-j-1}$$

$$= \frac{2^p - t^p}{2 - t} + t^p \sum_{j=0}^{k-p-1} 2^{k-p-j-1} = \frac{t^p - 2^p}{t - 2} + t^p(2^{k-p} - 1).$$

To derive the final ratio, divide by the cost of OPT, which is t^p (suppressing epsilons).

5   Conclusion

This work investigates the relationship between two forms of bicriteria analysis for online minimum bottleneck matching: the traditional one where resource augmentation comes in the form of added resources, and the one we propose here, where we are effectively allowing for some "degradation of service" in exchange for being online. Since the service provider is able to decline a number of requests upon arrival, this model may be more relevant to some practitioners than having to service all requests, but needing to supplement your resources for doing so. In addition, algorithms that perform well against an adversary with half the servers per server location, may depend on the assumption that there are at least two servers per server location, and may not readily translate to the native online matching setting. It is also interesting to consider the implications of the extra power afforded under the SoS(p) model, namely the responsibility to wield it wisely. While the greater freedom allowed under the SoS(p) model is appealing, more limited freedom can have the benefit of restricting the number of poor decisions a greedy algorithm can make. This may make designing good algorithms for the SoS(p) model more challenging.

  We showed in this work that the most natural greedy algorithm, GRINN(t), has a competitive ratio that is exponential in the number of requests assigned, and that this ratio is tight, implying that each free pass improves the performance ratio of GRINN(t) by a factor of 2. We proposed an improvement to this algorithm, GRIN*(t), and show that it does better under certain circumstances. Future work includes developing additional algorithms that may be constant-SoS(p)-competitive. This work provides further evidence for the fact that the bottleneck objective for online matching is quite different and far more elusive than the utilitarian total-cost objective. We believe that with the internet-induced trend toward decentralization, systems with multiple autonomous agents, and pervasive game theoretic concerns about fairness, a better understanding of optimizing for egalitarian objectives will become essential, and developing benchmarks of practical relevance is one path to gaining such an understanding.

References

1. Anthony, B.M., Chung, C.: Online bottleneck matching. J. Comb. Optim. **27**(1), 100–114 (2014)
2. Chung, C., Pruhs, K., Uthaisombut, P.: The online transportation problem: on the exponential boost of one extra server. In: Laber, E.S., Bornstein, C., Nogueira, L.T., Faria, L. (eds.) LATIN 2008. LNCS, vol. 4957, pp. 228–239. Springer, Heidelberg (2008)
3. Devanur, N.R., Hayes, T.P.: The adwords problem: online keyword matching with budgeted bidders under random permutations. In: Proceedings of the 10th ACM Conference on Electronic Commerce, EC '09, pp. 71–78 (2009). ISBN: 978-1-60558-458-4
4. Fernandes, C.G., Schouery, R.C.: Second-price ad auctions with binary bids and markets with good competition. Theor. Comput. Sci. **540–541**, 103–114 (2014). ISSN: 0304–3975
5. Fuchs, B., Hochstättler, W., et al.: Online matching on a line. Theor. Comput. Sci. **332**(1–3), 251–264 (2005). ISSN: 0304–3975
6. Gabow, H.N., Tarjan, R.E.: Algorithms for two bottleneck optimization problems. J. Algorithms **9**(3), 411–417 (1988)
7. Garfinkel, R.S.: An improved algorithm for the bottleneck assignment problem. Oper. Res. **19**(7), 1747–1751 (1971)
8. Goel, G., Mehta, A.: Online budgeted matching in random input models with applications to adwords. In: Proceedings of the Nineteenth Annual ACM-SIAM Symposium on Discrete Algorithms, SODA '08, pp. 982–991 (2008)
9. Goldberg, A.V., Hartline, J.D., et al.: Competitive auctions and digital goods. In: Proceedings of the 12th Annual ACM-SIAM Symposium on Discrete Algorithms, SODA '01, pp. 735–744 (2001). ISBN: 0-89871-490-7
10. Gu, A., Gupta, A., et al.: The power of deferral: maintaining a constant-competitive Steiner tree online. In Proceedings of the 45th Annual ACM Symposium on Theory of Computing, pp. 525–534. ACM (2013)
11. Gupta, A., Kumar, A., et al.: Maintaining assignments online: matching, scheduling, and flows. In: Proceedings of the 25th Annual ACM-SIAM Symposium on Discrete Algorithms, SODA 2014, pp. 468–479 (2014)
12. Hartline, J.D., Roughgarden, T.: Simple versus optimal mechanisms. In: ACM Conference on Electronic Commerce, pp. 225–234 (2009)
13. Idury, R., Schaffer, A.: A better lower bound for on-line bottleneck matching (manuscript, 1992)

14. Kalyanasundaram, B., Pruhs, K.: Online weighted matching. J. Algorithms **14**(3), 478–488 (1993)
15. Kalyanasundaram, B., Pruhs, K.: The online transportation problem. SIAM J. Discrete Math. **13**(3), 370–383 (2000)
16. Kalyanasundaram, B., Pruhs, K.: Speed is as powerful as clairvoyance. J. ACM **47**, 617–643 (2000). ISSN: 0004–5411
17. Kalyanasundaram, B., Pruhs, K.R.: An optimal deterministic algorithm for online b-matching. Theor. Comput. Sci. **233**(1), 319–325 (2000)
18. Khuller, S., Mitchell, S.G., et al.: On-line algorithms for weighted bipartite matching and stable marriages. Theor. Comput. Sci. **127**, 255–267 (1994). ISSN: 0304–3975
19. Megow, N., Skutella, M., Verschae, J., Wiese, A.: The power of recourse for online MST and TSP. In: Czumaj, A., Mehlhorn, K., Pitts, A., Wattenhofer, R. (eds.) ICALP 2012, Part I. LNCS, vol. 7391, pp. 689–700. Springer, Heidelberg (2012)
20. Mehta, A., Saberi, A., et al.: Adwords and generalized online matching. J. ACM **54**(5) (2007). ISSN: 0004–5411
21. Phillips, C.A., Stein, C., et al.: Optimal time-critical scheduling via resource augmentation. Algorithmica **32**(2), 163–200 (2002)
22. Roughgarden, T., Tardos, É.: How bad is selfish routing? J. ACM **49**(2), 236–259 (2002)

The Generalized 3-Edge-Connectivity
of Lexicographic Product Graphs

Xueliang Li[✉], Jun Yue, and Yan Zhao

Center for Combinatorics and LPMC-TJKLC, Nankai University,
Tianjin 300071, China
lxl@nankai.edu.cn, yuejun06@126.com, zhaoyan2010@mail.nankai.edu.cn

Abstract. The generalized k-edge-connectivity $\lambda_k(G)$ of a graph G is a natural generalization of the concept of edge-connectivity. The generalized edge-connectivity has many applications in networks. The lexicographic product of two graphs G and H, denoted by $G \circ H$, is an important method to construct large graphs from small ones. In this paper, we mainly study the generalized 3-edge-connectivity of $G \circ H$, and get lower and upper bounds of $\lambda_3(G \circ H)$. An example is given to show that all bounds are sharp.

Keywords: Edge-disjoint paths · Edge-connectivity · Steiner tree · Edge-disjoint steiner trees · Generalized edge-connectivity

1   Introduction

All graphs considered in this paper are simple, finite and undirected. We follow the terminology and notation of Bondy and Murty [3]. For a graph G, the *local edge-connectivity* between two distinct vertices u and v, denoted by $\lambda(u, v)$, is the maximum number of pairwise edge-disjoint uv-paths. A nontrivial graph G is *k-edge-connected* if $\lambda(u, v) \geq k$ for any two distinct vertices u and v of G. The *edge-connectivity* $\lambda(G)$ of a graph G is the maximum value of k for which G is k-edge-connected.

Naturally, the concept of edge-connectivity can be extended to a new concept, the generalized k-edge-connectivity, which was introduced by Li et al. [22]. For a graph $G = (V, E)$ and a set $S \subseteq V$ of at least two vertices, *a Steiner tree connecting S* (or simply, *an S-tree*) is a such subgraph $T = (V', E')$ of G that is a tree with $S \subseteq V'$. Two S-trees T and T' are said to be edge-disjoint if $E(T) \cap E(T') = \emptyset$. The *generalized local edge-connectivity* $\lambda(S)$ is the maximum number of pairwise edge-disjoint Steiner trees connecting S. For an integer k with $2 \leq k \leq n$, the *generalized k-edge-connectivity* $\lambda_k(G)$ of G is defined as $\lambda_k(G) = \min\{\lambda(S) \mid S \subseteq V(G), |S| = k\}$. Obviously, $\lambda_2(G) = \lambda(G)$. Set $\lambda_k(G) = 0$ if G is disconnected. Similarly, the concept of the generalized k-connectivity was introduced by Hager in [11] and it is also studied in [5].

Supported by NSFC No. 11371205 and PCSIRT.

© Springer International Publishing Switzerland 2014
Z. Zhang et al. (Eds.): COCOA 2014, LNCS 8881, pp. 412–425, 2014.
DOI: 10.1007/978-3-319-12691-3_31

We refer to [17–19,22,24,30] for some known results of the generalized connectivity and edge-connectivity.

The generalized edge-connectivity has a close relation to an important problem, the *Steiner tree packing problem*, which asks for finding a set of maximum number of edge-disjoint S-trees in a given graph G where $S \subseteq V(G)$, see [9,31]. An extreme of Steiner tree packing problem is the *Spanning tree packing problem* where $S = V(G)$. For any graph G, the *spanning tree packing number* or *STP number*, is the maximum number of edge-disjoint spanning trees contained in G. For the *STP* number, we refer to [1,25,26]. The difference between the Steiner tree packing problem and the generalized edge-connectivity is as follows: the former problem studies local properties of graphs since S is given beforehand, while the latter problem focuses on global properties of graphs since S runs over all k-subsets of $V(G)$.

The generalized edge-connectivity and the Steiner tree packing problem have applications in $VLSI$ circuit design, see [9,27]. In this application, a Steiner tree is needed to share an electronic signal by a set of terminal nodes. A Steiner tree is also used in computer communication networks and optical wireless communication networks, see [6,7]. Another application arises in the Internet Domain. Suppose that a given graph G represents a network. We select arbitrary k vertices as nodes. Suppose one of the nodes in G is a *broadcaster* and all other nodes are *users*. The broadcaster wants to broadcast as many streams of movies as possible, so that the users have the maximum number of choices. Each stream of movie is broadcasted via a tree connecting all the users and the broadcaster. So, in essence we need to find the maximum number of Steiner trees connecting all the users and the broadcaster, namely, we want to get $\lambda(S)$, where S is the selected k nodes. Clearly, it is a Steiner tree packing problem. Furthermore, if we want to know whether for any k nodes the network G has above properties, then we need to compute $\lambda_k(G) = \min\{\lambda(S)\}$ in order to prescribe the reliability and the security of the network.

From a theoretical perspective, both extremes of the generalized edge-connectivity problem are fundamental theorems in combinatorics. One extreme is when we have two terminals. In this case edge-disjoint trees are just edge-disjoint paths between the two terminals, and so the problem becomes the well-known edge version of Menger theorem. The other extreme is when all the vertices are terminals. In this case edge-disjoint trees are just spanning trees of the graph, and the problem becomes the classical Nash-Williams-Tutte theorem, see [23,29].

Graph product is an important method to construct large graphs from small ones. So it has many applications in the design and analysis of networks, see [9,14,15]. The lexicographic product (or composition), Cartesian product, strong product and the direct product are the main four standard products of graphs. More information about the (edge-) connectivity of these four product graphs can be found in [4,8,10,12,13,16,32]. The generalized 3-edge-connectivity of Cartesian product graphs was studied and the lower bound is given in [28]. In this paper, we study the generalized 3-edge-connectivity of lexicographic product graphs and provide both sharp lower and upper bounds.

Theorem 1. *Let G and H be two non-trivial graphs such that G is connected. Then $\lambda_3(H) + \lambda_3(G)|V(H)| \leq \lambda_3(G \circ H) \leq \min\left\{\left\lfloor \frac{4\lambda_3(G)+2}{3} \right\rfloor |V(H)|^2, \delta(H) + \delta(G)|V(H)| \right\}$. Moreover, the lower and upper bounds are sharp.*

Note that the vertex version, the generalized 3-connectivity of Cartesian product and lexicographic product graphs, was studied in [16, 20]. The results there are quite different from ours.

2   Preliminary and Notation

Let $G = (V, E)$ be a graph and S be an s-subset of V. $G[S]$ denotes the induced subgraph of G on S and $\mathcal{E}^{|S|}$ denotes the empty graph on S, that is, the union of s isolated vertices. *Connect x to S* is to join x to each vertex of S for a vertex x outside S. Given two sets X, Y of vertices, we call a path P an *XY-path* if the end-vertices of P are in X and Y, respectively, and all inner vertices are in neither X nor Y. If u and v are two vertices on a path P, uPv will denote the segment of P from u to v. Two distinct paths are *edge-disjoint* if they have no edges in common; *internally disjoint* if they have no internal vertices in common; *vertex-disjoint* if they have no vertices in common. For $X = \{x_1, x_2, \cdots, x_k\}$ and $Y = \{y_1, y_2, \cdots, y_k\}$, an *XY-linkage* is defined as a set Q of k vertex-disjoint XY-paths $x_i P_i y_i$, $1 \leq i \leq k$.

Let $G = (V_1, E_1)$ and $H = (V_2, E_2)$. The *lexicographic product (or composition)* $G \circ H$ of G and H is defined as follows: $V(G \circ H) = V_1 \times V_2$, two vertices (u, v) and (u', v') are adjacent if and only if either $uu' \in E_1$ or $u = u'$, $vv' \in E_2$. In other words, $G \circ H$ is obtained by substituting a copy $H(u)$ of H for every vertex u of G and joining each vertex of $H(u)$ with every vertex of $H(u')$ if $uu' \in E_1$. The vertex set $G(v) = \{(u, v)|u \in V_1\}$ for some fixed vertex v of H is called a layer of graph G or simply a *G-layer*. Analogously, we define the *H-layer* with respect to a vertex u of G and denote it by $H(u)$. It is not hard to see that any *G-layer* induces a subgraph of $G \circ H$ that is isomorphic to G and any *H-layer* induces a subgraph of $G \circ H$ that is isomorphic to H. For any $u, u' \in V(G)$ and $v, v' \in V(H)$, $(u, v), (u, v') \in V(H(u))$, $(u', v), (u', v') \in V(H(u'))$, $(u, v), (u', v) \in V(G(v))$, $(u, v'), (u', v') \in V(G(v'))$. We view (u, v') and (u', v) as *the vertices corresponding to* (u, v) in $G(v')$ and $H(u')$, respectively. Similarly, we can define the path and tree corresponding to some path and tree, respectively. The edge $(u, v)(u', v')$ is called a *first-type* edge if $uu' \in E_1$ and $v = v'$; a *second-type* edge if $vv' \in E_2$ and $u = u'$; a *third-type* edge if $uu' \in E_1$ and $v \neq v'$. For a subset W of $V(G)$ with $W = \{u_1, \cdots, u_t\}$, we denote $H(W) = H(u_1) \cup \cdots \cup H(u_t)$. We use K_W to denote a subgraph of $G \circ H$, where $V(K_W) = V(G[W] \circ H)$, $E(K_W) = E(G[W] \circ H) \setminus E(H(W))$, namely, the end-vertices of an edge of K_W are in different H-layers.

Unlike the other products, the lexicographic product does not satisfy the commutative law, that is, $G \circ H$ could not be isomorphic to $H \circ G$. By a simple observation, $G \circ H$ is connected if and only if G is connected. Moreover, $\delta(G \circ H) = \delta(G)|V(H)| + \delta(H)$.

Let $G = (V, E)$ be a connected graph, $S = \{x, y, z\} \subseteq V$, and T be an S-tree. We call T a *type I S-tree* if it is just a path whose end-vertices belong to S; a *type II S-tree* if it has exactly three leaves x, y, z. Note that each vertex in a type I S-tree has degree two except the two end-vertices in S. If T is of type II, every vertex in $T \setminus S$ has degree two except one vertex of degree three. By deleting some vertices and edges of an S-tree T, it is easy to check that T is of type I or II. Because our aim is to get as many S-trees as possible, in this paper, each S-tree is of type I or II. Therefore, we get the following proposition.

Proposition 1. *Let $G = (V, E)$ be a graph with $\lambda_3(G) = k \geq 2$, $S = \{x, y, z\} \subseteq V$. Then there exist $k - 2$ edge-disjoint S-trees $T_1, \cdots, T_{k-2}$ such that $E(T_i) \cap E(G[S]) = \emptyset$ where $1 \leq i \leq k - 2$.*

Proof. By the definition of an S-tree, we know that $|E(T_i) \cap E(G[S])| \leq 2$ and $|\{T_i \mid E(T_i) \cap E(G[S]) \neq \emptyset\}| \leq 3$. Let $T_1, \cdots, T_k$ be k edge-disjoint S-trees. If $|\{T_i \mid E(T_i) \cap E(G[S]) \neq \emptyset\}| \leq 2$, we are done. Thus, it remains to consider the case when $G[S]$ is a triangle. Without loss of generality, assume that $|\{T_i \mid E(T_i) \cap E(G[S]) \neq \emptyset\}| = 3$ and $E(T_i) \cap E(G[S]) \neq \emptyset$, where $i = 1, 2, 3$. Then T_1, T_2, T_3 have the structures F_1 or F_2 shown in Fig. 1. Furthermore, we can obtain T_1', T_2', T_3' from T_1, T_2, T_3 such that $E(T_i') \cap E(G[S]) = \emptyset$. See figures F_1' and F_2' in Fig. 1, where the S-tree T_1' is shown by gray lines. Thus $T_1', T_4, \cdots, T_k$ are our desired $k - 2$ edge-disjoint S-trees.

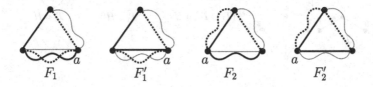

$$F_1 \qquad F_1' \qquad F_2 \qquad F_2'$$

Fig. 1. Three S-trees of type I.

Li et al. [21,22] got the following results which will be useful for our proof.

Observation 1. *[22] For any graph G of order n, $\lambda_k(G) \leq \lambda(G)$. Moreover, the upper bound is tight.*

Observation 2. *[22] If G is a connected graph, then $\lambda_k(G) \leq \delta(G)$. Moreover, the upper bound is tight.*

Proposition 2. *[21] Let G be a connected graph of order n with minimum degree δ. If there are two adjacent vertices of degree δ, then $\lambda_k(G) \leq \delta - 1$ for $3 \leq k \leq n$. Moreover, the upper bound is sharp.*

From Proposition 2, it is easy to get the following observation.

Observation 3. *Let G be a connected graph with $\lambda_3(G) = k$, and x, y be two adjacent vertices of G. Then $d_G(x) \geq k + 1$ or $d_G(y) \geq k + 1$.*

Example 1. Let G be a path of length two and H be a complete graph of order four, and T_1, T_2 be two edge-disjoint S-trees in H, where $S = \{x, y, z\} \subseteq V(H)$. The structure of $G \circ (T_1 \cup T_2)$ is shown as F_a in Fig. 2, where the edges of a complete bipartite graph is simplified by bold black crossing edges. Note that $E(G \circ T_1) \cap E(G \circ T_2) = E(G \circ \mathcal{E}^{|S|})$.

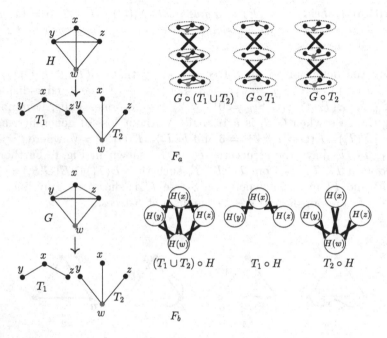

Fig. 2. The structures of $G \circ (T_1 \cup T_2)$ and $(T_1 \cup T_2) \circ H$.

Remark 1. Two edge-disjoint S-trees T_1, T_2 in H may have other vertices in common except S. If $V(T_1) \cap V(T_2) = W$, then $E(G \circ T_1) \cap E(G \circ T_2) = E(G \circ \mathcal{E}^{|W|})$.

Example 2. Let G be a complete graph of order four and H be an arbitrary graph, and T_1, T_2 be two edge-disjoint S-trees in G, where $S = \{x, y, z\} \subseteq V(G)$. The structure of $(T_1 \cup T_2) \circ H$ is shown as F_b in Fig. 2 and $E(T_1 \circ H) \cap E(T_2 \circ H) = E(H(S))$.

Remark 2. Two edge-disjoint S-trees T_1, T_2 in G may have other vertices in common except S. If $V(T_1) \cap V(T_2) = W$, then $E(T_1 \circ H) \cap E(T_2 \circ H) = E(H(W))$.

3   Lower Bound of $\lambda_3(G \circ H)$

In this section, we mainly prove the following theorem.

Theorem 2. *Let G and H be two non-trivial graphs such that G is connected. Then $\lambda_3(G \circ H) \geq \lambda_3(H) + \lambda_3(G)|V(H)|$. Moreover, the lower bound is sharp.*

By the following corollary, we know that the bound of the above theorem is sharp.

Corollary 1. $\lambda_3(P_s \circ P_t) = t + 1$.

Proof. By Theorem 2, $\lambda_3(P_s \circ P_t) \geq t + 1$. On the other hand, by Observation 2, $\lambda_3(P_s \circ P_t) \leq \delta(P_s \circ P_t) = t + 1$. Thus $\lambda_3(P_s \circ P_t) = t + 1$.

Let G be a graph with $V(G) = \{u_1, u_2, \cdots, u_{n_1}\}$ and $\lambda_3(G) = r_1$, and let H be a graph with $V(H) = \{v_1, v_2, \cdots, v_{n_2}\}$ and $\lambda_3(H) = r_2$. Set $S = \{x, y, z\} \subseteq V(G \circ H)$. Firstly, we give the sketch of the proof of Theorem 2. In total, the desired $r_2 + r_1 n_2$ S-trees are obtained on two stages: r_2 edge-disjoint S-trees by first-type and second-type edges on Stage I and $r_1 n_2$ edge-disjoint S-trees by the remaining first-type edges and the third-type edges on Stage II. Note that if H is disconnected, then $\lambda_3(H) = 0$ as defined, thus we omit Stage I immediately. Next we shall prove Theorem 2 by a series of lemmas according to the position of x, y, z in $G \circ H$.

Lemma 1. *If x, y, z belong to the same H-layer, then there exist $r_2 + r_1 n_2$ edge-disjoint S-trees.*

Proof. Without loss of generality, assume that $x, y, z \in H(u_1)$, $x = (u_1, v_1)$, $y = (u_1, v_2)$ and $z = (u_1, v_3)$. On Stage I, since $\lambda_3(H) = r_2$, there are r_2 edge-disjoint S-trees in $H(u_1)$. On Stage II, by Observation 2, u_1 has r_1 neighbors in G, say $\beta_1, \beta_2, \cdots, \beta_{r_1}$. Thus $T_{ij}^* = x(\beta_i, v_j) \cup y(\beta_i, v_j) \cup z(\beta_i, v_j)$ $(1 \leq i \leq r_1$ and $1 \leq j \leq n_2)$ are $r_1 n_2$ S-trees. These $r_2 + r_1 n_2$ S-trees are obviously edge-disjoint, as desired.

Lemma 2. *If exactly two of x, y and z belong to the same H-layer, then there exist $r_2 + r_1 n_2$ edge-disjoint S-trees.*

Proof. Assume that $x, y \in H(u_1)$, $z \in H(u_2)$. Let x'' and y'' be the vertices in $H(u_2)$ corresponding to x and y, and z' be the vertex in $H(u_1)$ corresponding to z, respectively. Consider the following two cases.

Case 1. $z' \in \{x, y\}$.

Without loss of generality, assume that $z' = x$, $x = (u_1, v_1)$, $y = (u_1, v_2)$ and $z = (u_2, v_1)$.

By Observation 1, there are r_2 edge-disjoint $v_1 v_2$-paths $P_1, P_2, \cdots, P_{r_2}$ in H such that $\ell(P_1) \leq \ell(P_2) \leq \cdots \leq \ell(P_{r_2})$. Denote the neighbor of v_1 in P_i by α_i $(1 \leq i \leq r_2)$. Set $D = \{\alpha_1, \alpha_2, \cdots, \alpha_{r_2}\}$. Notice that $\alpha_p \neq \alpha_q$ if $p \neq q$. Similarly, there are r_1 edge-disjoint u_1u_2-paths $Q_1, Q_2, \cdots, Q_{r_1}$ in G such that $\ell(Q_1) \leq \ell(Q_2) \leq \cdots \leq \ell(Q_{r_1})$. For each i with $1 \leq i \leq r_1$, set $Q_i = u_1\beta_{i,1}\beta_{i,2} \cdots \beta_{i,t_i-1}u_2$ and $\ell(Q_i) = t_i$. Also, note that $\beta_{p,1} \neq \beta_{q,1}$ if $p \neq q$.

On Stage I, the desired r_2 S-trees are obtained associated with the longest u_1u_2-path Q_{r_1}. If v_1 and v_2 are nonadjacent in H, then $T_i^* = P_i(u_1) \cup Q_{r_1}(\alpha_i) \cup z(u_2, \alpha_i)$ ($1 \leq i \leq r_2$) are r_2 S-trees as shown in Fig. 3(a), where $P_i(u_1)$ is the path in $H(u_1)$ corresponding to P_i in H, and $Q_{r_1}(\alpha_i)$ is the path in $G(\alpha_i)$ corresponding to Q_{r_1} in G. Now v_1 and v_2 are adjacent in H, that is, $P_1 = v_1v_2$ and $(u_1, \alpha_1) = y$. It follows from Observation 3 that $d_H(v_1) \geq r_2 + 1$ or $d_H(v_2) \geq r_2 + 1$, without loss of generality, say $d_H(v_1) \geq r_2 + 1$. For P_1, $T_1^* = xy \cup x(u_1, \alpha_{r_2+1}) \cup Q_{r_1}(\alpha_{r_2+1}) \cup z(u_2, \alpha_{r_2+1})$ is an S-tree, where $\alpha_{r_2+1} \notin D$, α_{r_2+1} is a neighbor of v_1 in H, and $Q_{r_1}(\alpha_{r_2+1})$ is the path in $G(\alpha_{r_2+1})$ corresponding to Q_{r_1}, see Fig. 3(b). For P_i ($2 \leq i \leq r_2$), set $T_i^* = P_i(u_1) \cup Q_{r_1}(\alpha_i) \cup z(u_2, \alpha_i)$. It is easy to see that these r_2 S-trees are edge-disjoint. The case that $d_H(v_2) \geq r_2+1$ can be proved similarly.

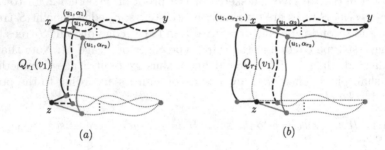

(a)   (b)

Fig. 3. The r_2 edge-disjoint S-trees where the edges of an S-tree are shown by the same type of lines.

Up to now, we should remark that the first-type edges incident with x and y in $G \circ H$ are not used whether or not v_1 and v_2 are adjacent in H. Since a vertex in $V(H) \setminus \{v_1, v_2\}$ may belong to more than one v_1v_2-path, we make use of either the r_2 neighbors of v_1 or the r_2 neighbors of v_2 to get our desired r_2 edge-disjoint S-trees.

Define a new graph $(G \circ H)^*$ from $G \circ H$ by deleting the edges of r_2 S-trees on Stage I. On Stage II, with the aid of Q_i ($1 \leq i \leq r_1$), we successively construct r_1n_2 S-trees in $(G \circ H)^*$ in non-decreasing order of the length of Q_i. We distinguish two subcases by the length t_1 of Q_1.

Subcase 1.1. $t_1 \geq 2$.

Recall that $Q_1 = u_1\beta_{1,1}\beta_{1,2} \cdots \beta_{1,t_1-1}u_2$. We will obtain n_2 internally disjoint xy-paths $A_1, A_2, \cdots, A_{n_2}$ in $K_{u_1,\beta_{1,1}}$, and a $V(H(\beta_{1,1}))V(H(\beta_{1,t_1-1}))$-linkage $B_1, B_2, \cdots, B_{n_2}$ by third-type edges associated with $\beta_{1,1}Q_1\beta_{1,t_1-1}$. Thus, $T_i^* = A_i \cup B_i \cup (\beta_{1,t_1-1}, v_i)z$ are n_2 edge-disjoint S-trees, where the subscript i ($1 \leq i \leq n_2$) of v_i is expressed module n_2 as one of $1, 2, \cdots, n_2$. Indeed, this can be done.

Set $A_i = x(\beta_{1,1}, v_i)y$ for $1 \leq i \leq n_2$. If $t_1 = 2$, then $B_i = \emptyset$. If $t_1 \geq 3$, then $B_i = (\beta_{1,1}, v_i)(\beta_{1,2}, v_{i+1})(\beta_{1,3}, v_i)(\beta_{1,4}, v_{i+1}) \cdots (\beta_{1,t_1-1}, v_i)$ and $T_i^* = A_i \cup B_i \cup (\beta_{1,t_1-1}, v_i)z$ when t_1 is even; $B_i = (\beta_{1,1}, v_i)(\beta_{1,2}, v_{i+1})(\beta_{1,3}, v_i)(\beta_{1,4}, v_{i+1}) \cdots (\beta_{1,t_1-1}, v_{i+1})$ and $T_i^* = A_i \cup B_i \cup (\beta_{1,t_1-1}, v_{i+1})z$ when t_1 is odd. For example, let $n_2 = 4$. Then 4 edge-disjoint S-trees are shown in Fig. 4 when $t_1 = 2$, $t_1 = 3$ and $t_1 = 4$, respectively.

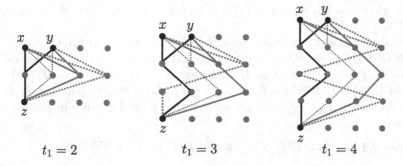

$$t_1 = 2 \qquad t_1 = 3 \qquad t_1 = 4$$

Fig. 4. The 4 edge-disjoint S-trees where the edges of an S-tree are shown by the same type of lines.

Subcase 1.2. $t_1 = 1$ and $Q_1 = u_1 u_2$.

Since $\lambda_3(G) = r_1$, it follows from Observation 3 that $d_G(u_1) \geq r_1 + 1$ or $d_G(u_2) \geq r_1 + 1$.

If $d_G(u_1) \geq r_1 + 1$, then denote another neighbor of u_1 in G by $\beta_{r_1+1,1}$ except u_2 and $\beta_{i,1}$ ($2 \leq i \leq r_1$). We obtain n_2 edge-disjoint S-trees associated with Q_1 as follows. Let $T_1^* = (\beta_{r_1+1,1}, v_1)x \cup (\beta_{r_1+1,1}, v_1)y \cup xz$, $T_2^* = (\beta_{r_1+1,1}, v_2)x \cup (\beta_{r_1+1,1}, v_2)y \cup yz$, $T_i^* = (u_2, v_i)x \cup (u_2, v_i)y \cup (u_2, v_i)(u_1, v_{i+1}) \cup (u_1, v_{i+1})z$ for $3 \leq i \leq n_2 - 1$, $T_{n_2}^* = (u_2, v_{n_2})x \cup (u_2, v_{n_2})y \cup (u_2, v_{n_2})(u_1, v_3) \cup (u_1, v_3)z$; see Fig. 5(a).

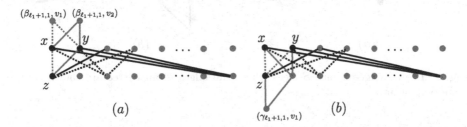

$$(a) \qquad\qquad (b)$$

Fig. 5. The n_2 edge-disjoint S-trees where the edges of an S-tree are shown by the same type of lines.

If $d_G(u_2) \geq r_1 + 1$, then denote another neighbor of u_2 in G by γ_{r_1+1} except u_1 and β_{i,t_i-1} ($2 \leq i \leq r_1$). For Q_1, set $T_1^* = xz \cup zy$, $T_2^* = xy'' \cup y''y \cup (\gamma_{r_1+1}, v_1)y'' \cup (\gamma_{r_1+1}, v_1)z$, $T_i^* = (u_2, v_i)x \cup (u_2, v_i)y \cup (u_2, v_i)(u_1, v_{i+1}) \cup (u_1, v_{i+1})z$ for

$3 \leq i \leq n_2 - 1$, and $T^*_{n_2} = (u_2, v_{n_2})x \cup (u_2, v_{n_2})y \cup (u_2, v_{n_2})(u_1, v_3) \cup (u_1, v_3)z$; see Fig. 5(b).

In both subcases, similar to Subcase 1.1, we are able to get n_2 edge-disjoint S-trees associated with Q_i $(2 \leq i \leq r_1)$, it follows that $n_2 r_1$ edge-disjoint S-trees are obtained, as desired.

Case 2. $z' \notin \{x, y\}$.

Assume that $x = (u_1, v_1)$, $y = (u_1, v_2)$ and $z = (u_2, v_3)$. Let $S' = \{v_1, v_2, v_3\}$ and $S'' = \{x, y, z'\}$.

By Observation 1, there are r_1 edge-disjoint $u_1 u_2$-paths $Q_1, Q_2, \cdots, Q_{r_1}$ in G such that $\ell(Q_1) \leq \ell(Q_2) \leq \cdots \leq \ell(Q_{r_1})$. By Proposition 1, r_2 edge-disjoint S'-trees $T_1, T_2, \cdots, T_{r_2}$ exist in H such that $0 \leq |\{T_i | E(T_i) \cap E(H[S'])\}| \leq 2$. Suppose $E(T_i) \cap E(H[S']) = \emptyset$ for $3 \leq i \leq r_2$. According to whether T_1 and T_2 share edges with $E(H[S'])$ or not, we get the desired S-trees in the following subcases.

Subcase 2.1. $E(T_1) \cap E(H[S']) = \emptyset$ and $E(T_2) \cap E(H[S']) = \emptyset$.

Denote the neighbor of v_3 in T_i by α_i where $1 \leq i \leq r_2$. On Stage I, let $T^*_i = T_i(u_1) \cup Q_{r_1}(\alpha_i) \cup z(u_2, \alpha_i)$, where $T_i(u_1)$ is the tree in $H(u_1)$ corresponding to T_i, $Q_{r_1}(\alpha_i)$ is the path in $G(\alpha_i)$ corresponding to Q_{r_1} for $1 \leq i \leq r_2$. On Stage II, if $\ell(Q_1) \geq 2$, then construct n_2 S-trees similar to Case 1; if $\ell(Q_1) = 1$, then either u_1 or u_2 has a neighbor which is not in each $u_1 u_2$-path Q_i in G. Thus n_2 S-trees associated with Q_1 are shown in Fig. 6 (u_1 has another neighbor in G in Fig. 6(a) and u_2 has another neighbor in G in Fig. 6(b)). Similar to Case 1, we obtain n_2 S-trees associated with Q_i for $2 \leq i \leq r_1$, thus there exist $r_2 + r_1 n_2$ edge-disjoint S-trees, as desired.

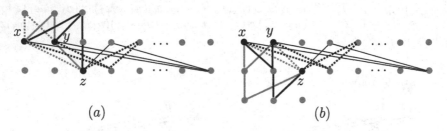

(a)   (b)

Fig. 6. The n_2 edge-disjoint S-trees where the edges of an S-tree are shown by the same type of lines.

Subcase 2.2. $E(T_1) \cap E(H[S']) \neq \emptyset$ and $E(T_2) \cap E(H[S']) = \emptyset$.

Suppose $|E(T_1) \cap E(H[S'])| = 1$ and $E(T_2) \cap E(H[S']) = \emptyset$. Furthermore, suppose $E(T_1) \cap E(H[S']) = v_1 v_2$ and $d_{T_1}(v_2) = 2$ (the other possibilities can be proved similarly). For $1 \leq i \leq r_2$, denote the neighbor of v_3 in T_i by α_i. Then we are able to obtain r_2 S-trees with the aid of α_i on Stage I and $r_1 n_2$ S-trees on Stage II similar to Subcase 2.1. It remains to consider the case that $|E(T_1) \cap E(H[S'])| = 2$

and $E(T_2) \cap E(H[S']) = \emptyset$. On Stage I, if $d_{Q_{r_1}}(u_1, u_2) \geq 2$ and $d_{T_1}(v_2) = 2$ or $d_{Q_{r_1}}(u_1, u_2) \geq 2$ and $d_{T_1}(v_3) = 2$, then an S-tree T_1^* associated with T_1 has the structure as shown in Fig. 7, where $\bar{x}$ is the neighbor of x'' in $Q_{r_1}(v_1)$; if $d_{Q_{r_1}}(u_1, u_2) = 1$, then $T_1^* = xyy''z$ (when u_1 has another neighbor outside Q_i) or $T_1^* = xyz'z$ (when u_2 has another neighbor outside Q_i). We obtain other $r_2 + r_1 n_2 - 1$ S-trees similar to Subcase 2.1. Thus there exist $r_2 + r_1 n_2$ edge-disjoint S-trees, as desired.

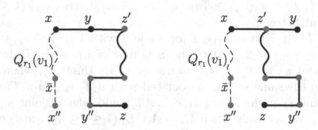

Fig. 7. The solid lines stand for the edges of the S-tree.

Subcase 2.3. $E(T_1) \cap E(H[S']) \neq \emptyset$ and $E(T_2) \cap E(H[S']) \neq \emptyset$.

Without loss of generality, suppose $|E(T_2) \cap E(H[S'])| = 1$. If $|E(T_1) \cap E(H[S'])| = 1$, then assume that the two S'-trees T_1 and T_2 have the structure as one of F_3, F_4, F_5, F_6 in Fig. 8, where v_2 is marked. For $1 \leq i \leq r_2$, denote the neighbor of v_2 in $T_i \setminus \{v_1, v_3\}$ by α_i and construct $r_2 + r_1 n_2$ S-trees similar to Subcase 2.1. So $|E(T_1) \cap E(H[S'])| = 2$, and then T_1 and T_2 have the structure F_7, where T_1 is shown in Fig. 8 by dotted lines. For $2 \leq i \leq r_2$, denote the neighbor of v_2 in $T_i \setminus \{v_1, v_3\}$ by α_i. Construct an S-tree T_1^* similar to Subcase 2.2 and other $r_2 + r_1 n_2 - 1$ S-trees similar to Subcase 2.1. Thus, there exist $r_2 + r_1 n_2$ edge-disjoint S-trees, as desired.

Fig. 8. Two S'-trees of type I.

Lemma 3. *If x, y, z belong to distinct H-layers, then there exist $r_2 + r_1 n_2$ edge-disjoint S-trees.*

Proof. Assume that $x \in H(u_1)$, $y \in H(u_2)$ and $z \in H(u_3)$. Let y', z' be the vertex corresponding to y, z in $H(u_1)$, x'', z'' be the vertex corresponding to x, z

in $H(u_2)$, and x''', y''' be the vertex corresponding to x, y in $H(u_3)$, respectively. We distinguish the following three cases.

Case 1. x, y, z belong to the same G-layer.

We may assume that $x = (u_1, v_1)$, $y = (u_2, v_1)$, $z = (u_3, v_1)$. It is easily seen that there are r_2 neighbors of v_1 in H, say $\alpha_1, \alpha_2, \cdots, \alpha_{r_2}$, and r_1 edge-disjoint $\{u_1, u_2, u_3\}$-trees $T_1, T_2, \cdots, T_{r_1}$ in G. For a tree T_i in G, set by $T_i(\alpha_j)$ the corresponding tree in $G(\alpha_j)$ for $1 \leq i \leq r_1$, $1 \leq j \leq r_2$.

On Stage I, $T_j^* = T_1(\alpha_j) \cup x(u_1, \alpha_j) \cup y(u_2, \alpha_j) \cup z(u_3, \alpha_j)$ $(1 \leq j \leq r_2)$ are r_2 edge-disjoint S-trees.

On Stage II, if T_j is of type I for some j with $1 \leq j \leq r_1$, then we may assume that $d_{T_j}(u_2) = 2$. Denote the neighbor of u_1, u_3 in T_j by η_j, γ_j and the neighbors of u_2 by β_j, $\bar{\beta}_j$ (β_j is nearer to u_1 than $\bar{\beta}_j$), where β_j, η_j and $\bar{\beta}_j$, γ_j may be the same vertex. Associated with $u_1 T_j u_2$ and $u_2 T_j u_3$, there are n_2 edge-disjoint xy-paths $A = \{A_1, \cdots, A_{n_2}\}$ and edge-disjoint yz-paths $B = \{B_1, \cdots, B_{n_2}\}$, respectively. Then $T_{ij}^* = A_i \cup B_i$ $(1 \leq i \leq n_2)$ are n_2 edge-disjoint S-trees. Indeed, this can be done. We will only provide the construction of A according to $d_{T_j}(u_1, u_2)$, since the construction of B is similar to that of A. If $d_{T_j}(u_1, u_2) = 1$, then set $A_1 = xy$, $A_i = x(u_2, v_i)(u_1, v_{i+1})y$ for $2 \leq i \leq n_2 - 1$, and $A_{n_2} = x(u_2, v_{n_2})(u_1, v_2)y$; if $d_{T_j}(u_1, u_2) = 2$, then set $A_i = x(\eta_j, v_i)y$ for $1 \leq i \leq n_2$. It remains to consider the case that $d_{T_j}(u_1, u_2) \geq 3$. Since there is a $V(H(\eta_j))V(H(\beta_j))$-linkage $D_1, D_2, \cdots, D_{n_2}$ by third-type edges of $G \circ H$ associated with $\eta_j T_j \beta_j$, it follows that $A_i = x(\eta_j, v_i) \cup D_i \cup (\beta_j, v_i)y$, where the subscript i $(1 \leq i \leq n_2)$ of v_i is expressed module n_2 as one of $1, 2, \cdots, n_2$. It remains to consider the case that T_j is of type II. Denote the neighbor of u_1, u_2, u_3 in T_j by η_j, β_j, γ_j and the only one three-degree vertex in T_j by w_j (η_j, β_j, γ_j and w_j may be the same vertex). We find a $V(H(\eta_j))V(H(\beta_j))$-linkage and a $V(H(\gamma_j))V(H(w_j))$-linkage respectively by third-type edges of $G \circ H$, and connect x, y, z respectively to $V(H(\eta_j))$, $V(H(\beta_j))$ and $V(H(\gamma_j))$. Thus, n_2 edge-disjoint S-trees are obtained associated with T_j. Since $1 \leq j \leq r_1$, it follows that $r_1 n_2$ edge-disjoint S-trees are obtained on Stage II, as desired.

Case 2.   Exactly two of x, y, z belong to the same G-layer.

We only consider the case $x = y'$ (the other cases $x = z'$ or $y' = z'$ can be proved by similar arguments). Assume that $x = (u_1, v_1)$, $y = (u_2, v_1)$ and $z = (u_3, v_2)$. Since $\lambda_3(H) = r_2$, there exist r_2 edge-disjoint $v_1 v_2$-paths $P_1, P_2, \cdots, P_{r_2}$ in H such that $\ell(P_1) \leq \ell(P_2) \leq \cdots \leq \ell(P_{r_2})$. For $1 \leq i \leq r_2$, denote the neighbor of v_1 and v_2 in P_i by α_i and β_i, respectively, and denote by $P_i(u_3)$ in $H(u_3)$ corresponding to P_i. Since $\lambda_3(G) = r_1$, there are r_1 edge-disjoint $\{u_1, u_2, u_3\}$-trees $T_1, T_2, \cdots, T_{r_1}$ in G.

On Stage I, if $\ell(P_1) \geq 2$, then set $T_i^* = x(u_1, \alpha_i) \cup y(u_2, \alpha_i) \cup z P_i(u_3)(u_3, \alpha_i) \cup T_1(\alpha_i)$ for $1 \leq i \leq r_2$. Otherwise, $\ell(P_1) = 1$, that is, v_1 is adjacent to v_2. Then $d_H(v_1) \geq r_2 + 1$ or $d_H(v_2) \geq r_2 + 1$. If $d_H(v_1) \geq r_2 + 1$, then $T_1^* = \{x(u_1, \alpha_{r_2+1}), y(u_2, \alpha_{r_2+1}), z x''', x'''(u_3, \alpha_{r_2+1})\} \cup T_1(\alpha_{r_2+1})$, where α_{r_2+1} is another neighbor of v_1 except α_i $(1 \leq i \leq r_2)$. If $d_H(v_2) \geq r_2 + 1$, then

$T_1^* = \{xz', z'(u_1, \beta_{r_2+1}), yz'', z''(u_2, \beta_{r_2+1}), z(u_3, \beta_{r_2+1})\} \cup T_1(\beta_{r_2+1})$, where β_{r_2+1} is another neighbor of v_1 except β_i ($1 \le i \le r_2$).

By similar arguments as in Case 1 of Lemma 3, $r_1 n_2$ edge-disjoint S-trees can be obtained on Stage II.

Case 3. x, y, z belong to different G-layers.

Assume that $x = (u_1, v_1)$, $y = (u_2, v_2)$ and $z = (u_3, v_3)$. Let $S' = \{v_1, v_2, v_3\}$ and $S'' = \{u_1, u_2, u_3\}$.

Since $\lambda_3(H) = r_2$, there are r_2 edge-disjoint S'-trees $T_1, T_2, \cdots, T_{r_2}$ in H. For $1 \le i \le r_2$, denote by α_i the vertex in T_i adjacent to a vertex in S', say v_1, and $\ell(T_i)$ denotes the size of T_i. Similarly, there are r_1 edge-disjoint S''-trees $T_1', T_2', \cdots, T_{r_1}'$ in G.

On Stage I, if $\ell(T_i) \ge 3$ for each i with $1 \le i \le r_2$, then let $T_i^* = x(u_1, \alpha_i) \cup yT_i(u_2)(u_2, \alpha_i) \cup zT_i(u_3)(u_3, \alpha_i) \cup T_1'(\alpha_i)$. Otherwise, similar to Case 2 of Lemma 2, the most difficult case is that there is an S'-tree of size two. Suppose $\ell(T_1) = 2$ and $d_{T_1}(v_2) = 2$. Thus T_1^* has three structures as shown in Fig. 9 where T_1' is of type II in Fig. 9(a), T_1' is of type I and $d_{T_1'}(u_1) = 2$ in Fig. 9(b) and T_1' is of type I and $d_{T_1'}(u_1) = 1$ in Fig. 9(c).

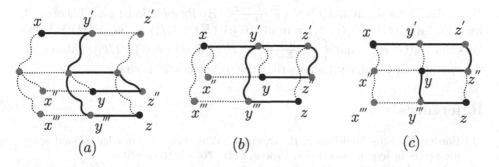

Fig. 9. The S-tree with the aid of T_1' shown by the solid lines.

On Stage II, $r_1 n_2$ edge-disjoint S-trees are obtained by similar arguments as in Case 1 of Lemma 3.

In each case, we obtain $r_2 + r_1 n_2$ S-trees, and it is easily seen that these S-trees are edge-disjoint, as desired.

From the above three lemmas, Theorem 2 follows immediately.

4   Upper Bound of $\lambda_3(G \circ H)$

In this section, we give an upper bound of the generalized 3-edge-connectivity of the lexicographic product of two graphs.

Yang and Xu [33] investigated the classical edge-connectivity of the lexicographic product of two graphs.

Theorem 3. [33] *Let G and H be two non-trivial graphs such that G is connected. Then*

$$\lambda(G \circ H) = \min\{\lambda(G)|V(H)|^2, \delta(H) + \delta(G)|V(H)|\}.$$

In [22], the sharp lower bound of the generalized 3-edge-connectivity of a graph is given as follows.

Proposition 3. *[22] Let G be a connected graph with n vertices. For every two integers s and r with $s \geq 0$ and $r \in \{0, 1, 2, 3\}$, if $\lambda(G) = 4s + r$, then $\lambda_3(G) \geq 3s + \lceil \frac{r}{2} \rceil$. Moreover, the lower bound is sharp. We simply write $\lambda_3(G) \geq \frac{3\lambda(G)-2}{4}$.*

From the above two results, we get the following upper bound of $\lambda_3(G \circ H)$.

Theorem 4. *Let G and H be two non-trivial graphs such that G is connected. Then*

$$\lambda_3(G \circ H) \leq \min\left\{ \left\lfloor \frac{4\lambda_3(G) + 2}{3} \right\rfloor |V(H)|^2, \delta(H) + \delta(G)|V(H)| \right\}.$$

Moreover, the upper bound is sharp.

Proof. By Proposition 3, $\lambda(G) \leq \lfloor \frac{4\lambda_3(G)+2}{3} \rfloor$. By Proposition 1 and Theorem 3, we have $\lambda_3(G \circ H) \leq \lambda(G \circ H) = \min\{\lambda(G)|V(H)|^2, \delta(H) + \delta(G)|V(H)|\}$. It follows that $\lambda_3(G \circ H) \leq \min\left\{ \left\lfloor \frac{4\lambda_3(G)+2}{3} \right\rfloor |V(H)|^2, \delta(H) + \delta(G)|V(H)| \right\}$. Moreover, the example in Corollary 1 shows that the upper bound is sharp.

References

1. Barden, B., Libeskind-Hadas, R., Davis, J., Williams, W.: On edge-disjoint spanning trees in hypercubes. Infor. Proces. Lett. **70**, 13–16 (1999)
2. Beineke, L.W., Wilson, R.J.: Topics in Structural Graph Theory. Cambrige University Press, Cambrige (2013)
3. Bondy, J.A., Murty, U.S.R.: Graph Theory. GTM 244. Springer, Berlin (2008)
4. Brešar, B., Špacapan, S.: Edge connectivity of strong products of graphs. Discuss. Math. Graph Theory **27**, 333–343 (2007)
5. Chartrand, G., Okamoto, F., Zhang, P.: Rainbow trees in graphs and generalized connectivity. Networks **55**(4), 360–367 (2010)
6. Cheng, X., Du, D.: Steiner Trees in Industry. Kluwer Academic Publisher, Dordrecht (2001)
7. Du, D., Hu, X.: Steiner Tree Problems in Computer Communication Networks. World Scientific, River Edge (2008)
8. Feng, M., Xu, M., Wang, K.: Identifying codes of lexicographic product of graphs. Electron. J. Comb. **19**(4), 56–63 (2012)
9. Grötschel, M.: The Steiner tree packing problem in $VLSI$ design. Math. Program. **78**, 265–281 (1997)
10. Hammack, R., Imrich, W., Klavžar, S.: Handbook of Product Graphs, 2nd edn. CRC Press, Boca Raton (2011)
11. Hager, M.: Pendant tree-connectivity. J. Combin. Theory **38**, 179–189 (1985)

12. Imrich, W., Klavžar, S.: Product Graphs: Structure and Recognition. Wiley, New York (2000)
13. Klavžar, S., Špacapan, S.: On the edge-connectivity of Cartesian product graphs. Asian-Europ. J. Math. **1**, 93–98 (2008)
14. Ku, S., Wang, B., Hung, T.: Constructing edge-disjoint spanning trees in product networks. IEEE Trans. Parallel Distrib. Syst. **14**(3), 213–221 (2003)
15. Li, F., Xu, Z., Zhao, H., Wang, W.: On the number of spanning trees of the lexicographic product of networks. Sci. China Ser. F **42**, 949–959 (2012)
16. Li, H., Li, X., Sun, Y.: The generalied 3-connectivity of Cartesian product graphs. Discrete Math. Theor. Comput. Sci. **14**(1), 43–54 (2012)
17. Li, H., Li, X., Mao, Y.: On extremal graphs with at most two internally disjoint Steiner trees connecting any three vertices. Bull. Malays. Math. Sci. Soc. **37**(2,3), 747–756 (2014)
18. Li, S., Li, X., Zhou, W.: Sharp bounds for the generalized connectivity $\kappa_3(G)$. Discrete Math. **310**, 2147–2163 (2010)
19. Li, S., Li, W., Li, X.: The generalized connectivity of complete equipartition 3-partite graphs. Bull. Malays. Math. Sci. Soc. **37**(1,2), 103–121 (2014)
20. Li, X., Mao, Y.: The generalized 3-connectivity of lexicographic product graphs. Discrete Math. Theor. Comput. Sci. **16**(1), 339–354 (2014)
21. Li, X., Mao, Y.: The minimal size of a graph with given generalized 3-edge-connectivity. Accepted for publication in Ars Comb
22. Li, X., Mao, Y., Sun, Y.: On the generalized (edge-)connectivity of graphs. Austral. J. Comb. **58**, 304–319 (2014)
23. Nash Williams, CStJA: Edge-disjonint spanning trees of finite graphs. J. London Math. Soc. **36**, 445–450 (1961)
24. Oellermann, O.R.: Connectivity and edge-connectivity in graphs: a survey. Cong. Numer. **116**, 231–252 (1996)
25. Ozeki, K., Yamashita, T.: Spanning trees: a survey. Graphs Combin. **27**(1), 1–26 (2011)
26. Palmer, E.: On the spanning tree packing number of a graph: a survey. Discrete Math. **230**, 13–21 (2001)
27. Sherwani, N.: Algorithms for *VLSI* Physical Design Automation, 3rd edn. Kluwer Academic Publishers, London (1999)
28. Sun, Y.: Generalized 3-edge-connectivity of Cartesian product graphs. Accepted for publication in Czech. Math. J.
29. Tutte, W.: On the problem of decomposing a graph into n connected factors. J. London Math. Soc. **36**, 221–230 (1961)
30. Volkmann, L.: Edge connectivity in p-partite graphs. J. Graph Theory **13**(1), 1–6 (1989)
31. West, D., Wu, H.: Packing Steiner trees and S-connectors in graphs. J. Comb. Theory Ser. B **102**, 186–205 (2012)
32. Xu, J., Yang, C.: Connectivity of Cartesian product graphs. Discrete Math. **306**, 159–165 (2006)
33. Yang, C., Xu, J.: Connectivity of lexicographic product and direct product of graphs. Ars Comb. **111**, 3–12 (2013)

Applied Optimization

Integer Programming Methods for Special College Admissions Problems

Péter Biró[1] (✉) and Iain McBride[2]

[1] Research Centre for Economic and Regional Studies, Institute of Economics, Hungarian Academy of Sciences, Budaörsi út 45, Budapest 1112, Hungary
peter.biro@krtk.mta.hu

[2] School of Computing Science, University of Glasgow, Sir Alwyn Williams Building, Glasgow G12 8QQ, UK
i.mcbride.1@research.gla.ac.uk

Abstract. We develop Integer Programming (IP) solutions for some special college admission problems arising from the Hungarian higher education admission scheme. We focus on four special features, namely the solution concept of stable score-limits, the presence of lower and common quotas, and paired applications. We note that each of the latter three special feature makes the college admissions problem NP-hard to solve. Currently, a heuristic based on the Gale-Shapley algorithm is being used in the application. The IP methods that we propose are not only interesting theoretically, but may also serve as an alternative solution concept for this practical application, and other similar applications.

Keywords: College admissions problem · Integer programming · Stable score-limits · Lower quotas · Common quotas · Couples

1 Introduction

Gale and Shapley [13] introduced and solved the college admissions problem, which generated a broad interdisciplinary research field in mathematics, computer science, game theory and economics[1]. The Hungarian higher education admission scheme is also based on the Gale-Shapley algorithm, but it is extended with a number of heuristics since the model contains some special features. In this paper we will study the possibility of modelling these special features with integer programming techniques.

In the Hungarian higher education matching scheme (see a detailed description in [5,27]), the students apply for programmes. However, for simplicity,

P. Biró: Visiting faculty at the Economics Department, Stanford University in year 2014. Supported by the Hungarian Academy of Sciences under its Momentum Programme (LD-004/2010) and also by OTKA grant no. K108673.

L. McBride: Supported by a SICSA Prize PhD Studentship.

[1] The 2012 Nobel-Prize in Economic Sciences has been awarded to Alvin Roth and Lloyd Shapley for the theory of stable allocations and the practice of market design.

© Springer International Publishing Switzerland 2014
Z. Zhang et al. (Eds.): COCOA 2014, LNCS 8881, pp. 429–443, 2014.
DOI: 10.1007/978-3-319-12691-3_32

we will refer to the programmes as colleges in our models. The first special feature of the application is the presence of ties, and the solution concept of stable score-limits. According to the Hungarian admission policy, when two applicants have the same score at a programme then they should either both be accepted or rejected by that programme. The solution of stable score-limits ensures that no quota is violated, since essentially the last group of students with the same score that would cause a quota violation is always rejected. A set of stable score-limits always exists, and a student-optimal solution can be found efficiently by an extension of the Gale-Shapley algorithm, as shown in [8]. This method is the basis of the heuristic used in the Hungarian application.

The second and third special features studied in this paper are the lower and common quotas. A university may set not just an upper quota for the number of admissible students for a programme, but also a lower quota. A violation of this lower quota would imply the cancellation of the programme. Furthermore, a common upper quota may be also introduced for a set of programmes, to limit the number of students admitted to a faculty, at a university or nationwide with regard to the state-financed seats in a particular subject. These concepts were studied in [6], where the authors showed that each of these special features makes the college admission problem NP-hard, as present in the Hungarian application. Finally, students can apply for pairs of programmes in case of teacher studies. This possibility was reintroduced to the scheme in 2010. This problem is closely related to the Hospitals/Residents problem with Couples, where couples may apply for pairs of positions. The latter problem is also known to be NP-hard [20], even for so-called consistent preferences [18], and even for a specific setting present in Scotland [7] where hospitals have common rankings.

The polytope of stable matchings was described in a number of papers for the stable marriage problem [22,24], and for the college admissions problem [4,11,25]. For these classical models, since the extremal point of the polytopes are integral, and thus correspond to stable matchings, one could always use a linear programming solver to compute stable solutions (although the computation of solutions can also be done efficiently using the Gale-Shapley algorithm).

However, by introducing even just one special feature, the existence of a stable matching can no longer be guaranteed, and the problem of finding a stable solution can even be NP-hard. In such cases it may be worth investigating integer programming techniques for solving these problems in theory and in practice as well. To the best of our knowledge there have been only two recent studies of this kind so far. In the first study Kwanashie and Manlove [17] investigated the problem of finding a maximum size weakly stable matchings for college admissions problems with ties, a problem known to be NP-hard, and motivated by the Scottish resident allocation scheme. In the other paper [10] the above mentioned matching with couples problem has been studied. As we have already mentioned, one of the four special features of the Hungarian higher education scheme, namely the presence of paired applications, has a close connection to the problem of couples. However, the remaining three special features studied in our paper, the stable score-limits, the lower and common quotas have not been investigated from this perspective.

It is interesting to note that whilst we are not aware of any large scale application for two-sided stable matching markets, except two minor examples[2], integer programming is the standard technique used for kidney exchange programs
[1,19,23].

Finally, we would like to highlight that the models and solution techniques
presented in this college admission context may well be useful for other applications too. Two important other applications are immediately apparent. Firstly,
controlled school choice [2], where the policy makers might want to improve the
social-ethnic diversity of the schools by setting different quotas for some types
of students. The other example is the resident allocation program, as used in
Japan [16], where both lower and upper quotas can be requested as regional
caps to ensure a better coverage in health care services in all geographic areas
with regard to each medical specialty.

In Sect. 2 we describe a basic model for the classical College Admissions
problem that will be the basis of our extended models. In Sect. 3 we consider the
College Admissions problem with ties and describe two integer linear programs
for finding a stable set of score-limits. The first model uses the objective function
to achieve stability and thus leads to the student-optimal stable set of score-
limits. The second model describes all the stable sets of score-limits using an
extended IP model. In Sect. 4 we formulate an IP for describing the College
Admissions problem with lower quotas, we provide some useful lemmas that can
speed up a solution. In Sect. 5 we study the College Admissions problem with
common quotas and we give an integer programming model that describes the
set of stable solutions.

2   A Model for the Classical College Admissions Problem

Our basic model is an extension of the Rothblum [24] model (analysed also in
[22]). This model has been described in [4].

Let $A = \{a_1, \ldots a_n\}$ be the set of applicants and $C = \{c_1, \ldots, c_m\}$ the set of
colleges. Let u_j denote the upper quota of college c_j. Regarding the preferences
and priorities, let r_{ij} denote the rank of college c_j in a_i's preference list, meaning
that a_i prefers c_j to c_k if $r_{ij} < r_{ik}$. Let s_{ij} be an integer representing the score
of a_i at college c_j, meaning that a_i has priority over a_k at college c_j if $s_{ij} > s_{kj}$,
where $\bar{s}$ is the maximum possible score. We denote the set of applications by E.

We introduce binary variables $x_{ij} \in \{0, 1\}$ for each application coming from
a_i to c_j, as a characteristic function of the matching, where $x_{ij} = 1$ corresponds

---

[2] In a famous study Roth [21] analysed the nature and the long term success of a
dozen resident allocation schemes established in the UK in the late seventies. He
found that two schemes produced stable outcomes and both of them remained in
use. From the remaining six ones, that did not always produce stable matchings, four
were eventually abandoned. The two programs that were not always produced stable
solutions but yet remained used were based on linear programming techniques and
has been operated for the two smallest market. Ünver [26] studied these programs
and the possible reasons of their survival in detail.

to the case when a_i is assigned to c_j. The feasibility of a matching can be ensured with the following two sets of constraints.

$$\sum_{j:(a_i,c_j)\in E} x_{ij} \leq 1 \text{ for each } a_i \in A \tag{1}$$

$$\sum_{i:(a_i,c_j)\in E} x_{ij} \leq u_j \text{ for each } c_j \in C \tag{2}$$

Here, (1) implies that no applicant can be assigned to more than one college, and (2) implies that the upper quotas of the colleges are respected.

One way to enforce the stability of a feasible matching is by the following constraint.

$$\left(\sum_{k:r_{ik}\leq r_{ij}} x_{ik}\right) \cdot u_j + \sum_{h:(a_h,c_j)\in E, s_{hj}>s_{ij}} x_{hj} \geq u_j \text{ for each } (a_i,c_j) \in E \tag{3}$$

Note that for each $(a_i, c_j) \in E$, if a_i is matched to c_j or to a more preferred college than the first term provides the satisfaction of the inequality. Otherwise, when the first term is zero, then the second term is greater than or equal to the right hand side if and only if the places at c_j are filled with applicants with higher scores.

Remark 1: When we have ties in the priorities (due to equal scores), then the following modified stability constraints, together with the feasibility constraints (1) and (2), lead to *weakly stable* matchings (in a model also known as Hospitals/Residents problem with Ties).

$$\left(\sum_{k:r_{ik}\leq r_{ij}} x_{ik}\right) \cdot u_j + \sum_{h:(a_h,c_j)\in E, s_{hj}\geq s_{ij}} x_{hj} \geq u_j \text{ for each } (a_i,c_j) \in E \tag{4}$$

Note that weakly stable matchings can have different sizes and the problem of finding a maximum size weakly stable matching is NP-hard (although some good approximation results exist, e.g. in [15]). See more about this problem, and its solutions by IP techniques in the recent paper of Kwanashie and Manlove [17].

Remark 2: In the absence of ties, we can get an applicant-optimal (resp. an applicant-pessimal) stable solution by setting the objective function of the IP as the minimum (resp. maximum) of the following term:

$$\sum_{(a_i,c_j)\in E} r_{ij} \cdot x_{ij}$$

We note that these extreme solutions can be obtained with the two versions of Gale and Shapley's deferred acceptance algorithm in linear time [13].

Remark 3: Baïou and Balinski [4] proposed an alternative model to describe the stable admission polytope, since the above simple integer program may admit fractional solutions as extreme point. See also Sethuraman et al. [25] about the alternative model. Fleiner [11] provided a different description for the stable admission polytope.

3   Stable Score-Limits

The use of score-limits (or cutoff scores) is very common in college admission systems. The applicants have a score at each place to which they are applying and they are ranked according to these scores by the colleges. The solution is announced in terms of score-limits, each college (or a central coordinator) announces the score of the last admitted student, and each student is then admitted to her most preferred place on her preference list where she achieved the score-limit. See more about the Irish, Hungarian, Spanish and Turkish applications in [8]. The score-limits can be seen as a kind of dual solution of a matching, or prices in a competitive equilibrium. Azevedo and Leshno [3] analysed this phenomenon in detail.

In this section we first develop a basic model for the classical College Admissions problem by using score-limits. Then we discuss the case when ties can appear due to students applying to a place with the same score, as happens in Hungary. We show how this extended setting can be described with a similar IP both with and without the use of an objective function.

3.1   Stable Score-Limits with No Ties

If we are given a stable matching for a College Admissions problem then we can define a stable set of score-limits by keeping the following requirements. Each student must meet the score-limit of the college where she is admitted, and no student meets the score-limit of a college that rejected her. These two requirements imply that every student is admitted to the best place where she achieved the score-limit. Finally, to ensure that no student is rejected if a quota was not filled we shall require the score-limit of each unfilled college to be minimal.

To describe this solution concept with an integer program, we introduce new variables for the score-limits. Let t_j be the score-limit at college c_j, where $0 \leq t_j \leq \bar{s} + 1$. The feasibility constraints (1) and (2) remain the same, we only need to link the score-limits to the matching and establish the new stability conditions as follows.

$$t_j \leq (1 - x_{ij}) \cdot (\bar{s} + 1) + s_{ij} \text{ for each } (a_i, c_j) \in E \qquad (5)$$

and

$$s_{ij} + 1 \leq t_j + \left(\sum_{k : r_{ik} \leq r_{ij}} x_{ik} \right) \cdot (\bar{s} + 1) \text{ for each } (a_i, c_j) \in E \qquad (6)$$

Here, (5) implies that if an applicant is admitted to a college then she achieved the score-limit of that college. The other constraint, (6), ensures that if applicant a_i is not admitted to c_j then either her score at c_j is lower than the score-limit, t_j, or she is admitted to a college that she preferred.

Finally, we need to ensure that each college that could not fill its quota has a minimal score-limit. We introduce an indicator variable $f_j \in \{0, 1\}$ for each college c_j which is equal to zero if the college is unfilled, by using the following constraint.

$$f_j \cdot u_j \leq \sum_{i:(a_i, c_j) \in E} x_{ij} \text{ for each } c_j \in C \tag{7}$$

Then the following constraint ensures that if a college is unfilled then its score-limit is minimal.

$$t_j \leq f_j(\bar{s} + 1) \text{ for each } c_j \in C \tag{8}$$

We summarise the above statements in the following Theorem.

Theorem 1. *The stable matchings and the related stable sets of score-limits of a College Admissions problem correspond to the solutions of the integer linear program consisting of the feasibility conditions* (1), (2), *the stability conditions* (5), (6) *and conditions* (7), (8).

Proof. A matching is feasible if and only if the corresponding solution satisfy the feasibility constraints. Condition (5) implies that if an applicant is admitted to a college then she achieved the score-limit of that college, and (6) implies that she has not achieved the score-limit of any college that she prefers to her assignment. Therefore, these two conditions are satisfied if and only if every applicant is admitted to the best place in her list where she achieved the score-limit. Finally, (7) and (8) ensure that no college can have positive score-limit, and therefore no college can reject any applicant, if its quota is not filled.   □

3.2   Stable Score-Limits with Ties

The problem with ties has been defined in [5] and studied in [8,12]. Ties can appear, as the scores of the applicants might be equal at a college, and these ties are never broken in the Hungarian application. Therefore a group of students with the same score are either all accepted or all rejected. In the Hungarian application the upper quotas are always satisfied, and the stability is defined with score-limits (cutoff scores) as follows. For a set of score-limits, each applicant is admitted to the first place in her list where she achieves the score-limit. A set of score-limits is feasible if no quota is violated. It is stable if no score-limit can be lowered at any college without violating its quota, while keeping the other score-limits unchanged. When no ties occur then this definition is equivalent to the original Gale-Shapley one. A stable set of score-limits always exists, and may be found using a generalised Gale-Shapley algorithm. Moreover the applicant proposing version leads to an applicant-optimal solution, where each of the score-limit at each college is as small as possible, and a similar statement applies for the college proposing-version (see [5, 8, 12] for details).

Here we present an IP formulation to find an applicant-optimal set of score-limits with respect to the Hungarian version (where no upper quota may be violated) this is called *H-stable set of score-limits* in [8]. The feasibility constraints (1) and (2) remain the same as in the previous model, and also the two requirements regarding the score-limits, expressed in constraints (5), (6). However, we cannot require the unfilled colleges to have minimal score-limits in this model, since an unfilled seat might be created by a tie. We describe two possible solutions for this problem. The first is by the following simple objective function.

$$\min \sum_{j=1\ldots m} t_j \tag{9}$$

The above objective function is sufficient to ensure the stability condition, that is no college can decrease its score limit without violating its quota, supposing that the other score-limits remain the same. To summarise, we state the correctness of the integer program as follows, the proof is in the extended version of our paper [9].

Theorem 2. *Feasibility conditions* (1), (2) *and stability conditions* (5), (6) *together with the objective function* (9) *comprise an integer linear program such that the optimal solution of this IP corresponds to the applicant-optimal stable set of score-limits.*

When we want to describe all sets of stable score-limits (and not only the applicant-optimal one) then we have to replace the objective function with some additional conditions. We can do that by introducing some new variables, as described in the next subsection.

3.3 Stable Score-Limits with Ties and Free Objective Function

First, we introduce a binary variable y_j for each college c_j that is equal to one when t_j is positive (i.e. when there are some applicants rejected from c_j), and otherwise it is zero.

$$t_j \leq (\bar{s} + 1)y_j \text{ for each } c_j \in C \tag{10}$$

Then, for each application (a_i, c_j), we define a new binary variable, d_{ij}, that can be equal to one if a_i desires c_j compared to her actual match and a_i would meet the admission criteria at c_j if the score-limit at c_j was decreased by one, where m denotes the number of colleges.

$$\sum_{r_{ik} \geq r_{ij}} d_{ik} \leq (1 - x_{ij})m \text{ for each } (a_i, c_j) \in E \tag{11}$$

$$t_j - 1 \leq (1 - d_{ij})\bar{s} + s_{ij} \text{ for each } (a_i, c_j) \in E \tag{12}$$

With the help of the new variables, we can now describe the stability condition of the score-limits as follows.

$$(u_j + 1)(1 - y_j) + \sum_{i:(a_i,c_j)\in E} (x_{ij} + d_{ij}) \geq u_j + 1 \text{ for each } c_j \in C \qquad (13)$$

Theorem 3. *Feasibility conditions* (1), (2), *stability conditions* (5) *and* (6) *and conditions* (10), (11), (12) *for the new variables* d_{ij}, *together with a new stability condition* (13) *comprise an integer linear program such that each feasible integer solution corresponds to a stable set of score-limits.*

Proof. Again, the feasibility conditions ensure that the corresponding matching is feasible, if and only if the assignment of values to the variables in the ILP is feasible. Similarly to the previous model, (5) implies that if an applicant is admitted to a college then she achieved the score-limit of that college, and (6) implies that each applicant is admitted to the best available place in her list, if admitted somewhere. Now we shall prove that the remaining four sets of conditions are satisfied if and only if the set of score-limits is stable, i.e. when no college can decrease its score-limit without violating its quota. Suppose first that we have a stable set of score-limits. We assign values to all variables in the IP model appropriately and we prove that the constraints are satisfied. So let t_j be the score-limit at college c_j and we set y_j to be one if t_j is positive. Let d_{ij} be equal to one for each applicant a_i who would prefer to be matched to college c_j than her current partner and who would also meet the score-limit at c_j if it was decreased by one (i.e. $s_{ij} = t_j - 1$), and we set all the other d_{ij} variables to be zero. When doing so, we satisfy conditions (11) and (12), since (11) is satisfied when no d_{ik} is equal to one if a_i prefers her current match c_j to c_k, and (12) is satisfied if a_i meets the score-limit at c_j if it is decreased by one. The stability of the set of score-limits means that no college (with a positive score-limit) can decrease its score-limit without violating its quota. This means that if a college has a positive score-limit, so $y_j = 1$, then if it decreased its score-limit by one then the new students admitted were exactly those for whom the corresponding variable d_{ij} is equal to one. The violation of the quota implies that (13) must be satisfied. To prove the converse, let us suppose that we have an assignment of values to the variables in our IP model such that all the constraints in our IP model are satisfied. We shall prove that the set of score-limits as defined by variables t_j is stable, that is, for each college c_j either its score-limit is zero or the decrease of its score-limit would cause a quota violation. When t_j is positive then y_j must be zero, so $\sum_{i:(a_i,c_j)\in E}(x_{ij} + d_{ij}) \geq u_j + 1$. Since d_{ij} can be one only if a_i both desire and deserves c_j when t_j is decreased to $t_j - 1$, this means that the quota at c_j would be indeed violated when the score-limit would be decreased by one. (Note that y_j does not necessarily have to be zero when t_j is zero and d_{ij} does not necessarily have to be one when a_i both deserves and desires c_j.) □

Therefore now we can compute both the student-optimal and student-pessimal stable score-limits, by setting the objective function as described in Remark 2.

These extremal solutions can be also computed efficiently by the two generalised versions of the Gale-Shapley algorithm, as shown in [8].

We note that in [8] there was another stability definition, the so-called L-stability, that is based on a more relaxed admission policy, namely when the last group of student with the same score with whom the quota would be violated are always accepted. In this paper we are focusing on the setting that is present in the Hungarian application, so we do not deal with L-stability, but it would be possible to describe an IP model for that version as well in a similar fashion.

4 Lower Quotas

In this section we extend the classical College Admissions problem with the possibility of having lower quotas set for the colleges. After developing an integer program for finding a stable solution for this problem we describe the current heuristic used in Hungary and we also provide some Lemmas that can speed up the solution of the IP. Finally we discuss the possibility of having lower quotas for sets of colleges.

4.1 College Admissions Problem with Lower Quotas

This problem has been defined in [6]. In addition to the College Admissions model, here we have lower quotas as well. Let l_j be the lower quota of college c_j. In a feasible solution a college can either be closed (in which case there is no student assigned to there), or open, when the number of students admitted must be between its lower and upper quotas. To describe this feasibility requirement, besides keeping (1), we modify (2) as follows. We introduce a new binary variable, $o_j \in \{0,1\}$ for each college c_j, where $o_j = 1$ corresponds to the case when the college is open.

$$o_j \cdot l_j \leq \sum_{i:(a_i,c_j)\in E} x_{ij} \leq o_j \cdot u_j \text{ for each } c_j \in C \qquad (14)$$

The above set of constraints together with (1) ensure the feasibility of the matching.

The stability of a solution requires the lack of traditional blocking pairs for open colleges, and the lack of blocking groups for closed colleges. The latter means that there cannot be at least as many unsatisfied students (unassigned or assigned to a less preferred place) at a college as the lower quota of that college. The stability conditions can be enforced with the following conditions.

$$\left(\sum_{k:r_{ik}\leq r_{ij}} x_{ik} \right) \cdot u_j + \sum_{h:(a_h,c_j)\in E, s_{hj}>s_{ij}} x_{hj} \geq o_j \cdot u_j \text{ for each } (a_i,c_j) \in E \quad (15)$$

$$\sum_{i:(a_i,c_j)\in E} \left[1 - \sum_{k:r_{ik}<r_{ij}} x_{ik} \right] \leq (1-o_j) \cdot (l_j - 1) + o_j \cdot n \text{ for each } c_j \in C \quad (16)$$

The first condition implies the usual stability for open colleges, whilst the second condition implies the group-stability for closed colleges. Below we give a formal definition for the College Admission problem with lower quota and a proof of the above description in the following Theorem.

Theorem 4. *The feasibility conditions* (1) *and* (14) *together with the stability conditions* (15) *and* (16) *form an integer program such that its solutions correspond to the stable matchings of a college admissions problem with lower quotas.*

Proof. A solution of the IP satisfies the feasibility conditions (1) and (14) if and only if the corresponding matching is feasible, i.e., no student is admitted to more than one college and the lower and upper quotas are respected in each open college. Regarding stability, condition (15) is redundant if the college is closed and implies the pairwise stability condition for any open college. Condition (16) is redundant for open colleges, and enforces group-stability for any closed college. To show the latter we shall see that the right hand side of the constraint is $l_j - 1$ for a closed college c_j and on the left hand side those applicants of c_j are counted who are not admitted to any preferred place according to their preferences, so these are the students who would be happy if c_j would be open and admit them them.   □

4.2 Heuristics

The problem of finding a stable matching for the college admissions problem with lower quotas is proved to be NP-hard [6]. In the Hungarian application, where lower quotas can be set for any programme, the following heuristic is used with regard to this special feature. First, the applicant-proposing Gale-Shapley algorithm produces a stable matching where some lower quotas might be violated. The heuristic closes one programme, where the ratio of the number of students admitted and lower quota is minimal, and then the applicant-proposing Gale-Shapley algorithm continues by letting the rejected students (whose assigned programme has just been cancelled) apply to their next choices. This heuristic runs in linear time in the number of applications, and it produces the applicant-optimal stable matching for the remaining open colleges. However, as it was illustrated in [6], this heuristic can easily produce unstable outcomes even when the problem is solvable. With the IP technique, however, the IP model should guarantee to find a stable solution, whenever it exists. The following lemmas can help in speeding up the solver. The proof of these lemmas can be found it the extended version of our paper [9].

Lemma 1. *Let I be in instance of College Admission problem and I' be a reduced market where a college is missing. Then the number of students admitted to any college in I' must be at least as many as the number of students admitted in I. Moreover, when we compare the student-optimal (resp. college optimal) stable matchings in I and I' then each student gets admitted to a college at least as good in I then in I'.*

Lemma 2. *The colleges that reach their lower quotas in the stable solutions of a College Admissions problem with no lower quotas must be open in every stable solution where lower quotas are respected.*

Lemma 3. *Suppose that X is the set of colleges that do not reach their lower quotas in the stable solutions with no lower quotas. Given a college c_j of X, if all the colleges in X but c_j are closed and c_j still does not achieve its lower quota then c_j must be closed in any stable solution with lower quotas.*

5   Common Quotas

This problem has also been defined in [6]. For each set of colleges $C_p \subseteq C$ the coordinator of the scheme may set a common upper quota, u_p, meaning that the total number of students admitted to colleges in C_p cannot exceed this quota. Therefore, the set of feasibility constraints, (1) and (2), has to be extended with some new constraints enforcing the common quotas, as follows.

$$\sum_{(a_i,c_j)\in E, c_j \in C_p} x_{ij} \leq u_p \text{ for each } C_p \subseteq C \qquad (17)$$

Regarding stability, first of all we have to suppose that any two colleges, c_j and c_k, that belong to a set of colleges C_p with a common quota must rank their applicants in the same way. In particular, in the Hungarian application any student a_i has the same score at such colleges (i.e. programmes in Hungary) with common quota, so $s_{ij} = s_{ik}$ holds. (In a more general model, we should suppose to have a specific scoring for each set of colleges with common quota, which is in agreement with the individual scorings of the colleges belonging to this set. For instance, we could have a score s_{ij}^p for each application associated to a set of colleges C_p with a common quota such that $s_{ij} > s_{lj}$ implies $s_{ij}^p > s_{lj}^p$).

In this setting stability means that if a student a_i is not admitted to a college c_j or to any better college of her preference then either c_j must have filled its quota with better students or there is a set of colleges C_p, such that $c_j \in C_p$ and all the u_p places in C_p have been filled with better students than a_i. Biró et al. [6] showed that if the sets of colleges with common upper quotas is nested, i.e., when $C_p \cap C_q \neq \emptyset$ implies either $C_p \subset C_q$ or $C_p \supset C_q$, then a stable matching always exists. Moreover, a stable matching can be found efficiently by the generalised Gale-Shapley algorithm and there are applicant and college optimal solutions. However, if the set system is not nested then stable solution may not exist and the problem of finding a stable matching is NP-hard. Interestingly, the Hungarian application involved nested set systems until 2007 when a legislative change modified the structure of the underlying model and made the set system non-nested, with the possibility of having no stable solution and also making the problem computationally hard.

Here, we show that we can express this stability condition with the use of score-limits, in a similar fashion to the method we described in Sect. 2. However, here we need to assume that there are no ties. We set a score-limit t_p for each set of colleges C_p with common quota, which is less than or equal to the score of the weakest admitted student if the common quota is filled, and 0 if the common quota is unfilled in the matching. When describing the model in this way, stability implies that if a student a_i is admitted to college c_j then $s_{ij} \geq t_j$ and also $s_{ij} \geq t_p$ for any set of colleges C_p with common quota where C_p includes c_j. Furthermore, if a_i is not admitted to c_j or to any better college of her preference then it must be the case that either $s_{ij} < t_j$ or $s_{ij} < t_p$ for some set of colleges C_p with common quota where C_p contains c_j. These conditions can be formalised with the following set of conditions, where q_j denotes the number of sets of colleges with common quota involving college c_j, $\{c_j\}$ also being one of them.

$$t_p \leq (1 - x_{ij}) \cdot (\bar{s} + 1) + s_{ij} \text{ for each } (a_i, c_j) \in E \text{ and } c_j \in C_p \qquad (18)$$

and

$$s_{ij} + 1 \leq t_p + \left(\sum_{k:r_{ik} \leq r_{ij}} x_{ik} + y_i^p \right) \cdot (\bar{s} + 1) \text{ for each } (a_i, c_j) \in E \text{ and } c_j \in C_p \quad (19)$$

with

$$\sum_{p:c_j \in C_p} y_i^p \leq q_j - 1 \text{ for each } (a_i, c_j) \in E \qquad (20)$$

where $y_i^p \in \{0, 1\}$ is a binary variable. These conditions are needed to establish the links between a matching and the corresponding score-limits. However, for stability we also have to ensure that the score-limits are minimal. In case of strict preferences (i.e., when no two students have the same score at colleges belonging to a set of colleges with a common quota), we can ensure the minimality of the score-limits with the following conditions.

Again, we introduce an indicator variable f_p for each set of colleges C_p which is equal to zero if the common quota of these colleges is unfilled, by using the following constraints.

$$f_p \cdot u_p \leq \sum_{i:(a_i,c_j) \in E, c_j \in C_p} x_{ij} \text{ for each } C_p \subseteq C \qquad (21)$$

Then we ensure that if a college or a set of colleges with a common quota is unfilled then its score-limit is zero.

$$t_p \leq f_p(\bar{s} + 1) \text{ for each } C_p \subseteq C \qquad (22)$$

We describe and prove the correctness of the IP model in the following Theorem.

Theorem 5. *Feasibility conditions (1), (2) and (17), with the stability conditions (18), (19) and (20), together with (21) and (22) describe an integer program such that its solutions correspond to stable matching for the College Admissions problem with common quotas.*

Proof. To see the correctness, we have to note first that the matching is feasible if and only if the feasibility constraints (1), (2) and (17) are satisfied, and conditions (21) and (22) are satisfied if and only if the score-limit of any unfilled college or set of colleges is zero.

Now, suppose first that we have a stable solution and we show that all the stability conditions can be satisfied by setting the variables appropriately. When a common quota C_p is filled we set t_p to be equal to the last admitted applicant at any college included in C_p. This ensures that the first set of conditions (18) is satisfied. Let $y_i^p = 0$ if a_i does not meet t_p at any college $c_j \in C_p$ where she applied to, and $y_i^p = 1$ otherwise, with the exception of set $\{c_j\}$, where we set $y_i^j = 0$ if i is admitted to c_j or to a better place. The stability of the matching then implies (20). Finally, let us consider an application (a_i, c_j) where $c_j \in C_p$. If a_i is admitted to c_j or to a better college then the corresponding constraint (19) is satisfied, irrespective of the value y_i^p. Otherwise, suppose that a_i is not admitted to c_j or to any better place. If a_i does not meet the score-limit t_p then (19) is satisfied, obviously, and it is also satisfied when she meets t_p, since $y_i^p = 1$ in that case.

Conversely, suppose that we have a solution for the IP model, and we will show that this ensures the stability of the corresponding matching. If (a_i, c_j) is in the matching then constraints (18) imply that a_i achieves the score-limit of c_j and also the score-limit of every set of colleges with common quota containing c_j. Finally, suppose that a_i is not admitted to c_j or to any better college of her preference. Since one of the additional variables of form y_i^l must be zero, say y_i^p, the corresponding constraint (19) implies that the set of colleges C_p containing c_j has a score-limit t_p greater than s_{ij}. So the matching is indeed stable.   □

Finally we note that if we have ties then we shall use an objective function that minimises the sum of the score-limits or an extended model, similar to the ones described in Sect. 2.

Further Notes

In the Hungarian application students can apply for pairs of programmes in case of teachers' studies, e.g. when they want to become a teacher in both maths and physics. The case of paired applications can be modeled in a similar way as the case of common quotas. Each pair of colleges can be seen as an artificial college, and the upper quotas of the original colleges will become common quotas. For completeness, we describe the IP model for college admissions with paired application in the extended version of our paper [9].

The combination of the four special features result in interesting challenges in two ways. First, we need to define appropriate stability criteria when both lower and upper quotas are present. Second, we need to combine the separate integer programmes into a single programme that would result in a suitable solution for the real application. We describe some solutions and particular difficulties related to issue in the extended version of our paper [9].

It would be also important to know whether these IP formulations may be solved within a realistic timescale for such a large scale application as the Hungarian higher education matching schemes, with around 100000 applicants. Our plan is to conduct experiments on real data we have access to, which comes from the 2008 match run of the Hungarian higher education scheme.

Furthermore, one could try to solve our special college admissions problems with other approaches, e.g. with different integer programmes. Finally, our models may be useful in other applications as well, such as controlled school choice (see e.g. [2]), resident allocation with distributional constraints (see e.g. [16]), or for finding stable solutions with additional restrictions, such as matchings with no Pareto-improving swaps [14].

References

1. Abraham, D., Blum, A., Sandholm, T.: Clearing algorithms for barter-exchange markets: enabling nationwide kidney exchanges. In: Proceedings of ACM-EC (2007)
2. http://www.sciencedirect.com/science/article/pii/S0022053114000301
3. Azavedo, E.M., Leshno, J.D.: A supply and demand framework for two-sided matching markets. Working paper (2012)
4. Baïou, M., Balinski, M.: The stable admissions polytope. Math. Program. **87**(3, Ser. A), 427–439 (2000)
5. Biró, P.: Student admissions in Hungary as Gale and Shapley envisaged. Technical report, no. TR-2008-291 of the CS Department of Glasgow University (2008)
6. Biró, P., Fleiner, T., Irving, R.W., Manlove, D.F.: The college admissions problem with lower and common quotas. Theor. Comput. Sci. **411**, 3136–3153 (2010)
7. Biró, P., Irving, R.W., Schlotter, I.: Stable matching with couples - an empirical study. ACM J. Exp. Algorithmics **16**, 12:1 (2011). (Article No.: 1.2)
8. Biró, P., Kiselgof, S.: College admissions with stable score-limits. Central Eur. J. Oper. Res. (2014, To appear)
9. Biró, P., McBride, I.: Integer programming methods for special college admissions problems. Technical report no. 1408.6878, Computing Research Repository, Cornell University Library (2014)
10. Biró, P., Manlove, D.F., McBride, I.: The hospitals/residents problem with couples: complexity and integer programming models. In: Gudmundsson, J., Katajainen, J. (eds.) SEA 2014. LNCS, vol. 8504, pp. 10–21. Springer, Heidelberg (2014)
11. Fleiner, T.: On the stable b-matching polytope. Math. Soc. Sci. **46**, 149–158 (2003)
12. Fleiner, T., Jankó, Z.: Choice function-based two-sided markets: stability, lattice property, path independence and algorithms. Algorithms **7**(1), 32–59 (2014)
13. Gale, D., Shapley, L.S.: College admissions and the stability of marriage. Am. Math. Monthly **69**(1), 9–15 (1962)
14. Irving, R.W.: Stable matching problems with exchange restrictions. J. Comb. Optim. **16**, 344–360 (2008)
15. Irving, R.W., Manlove, D.F.: Approximation algorithms for hard variants of the stable marriage and hospitals/residents problems. J. Comb. Optim. **16**, 279–292 (2008)
16. Kamada, Y., Kojima, F.: Stability and strategy-proofness for matching with constraints: a problem in the Japanese medical match and its solution. Am. Econ. Rev. (P&P) **102**(3), 366–370 (2012)

17. Kwanashie, A., Manlove, D.F.: An integer programming approach to the hospitals/residents problem with ties. In: Huisman, D., Louwerse, I., Wagelmans, A.P.M. (eds.) Operations Research Proceedings 2013. Operations Research Proceedings, pp. 263–269. Springer, Heidelberg (2014)

18. McDermid, E.J., Manlove, D.F.: Keeping partners together: algorithmic results for the hospitals/residents problem with couples. J. Comb. Optim. **19**, 279–303 (2012)

19. Manlove, D.F., O'Malley, G.: Paired and altruistic kidney donation in the UK: algorithms and experimentation. In: Klasing, R. (ed.) SEA 2012. LNCS, vol. 7276, pp. 271–282. Springer, Heidelberg (2012)

20. Ronn, E.: NP-complete stable matching problems. J. Algorithms **11**, 285–304 (1990)

21. Roth, A.E.: A natural experiment in the organization of entry-level labor markets: regional markets for new physicians and surgeons in the United Kingdom. Am. Econ. Rev. **81**, 415–440 (1991)

22. Roth, A.E., Rothblum, U.G., Vande Vate, J.H.: Stable matchings, optimal assignments, and linear programming. Math. Oper. Res. **18**(4), 803–828 (1993)

23. Roth, A.E., Sönmez, T., Ünver, M.U.: Efficient kidney exchange: coincidence of wants in markets with compatibility-based preferences. Am. Econ. Rev. **97**(3), 828–851 (2007)

24. Rothblum, U.G.: Characterization of stable matchings as extreme points of a polytope. Math. Program. **54**(1, Ser. A), 57–67 (1992)

25. Sethuraman, J., Teo, C.-P., Qian, L.: Many-to-one stable matching: geometry and fairness. Math. Oper. Res. **31**(3), 581–596 (2006)

26. Ünver, M.U.: Backward unraveling over time: the evolution of strategic behavior in the entry-level British medical labor markets. J. Econ. Dyn. Control **25**, 1039–1080 (2001)

27. Biró, P.: University admission practices - Hungary. matching-in-practice.eu

On the Width of Ordered Binary Decision Diagrams

Beate Bollig[✉]

LS2 Informatik, TU Dortmund, Dortmund, Germany
beate.bollig@tu-dortmund.de

Abstract. Ordered binary decision diagrams (OBDDs) are a popular data structure for Boolean functions. One of its complexity measures is the width which has been investigated in several areas in computer science like machine learning, property testing, and the design and analysis of implicit graph algorithms. Maybe the most important issue of OBDDs is the possibility to choose the variable ordering and for a given function the width of an OBDD is very sensitive to this choice. The main result of the paper is the proof that the width minimization problem is NP-hard. Furthermore, two basic problems in the design and analysis of implicit graph algorithms are reinvestigated and known upper bounds on their complexity that depend on the width of the input OBDDs are improved.

1 Introduction

Ordered binary decision diagrams (OBDDs) are very restricted branching programs, a model well-known in complexity theory for space bounded computations. They are a popular data structure for Boolean functions [7]. Among the many areas of applications are verification, model checking, and computer aided design (for a survey see, e.g., [21]). To define the width of an OBDD, complete OBDDs are considered where on all computation paths all variables have to be tested. (For the formal definitions see Sect. 2.) Maybe the most important issue of OBDDs is the possibility to choose the variable ordering since the size and the width of an OBDD representing a function f heavily depend on this choice. For example for the most significant bit of binary addition the width may vary between constant and exponential size. Therefore, in applications it is an important issue to pick a good variable ordering.

The width of ordered binary decision diagrams has been part of investigations in several areas in computer science. For example, OBDDs of bounded width have been studied in the machine learning context rather extensively. In particular, the influence of the width on the difficulty of the corresponding learning problem has been analyzed (see, e.g., [8]). It has been shown that OBDDs of width 2 are PAC-learnable while OBDDs of width at least 3 are as hard to learn as DNF formulas. Moreover, also in complexity theory the width of OBDDs has been investigated, e.g., in property testing. Newman has presented a property testing algorithm

---

The author is supported by DFG project BO 2755/1-2.

Z. Zhang et al. (Eds.): COCOA 2014, LNCS 8881, pp. 444–458, 2014.
DOI: 10.1007/978-3-319-12691-3_33

for any property decidable by an ordered binary decision diagram of constant width [13]. The performance of his algorithm, i.e., the number of queries, depends highly on the width of the input OBDD. Lower and upper bounds have been shown for testing functions for the property of being computable by an ordered binary decision diagram of small width, i.e., given oracle access to a Boolean function f testing whether f can be represented by an OBDD of small width or is in a certain sense far from any such function [6,11,14]. Here, the computational complexity to find an optimal variable ordering such that the corresponding OBDD width for the representation of a given function is minimal is investigated. It is known that the computation of an optimal variable ordering with respect to the OBDD size is NP-hard [5]. Only recently, an NP-hardness proof for the computation of an optimal variable ordering for the size of complete OBDDs for a given function has been presented [2]. Variable orderings that are optimal with respect to the size are not necessarily optimal for the width of OBDDs. An example is the multiplexer or direct storage access function. Furthermore, an optimal variable ordering with respect to the width is not necessarily optimal with respect to the size of complete OBDDs (see Sect. 3). Here, using a new reduction it is shown that the computation of the minimal OBDD width is NP-hard. For this result we use some of the ideas presented in [2] but the reduction and the proof are different because of our different objective function. Anyway, the new reduction underlines the robustness of the construction in [2].

Another application of OBDDs is in the design and analysis of implicit algorithms. Given the rapid growth of application-based networks, the design and analysis of graph algorithms is faced with the challenge to deal with very large graphs. An heuristic approach are implicit OBDD-based graph algorithms where vertices of an input graph are binary encoded and the edge set of the input graph is represented by its characteristic function (see, e.g., [1,3,12] for some implicit OBDD-based algorithms for classical graph problems like minimal spanning tree, maximum matching, and maximum flow in 0-1 networks). The runtime and space requirement of implicit OBDD-based algorithms do not only depend on the number of operations but also on the size of the OBDDs involved. Unfortunately, analyzing the runtime and the space requirement is often not easy because of the missing structure of intermediate OBDDs. Nevertheless, in [16,22] it has been shown that some implicit OBDD-based algorithms for the maximum flow problem in 0-1 networks and for topological sorting have an overall polylogarithmic runtime with respect to $|V|$, the number of nodes in an input graph, on special instances, where the input OBDDs have small width. Moreover, Sawitzki has presented an implicit OBDD-based algorithm for the all-pairs shortest-paths problem, which works on loopless directed graphs where the edges weights are strictly positive [17]. He has analyzed the OBDD sizes occurring during the run of this algorithm and has proved that his algorithm has polylogarithmic runtime and space usage with respect to $|V|$ and the maximal weight on the edges if the width of the input OBDD is very restricted. To understand better the behavior of OBDD-based graph algorithms, in this paper two basic problems are reinvestigated. For argument reordering, important to reverse the direction of directed edges in an OBDD-represented graph, and for quantification over a set of variables, which is most of the time the most time consuming step in implicit algorithms, known upper bounds on their resource requirements, that depend on the input OBDD width, presented in [18,19] are improved. For the result on quantification we use knowledge from automata theory since complete OBDDs are closely related to nonuniform finite automata for Boolean languages L, where $L \subseteq \{0,1\}^n$ (see, e.g., Sect. 3.2 in [21]). We demonstrate that the well-known powerset construction for the transformation of nondeterministic finite automata into deterministic ones can be adapted. As a result the known upper bounds on the resource requirements can be slightly improved and the proof structure can be simplified.

The rest of the paper is organized as follows. In Sect. 2 we define some notation and review some basics concerning OBDDs, implicit graph representations, and functional operations. Section 3 contains the main result of the paper and is dedicated to the width minimization problem and its NP-completeness proof. Finally, argument reordering and quantification over a set of variables are revisited in Sect. 4.

2   Preliminaries

We briefly recall the main notions concerning OBDDs.

On (complete) ordered binary decision diagrams. OBDDs are a popular dynamic data structure in areas working with Boolean functions, like circuit verification or model checking.

Definition 1. *Let $X_n = \{x_1, \ldots, x_n\}$ be a set of Boolean variables. A variable ordering π on X_n is given by a permutation on $\{1, \ldots, n\}$ leading to the ordered list $x_{\pi(1)}, \ldots, x_{\pi(n)}$ of the variables. A π-OBDD on X_n is a directed acyclic graph $G = (V, E)$ whose sinks are labeled by the Boolean constants 0 and 1 and whose non-sink (or decision) nodes are labeled by Boolean variables from X_n. Each decision node has two outgoing edges, one labeled by 0 and the other by 1. The edges between decision nodes have to respect the variable ordering π, i.e., if an edge leads from an x_i-node to an x_j-node, then $\pi^{-1}(i) < \pi^{-1}(j)$ (x_i precedes x_j in $x_{\pi(1)}, \ldots, x_{\pi(n)}$). At each node v a Boolean function $f_v \in B_n$, i.e., $\{0,1\}^n \to \{0,1\}$, is represented. A c-sink represents the constant function c. If f_{v_0} and f_{v_1} are the functions at the 0- or 1-successor of v, resp., and v is labeled by x_i, f_v is defined by Shannon's decomposition rule $f_v(a) := \overline{a_i} f_{v_0}(a) \vee a_i f_{v_1}(a)$. The size of a π-OBDD G, denoted by $|G|$, is equal to the number of its decision nodes. A π-OBDD of minimal size for a given function f and a fixed variable ordering π is unique up to isomorphism. A π-OBDD for a function f is called reduced if it is the minimal π-OBDD for f. The π-OBDD size of a function f, denoted by π-OBDD(f), is the size of the reduced π-OBDD representing f. An OBDD is a π-OBDD for an arbitrary variable ordering π. The OBDD size of f is the minimum of all π-OBDD(f).*

A variable ordering is called a *natural variable ordering* if π is the identity $1, 2, \ldots, n$. Obviously a variable ordering π can be identified with the corresponding ordering $x_{\pi(1)}, \ldots, x_{\pi(n)}$ of the variables if the meaning is clear from the context. Since (ordered) binary decision diagrams are a nonuniform model of computation, usually sequences of binary decision diagrams $(G_n)_{n \in \mathbb{N}}$ representing a sequence of Boolean functions $(f_n)_{n \in \mathbb{N}}$ with respect to sequences of variable orderings $(\pi_n)_{n \in \mathbb{N}}$ are considered, where f_n is a Boolean function on n variables. In the following we simplify the notation because the meaning is clear from the context.

In order to define the width of a π-OBDD complete OBDDs are introduced. Here, there are only edges between nodes labeled by neighboring variables, i.e., if an edge leads from an x_i-node to an x_j-node, then $\pi^{-1}(i) = \pi^{-1}(j) - 1$.

Definition 2. *An OBDD on X_n is complete if all paths from the source to one of the sinks have length n. The width of a complete OBDD is the maximal number of nodes labeled by the same variable. A complete π-OBDD of minimal size for a given function f and a fixed variable ordering π is unique up to isomorphism. A π-OBDD for a function f is called* quasi-reduced *if it is the minimal complete π-OBDD for f. The complete π-OBDD size of a function f, denoted by π-QOBDD(f), is the size of the quasi-reduced π-OBDD representing f. The π-OBDD width of a function f, denoted by π-OBDD$_w(f)$, is the width of the quasi-reduced π-OBDD representing f. A complete OBDD, or QOBDD for short, is a complete π-OBDD for an arbitrary variable ordering π. The complete OBDD size or QOBDD size of f is the minimum of all π-QOBDD(f). The OBDD width of f is the minimum of all π-OBDD$_w(f)$.*

Complete OBDDs with respect to natural variable orderings differ from nonuniform deterministic finite automata only in the minor aspect that variables have still to be tested if the corresponding subfunction is the constant function 0.

Definition 3. *A level p, $1 \leq p \leq n$, in a complete π-OBDD G contains all nodes in G labeled by the variable at position p in the variable ordering π.*

Let f be a Boolean function on the variables $x_1, \ldots, x_n$. The *subfunction* $f_{|x_i=c}$, $1 \leq i \leq n$ and $c \in \{0, 1\}$, is defined as $f(x_1, \ldots, x_{i-1}, c, x_{i+1}, \ldots, x_n)$. A function f *depends essentially* on a Boolean variable z if $f_{|z=0} \neq f_{|z=1}$. The size of the (quasi-)reduced π-OBDD representing f is described by the following result.

Proposition 1. *([20]) The number of $x_{\pi(i)}$-nodes in the quasi-reduced (reduced) π-OBDD for f is equal to the number of different subfunctions*

$$f_{|x_{\pi(1)}=a_1, \ldots, x_{\pi(i-1)}=a_{i-1}} \text{ (that essentially depend on } x_{\pi(i)}),$$

where $a_1, \ldots, a_{i-1} \in \{0, 1\}$.

In a reduced OBDD each node encodes a different function, whereas in a quasi-reduced OBDD each node labeled by the same variable represents a different function.

Nondeterminism is an important concept in computer science. One way to define nondeterministic π-OBDDs is the following.

Definition 4. *A nondeterministic π-OBDD on a variable set $X_n = \{x_1, \ldots, x_n\}$ is a π-OBDD on X_n with some additional unlabeled nodes, called nondeterministic nodes. An input $b \in \{0,1\}^n$ activates all edges consistent with b, i.e., the edges labeled by b_i which leave nodes labeled by x_i. A computation path for an input b in a (nondeterministic) OBDD G is a path of edges activated by the input b which leads from the source to a sink. A computation path for an input b which leads to the 1-sink is called* accepting path *for b. G represents the function $f \in B_n$ for which $f(b) = 1$ iff there exists an accepting path for the input b.*

On OBDD-based graph representations and functional operations. To represent large but highly regular graphs by means of data structures smaller than adjacency matrices or adjacency lists seems to be a natural idea. Boolean encodings for the vertices can be used to characterize sets of vertices or edges by their characteristic Boolean functions, and data structures for Boolean functions can be used to represent and manipulate the input graphs. Let $G = (V, E)$ be a graph with N vertices $v_0, \ldots v_{N-1}$ and $|z|_2 := \sum_{i=0}^{n-1} z_i 2^i$, where $z = (z_0, \ldots, z_{n-1}) \in \{0,1\}^n$ and $n = \lceil \log N \rceil$. Now, E can be represented by an OBDD for its characteristic function, where $x, y \in \{0,1\}^n$ and

$$\chi_E(x, y) = 1 \Leftrightarrow (|x|_2, |y|_2 < N) \wedge (v_{|x|_2}, v_{|y|_2}) \in E.$$

Undirected edges are represented by symmetric directed ones.

Implicit OBDD-based graph algorithms have to solve problems on a given graph instance by efficient functional operations offered by the OBDD data structure. In the following we briefly describe three important functional operations and the corresponding time and additional space requirements for OBDDs (for a detailed discussion see, e.g., Sect. 3.3 in [21]). Let f and g be Boolean functions in B_n on the variable set $X_n = \{x_1, \ldots, x_n\}$ and G_f and G_g be π-OBDDs for the representations of f and g, respectively.

- *Replacement by constant:* Given G_f, an index $i \in \{1, \ldots, n\}$, and a Boolean constant $c_i \in \{0, 1\}$, compute a π-OBDD for the subfunction $f_{|x_i = c_i}$. This can be done in time $\mathcal{O}(|G_f|)$ and the π-OBDD for $f_{|x_i = c_i}$ is not larger than G_f.
- *Synthesis:* Given G_f and G_g and a binary Boolean operation $\otimes \in B_2$, compute a π-OBDD G_h for the function $h \in B_n$ defined as $h := f \otimes g$. This can be done in time and space $\mathcal{O}(|G_f| \cdot |G_g|)$ and the size of G_h is bounded above by $\mathcal{O}(|G_f| \cdot |G_g|)$.

  Sometimes the following estimation leads to better upper bounds. Let G_h^* be the graph that consists of the nodes in the product graph of G_f and G_g reachable from the node representing the function h. The computation of G_h can be done in time $\mathcal{O}(|G_h^*| \log |G_h^*|)$ and space $\mathcal{O}(|G_h^*|)$.
- *Quantification:* Given G_f, an index $i \in \{1, \ldots, n\}$, and a quantifier $Q \in \{\exists, \forall\}$, compute a π-OBDD G_h for the function $h \in B_n$ defined as $h := (Qx_i)f$, where $(\exists x_i)f := f_{|x_i=0} \vee f_{|x_i=1}$ and $(\forall x_i)f := f_{|x_i=0} \wedge f_{|x_i=1}$.

The computation of G_h can be realized by two replacements of constants and a synthesis operation. This can be done in time and space $\mathcal{O}(|G_f|^2)$.

Since complete OBDDs are closely related to nonuniform finite automata, knowledge from automata theory can sometimes be used in the design and analysis of OBDD-based algorithms. Here, if G_f and G_g are complete π-OBDDs of maximal width w, we can apply the synthesis algorithm for finite automata. If we take into account that OBDDs are acyclic, the synthesis can be done in time and space $\mathcal{O}(w^2 \cdot n)$.

3   On the Minimization of the OBDD Width

In this section we investigate the minimization of the OBDD width. First, we prove that for a given Boolean function an optimal variable ordering with respect to the OBDD width is not necessarily optimal for the complete OBDD size. Our auxiliary function g_4 is defined on the x-variables $x_1, \ldots, x_4$ and $g_4(x) := (x_1 \wedge x_2) \vee (x_3 \wedge x_4)$. Let f_n be an arbitrary Boolean function defined on the y-variables $y_1, \ldots, y_n$ whose OBDD width is at least 4. Obviously such a function exists. Let π be an optimal variable ordering with respect to the OBDD width of f_n. Now, we define a new Boolean function as conjunction of f_n and g_4. The function h_n is defined on the variables $x_1, \ldots, x_4, y_1, \ldots y_n$, and $h_n(x, y) := f_n(y) \wedge g_4(x)$. Let π' be $y_{\pi(1)}, \ldots, y_{\pi(n)}, x_1, x_2, x_3, x_4$ and let π'' be $y_{\pi(1)}, \ldots, y_{\pi(n)}, x_1, x_3, x_2, x_4$. It is not difficult to prove that the variable orderings π' and π'' are both optimal with respect to the OBDD width of g_n but π'-QOBDD(h_n) is smaller than π''-QOBDD(h_n). Therefore, the variable ordering π'' is optimal with respect to the OBDD width but not with respect to the size of complete OBDDs for the function h_n. Figure 1 shows the quasi-reduced sub-OBDDs for g_4.

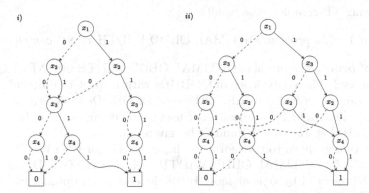

Fig. 1. Quasi-reduced OBDDs with respect to the variable orderings (i) x_1, x_2, x_3, x_4 and (ii) x_1, x_3, x_2, x_4 for the function $g_4(x) = (x_1 \wedge x_2) \vee (x_3 \wedge x_4)$.

Note, our counterexample can also be used to show that optimal variable orderings with respect to the width are not necessarily optimal regarding the

OBDD size and it can also be used with respect to alternative definitions of the OBDD width (see the remark at the end of this section).

Now, we start to investigate the complexity of the OBDD width minimization by the definition of the well-known NP-hard cutwidth minimization problem (CUTWIDTH for short) [10]. Given a graph H and a positive integer C the question is whether there is a linear ordering of its vertices such that any line inserted between two consecutive vertices in the ordering cuts at most C edges of the input graph.

Definition 5 (CUTWIDTH). *Given an undirected graph $H = (V, E)$, $n :=$ $|V|$, and an injective mapping $s : V \to \{1, 2, \ldots, n\}$, the cutwidth of a vertex $v \in V$ with respect to s, denote $\mathrm{CW}_s(v)$ is the number of edges $(u, w) \in E$ satisfying $s(u) \leq s(v) < s(w)$. The cutwidth of H with respect to s is the maximum cutwidth of its vertices: $\mathrm{CW}_s(H) := \max_{v \in V} \mathrm{CW}_s(v)$. The answer to the problem CUTWIDTH is yes iff for a given input graph $H = (V, E)$ and a bound C there is an injective mapping $s : V \to \{1, 2, \ldots, n\}$ such that $\mathrm{CW}_s(H)$ is at most C.*

It is not difficult to see that for the definition of $\mathrm{CW}_s(H)$ it does not matter whether $\mathrm{CW}_s(v)$ is the number of edges $(u, w) \in E$ satisfying $s(u) \leq s(v) < s(w)$ or $s(u) < s(v) \leq s(w)$. In the following the problem OPTIMAL OBDD WIDTH is investigated.

Definition 6 (OPTIMAL OBDD WIDTH). *Given a QOBDD G and a bound W, the answer to the problem OPTIMAL OBDD WIDTH is yes iff the function represented by G can be represented by a QOBDD (respecting an arbitrary variable ordering) whose width is at most W.*

Note that the problem OPTIMAL OBDD WIDTH is defined via QOBDDs because of the standard definition of the OBDD width. It does not matter for the following NP-completeness result.

Theorem 1. *The problem OPTIMAL OBDD WIDTH is NP-complete.*

Sketch of proof. The problem OPTIMAL OBDD WIDTH is in NP. A QOBDD can be guessed. The equivalence of QOBDDs with respect to different variable orderings can be verified similarly to the case for OBDDs in deterministic polynomial time [9]. Our NP-hardness proof uses a polynomial time reduction from CUTWIDTH. Let $H = (V, E)$ and C be given and $m := |E|$. For the polynomial reduction we have to transform the input (H, C) for CUTWIDTH into an input (G, W) for OPTIMAL OBDD WIDTH such that the OBDD width of the function represented by G is at most W iff the cost of an optimal cut for H is at most C.

Only recently it has been shown that the problem OPTIMAL QOBDD which decides whether given a QOBDD G and a size bound s the function represented by G can be represented by a QOBDD with at most s nodes is NP-complete [2]. The NP-hardness has been shown by a polynomial reduction from the well-known NP-complete problem Optimal Linear Arrangement. Here, we use some of the ideas of this reduction but our proofs are different because of our different objective function.

In the following we identify the vertex set V of the input H with the set $\{1, 2, \ldots, n\}$. As a result an injective mapping $s : V \to \{1, 2, \ldots, n\}$ can be identified by a permutation on $\{1, 2, \ldots, n\}$. Now, for the kth edge $\{i, j\} \in E, 1 \leq k \leq m$ and $i, j \in \{1, 2, \ldots, n\}$, we introduce an edge function $f_k(v_1, v_2, \ldots, v_n) = (v_i \vee v_j)$ (see also [2]). Figure 2 shows an example of a quasi-reduced OBDD representing an edge function.

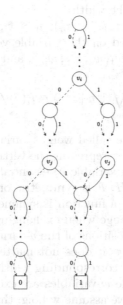

Fig. 2. A quasi-reduced OBDD for the edge function $f_k(v) = v_i \vee v_j$ with respect to a variable ordering π where $\pi^{-1}(i) < \pi^{-1}(j)$.

The intuition for the definition of the edge function is the following. If we ignore the nodes representing the constant function 0 in a quasi-reduced π-OBDD for the edge function f_k for now, there are two nodes labeled by a variable v_{i_1} iff $\pi^{-1}(i) < \pi^{-1}(i_1) \leq \pi^{-1}(j)$. For all other v-variables the number of nodes is 1. Now, if we consider the quasi-reduced π-OBDDs for all edge functions of an input graph H without mergings between the different representations for the edge functions and ignoring the nodes representing the constant function 0, the maximal number of nodes labeled by the same variable is equal to $m + \mathrm{CW}_\pi(H)$. Here, we use the observation above that for the definition of $\mathrm{CW}_\pi(H)$ it does not matter whether $\mathrm{CW}_s(v)$ is the number of edges (u, w) in the input graph H satisfying $s(u) \leq s(v) < s(w)$ or $s(u) < s(v) \leq s(w)$.

We have still several problems to solve. First, in order to obtain a single Boolean function the edge functions have to be combined to one function. Second, we have to make sure that representations for different edge functions do share the nodes representing the constant function 0 but no other nodes. Finally, the part of the OBDD for the representation of the edge functions should define the width of the ordered binary decision diagram. The solution is to frame the edge functions in an appropriate way. For the framing we use counting functions (see also [2]). Intuitively, we will show that there exists a variable ordering which is optimal with respect to the OBDD width where the variables that represent the vertices of the input graph for CUTWIDTH are tested in the middle. Therefore, in the corresponding QOBDD the representations for the edge functions cannot share BDD nodes. Furthermore, we guarantee that a level that corresponds to one of these variables defines the width.

For a variable vector $z = (z_1, \ldots, z_n)$, $n \in \mathbb{N}$, let $\| z \|$ be $\sum_{i=1}^{n} z_i$. The function $F \in B^{2m+n}$ is defined on the variable vectors $u = (u_1, u_2, \ldots, u_m)$, $v = (v_1, v_2, \ldots, v_n)$, and $w = (w_1, w_2, \ldots, w_m)$, and

$$F(u, v, w) := \bigvee_{i=1}^{m} (\| u \| = i) \wedge f_i(v) \wedge (\| w \| = i).$$

The u- and the w-variables are called weight variables and the v-variables are called vertex variables since they represent the vertices of the input graph H for CUTWIDTH. We call a vertex variable v_j essential for an edge function f_i iff j is incident to the ith edge in H. F is symmetric on the u-variables and on the w-variables, respectively. Here, a function is symmetric on two variables x_i and x_j if the function does not change when exchanging the variables x_i and x_j. If $P = \{i_1, \ldots, i_m\}$ is the set of positions of the u-variables (or w-variables, respectively) in a variable ordering π, it does not matter which u-variable is tested on which position in P for the corresponding width of a QOBDD for F. Moreover, the roles of the u-and the w-variables are exchangeable, therefore, in the remaining part of the section we assume w.l.o.g. that the u-variables are tested in the ordering $u_1, u_2, \ldots, u_m$ and the w-variables in the ordering $w_1, w_2, \ldots, w_m$ and u_1 is the first variable of all u- and w-variables. Our transformation computes the (quasi-reduced) OBDD representing F with respect to the ordering $u_1, u_2, \ldots, u_m, v_1, v_2, \ldots, v_n, w_1, w_2, \ldots, w_m$ in polynomial time.

A *sandwich variable ordering* is a variable ordering where the v-variables are tested between the u- and the w-variables, and all u-variables as well as all w-variables are tested consecutively.

The following lemma is not difficult to prove.

Lemma 1. *Let π be a sandwich variable ordering and let π' be the subordering of π on the v-variables. Then π-$OBDD_w$ of $F(u, v, w)$ is $m + 1 + CW_{\pi'}(H)$.*

We are now able to define the bound W in our reduction: $W := m + 1 + C$. In order to prove the correctness of our reduction we have to show that the input graph H for CUTWIDTH has a minimum cut whose cost is bounded by C iff F can be represented by a QOBDD whose width is at most W. Using Lemma 1 the only-if-part is easy. The if-part of the correctness proof is the tricky one. By our considerations above it remains to prove that some variable ordering of

$F = (u, v, w)$ which is optimal with respect to the width is a sandwich variable ordering.

The proof structure is to change a given (optimal) variable ordering π in three phases until it is a sandwich variable ordering. This is similar as in [2] but the proofs are different. In all phases we do not change the ordering among the u-variables, among the v-variables, and among the w-variables, respectively. We know that the roles of the u- and the w-variables are symmetric, therefore, we assume w.l.o.g. that the first weight variable is a u-variable. First, we ensure that all u-variables are tested before all w-variables. We do this by exchanging the positions of the first w-variable in the variable ordering and the following u-variable without increasing the width of the corresponding QOBDD. Since the procedure can be iterated, we are done. Figure 3 illustrates the changement in the variable ordering after one step.

Lemma 2. *Let π be a variable ordering on the u-, v-, and w-variables and let i_k be the position of the variable u_k and let j_k be the position of the variable w_k, $1 \leq j \leq m$. Furthermore, let j_1 be between i_l and i_{l+1}, $l \in \{1, \ldots, m - 1\}$. Let π' be the variable ordering where the variable w_1 is at position i_{l+1} and u_{l+1} is at position j_1 and all other variables are ordered according to π. Then $\pi'\text{-OBDD}_w(F)$ is not larger than $\pi\text{-OBDD}_w(F)$.*

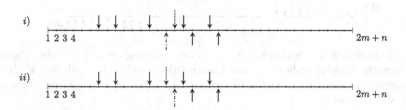

Fig. 3. A simplified description how the variable orderings π and π' differ. The arrows pointing down indicate the positions of the u-variables and the arrows pointing up the positions of the w-variables. Figure (i) symbolizes the variable ordering π, Figure (ii) the ordering π'.

Next, we change the variable ordering in such a way that the u-variables are tested in the beginning. Figure 4 shows how the variable orderings differ.

Lemma 3. *Let π be a variable ordering on the u-, v-, and w-variables where all u-variables are before the w-variable. Let π' be the variable ordering that starts with the u-variables followed by the remaining variables in the same order as in π. Then $\pi'\text{-OBDD}_w(F)$ is not larger than $\pi\text{-OBDD}_w(F)$.*

Finally, we modify the variable ordering such that the w-variables are tested in the end. Figure 5 shows how the variable orderings differ.

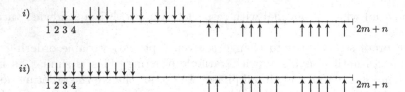

Fig. 4. A simplified description how the variable orderings π and π' differ. Figure (i) symbolizes the variable ordering π and Figure (ii) illustrates the ordering π'. In both orderings the u-variables are before the w-variables, furthermore in π' all u-variables are in the beginning of the ordering.

Lemma 4. *Let π be a variable ordering on the u-, v-, and w-variables that starts with the u-variables. Let π' be the variable ordering where the u-variables are in the beginning of the ordering, the v-variables are ordered in the same suborder as in π, and the w-variables are the last variables in the ordering. Then π'-$OBDD_w(F)$ is not larger than π-$OBDD_w(F)$.*

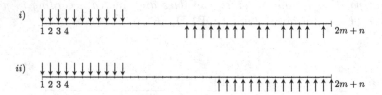

Fig. 5. A simplified description how the variable orderings π and π' differ. Figure (i) symbolizes the variable ordering π and Figure (ii) illustrates the ordering π'. In both orderings the u-variables are in the beginning. Moreover, in π' the w-variables are in the end of the ordering.

Since each phase does not increase the width of the BDD representation, we are done and by Lemmas 2–4 we have shown the following result.

Corollary 1. *There exists a sandwich variable ordering that is optimal with respect to the width of an OBDD for $F(u, v, w)$.*

Remark. Note that throughout the paper the width of an OBDD is defined via the complete or leveled model as is often the case in the literature (see, e.g., [6,8,11,13,14]). One might argue that in applications often reduced OBDDs are used and therefore another definition of the width like the maximal number of nodes labeled by the same variable or the maximal number of nodes with the same distance from the source would be more appropriate. However, in this case our NP-completeness proof can be adapted by a modification of the edge functions from $f(v) = v_i \vee v_j$ into

$$f(v) = (v_i \oplus v_j) \wedge \bigwedge_{\substack{1 \le \ell \le n \\ \ell \notin \{i,j\}}} v_\ell.$$

4   Revisiting Basic Problems for OBDD-based Graph Algorithms

In this section we reconsider two basic problems in the implicit setting and improve upper bounds on their resource requirements.

On argument reordering in the implicit setting. To solve a graph problem, it is sometimes useful to reverse the edges of a given graph, e.g., for maximum flow problems. Therefore, the following operation is defined (see, e.g., [18]).

Definition 7. *Let ρ be a permutation on $\{1, \ldots, k\}$ and $f \in B_{kn}$ be defined on k Boolean variable vectors $x^{(1)}, \ldots, x^{(k)}$ each of length n. The argument reordering $\mathcal{R}_\rho(f) \in B_{kn}$ with respect to ρ is $\mathcal{R}_\rho(f)(x^{(1)}, \ldots, x^{(k)}) = f(x^{(\rho(1))}, \ldots, x^{(\rho(k))})$.*

Given a π-OBDD G_f representing f, a π-OBDD for the function $\mathcal{R}_\rho(f)$ can be computed by renaming the variables in an appropriate way followed by at most $(k-1)n$ so-called jump-up operations, where a variable jumps to another position in the variable ordering.

A *k-interleaved variable ordering* on the k variable vectors $x^{(i)} = (x_1^{(i)}, \ldots, x_n^{(i)})$, $1 \leq i \leq k$, denoted by $\pi_{k,n}^\tau$ is defined in the following way:

$$\pi_{k,n}^\tau = \left(x_{\tau(1)}^{(1)}, \ldots, x_{\tau(1)}^{(k)}, x_{\tau(2)}^{(1)}, \ldots, x_{\tau(2)}^{(k)}, \ldots, x_{\tau(n)}^{(k)} \right),$$

where τ is a permutation on $\{1, \ldots, n\}$. In the design and analysis of implicit graph algorithms it is very common to use interleaved variable orderings mainly because of two reasons. The first one is that some Boolean functions that are useful in the implicit setting can be represented in small size according to such an ordering. Moreover, some interleaved variable orderings belong to the class of optimal variable orderings (with respect to the size of the representation) for important functions like the equality function $\mathrm{EQ}_n(x, y)$, the greater than function $\mathrm{GT}_n(x, y)$, or the most significant bit of binary addition $\mathrm{ADD}_{n,n}(x, y)$. The second reason why interleaved variable orderings are used in the implicit setting is that an argument reordering operation does not lead to a blow-up in the representation size (asymptotically with respect to n).

Lemma 5. *Let ρ be a permutation on $\{1, \ldots, k\}$ and let τ be a permutation on $\{1, \ldots, n\}$. Furthermore, let $f \in B_{kn}$ be defined on k Boolean variable vectors $x^{(1)}, \ldots, x^{(k)}$ each of length n, $\pi_{k,n}^\tau$ be a k-interleaved variable ordering, and G_f be a complete $\pi_{k,n}^\tau$-OBDD of width w representing f. The width of the quasi-reduced $\pi_{k,n}^\tau$-OBDD for the function $\mathcal{R}_\rho(f)$ obtained from the function f by an argument reordering according to τ is at most $2^{k-1}w$.*

Proof. Given the complete $\pi_{k,n}^\tau$-OBDD G_f representing f, a $\pi_{k,n}^\tau$-OBDD for the function $\mathcal{R}_\rho(f)$ can be computed by renaming the variables according to ρ followed by a reconstruction of a complete $\pi_{k,n}^\tau$-OBDD for $\mathcal{R}_\rho(f)$. There are n variable blocks, where the variable block j consists of $x_{\tau(j)}^{(1)}, \ldots, x_{\tau(j)}^{(k)}$, $1 \leq j \leq n$. Using Proposition 1 we know that each of these blocks can be handled separately without any affection on the number of nodes labeled by a variable which does not belong to the chosen variable block. Therefore, it is sufficient to consider only one variable block. In the following we assume that the blocks are processed bottom-up. We consider the variable block j. First, the variable $x_{\tau(j)}^{(1)}$ jumps to the first position with respect to the variables $x_{\tau(j)}^{(1)}, \ldots, x_{\tau(j)}^{(k)}$ of the variable block j and the given OBDD is reconstructed accordingly, then the next variable $x_{\tau(j)}^{(2)}$ jumps to the second position, and so on. Such a *jump-up* operation can be obtained by a synthesis operation of $x_{\tau(j)}^{(1)}$ and the OBDDs G_0 and G_1 obtained from the given OBDD by a replacement of $x_{\tau(j)}^{(1)}$ to the constant 0 and 1, respectively (see, e.g., [4]). Using Proposition 1 again, we can conclude that the number of nodes labeled by a variable at a position between the old and the new position of the jumping variable has at most doubled because the number of the different relevant subfunctions has at most doubled. (Similar consideration can be found in the proof of Theorem 2 in [4].) To bound the new number of $x_{\tau(j)}^{(1)}$-nodes, we need another simple argument. For each $x_{\tau(j)}^{(1)}$-node, there exist at least one incoming edge. Since we consider complete OBDDs, we can conclude that there can only be at most $2w$ edges leading into a node labeled by $x_{\tau(j)}^{(1)}$. There are at most $k - 1$ jump-up operations to reestablish the variable ordering $\pi_{k,n}^\tau$ for this part of the OBDD. Therefore, we can conclude that the width of the new OBDD is at most $2^{k-1}w$.   □

This result is an improvement of Lemma 3 in [19] where an upper bound of $3^k w$ on the width of the new OBDD has been shown. Note that already in [17] a better upper bound has been presented but the proof was not correct.

For the resource requirements we know that a jump-up operation can be done in time $\mathcal{O}(|F| \log |F|)$ and space $\mathcal{O}(|F|)$ for the output-OBDD F (using Theorem 1 in [15]). Furthermore, we can generalize the result for shared binary decision diagrams, a generalized OBDD model for multiple output Boolean functions. Then, dealing with all the variable blocks simultaneously, we can prove the following.

Proposition 2. *Let ρ be a permutation on $\{1, \ldots, k\}$ and let τ be a permutation on $\{1, \ldots, n\}$. Furthermore, let $f \in B_{kn}$ be defined on k Boolean variable vectors $x^{(1)}, \ldots, x^{(k)}$ each of length n, $\pi_{k,n}^\tau$ be a k-interleaved variable ordering, and G_f be a complete $\pi_{k,n}^\tau$-OBDD of width w representing f. A $\pi_{k,n}^\tau$-OBDD for the argument reordering $\mathcal{R}_\rho(f)$ can be constructed in time $\mathcal{O}(2^k wkn \log(2^k wkn))$ and space $\mathcal{O}(2^k wkn)$.*

On the quantification over a set of variables in the implicit setting. The quantification not only over one variable but over a variable vector is a very common operation in an implicit graph algorithm. Sawitzki has investigated the time and space requirement of the quantification over a set of variables with respect to the width of the given complete OBDD.

Lemma 6 (Lemma 2 in[18]). *Let $f \in B_n$ be a function on the variable set $X_n = \{x_1, \ldots, x_n\}$, $X \subseteq X_n$, $Q \in \{\exists, \forall\}$, and G_f be a complete π-OBDD representing* *f with respect to a variable ordering π. The function f' is defined by $(QX)f$, which means the function obtained from f by quantifying all variables in X. The width of the quasi-reduced π-OBDD $G_{f'}$ for f' is bounded by 2^w, where w is the width of G_f. $G_{f'}$ can be computed in time $\mathcal{O}(|X|n2^{2w}\log(n2^{2w}))$ and space $\mathcal{O}(n2^{2w})$.*

The proof structure for Lemma 6 is simple. First, it has been shown that the width of any intermediate OBDD is upper bounded by 2^w. Since $|X|$ synthesis operations are sufficient for the quantification over $|X|$ variables, the result has been obtained by using well-known bounds for one synthesis step (see Sect. 2). Our aim is to improve the upper bounds and to simplify the proof by using knowledge from automata theory. W.l.o.g. we assume that Q is $\exists$. (For the $\forall$-quantification the proof is similar using a model of OBDDs with $\wedge$-nondeterminism.) All nodes labeled by a variable in the set X in the given π-OBDD are changed into nondeterministic nodes. It is easy to see that the result is a nondeterministic π-OBDD representing the function $f' := (QX)f$. A variable z is tested on a path if there exist a node on the path labeled by z. Using the well-known powerset construction from automata theory and taking into account that OBDDs are acyclic and on every path from the source to the sinks the same set of variables $X_n \setminus X$ is tested, we convert the nondeterministic π-OBDD into a deterministic one for f'. It is easy to see that the width of the resulting OBDD $G_{f'}$ is at most 2^w. To construct $G_{f'}$ we use an array of dimension $2^w \times (|X_n| - |X|)$ to know which node has already been constructed. Each successor of a node $(v_{i_1}, v_{i_2}, \ldots, v_{i_k})$, where $v_{i_1}, v_{i_2}, \ldots, v_{i_k}$ are different nodes in the nondeterministic OBDD labeled by the same variable and $k \leq w$, can be determined by the union of maximal w nodes. Therefore, altogether we have proved the following result.

Proposition 3. *Let $f \in B_n$ be a function on the variable set $X_n = \{x_1, \ldots, x_n\}$, $X \subseteq X_n$, $Q \in \{\exists, \forall\}$, and G_f be a complete π-OBDD representing f with respect to a variable ordering π. The function f' is defined by $(QX)f$, which means the function obtained from f by quantifying all variables in X. The width of the quasi-reduced π-OBDD $G_{f'}$ for f' is bounded by 2^w, where w is the width of G_f. $G_{f'}$ can be computed in time $\mathcal{O}(n2^w w)$ and space $\mathcal{O}(n2^w)$.*

Acknowledgment. The author would like to thank the referees for comments which helped to improve the presentation of the paper.

References

1. Bollig, B.: On symbolic OBDD-based algorithms for the minimum spanning tree problem. Theor. Comput. Sci. **447**, 2–12 (2012)
2. Bollig, B.: On the complexity of some ordering problems. In: Csuhaj-Varjú, E., Dietzfelbinger, M., Ésik, Z. (eds.) MFCS 2014, Part II. LNCS, vol. 8635, pp. 118–129. Springer, Heidelberg (2014)
3. Bollig, B., Gillé, M., Pröger, T.: Implicit computation of maximum bipartite matchings by sublinear functional operations. Theor. Comput. Sci. (2014). doi:10.1016/j.tcs.2014.07.020

4. Bollig, B., Löbbing, M., Wegener, I.: On the effect of local changes in the variable ordering of ordered decision diagrams. Inf. Process. Lett. **59**, 233–239 (1996)
5. Bollig, B., Wegener, I.: Improving the variable ordering of OBDDs is NP-complete. IEEE Trans. Comput. **45**(9), 993–1002 (1996)
6. Brody, J., Matulef, K., Wu, C.: Lower bounds for testing computability by small width OBDDs. In: Ogihara, M., Tarui, J. (eds.) TAMC 2011. LNCS, vol. 6648, pp. 320–331. Springer, Heidelberg (2011)
7. Bryant, R.: Graph-based algorithms for boolean function manipulation. IEEE Trans. Comput. **35**(8), 677–691 (1986)
8. Ergün, F., Kumar, R., Rubinfeld, R.: On learning bounded-width branching programs. In: Mass, W. (ed.) COLT, pp. 361–368. ACM (1995)
9. Fortune, F., Hopcroft, J., Schmidt, E.M.: The complexity of equivalence and containment for free single variable program schemes. In: Ausiello, G., Böhm, C. (eds.) ICALP 1978. LNCS, vol. 62, pp. 227–240. Springer, Heidelberg (1978)
10. Gavril, F.: Some NP-complete problems on graphs. In: 11th Conference on Information Science and Systems, pp. 91–95 (1977)
11. Goldreich, O.: On testing computability by small width OBDDs. In: Serna, M., Shaltiel, R., Jansen, K., Rolim, J. (eds.) APPROX 2010. LNCS, vol. 6302, pp. 574–587. Springer, Heidelberg (2010)
12. Hachtel, G., Somenzi, F.: A symbolic algorithm for maximum flow in 0-1 networks. Form. Meth. Syst. Des. **10**, 207–219 (1997)
13. Newman, I.: Testing membership in languages that have small width branching programs. SIAM J. Comput. **31**(5), 1557–1570 (2002)
14. Ron, D., Tsur, G.: Testing computability by width-two OBDDs. Theor. Comput. Sci. **420**, 64–79 (2012)
15. Savický, P., Wegener, I.: Efficient algorithms for the transformation between different types of binary decision diagrams. Acta Informatica **34**(4), 245–256 (1997)
16. Sawitzki, D.: Implicit flow maximization by iterative squaring. In: Van Emde Boas, P., Pokorný, J., Bieliková, M., Štuller, J. (eds.) SOFSEM 2004. LNCS, vol. 2932, pp. 301–313. Springer, Heidelberg (2004)
17. Sawitzki, D.: A symbolic approach to the all-pairs shortest-paths problem. In: Hromkovič, J., Nagl, M., Westfechtel, B. (eds.) WG 2004. LNCS, vol. 3353, pp. 154–167. Springer, Heidelberg (2004)
18. Sawitzki, D.: The complexity of problems on implicitly represented inputs. In: Wiedermann, J., Tel, G., Pokorný, J., Bieliková, M., Štuller, J. (eds.) SOFSEM 2006. LNCS, vol. 3831, pp. 471–482. Springer, Heidelberg (2006)
19. Sawitzki, D.: Exponential lower bounds on the space complexity of OBDD-based graph algorithms. In: Correa, J.R., Hevia, A., Kiwi, M. (eds.) LATIN 2006. LNCS, vol. 3887, pp. 781–792. Springer, Heidelberg (2006)
20. Sieling, D., Wegener, I.: NC-algorithms for operations on binary decision diagrams. Parallel Process. Lett. **3**, 3–12 (1993)
21. Wegener, I.: Branching programs and binary decision diagrams: theory and applications. SIAM, Philadelphia (2000)
22. Woelfel, P.: Symbolic topological sorting with OBDDs. J. Discrete Algorithm **4**(1), 51–71 (2006)

Tight Analysis of Priority Queuing
for Egress Traffic

Jun Kawahara[1], Koji M. Kobayashi[2(✉)], and Tomotaka Maeda[3]

[1] Nara Institute of Science and Technology, Nara, Japan
jkawahara@is.naist.jp
[2] National Institute of Informatics, Chiyoda, Japan
kobaya@nii.ac.jp
[3] Academic Center for Computing and Media Studies,
Kyoto University, Kyoto, Japan
tomo@net.ist.i.kyoto-u.ac.jp

Abstract. Recently, the problems of evaluating performances of switches and routers have been formulated as online problems, and a great amount of results have been presented. In this paper, we focus on managing outgoing packets (called *egress traffic*) on switches that support Quality of Service (QoS), and analyze the performance of one of the most fundamental scheduling policies *Priority Queuing* (*PQ*) using competitive analysis. We formulate the problem of managing egress queues as follows: An output interface is equipped with m queues, each of which has a buffer of size B. The size of a packet is unit, and each buffer can store up to B packets simultaneously. Each packet is associated with one of m priority values α_j ($1 \leq j \leq m$), where $\alpha_1 \leq \alpha_2 \leq \cdots \leq \alpha_m$, $\alpha_1 = 1$, and $\alpha_m = \alpha$ and the task of an online algorithm is to select one of m queues at each scheduling step. The purpose of this problem is to maximize the sum of the values of the scheduled packets.

For any B and any m, we show that the competitive ratio of PQ is exactly $2 - \min_{x \in [1, m-1]} \{ \frac{\alpha_{x+1}}{\sum_{j=1}^{x+1} \alpha_j} \}$. That is, we conduct a complete analysis of the performance of PQ using worst case analysis. Moreover, we show that no deterministic online algorithm can have a competitive ratio smaller than $1 + \frac{\alpha^3 + \alpha^2 + \alpha}{\alpha^4 + 4\alpha^3 + 3\alpha^2 + 4\alpha + 1}$.

1 Introduction

In recent years, the Internet has provided a rich variety of applications, such as teleconferencing, video streaming, IP telephone, mainly thanks to the rapid growth of the broadband technology. To enjoy such services, the demand for the Quality of Service (QoS) guarantee is crucial. For example, usually there is little requirement for downloading programs or picture images, whereas real-time services, such as distance meeting, require constant-rate packet transmission. One possible way of supporting QoS is differentiated services (Diffserv) [15]. In DiffServ, a value is assigned to each packet according to the importance of the packet. Then, switches that support QoS (QoS switches) decide the order of

© Springer International Publishing Switzerland 2014
Z. Zhang et al. (Eds.): COCOA 2014, LNCS 8881, pp. 459–473, 2014.
DOI: 10.1007/978-3-319-12691-3_34

packets to be processed, based on the value of packets. In such a mechanism, one of the main issues in designing algorithms is how to treat packets depending on the priority in buffering or scheduling. This kind of problems was recently modeled as an *online problem*, and the *competitive analysis* [16,39] of algorithms has been done.

Aiello et al. [1] was the first to attempt this study, in which they considered a model with only one First In First Out (FIFO) queue. This model mainly focuses on the buffer management issue of the input port of QoS switches: There is one FIFO queue of size B, meaning that it can store up to B packets. An input is a sequence of events. An event is either an *arrival event*, at which a packet with a specified priority value arrives, or a *scheduling event*, at which the packet at the head of the queue will be transmitted. The task of an online (buffer management) algorithm is to decide, when a packet arrives at an arrival event, whether to accept or to reject it (in order to keep a room for future packets with higher priority). The purpose of the problem is to maximize the sum of the values of the transmitted packets. Aiello et al. analyzed the competitiveness of the Greedy Policy, the Round Robin Policy, the Fixed Partition Policy, etc.

After the publication of this seminal paper, more and more complicated models have been introduced and studied, some of which are as follows: Azar et al. [9] considered the *multi-queue switch model*, which formulates the buffering problem of one input port of the switch. In this problem, an input port has N input buffers connected to a common output buffer. The task of an online algorithm is now not only buffer management but also scheduling. At each scheduling event, an algorithm selects one of N input buffers, and the packet at the head of the selected buffer is transmitted to the inside of the switch through the output buffer. There are some formulations that model not only one port but the entire switch. For example, Kesselman et al. [29] introduced the *Combined Input and Output Queue (CIOQ) switch model*. In this model, a switch consists of N input ports and N output ports, where each port has a buffer. At an *arrival phase*, a packet (with the specified destination output port) arrives at an input port. The task of an online algorithm is buffer management as mentioned before. At a *transmission phase*, all the packets at the top of the nonempty buffers of output ports are transmitted. Hence, there is no task of an online algorithm. At a *scheduling phase*, packets at the top of the buffers of input ports are transmitted to the buffers of the output ports. Here, an online algorithm computes a matching between input ports and output ports. According to this matching, the packets in the input ports will be transmitted to the corresponding output ports. Kesselman et al. [32] considered the *crossbar switch model*, which models the scheduling phase of the CIOQ switch model more in detail. In this model, there is also a buffer for each pair of an input port and an output port. Thus, there arises another buffer management problem at scheduling phases.

In some real implementation (e.g., [17]), additional buffers are equipped with each output port of a QoS switch to control the outgoing packets (called *egress traffic*). Assume that there are m priority values of packets $\alpha_1, \alpha_2, \ldots, \alpha_m$ such that $\alpha_1 \leq \alpha_2 \leq \cdots \leq \alpha_m$. Then, m FIFO queues $Q^{(1)}, Q^{(2)}, \ldots, Q^{(m)}$ are introduced for each output port, and a packet with the value α_i arriving at this output port is stored in the queue $Q^{(i)}$. Usually, this buffering policy is greedy, namely, when a packet arrives, it is rejected if the corresponding queue is full, and accepted otherwise. The task of an algorithm is to decide which queue to transmit a packet at each scheduling event.

Several practical algorithms, such as Priority Queuing (PQ), Weighted Round-Robin (WRR) [24], and Weighted Fair Queuing (WFQ) [20], are currently implemented in network switches. PQ is the most fundamental algorithm, which selects the highest priority non-empty queue. This policy is implemented in many switches by default. (e.g., Cisco's Catalyst 2955 series [18]) In the WRR algorithm, queues are selected according to the round robin policy based on the weight of packets corresponding to queues, i.e., the rate of selecting $Q^{(i)}$ in one round is proportional to α_i for each i. This algorithm is implemented in Cisco's Catalyst 2955 series [18] and so on. In the WFQ algorithm, length of packets, as well as the priority values, are taken into consideration so that shorter packets are more likely to be scheduled. This algorithm is implemented in Cisco's Catalyst 6500 series [19] and so on.

In spite of intensive studies on online buffer management and scheduling algorithms, to the best of our knowledge, there have been no research on the egress traffic control, which we focus on in this paper. Our purpose is to evaluate the performances of actual scheduling algorithms for egress queues.

Our Results. We formulate this problem as an online problem, and provide a tight analysis of the performance of PQ using competitive analysis. Specifically, for any B, we show that the competitive ratio of PQ is exactly $2 - \min_{x \in [1,m-1]} \{ \frac{\alpha_{x+1}}{\sum_{j=1}^{x+1} \alpha_j} \}$. PQ is trivial to implement, and has a lower computational load than the other policies, such as WRR and WFQ. Hence, it is meaningful to analyze the exact performance of PQ. Moreover, we present a lower bound of $1 + \frac{\alpha^3 + \alpha^2 + \alpha}{\alpha^4 + 4\alpha^3 + 3\alpha^2 + 4\alpha + 1}$ on the competitive ratio of any deterministic algorithm.

Related Work. Independently of our work, Al-Bawani and Souza [2] have recently considered much the same model. PQ is called the greedy algorithm in their paper. They consider the case where $0 < \alpha_1 < \alpha_2 < \cdots < \alpha_m$. Also, they assume that for any $j(\in [1, m])$, the jth queue can store at most $B_j(\in [1, B])$ packets at a time. In the case of $B_j = B$, that is, in the same setting as ours, they showed that the competitive ratio of PQ is at most $2 - \min_{j \in [1,m-1]} \{ \frac{\alpha_{j+1} - \alpha_j}{\alpha_{j+1}} \}$ for any m and B. When comparing our result and their upper bound, we have $2 - \min_{x \in [1,m-1]} \{ \frac{\alpha_{x+1}}{\sum_{j=1}^{x+1} \alpha_j} \} < 2 - \min_{j \in [1,m-1]} \{ \frac{\alpha_{j+1} - \alpha_j}{\alpha_{j+1}} \}$ by elementary calculation (see [25]). Note that $2 - \min_{j \in [1,m-1]} \{ \frac{\alpha_{j+1} - \alpha_j}{\alpha_{j+1}} \}$ is equal to 2 when there exists some z such that $\alpha_{z+1} = \alpha_z$. In general practical switches, the sizes of any two egress queues attached to the same output port are equivalent by default. Since we focus on evaluating the performance of algorithms in a more practical setting (which might be less generalized), we assume that the size of each queue is B. Moreover, our analysis in this paper does not depend on the maximum numbers of packets stored in buffers, and instead it depends on whether buffers are full of packets. Thus, the exact competitive ratio of PQ would be derived for the setting where for any j, the size of the jth queue is B_j in the same way as this paper. (If we apply our method in their setting, Lemma 6 in Sect. 3.3 has to be fixed slightly. However the competitive ratio obtained in this setting seems to be a more complicated value including some mins or maxes.)

As mentioned earlier, there are a lot of studies concentrating on evaluating performances of functions of switches and routers, such as queue management and packet scheduling. The most basic one is the model consisting of single FIFO queue by Aiello et al. [1] mentioned above. In their model, each packet can take one of two values 1 or $\alpha(> 1)$. Andelman et al. [7] generalized the values of packets to any value between 1 and α. Another generalization is to allow *preemption*, namely, one may drop a packet that is already stored in a queue. Results of the competitiveness on this model are given in [1, 5–7, 21, 26, 28]. The multi-queue switch model [9, 11, 36] consists of m FIFO queues. In this model, the task of an algorithm is to manage its buffers and to schedule packets. The problem of designing only a scheduling algorithm in multi-queue switches is considered in [4, 8, 13, 14, 35]. Moreover, Albers and Jacobs [3] performed an experimental study for the first time on several online scheduling algorithms for this model. Also, the overall performance of several switches, such as shared-memory switches [23, 27, 34], CIOQ switches [10, 29, 30, 33], and crossbar switches [31, 32], are extensively studied.

Fleischer and Koga [38] and Bar-Noy et al. [12] studied the online problem of minimizing the length of the longest queue in a switch, in which the size of each queue is unbounded. In [12, 38], they showed that the competitive ratio of any online algorithm is $\Omega(\log m)$, where m is the number of queues in a switch. Fleischer and Koga [38] presented a lower bound of $\Omega(m)$ for the round robin policy. In addition, in [12, 38], the competitive ratio of a greedy algorithm called Longest Queue First is $O(\log m)$. Recently, Kogan et al. [37] studied a multi-queue switch where packets with different required processing times arrive. (In the other settings mentioned above, the required processing times of all packets are equivalent.)

2   Model Description

In this section, we formally define the problem studied in this paper. Our model consists of m queues, each with a buffer of size B. The size of a packet is unit, which means that each buffer can store up to B packets simultaneously. Each packet is associated with one of m values α_i ($1 \le i \le m$), which represents the priority of this packet where a packet with larger value is of higher priority. Without loss of generality, we assume that $\alpha_1 = 1$, $\alpha_m = \alpha$, and $\alpha_1 \le \alpha_2 \le \cdots \le \alpha_m$. The ith queue is denoted $Q^{(i)}$ and is also associated with its priority value α_i. An arriving packet with the value α_i is stored in $Q^{(i)}$.

An input for this model is a sequence of *events*. Each event is an *arrival event* or a *scheduling event*. At an arrival event, a packet arrives at one of m

queues, and the packet is *accepted* to the buffer when the corresponding queue has free space. Otherwise, it is *rejected*. If a packet is accepted, it is stored at the tail of the corresponding queue. At a scheduling event, an online algorithm selects one non-empty queue and transmits the packet at the head of the selected queue. We assume that any input contains enough scheduling events to transmit all the arriving packets in it. That is, any algorithm can certainly transmit a packet stored in its queue. Note that this assumption is common in the buffer management problem. (See e.g. [22].) The *gain* of an algorithm is the sum of the values of transmitted packets. Our goal is to maximize it. The gain of an algorithm ALG for an input σ is denoted by $V_{ALG}(\sigma)$. If $V_{ALG}(\sigma) \geq V_{OPT}(\sigma)/c$ for an arbitrary input σ, we say that ALG is *c-competitive*, where OPT is an optimal offline algorithm for σ.

3   Analysis of Priority Queuing

3.1   Priority Queuing

PQ is a greedy algorithm. At a scheduling event, PQ selects the non-empty queue with the largest index. For analysis, we assume that OPT does not reject an arriving packet. This assumption does not affect the analysis of the competitive ratio. (See the full version of this paper [25].)

3.2   Overview of the Analysis

We define an *extra packet* as a packet which is accepted by OPT but rejected by PQ. In the following analysis, we evaluate the sum of the values of extra packets to obtain the competitive ratio of PQ. We introduce some notation for our analysis. For any input σ, $k_j(\sigma)$ denotes the number of extra packets arriving at $Q^{(j)}$ when treating σ. We call a queue at which at least one extra packet arrives a *good queue* when treating σ. $n(\sigma)$ denotes the number of good queues for σ. Moreover, for any input σ and any $i(\in [1, n(\sigma)])$, $q_i(\sigma)$ denotes the good queue with the ith minimum index. That is, $1 \leq q_1(\sigma) < q_2(\sigma) < \cdots < q_{n(\sigma)}(\sigma) \leq m$. Also, we define $q_{n(\sigma)+1}(\sigma) = m$. In addition, for any input σ, $s_j(\sigma)$ denotes the number of packets which PQ transmits from $Q^{(j)}$. We drop the input σ from the notation when it is clear. Then, $V_{PQ}(\sigma) = \sum_{j=1}^{m} \alpha_j s_j$, and $V_{OPT}(\sigma) = V_{PQ}(\sigma) + \sum_{i=1}^{n} \alpha_{q_i} k_{q_i}$. (The equality follows from the lemma proven in the full version of this paper [25].)

First, we show that $k_m = 0$, that is, $q_n + 1 \leq m$, in Lemma 1. We will gradually construct some input set $\mathcal{S}^*$ (defined below) from Lemmas 3 to 8 using some adversarial strategies against PQ. Moreover, in Lemma 9, we prove that the set $\mathcal{S}^*$ includes an input σ such that the ratio $\frac{V_{OPT}(\sigma)}{V_{PQ}(\sigma)}$ is maximized. That is, we show that there exists an input σ^* in the set $\mathcal{S}^*$ to get the competitive ratio of PQ in the lemma. More formally, we define the set $\mathcal{S}^*$ of the inputs σ' satisfying the following five conditions: (i) for any $i(\in [1, n(\sigma') - 1])$, $q_i(\sigma') + 1 = q_{i+1}(\sigma')$, (ii) for any $i(\in [1, n(\sigma')])$, $k_{q_i(\sigma')}(\sigma') = B$, (iii) for any $j(\in [q_1(\sigma'), q_{n(\sigma')}(\sigma')+1])$,

$s_j(\sigma') = B$, (iv) for any $j(\in [1, q_1(\sigma') - 1])$, $s_j(\sigma') = 0$ if $q_1(\sigma') - 1 \geq 1$, and (v) for any $j(\in [q_{n(\sigma')}(\sigma') + 2, m])$, $s_j(\sigma') = 0$ if $q_{n(\sigma')}(\sigma') + 2 \leq m$. Then, we show that there exists an input $\sigma^* \in \mathcal{S}^*$ such that $\max_{\sigma''}\{\frac{V_{OPT}(\sigma'')}{V_{PQ}(\sigma'')}\} = \frac{V_{OPT}(\sigma^*)}{V_{PQ}(\sigma^*)}$ in Lemma 9.

By the above lemmas, we can obtain the competitive ratio of PQ as follows: For ease of presentation, we write $s_i(\sigma^*)$, $n(\sigma^*)$, $q_i(\sigma^*)$ and $k_i(\sigma^*)$ as s_i^*, n^*, q_i^* and k_i^*, respectively. Thus, $\frac{V_{OPT}(\sigma^*)}{V_{PQ}(\sigma^*)} = \frac{V_{PQ}(\sigma^*) + \sum_{i=1}^{n^*} \alpha_{q_i^*} k_{q_i^*}^*}{V_{PQ}(\sigma^*)} = 1 + \frac{B \sum_{j=q_1^*}^{q_{n^*}^*} \alpha_j}{B \sum_{j=q_1^*}^{q_{n^*}^*+1} \alpha_j} \leq$

$1 + \frac{\sum_{j=1}^{q_{n^*}^*} \alpha_j}{\sum_{j=1}^{q_{n^*}^*+1} \alpha_j} = 2 - \frac{\alpha_{q_{n^*}^*+1}}{\sum_{j=1}^{q_{n^*}^*+1} \alpha_j}$. The last inequality follows from $\frac{\sum_{j=x-1}^{y} \alpha_j}{\sum_{j=x-1}^{y+1} \alpha_j} - $

$\frac{\sum_{j=x}^{y} \alpha_j}{\sum_{j=x}^{y+1} \alpha_j} = (\sum_{j=x-1}^{y} \alpha_j \sum_{j=x}^{y+1} \alpha_j - \sum_{j=x}^{y} \alpha_j \sum_{j=x-1}^{y+1} \alpha_j)/(\sum_{j=x-1}^{y+1} \alpha_j \sum_{j=x}^{y+1} \alpha_j) = $

$(\alpha_{x-1}\alpha_{y+1})/(\sum_{j=x-1}^{y+1} \alpha_j \sum_{j=x}^{y+1} \alpha_j) > 0$. This gives an upper bound on the competitive ratio of PQ.

On the other hand, we show that there exists some input $\hat{\sigma}$ such that $\frac{V_{OPT}(\hat{\sigma})}{V_{PQ}(\hat{\sigma})} = 2 - \min_{x \in [1, m-1]}\{\frac{\alpha_{x+1}}{\sum_{j=1}^{x+1} \alpha_j}\}$ in Lemma 10, which presents a lower bound for PQ. Therefore, we have the following theorem:

Theorem 1. *The competitive ratio of PQ is exactly* $2 - \min_{x \in [1, m-1]}\{\frac{\alpha_{x+1}}{\sum_{j=1}^{x+1} \alpha_j}\}$.

3.3   Competitive Analysis of PQ

For ease of presentation, an *event time* denotes a moment when an event happens, and any other moment is called a *non-event time*. We assign index numbers 1 through B to each position of a queue from the head to the tail in increasing order. The jth position of $Q^{(i)}$ is called the jth *cell*. For any non-event time t, suppose that the jth cell in $Q^{(i)}$ of PQ holds a packet at t but the jth cell c in $Q^{(i)}$ of OPT does not at t. Then, we call c a *free* cell at t. Note that any extra packet is accepted at a free cell. For any non-event time t, let $h_{ALG}^{(j)}(t)$ denote the number of packets which an algorithm ALG stores in $Q^{(j)}$ at t. We first prove the following lemma. (The lemma is similar to Lemma 2.3 in [2].) Due to page limitations, we omit most of the proofs of the following lemmas. They are included in the full version of this paper [25].

Lemma 1. $k_m = 0$.

Next, in order to evaluate the total number of extra packets accepted at each $Q^{(q_i)}$ ($i \in [1, n]$), we construct some matching between extra packets and PQ's packets according to the matching routine defined later. (Note that evaluating the number of extra packets is related to the property (ii) of $\mathcal{S}^*$.) Suppose that extra packet p is matched with PQ's packet p' such that p and p' are transmitted from $Q^{(i)}$ and $Q^{(i')}$, respectively. Then, the routine constructs this matching where $i < i'$. Let us explain how to construct the matching. We match extra packet one by one with time. However, it is difficult to match an extra packet with PQ's packet in a direct way. Thus, the matching is formed in two stages. That is, at first, for any free cell c, we match c with some PQ's packet p when c becomes free at an event time. At a later time, we rematch the extra packet p' accepted into c with p at an event time when OPT accepts p'.

In order to realize such matching, we first verify a change in the number of free cells at each event before introducing our matching routine. We give some definitions for that reason. For any event time t, $t-$ denotes the non-event time before t and after the previous event time. Also, $t+$ denotes the non-event time after t and before the next event time. Let $f^{(j)}(t)$ denote the number of free cells in $Q^{(j)}$ at a non-event time t, that is, $f^{(j)}(t) = \max\{h_{PQ}^{(j)}(t) - h_{OPT}^{(j)}(t), 0\}$. Note that OPT does not reject any packet by our assumption (We detail the assumption in the full version of this paper [25]). Thus, for any non-event time t, $\sum_{j=1}^{m} h_{OPT}^{(j)}(t) > 0$ if $\sum_{j=1}^{m} h_{PQ}^{(j)}(t) > 0$.

Arrival event: Let p be the packet arriving at $Q^{(x)}$ at an event time t.

Case A1: Both PQ and OPT accept p, and $h_{PQ}^{(x)}(t-) - h_{OPT}^{(x)}(t-) > 0$:
Since $h_{PQ}^{(x)}(t+) = h_{PQ}^{(x)}(t-) + 1$ and $h_{OPT}^{(x)}(t+) = h_{OPT}^{(x)}(t-) + 1$, $h_{PQ}^{(x)}(t+) - h_{OPT}^{(x)}(t+) > 0$. Thus, the $(h_{PQ}^{(x)}(t-) + 1)$st cell of $Q^{(x)}$ becomes free in place of the $(h_{OPT}^{(x)}(t-) + 1)$st cell of $Q^{(x)}$. Hence $f^{(x)}(t+) = f^{(x)}(t-)$.

Case A2: Both PQ and OPT accept p, and $h_{PQ}^{(x)}(t-) - h_{OPT}^{(x)}(t-) \leq 0$:
Since $h_{PQ}^{(x)}(t+) = h_{PQ}^{(x)}(t-) + 1$ and $h_{OPT}^{(x)}(t+) = h_{OPT}^{(x)}(t-) + 1$, $h_{PQ}^{(x)}(t+) - h_{OPT}^{(x)}(t+) \leq 0$. Since the states of all the free cells do not change before and after t, $f^{(x)}(t+) = f^{(x)}(t-)$.

Case A3: PQ rejects p, but OPT accepts p: p is an extra packet since only OPT accepts p. p is accepted into the $(h_{OPT}^{(x)}(t-) + 1)$st cell, which is free at $t-$, of $Q^{(x)}$. $h_{PQ}^{(x)}(t+) = h_{PQ}^{(x)}(t-) = B$, and $h_{OPT}^{(x)}(t+) = h_{OPT}^{(x)}(t-) + 1$, which means that $f^{(x)}(t+) = f^{(x)}(t-) - 1$.

Scheduling event: If PQ (OPT, respectively) has at least one non-empty queue, suppose that PQ (OPT, respectively) transmits a packet from $Q^{(y)}$ ($Q^{(z)}$, respectively) at t.

Case S: $\sum_{j=1}^{m} h_{PQ}^{(j)}(t-) > 0$ and $\sum_{j=1}^{m} h_{OPT}^{(j)}(t-) > 0$:

Case S1: $y = z$:

Case S1.1: $h_{PQ}^{(y)}(t-) - h_{OPT}^{(y)}(t-) > 0$:

Since $h_{PQ}^{(y)}(t+) = h_{PQ}^{(y)}(t-) - 1$ and $h_{OPT}^{(y)}(t+) = h_{OPT}^{(y)}(t-) - 1$, $h_{PQ}^{(y)}(t+) - h_{OPT}^{(y)}(t+) > 0$ holds. Thus, the $h_{OPT}^{(y)}(t-)$th cell of $Q^{(y)}$ becomes free in place of the $h_{PQ}^{(y)}(t-)$th cell of $Q^{(y)}$. Hence $f^{(y)}(t+) = f^{(y)}(t-)$.

Case S1.2: $h_{PQ}^{(y)}(t-) - h_{OPT}^{(y)}(t-) \leq 0$:

Since $h_{PQ}^{(y)}(t+) = h_{PQ}^{(y)}(t-) - 1$ and $h_{OPT}^{(y)}(t+) = h_{OPT}^{(y)}(t-) - 1$ hold, $h_{PQ}^{(y)}(t+) - h_{OPT}^{(y)}(t+) \leq 0$. Hence the states of all the free cells do not change before and after t.

Case S2: $y > z$:

Case S2.1: $h_{PQ}^{(z)}(t-) - h_{OPT}^{(z)}(t-) < 0$:

Since $h_{PQ}^{(z)}(t+) = h_{PQ}^{(z)}(t-)$ and $h_{OPT}^{(z)}(t+) = h_{OPT}^{(z)}(t-) - 1$, $h_{PQ}^{(z)}(t+) \leq h_{OPT}^{(z)}(t+)$. Thus, the states of all the free cells of $Q^{(z)}$ do not change before and after t.

Case S2.1.1: $h_{PQ}^{(y)}(t-) - h_{OPT}^{(y)}(t-) > 0$:

Since $h_{PQ}^{(y)}(t+) = h_{PQ}^{(y)}(t-) - 1$ and $h_{OPT}^{(y)}(t+) = h_{OPT}^{(y)}(t-)$, $f^{(y)}(t+) = f^{(y)}(t-) - 1$ holds.

Case S2.1.2: $h_{PQ}^{(y)}(t-) - h_{OPT}^{(y)}(t-) \leq 0$:

Since $h_{PQ}^{(y)}(t+) = h_{PQ}^{(y)}(t-) - 1$ and $h_{OPT}^{(y)}(t+) = h_{OPT}^{(y)}(t-)$, $h_{PQ}^{(y)}(t+) < h_{OPT}^{(y)}(t+)$. Hence, the states of all the free cells of $Q^{(y)}$ do not change before and after t.

Case S2.2: $h_{PQ}^{(z)}(t-) - h_{OPT}^{(z)}(t-) \geq 0$:

$h_{PQ}^{(z)}(t+) = h_{PQ}^{(z)}(t-)$ and $h_{OPT}^{(z)}(t+) = h_{OPT}^{(z)}(t-) - 1$. Thus, the $h_{OPT}^{(z)}(t-)$th cell of $Q^{(z)}$ becomes free, which means that $f^{(z)}(t+) = f^{(z)}(t-) + 1$ holds.

Case S2.2.1: $h_{PQ}^{(y)}(t-) - h_{OPT}^{(y)}(t-) > 0$:

Since $h_{PQ}^{(y)}(t+) = h_{PQ}^{(y)}(t-) - 1$ and $h_{OPT}^{(y)}(t+) = h_{OPT}^{(y)}(t-)$, $f^{(y)}(t+) = f^{(y)}(t-) - 1$.

Case S2.2.2: $h_{PQ}^{(y)}(t-) - h_{OPT}^{(y)}(t-) \leq 0$:

Since $h_{PQ}^{(y)}(t+) = h_{PQ}^{(y)}(t-) - 1$ and $h_{OPT}^{(y)}(t+) = h_{OPT}^{(y)}(t-)$, $h_{PQ}^{(y)}(t+) < h_{OPT}^{(y)}(t+)$, which means that the states of all the free cells of $Q^{(y)}$ do not change before and after t.

Case S3: $y < z$:

Since $h_{PQ}^{(z)}(t+) = h_{PQ}^{(z)}(t-) = 0$ by the definition of PQ, no new free cell arises in $Q^{(z)}$.

Case S3.1: $h_{PQ}^{(y)}(t-) - h_{OPT}^{(y)}(t-) > 0$:

Since $h_{PQ}^{(y)}(t+) = h_{PQ}^{(y)}(t-) - 1$ and $h_{OPT}^{(y)}(t+) = h_{OPT}^{(y)}(t-)$, $f^{(y)}(t+) = f^{(y)}(t-) - 1$ holds.

Case S3.2: $h_{PQ}^{(y)}(t-) - h_{OPT}^{(y)}(t-) \leq 0$:

Since $h_{PQ}^{(y)}(t+) = h_{PQ}^{(y)}(t-)-1$ and $h_{OPT}^{(y)}(t+) = h_{OPT}^{(y)}(t-)$, $h_{PQ}^{(y)}(t+) < h_{OPT}^{(y)}(t+)$ holds. Hence, the states of all the free cells of $Q^{(y)}$ do not change before and after t.

Case S̄: $\sum_{j=1}^{m} h_{PQ}^{(j)}(t-) = 0$ and $\sum_{j=1}^{m} h_{OPT}^{(j)}(t-) > 0$:

Since the buffer of PQ is empty, there does not exist any free cell in it.

Based on a change in the state of free cells, we match each extra packet with a packet transmitted by PQ according to the matching routine in Table 1. (All the names of the cases in the routine correspond to the names of cases in the above sketch about free cells.) We outline the matching routine. Roughly speaking, the routine either adds a new edge to a tentative matching if a new free cell arises (Cases A1, S1.1, S2.2), or fixes some edge if OPT accepts an extra packet (Case A3), while keeping edges constructed before. In the other cases (Cases A2, S1.2, S2.1, S3, S̄), the routine does nothing. Specifically, both OPT and PQ accept arriving packets at the same queue in Case A1, and they transmit packets from the same queue in Case S1.1. Since the total numbers of free cells do not change in these cases but the states of free cells do, the routine updates an edge in a tentative matching, namely removes an edge between PQ's packet p and a cell that became non-free and adds a new edge between p and a new free cell. When the routine executes Case S2.2, the queue where OPT transmits a packet is different from that of PQ. By the conditions of the numbers of packets in their queues and so on (see the condition of Case S2.2), a cell of OPT's queue becomes free. The routine matches the cell with the packet transmitted by PQ at this event. In Case A3, an extra packet is accepted into a free cell c. Since c has been already matched with some PQ's packet p', which can be proven inductively in Lemma 2, the routine replaces the partner of p' from c to p. Once an extra packet is matched, the partner of the packet never changes.

We give some definitions. For any packet p, $g(p)$ denotes the index of the queue at which p arrives. Also, for any cell c, $g(c)$ denotes the index of the queue including c. We now show the feasibility of the routine.

Lemma 2. *For any non-event time t', and any extra packet p which arrives before t', there exists some packet p' such that PQ transmits p' before t', $g(p) < g(p')$ and p is matched with p' at t'. Moreover, for any free cell c at t', there exists some packet p'' such that PQ transmits p'' before t', $g(c) < g(p'')$, and c is matched with p'' at t'.*

Proof. The proof is by induction on the event time. The base case is clear. Let t be any event time. We assume that the statement is true at $t-$, and prove that it is true at $t+$.

First, we discuss the case where the routine executes Case A1 or S1.1 at t. Let c be the cell which becomes free at t. Also, let c' be the cell which is free at $t-$ and not free at $t+$. By the induction hypothesis, a packet p which is transmitted by PQ before $t-$ is matched with c' at $t-$. Then, the routine unmatches p, and matches p with c by the definitions of Cases A1 and S1.1. $g(c) = g(c')$ clearly holds. Also, since $g(c') < g(p)$ by the induction hypothesis, the statement is true at $t+$.

Next, we consider the case where the routine executes Case A3 at t. Let p' be the extra packet accepted by OPT at t. Also, let c be the free cell into which OPT accepts p' at t. By the induction hypothesis, a packet p which is transmitted by PQ before $t-$ is matched with c at $t-$. Then, by the definition of Case A3, the routine unmatches p, and matches p with p'. $g(c) = g(p')$ holds by definition. In addition, $g(c) < g(p)$ by the induction hypothesis. Thus, $g(p') < g(p)$, which means that the statement holds at $t+$.

Third, we investigate the case where the routine executes Case S2.2 at t. Suppose that PQ transmits a packet p at t, and the new free cell c arises at t.

Table 1. Matching routine

Matching routine: Let t be an event time.

Arrival event: Suppose that the packet p arrives at $Q^{(x)}$ at t. Execute one of the following three cases at t.

Case A1: Both PQ and OPT accept p, and $h_{PQ}^{(x)}(t-) - h_{OPT}^{(x)}(t-) > 0$:

Let c be OPT's $(h_{OPT}^{(x)}(t-) + 1)$st cell of $Q^{(x)}$, which is free at $t-$ but not at $t+$. Let c' be OPT's $(h_{PQ}^{(x)}(t-) + 1)$st cell which is not free at $t-$ but is free at $t+$. There exists the packet q matched with c at $t-$. (The existence of such q is guaranteed by Lemma 2.) Change the matching partner of q from c to c'.

Case A2: Both PQ and OPT accept p, and $h_{PQ}^{(x)}(t-) - h_{OPT}^{(x)}(t-) \leq 0$:

Do nothing.

Case A3: PQ rejects p, but OPT accepts p:

Let c be OPT's $(h_{OPT}^{(x)}(t-) + 1)$st cell of $Q^{(x)}$, that is, the cell to which the extra packet p is now stored. Note that c is free at $t-$ but is not at $t+$. There exists the packet q matched with c at $t-$. (See Lemma 2.) Change the partner of q from c to p.

Scheduling event: If PQ (OPT, respectively) has at least one non-empty queue at $t-$, suppose that PQ (OPT, respectively) transmits a packet from $Q^{(y)}$ ($Q^{(z)}$, respectively) at t. Execute one of the following three cases at t.

Case S1.1: $\sum_{j=1}^{m} h_{PQ}^{(j)}(t-) > 0$, $\sum_{j=1}^{m} h_{OPT}^{(j)}(t-) > 0$, $y = z$, and $h_{PQ}^{(y)}(t-) - h_{OPT}^{(y)}(t-) > 0$:

Let c be OPT's $h_{PQ}^{(y)}(t-)$th cell of $Q^{(y)}$, which is free at $t-$ but is not free at $t+$. Let c' be OPT's $h_{OPT}^{(y)}(t-)$th cell of $Q^{(y)}$, which is not free at $t-$ but is free at $t+$. There exists the packet q matched with c at $t-$. (See Lemma 2.) Change the matching partner of q from c to c'.

Case S2.2: $\sum_{j=1}^{m} h_{PQ}^{(j)}(t-) > 0$, $\sum_{j=1}^{m} h_{OPT}^{(j)}(t-) > 0$, $y > z$, and $h_{PQ}^{(z)}(t-) - h_{OPT}^{(z)}(t-) \geq 0$:

Let c be OPT's $h_{OPT}^{(z)}(t-)$th cell of $Q^{(z)}$, which becomes free at $t+$. Since the packet p transmitted from $Q^{(y)}$ by PQ is not matched with anything (see Lemma 2), match p with c.

Otherwise (Cases S1.2, S2.1, S3, $\bar{S}$): Do nothing.

By the induction hypothesis, any PQ's packet which is matched with a free cell or an extra packet is transmitted before t. Hence, p is not matched with anything at $t-$. Thus, the routine can match p with c at t. Moreover, $g(c) < g(p)$ by the condition of Case S2.2. By the induction hypothesis, the statement is true at $t+$.

In the other cases, a new matching does not arise. Therefore, the statement is clear by the induction hypothesis, which completes the proof.

In the next lemma, we obtain part of the properties of the set S^*.

Lemma 3. *Let σ be an input such that for some $u(\in [1,m])$, $s_u(\sigma) > B$. Then, there exists an input $\hat{\sigma}$ such that for each $j(\in [1,m])$, $s_j(\hat{\sigma}) \leq B$, and $\frac{V_{OPT}(\sigma)}{V_{PQ}(\sigma)} < \frac{V_{OPT}(\hat{\sigma})}{V_{PQ}(\hat{\sigma})}$.*

We give the notation. S_1 denotes the set of inputs σ such that for any $j(\in [1,m])$, $s_j(\sigma) \leq B$. In what follows, we analyze only inputs in S_1 by Lemma 3. Next, we evaluate the number of extra packets arriving at each good queue using Lemma 2.

Lemma 4. *For any $x(\in [1,n])$, $\sum_{i=x}^{n} k_{q_i} \leq \sum_{j=q_x+1}^{m} s_j$.*

Now we gradually gain all the properties of S^* in the following lemmas while proving S^* contains inputs σ such that $\frac{V_{OPT}(\sigma)}{V_{PQ}(\sigma)}$ is maximized. Specifically, for $i = 1, \ldots, 4$, we construct some subset S_{i+1} from the set S_i in each of the following lemmas, and eventually we can gain S^* from S_5. (We have already obtained S_1 in Lemma 3.) It is difficult to show all the properties of S^* in one lemma, and thus we progressively give the definitions of the S_{i+1} that has more restrictive properties than S_i.

Next in Lemma 5, we discuss the condition of events where the number of extra packets accepted into a good queue $Q^{(q_i)}$ ($i \in [1,n]$) is maximized, and show that it is true when $k_{q_i} = \sum_{j=q_i+1}^{q_{i+1}} s_j$. Throughout the proofs of all the following lemmas, we drop σ from $s_j(\sigma)$, $n(\sigma)$, $q_i(\sigma)$ and $k_j(\sigma)$.

Lemma 5. *For any input $\sigma \in S_1$, there exists an input $\hat{\sigma}(\in S_1)$ such that (i) for any $i(\in [1, n(\hat{\sigma})])$, $k_{q_i(\hat{\sigma})}(\hat{\sigma}) = \sum_{j=q_i(\hat{\sigma})+1}^{q_{i+1}(\hat{\sigma})} s_j(\hat{\sigma})$, (ii) for any $j(\in [1, q_1(\hat{\sigma}) - 1])$, $s_j(\hat{\sigma}) = 0$ if $q_1(\hat{\sigma}) - 1 \geq 1$, and (iii) $\frac{V_{OPT}(\sigma)}{V_{PQ}(\sigma)} \leq \frac{V_{OPT}(\hat{\sigma})}{V_{PQ}(\hat{\sigma})}$.*

In light of the above lemma, we introduce the next set of inputs. S_2 denotes the set of inputs $\sigma(\in S_1)$ satisfying the following conditions: (i) for any $i(\in [1,n])$, $k_{q_i} = \sum_{j=q_i+1}^{q_{i+1}} s_j$, (ii) for any $j(\in [q_1, m])$, $s_j \leq B$, and (iii) for any $j(\in [1, q_1-1])$, $s_j = 0$ if $q_1 - 1 \geq 1$.

Lemma 6. *Let $\sigma(\in S_2)$ be an input such that for some $z(\leq n(\sigma)-1)$, $q_z(\sigma)+1 < q_{z+1}(\sigma)$. Then, there exists an input $\hat{\sigma}(\in S_2)$ such that (i) for each $i(\in [1, n(\hat{\sigma}) - 1])$, $q_i(\hat{\sigma}) + 1 = q_{i+1}(\hat{\sigma})$ and $k_{q_i(\hat{\sigma})}(\hat{\sigma}) = B$, and (ii) $\frac{V_{OPT}(\sigma)}{V_{PQ}(\sigma)} \leq \frac{V_{OPT}(\hat{\sigma})}{V_{PQ}(\hat{\sigma})}$.*

We define the set S_3 of inputs. S_3 denotes the set of inputs $\sigma(\in S_2)$ such that (i) for each $i(\in [1, n-1])$, $q_i + 1 = q_{i+1}$, (ii) for each $i(\in [1, n-1])$, $k_{q_i} = B$, (iii) for each $j(\in [q_1, q_n])$, $s_j = B$, (iv) for any $j(\in [1, q_1 - 1])$, $s_j = 0$ if $q_1 - 1 \geq 1$, and (v) for each $j(\in [q_n + 1, m])$, $s_j \leq B$. (By Lemma 1, $q_n + 1 \leq m$.)

Lemma 7. *For any input $\sigma(\in \mathcal{S}_3)$, there exists an input $\sigma'(\in \mathcal{S}_3)$ such that (i)* $s_{q_{n(\sigma)}(\sigma)+u+1}(\sigma') = (\sum_{j=q_{n(\sigma)}(\sigma)+1}^{m} s_j(\sigma)) - uB$, *where* $u = \lfloor \frac{\sum_{j=q_{n(\sigma)}(\sigma)+1}^{m} s_j(\sigma)}{B} \rfloor$, *and for any* $j(\in [q_{n(\sigma)}(\sigma), q_{n(\sigma)}(\sigma) + u])$, $s_j(\sigma') = B$, *and (ii)* $\frac{V_{OPT}(\sigma)}{V_{PQ}(\sigma)} \leq \frac{V_{OPT}(\sigma')}{V_{PQ}(\sigma')}$.

We next introduce the set $\mathcal{S}_4$ of inputs. Let $\mathcal{S}_4$ denote the set of inputs $\sigma(\in \mathcal{S}_3)$ satisfying the following five conditions: (i) for each $i(\in [1, n-1])$, $q_i + 1 = q_{i+1}$, (ii) for each $i(\in [1, n-1])$, $k_{q_i} = B$, (iii) for each $j(\in [q_1, q_n])$, $s_j = B$, (iv) for any $j(\in [1, q_1 - 1])$, $s_j = 0$ if $q_1 - 1 \geq 1$, and (v) there exists some u such that $0 \leq u \leq m - q_n - 1$. Also, for any $j(\in [q_n, q_n + u])$, $s_j = B$, $B \geq s_{q_n+u+1} \geq 1$, and for any $j(\in [q_n + u + 2, m])$, $s_j = 0$ if $q_n + u + 2 \leq m$.

Lemma 8. *Let $\sigma(\in \mathcal{S}_4)$ be an input such that $q_{n(\sigma)}(\sigma)+2 \leq m$, $s_{q_{n(\sigma)}(\sigma)+1}(\sigma) = B$, and $\sum_{j=q_{n(\sigma)}(\sigma)+2}^{m} s_j(\sigma) > 0$.*
Then, there exists an input $\hat{\sigma}(\in \mathcal{S}_4)$ such that (i) $n(\hat{\sigma}) = n(\sigma) + 1$, (ii) for each $i(\in [1, n(\hat{\sigma}) - 1])$, $q_i(\hat{\sigma}) = q_i(\sigma)$, and $q_{n(\hat{\sigma})}(\hat{\sigma}) = q_{n(\sigma)}(\sigma) + 1$, and (iii) $\frac{V_{OPT}(\sigma)}{V_{PQ}(\sigma)} \leq \frac{V_{OPT}(\hat{\sigma})}{V_{PQ}(\hat{\sigma})}$.

$\mathcal{S}_5$ denotes the set of inputs $\sigma(\in \mathcal{S}_4)$ satisfying the following six conditions: (i) for each $i(\in [1, n-1])$, $q_i + 1 = q_{i+1}$, (ii) for each $i(\in [1, n-1])$, $k_{q_i} = B$, (iii) for each $j(\in [q_1, q_n])$, $s_j = B$, (iv) for any $j(\in [1, q_1 - 1])$, $s_j = 0$ holds if $q_1 - 1 \geq 1$, (v) $k_{q_n} = s_{q_n+1}$ (By Lemma 1, $q_n + 1 \leq m$.) and $1 \leq s_{q_n+1} \leq B$, and (vi) for any $j(\in [q_n + 2, m])$, $s_j = 0$ holds if $q_n + 2 \leq m$.

Lemma 9. *For any input $\sigma(\in \mathcal{S}_5)$, there exists an input $\hat{\sigma}(\in \mathcal{S}_5)$ such that (i) $s_{q_{n(\hat{\sigma})}(\hat{\sigma})+1}(\hat{\sigma}) = B$, and (ii) $\frac{V_{OPT}(\sigma)}{V_{PQ}(\sigma)} \leq \frac{V_{OPT}(\hat{\sigma})}{V_{PQ}(\hat{\sigma})}$.*
That is, there exists an input $\sigma^* \in \mathcal{S}^*$ such that $\max_{\sigma'}\{\frac{V_{OPT}(\sigma')}{V_{PQ}(\sigma')}\} = \frac{V_{OPT}(\sigma^*)}{V_{PQ}(\sigma^*)}$.

Proof. Since $\sigma \in \mathcal{S}_5$ holds, $\frac{V_{OPT}(\sigma)}{V_{PQ}(\sigma)} = \frac{V_{PQ}(\sigma)+\sum_{i=1}^{n} \alpha_{q_i} k_{q_i}}{V_{PQ}(\sigma)} \leq 1 + \frac{B(\sum_{j=q_1}^{q_n-1} \alpha_j)+\alpha_{q_n} s_{q_n+1}}{\sum_{j=q_1}^{q_n+1} \alpha_j s_j} \leq 1 + \frac{B(\sum_{j=q_1}^{q_n-1} \alpha_j)+\alpha_{q_n} s_{q_n+1}}{B(\sum_{j=q_1}^{q_n} \alpha_j)+\alpha_{q_n+1} s_{q_n+1}}$, which we define as $x(s_{q_n+1})$.

Let $\sigma_1, \sigma_2 \in \mathcal{S}_5$ be any inputs such that (i) $n = n(\sigma_2) = n(\sigma_1)+1$, (ii) for any $i(\in [1, n-1])$, $q_i = q_i(\sigma_1) = q_i(\sigma_2)$, (iii) $q_n = q_n(\sigma_2)$, and (iv) $s_{q_{n-1}+1}(\sigma_1) = B$ and $s_{q_n+1}(\sigma_2) = B$. Then, since $x(s_{q_n+1})$ is monotone (increasing or decreasing) as s_{q_n+1} increases, $\frac{V_{OPT}(\sigma)}{V_{PQ}(\sigma)} \leq \max\{\frac{V_{OPT}(\sigma_1)}{V_{PQ}(\sigma_1)}, \frac{V_{OPT}(\sigma_2)}{V_{PQ}(\sigma_2)}\}$. Therefore, let $\hat{\sigma}$ be the input such that $\hat{\sigma} \in \arg\max\{\frac{V_{OPT}(\sigma_1)}{V_{PQ}(\sigma_1)}, \frac{V_{OPT}(\sigma_2)}{V_{PQ}(\sigma_2)}\}$, which means that the statement is true.

Lemma 10. *The competitive ratio of PQ is at least $2 - \min_{x \in [1, m-1]}\{\frac{\alpha_{x+1}}{\sum_{j=1}^{x+1} \alpha_j}\}$.*

Proof. Consider the following input σ. Define $m' \in \arg\min_{x \in [1, m-1]}\{\frac{\alpha_{x+1}}{\sum_{j=1}^{x+1} \alpha_j}\}$. Initially, $(m' + 1)B$ arrival events happen such that B packets arrive at $Q^{(1)}$ to $Q^{(m'+1)}$. Then, for $k = 1, 2, \ldots, m'$, the kth round consists of B scheduling events followed by B arrival events in which all the B packets arrive at $Q^{(m'-k+1)}$.

For σ, PQ transmits B packets from $Q^{(m'-k+2)}$ at the kth round. As a result, PQ cannot accept arriving packets in the $(k+1)$st round. Hence, $V_{PQ}(\sigma) = B \sum_{j=1}^{m'+1} \alpha_j$ holds. On the other hand, OPT transmits B packets from $Q^{(m'-k+1)}$ at the kth round, and hence can accept all the arriving packets. Thus, $V_{OPT}(\sigma) = 2B \sum_{j=1}^{m'} \alpha_j + B\alpha_{m'+1}$. Therefore, $\frac{V_{OPT}(\sigma)}{V_{PQ}(\sigma)} = \frac{2\sum_{j=1}^{m'} \alpha_j + \alpha_{m'+1}}{\sum_{j=1}^{m'+1} \alpha_j} = 2 - \frac{\alpha_{m'+1}}{\sum_{j=1}^{m'+1} \alpha_j}$.

(It is easy to see that $\sigma \in \mathcal{S}_5$.)

4   Lower Bound for Deterministic Algorithms

In this section, we show a lower bound for any deterministic algorithm. We make an assumption that is well-known to have no effect on the analysis of the competitive ratio. We consider only online algorithms that transmit a packet at a scheduling event whenever their buffers are not empty. (Such algorithms are called *work-conserving*. See e.g. [9].)

Theorem 2. *No deterministic online algorithm can achieve a competitive ratio smaller than* $1 + \frac{\alpha^3 + \alpha^2 + \alpha}{\alpha^4 + 4\alpha^3 + 3\alpha^2 + 4\alpha + 1}$.

5   Concluding Remarks

A lot of packets used by multimedia applications arrive in a QoS switch at a burst, and managing queues to store outgoing packets (egress traffic) can become a bottleneck. In this paper, we have formulated the problem of controlling egress traffic, and analyzed Priority Queuing policies (PQ) using competitive analysis. We have shown that the competitive ratio of PQ is exactly $2 - \min_{x \in [1,m-1]}\{\frac{\alpha_{x+1}}{\sum_{j=1}^{x+1} \alpha_j}\}$. Moreover, we have shown that there is no $1 + \frac{\alpha^3 + \alpha^2 + \alpha}{\alpha^4 + 4\alpha^3 + 3\alpha^2 + 4\alpha + 1}$-competitive deterministic algorithm.

We present some open questions as follows: (i) What is the competitive ratio of other practical policies, such as WRR? (ii) We consider the case where the size of each packet is one, namely fixed. In the setting where packets with variable sizes arrive, what is the competitive ratio of PQ or other policies? (iii) We are interested in comparing our results with experimental results using measured data in QoS switches. (iv) The goal was to maximize the sum of the values of the transmitted packets in this paper, which is generally used for the online buffer management problems. However, this may not be able to evaluate the actual performance of practical scheduling algorithms correctly. (We showed that the worst scenario for PQ is extreme in this paper.) What if another objective function (e.g., fairness) is used for evaluating the performance of a scheduling algorithm? (v) An obvious open question is to close the gap between the competitive ratio of PQ and our lower bound for any deterministic algorithm.

Acknowledgments. We would like to deeply thank Associate Professor Shuichi Miyazaki for a lot of advice on an earlier version of this paper. This work was supported by JSPS KAKENHI Grant Number 26730008 and Cyber Physical System Integrated IT Platform project.

References

1. Aiello, W., Mansour, Y., Rajagopalan, S., Rosén, A.: Competitive queue policies for differentiated services. J. Algorithms **55**(2), 113–141 (2005)
2. Al-Bawani, K., Souza, A.: Buffer overflow management with class segregation. Inf. Process. Lett. **113**(4), 145–150 (2013)
3. Albers, S., Jacobs, T.: An experimental study of new and known online packet buffering algorithms. Algorithmica **57**(4), 725–746 (2010)
4. Albers, S., Schmidt, M.: On the performance of greedy algorithms in packet buffering. SIAM J. Comput. **35**(2), 278–304 (2005)
5. Andelman, N.: Randomized queue management for DiffServ. In: Proceedings of the 17th ACM Symposium on Parallel Algorithms and Architectures, pp. 1–10 (2005)
6. Andelman, N., Mansour, Y.: Competitive management of non-preemptive queues with multiple values. In: Fich, F.E. (ed.) DISC 2003. LNCS, vol. 2848, pp. 166–180. Springer, Heidelberg (2003)
7. Andelman, N., Mansour, Y., Zhu, A.: Competitive queueing policies for QoS switches. In: Proceedings of the 14th ACM-SIAM Symposium on Discrete Algorithms, pp. 761–770 (2003)
8. Azar, Y., Litichevskey, A.: Maximizing throughput in multi-queue switches. Algorithmica **45**(1), 69–90 (2006)
9. Azar, Y., Richter, Y.: Management of multi-queue switches in QoS networks. Algorithmica **43**(1–2), 81–96 (2005)
10. Azar, Y., Richter, Y.: An improved algorithm for CIOQ switches. ACM Trans. Algorithms **2**(2), 282–295 (2006)
11. Azar, Y., Richter, Y.: The zero-one principle for switching networks. In: Proceedings of the 36th ACM Symposium on Theory of Computing, pp. 64–71 (2004)
12. Bar-Noy, A., Freund, A., Landa, S., Naor, J.: Competitive on-line switching policies. Algorithmica **36**(3), 225–247 (2003)
13. Bienkowski, M., Mądry, A.: Geometric aspects of online packet buffering: an optimal randomized algorithm for two buffers. In: Laber, E.S., Bornstein, C., Nogueira, L.T., Faria, L. (eds.) LATIN 2008. LNCS, vol. 4957, pp. 252–263. Springer, Heidelberg (2008)
14. Bienkowski, M.: An optimal lower bound for buffer management in multi-queue switches. Algorithmica **68**(2), 426–447 (2014)
15. Blanke, S., Black, D., Carlson, M., Davies, E., Wang, Z., Weiss, W.: An architecture for differentiated services. RFC2475, IETF, December 1998
16. Borodin, A., El-Yaniv, R.: Online Computation and Competitive Analysis. Cambridge University Press, Cambridge (1998)
17. Cisco Systems Inc. Campus QoS Design (2014). http://www.cisco.com/en/US/docs/solutions/Enterprise/WAN_and_MAN/QoS_SRND/QoSDesign.html
18. Cisco Systems Inc. Cisco Catalyst 2955 series switches data sheets (2014). http://www.cisco.com/en/US/products/hw/switches/ps628/products_data_sheets_list.html
19. Cisco Systems Inc. Cisco Catalyst 6500 series switches data sheets (2014). http://www.cisco.com/en/US/products/hw/switches/ps708/products_data_sheets_list.html
20. Demers, A., Keshav, S., Shenker, S.: Analysis and simulation of a fair queueing algorithm. J. Internetworking Res. Exper. **1**(1), 3–26 (1990)
21. Englert, M., Westermann, M.: Lower and upper bounds on FIFO buffer management in QoS switches. Algorithmica **53**(4), 523–548 (2009)

22. Goldwasser, M.: A survey of buffer management policies for packet switches. ACM SIGACT News **41**(1), 100–128 (2010)
23. Hahne, E., Kesselman, A., Mansour, Y.: Competitive buffer management for shared-memory switches. In: Proceedings of the 13th ACM Symposium on Parallel Algorithms and Architectures, pp. 53–58 (2001)
24. Katevenis, M., Sidiropopulos, S., Courcoubetis, C.: Weighted round-robin cell multiplexing in a general-purpose ATM switch chip. IEEE J. Sel. Area Commun. **9**(8), 1265–1279 (1991)
25. Kawahara, J., Kobayashi, K.M., Maeda, T.: Tight analysis of priority queuing policy for egress traffic (2012). arXiv:1207.5959 [cs.DS]
26. Kesselman, A., Lotker, Z., Mansour, Y., Patt-Shamir, B., Schieber, B., Sviridenko, M.: Buffer overflow management in QoS switches. SIAM J. Comput. **33**(3), 563–583 (2004)
27. Kesselman, A., Mansour, Y.: Harmonic buffer management policy for shared memory switches. Theoret. Comput. Sci. **324**(2–3), 161–182 (2004)
28. Kesselman, A., Mansour, Y., van Stee, R.: Improved competitive guarantees for QoS buffering. Algorithimica **43**(1–2), 63–80 (2005)
29. Kesselman, A., Rosén, A.: Scheduling policies for CIOQ switches. J. Algorithms **60**(1), 60–83 (2006)
30. Kesselman, A., Rosén, A.: Controlling CIOQ switches with priority queuing and in multistage interconnection networks. J. Interconnection Netw. **9**(1/2), 53–72 (2008)
31. Kesselman, A., Kogan, K., Segal, M.: Packet mode and QoS algorithms for buffered crossbar switches with FIFO queuing. Distrib. Comput. **23**(3), 163–175 (2010)
32. Kesselman, A., Kogan, K., Segal, M.: Best effort and priority queuing policies for buffered crossbar switches. Chicago J. Theor. Sci. 1–14 (2012)
33. Kesselman, A., Kogan, K., Segal, M.: Improved competitive performance bounds for CIOQ switches. Algorithmica **63**(1–2), 411–424 (2012)
34. Kobayashi, K., Miyazaki, S., Okabe, Y.: A tight bound on online buffer management for two-port shared-memory switches. In: Proceedings of the 19th ACM Symposium on Parallel Algorithms and Architectures, pp. 358–364 (2007)
35. Kobayashi, K., Miyazaki, S., Okabe, Y.: A tight upper bound on online buffer management for multi-queue switches with bicodal buffers. IEICE Trans. Fund. Electron. Commun. Comput. Sci. **E91–D**(12), 2757–2769 (2008)
36. Kobayashi, K., Miyazaki, S., Okabe, Y.: Competitive buffer management for multi-queue switches in QoS networks using packet buffering algorithms. In: Proceedings of the 21st ACM Symposium on Parallel Algorithms and Architectures, pp. 328–336 (2009)
37. Kogan, K., Lopez-Ortiz, A., Nikolenko, S., Sirotkin, A.: Multi-queued network processors for packets with heterogeneous processing requirements. In: Proceedings of the 5th International Conference on Communication Systems and Networks, pp. 1–10 (2013)
38. Fleischer, R., Koga, H.: Balanced scheduling toward loss-free packet queuing and delay fairness. Algorithmica **38**(2), 363–376 (2004)
39. Sleator, D., Tarjan, R.: Amortized efficiency of list update and paging rules. Commun. ACM **28**(2), 202–208 (1985)

Optimally Bracing Grid Frameworks with Holes

Yoshihiko Ito[1], Yuki Kobayashi[1]($\boxtimes$), Yuya Higashikawa[1], Naoki Katoh[1],
Sheung-Hung Poon[2], and Maria Saumell[3]

[1] Department of Architecture and Architectural Engineering,
Kyoto University, Kyoto, Japan
{as-ito-y,as-kobayashi,as.higashikawa,naoki}@archi.kyoto-u.ac.jp
[2] Department of Computer Science, National Tsing Hua University, Hsinchu, Taiwan
spoon@cs.nthu.edu.tw
[3] Department of Mathematics and European Centre of Excellence NTIS,
University of West Bohemia, Pilsen, Czech Republic
saumell@kma.zcu.cz

Abstract. We consider the bracing problem of a square grid framework possibly with holes and present an efficient algorithm for making the framework infinitesimally rigid by augmenting it with the minimum number of diagonal braces. This number of braces matches the lower bound given by Gáspár, Radics and Recski [2]. Our contribution extends the famous result on bracing the rectangular grid framework by Bolker and Crapo [1].

Keywords: Combinatorial rigidity · Bar-joint framework · Square grid framework · Bracing

1 Introduction

Bolker and Crapo gave a necessary and sufficient condition in their seminal paper [1] for an $m \times n$ square grid framework with some diagonal braces of unit grid squares to be infinitesimally rigid. They defined a bipartite graph corresponding to the square grid framework with some diagonal braces such that the infinitesimal rigidity of the framework can be tested by checking the connectivity of the graph. In particular, the minimum number of diagonal braces that are necessary and sufficient to make the $m \times n$ square grid framework infinitesimally rigid is $m + n - 1$ (see Fig. 1).

Yuya Higashikawa: Supported by JSPS Grant-in-Aid for JSPS Fellows (26 · 4042).
Naoki Katoh: Supported by JSPS Grant-in-Aid for Scientific Research(A) (25240004).
Sheung-Hung Poon: Supported in part by grant NSC 100–2628-E-007–020-MY3 in Taiwan, R.O.C.
Maria Saumell: Supported by the project NEXLIZ - CZ.1.07/2.3.00/30.0038, which is co-financed by the European Social Fund and the state budget of the Czech Republic.

© Springer International Publishing Switzerland 2014
Z. Zhang et al. (Eds.): COCOA 2014, LNCS 8881, pp. 474–489, 2014.
DOI: 10.1007/978-3-319-12691-3_35

Radics and Recski [5] studied the case with holes where the outer boundary of the square grid framework is a simple rectilinear polygon, and long diagonal bars as well as cables can be used. They showed a lower bound for the number of diagonal bars and cables required to make the framework rigid; this bound matches the one for the case where only short braces (diagonal edges of unit grid squares) are allowed. However, in this case, they noted that the characterization based on a bipartite graph is no longer valid.

Gáspár, Radics and Recski [2] studied the case with holes where the outer boundary is rectangular, and derived a necessary and sufficient condition in terms of the rank of a certain matrix for an $m \times n$ square grid framework with some diagonal braces of unit grid squares to be infinitesimally rigid. The advantage of using the matrix introduced by [2] is that the matrix size is much reduced compared with using the original rigidity matrix, which helps to substantially reduce the running time for checking the rigidity. The paper [2] mentioned that the result can be generalized to the case where the outer boundary is a rectilinear simple polygon. However, the details were not given.

Although not clearly stated, the paper [5] mentioned that the above necessary and sufficient condition implies a lower bound on the number of diagonal bars and/or cables that are necessary to make the framework infinitesimally rigid, which is stated as

$$(\#\{\text{row and column segments}\}) - 2(\#\{\text{holes}\}) - 1 \qquad (1)$$

where row and column segments will be defined later.

In this paper, we consider the case where short diagonal bars connecting opposite corners of a unit square (simply called *braces*) are used, and generalize the previous results in the following three directions.

1. We consider the case where there is no hole, but the outer boundary of the square grid polygon is a general rectilinear polygon. For this case, we shall give a characterization based on a bipartite graph which is the same as the one in [1].
2. For the case with holes, we observe that it is not possible to generalize the characterization in [1]. However, for this case we shall show a lower bound on the number of braces required to make the square grid framework infinitesimally rigid which matches the one obtained in [2].
3. We shall propose an algorithm for bracing a square grid framework using the minimum number of braces and matching the lower bound.

The motivation of the work by Bolker and Crapo [1] was finding a way to add braces to a one-story building to make it rigid. In view of this, it is a very important issue to investigate the bracing problem of grid frameworks with holes because walls and/or ceilings may have holes such as windows.

The rest of this paper is organized as follows. Section 2 introduces the necessary definitions and notations. Section 3 considers the case of a square grid framework with no holes such that the outer boundary is a general rectilinear polygon, and gives a characterization based on a bipartite graph which generalizes the result by [1]. Sections 4 and 5 consider the case of a square grid framework with holes such that the outer boundary is a general rectilinear polygon. In Sect. 4, we first show that the characterization using a bipartite graph is no longer possible, and then we give a lower bound on the number of braces required to make the framework rigid. Section 5 proposes an algorithm that adds the minimum number of braces required to make the framework rigid.

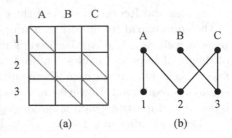

(a)   (b)

Fig. 1. A braced $m \times n$ square grid framework and the corresponding bipartite graph.

2   Preliminaries

We define a *grid framework* as a connected two-dimensional bar-joint framework which can be viewed as a union of unit grid squares. More formally, suppose we are given a rectilinear simple polygon P with holes $H_1, H_2, \ldots, H_h$ such that all vertices are located at the integer grid. We assume that (i) the area of every hole is at least two, (ii) the outer face of P and any hole H_j do not share any vertex, and (iii) any two distinct holes do not share any vertex. Let B_0 denote the outer boundary of P (i.e., the enclosing cycle of P) and let B_i (for $i = 1, 2, \ldots, h$) denote the boundary of H_i. Since all vertices are on integer grid points, the interior of P is decomposed into unit grid squares.

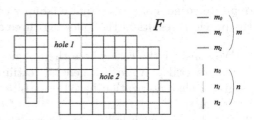

Fig. 2. A grid framework (for $m_0 = 15$, $m_1 = 3$, $m_2 = 3$, $m = 21$, $n_0 = 12$, $n_1 = 4$, $n_2 = 3$, $n = 19$, $h = 2$).

We add all such unit grid squares (i.e., their vertices and edges) to the interior of P. The resulting framework is a grid framework which is denoted by F (see Fig. 2). We regard it as a bar-joint framework in which every edge is a rigid bar and every vertex is a universal joint (free joint) (see [7]). Let V and E denote the set of vertices and edges of F, respectively. Let S denote the set of squares of F. For any grid framework X, let V_X and E_X be the set of vertices and edges of X, respectively.

We define the set of *upper boundary squares*, denoted by S_u, as the set of squares s of S such that the upper edge of s is not shared with any other square

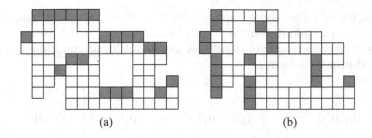

(a)   (b)

Fig. 3. (a) Upper boundary squares, and (b) left boundary squares of the grid framework of Fig. 2.

of S (see Fig. 3(a)). Let $m = |S_u|$. Similarly, we define the set of *left boundary squares* S_l, and let $n = |S_l|$ (see Fig. 3(b)). For i with $0 \leq i \leq h$, let S_{ui} be the set of upper boundary squares whose upper edge is on the hole boundary B_i. Let $m_i = |S_{ui}|$ for $0 \leq i \leq h$. From our assumption, the sets S_{ui} $(0 \leq i \leq h)$ are mutually disjoint, and thus $m = \sum_{i=0}^{h} m_i$ holds. For i with $0 \leq i \leq h$, , let S_{li} be the set of left boundary squares whose left edges are on B_i. Let $n_i = |S_{li}|$ for $0 \leq i \leq h$. Notice that the sets S_{li} $(0 \leq i \leq h)$ are mutually disjoint, and thus $n = \sum_{i=0}^{h} n_i$ holds.

In what follows, when we add a brace to a unit grid square, it is placed so that it connects the upper left and lower right corners of the square. This restriction does not change the rigidity of the framework as will be discussed in Subsect. 3.1.

We shall give some definitions and facts related to the rigidity of bar-joint frameworks. A bar-joint framework in the plane is represented as a pair $(G, \mathbf{p})$ of a graph $G = (V, E)$ and a mapping $\mathbf{p}$ from V to $\mathbb{R}^2$. A bar corresponds to an edge $e = (u, v) \in E$ whose length is given by $\|\mathbf{p}(u) - \mathbf{p}(v)\|$ (the distance between u and v). Since we consider that bars are rigid, $\|\mathbf{p}(u) - \mathbf{p}(v)\|$ does not change under any deformation of the framework. When considering a smooth deformation of the framework, we express $\mathbf{p}(u)$ as a continuously differentiable function of a parameter t, i.e., $\mathbf{p}_t(u)$. Here we assume $\mathbf{p}_0(v) = \mathbf{p}(v)$ for all $v \in V$. Since the edge length remains the same under the deformation, we have

$$(\mathbf{p}_t(u) - \mathbf{p}_t(v)) \cdot (\dot{\mathbf{p}}_t(u) - \dot{\mathbf{p}}_t(v)) = 0, \forall (u, v) \in E. \tag{2}$$

$\dot{\mathbf{p}}_0(v)$ can be regarded as a velocity vector (i.e., an infinitesimal motion) of the joint v at $t = 0$, and is simply denoted by $\dot{\mathbf{p}}(v) = (\dot{x}(v), \dot{y}(v))$. Letting $\mathbf{u}(v) = \dot{\mathbf{p}}_0(v)$, the Eq. (2) at $t = 0$ can be rewritten as

$$(\mathbf{p}(u) - \mathbf{p}(v)) \cdot (\mathbf{u}(u) - \mathbf{u}(v)) = 0, \forall (u, v) \in E. \tag{3}$$

A linear combination of vectors $\mathbf{u}_x, \mathbf{u}_y, \mathbf{u}_r$ defined by

$$\begin{aligned} \mathbf{u}_x &= [1 \, 0 \ldots 1 \, 0] \\ \mathbf{u}_y &= [0 \, 1 \ldots 0 \, 1] \\ \mathbf{u}_r &= [-y(v_1) \, x(v_1) \ldots -y(v_{|V|}) \, x(v_{|V|})] \end{aligned} \tag{4}$$

is called a *trivial infinitesimal motion*. When a system of linear equations (3) admits only trivial infinitesimal motions, the framework is *infinitesimally rigid*. Otherwise it is *infinitesimally flexible*. A system of linear equations (3) can be represented in matrix form as

$$
R_G(\mathbf{p})\mathbf{u}^\top = \begin{bmatrix} \vdots & \ddots & \vdots & \cdots & \vdots & \ddots & \vdots \\ 0 & \cdots & (\mathbf{p}_i - \mathbf{p}_j) & \cdots & (\mathbf{p}_j - \mathbf{p}_i) & \cdots & 0 \\ \vdots & \ddots & \vdots & \cdots & \vdots & \ddots & \vdots \end{bmatrix} \mathbf{u}^\top = \mathbf{0}^\top. \tag{5}
$$

Here $R_G(\mathbf{p})$ is called the *rigidity matrix*. In what follows, unless confusion occurs, "infinitesimally rigid" is simply called "rigid". The kernel dimension of $R_G(\mathbf{p})$ (denoted by dim ker $R_G(\mathbf{p})$) represents the degree of freedom of the framework $(G, \mathbf{p})$. If dim ker $R_G(\mathbf{p}) = 3$ or, in other words, rank $R_G(\mathbf{p}) = 2|V| - 3$, then the framework allows only trivial motions, namely, it is rigid.

Definition 1. *[3] A Henneberg 1 operation consists of creating a graph G_{k+1} of $k + 1$ vertices from a graph G_k of k vertices by adding a new vertex w, and connect it to two existing vertices u and v.*

Proposition 1. *([7]) Suppose that we are given a rigid framework $(G_k = (V_k, E_k), \mathbf{p})$ of k vertices. Choose a new vertex v_{k+1} and two existing vertices u and v such that $\mathbf{p}'(v_{k+1})$, $\mathbf{p}'(u)$ and $\mathbf{p}'(v)$ are not collinear, where $\mathbf{p}'$ represents a mapping from $V_K \cup \{v_{k+1}\}$ to $\mathbb{R}^2$ such that $\mathbf{p}(v) = \mathbf{p}'(v)$ for $v \in V_k$. Then applying the Henneberg 1 operation that adds v_{k+1} together with new edges (v_{k+1}, u) and (v_{k+1}, v) creates a rigid framework $(G_{k+1} = (V_k \cup \{v_{k+1}\}, E_{k+1}), \mathbf{p}')$ of $k+1$ vertices.*

Now let us consider the question of making F rigid by adding braces. Here a brace is defined as a diagonal edge of a unit grid square i.e., the length of any brace is $\sqrt{2}$. We do not consider longer braces. The following theorem by Maxwell [4] states a necessary condition for a given framework to be rigid.

Theorem 1 (Maxwell [4]). *Given a 2-dimensional bar-joint framework, let v and e be the number of vertices and edges of the framework, respectively. Then, if the framework is rigid, e satisfies the following inequality:*

$$
e \geq 2v - 3. \tag{6}
$$

3   Case for Frameworks Without Holes

In this section, we assume that the number of holes is $h = 0$. In this case, we show a necessary and sufficient condition to make a grid framework rigid which generalizes the theorem in [1] in the sense that the theorem still holds for a general rectilinear simple polygon.

3.1 Brace Matrix

The basic idea of this subsection is based on the paper [1]. A maximal sequence
of consecutively connected edges in a vertical or horizontal direction is called a
vertical or a *horizontal segment*, respectively. Notice that the number of vertical
and horizontal segments of F are $m+1$ and $n+1$, respectively, where m and n are
the number of upper boundary squares and left boundary squares, respectively
(see the beginning of Sect. 2).

Suppose that the ith horizontal segment consists of points $p_1 = (x_{i_1}, y), p_2 =
(x_{i_2}, y), \ldots, p_l = (x_{i_l}, y)$. From Eq. (2) and the fact that $x_{i_{k+1}} - x_{i_k} = 1$ for
$k = 1, 2, \ldots, l-1$, we obtain that

$$\dot{x}_{i_1} = \dot{x}_{i_2} = \cdots = \dot{x}_{i_l}. \tag{7}$$

Similarly, for a vertical segment consisting of points $p_1 = (x, y_{j_1}), p_2 = (x, y_{j_2}),
\ldots, p_v = (x, y_{j_v})$, it follows that

$$\dot{y}_{j_1} = \dot{y}_{j_2} = \cdots = \dot{y}_{j_v}. \tag{8}$$

Let $\dot{a}_i$ and $\dot{b}_j$ denote the values of (7) and (8), respectively. We then associate $\dot{a}_i$
$(i = 1, 2, \ldots, n+1)$ and $\dot{b}_j$ $(j = 1, 2, \ldots, m+1)$ with the horizontal and vertical
segments, respectively. Then $\dot{a}_i$ denotes the infinitesimal horizontal motion of
the ith horizontal segment, and $\dot{b}_j$ denotes the infinitesimal vertical motion of
the jth vertical segment (see Fig. 4(a)).

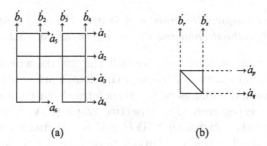

(a)   (b)

Fig. 4. (a) Illustration of a framework and its infinitesimal motion of horizontal and
vertical segments, where $m = 3$, $n = 5$, and the number of vertical and horizontal
segments is $m + 1 = 4$, and $n + 1 = 6$, respectively. (b) An infinitesimal motion of a
unit square.

Now let us consider a unit grid square which is surrounded by the p-th and
q-th horizontal segments and the r-th and s-th vertical segments (see Fig. 4(b)).
If the square is braced along the diagonal from the upper left corner to the lower
right one, the velocity vectors (i.e., infinitesimal motions) at the ends of the
brace are $(\dot{a}_p, \dot{b}_r)$ and $(\dot{a}_q, \dot{b}_s)$. By (2), these vectors satisfy $\dot{a}_p - \dot{b}_r = \dot{a}_q - \dot{b}_s$,
i.e., $\dot{a}_p + \dot{b}_s - \dot{a}_q - \dot{b}_r = 0$. If the square is braced along the other diagonal, then

we have $\dot{a}_p + \dot{b}_s = \dot{a}_q + \dot{b}_r$, i.e., $\dot{a}_p + \dot{b}_s - \dot{a}_q - \dot{b}_r = 0$. Thus the two possible cross-braces yield the same equation.

Now let Br be a set of braces added to the framework F. Then we have $|Br|$ linear equations which form the system of linear equations (rigidity matrix) expressed by

$$\begin{pmatrix} -1 & 1 & 0 & \cdots\cdots & 1 & -1 & 0 & \cdots\cdots\cdots \\ -1 & 1 & 0 & \cdots\cdots & 0 & 1 & -1 & 0 & \cdots\cdots \\ -1 & 1 & 0 & \cdots\cdots & 0 & 0 & 1 & -1 & 0 & \cdots \\ \vdots & \vdots & \vdots & \vdots & \vdots & \vdots & \vdots & \vdots & \vdots & \vdots \\ \cdots\cdots\cdots & -1 & 1 & \cdots\cdots\cdots\cdots & -1 & 1 \end{pmatrix} \cdot \begin{pmatrix} \dot{b}_1 \\ \dot{b}_2 \\ \vdots \\ \dot{b}_{m+1} \\ \dot{a}_1 \\ \dot{a}_2 \\ \vdots \\ \dot{a}_{n+1} \end{pmatrix} = 0. \qquad (9)$$

This matrix is called the *brace matrix* of $F \cup Br$.

3.2 Transformation

By extending the result of Bolker and Crapo [1], we can prove that the rigidity of a square grid framework can be tested by checking the connectedness of some bipartite graph. This bipartite graph is defined similarly as in [1]. We shall first transform the brace matrix of $F \cup Br$ so that every row has two nonzero entries. In fact, we show the following.

Claim 1. *We can transform the brace matrix of $F \cup Br$ into the incidence matrix of a bipartite graph without changing the independence of each row.*

The proof of this claim is omitted. We call the matrix which is obtained by Claim 1 the *transformed matrix*. Although the details are omitted, it holds that the number of rows and columns of the transformed matrix are the number of braces and $m + n$, respectively. The bipartite graph (U, V, E) corresponding to the transformed matrix satisfies that (i) $U \cup V$ is a vertex set with $U \cap V = \emptyset$, (ii) U and V correspond to the set of upper boundary squares and left boundary squares, respectively, and (iii) E is an edge set which corresponds to the set of braces.

For each square $s \in F$, let s_u (resp. s_l) be the upper boundary square which is above (resp. left of) s and is closest to s. If s is braced, then the edge e corresponding to s is the one that connects the vertex in U corresponding to s_u and the one in V corresponding to s_l.

3.3 Necessary and Sufficient Condition to Make a Framework Rigid

As observed in [1], it suffices to look at the adjacency matrix of a bipartite graph. By the argument of the previous subsection and by [1], we have the following theorem.

Theorem 2. *In the case of $h = 0$, $F \cup Br$ is minimally rigid if and only if the bipartite graph corresponding to the transformed matrix of $F \cup Br$ is a spanning tree.*

4 Lower Bound on the Number of Braces Required

In this section, we shall show a lower bound on the number of braces required to make a grid framework with holes rigid. In [2], Gáspár et al. mentioned the same lower bound, but did not give a rigorous proof. Furthermore, our proof is based on a combinatorial argument which seems to be different from the one in [2].

We consider the problem of making X rigid by adding braces. Let Br_X be a set of braces whose addition to X makes it rigid, and let $X \cup Br_X$ be the framework resulting from adding Br_X to X. We show the following theorem.

Theorem 3. *For any framework F with m upper boundary squares, n left boundary squares, and h holes, the minimum number of braces which make F rigid is at least $m + n - 2h - 1$.*

Proof. We define W as the framework without holes which has the same outer boundary as F (see Fig. 5(b)). For i with $1 \leq i \leq h$, let $V_i \subset V_W$ be the set of vertices which belong to the interior of the ith hole H_i of F (i.e., $V_i \cap V_F = \emptyset$), and let $E_i \subset E_W$ be the set of edges which belong to the the interior of the ith hole of F (i.e., $E_i \cap E_F = \emptyset$).

Applying the Henneberg 1 operation, we augment a rigid framework $F \cup Br_F$ by adding vertices of V_i together with edges of E_i, as much as possible starting from all upper left corners of H_i. Let $I \cup Br_F$ be the resulting framework, and let I be the framework obtained by removing Br_F. Since $F \cup Br_F$ is rigid, $I \cup Br_F$ is also rigid. It is not difficult to see that $I \cup Br_F$ contains all vertices in $\cup_{i=1}^{h} V_i$.

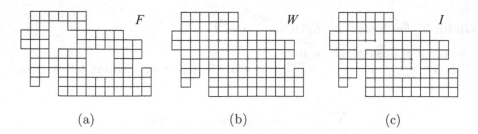

$$F \qquad\qquad W \qquad\qquad I$$

(a)   (b)   (c)

Fig. 5. Illustration of (a) F, (b) W, and (c) I.

Since $I \cup Br_F$ is rigid, the following inequality holds by Maxwell's condition (6).

$$|Br_F| \geq 2|V_I| - 3 - |E_I|. \tag{10}$$

By Theorem 2, we can make W rigid by adding a set of braces Br_W whose cardinality matches the lower bound in Maxwell's condition. Then,

$$|Br_W| = 2|V_W| - 3 - |E_W|.\tag{11}$$

W has m_0 upper boundary squares, n_0 left boundary squares, and no holes. Thus,

$$|Br_W| = m_0 + n_0 - 1.\tag{12}$$

By (11) and (12),

$$2|V_W| - 3 = |E_W| + m_0 + n_0 - 1.\tag{13}$$

Now we plug $V_I = V_W$ into equation (10) and obtain

$$|Br_F| \geq 2|V_W| - 3 - |E_I|.\tag{14}$$

By (13) and (14),

$$|Br_F| \geq |E_W| - |E_I| + m_0 + n_0 - 1.\tag{15}$$

We define E_i^* as the subset of $E_W \setminus E_I$ in the ith hole (see Fig. 6). Then, we have:

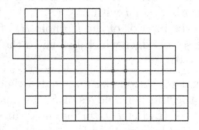

Fig. 6. Illustration of E_i^* (blue edges).

Claim 2. *For $1 \leq i \leq h$, $|E_i^*| = m_i + n_i - 2$.*

The proof of this claim is omitted.

Since $\sum_{i=0}^{h} m_i = m$ and $\sum_{i=0}^{h} n_i = n$, the following equation follows from Claim 2:

$$\sum_{i=1}^{h} |E_i^*| = m + n - m_0 - n_0 - 2h.\tag{16}$$

By $\sum_{i=1}^{h} |E_i^*| = |E_W| - |E_I|$,

$$|E_W| - |E_I| = m + n - m_0 - n_0 - 2h.\tag{17}$$

By (15) and (17), the following inequality follows:

$$|Br_F| \geq m + n - 2h - 1.\tag{18}$$

This completes the proof.   □

5   Algorithm

In this section, we first propose an algorithm for adding the minimum number of braces necessary to make a given square grid framework possibly with holes rigid. Then we prove the correctness of the algorithm. Our algorithm is called BULS (Bracing Upper and Left Squares) and consists of three steps described as follows:

Algorithm BULS
Input : A square grid framework F
Step 1: Add braces to the squares in $S_u \cup S_l$ (Fig. 7(a)).
Step 2: Add braces to the squares $s \notin S_u \cup S_l$ such that s shares the upper edge with a square in S_l and the left edge with a square in S_u (blue braces in Fig. 7(b)). (Notice that the upper left corner of a selected square coincides with either an upper left concave corner of the boundary B_0 or a lower right convex corner of the boundary B_i, for some $1 \le i \le h$.)
Step 3: Remove three braces per hole as follows (Fig. 7(c)). The boundary of each hole H_i with $1 \le i \le h$ has at least one vertex that is a lower right convex corner. If there is more than one, choose the rightmost one (ties are broken by giving the highest priority to the one with the smallest y-coordinate. Let v_i^* be such vertex. We remove all braces of the three squares which share v_i^*.

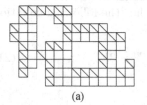
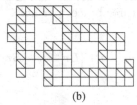
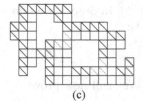

(a)   (b)   (c)

Fig. 7. (a) **Step 1:** Add braces to squares in $S_u \cup S_l$. (b) **Step 2:** Add new braces (colored in blue) to the squares whose upper left corner belongs to $\cup_{i=1}^h B_i$. (c) **Step 3:** Remove three braces per hole.

Let F^{BULS} denote the framework obtained by applying BULS to F. Then, we shall prove the following two lemmas:

Lemma 1. $|Br^{BULS}| = m + n - 2h - 1$.

Lemma 2. F^{BULS} is rigid.

By these two lemmas, we prove that Algorithm BULS makes any square grid framework rigid by adding $m + n - 2h - 1$ braces. By Theorem 3, we have the following theorem:

Theorem 4. *Algorithm BULS adds the minimum number of braces to make any square grid framework rigid.*

The following two subsections give the proofs of Lemmas 1 and 2, respectively.

5.1 Proof of Lemma 1

In Step 1 of BULS, let Br^{ulv} be the brace set added to the squares in $S_u \cap S_l$. Let $Br_0^{ulv} \subset Br^{ulv}$ be the brace set added to squares in $S_u \cap S_l$ whose upper left corners are on the outer face, and Br_i^{ulv} $(1 \leq i \leq h)$ be the brace set added to squares in $S_u \cap S_l$ whose upper left corners are on the boundary of the ith hole. Notice that Br_i^{ulv} $(0 \leq i \leq h)$ are mutually disjoint, and thus the following equation holds:

$$|Br^{ulv}| = |Br_0^{ulv}| + |Br_1^{ulv}| + \cdots + |Br_h^{ulv}|. \tag{19}$$

The number of braces added in Step 1 is thus $m + n - |Br^{ulv}|$.

In Step 2, let S_0^{lrv} denote the set of squares whose upper left corner coincide with an upper left concave corner of the boundary B_0. Let Br_0^{ulc} be the set of braces added to these squares. Let S_i^{lrv} $(1 \leq i \leq h)$ denote the set of squares whose upper left corner coincide with a lower right convex corner of the boundary B_i. We added new braces to each square $s \in S_i^{lrv}$. Let Br_i^{ulc} be the set of such braces, and let $Br^{ulc} = \cup_{i=0}^h Br_i^{ulc}$. Notice that Br_i^{ulc} with $0 \leq i \leq h$ are mutually disjoint and thus the following equation holds:

$$|Br^{ulc}| = |Br_0^{ulc}| + |Br_1^{ulc}| + \cdots + |Br_h^{ulc}|. \tag{20}$$

Then after Step 2, the number of braces added is equal to $m + n + |Br^{ulc}| - |Br^{ulv}|$. In Step 3, we remove three braces per hole. Thus the following equation holds:

$$|Br^{BULS}| = m + n + |Br^{ulc}| - |Br^{ulv}| - 3h. \tag{21}$$

We shall now prove

$$|Br^{ulc}| - |Br^{ulv}| = h - 1, \tag{22}$$

which completes the proof. This is derived by proving two equalities.

$$|Br_0^{ulc}| = |Br_0^{ulv}| - 1, \text{and} \tag{23}$$

$$|Br_i^{ulc}| = |Br_i^{ulv}| + 1, \ i = 1, \ldots, h. \tag{24}$$

Let R be a general rectilinear simple polygon, and V_R be the set of corner vertices of R. Let V_R^{ulv} denote the set of upper left convex corners, and V_R^{ulc} denote the set of upper left concave corners (Fig. 8). Then, we have the following claim.

Claim 3. $|V_R^{ulc}| = |V_R^{ulv}| - 1$.

Proof. We can prove this claim by induction on the number of corners. We omit the details. □

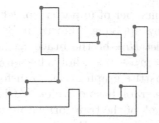

Fig. 8. The red vertices are upper left convex corners, and the blue vertices are upper left concave corners.

Let V_{B_0} be the corner vertex set on the outer boundary B_0. By Claim 3, we have

$$|V_{B_0}^{ulc}| = |V_{B_0}^{ulv}| - 1. \tag{25}$$

A square which has a brace in Br_0^{ulv} has a vertex in $V_{B_0}^{ulv}$, so we have $|V_{B_0}^{ulv}| = |Br_0^{ulv}|$. A square which has a brace in Br_0^{ulc} has a vertex in $V_{B_0}^{ulc}$, so we have $|V_{B_0}^{ulc}| = |Br_0^{ulc}|$. Then, by (25), the Eq. (23) holds.

By applying a similar argument to the boundary of holes, we can derive (24) (the details are omitted). By (23), and (24),

$$\sum_{i=0}^{h} |Br_i^{ulc}| = \sum_{i=0}^{h} |Br_i^{ulv}| + h - 1. \tag{26}$$

By (19), (20), and (26),

$$|Br^{ulc}| - |Br^{ulv}| = h - 1. \tag{27}$$

By (21) and (27),

$$|Br^{BULS}| = m + n - 2h - 1. \tag{28}$$

5.2   Proof of Lemma 2

In this subsection, we prove F^{BULS} is rigid. The proof is done by induction on h, i.e., the number of holes in a square grid framework F, as follows:

Basis : Algorithm BULS correctly adds the minimum number of braces necessary to make rigid a given square grid framework with $h = 0$.

The correctness of Algorithm BULS for $h = 0$ is based on the following claim.

Claim 4. If $h = 0$, the associated bipartite graph of the transformed matrix of F^{BULS} is acyclic.

The proof of this claim is omitted.

Let m_F (resp. n_F) be the number of upper (resp. left) boundary squares of F. F^{BULS} can be obtained by applying Algorithm BULS to the corresponding framework without holes: let Br_F be the brace set added onto F by the algorithm. By Eq. (28), $|Br_F| = m_F + n_F - 1$ holds because F has no holes. Thus, the number of edges of the bipartite graph of the transformed matrix of F^{BULS} is $m_F + n_F - 1$. This bipartite graph has no cycles by Claim 4. Since the number of vertices of the bipartite graph of the transformed matrix of F^{BULS} is $m_F + n_F$, the bipartite graph is a spanning tree. By Theorem 2, F^{BULS} is rigid.

Induction Step: Suppose that Algorithm BULS correctly adds the minimum number of braces necessary to make rigid a given square grid framework with $h = k$. We then prove that Algorithm BULS also correctly adds the minimum number of braces necessary to make rigid a given square grid framework with $h = k + 1$.

Now suppose that we are given a square grid framework F with $h = k+1$. We apply Algorithm $BULS$ to F, and let F^{BLUS} be the obtained framework. We introduce an operation called *Reduce-Holes* which removes a part of F^{BLUS} and decreases the number of holes by one. First, we need to define some notation.

Recall that we defined in Step 3 of Algorithm BULS v_i^* as a vertex that is a lower right convex corner in H_i for $i = 1, 2, \ldots, h$. Let $V^* = \{v_1^*, v_2^*, \ldots, v_h^*\}$. We define v^F as the lowest vertex among the rightmost vertices in V^*. Let u_1^F (resp. l_1^F) be the vertex adjacent to v^F on the upper side (resp. the left side). Let s_h^F (resp. s_v^F) be the horizontal (resp. vertical) segment passing through u_1^F (resp. l_1^F). Let $x(v)$ (resp. $y(v)$) be the x-coordinate (resp. y-coordinate) of a vertex v. We define $u_2^F = \arg\min\{x(v)|v \in B_0 \cap s_h^F, x(v) > x(u_1^F)\}$, $l_2^F = \arg\max\{y(v)|v \in B_0 \cap s_v^F, y(v) < y(l_1^F)\}$. We define c_h^F (resp. c_v^F) as the segment connecting u_1^F and u_2^F (resp. l_1^F and l_2^F). Let V_h^F (resp. V_v^F) be the set of vertices which are adjacent to vertices in c_h^F except u_1^F on the lower side (resp. c_v^F except l_1^F on the right side) (see Fig. 9).

Then, the operation Reduce-Holes consists of three steps: (i) Remove all vertical edges (resp. the horizontal edges) which are incident to vertices in c_h^F on the lower side (resp. c_v^F on the right side). (ii) Remove the upper (resp. left) edge and the diagonal edge of the unit square whose upper left corner is u_2^F (resp. l_2^F). (iii) Remove the connected part containing v^F.

In this subsection, we rename F^{BULS} as F_{k+1}. We define F_k as the resulting framework which is obtained by applying Reduce-Hole to F_{k+1} (see Fig. 9). Note that F_k has k holes, and that F_k is rigid because the placement of braces in F_k is exactly what Algorithm BULS does to the framework with k holes.

Thus, we complete the proof by proving the following claim.

Claim 5. *If F_k is rigid, F_{k+1} is rigid.*

Proof. We suppose that F_k is rigid. We add $v^{F_{k+1}}$ and its incident edges $v^{F_{k+1}} u_1^{F_{k+1}}$ and $v^{F_{k+1}} l_1^{F_{k+1}}$ by using Henneberg 1, and the resulting framework is rigid. After this, we add all vertices in $V_h^{F_{k+1}} \cup V_v^{F_{k+1}}$ by using Henneberg

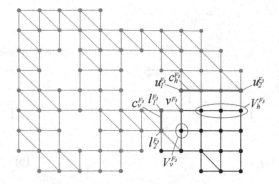

Fig. 9. An example of F_3. The blue part is F_2. The green edges are removed in Step (i) or (ii) of the operation Reduce-Holes.

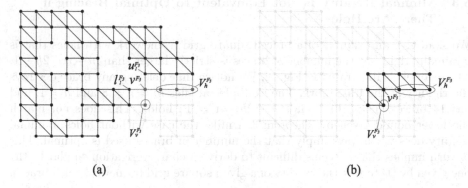

(a)   (b)

Fig. 10. An example of (a) F_2' and (b) F_2''.

1 repeatedly. We define the resulting framework as F_k' (see Fig. 10(a)). Using Proposition 1 and the fact that F_k is rigid, we obtain that F_k' is rigid.

Now we consider another framework which consists of all squares of which the lower right vertices are in $\{v^{F_{k+1}}\} \cup V_h^{F_{k+1}} \cup V_v^{F_{k+1}}$. Let F_k'' be the framework obtained by bracing all squares of the above framework (see Fig. 10(b)). Note that F_k'' is triangulated, and thus rigid.

Next, we replace F_k' by F_k'' in F_{k+1}. After this replacement, we call the resulting framework F_{k+1}^* (see Fig. 11(b)). Then, since both F_k' and F_k'' are rigid subgraphs of F_{k+1} and F_{k+1}^*, respectively, F_{k+1} is rigid if and only if F_{k+1}^* is rigid. We here notice that F_{k+1}^* has no holes and that F_{k+1}^* is obtained by applying Algorithm BULS to the framework obtained by removing braces from F_{k+1}^*. Thus F_{k+1}^* is rigid as shown in the proof for the basis at the beginning of this subsection. □

Therefore, by induction on h, F^{BULS} is rigid. □

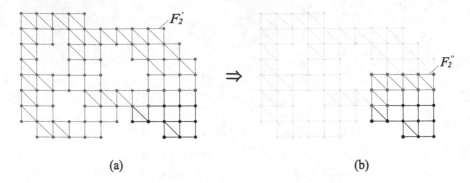

Fig. 11. An example of (a)F_3, and (b)F_3^*.

5.3 Minimal Rigidity Is Not Equivalent to Optimal Bracing if There Are Holes

We shall give an example of a braced square grid framework with holes that is minimally rigid but the number of braces is strictly larger than $m + n - 2h - 1$. The example is illustrated in Fig. 12. We notice that deleting any brace destroys the rigidity of the framework. The example satisfies $m = n = 8$ and $h = 1$, and has 14 braces. Thus, $14 > m + n - 2h - 1 = 13$ holds. This does not match the lower bound given by Theorem 3. Unlike the case without holes, minimal rigidity does not always imply that the number of braces used is optimal. This in turn implies that it seems difficult to derive a characterization similar to the one given by [1] to test the rigidity of a given square grid framework with braces.

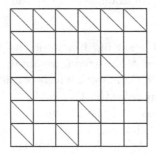

Fig. 12. Illustration of a braced square grid framework with one hole that is minimally rigid but does not satisfy the condition in Theorem 3.

6   Conclusion

This paper considers the problem of bracing a square grid framework with holes to make it minimally rigid. The paper first studies the case with no holes and establishes a theorem that generalizes the result by [1]. It then considers the case with holes. The paper presents a lower bound on the number of braces required to make a given grid framework minimally rigid, and then proposes an algorithm that adds a set of braces whose cardinality matches the lower bound. However, the number of braces used to obtain a minimally rigid square grid framework does not always match the lower bound. This means that it seems difficult to derive a characterization in terms of a bipartite graph as in [1] to test the rigidity of a given square grid framework with braces. It is open whether a combinatorial characterization is possible.

References

1. Bolker, E.D., Crapo, H.: Bracing Rectangular Frameworks. I. SIAM J. Appl. Math. **36**(3), 473–490 (1979)
2. Gáspár, Z., Radics, N., Recski, A.: Rigidity of Square Grids with Holes. Comput. Assist. Mech. Eng. Sci. **6**(3–4), 329–335 (1999)
3. Henneberg, L.: Die Graphische Statik der Starren System, Leipzig 1911, Johnson Reprint 1968
4. Maxwell, J.C.: On the Calculation of the Equilibrium and Stiffness of Frames. Phil. Mag. **27**, 294–299 (1864)
5. Radics, N., Recski, A.: Applications of Combinatorics to Statics - Rigidity of Grids. Discrete Appl. Math. **123**, 473–485 (2002)
6. Bar-Yehuda, R., Ben-Hanoch, E.: A Linear-Time Algorithm for Covering Simple Polygons with Similar Rectangles. Int. J. Comput. Geom. Appl. **6**(1), 79–102 (1996)
7. Whiteley, W.: Some Matroids from Discrete Applied Geometry. AMS Contemp. Math. **197**, 171–313 (1996)

Top-K Query Retrieval of Combinations with Sum-of-Subsets Ranking

Subhashis Majumder[1], Biswajit Sanyal[2]([✉]), Prosenjit Gupta[1], Soumik Sinha[3], Shiladitya Pande[4], and Wing-Kai Hon[5]

[1] Department of Computer Science and Engineering,
Heritage Institute of Technology, Kolkata 700 107, West Bengal, India
[2] Department of Information Technology, Government College of Engineering
and Textile Technology, Hooghly, Serampore 712 201, West Bengal, India
biswajit_sanyal@yahoo.co.in
[3] Department of Computer Science and Engineering, IIT Kanpur, Kanpur, India
[4] Department of Computer Science and Engineering, IIT Madras, Chennai, India
[5] National Tsing Hua University, Hsinchu, Taiwan

Abstract. *Top-k* query processing is an important building block for ranked retrieval, with applications ranging from text and data integration to distributed aggregation of network logs and sensor data. *Top-k* queries return a ranked set of the k best data objects selected on the basis of the ranks (scores) of the objects, assigned by any ranking (scoring) function. In this paper, we consider a problem on generation of combinations. Given a set S of n real numbers and an integer $r \leq n$, we consider the $\binom{n}{r}$ different *r-combinations* of the elements of S. Let all these $\binom{n}{r}$ combinations be indicated by the set $\mathcal{C} = \{C_1, C_2, \ldots, C_{\binom{n}{r}}\}$. From this set $\mathcal{C}$, given any positive integer $k \leq \binom{n}{r}$, our goal is to generate the k best combinations (*top-k* combinations) ranked on the basis of some aggregation function F. We consider *Summation* as the aggregation function. This calculates the sum of all the r real numbers in any combination C_i, and the ranking criterion is that a combination C_i is ranked higher than a combination C_j, if the sum of the constituent numbers of C_i is larger than that of C_j. For any given n and r ($\leq n$), we build a *metadata structure* G (basically a DAG), even before S and k are known. We can later use G to report the *top-k* combinations efficiently when S is available. We further present an alternative incremental method, where we generate only the required portions of G on demand, instead of constructing the whole G explicitly. It helps us to save the time and space overhead of the preprocessing phase.

1 Introduction

Nowadays, in many application domains, end users are more interested in knowing the most important query answers rather than merely retrieving a list of all the data items that satisfy a query, i.e., a manageable summary of the query results in the potentially huge answer space. Generating such a summary often requires applying aggregation functions to the query results. One of the simplest

© Springer International Publishing Switzerland 2014
Z. Zhang et al. (Eds.): COCOA 2014, LNCS 8881, pp. 490–505, 2014.
DOI: 10.1007/978-3-319-12691-3_36

aggregation function is reporting the k best items among the ones that satisfy the query. This has led to the formulation of the so called *top-k* problems [3,4].

In mathematics, a *combination* is a way of selecting several things from a larger group, where (unlike permutations) order does not matter. A *r-combination* of a set S is a subset of r distinct elements of S. If S has n elements, the number of *r-combinations* is represented by $\binom{n}{r}$. Given a set of n real numbers and an integer $r \leq n$, we consider $\binom{n}{r}$ different *r-combinations* of S. Let all these $\binom{n}{r}$ combinations be indicated by the set $C = \{C_1, C_2, \ldots, C_{\binom{n}{r}}\}$. From this set C, our goal is to generate the k best combinations (*top-k* combinations) for any input value k, ranked on the basis of some aggregation function F. Let combination C be represented as $C = (a_1, a_2, \ldots, a_r)$ with $F(C) = \sum_{i=1}^{r} a_i$. In this paper, we consider Summation as the aggregation function that calculates the sum of the r real numbers in any combination C_i, and the ranking criterion is that a combination C_i is ranked higher than a combination C_j if the sum of the constituent numbers of C_i is larger than that of C_j. Here, rank is unique even if sums are same, ties must be broken by some criterion or even arbitrarily. We have $rank_F(C) =$ rank of combination C according to ranking function F.

1.1 Past Work

The *top-k* problem has been well studied in different domains like information retrieval, multimedia [2], text and data integration, business analytics, computational geometry [1,5,7], set theory [9] and so on. It is also quite popular in database related queries where retrieval of *top-k* combinations over a join of ranked list is a hot research topic of today [8]. Recently Lu et al. [6] studied *top-k,m* queries, where they considered a set of groups each containing a set of attributes, each of which was associated with a ranked list of tuples, with Id and score. They reported the *top-k* combinations of attributes according to the corresponding *top-m* tuples with matching Ids. However, we solve a different problem and to the best of our knowledge, no other results are known till date for our problem.

2 Our Contribution

For any input set S of n real numbers and for any input value r, it is possible to generate *top-k r-combinations* by sorting all the $\binom{n}{r}$ combinations on the basis of the corresponding aggregation function. However, such a straightforward approach will use up large processing time and space. The concept of using a preprocessing step is quite popular in the domain of *top-k* problems, so that the time required to process the actual query is less. Though preprocessing takes some time, it is justified as the preprocessing step normally happens only once, usually at the beginning of an algorithm, while the queries are executed several times. As a result, for many algorithms it has become a common practice to report the time complexity of preprocessing step and the query step separately.

In our problem, the inputs are the set S of n real numbers, the integer r and the integer k. The main novelty of our technique is that preprocessing can start even before the actual numbers of set S are available. If we just know the values of n and r, we can build a *metadata structure* G (basically a DAG) entirely offline, that can save considerable amount of computation when the actual numbers become available. We call G a metadata structure, as it uses the relative positioning of the n numbers on the real line instead of the actual numbers themselves. With the help of G generated in the preprocessing phase, we can then generate the *top-k r-combinations* quite efficiently.

The paper is organized as follows. In Sect. 3, we describe the DAG G and show how it can be used in conjunction with a max-heap structure H to obtain the *top-k* combinations efficiently. In Sect. 4, we describe how we can build the DAG for $\binom{m}{r}$ using the DAG for $\binom{n}{r}$, where $m > n$, so that if the DAG for $\binom{n}{r}$ is available, we can reuse some of the computations. In Sect. 5, first we present some experimental results and then we briefly explain an alternative idea of implicitly constructing only the required portions of the metadata structure on demand. Finally in the last section, we conclude citing some directions for future research. Due to lack of space, we omit some proofs and many details.

3   Solution Methodology

If we want to select r numbers from a set of n distinct numbers such that their summation is as large as possible, the answer is quite obvious, we have to select the r largest numbers from the set. In other words, if we have the n numbers sorted in a list in non-increasing order, all we have to do is pick up the first r numbers from that list and that solves our *top-k* problem for $k = 1$. While doing so, we do not even need to know what exactly the numbers are. For $k > 1$ too, we can use some similar facts to do some computation without even knowing what the actual numbers are. To perform these computations, all we need to know are the values of n and r, and we store the preprocessing knowledge gathered from those computations in the DAG G. Afterwards, whatever be the values of the set of n numbers, we can successfully use the knowledge stored in G to obtain the *top-k r-combinations* efficiently.

In this section, we first discuss how to build the DAG G, given n and r, and then we show how to use G with a max-heap structure H, to obtain the *top-k r-combinations*, once the n numbers are known.

3.1   Construction of Metadata Structure

Let us assume that the given n numbers are kept in a list sorted in non-increasing order and also let $P = \{1, 2, \ldots, n\}$ be the set of positions of the numbers in the list. Now, we can view a particular *r-combination* as a set of r distinct positions chosen from P. We build a layered DAG, $G = (V, E)$, where each node $v \in V$ contains the information of r positions related to a *r-combination* and a directed edge from node V_i to node V_j will denote that the *r-combination* corresponding to parent node V_i is better than the one corresponding to the child node V_j, and hence V_i will occupy a higher position compared to that of V_j in the *top-k* list.

Definition 1. *Let V_i and V_j be two nodes of the DAG G. $V_i \succeq V_j$ if the sum of r numbers belonging to the r-combination corresponding to V_i is greater than or equal to the sum of r numbers corresponding to V_j.*

Initially, the input set S is considered as sorted (though set S is not physically available at that time) from left to right in non-increasing order, and so the first r positions from the left gives the best (*top-1*) *r-combination*. This *top-1 r-combination* is considered as the root node of the DAG G. In order to generate the other nodes of G, we consider the relative positions of r numbers from set S, in the available positions from the set $P = \{1, 2, \ldots, n\}$.

Definition 2. *Let C be a r-combination such that its r numbers are at positions $p_1, p_2, \ldots, p_r, 1 \le p_i \le n$. Let C' be another r-combination whose r numbers are at positions $p'_1, p'_2, \ldots, p'_r, 1 \le p'_i \le n$, such that for some j, $p'_j = p_j + 1$ and $p'_i = p_i$ for $i \ne j, 1 \le i \le r$. Then we say that C' has one more **shift** than C.*

In DAG G, for each node we store the information of r positions related to a *r-combination* by an array of flag bits F, termed as its shift information. The array F, contains $n - r$ number of $0s$ and r number of $1s$ in r different positions like $p_1, p_2, \ldots, p_r$, for all $i, 1 \le p_i \le n$, those represent the r different positions of the r real numbers of the set S for the r-combination corresponding to the DAG node (S is considered as sorted from left to right in non-increasing order and not physically available at that time). Figure 1 shows the DAG for $\binom{6}{3}$, where each node stores its shift information in an array of flag bits of size 6, containing 3 number of $1s$ and 3 number of $0s$. The root node of the DAG of Fig. 1, stores the shift information (111000) that corresponds to a $r - combination$, containing the first three numbers (from the positions 1, 2 and 3) of the set S.

We categorize the nodes in terms of the total number of shifts required for all its r numbers from the shift information of the root node. Hence if the corresponding shift information needs m shifts, then the node will be in the category V_m. Hence, by this definition, the root node will be in category V_0 and that is the only node in this category. Further, in a category V_m, if there are t different variations, we write them as $V_m = \{V_{m1}, V_{m2}, \ldots, V_{mt}\}$. As for example, in Fig. 1, variation V_{32} is in category V_3, as the total number of shifts required to create its shift information (101010) is 3. Note that the 3rd 1 has suffered two shifts and the 2nd 1 has suffered one shift from their original positions in the shift information (111000) of the root node. Finally, a directed edge is drawn from a node $V_{mi} \in V_m$ of m shifts to some other node $V_{(m+1)j} \in V_{m+1}$ of $(m+1)$ shifts iff $V_{mi} \succeq V_{(m+1)j}$.

Considering the position information of the r numbers in their respective combinations, without even considering the actual r numbers, from Fig. 1 ($r = 3$) we can straightaway conclude that $V_{01} \succeq V_{11}$. This is because the first $r - 1$ numbers in both cases are same and the rth number of V_{01} is greater than

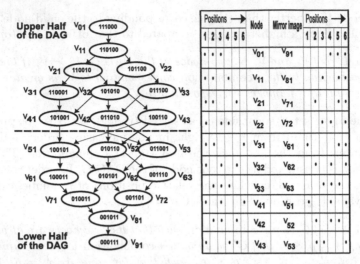

V_{fs} - the first index f represents the no. of shifts and the second index s represents the variation

Fig. 1. DAG G for $\binom{6}{3}$ with shift information of all the nodes

or equal to that of the rth number of V_{11}, as the numbers are sorted in non-increasing order. Similarly we can argue that $V_{11} \succeq V_{21}$, $V_{11} \succeq V_{22}$, $V_{21} \succeq V_{31}$ and $V_{21} \succeq V_{32}$, $V_{22} \succeq V_{32}$ and $V_{22} \succeq V_{33}$ and so on. Hence in Fig. 1, we have drawn an edge from V_{01} to V_{11}, V_{11} to V_{21} and V_{22}, V_{21} to V_{31} and V_{32} and likewise to construct the whole DAG. Just for illustration, whatever be the actual numbers are, $V_{22} \succeq V_{32}$, as in both cases 2 of the 3 numbers are exactly the same, and the smallest number in the former case can never be smaller than the smallest number in the latter case because of their positions in the original sorted list. From the above discussion, it becomes obvious that a number of edges can be drawn in G, without knowing the actual set S of n real numbers. Note that we draw an edge from V_{pi} to V_{qj}, only if for all possible choices of n numbers, $V_{pi} \succeq V_{qj}$ holds. Formally, let P be the combination corresponding to V_{pi} with its r numbers in r different positions like $p_1, p_2, \ldots, p_r$, for all $i, 1 \le p_i \le n$. Likewise, let Q be the combination corresponding to V_{qj} with its r numbers in r different positions like $q_1, q_2, \ldots, q_r$, for all $i, 1 \le q_i \le n$. We draw an edge from node V_{pi} to node V_{qj}, if $p_i \le q_i$ for all $1 \le i \le r$ and $q = p + 1$, i.e., edges exist only between consecutive levels. We state below several results on the properties of the DAG G.

Lemma 1. *In G, not all pairs of nodes (a, b), where $a \in V_m$ and $b \in V_{m+1}$, can be connected by directed edges.*

Proof. A simple counter-example establishes this result. In Fig. 1, note that no edge exists between V_{21} and V_{33}. It is actually not possible to conclude whether $V_{21} \succeq V_{23}$ or $V_{23} \succeq V_{21}$, without knowing the actual numbers. Let the 6 numbers be $16, 14, 12, 10, 8$ and 2. Now in both configurations the 2nd number appears, so its value does not matter in this case. However, in one case the 1st number and the 5th number appear whose sum is $16 + 8 = 24$. In the other case, the 3rd number and the 4th number appear. Their sum is $12 + 10 = 22$. Hence for these set of numbers node $V_{21} \succeq V_{33}$. However, if the original numbers were $16, 14, 12, 10, 4$ and 2, the reverse thing would have happened, as $16 + 4 < 12 + 10$. Hence the result.

Lemma 2. *There are no edges between the nodes of the same category.*

Proof. Note that two nodes are in the same category (at the same level) in the graph G, iff the total number of shifts with respect to root node undergone by the numbers in those nodes are exactly the same. Consider a combination P with its r numbers in r different positions like $p_1, p_2, \ldots, p_r$, $1 \leq p_i \leq n$, $1 \leq i \leq r$ and also combination Q with its r numbers at $q_1, q_2, \ldots, q_r$, $1 \leq q_i \leq n$ in the same category. Then we can write $\sum_{i=1}^{r} p_i = \sum_{i=1}^{r} q_i = \sum_{i=1}^{r} i + S$, where S is the shift of each of these combinations from the top combination stored in root node, the position of whose i^{th} number is i. Now a directed edge can exist from P to Q, iff $p_i \leq q_i$ for all $1 \leq i \leq r$, which can be true iff $p_i = q_i$, $1 \leq i \leq r$, i.e., the two combinations are exactly the same. For the same reason, there cannot exist any edge from a lower category (level) to a higher category (level).

Lemma 3. *The $\succeq$ relationship is transitive.*

Proof. This follows from the transitivity of the relation $\geq$ on the set of real numbers.

From the above lemma, it follows that the directed edges between consecutive levels are enough to capture all the relevant order information and we do not need edges from nodes of level V_m to nodes of level V_{m+j}, where $j > 1$.

Lemma 4. *There is a horizontal line of symmetry in the DAG G.*

Proof. This from the symmetry of the shift information of the nodes in G.

We illustrate the symmetry of the DAG G in Fig. 1. Note that we can replicate the lower half of the DAG G from the upper half, which can also save considerable amount of preprocessing time.

3.2   Generation of Different Variations Within the Same Category

In order to generate all the nodes of the DAG $G_{\binom{n}{r}}$, we need to change the relative positions of r numbers from the set S, in the available position set $P = \{1, 2, \ldots, n\}$, termed as shifts. It is easy to observe that a maximum of $r(n - r)$ shifts are possible, when all the r numbers will occupy the r rightmost positions from the position set. Hence, $G_{\binom{n}{r}}$ has $1 + r(n - r)$ levels in total including level 0, where at level i, each node corresponds to a combination of i number of shifts.

Now for any number of shifts p we need to generate different variations $V_p = \{V_{p1}, V_{p2}, \ldots, V_{pt}\}$, where each variation V_{pi} contributes one single node to the DAG at level p. We briefly explain below how to enlist the different variations that are present at the same category.

Let us try to find the different variations when the total number of shifts is 8 for a case when $n = 10$ and $r = 5$. Then we need an algorithm to find the ways in which 8 can be written as a sum of positive integers, i.e. the integer partitions of 8. Note that the components of the partitions will then signify the number of right shifts for the individual numbers.

Definition 3. *A partition of n is a sequence of non-negative integers $a_1 \geq a_2 \geq \ldots \geq a_t$ such that $n = a_1 + a_2 + \ldots + a_t$.*

For example, one partition of 7 can be $a_1 = a_2 = 3$, $a_3 = 1$ and $a_4 = a_5 = \cdots = 0$. The number of non-zero terms is called the number of parts, and the zero terms are usually suppressed. So we write $7 = 3+3+1$, or simply 331 when the context is clear. We use an efficient algorithm by Yamanaka et al. [10] to generate all the partitions of a positive integer n. We generate the partitions of a number by using a tree structure.

Definition 4. *The partition tree of a number n is a tree, where each node represents a partition of n. The root node contains the partition n itself. The right child of a node is obtained by decreasing the leftmost element by 1 and increasing the rightmost element by 1 and the left child of a node is obtained by decreasing the leftmost element by 1 and including a new 1 to the right.*

In Fig. 2, we show the partition tree of 8, from which we get all the partitions of 8. Then we apply the shifts corresponding to each partition to the root combination. For example, for the partition 332 we apply 3 shifts to the last number, i.e., r^{th} number, 3 shifts to the $(r-1)^{th}$ number, 2 shift to the $(r-2)^{th}$ number and 0 shifts to the rest. It can be trivially observed that out of the r numbers, we can never give a larger shift to a number L than a number R, if the original position of L was to the left of R. Now, consider the following two cases:

Case 1: The number of shifts for a particular number is greater than $n - r = 10 - 5 = 5$.

Case 2: The total number of shifts (total number of parts in the partition) is greater than $r = 5$.

As an example of Case 1, consider the root partition for $p = 8$, 8 shifts cannot be applied to the 5th number as the maximum number of shifts that can be applied is $n - r = 10 - 5 = 5$. Similarly, partitions $71, 611, 62$ will also lead to invalid combinations. These partitions are discarded.

As an example of Case 2, consider the partition 311111, where number of parts $= 6$. This partition will also lead to an invalid combination because there are only 5 numbers and we can apply shifts to only the corresponding 5 dots in

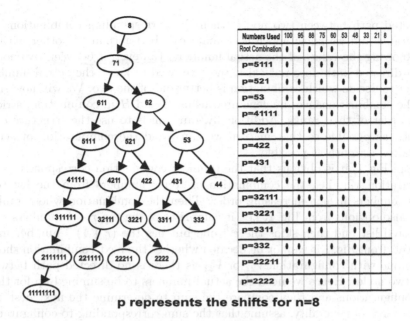

p represents the shifts for n=8

Fig. 2. Partition tree for n = 8 with shift information of all the nodes

the table. So, we have to discard all the partitions where no. of parts $>r$, i.e., no. of parts >5 in the example considered.

All partitions at and below level 3 are valid from that point of view. The partitions of level 3 start with 5 (5111, 521 and 53). The partitions of level 4 start with 4 and so on.

Following the above logic, we form the different variations for shifts = 8, where $n = 10$ and $r = 5$, and this is being illustrated in Fig. 2.

In this way, we continue finding the variations when the total no. of shifts = $9, 10, 11, \ldots, 25$, for the example considered. We just illustrate once more the maximum number of shifts that can be applied to the root combination. In the example considered above, the maximum number of shifts that can be applied to the 5th dot of the root combination is $n - r = 5$. The maximum no. of shifts that can be applied to the 4th dot is again 5. Similarly, the maximum no. of shifts that can be applied to each dot is 5, making the total no. of shifts $= r(n - r) = 5 * (10 - 5) = 25$.

We can summarize the procedure to generate all the possible variations from partition trees as follows. For all shifts from 1 to $r(n - r)$, do

(1) Generate the partition tree corresponding to the total number of shifts.
(2) Create nodes with shift information by selecting the valid partitions.

3.3 Generating the Combinations Using *G* and a Max-Heap

Here, we use a max-heap data structure H to obtain the *top-k r-combinations* from all the possible $\binom{n}{r}$ combinations. We have noted that in G, if there is a directed path between two nodes, then the corresponding combinations are comparable and we know which combination is better than the other without even knowing the values of the actual numbers. However, if between two nodes, such a directed path does not exist, then we need to know the actual numbers in order to decide which combination is better out of the two. We will now show how the combinations can be generated using G and H in conjunction, strictly in the order of their ranks. Additionally, our aim is to use the structure of G as much as possible so that in turn we can avoid the computation of actual summation as much as possible.

From Fig. 1, it is obvious that the topmost combination corresponds to the configuration V_{01}, i.e., the combination of top r numbers from the list of n sorted numbers in non-increasing order. Then the combination whose rank is 2nd is also obvious from Fig. 1, i.e., it comprises of the top $r - 1$ numbers from the sorted list and then skips the rth one but has the $(r + 1)^{th}$ number in it. However, from Fig. 1 it cannot be decided whether the next combination should be the one with configuration V_{21} or V_{22} as there is no directed path between these two nodes. Hence we need the actual numbers to be summed up for these two configurations and then compare the sum to determine the next best one. Without loss of generality, assume that the sum corresponding to configuration V_{21} is higher. Then going further down one level in Fig. 1, we can conclude node $V_{21} \succeq V_{31}$, but G does not help us to further decide, whether node $V_{31} \succeq v_{22}$ or not as there is no directed path between the two. Hence we devise the following method using a max-heap H to compare the combinations, as and when they are getting generated to resolve the ties that are left undecided from G.

Initially we place the root node of the DAG G as the only element in the heap. Then, in order to generate the next combination at each step, we remove the top element of H, which is in fact the next combination and then insert all its children from G into H. While they get inserted into H, the key is the sum of the numbers in that configuration and hence the heap gets automatically restructured, so that the next best combination always stays at the top for the next removal. Note that there is a minor detail here that needs to be taken care of. Some of the children of a particular node x may be already present in the heap H, as x may be a one-shift of more than one nodes in G, i.e., there may be incoming edges to x in G from multiple nodes say p and q in G. Suppose q is getting extracted from H. Hence, x being a child of q should enter H, but it may have already entered H when p got extracted. To avoid this problem of duplication, we insert x into H only when we extract that parent of x, which comes first in the lexicographic order of the subsets containing the r positions corresponding to the r-combinations related to all the parents of x. We illustrate this with a small example. Suppose r = 5, and the currently extracted node z represents the subset of positions $\{1, 2, 3, 5, 9\}$. Then such a node z would initiate the insertion of nodes corresponding to the subset of positions $\{1, 2, 4, 5, 9\}$ and $\{1, 2, 3, 6, 9\}$, but would not insert $\{1, 2, 3, 5, 10\}$. $\{1, 2, 3, 5, 10\}$ would rather be inserted when a node representing the subset $\{1, 2, 3, 4, 10\}$ will get extracted. Note that the delayed entry of node x is justified, as all the

---

Algorithm 1. Top-k-Combinations-with-Preprocessing(n: integer, r: integer, k: integer)

1: Sort the set S of n real numbers in non-increasing order and let the sorted list be
   $S_s = \{a_1, a_2, \ldots, a_n\}$, where the number a_p is at position p, $1 \leq p \leq n$;
2: $R \leftarrow$ Retrieve the root node from the DAG G;
3: $F \leftarrow R.shiftInfo$; //$F = \{F_1, F_2, \ldots, F_n\}$, where $F_j = 1$ for $j = 1, 2, 3, \ldots, r$
   and $F_j = 0$ for $j = r+1, r+2, \ldots, n$;
4: $R.sumCost \leftarrow \sum\limits_{j=1}^{n} F_j \times a_j$;
5: Insert R into MaxHeap;
6: **for** $p = 1$ to k **do**
7:     $R \leftarrow$ Extract the root node from MaxHeap;
8:     **for** $i = 1$ to n **do**
9:         $C \leftarrow \phi$;
10:        **if** $R.shiftInfo[i] = 1$ **then**
11:            $C \leftarrow C \cup \{a_i\}$; //construct the next best combination
12:        **end if**
13:    **end for**
14:    Print C; //report the next best combination
15:    Retrieve listChildDAG(R); //get child nodes of R from G
16:    **for** each node I in listChildDAG(R) **do**
17:        $I.sumCost \leftarrow \sum\limits_{i=1}^{n} I.shiftInfo[i] \times a_i$; //calculate total cost of the node I
18:        Insert I into MaxHeap;
19:    **end for**
20: **end for**

---

parents of x have to be extracted from H before x is extracted, which follows easily from Definition 2. Note that for all the child nodes to be inserted while extracting a particular node, such an extra checking can be done in a total of $O(r)$ time. The above process is performed k times until all *top-k* combinations are generated as output. The total details is given in Algorithm 1.

3.4 Computational Complexity

In our solution methodology, we at first built a metadata structure G (basically a DAG) as part of preprocessing and later on in the query step using Algorithm 1, we use this data structure G to report the k best combinations efficiently. In this section, we state the time and space complexities of the preprocessing step and the query step separately.

Lemma 5. *The overall space complexity of preprocessing is $O(rT_{node})$, where T_{node} is the total number of nodes of the DAG G.*

Proof. Let the total number of nodes of the DAG G be T_{node}. For each node of G, we have kept an array of n bits as shift information. Hence the overall space requirement is $O(nT_{node})$. However, storing the shift information as an array of r numbers instead of n bits will reduce the space requirement to $O(rT_{node})$.

Table 1. Information of the DAG G with varying values of n and r

n	r	Maximum number of nodes at any level	Number of Levels in the DAG, $1+r(n-r)$	Total Number of Nodes in DAG	Actual value of $\sum_{p=0}^{r(n-r)} |N(p)|$
10	5	20	26	252	9296
10	7	10	22	120	3506
15	5	141	51	3003	1295971
15	9	227	55	5005	2533589
20	5	521	76	15504	61537395
20	8	3788	97	125970	999764335

In Sect. 3.2, we explained the construction methodology of any level $p \in [0 \ldots r(n-r)]$ of the DAG G, in detail from any partition tree for integer p. Let the set of nodes at level p of the DAG G be $V(p)$ and the set of nodes in the partition tree of integer p be $N(p)$. In Sect. 3.2, we described how to construct $|V(p)|$ number of nodes at level p of the DAG G from the total number of nodes, $|N(p)|$ of any partition tree of integer p. In Sect. 3.2, we further showed that $|V(p)| \leq |N(p)|$ and for most of the cases $|V(p)| << |N(p)|$. The DAG G has in total $1 + r(n-r)$ levels. So total number of nodes of the DAG G,
$$T_{node} = \sum_{p=0}^{r(n-r)} |V(p)| < \sum_{p=0}^{r(n-r)} |N(p)|.$$
In reality, the value of $\sum_{p=0}^{r(n-r)} |N(p)|$ is much higher compared to the value of T_{node}. We have implemented the DAG G with varying values of n and r and reported the information of G in Table 1. The 5th column of Table 1 represents the total number of nodes, T_{node} of the DAG G. In the last column, we report the actual value of $\sum_{p=0}^{r(n-r)} |N(p)|$, which is, as expected much higher than the value of T_{node}.

Lemma 6. *The overall time complexity of preprocessing is $O(nT_{node})$.*

Proof. In the DAG G, any node V_{mi} at level m, $0 \leq m \leq r(n-r)-2$, has one less shift than all its child nodes $V_{(m+1)j}$ at level $(m+1)$, where any parent node V_{mi} is connected to at most r child nodes $V_{(m+1)j}$. Generation of each child node takes $O(n)$ time, so overall time required to construct G is $O(nT_{node})$.

There is a horizontal line of symmetry in the DAG G as depicted in Fig. 1 and replicating the lower half of G from the upper half can also save considerable amount of preprocessing time and space.

Lemma 7. *The overall space complexity of Algorithm 1 (query step) is $O(kr^2)$.*

Proof. After each extraction from the heap H, we need to insert at most r nodes into H from the DAG G. Hence, the total number of nodes to be inserted into H is bounded by $O(kr)$. However, in practice this number is far less, as reported in Sect. 5. Note that for each node, we kept an array of n bits as shift information. Hence the overall space requirement is $O(krn)$. However, storing the shift information as an array of r numbers instead of n bits will reduce the space requirement to $O(kr^2)$.

Lemma 8. *The time complexity of Algorithm 1 (query step) is* $O(kr \log kr)$.

Proof. We start with only one node (*top*-1 combination) in the heap H. The generation of each new combination will take $O(n)$ time, which may be reduced to $O(r)$ by changing the storage procedure of shift information. Also, after each extraction from H, at most r new nodes need to be inserted into H. Hence, after generating the pth best combination, H can hold at most pr nodes, so inserting at most r new nodes will take $O(r \log pr)$ time. So, the overall time requirement is,

$$T(k) = \sum_{p=1}^{k-1} O(r + r \log pr) = \sum_{p=1}^{k-1} O(r \log pr) = r(k-1)O(\log kr) = O(kr \log kr).$$

4 DAG for $\binom{m}{r}$ from that of $\binom{n}{r}$, where $m > n$

We state below some relationships between $G_{\binom{m}{r}}$, the DAG for $\binom{m}{r}$ and $G_{\binom{n}{r}}$, the DAG for $\binom{n}{r}$, where $m > n$. We can then construct $G_{\binom{m}{r}}$ by adding some additional nodes and edges to $G_{\binom{n}{r}}$, thus saving some computation and hence preprocessing time. We assume that $G_{\binom{n}{r}}$ is available from some previous computation. Note that $G_{\binom{m}{r}}$ will have $1 + r(m - r)$ levels in total.

Case 1: Up to level $n - r$, $G_{\binom{m}{r}}$ is isomorphic to $G_{\binom{n}{r}}$ (just $m - n$ extra 0s need to be padded to the right of the shift information).

Case 2: For any level i of $G_{\binom{m}{r}}$, where $i \in [(n - r + 1) \ldots (m - r)]$, all the nodes of $G_{\binom{n}{r}}$ that appear above level L of the partition tree [10] of i, where $L = i - (n - r)$, are added at level i of $G_{\binom{m}{r}}$.

Case 3: For any level $i > m - r$ of $G_{\binom{m}{r}}$, all the nodes of $G_{\binom{n}{r}}$ that appear above level L of the partition tree of i, where $L = i - (n - r)$, are not added at level i of $G_{\binom{m}{r}}$, rather if any element of a partition is greater than $m - r$, its corresponding node is excluded.

Case 4: For any level $i > r(n - r)$, nodes of the $G_{\binom{m}{r}}$ cannot be generated from $G_{\binom{n}{r}}$ because $G_{\binom{n}{r}}$ itself has $r(n - r)$ levels only.

5 Experimental Results

We have implemented the solution methodology of Sect. 3 and ran experiments with varying values of n, r and k, where we chose the n real numbers randomly.

Table 2. Performance of two methods with varying values of n, r and k

1	2	3	4	5	6	7		9
n	r	k	Maximum heap size	Total entry	Total heap entry/k	*Top-k* combination with preprocessing		*Top-k* combination on demand
						Pre-processing time (in sec)	Query time (in sec)	Time (in sec)
10	5	252	30	252	1	0.157	0.078	0.175
15	5	500	191	682	1.364	1.688	0.156	0.356
15	10	1000	234	1234	1.234	2.407	0.594	1.216
20	5	2000	496	2493	1.246	9.61	0.687	1.588
20	6	3000	959	3958	1.319	32.627	1.203	2.88
20	7	6000	1885	7844	1.307	94.099	3.157	8.87
20	8	12000	3451	15450	1.287	208.84	7.969	26.736
20	9	24000	6008	30008	1.250	-	-	82.86
20	10	5000	2011	7006	1.401	-	-	9.275
40	5	10000	3147	13145	1.314	2999.907	4.584	16.377
45	5	20000	4827	24819	1.241	10218.046	8.965	43.92
100	50	10000	7195	17195	1.719	-	-	103.103
200	100	40000	22197	62191	1.554	-	-	1068.561
400	200	100000	49051	149051	1.490	-	-	8991.857

We ran the experiments in a Core 2 Duo Pentium IV machine with 4 GB RAM and reported the performance in Table 2.

Note that as we go on generating the *top-k* combinations, the heap size waxes and wanes depending on the distribution of those n numbers being considered. The 4th column of Table 2 represents the maximum number of numbers simultaneously present in the heap. In the next column, we report the total number of heap entries, i.e., how many combinations were evaluated in all in order to report the *top-k* combinations. Note that this value will always be between k and $\binom{n}{r}$. The closer it is towards k, the more it shows that our method is effective. The 6th Column shows that the total number of entries remained within $2k$.

In the 7th column we report the time needed to construct the DAG and naturally it is high as the time required is exponential. Since it is a part of preprocessing, it is not a major concern. In the 8th column, we report the execution time of Algorithm 1, which is very low, as expected. However, when the values of n, r and k were sufficiently large, the preprocessing phase ran for several hours and was forcibly terminated and hence in some of the rows of column 7 there are blank entries. This gave us the impetus for exploring some alternative method.

5.1   Generating Combinations Without Explicit Construction of G

In order to reduce the burden of total DAG construction, we resorted to constructing the levels of the DAG G on demand, as and when needed, keeping the

---

Algorithm 2. Top-k-Combinations-on-Demand(n: integer, r: integer, k: integer)

1: Sort the set S of n real numbers in non-increasing order and let the sorted list be
   $S_s = \{a_1, a_2, \ldots, a_n\}$, where the number a_p is at position p, $1 \leq p \leq n$;
2: Make a list of flag bits $F \leftarrow \{F_1, F_2, \ldots, F_n\}$, where $F_j \leftarrow 1$ for $j = 1, 2, 3, \ldots, r$
   and $F_j \leftarrow 0$ for $j = r + 1, r + 2, \ldots, n$;
3: Construct the root node T of the DAG G, where $T.shiftInfo \leftarrow F$ and
   $T.sumCost \leftarrow \sum_{j=1}^{n} F_j \times a_j$;
4: Insert T into MaxHeap;
5: **for** $p = 1$ to k **do**
6:    $R \leftarrow$ Extract the root node from MaxHeap;
7:    **for** $i = 1$ to n **do**
8:       $C \leftarrow \phi$;
9:       **if** $R.shiftInfo[i] = 1$ **then**
10:          $C \leftarrow C \cup \{a_i\}$; //construct the next best combination
11:      **end if**
12:   **end for**
13:   Print C; //report the next best combination
14:   Construct listChildDAG(R); //get child nodes of R from the construction
      method of G
15:   **for** each node I in listChildDAG(R) **do**
16:      $I.sumCost \leftarrow \sum_{i=1}^{n} I.shiftInfo[i] \times a_i$; //calculate total cost of the node I
17:      Insert I into MaxHeap;
18:   **end for**
19: **end for**

---

construction idea of G in the mind. We had made a few changes in Algorithm 1 in lines numbered 2 and 15, where we generate the nodes of the DAG G on the fly instead of retrieving them from the DAG. This is not difficult, since we can always construct the shift information of the child nodes from the shift information of their parent, whch is just 1 more. As in Algorithm 1, we perform similar checks in $O(r)$ time while inserting the child nodes into H, in order to avoid the duplication of the nodes that are already present in H. The total process is described in Algorithm 2. We have implemented this alternative approach also and report the performance in the last column of Table 2. As expected, the times reported in this column are higher compared to the corresponding values in the previous column, as there is no preprocessing phase in this approach. However, the remarkable fact is that this gets rid of the entire preprocessing overhead (both space and time) in exchange for a slight increase in query time. It can now be used for any practical case, however high the values of n and r may be, as long as k is of reasonable size.

Computational Complexity

Lemma 9. *The overall space complexity of Algorithm 2 is $O(kr^2)$.*

Proof. Follows from the proof of Lemma 7.

Lemma 10. *The overall time complexity of Algorithm 2 is $O(n \log n + kr \log k)$.*

Proof. The initial sorting takes $O(n \log n)$ time. For *top*-1 combination, max-heap H needs to be constructed with only the root node of G, taking $O(r)$ time. So time requirement for reporting the *top*-1 combination is $T(1) = O(r)$. At each step of generating the next best combination, we need to consider the previous extraction R from the max-heap H and insert at most r of its child nodes from the DAG G into H. Before inserting any node into H, we calculate its sum of numbers from its shift information, so that the next best combination always stays at the top for the next removal. Here we do not construct the DAG G explicitly but from the knowledge of the shift mechanism, we figure out which child nodes have to be inserted in the heap H. When, we are considering the pth best combination, heap H already holds $O(pr)$ nodes, so inserting r new nodes will take $O(r \log pr)$ time. When we insert the children of an extracted node, all the children (not each one) can be implicitly constructed in $O(r)$ time, by scanning the positions of the r entries of the extracted node. So, the time requirement for extracting k elements is, $T(k) = \sum_{p=1}^{k-1} O(r + r \log pr) = O(\sum_{p=1}^{k-1} (r + r \log pr)) = O((k-1)r) + O(r(k-1) \log kr) = O(kr + kr \log kr)$. The running time can be slightly improved to $O(kr + kr \log k) = O(kr \log k)$, if we maintain a min-heap H' together with H. The purpose of the min-heap is to identify the minimum element, whenever the size of H exceeds k, so that we can remove this element immediately, and always keep the size of H', as well as H to be at most k. Consequently, any heap operation will be performed in $O(\log k)$ time. Hence overall running time of Algortihm 2 is $O(n \log n + kr \log k)$.

6   Conclusions and Future Research

We have proposed two efficient algorithms for generating the *top-k r-combinations* of a set of n real numbers using a novel metadata structure. Solving the problem for aggregation functions other than summation, probabilistic analysis of the heap size and finding other applications of the metadata structure remain possible directions for future research.

References

1. Afshani, P., Brodal, G.S., Zeh, N.: Ordered and unordered *Top-k* range reporting in large data sets. In: 2011 SODA, pp. 390–400 (2011)
2. Chaudhuri, S., Gravano, L., Marian, A.: Optimizing *top-k* selection queries over multimedia repositories. IEEE Trans. Knowl. Data Eng. **16**(8), 992–1009 (2004)
3. Chen, J., Kanj, I.A., Meng, J., Xia, G., Zhang, F.: Parameterized *top-k* algorithms. Theor. Comput. Sci. **470**, 105–119 (2013)
4. Chen, J., Yi, K.: A dynamic data structure for *top-k* queries on uncertain data. Theor. Comput. Sci. **407**(1–3), 310–317 (2008)

5. Karpinski, M., Nekrich, Y.: *Top-k* color queries for document retrieval. In: 2011 SODA, pp. 401–411 (2011)
6. Lu, J., Senellart, P., Lin, C., Du, X., Wang, S., Chen, X.: Optimal top-k generation of attribute combinations based on ranked lists. In: 2012 SIGMOD, pp. 409–420 (2012)
7. Rahul, S., Gupta, P., Janardan, R., Rajan, K.S.: Efficient top-k queries for orthogonal ranges. In: Katoh, N., Kumar, A. (eds.) WALCOM 2011. LNCS, vol. 6552, pp. 110–121. Springer, Heidelberg (2011)
8. Suzuki, T., Takasu, A., Adachi, J.: *Top-k* query processing for combinatorial objects using Euclidean distance. In: IDEAS 11 Proceedings of the 15th Symposium on International Database Engineering and Applications, pp. 209–213 (2011)
9. Vajnovszki, V.: A loopless algorithm for generating the permutations of a multiset. Theor. Comput. Sci. **307**(2), 415–431 (2003)
10. Yamanaka, K., Kawano, S., Kikuchi, Y., Nakano, S.: Constant time generation of integer partitions. IEICE Trans. Fundam. **E90–A**(5), 888–895 (2007)

Efficient Group Testing Algorithms with a Constrained Number of Positive Responses

Responses

Annalisa De Bonis$^{(\boxtimes)}$

Diparitmento di Informatica, Università di Salerno, Fisciano, SA, Italy
debonis@dia.unisa.it

Abstract. *Group testing* is a well known search problem that consists in detecting the defective members of a set of objects O by performing tests on properly chosen subsets (*pools*) of the given set O. In classical group testing the goal is to find all defectives by using as few tests as possible. We consider a variant of classical group testing in which one is concerned not only with minimizing the total number of tests but aims also at reducing the number of tests involving defective elements. The rationale behind this search model is that in many practical applications the devices used for the tests are subject to deterioration due to exposure to or interaction with the defective elements. As an example, consider the leak testing procedures aimed at guaranteeing safety of sealed radioactive sources. Personnel involved in these procedures are at risk of being exposed to radiation whenever a leak in the tested sources is present. The number of positive tests admitted by a leak testing procedure should depend on the dose of radiation which is judged to be of no danger for the health. Obviously, the total number of tests should also be taken into account in order to reduce the costs and the work load of the safety personnel. In this paper we consider both adaptive and non adaptive group testing and for both scenarios we provide almost matching upper and lower bounds on the number of "yes" responses that must be admitted by any strategy performing at most a certain number t of tests. The lower bound for the non adaptive case follows from the upper bound on the optimal size of a variant of d-cover free families introduced in this paper, which we believe may be of interest also in other contexts.

1 Introduction and Contributions

Group testing is a well known search paradigm that consists in detecting the defective members of a set of objects O by performing tests on properly chosen subsets (*pools*) of the given set O. A test yields a "yes" response if the tested pool contains one or more defective elements, and a "no" response otherwise. The goal is to find all defectives by using as few tests as possible. Group testing origins date back to World War II when it was introduced as a possible technique for mass blood testing [12]. Since then group testing has found applications in a wide variety of situations ranging from conflict resolution algorithms for

© Springer International Publishing Switzerland 2014
Z. Zhang et al. (Eds.): COCOA 2014, LNCS 8881, pp. 506–521, 2014.
DOI: 10.1007/978-3-319-12691-3_37

multiple-access systems [10,29], fault diagnosis in optical networks [17], quality control in product testing [25], failure detection in wireless sensor networks [22], data compression [18], and many others. Among the modern applications of group testing, some of the most important are related to the field of molecular biology, where group testing is especially employed in the design of screening experiments. Du and Hwang [14] provide an extensive coverage of the most relevant applications of group testing in this area.

The different contexts to which group testing applies often call for variations of the classical research model that best adapt to the characteristics of the problems. These variants concern the test model [3,4,7,10,11], the number of pursued defective elements [1,8], as well as the structure of the test groups [5,8,28].

In this paper, we consider a variant of the classical model in which one is concerned not only with minimizing the total number of tests but aims also at reducing the number of tests involving defective elements. Therefore, the test groups should be structured so as to reduce the number of groups intersecting the set of defectives. The rationale behind this search model is that in many practical applications the devices used for the tests are subject to deterioration due to exposure to or interaction with the defective elements. In some contexts, the positive groups may even represent a risk for the safety of the persons that perform the tests. An example of such applications are leak testing procedures aimed at guaranteeing the safety of sealed radioactive sources [26,27]. Radioactive sources are widely used in medical, industrial and agricultural applications, as well as in scientific research. Sealed sources are small metal containers in which radioactive material is sealed. As long as the sealed sources are handled correctly and the enclosing capsules are intact, they do not represent a health hazard. According to the radiation safety standards, sealed radioactive sources should be tested at regular intervals in order to verify the integrity of the capsules. Leak testing procedures are crucial in preventing contamination of facilities and personnel due to the escape of radioactive material. However, these procedures put the safety personnel at the risk of being exposed to radiation whenever a leak in the tested sources is present. Commonly, when not used, the sources are stored in lead-shielded drawers. In order to be tested for leakage, sources are removed one at time from the storage area and wiped with absorbent paper or a cotton swab held by a long pair of forceps. The wipe sample is then analyzed for radioactive contamination. An alternative procedure consists in testing the sources in groups. To this aim, the sources are not removed from the shielded storage drawer and a wipe sample is taken from the upper surface of the storage drawer. If the sample is contaminated then at least one source in the tested storage drawer is leaking; otherwise all sources in the drawer are intact. This idea suggests the use of group testing in leak testing procedures. Since leak testing procedures expose to risk the personnel that perform the tests on contaminated wipe samples, the number of positive tests admitted by the group testing procedure should depend on the dose of radiation which is judged to be of no danger for the health. Obviously, the total number of tests should also be taken into account in order to reduce the costs and the work load of the safety personnel. Trivially, the procedure that tests all elements individually attains the minimum number of positive responses, which is equal to the number of defectives in the input set. While this procedure may be an option when the danger implied by testing positive samples is extremely high, many practical applications call for procedures that can be tuned to obtain the desired tradeoff between the number of "admissible" positive responses and the total number of tests.

We consider both *adaptive* and *non adaptive* group testing procedures. In adaptive group testing, at each step the algorithm decides which group to test by observing the responses of the previous tests. For classical group testing, there exist adaptive strategies that achieve the information theoretic lower bound $\Omega(d\log(n/d))$, where n is the total number of elements and d is the upper bound on the number of defectives. However, in many practical scenarios adaptive strategies are useless due to the fact that assembling the groups for the tests may be very time consuming and that some kind of group tests may take long time to give a response. In such applications, it is preferable to use non adaptive strategies, i.e., strategies in which all tests are decided in advance and can be performed in parallel. Non adaptive group testing strategies are much more costly than adaptive algorithms. Indeed, the minimum number of tests used by these procedures is equal to the minimum length of certain combinatorial structures known under the name of *cover free families* (or equivalently, *superimposed codes*) [15,16,20]. The known bounds for these combinatorial structures imply that the number of tests of any non adaptive group testing algorithm is lower bounded by $\Omega((d^2/\log d)\log n)$ and that there exist non adaptive group testing algorithms that use $O(d^2\log n)$ tests.

In both the adaptive and the non adaptive scenarios, we present almost matching upper and lower bounds on the number of "yes" responses that must be admitted by any strategy performing at most a certain number t of tests. As expected, in these bounds the number of positive responses decreases with the total number of tests. In Sect. 2, we present the lower bound for the adaptive case and give an algorithm that almost achieves this bound. In Sect. 3, we consider the non adaptive scenario. In that section we introduce a variant of cover free families in which the union of any d members is required to be smaller than a certain parameter. The lower bound for the non adaptive case follows from the upper bound on the optimal size of these families. For the non adaptive case we present two algorithms both achieving the same asymptotical bound. An interesting feature of one of these constructions consists in being an *explicit* construction, in that there exists an efficient algorithm to design the underlying combinatorial structure.

2   The Adaptive Case

In this section we deal with the case when tests are performed adaptively by looking at the feedbacks of already performed tests. For the purpose of our analysis, we need to introduce the following definition.

Definition 1. *Let t, n, d be positive integers with $n \geq d \geq 1$, and let O be a set of n elements containing at most d defective elements. Moreover, let $\mathcal{A}$ be a group testing strategy that finds all defective items in O by at most t tests. We denote by $y_{\mathcal{A}}(d, n, t)$ the maximum number of positive responses that occurs during the search process performed by $\mathcal{A}$, where the maximum is taken over all possible subsets of up to d defectives. The minimum value of $y_{\mathcal{A}}(d, n, t)$ is denoted by $y(d, n, t)$, where the minimum is taken over all group testing algorithms that use at most t tests to find all defectives in O.*

Notice that $y(d, n, t)$ represents the minimum number of positive responses that must be admitted in order to find up to d defectives in a set of n elements by at most t tests. The following lemma is quite straightforward.

Lemma 1. *Let t, n, d be positive integers with $n \geq d \geq 1$. Then, $y(d, n, t) \geq d$.*

Proof. Suppose by contradiction that $y(d, n, t) < d$. Then, there would be at least one defective element which either is never tested or appears only in groups that contain also other defective elements. In both cases, the algorithm could not decide whether this element is defective or not. This is due to the fact that we are considering the case when d is not the exact number of defectives but an upper bound on it. ☐

In order to derive a lower bound on $y(d, n, t)$, we describe the search process by a binary tree where each internal node corresponds to a test and each leaf to one of the possible outcomes of the algorithm. For each internal node, its left branch is labelled with 0 and corresponds to a negative response, while its right branch is labelled with 1 and corresponds to a positive response. A path from the root to a leaf x represents the sequence of tests performed by the algorithm when the set of defective items is the one associated with x. Obviously, for an input set of size n that contains d defective elements, a group testing strategy is successful if and only if the corresponding tree has $\binom{n}{d}$ leaves. Let us denote by y the maximum number of "yes" responses in the whole sequence of test responses. Each root-to-leaf path can be represented by the binary vector whose entries are the labels of the branches along the path taken in the order they are encountered starting from the root. Since each path that starts from the root and ends in a leaf must contain at most y branches labelled with 1, then the number of such binary vectors is smaller than or equal to $\sum_{i=1}^{y} \binom{t}{i}$. As a matter of fact, by Lemma 1, the sum should start at $i = d$. However, this would not affect significantly the analysis. Since the number of leaves cannot be larger than the upper bound on the number of root-to-leaf paths, then it holds

$$\sum_{i=1}^{y} \binom{t}{i} \geq \binom{n}{d}. \tag{1}$$

The above bound obviously holds also in the case when d is an upper bound on the number of defective elements since in that case the number of possible outcomes is $\sum_{k=1}^{d} \binom{n}{k} \geq \binom{n}{d}$.

Inequality (1) allows to derive a lower bound on $y(d, n, t)$. In order to obtain the desired bound, we make use of the following well known inequalities on the binomial coefficient

$$\left(\frac{N}{m}\right)^m \leq \binom{N}{m} \leq \left(\frac{eN}{m}\right)^m, \tag{2}$$

where e denotes the Neper's constant $e = 2,71828\ldots$.

In the following, unless specified differently, all logarithms are in base 2.

Theorem 1. *Let t, n, d be positive integers with $n \geq d \geq 1$. It holds*

$$y(d, n, t) > \max\left\{d, \frac{d\log(n/d)}{\log t} + \frac{\log(t-1)}{\log t} - 1\right\}.$$

Proof. Let y denote the maximum number of positive responses admitted by an adaptive group testing algorithm that uses at most t tests to find up to d defectives. By applying the upper bound in (2) to $\binom{t}{i}$, for $i \geq 3$, we have that

$$\sum_{i=1}^{y} \binom{t}{i} \leq t + \frac{(t-1)t}{2} + \sum_{i=3}^{y} \left(\frac{et}{i}\right)^i. \tag{3}$$

The righthand side of (3) is smaller than

$$t + \frac{(t-1)t}{2} + \sum_{i=3}^{y} t^i = t + \frac{(t-1)t}{2} + \left(\frac{t^{y+1}-1}{t-1} - 1 - t - t^2\right),$$

from which it follows

$$\sum_{i=1}^{y} \binom{t}{i} \leq \frac{t^{y+1}-1}{t-1}. \tag{4}$$

On the other hand, the lower bound on the binomial coefficients in (2) implies

$$\binom{n}{d} \geq \left(\frac{n}{d}\right)^d. \tag{5}$$

The lower bound in the statement of the theorem follows immediately from Lemma 1 and inequalities (1), (4), and (5). □

Now we present an algorithm that almost attains the lower bound of Theorem 1. The algorithm is designed after Li's stage group testing algorithm [21]. While Li's analysis aims at minimizing the total number of tests, our algorithm performs a number of tests that depends on the number of positive responses admitted by the algorithm.

The algorithm works as follows. The tests are organized in stages in such a way that each stage tests a collection of disjoint subsets that form a partition of the search space. At stage i the search space is partitioned into g_i groups, $g_i - 1$ of which have size k_i, while the remaining one might have size smaller than k_i. The elements in the subsets that test negative are discarded, while those in the subsets that test positive are grouped together to form the new search space. Therefore, in the first stage the input set is partitioned into g_1 subsets, $g_1 - 1$ of which have size k_1, while the remaining one might have size smaller than k_1. The elements in the subsets that test positive are grouped together and further searched in stage 2. At stage 2 the new search space is partitioned into g_2 groups, $g_2 - 1$ of which have size k_2, while the remaining one might have size smaller than k_2. The elements in the groups that test positive in stage 2 are further searched in stage 3, and so on. Notice that the tests in each stage can be performed in parallel. Let f denote the total number of stages. Notice that in stage i, $i = 1, \ldots, f$, the defective elements are contained in at most d of the g_i groups and therefore, after this stage, the search space consists of at most dk_i elements. The algorithm is successful if and only if after stage f the search space contains only the defective elements. This is insured by setting $k_f = 1$.

Let us ignore for the moment the integral constraints. The total number of tests performed by the algorithm is

$$t \leq \sum_{i=1}^{f} g_i = \frac{n}{k_1} + \frac{dk_1}{k_2} + \frac{dk_2}{k_3} + \ldots + \frac{dk_{f-2}}{k_{f-1}} + dk_{f-1}. \tag{6}$$

As observed before, in each stage at most d groups test positive and consequently, the total number of positive responses is upper bounded by fd. Obviously, the minimum is attained for $f = 1$, i.e., in the case when the algorithm consists in a single stage that tests each element individually. Therefore, it trivially holds $y(d, n, n) = d$.

If we fix the number of stages f, the values of the k_i's does not affect the upper bound on the number of positive responses (as far as $g_i = \frac{dk_{i-1}}{k_i} \geq d$, i.e., $k_{i-1} \geq k_i$). Therefore, we choose the values of $k_1, \ldots, k_{f-1}$ which minimize the upper bound on t. The minimum value of the righthand side of (6) is attained for $k_i^* = \left(\frac{n}{d}\right)^{\frac{f-i}{f}}$, $i = 1, \ldots, f - 1$. We set $g_1 = \left\lceil \frac{n}{k_1^*} \right\rceil$ and $g_i = d \left\lceil \frac{k_i^*}{k_{i+1}^*} \right\rceil$, for $i = 2, \ldots, f$. Therefore, in each stage, the number of tests is at most $d \left\lceil \left(\frac{n}{d}\right)^{\frac{1}{f}} \right\rceil$, and consequently, the total number of tests is

$$t < fd \left(\frac{n}{d}\right)^{\frac{1}{f}} + fd.$$

The above upper bound on t implies

$$f < \frac{\log(\frac{n}{d})}{\log(\frac{t}{fd} - 1)} \tag{7}$$

Since the maximum number of positive responses is fd, we set $y_A(d, n, t) = fd$ and have that inequality (7) implies the following theorem.

Theorem 2. *Let t, n, d be positive integers with $n \geq d \geq 1$. There exists a group testing strategy A for which*

$$y_A(d, n, t) < \frac{d \log(\frac{n}{d})}{\log(\frac{t}{y_A(d,n,t)} - 1)}.$$

3   The Non Adaptive Case

A group testing algorithm is said to be *non adaptive* if all tests must be decided beforehand without looking at the responses of previous tests. In this section, we derive upper and lower bounds on the minimum number of positive responses that a group testing algorithm must admit when all tests are decided non adaptively. We denote by $\tilde{y}(n, d, t)$ the minimum value of $y_{\mathcal{A}}(n, d, t)$ over all non adaptive strategies $\mathcal{A}$ that find up to d defective elements in an input set of size n by at most t tests.

There exists a correspondence between non adaptive group testing algorithms for input sets of size n and families of n subsets. Indeed, given a family $\mathcal{F} = \{F_1, \ldots, F_n\}$ with $F_i \subseteq [t]$, we design a non adaptive group testing strategy as follows. We denote the elements in the input set by the integers in $\{1, \ldots, n\}$ and for $i = 1, \ldots, t$, define the group $T_i = \{j : i \in F_j\}$. Obviously, $T_1, \ldots, T_t$ can be tested in parallel and therefore the resulting algorithm is non adaptive. Conversely, given a non adaptive group testing strategy for an input set of size n that tests $T_1, \ldots, T_t$, we define a family $\mathcal{F} = \{F_1, \ldots, F_n\}$ by setting $F_j = \{i \in [t] : j \in T_i\}$, for $j = 1, \ldots, n$. Equivalently, any non adaptive group testing algorithm for an input set of size n that performs t tests corresponds to a binary code of length t and size n. This is due to the fact that any family of size n on the ground set $[t]$ can be represented by the binary code of length t whose codewords are the characteristic vectors of the members of the family. Given such a binary code $\mathcal{C} = \{c_1, \ldots, c_n\}$, one has that j belongs to the pool T_i if and only if c_j has the i-th entry equal to 1.

It is well known that a non adaptive group testing strategy is successful if and only if the corresponding family is $\bar{d}$-*separable*, i.e., a family in which the unions of up to d members are pairwise distinct [13,14]. To see this, let us represent the test responses by a binary vector whose i-th entry is equal to 1 if and only if T_i tests positive. We call this vector the *response vector*. Notice that the response vector is the characteristic vector of the union of the members of the family associated with the defective elements. In the binary code representation, this is equivalent to saying that the response vector is the OR of the codewords associated with the defective elements. Therefore, the set of the defective elements is univocally identified if and only if the union of up to d members of the family are pairwise distinct. It is immediate to see that this condition holds if and only if the family is $\bar{d}$-separable. The reader is referred to [13,14] for a detailed account on these issues.

In spite of the equivalence between separable families and non adaptive group testing strategies, typically in the literature the design of non adaptive algorithms is based on families satisfying a slightly stronger property that allows for a more efficient decoding algorithm to obtain the set of defectives from the test responses. These families satisfy the property that no member of the family is contained in the union of any other d members. Families with this property are called d-*cover free* families [16], whereas the corresponding binary codes are said to be d-*superimposed* or d-*disjunct* [9,13–15,20]. Such codes have the property that for each codeword c and any other d codewords $c_{j_1}, \ldots, c_{j_d}$ there exists an index i such that c has the i-th entry equal to 1, whereas all of $c_{j_1}, \ldots, c_{j_d}$ have the i-th entry equal to 0. We say that c is not *covered* by the OR of $c_{j_1}, \ldots, c_{j_d}$. A consequence of this property is that any codeword associated with a regular (e.g., non defective) element is not covered by the response vector. Therefore, it is possible to recover the set of the defective elements by simply comparing the response vector with each codeword. On the other hand, if we use an algorithm based on a $\bar{d}$-separable family then, in order to obtain the set of the defective elements, we need to examine all subsets of up to d codewords.

In the following, we recall the formal definition of $\bar{d}$-separable and d-cover free families and introduce a new variant of those families that can be used to derive upper and lower bounds for the group testing problem we are considering.

Let $\mathcal{F} = \{F_1, \ldots, F_n\}$ be a family of subsets of $[t] = \{1, \ldots, t\}$. We will refer to the set $[t]$ as the *ground set* of the family. For a positive integer $k \leq t$, a family $\mathcal{F} = \{F_1, \ldots, F_n\}$ of subsets of $[t]$ is said k-*uniform*, if $|F_i| = k$, for $i = 1, \ldots, n$.

Definition 2. *Let d and t be positive integers. A family $\mathcal{F}$ on the ground set $[t]$ is said to be $\bar{d}$-separable, if the unions of up to d members of the family are pairwise distinct, whereas it is said to be d-cover free if any member of the family is not contained in the union of any other d members of the family.*

Given a family $\mathcal{F} = \{F_1, \ldots, F_n\}$, the corresponding group testing algorithm must admit a number of positive responses which is as large as the size of the largest union of up to d members of the family. Indeed, let $j_1, \ldots, j_m$, with $m \leq d$, be the defective elements. A group T_i intersects $\{j_1, \ldots, j_m\}$ if and only if $i \in F_{j_1} \cup \ldots \cup F_{j_m}$. Therefore, the number of positive responses is equal to $|F_{j_1} \cup \ldots \cup F_{j_m}|$. By the above argument, a group testing strategy that uses t tests and admits at most s positive responses is equivalent to the following notion of $\cup_{\leq s}$ $\bar{d}$-separable family.

Definition 3. *Let d, s, and t be positive integers. We say that a family $\mathcal{F}$ on the ground set $[t]$ is a $\cup_{\leq s}$ $\bar{d}$-separable family if $\mathcal{F}$ is $\bar{d}$-separable and the union of any d members of $\mathcal{F}$ has size at most s. The maximal cardinality of a $\cup_{\leq s}$ $\bar{d}$-separable family on the ground set $[t]$ is denoted by $n_{sep}(d, \cup_{\leq s}, t)$.*

Similarly to what happens in classical group testing, cover free families allow to decode the response vector much more efficiently. Therefore, we introduce the following definition.

Definition 4. *Let d, s, and t be positive integers. We say that a family $\mathcal{F}$ on the ground set $[t]$ is a $\cup_{\leq s}$ d-cover free family if $\mathcal{F}$ is d-cover free and the union of any d members of $\mathcal{F}$ has size at most s. The maximal cardinality of a $\cup_{\leq s}$ d-cover free family on the ground set $[t]$ is denoted by $n_{cf}(d, \cup_{\leq s}, t)$.*

It is immediate to see that $\cup_{\leq s}$ d-cover free families are $\cup_{\leq s}$ $\bar{d}$-separable families, and consequently, existential results for the former families apply also to the latter families. The following theorem shows that upper bounds on the maximum cardinality of $\cup_{\leq s}$ $(d-1)$-cover free families can be used to derive upper bounds on the maximum size of $\cup_{\leq s}$ $\bar{d}$-separable families.

Theorem 3. *Let d, s, and t be positive integers. Any $\cup_{\leq s}$ $\bar{d}$-separable family is $\cup_{\leq s}$ $(d-1)$-cover free.*

Proof. First we show that any $\bar{d}$-separable family is a $(d-1)$-cover free family. This relation was noted by Kautz and Slingleton [20] and is quite simple to see. Indeed, suppose by contradiction that a $\bar{d}$-separable family is not $(d-1)$-cover free. As a consequence, there exist d members of the family $F_1, F_2, \ldots, F_d$ such that $F_d \subseteq F_1 \cup \ldots \cup F_{d-1}$, and therefore, it holds $\bigcup_{i=1}^{d} F_i = \bigcup_{i=1}^{d-1} F_i$ thus contradicting the fact that the family is $\bar{d}$-separable. Moreover, it holds $|\bigcup_{i=1}^{d-1} F_i| < |\bigcup_{i=1}^{d} F_i| \leq s$, thus proving that the family is $\cup_{\leq s}$ $(d-1)$-cover free. $\qquad\square$

3.1 Negative Results

Theorem 4. *Let d, s and t be positive integers. The maximal size of a $\cup_{\leq s}$ d-cover free family on the ground set $[t]$ is*

$$n_{cf}(d, \cup_{\leq s}, t) \leq \left(\frac{etd^2}{4s}\right)^{\left\lceil \frac{4s}{d^2} \right\rceil} + \lceil d/2 \rceil - 1,$$

where e denotes the Neper's constant $e = 2,71828\ldots$.

Proof. Let $\mathcal{F}$ be a $\cup_{\leq s}$ d-cover free family on the ground set $[t]$. Let us define the sets $G_1, \ldots, G_{\lceil d/2 \rceil - 1}$ as follows. We set G_1 to be the largest member of $\mathcal{F}$ and, for each $i = 2, \ldots, \lceil d/2 \rceil - 1$, G_i to be the largest set in $\{F \setminus \bigcup_{j=1}^{i-1} G_j : F \in \mathcal{F} \setminus \{G_1, \ldots, G_{i-1}\}\}$. In other words, after choosing G_1 as the largest member of the family, we remove the elements of G_1 from all members of $\mathcal{F} \setminus \{G_1\}$ and set G_2 to be the largest of the resulting sets. Then, we remove the elements of G_2 from all unselected sets and set G_3 to be the largest of the sets of the form $F \setminus (G_1 \cup G_2)$, for $F \in \mathcal{F} \setminus \{G_1, G_2\}$, and so on until $\lceil d/2 \rceil - 1$ sets are selected. Let $\mathcal{F}'$ be the family obtained by removing the elements of $G_1, \ldots, G_{\lceil d/2 \rceil - 1}$ from all members of $\mathcal{F} \setminus \{G_1, \ldots, G_{\lceil d/2 \rceil} - 1\}$, i.e., $\mathcal{F}' = \{F \setminus \bigcup_{j=1}^{\lceil d/2 \rceil - 1} G_j : F \in \mathcal{F} \setminus \{G_1, \ldots, G_{\lceil d/2 \rceil - 1}\}\}$. It is possible to see that $\mathcal{F}'$ is a $(\lfloor d/2 \rfloor + 1)$-cover free family. Suppose by contradiction that $\mathcal{F}'$ is not $(\lfloor d/2 \rfloor + 1)$-cover free. By this assumption, there are $\lfloor d/2 \rfloor + 2$ sets $F_1', F_2', \ldots, F_{\lfloor d/2 \rfloor + 2}' \in \mathcal{F}'$ such that $F_{\lfloor d/2 \rfloor + 2}' \subseteq F_1' \cup \ldots \cup F_{\lfloor d/2 \rfloor + 1}'$. Since for $i = 1, \ldots, \lfloor d/2 \rfloor + 2$, it is $F_i' = F_i \setminus \bigcup_{j=1}^{\lceil d/2 \rceil - 1} G_j$ for some set $F_i \in \mathcal{F} \setminus \{G_1, \ldots, G_{\lceil d/2 \rceil - 1}\}$, it holds $F_{\lfloor d/2 \rfloor + 2} \subseteq F_1 \cup \ldots \cup F_{\lfloor d/2 \rfloor + 1} \cup G_1 \cup \ldots \cup G_{\lceil d/2 \rceil - 1}$, thus contradicting the fact that $\mathcal{F}$ is d-cover free. By the same argument as above it is possible to see that all sets in $\mathcal{F}' \cup \{G_1, \ldots, G_{\lceil d/2 \rceil - 1}\}$ are non-empty and pairwise distinct.

In the following we derive an upper bound on the cardinality of $\mathcal{F}'$. Notice that $G_1, \ldots, G_{\lceil d/2 \rceil - 1}$ are pairwise disjoint and that $|G_1| \geq |G_2| \geq \ldots \geq |G_{\lceil d/2 \rceil - 1}|$. Moreover, it holds $G_i \cap F' = \emptyset$ and $|G_i| \geq |F'|$, for any $i = 1, \ldots, \lceil d/2 \rceil - 1$ and $F' \in \mathcal{F}$. Therefore, for any member $F' \in \mathcal{F}'$, one has that

$$s \geq \left| \bigcup_{i=1}^{\lceil d/2 \rceil - 1} G_i \cup F' \right| = \sum_{i=1}^{\lceil d/2 \rceil - 1} |G_i| + |F'| \geq \lceil d/2 \rceil |F'|. \tag{8}$$

It follows that $|F'| \leq \lfloor \frac{2s}{d} \rfloor$. Since F' is an arbitrary member of $\mathcal{F}'$ then inequality (8) holds for any member F' of $\mathcal{F}'$.

As observed in [2,24], any member F of an r-cover free family contains a subset of size $\lceil |F|/r \rceil$ which is not contained in any other member of the family. Since $\mathcal{F}'$ is $(\lfloor d/2 \rfloor + 1)$-cover free, then any member $F' \in \mathcal{F}'$ contains a subset of size at most $\lceil \frac{2|F'|}{d} \rceil \leq \lceil \frac{4s}{d^2} \rceil$ which is not contained in any other member of the family. Hence, the size of $\mathcal{F}'$ is upper bounded by the number of distinct subsets of $[t]$ of size $\lceil \frac{4s}{d^2} \rceil$. Therefore,

$$|\mathcal{F}'| \leq \binom{t}{\lceil \frac{4s}{d^2} \rceil}. \tag{9}$$

By the upper bound on the binomial coefficient in (2), it follows

$$|\mathcal{F}'| \leq \left(\frac{etd^2}{4s} \right)^{\lceil \frac{4s}{d^2} \rceil}, \tag{10}$$

where e denotes the Neper's constant $e = 2,71828\ldots$.

The theorem is a consequence of the above inequality and of $|\mathcal{F}|$ being equal to $|\mathcal{F}'| + \lceil d/2 \rceil - 1$.   □

If we relax the constraint on the cardinality of the unions of d members of the familiy by setting $s = N$ in the upper bound of Theorem 4, we obtain the best known lower bound $\Omega(\frac{d^2}{\log d} \log n)$ on the minimum cardinality of the ground set of a d-cover free family of size n.

The following corollary is a consequence of Theorems 3 and 4.

Corollary 1. *Let d, s and t be positive integers. The maximal size of a $\cup_{\leq s}$ $\bar{d}$-separable family on the ground set $[t]$ is*

$$n_{sep}(d, \cup_{\leq s}, t) \leq \left(\frac{et(d-1)^2}{4s} \right)^{\lceil \frac{4s}{(d-1)^2} \rceil} + \left\lceil \frac{d-1}{2} \right\rceil - 1,$$

where e denotes the Neper's constant $e = 2,71828\ldots$.

Lemma 1 and Corollary 1 imply the following lower bound on $\tilde{y}(d,n,t)$.

Theorem 5. *Let t, n, d be positive integers with $n \geq d \geq 2$. It holds*

$$\tilde{y}(d,n,t) \geq \max \left\{ d, \frac{(d-1)^2}{4} \left(\frac{\log(n - d/2 + 1/2)}{\log(\frac{et(d-1)^2}{4\tilde{y}(d,n,t)})} - 1 \right) \right\}.$$

3.2   Two Almost Optimal Non Adaptive Algorithms

In this section we present two algorithms that asymptotically get very close to the lower bound of Theorem 5. The first one is based on a construction of d-cover families very similar to that of Hwang and Sós [19], while the second one exploits a breakthrough result of [23]. We remark that the results in this section translate into lower bounds on the size of $\cup_{\leq s}$ d-cover free families which are very close to the upper bound of Theorem 4. The underlying combinatorial structures of both algorithms consist of families in which any two members share at most a certain number λ of elements. The following simple lemma will be used in the analysis of both algorithms. It can be easily proved by induction.

Lemma 2. *Let d and λ be two positive integers and let $\mathcal{F}$ be a family of sets with $|\mathcal{F}| \geq d$ and such that any two members $F_1, F_2 \in \mathcal{F}$ intersect in at most λ elements. Then, for any d members $F_1, \ldots, F_d$ of $\mathcal{F}$, it holds $|\bigcup_{i=1}^d F_i| \geq \sum_{i=1}^d |F_i| - \frac{1}{2}d(d-1)\lambda$.*

The First Algorithm. This construction builds a k-uniform family in which any two members intersect in at most $\lfloor \frac{k-1}{d} \rfloor$ elements. In any such family, the union of d members shares at most $d\lfloor \frac{k-1}{d} \rfloor < k$ elements with any other member of the family, and consequently, the union of any d members cannot contain any other member. The idea is to start with the collection S_k of all subsets of k elements of $[t]$, where the relationship between k and t is going to be established later. Let us assume that k is a multiple of d, i.e., $k = vd$, for some integer constant $v > 1$. The algorithm builds a family $\{F_1, \ldots, F_n\}$ whose members intersect in less than $v = \frac{k}{d}$ elements, as follows. At step 1, the algorithm selects F_1 as an arbitrary subset of S_k and discards from S_k all subsets that intersect F_1 in at least v elements. At step 2, it chooses F_2 arbitrarily among the k-subsets that have not been discarded at step 1, and gets rid of all k-subsets that intersect F_2 in at least v elements. At a generic step i, it chooses F_i arbitrarily among the k-subsets that have not been discarded at the previous steps, and remove all k-subsets that intersect F_i in at least v elements. In such a way, the algorithm selects only subsets whose pairwise intersections have size at most $v - 1$. The cardinality n of the resulting family depends on the number of steps that the algorithm can perform before no subset is left in S_k.

Our analysis follows in part that of Hwang and Sòs but we do not fix the parameter k since we aim at deriving an upper bound on kd which is an upper bound on the size of the union of any d members of the family. We remark that kd is not a loose estimate of the size of this union. Indeed, any two members of the family intersect in at most $v-1$ elements, and therefore, by Lemma 2 it holds $|\bigcup_{i=1}^d F_{j_i}| \geq \sum_{i=1}^d |F_{j_i}| - \frac{1}{2}d(d-1)(v-1)$. Since the family is k-uniform and $v = \frac{k}{d}$, the above inequality implies $|\bigcup_{i=1}^d F_{j_i}| \geq kd - \frac{1}{2}(k/d-1)d(d-1) \geq kd/2+k/2$. Notice that at each step the algorithm discards at most $\sum_{i=v}^k \binom{k}{i}\binom{t-k}{k-i}$ subsets from S_k, and consequently the number of sets selected by the algorithm is

$$n \geq \frac{\binom{t}{k}}{\sum_{i=v}^k \binom{k}{i}\binom{t-k}{k-i}}. \tag{11}$$

Let $b_i = \sum_{i=v}^{k} \binom{k}{i}\binom{t-k}{k-i}$ and let us estimate the ratio $\frac{b_{i+1}}{b_i}$. One has

$$\frac{b_{i+1}}{b_i} = \frac{(k-i)^2}{(i+1)(t-2k+i+1)} \tag{12}$$

The righthand side of (12) decreases with i and therefore, for $q < v \leq i$, it holds

$$\frac{b_{i+1}}{b_i} \leq \frac{(k-q)^2}{(1+q)(t-2k+q+1)} \leq \frac{(k-q)^2}{q(t-2k+q)}. \tag{13}$$

Let us set $a = \frac{(k-q)^2}{q(t-2k+q)}$. From inequality (13) it follows that $b_{i+1} < b_q a^{i-q+1}$ which, along with inequality (11), implies $n \geq \frac{\binom{t}{k}}{b_q a^{-q}\sum_{i=v}^{k} a^i} = \frac{\binom{t}{k}(1-a)}{b_q a^{v-q}(1-a^{k-v+1})}$.

Since $b_q < \sum_{i=1}^{t} b_i = \binom{t}{k}$, then the above inequality implies

$$n > \frac{(1-a)}{a^{v-q}(1-a^{k-v+1})} \geq \frac{(1-a)}{a^{v-q}}. \tag{14}$$

Let us set $q = pv$, with $0 < p < 1$ being a constant, in the expression of a. It holds

$$a = \frac{(k-pv)^2}{pv(t-2k+pv)} = \frac{(d-p)^2}{p(t/v-2k/v+p)} = \frac{k(d-p)^2}{p(td-2kd+kp)} \leq \frac{dk}{p(t-2k)}.$$

If we assume $t \geq 2k(\frac{d}{p}+1)$, then it holds $a \leq \frac{1}{2}$, and consequently, inequality (14) implies

$$n > \frac{1}{a^{v-q-1}} > \left(\frac{p(t-2k)}{dk}\right)^{\frac{k}{d}(1-p)-1}.$$

Applying the base 2 logarithm, we obtain $\log n > (\frac{k}{d}(1-p)-1)\log\left(\frac{p(t-2k)}{dk}\right)$, from which

$$k \leq \frac{d}{1-p}\left(\frac{\log n}{\log\left(\frac{p(t-2k)}{dk}\right)}+1\right). \tag{15}$$

Since kd is an upper bound on the cardinality of the union of any d members of the family then it follows

$$\tilde{y}_A(d,n,t) = O\left(\frac{d^2\log n}{\log\left(\frac{t-2k}{dk}\right)}\right).$$

On the other hand, we proved that, for any d members $F_{j_1},\ldots,F_{j_d}$ of the family, it holds $|\bigcup_{i=1}^{d} F_{j_i}| \geq kd/2 + k/2$. This implies $dk \leq 2|\bigcup_{i=1}^{d} F_{j_i}| - k$, and consequently,

$$\tilde{y}_A(d,n,t) = O\left(\frac{d^2\log n}{\log\left(\frac{t-2k}{2\tilde{y}_A(d,n,t)-k}\right)}\right). \tag{16}$$

Since we are assuming $t \geq 2k(\frac{d}{p}+1)$, then $t-2k \geq t(1-\frac{p}{d+p}) > t(\frac{d-1}{d})$. Hence, inequality (16) implies the following theorem.

Theorem 6. *Let t, n, d be positive integers with $n \geq d \geq 1$. There exists a non adaptive group testing strategy $\mathcal{A}$ for which*

$$\tilde{y}_{\mathcal{A}}(d, n, t) = O \left(\frac{d^2 \log n}{\log \left(\frac{t}{\tilde{y}_{\mathcal{A}}(d,n,t)} \right)} \right).$$

The Second Algorithm. In this section we present a second construction of $\cup_{\leq s}$ d-cover free families that obtains the same bound of Theorem 6. An interesting feature of this construction is that it is an explicit construction. It is based on a breakthrough result by Porat and Rothschild [23] which provides the first deterministic explicit construction of error correcting codes meeting the Gilbert-Varshamov bound. In fact, the result in [23] provides a construction for $[m, k, \delta m]_q$-linear codes. We recall that an $[m, k, \delta m]_q$-linear code is a q-ary code over the alphabet $\mathbb{F}_q$ with length m, size $n = q^k$ and Hamming distance equal to δm. In the following, we denote by $H_q(p)$ the q-ary entropy function $H_q(p) = p \log_q \frac{q-1}{p} + (1-p) \log_q \frac{1}{1-p}$. Porat and Rothschild proved the following

Theorem 7. *[23] Let q be a prime power, m and k positive integers, and $\delta \in [0,1]$. If $k \leq (1 - H_q(\delta))m$, then it is possible to construct an $[m, k, \delta m]_q$-linear code in time $\Theta(mq^k)$.*

In [23], Porat and Rothschild show how to construct an (n, r)-strongly selective family [6] from a linear code with properly chosen parameters and then exploit the above mentioned theorem to construct in time $\Theta(rn \ln n)$ a linear code that can be reduced to an (n, r)-*strongly selective* family of size $\Theta(r^2 \ln n)$. We just mention that an (n, r)-strongly selective family is a combinatorial structure which is essentially equivalent to an $(r - 1)$-cover free family. The following theorem rephrases the result in [23] in terms of cover free families.

Theorem 8. *If there exists an $[m, k, \delta m]_q$-linear code then it is possible to construct an m-uniform $(\lceil \frac{1}{1-\delta} \rceil - 1)$-cover free family of size $n = q^k$ on the ground set $[mq]$.*

Proof. Given an $[m, k, \delta m]_q$-linear code $\mathcal{C} = \{\mathbf{c}_1, \ldots, \mathbf{c}_n\}$, let us define the family $\mathcal{F}$ as $\mathcal{F} = \{F(\mathbf{c}_1), \ldots, F(\mathbf{c}_n)\}$, where $F(\mathbf{c}_j) = \{f(i, a) : (i, a) \in [m] \times [q], \mathbf{c}_j[i] = a\}$, with f being an injection from $[m] \times [q]$ to $[mq]$. It is immediate to see that $\mathcal{F}$ is m-uniform in that for each index $i \in [m]$ there is a unique pair $(i, a) \in [m] \times [q]$ such that $\mathbf{c}_j[i] = a$. It is also possible to verify that any two members of $\mathcal{F}$ intersect in at most $m - \delta m$ elements. Indeed, for any two distinct words $\mathbf{c}_j, \mathbf{c}_\ell \in \mathcal{C}$ there are at least δm indices $i \in [m]$ such that $\mathbf{c}_\ell(i) \neq \mathbf{c}_j(i)$. This implies that there are at least δm pairs $(i, a) \in [m] \times [q]$ such that $f(i, a) \in F(\mathbf{c}_j)$ and $f(i, a) \notin F(\mathbf{c}_\ell)$, and consequently, $F(\mathbf{c}_j)$ and $F(\mathbf{c}_\ell)$ share at most $m - \delta m$ elements. It follows that the union of any $\lceil \frac{1}{1-\delta} \rceil - 1$ members of $\mathcal{F}$ shares at most $m - 1$ elements with any other member of the family, implying that $\mathcal{F}$ is $(\lceil \frac{1}{1-\delta} \rceil - 1)$-cover free. $\qquad \square$

Theorem 9. *Let t, n, d be positive integers with $n \geq d \geq 1$. There exists a non adaptive group testing strategy $\mathcal{A}$ for which*

$$\tilde{y}_{\mathcal{A}}(d, n, t) = \Theta\left(\frac{d^2 \ln n}{\ln\left(\frac{t}{\tilde{y}_{\mathcal{A}}(d,n,t)}\right)}\right).$$

The underlying family can be constructed in time $\Theta\left(\frac{dn \ln n}{\ln\left(\frac{t}{\tilde{y}_{\mathcal{A}}(d,n,t)}\right)}\right)$.

Proof. By Theorem 7 it is possible to construct an $[m, k, \delta m]_q$ linear code in time $\Theta(mq^k)$, where q is a prime power, m a positive integer, $\delta \in [0, 1]$ and $k = (1 - H_q(\delta))m$. Theorem 8 then implies that such a code can be transformed into an m-uniform $(\lceil \frac{1}{1-\delta} \rceil - 1)$-cover free family $\mathcal{F}$ on the ground set $[mq]$. Let us set $\delta = \frac{d}{d+1}$, and let $q > 2d + 1$. It holds

$$1 - H_q(\delta) = 1 - \left[\left(\frac{d}{d+1}\right) \log_q\left(\frac{(d+1)(q-1)}{d}\right) + \frac{1}{d+1} \log_q(d+1)\right]$$

$$= \frac{1}{(d+1)\ln q}\left[d\ln\left(\frac{q}{q-1}\right) - d\ln\left(\frac{d+1}{d}\right) + \ln\frac{q}{d+1}\right]. \quad (17)$$

We can exploit the well known relation $\ln\frac{z}{z-1} = \frac{1}{z} + o(\frac{1}{z})$, to estimate (17). Therefore, we get

$$1 - H_q(\delta) = \frac{1}{(d+1)\ln q}\left[\frac{d}{q} - \frac{d}{d+1} + \ln\left(\frac{q}{d+1}\right)\right] + o\left(\frac{1}{(d+1)\ln q}\right). \quad (18)$$

Since we are assuming $q > 2d+1$, then $c \cdot \ln\frac{q}{d+1} \leq \left[\left(\frac{d}{q}\right) - \left(\frac{d}{d+1}\right) + \ln\left(\frac{q}{d+1}\right)\right] < \ln\frac{q}{d+1}$, for any constant $c \leq 1/4$. Therefore,

$$1 - H_q(\delta) = \Theta\left(\left(\frac{1}{(d+1)\ln q}\right)\ln\left(\frac{q}{d+1}\right)\right). \quad (19)$$

It follows that

$$\log_q n = k = m(1 - H_q(\delta)) = \Theta\left(\frac{m}{(d+1)\ln q}\left(\ln\frac{q}{d+1}\right)\right). \quad (20)$$

By setting $s = dm$ and $t = mq$ in (20), we get

$$\ln n = \log_q n \ln q = \Theta\left(\frac{s}{d(d+1)}\left(\ln\frac{td}{s(d+1)}\right)\right) = \Theta\left(\frac{s}{d^2}\left(\ln\frac{t}{s}\right)\right). \quad (21)$$

The maximum number $\tilde{y}_{\mathcal{A}}(d, n, t)$ of positive responses admitted by the algorithm is equal to the maximum number of elements contained in the union of d members of the family. Since s is an upper bound on the size of the union of any d members of the family, one has that $\tilde{y}_{\mathcal{A}}(d, n, t) \leq s$. In the proof of Theorem 8 it has been shown that any two members of the family intersect in at most $m - \delta m$ elements. Hence, Lemma 2 implies $|\bigcup_{i=1}^{d} F_{j_i}| \geq md - \frac{1}{2}(m - \delta m)d(d-1) = s - s(d-1)/(2d+2) \geq s/2$, for any d members $F_{j_1}, \ldots, F_{j_d}$ of the family. Therefore, it holds $s/2 \leq \tilde{y}_{\mathcal{A}}(d, n, t) \leq s$, from which the theorem follows.   □

Acknowledgement. The author wishes to thank Prof. Ugo Vaccaro for the inspiring discussions.

References

1. Ahlswede, R., Deppe, C., Lebedev, V.: Threshold and majority group testing. In: Aydinian, H., Cicalese, F., Deppe, C. (eds.) Ahlswede Festschrift. LNCS, vol. 7777, pp. 488–508. Springer, Heidelberg (2013)
2. Alon, N., Asodi, V.: Learning a hidden subgraph. SIAM J. Discrete Math. **18**(4), 697–712 (2005)
3. Chen, H.B., De Bonis, A.: An almost optimal algorithm for generalized threshold group testing with inhibitors. J. Comp. Biol. **18**, 851–864 (2011)
4. Chin, F.Y.L., Leung, H.C.M., Yiu, S.M.: Non-adaptive complex group testing with multiple positive sets. Theoret. Comput. Sci. **505**, 11–18 (2013)
5. Cicalese, F., Damaschke, P., Vaccaro, U.: Optimal group testing strategies with interval queries and their application to splice site detection. Int. J. Bioinform. Res. Appl. **1**(4), 363–388 (2005)
6. Clementi, A.E.F., Monti, A., Silvestri, R.: Selective families, superimposed codes, and broadcasting on unknown radio networks. In: Twelfth Annual ACM-SIAM Symposium on Discrete Algorithms, pp. 709–718 (2001)
7. Damaschke, P.: Randomized group testing for mutually obscuring defectives. Inf. Process. Lett. **67**, 131–135 (1998)
8. Damaschke, P., Sheikh Muhammad, A., Triesch, E.: Two new perspectives on multi-stage group testing. Algorithmica **67**(3), 324–354 (2013)
9. De Bonis, A., Gąsieniec, L., Vaccaro, U.: Optimal two-stage algorithms for group testing problems. SIAM J. Comput. **34**(5), 1253–1270 (2005)
10. De Bonis, A., Vaccaro, U.: Constructions of generalized superimposed codes with applications to group testing and conflict resolution in multiple access channels. Theoret. Comput. Sci. **306**, 223–243 (2003)
11. De Bonis, A., Vaccaro, U.: Optimal algorithms for two group testing problems and new bounds on generalized superimposed codes. IEEE Trans. Inf. Theory **10**, 4673–4680 (2006)
12. Dorfman, R.: The detection of defective members of large populations. Ann. Math. Statist. **14**, 436–440 (1943)
13. Du, D.Z., Hwang, F.K.: Combinatorial Group Testing and Its Applications. World Scientific, River Edge (2000)
14. Du, D.Z., Hwang, F.K.: Pooling design and nonadaptive group testing. Series Appl. Math. (World Scientific), vol. 18 (2006)
15. Dyachkov, A.G., Rykov, V.V.: A survey of superimposed code theory. Probl. Control Inform. Theory **12**, 229–242 (1983)
16. Erdös, P., Frankl, P., Füredi, Z.: Families of finite sets in which no set is covered by the union of r others. Israel J. Math. **51**, 75–89 (1985)

17. Harvey, N.J.A., Patrascu, M., Wen, Y., Yekhanin, S., Chan, V.W.S.: Non-adaptive fault diagnosis for all-optical networks via combinatorial group testing on graphs. In: 26th IEEE International Conference on Computer Communications, pp. 697–705 (2007)

18. Hong, E.S., Ladner, R.E.: Group testing for image compression. IEEE Trans. Image Process. 11(8), 901–911 (2002)

19. Hwang, F.K., Sós, V.T.: Non adaptive hypergeometric group testing. Studia Sc. Math. Hung. 22, 257–263 (1987)

20. Kautz, W.H., Singleton, R.C.: Nonrandom binary superimposed codes. IEEE Trans. Inf. Theory 10, 363–377 (1964)

21. Li, C.H.: A sequential method for screening experimental variables. J. Amer. Statist. Assoc. 57, 455–477 (1962)

22. Lo, C., Liu, M., Lynch, J.P., Gilbert, A.C.: Efficient sensor fault detection using combinatorial group testing. In: 2013 IEEE International Conference on Distributed Computing in Sensor Systems, pp. 199–206 (2013)

23. Porat, E., Rothschild, A.: Explicit non adaptive combinatorial group testing schemes. IEEE Trans. Inf. Theory 57(12), 7982–7989 (2011)

24. Ruszinkó, M.: On the upper bound of the size of the r-cover-free families. J. Combin. Theory Ser. A 66, 302–310 (1994)

25. Sobel, M., Groll, P.A.: Group testing to eliminate efficiently all defectives in a binomial sample. Bell Syst. Tech. J. 38, 1179–1252 (1959)

26. Pasternack, B.S., Bohnin, D.E., Thomas, J.: Group-sequential leak-testing of sealed radium sources. Technometrics 18(1), 59–66 (1975)

27. Thomas, J., Pasternack, B.S., Vacirca, S.J., Thompson, D.L.: Application of group testing procedures in radiological health. Health Phys. 25, 259–266 (1973)

28. Wang, F., Du, H.D., Jia, X., Deng, P., Wu, W., MacCallum, D.: Non-unique probe selection and group testing. Theoret. Comput. Sci. 381, 29–32 (2007)

29. Wolf, J.: Born again group testing: multiaccess communications. IEEE Trans. Inf. Theory 31, 185–191 (1985)

Maximizing Revenues for On-Line Dial-a-Ride

Ananya Christman[1](✉) and William Forcier[2]

[1] Middlebury College, Middlebury, VT 05753, USA
achristman@middlebury.edu
[2] Abbott Laboratories, Lake Forest, IL 60045, USA
william.forcier@abbott.com

Abstract. We consider the On-Line Dial-a-Ride Problem, where a server fulfills requests that arrive over time. Each request has a source, destination, and release time. We study a variation of this problem where each request also has a revenue that the server earns for fulfilling the request. The goal is to serve requests within a time limit while maximizing the total revenue. We first prove that no deterministic online algorithm can be competitive unless the input graph is complete and edge weights are unit. We therefore focus on these graphs and present a 2-competitive algorithm for this problem. We also consider two variations of this problem: (1) the input graph is complete bipartite and (2) there is a single node that is the source for every request, and present a 1-competitive algorithm for the former and an optimal algorithm for the latter.

Keywords: Online algorithms · Dial-a-ride · Competitive analysis · Graphs

1 Introduction

In the On-Line Dial-a-Ride Problem (OLDARP), a server travels in some metric space to serve requests for rides. The server has a *capacity* that specifies the maximum number of requests it can serve at any time. The server starts at a designated location of the space, the *origin*, and moves along the space to serve requests. Requests arrive dynamically and each request specifies a *source*, which is the pick-up (or start) location of the ride, a *destination*, which is the delivery (or end) location, and the release time of the request, which is the earliest time the request may be served. For each request, the server must decide whether to serve the request and at what time, with the goal of meeting some optimality criterion. In many cases preemption is not allowed, so if the server decides to serve a request, it must do so until completion. On-Line Dial-a-Ride Problems have many practical applications in settings where a vehicle (or multiple vehicles) is dispatched to satisfy requests involving pick-up and delivery of people or goods. Important examples include ambulance routing, transportation for the elderly and disabled, taxi services, and courier services.

© Springer International Publishing Switzerland 2014
Z. Zhang et al. (Eds.): COCOA 2014, LNCS 8881, pp. 522–534, 2014.
DOI: 10.1007/978-3-319-12691-3_38

In the version of OLDARP that we consider, there is a global time limit such that requests must be served before this time, and each request has an associated revenue, which is the amount earned by the server for serving the request. In the context of the applications described above, the revenue can represent the priority levels of the requests. We assume that the server can serve at most one request at a time (in other words the server has unit capacity). Such is the case for ambulance routing and many taxi services. The goal is to serve requests within the time limit so as to maximize the total revenue. We assume preemption is not allowed, therefore serving a particular request may prevent the server from serving a higher revenue request that arrives later. We refer to this problem as ROLDARP (Revenue On-Line Dial-A-Ride Problem) and describe it formally in Sect. 3.

For ROLDARP, the metric space is modeled by a graph of nodes and weighted edges, where edge weights represent the travel times between nodes. We present a 2-competitive algorithm to solve this problem for complete graphs where edge weights are unit. While our algorithm applies to only these graphs, we show that no competitive algorithm exists if we relax either of these graph properties. Specifically, we prove that no deterministic online algorithm can be competitive for complete graphs with varying edge weights nor for non-complete graphs. We also consider a variation of this problem where the input graph is complete bipartite with unit-weight edges, and where every source is from the left-hand side and every destination is from the right-hand side. For these graphs, we present a 1-competitive algorithm. We analyze our algorithms based on their *competitive ratio*, i.e. the worst-case ratio between the revenue earned by the algorithm and the revenue earned by an optimal offline algorithm that knows all of the requests in advance. Although the competitive ratios of both algorithms include an additive constant equal to the last request served by the optimal algorithm, we show that no online algorithm can avoid this constant. Finally, we consider a version of ROLDARP where there is a single node that is the source of all requests and present an optimal algorithm for this problem.

The remainder of this paper is organized as follows. In Sect. 2 we describe several works related to our problem. In Sect. 3 we formally define ROLDARP and describe the variations: V-ROLDARP (where edge weights are varying), complete-bipartite ROLDARP, and single-source ROLDARP. In Sect. 4 we prove that no deterministic online algorithm can be competitive for V-ROLDARP. In Sect. 5 we present a 2-competitive algorithm for ROLDARP. In Sect. 6.1 we present a 1-competitive algorithm for complete bipartite ROLDARP and in Sect. 6.2 we present an optimal algorithm for single-source ROLDARP. Finally in Sect. 7 we summarize our results and discuss some possible extensions.

2   Related Work

Several variations of the On-Line Dial-a-Ride Problem have been studied in the past. The authors of [10] studied the problem for the unit metric space with two different objectives. One is to minimize the time to serve all requests and return to the origin (also known as *completion time*); the other is to minimize the average completion time of the requests (also known as *latency*). For minimizing completion time, they showed that any deterministic algorithm must have competitive ratio of at least 2 regardless of the server capacity. They presented algorithms for the cases of finite and infinite capacity with competitive ratios of 2.5 and 2, respectively. For minimizing latency, they proved that any algorithm must have a competitive ratio of at least 3. They presented a 15-competitive algorithm for this problem on the real line in which the server has infinite capacity.

The authors of [1] studied minimizing total completion time for OLDARP with multiple servers and capacity constraints and present a 2-competitive algorithm for this problem. For the version of the problem with one server and no capacity constraints, they presented a $1 + \sqrt{(1 + 8\rho)}/2$-competitive algorithm, where ρ is the approximation ratio of a related offline problem.

The work in [9] considered a modified version of OLDARP where at the release time of a request only the source location is revealed. The destination location is revealed only when the server arrives at the source. Such a setting is appropriate for applications such as elevator scheduling or ride scheduling for taxis. The authors proved that for minimizing completion time, when preemption is allowed (i.e. the server is allowed to halt a ride at any time and possibly proceed with it later) any deterministic algorithm must have competitive ratio of at least 3. They also gave a 3-competitive algorithm to solve this problem.

The work in [7] considered a version of OLDARP where each request consists of one or more locations, and precedence and capacity requirements for the locations. To serve a request the server must visit each location, while satisfying the requirements. The authors provided a non-polynomial 2-competitive algorithm for minimizing completion time.

The authors of [2] considered a version of the problem where each request consists of a single location and a release time. This is referred to as the On-Line Traveling Salesman Problem (OLTSP). The server starts at an origin and must serve all requests while minimizing the overall completion time. The authors studied this problem on the Euclidean space and proved that no online algorithm can be better than 2-competitive. They gave a 2.5-competitive non-polynomial time online algorithm and a 3-competitive polynomial time algorithm to solve this problem.

The authors of [3] also aimed to minimize completion times for OLTSP but considered an asymmetric network where the distance from one point to another may differ in the inverse direction. They considered two versions of the problem: *homing*, where the server is required to return to the origin after completing the requests, and *nomadic*, where there is no such requirement. They presented a non-polynomial $(3 + \sqrt{5})/2$-competitive algorithm for the homing version and proved that this is the best possible. For the nomadic version, they proved that no general online competitive algorithm can exist for this problem; instead the competitive ratio for any online algorithm must depend on the asymmetry of the space.

The work in [5] considers a version of OLTSP where edge costs change over time. They prove upper and lower bounds on the competitive ratio and provide a competitive ratio that is based on the minimum and maximum edge costs. However, their competitive algorithms require a solution to a related NP-hard problem.

The authors of [8] studied both OLDARP and OLTSP for the uniform metric space. Their objective is to minimize the maximum *flow time*, the difference between a request's release and service times. They proved that no competitive algorithm exists for OLDARP and gave a 2-competitive algorithm to solve OLTSP.

More recently, the authors of [6] considered a variation of OLTSP where each request also has a penalty (incurred if the request is rejected). The goal is to minimize the time to serve all accepted requests plus the sum of the penalties of rejected requests. They gave a 2-competitive algorithm to solve the problem on the real line and a 2.28-competitive algorithm on a general metric space.

The authors of [4] studied a variation of OLTSP where each request has both a penalty and a weight. The goal is to collect a specified quota of weights by satisfying a sufficient number of requests, while minimizing the total service time plus the penalties of rejected requests. They gave a 7/3-competitive algorithm for this problem on a general graph and proved lower and upper bounds of 1.89 and 2, respectively, on the real halfline.

To our knowledge, this is the first work that studies OLDARP with the goal of maximizing total revenue within a time limit. The related works of [1,7–10] study OLDARP, however none consider requests with revenues. Although in the work of [4], requests have revenues (their paper refers to them as "weights"), each request has only one location, whereas requests for our problem consist of both a source and a destination.

3   Problem Statement

In the basic form of the *On-Line Dial-a-Ride Problem* (OLDARP), a network server receives requests dynamically and each request has a source, destination, and release time. The server starts at a predefined origin location and serves a request by picking up at the source and delivering at the destination. The server can serve only one request at a time and cannot serve a request prior to its release time. We consider a version of the problem where there is a given time limit by which the server must serve all requests and every request has a revenue that the server earns for fulfilling the request. The goal is to serve requests within the time limit so as to maximize the total revenue.

We study competitive algorithms for this problem and use standard terminology from competitive analysis. An algorithm ON is considered *online* if it learns about a request only at its release time, whereas an algorithm is considered offline if it is aware of all requests at time 0 (i.e. the earliest time). We let OPT denote the optimal offline algorithm, as in the algorithm that given any input will earn the greatest revenue of any other algorithm on that input.

Given an input graph G, a sequence $\sigma = r_1, \ldots r_m$ of requests and an algorithm ALG, we denote $\text{ALG}(G, \sigma)$ as the total revenue earned by ALG from σ on G. We say that ON is *c-competitive* if there exists $c > 0, b \geq 0$ such that for all σ:

$$\text{OPT}(G, \sigma) \leq c \cdot \text{ON}(G, \sigma) + b \tag{1}$$

The input to the problem is an undirected complete graph $G = (V, E)$ where V is the set of vertices (or nodes) and $E = \{(u, v) : u, v \in V, u \neq v\}$ is the set of edges. For every edge $(u, v) \in E$, there is a weight $w_{u,v} > 0$. If for every edge $w_{u,v} = 1$ (i.e. the graph represents the unit metric space), we refer to the problem as ROLDARP. If edge weights are varying, we refer to the problem as V-ROLDARP. One node in the graph, o, is designated as the origin and is where the server is initially located (i.e. at time 0). The input also includes a time limit T and a sequence of requests, σ, that is dynamically issued to the server. Each request is of the form (s, d, t, r) where s is the source node, d is the destination, t is the time the request is released, and r is the revenue earned by the server for serving the request. To serve a request, the server must move from its current location x to s, then from s to d. The total time is equal to the length of the path from x to d. We assume the earliest time a request may be released is at $t = 0$. For each request, the server must decide whether to serve the request and if so, at what time. A request may not be served earlier than its release time and at most one request may be served at any given time. Once the server starts serving a request, it must serve the request until completion (i.e. preemption is not allowed). The goal for the server is to serve requests within the time limit so as to maximize the total earned revenue. As a preprocessing step, we can remove any edge (u, v) such that $w_{u,v} > T$, since no algorithm (either online or offline) can use this edge to serve a request.

We consider several variations of ROLDARP which we summarize below:

- original ROLDARP - The graph is a complete undirected graph where every edge has unit weight. Each request has a source, destination, release date, and revenue that is earned for serving the request. There is a global time limit before which requests must be served. The goal is to maximize the total revenue earned within the time limit.
- V-ROLDARP - Edges in the graph have varying weights.
- complete-bipartite ROLDARP - The graph is an undirected complete-bipartite graph where every source is from the left-hand side and every destination is from the right-hand side of the graph.
- single-source ROLDARP - One vertex is the source for every request.

4   Non-competitiveness of V-ROLDARP

We first consider V-ROLDARP. The input to this problem is a complete undirected graph G of $n \geq 2$ nodes where for every edge (u, v) there is a weight $w_{u,v} > 0$, and there are at least two distinct edges (u, v) and (x, y) where $w_{u,v} \neq w_{x,y}$. In this section, we prove that no deterministic online algorithm can be competitive for V-ROLDARP. Note that any connected graph can be converted to a complete graph such that the pairwise distance between nodes of both graphs is equivalent. Specifically, for two non-adjacent nodes i and j of a non-complete graph, we can create an edge (i, j) with weight equal to the distance between i and j. Therefore this proof also holds for connected non-complete graphs.

In Sect. 5 we consider ROLDARP and provide an algorithm that is 2-competitive when the additive constant b from Eq. (1) is the revenue of the last request served by OPT, v_{last}. In this section, we prove that no algorithm can be competitive for V-ROLDARP for any $b \leq v_{last}{}^1$). In particular, we first show that there is no value b independent of the input such that a deterministic online algorithm can be c-competitive. We then show that even if we allow b to depend on the input, no deterministic online algorithm can be c-competitive with $b \leq v_{last}$. Specifically, we show that an adversary can issue a request sequence σ such that for any $\alpha \geq 1, b \leq v_{last}$, $\alpha \cdot \text{ON}(G, \sigma) + b < \text{OPT}(G, \sigma)$ where G is any input graph as described above.

Theorem 1. *No deterministic online algorithm can be competitive for V-ROLDARP.*

Proof. We first show that there is no value b independent of the input such that a deterministic online algorithm can be competitive. In other words, the additive constant b in Eq. (1) must depend on the input (in particular, the last request served by OPT, v_{last}). Let $T \geq 2$; for all time units before $T - 1$, the adversary will release no requests. At time $T - 1$ a deterministic online algorithm must be at some node u of G. The adversary can release a request $(w, u, T - 1, v_{last})$, so the request sequence σ consists only of this request. This request cannot be served by the online algorithm, but it can be served by the optimal algorithm, so the online algorithm earns 0 while the optimal algorithm earns v_{last}. Since we can set $v_{last} = b + 1$, for any b, $\text{OPT}(G, \sigma) > \text{ON}(G, \sigma) + b$. This also shows that no deterministic online algorithm can be competitive when $b < v_{last}$.

We now show that even if we allow an additive constant of $b = v_{last}$, still no online algorithm can be competitive. Specifically we show that an adversary can issue a request sequence σ such that for any $\alpha \geq 1, b \leq v_{last}$, $\alpha \cdot \text{ON}(G, \sigma) + b < \text{OPT}(G, \sigma)$.

Let G denote a complete graph with three nodes s, x, and y, where s is the origin, $w_{s,x} = c$ for any $c \geq 3$, $w_{s,y} = 1$ and $w_{x,y} > c$, and let $T \geq 2c$ denote the time limit. The adversary will release two requests: $(s, x, T - 2c, v)$ and $(x, s, T - c, v)$ for $v > 0$. There are two cases for ON:

Case 1: ON serves neither of these requests.
Since there is enough time for OPT to serve both requests, $\text{OPT}(G, \sigma) = 2v$ while $\text{ON}(G, \sigma) = 0$. If $b = v_{last}$ we have $\alpha \cdot 0 + v_{last} = v \leq 2v$ for any α, so ON is not competitive.

---

[1] Note that since v_{last} can be at most the maximum allowed revenue, this proof shows non-competitiveness for b equal to all possible revenue values.

Case 2: ON serves at least one of these requests.

ON earns revenue at most $2v$. Suppose ON starts serving a request at time t. Then the adversary will release two requests $(s, y, t + 1, v^*)$ and $(y, s, t + 1, v^*)$ where $v^* = \alpha \cdot 2v + 1$. ON will not have enough time to serve either of these requests but OPT will serve both of these requests and earn revenue $2v^* = v^* + v_{last} = \alpha \cdot 2v + 1 + v_{last}$. For $b = v_{last}$ we have $\alpha \cdot 2v + v_{last} \leq \alpha \cdot 2v + 1 + v_{last}$ for any α, so ON is not competitive.

5 ROLDARP on a Complete Graph with Unit Edges

In this section, we provide an online algorithm, Greatest Revenue First (GRF) that is 2-competitive for ROLDARP. Specifically, for input graph G, given request sequence σ, if $\text{OPT}(G, \sigma)$ denotes the optimal revenue earned from σ, $\text{GRF}(G, \sigma)$ denotes the amount of revenue earned by GRF from σ, and v_{last} denotes the last revenue earned by OPT, we show:

$$\text{OPT}(G, \sigma) \leq 2 \cdot \text{GRF}(G, \sigma) + v_{last} \qquad (2)$$

We first show that for any graph G with $n \geq 2$ nodes and a time limit $T \geq 2$ no online algorithm can avoid the v_{last} additive constant of Eq. (2). In particular an adversary can generate a request sequence such that no online algorithm can serve the last request of the sequence.

At time $T - 1$ any online algorithm must be at some node in G. Consider two arbitrary nodes u and v. If the algorithm is at u the adversary can release a request from v to u. If the algorithm is not at u the adversary can release a request from u to v. In either case the online algorithm cannot satisfy the request while an optimal algorithm can.

Algorithm 1.1 describes the GRF algorithm.[2] The main idea is that for every time unit for which there is some unserved request, GRF either moves to the source location of the request with the highest revenue or serves a previously issued request with the highest revenue.

Theorem 2. *Greatest Revenue First is 2-competitive.*

Proof. We now prove Eq. 2 to show that GRF is 2-competitive. We assume without loss of generality, that a request is issued at every time unit[3]. To prove Eq. 2, we consider another algorithm MAX. We define MAX such that at every time unit

---

[2] We note that there are at least two enhancements that can improve the performance of GRF without improving the competitive ratio: (1) In steps 2 and 7, instead of simply moving to the request that earns the greatest revenue, the algorithm can serve a request if there is one available while performing this move. (2) In steps 3 and 8, the algorithm can check if a request with higher revenue has been released since the previous step and if so, serve this request instead of r.

[3] If at any time t, there is no request issued, we can generate a "dummy" request of the form $(s, d, t, 0)$, where s and d are nodes in the input graph, since neither GRF nor any optimal algorithm would accept this request.

Algorithm 1.1. Algorithm GRF. Input is complete graph G and time limit T.

1: **if** T is even **then**
2:    At every even time, determine which released request earns the greatest revenue and move to the source location of this request. Denote this request as r. If no unserved requests exist, do nothing until the next even time.
3:    At every odd time, serve request r (if it exists) from the previous step.
4: **end if**
5: **if** T is odd **then**
6:    At time 0, do nothing.
7:    At every odd time, determine which released request earns the greatest revenue and move to the source location of this request. Denote this request as r. If no unserved requests exist, do nothing until the next odd time.
8:    At every even time, serve request r (if it exists) from the previous step.
9: **end if**

except $T - 1$, MAX serves the request with the greatest revenue regardless of the source node of the request. At time $T - 1$, MAX does nothing. Note that MAX may not coincide with the request set of a feasible algorithm. In other words, given the input graph, request sequence, and time limit, MAX may fulfill a set of requests that *no* algorithm can fulfill. For example, suppose the origin is some node o. Suppose at time 0, a request with maximal revenue is released with source $s_0 \neq o$ and destination d_0. No algorithm can serve this request in the time slot from 0 to 1, but we assume that MAX does. Thus, by the construction of MAX, for any sequence of requests σ, the following equation holds:

$$\text{MAX}(G, \sigma) \geq \text{OPT}(G, \sigma) - v_{last} \qquad (3)$$

We will show that:

$$2 \cdot \text{GRF}(G, \sigma) \geq \text{MAX}(G, \sigma) \qquad (4)$$

A proof of (4) will immediately prove (2).

We now prove Eq. (4). For this proof we use the terminology "algorithm A *serves* (or *has served*) request r at time t" to indicate that A *begins* serving r at time t and *completes* serving r at time $t + 1$.

Let $r_0, r_1, r_2, \ldots r_{T-2}$ denote the requests served by MAX at times $0, 1, 2, \ldots$ $T - 2$ earning revenues $v_0, v_1, v_2, \ldots v_{T-2}$. We consider two cases based on the parity of T.

Case 1: T is even (Table 1 shows an example with $T = 6$).
At $t = 0$, GRF and MAX determine that v_0 is the greatest revenue. At $t = 0$ MAX fulfills v_0 and at $t = 1$ GRF fulfills v_0.

We now show that for every odd time $t \neq 1$, if MAX serves r_{t-1} and r_{t-2} at times $t - 1$ and $t - 2$, earning total revenue $v_{t-1} + v_{t-2}$, then at time t GRF earns revenue at least $\max\{v_{t-1}, v_{t-2}\}$. Note that when T is even GRF serves requests only at odd times. We show that at time $t - 1$, when GRF decides which request to serve at time t, both r_{t-1}, r_{t-2} are available requests (so GRF will serve the one with the higher revenue).

Consider r_{t-1} and r_{t-2}. Since MAX has served them at $t-1$ and $t-2$, they must have been released by times $t-1$ and $t-2$, respectively. The only way that they would not be available requests for GRF at time $t-1$ is if GRF has already served them. However, this is not possible because if GRF serves a request at some time τ, it must have been released by time $\tau - 1$, and therefore MAX must serve this request by time $\tau - 1$ at the latest.

Case 2: T is odd.
Omit the first paragraph and replace odd with even and even with odd in the proof for Case 1.

Now, let r'_t denote the request served by GRF at time t and let v'_t denote the revenue of r'_t. Then (assuming $v_t = 0$ for $t < 0$), for all times t in which GRF earns revenue:

$$v'_t \geq \max\{v_{t-1}, v_{t-2}\} \tag{5}$$

$$2 \cdot v'_t \geq v_{t-1} + v_{t-2} \tag{6}$$

Since Eq. (6) holds for all $r'_t \in \sigma$ served by GRF, we have:

$$2 \cdot \text{GRF}(G, \sigma) \geq \text{MAX}(G, \sigma) \tag{7}$$

Then from (3), we have

$$2 \cdot \text{GRF}(G, \sigma) \geq \text{OPT}(G, \sigma) - v_{last} \tag{8}$$

$$2 \cdot \text{GRF}(G, \sigma) + v_{last} \geq \text{OPT}(G, \sigma) \tag{9}$$

Table 1. An example of the revenues earned by GRF and MAX for $T = 6$ (GRF revenues given w.l.o.g.). GRF earns total revenue $v_{\text{GRF}} = v_0 + v_1 + v_3$ and MAX earns total revenue $v_{\text{MAX}} = v_0 + v_1 + v_2 + v_3 + v_4$. Since $v_1 \geq v_2$ and $v_3 \geq v_4$, we have $2 \cdot v_{\text{GRF}} \geq v_{\text{MAX}}$.

t	GRF	MAX
0	0	v_0
1	v_0	v_1
2	0	v_2
3	$v_1 = max(v_1, v_2)$	v_3
4	0	v_4
5	$v_3 = max(v_2, v_3, v_4)$	0

6   Variants of ROLDARP

6.1   Complete Bipartite ROLDARP

In this section, we consider ROLDARP for complete bipartite graphs, specifically where if V_1 and V_2 denote the two sets of nodes, then every source is in V_1 and every destination is in V_2. We prove that a modified version of Greatest Revenue First, BGRF (Bipartite Greatest Revenue First) is 1-competitive for this version of ROLDARP (see Algorithm 1.2).

Algorithm 1.2. Algorithm BGRF. Input is a complete bipartite graph G and time limit T.

1: **if** T is even **then**
2:　　At time 0, do nothing.
3:　　At time 1, move to any destination node (i.e. any node in V_2). Do nothing if already at a destination node.
4:　　At every even time, determine which released request earns the greatest revenue and move to the source location of this request. Denote this request as r. If no unserved requests exist, do nothing until the next even time.
5:　　At every odd time, serve request r (if it exists) from the previous step.
6: **end if**
7: **if** T is odd **then**
8:　　At time 0, move to any destination node (i.e. any node in V_2). Do nothing if already at a destination node.
9:　　At every odd time, determine which released request earns the greatest revenue and move to the source location of this request. Denote this request as r. If no unserved requests exist, do nothing until the next odd time.
10:　　At every even time, serve request r (if it exists) from the previous step.
11: **end if**

Proposition 1. *Algorithm* BGRF *is 1-competitive for Complete Bipartite ROLDARP.*

Proof. We will show that given request sequence σ and input graph G, if OPT(G, σ) denotes the optimal revenue earned from σ and BGRF(G, σ) denotes the amount of revenue earned by BGRF from σ, then:

$$\text{OPT}(G, \sigma) \leq \text{BGRF}(G, \sigma) + v_{last} \tag{10}$$

where v_{last} is the revenue of the last request served by OPT.

We now prove Eq. 10. We consider two cases based on the parity of T:

Case 1: T is odd.
As in Sect. 5, we consider another algorithm MAX such that for any optimal algorithm OPT, MAX serves all except the last request served by OPT, but in a different order. Specifically, for $t < T - 1$, when OPT serves a request at time t, MAX serves the request with the greatest revenue that has been released by time $t + 1$ that OPT eventually serves. In other words, out of all the requests released by $t + 1$ that OPT serves, MAX serves the one with the greatest revenue. Note that for any request sequence σ:

$$\text{MAX}(G, \sigma) = \text{OPT}(G, \sigma) - v_{last} \tag{11}$$

where v_{last} is the last revenue earned by OPT.

We show that every time MAX earns a revenue, BGRF earns at least as much revenue two time units later. We assume without loss of generality that there are enough requests such that at every time unit, MAX and BGRF have a request to serve. Note that any algorithm requires at least two time units to serve each request after the first request — the first time unit for moving to the source of the request and the second time unit for moving to the destination. Therefore MAX and BGRF serve requests at every other time unit. Specifically, MAX serves at every even t for $t < T - 1$ and BGRF serves at every even t except $t = 0$.

Assume at some time t MAX serves r^* and earns revenue v^*. We show by contradiction that r^* must be an available request for BGRF to serve at $t + 2$. Since r^* must have been released by $t + 1$, it must be available for BGRF to serve at $t + 2$ unless BGRF has already served it prior to $t + 1$. Suppose that BGRF served it prior to $t + 1$ at some time t'. Then r^* must have been the highest revenue request released by t'. But this implies that MAX would have served r^* by t' which is a contradiction since MAX serves r^* at time $t \geq t'$.

Thus by every odd time $t = 1, 3, 5, \ldots T - 2$, MAX has served one more request than BGRF and each request BGRF serves earns at least as much revenue as the request served by MAX two time units earlier. Finally, at $T - 1$, BGRF will serve the last request that MAX serves. So for any request sequence σ, $\mathrm{BGRF}(G, \sigma) \geq \mathrm{MAX}(G, \sigma)$. Combining this with Eq. 11 proves Eq. 10.

Case 2: T is even.
A few modifications to the odd case will prove the even case. For the even case, we consider a modified version of OPT, $\overline{\mathrm{OPT}}$, which we describe below.

First note that any optimal algorithm requires one time slot to serve the first request and two time slots to serve every additional request, so in total, an odd number of time slots. Therefore, if T is even, one time slot will serve no purpose so the algorithm can simply wait for more requests to be released during this time slot. The optimal choice is to use the first time slot (i.e. $t = 0$ to $t = 1$) to wait as this allows the maximum number of requests to be released from which the algorithm can choose. Therefore, any optimal algorithm that does not wait during the first time slot can be converted to an equivalent algorithm that does. Specifically if OPT is an optimal algorithm that waits after the first time slot, we can convert OPT to an equivalent algorithm $\overline{\mathrm{OPT}}$ by shifting the wait to the first time slot and then performing the remaining moves of OPT in the same order as OPT. Now both $\overline{\mathrm{OPT}}$ and BGRF wait during the first time slot. As in the odd case, we consider an algorithm MAX that serves the same requests as $\overline{\mathrm{OPT}}$ (instead of OPT). Now we can apply the proof for the odd case by simply replacing odd for even and even for odd.

6.2   Single-Source ROLDARP

In this section, we consider a modified version of ROLDARP where there is a single source node, S, which is the source of every request and within unit distance of every other node in the graph. We present an algorithm SGRF (Single Greatest Revenue First) that is optimal for this problem (see Algorithm 1.3).

Algorithm 1.3. Algorithm SGRF. Input is graph G, time limit T, and the source node S.

1: **if** T is even **then**
2:    At every odd time, serve the request with the greatest revenue.
3:    At every even time, move to S.
4: **end if**
5: **if** T is odd **then**
6:    At every even time, serve the request with the greatest revenue.
7:    At every odd time, move to S.
8: **end if**

Proposition 2. *Algorithm* SGRF *is equivalent to an optimal offline solution for Single-Source ROLDARP.*

Proof. We can assume without loss of generality, that the origin is S: since *any* algorithm will need to move to S to serve any request, an instance of the problem where the origin is not S is equivalent to an instance where the origin is S and T is decremented by 1.

We consider an optimal offline algorithm OPT and prove by way of contradiction that the set of requests served by SGRF earns as much revenue as the set of requests served by OPT.

Let R denote the set of requests served by OPT, where r_i denotes a request with revenue v_i. We consider a set R^* that is a rearrangement of the requests from R. Let r_j and r_k denote two requests served by OPT where both r_j and r_k were released by some time t. Then in R^*, r_j and r_k will be ordered such that the request with the higher revenue appears first, i.e. r_j followed by r_k if $v_j > v_k$ or r_k followed by r_j if $v_k \geq v_j$. In other words, R^* is the set of requests of R such that if any request in R can be swapped with another request with a higher revenue, then this swap occurs in R^*.

Note that since any algorithm must move back to S after completing a request, serving any request takes exactly two time units, so rearranging the requests in R as described will not affect the overall time required. Therefore every request in R also appears in R^* so the two sets earn equal amounts of revenue.

Now, suppose, by contradiction, that in R^* there is some request r with revenue v served by OPT at time t but not served by SGRF.

Case 1: T is even.
Suppose t is odd. Since at every odd time, SGRF serves the request with the highest revenue, at time t SGRF must earn revenue at least v. If t is even, then there are at least 2 more time units remaining (i.e. $T \geq t + 2$), so SGRF can serve r from $t + 1$ to $t + 2$; OPT would not be able to serve another request during this time slot as it would need to use this time to move back to s.

Case 2: T is odd.
Simply replace odd with even and even with odd in the proof for Case 1.

7   Conclusion

We studied ROLDARP and variations of this problem. For ROLDARP we proved that deterministic competitive algorithms do not exist for complete graphs with varying edge weights nor for non-complete graphs. For complete graphs with unit edge weights, we presented a 2-competitive algorithm to solve this problem. For ROLDARP on complete bipartite graphs, we presented a 1-competitive algorithm. Finally, for ROLDARP on graphs with a single source vertex, we presented an optimal online algorithm.

Obviously improving the competitive ratio for original ROLDARP or proving a lower bound of 2 would be interesting extensions of our work. Some other open problems involve incorporating modifications that would make the problem more reflective of realistic settings. For example, we can consider a server that can serve multiple requests at a time or a setting with multiple servers. We can also associate with each request a penalty that is incurred if the request is not served; one goal would be to earn a specified amount of revenue while minimizing the penalties of unserved requests.

References

1. Ascheuer, N., Krumke, S.O., Rambau, J.: Online dial-a-ride problems: minimizing the completion time. In: Reichel, H., Tison, S. (eds.) STACS 2000. LNCS, vol. 1770, pp. 639–650. Springer, Heidelberg (2000)
2. Ausiello, G., Feuerstein, E., Leonardi, S., Stougie, L., Talamo, M.: Algorithms for the on-line traveling salesman. Algorithmica 29(4), 560–581 (2001)
3. Ausiello, G., Bonifaci, V., Laura, L.: The on-line asymmetric traveling salesman problem. J. Discrete Algorithms 6(2), 290–298 (2008)
4. Ausiello, G., Bonifaci, V., Laura, L.: The on-line prize-collecting traveling salesman problem. Inf. Process. Lett. 107(6), 199–204 (2008)
5. Broden, B., Hammar, M., Nilsson, B.: Online and offline algorithms for the time-dependent TSP with time zones. Algorithmica 39(4), 299–319 (2004)
6. Jaillet, P., Lu, X.: Online traveling salesman problems with flexibility. Networks 58, 137–146 (2011)
7. Jaillet, P., Wagner, M.: Generalized online routing: new competitive ratios, resource augmentation and asymptotic analyses. Oper. Res. 56(3), 745–757 (2008)
8. Krumke, S.: On minimizing the maximum flow time in the online dial-a-ride problem. Networks 44, 41–46 (2004)
9. Lipmann, M., Lu, X., de Paepe, W.E., Sitters, R.A., Stougie, L.: On-line dial-a-ride problems under restricted information model. Algorithmica 40, 319–329 (2004)
10. Stougie, L., Feuerstein, E.: On-line single-server dial-a-ride problems. Theor. Comput. Sci. 268(1), 91–105 (2001)

CSoNet

Global Internet Connectedness: 2002–2011

Hyunjin Seo[1](✉) and Stuart Thorson[2]

[1] William Allen White School of Journalism and Mass Communications,
University of Kansas, Lawrence, USA
hseo@ku.edu
[2] The Maxwell School, Syracuse University, Syracuse, USA
thorson@syr.edu

Abstract. We examine the communications networks formed by direct international Internet links, weighted by bandwidth capacity, each year over the 2002–2011 period. While the 2011 network closely resembles that of 2002, the network has become more tightly interconnected over time. With countries as nodes, connectedness was measured by both changes in median degree and overall network density. We also considered networks formed by aggregating countries into United Nations (UN) *continent* and *region* categories as well as network communities identified through tightness of degree interconnection weighted by bandwidth. While relative connectedness as measured by percentage of bandwidth staying within UN geographic regions is decreasing, the percentage remaining within the continent has been fairly constant over the period. All of this must, of course, be understood in the context of enormous total international bandwidth growth between 2002 and 2011 at all levels of analysis.

1 Introduction

Internet-based information and communication technologies have brought about important changes in our society, particularly with respect to building and maintaining social relationships and producing and sharing information and knowledge. For example, individuals increasingly rely on social networking sites such as Facebook or Twitter to build and maintain relationships with friends, families, and others [10]. New digital technologies have also had a significant impact on global activism, as they afford rapid mobilization of citizens around the world [8,13,14]. These changes in the nature of interactions between actors in society are consequences of decentralized information and cooperative peer production characteristic of the networked information society [2,5,7,17].

Transnational Internet-based communication tools empower individuals in disparate parts of the world to collaborate in producing and sharing information. TeleGeography's annual surveys of Internet traffic and capacity show that overall international Internet bandwidth has increased from less than 1 Tbps in 2002 to about 55 Tbps in 2011. International Internet bandwidth refers to the amount of data that can be transferred over the Internet, across national borders, in a given amount of time, and thus it is a good indicator of transnational Internet traffic flows [3].

© Springer International Publishing Switzerland 2014
Z. Zhang et al. (Eds.): COCOA 2014, LNCS 8881, pp. 537–546, 2014.
DOI: 10.1007/978-3-319-12691-3_39

Previous research analyzing global Internet bandwidth between 2002 and 2011 found that global Internet connectedness has grown significantly over the period and the global Internet network has become denser [16]. In the aggregate, countries have become more tightly connected. Countries that were central in the 2002 global Internet network with a larger number of direct connections with other countries and with relatively large amounts of international bandwidth largely remained so in 2011.

While previous studies are helpful for understanding the global Internet network, there are several important questions that have not been empirically tested. For example, how, if at all, has the community structure of the global Internet network changed over the past decade? How has global Internet connectedness between regions and continents evolved? To what extent is the significance of geographical distance between countries receding as the Internet develops? There are very few studies examining the over time global distribution of Internet assets at country, regional, and continental levels, as most research has focused on identifying determinants of Internet diffusion or distribution [1,4,11]. We attempt to fill the gap in the literature by examining these and other issues in the area of global Internet connectedness.

2   Data and Method

We analyze international bandwidth data curated by TeleGeography [18] and purchased from them. International Internet bandwidth is an upper bound on direct country to country Internet traffic flow [3]. Using a combination of bandwidth metrics and centrality indicators, we examine bandwidth data over the 10-year period from 2002 to 2011. The raw dataset consisted of the English language names of all pairs of countries for which Telegeography had bandwidth data for the period together with the direct bandwidth between that pair for each of the 10 years.

To construct the networks reported here we first broke apart the country name pairs where a direct connection existed that year and for each country added the country, region, and continent ISO codes as used by the Statistics Division of the United Nations Secretariat. These data were read into R [15] and converted to edge lists for each year in the 2002–2011 period. This resulted, for each year, in R dataframes containing each dyad of countries with a direct bandwidth connection for that year together with the amount of bandwidth capacity shared by that connection. Countries with no reported direct connections were dropped from the analysis for that year. This resulted in 186 countries that shared bandwidth with at least one other country in 2002 and grew to 201 countries by 2011. The R package *igraph* [9] was used to build and analyze graphs from each of the edge lists. We report here on networks aggregated at the United Nations (UN) region and continent levels as well as the country level networks.

Internet connections transmit data in either direction and the resultant graphs are treated as undirected. In apportioning bandwidth to nodes (countries, regions, or continents) we adopted the convention of assigning one-half of dyad bandwidth to each node in the pair.

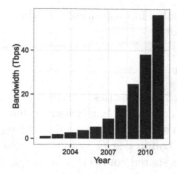

Fig. 1. International bandwidth capacity: 2002–2011

This procedure resulted in 30 network graphs–one for each year in the 2002–2011 period under study for countries, regions, and continents. We ask a number of questions of each of these networks. These include which countries are connected to many other countries and which to only a few (degree); which countries have a lot of bandwidth and which have a relatively small amount (bandwidth distribution); and which countries are more *central* in the network. We also discuss global network properties such as network size and density. Finally, and most importantly, each of these questions can be examined both for single time points and for possible patterned changes over time.

In the next section we describe some properties of these graphs in some detail especially as they relate to changes in the interconnectedness of the international Internet. All of this should be viewed against the background of the enormous growth of international Internet bandwidth since 2002 (see Fig. 1). All countries and regions have experienced significant bandwidth increases though, as will be shown in Sect. 3, the greatest growth in raw bandwidth was generally enjoyed by the countries with the largest bandwidth shares in 2002. While bandwidth rich countries have grown bandwidth richer, so have countries with less bandwidth. Understanding patterns of over time changes in bandwidth distribution requires careful examination of rates of increase at various points in the distributions.

3 Results

Changes in both compression technologies and uses of bandwidth suggest caution in interpreting raw bandwidth. As compression technology continues to improve, a fixed amount of data may require less bandwidth. At the same time, changing uses of bandwidth can result in shifts in demand. For example, as video compression technology improves, demand for transmitting video may increase thus requiring increased capacity. Furthermore, there are country specific policies which may result in different demand patterns. As an instance, at the time of this writing China is blocking access to Google and related sites. Finally, it is important to keep in mind that our analysis focuses solely on international bandwidth as measured by direct country to country connections. This makes sense given our focus on global interconnectedness but it also means that we are ignoring domestic bandwidth.

3.1   Country Analysis

The country level network has countries as nodes with edges between countries where there is a direct international bandwidth connection between them. The bandwidth associated with each edge is treated as the edge's weight. This results in a weighted undirected country-level graph for each of the 10 years of our dataset. Each country in a shared connection is assigned one-half of that edge's bandwidth. As reported above, the number of countries with at least one shared connection grew slightly from 186 in 2002 to 201 in 2011. The number of edges (direct connections) increased from 521 to 766. The median degree (number of direct connections) went from 2 ($mean = 5.57$) in 2002 to 4 ($mean = 7.62$) in 2011. This is reflected in the increase in graph density from .03 to .08. Over the period, the correlation (Spearman ρ) between country degree rank and country bandwidth rank ranged between .75 and .81 ($p < .001$ in all cases) indicating that high ranking bandwidth countries also tended to be high ranking degree countries. The United States had the highest eigenvector centrality score [6], weighted by bandwidth, in 2002 and 2003. Great Britain has the highest eigenvector score from 2004 to 2011.

It is worth emphasizing the heavy right tail of the bandwidth distribution in each year. Countries at or above the 50th percentile in international bandwidth held over 97 percent of the total international bandwidth and those in the top 10 percentile enjoyed around 83 percent of all network international bandwidth.

To visualize the 2002 and 2011 tails we first extracted the subnetworks consisting only of the degree tail countries and then used the Walktrap algorithm with bandwidth as edge weights as implemented in igraph to identify communities within the subnetwork. The basic idea here is that a community would be a densely connected subnetwork of the tail. The Walktrap approach to identifying communities takes random walks (here of length 4) from nodes and identifies communities as those networks that are easily reached by those walks. In our case we took into account the bandwidth of each connection in identifying the communities. We focused on countries ranking in the top 10 percentile as measured by degree. Twenty countries were in the top 10 percent (with regard to degree) in 2002 with the largest being the United States (degree = 124) and with United Arab Emirates, Australia, Taiwan, Portugal, and Malaysia as the smallest (degree = 12). In 2011 there were 23 countries in the top 10 percent. The largest was the United States (degree = 108). The smallest were Spain, Sweden, Turkey, Saudi Arabia, and Qatar (degree = 15).

Results are shown in Figs. 2 and 3. Here node diameters are roughly proportional to their degree in the subnetwork and the edge width is proportionate to the log of the bandwidth associated with that edge. The colors of the nodes reflect the community of which the node is identified as a member. Note also that the connections shown are only those within the subnetwork.

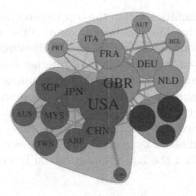

Fig. 2. Top 10 percentile degree communities: 2002 (Note: Vertex diameters are proportional to their degree and edge widths are proportional to log of edge bandwidth. Vertex labels are three character ISO country codes.)

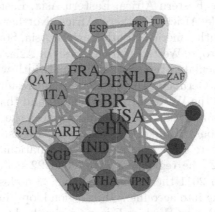

Fig. 3. Top 10 percentile degree communities: 2011 (Note: Vertex diameters are proportional to their degree and edge widths are proportional to log of edge bandwidth. Vertex labels are three character ISO country codes.)

Several things are interesting. First, communities, not surprisingly, reflect spatial geography. Countries near one another tend to be in the same community. In 2002 three communities were identified–Asia Pacific, Western Europe, and a Nordic one consisting of Russia, Sweden, and Norway. It is noteworthy that in 2002, the United States and Canada were identified as being in a community with mostly Asia Pacific counties. Also in that community was the United Arab Emirates. For 2011, we identify four communities: Asia Pacific, Western Europe, Nordic, and Middle East. The first three communities are similar to those found in 2002, but now there is also a largely Middle East community (United Arab Emirates, Qatar, and Saudi Arabia) together with the United States and South Africa.

This analysis of communities within the top 10 percentile tail degree countries provides important insight. The number of communities within the subnetwork increased from three to four between 2002 and 2011. While this finding is consistent with previous research showing growth of Internet connections in the Middle East [12,16], it also provides additional empirical support for the emergence of these Middle Eastern countries as an important community (using node degree as the indicator) within the upper 10 percentile tail of the global Internet. This may help us better understand political and social movements occurring in the region that may be facilitated by more tightly connected digital media-based collaborative networks with the larger international Internet.

3.2   Region and Continent Analysis

To construct regional networks, we associated each country with its UN region. This resulted in 22 nodes: Australia and New Zealand, Caribbean, Central America, Central Asia, Eastern Africa, Eastern Asia, Eastern Europe, Melanesia, Micronesia, Middle Africa, Northern Africa, Northern America, Northern Europe, Polynesia, South America, South-Eastern Asia, Southern Africa, Southern Asia, Southern Europe, Western Africa, Western Asia, and Western Europe. For each region we calculated the amount of bandwidth which remained in the region (for example, a Germany-Austria edge would result in bandwidth being assigned as internal to the Western Europe region) and that which went outside the region (the Germany-Sweden edge would connect the Western Europe and Northern Europe regions). Unlike the country networks, region networks can have loops in cases where some of a country's international bandwidth remains in the region. In 2002, the regional network had 22 nodes and was interconnected by 91 edges. In 2011 there were the same 22 nodes and now 110 edges. Network density, taking into account within region loops, increased slightly from .36 (2002) to .43 (2011). The Western Europe region had the highest eigenvector centrality score, weighted by bandwidth, in both 2002 and 2011. Not surprisingly, the regions are much more tightly connected than were the individual countries.

Figures 4 and 5 provide a visualization of the community structures identified for 2002 and 2011. In 2002 there was essentially one large community consisting of all the regions except for Micronesia and Polynesia. By 2011 two large communities had emerged. One was comprised of the Americas, most of Asia, and Australia and New Zealand. The other included the European regions, Africa, and South Asia.

Continent networks were constructed in an analogous manner to the region ones though, of course, using each country's continent as assigned by the UN. This resulted in five nodes: Africa, Americas, Asia, Europe, and Oceania with 13 edges in 2002 and 14 by 2011 with the addition of an Oceania-Europe edge. Not surprisingly, given the level of aggregation, network density was also high–going from .87 to .93. Also, as expected, no communities were identified at the continent level. Europe had the highest eigenvector centrality score, weighted by bandwidth, in both 2002 and 2011.

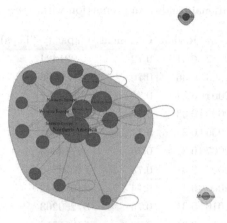

Fig. 4. Regional communities: 2002 (Note: Vertex diameters are proportional to their degree and edge widths are proportional to log of edge bandwidth.)

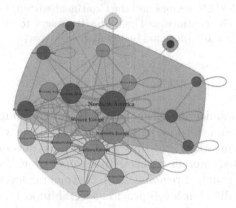

Fig. 5. Regional communities: 2011 (Note: Vertex diameters are proportional to their degree and edge widths are proportional to log of edge bandwidth.)

More interesting results can be seen in Table 1. The column labeled *Region* shows the proportion of total bandwidth for that region that went outside the region. Note that this proportion *decreased* from .41 to .31 over the period. This suggests that as total bandwidth increased, an increasing proportion of it was being allocated to connections outside the region. This is consistent with increasing interconnectedness. In contrast, reflected in the *Continent* column, at the continent level the proportion of bandwidth staying within the continent actually increased slightly. While, on average, countries are increasingly directing bandwidth outside their local region, that bandwidth is generally remaining on the same continent. Interestingly the Americas (.49 in 2002 to .58 in 2011) and Asia (.39 in 2002 to .47 in 2011) had the largest increase in the proportion of their continent's international bandwidth being directed within the continent. Europe remained fairly constant going from .85 to .87 over the period.

Table 1. International bandwidth proportion within region and continent

Year	Region	Continent	Capacity (Tbps)
2002	0.41	0.72	931424
2003	0.36	0.70	1759866
2004	0.34	0.71	2512052
2005	0.33	0.70	3548382
2006	0.33	0.71	5133354
2007	0.31	0.70	8714300
2008	0.31	0.73	14654850
2009	0.31	0.74	24082842
2010	0.31	0.75	37269263
2011	0.31	0.76	54855718

Note: The Region column shows the yearly proportion of international bandwidth with the U.N. regions and the Continent within the U.N. continents. The Capacity refers to the total international bandwidth each year.

4   Conclusion

Information and communication technologies (ICTs) have implications for political, economic, social, and cultural relationships around the world. To help better understand these implications and thus contribute to developing relevant policies, we analyzed global Internet connectedness, operationalized by international bandwidth links, at country, regional, and continental levels. Taken together, our analyses at these different levels provide several important results. First, our regional findings show that the proportion of international bandwidth allocated to connections outside the region has increased between 2002 and 2011 indicating increased Internet-based communications across regions compared with those within regions. In comparison, the proportion of international bandwidth allocated to connections outside the continent has slightly decreased. The greatest increased in global interconnectivity occurred within continents rather than between them for two of the bandwidth richest continents, the Americas and Asia. Europe, the most central (as measured by weighted eigenvector centrality), showed a constant proportion of bandwidth directed within the continent over the period.

Of particular interest is the emergence of two major regional communities by 2011–one Eurocentric and the other centered on the Northern America and Eastern Asia regions. This is consistent with United States policy pronouncements indicating a tilt toward Asia as well as policies in Eastern Asian countries including China, South Korea, and Japan directed at projecting a regional or global presence. Given the lead time involved in adding significant bandwidth, this may provide further support for the notion that shifting bandwidth patterns can serve as leading indicators of deeper socio-political changes.

While we identified some interesting changes in both bandwidth distribution and community structure, these changes took place largely within continental boundaries, with the exception of several Middle Eastern countries. More generally, the underlying network structure became a bit more complex and interconnected, as the international Internet saw large annual bandwidth capacity increases between 2002 and 2011.

With its longitudinal network analysis, this study makes theoretical and methodological contributions relevant to both scholarly and policy communities in the areas of international communication networks and global digital divide. Our study is one of the few that looks at changes in Internet distributions at country, regional and continental levels for a long period of time and should inform studies looking at determinants of global Internet growth and diffusion [1,11]. Use of mathematical concepts from network analysis to a communication phenomenon helps broaden social science approaches. Moreover, the results can help policymakers better identify appropriate strategies for global communication issues such as digital divide.

Acknowledgments. This research was partly funded by the University of Kansas (NFGRF 2302269) and the Maxwell School of Syracuse University.

References

1. Andrés, L., Cuberes, D., Diouf, M., Serebrisky, T.: The diffusion of the internet: a cross-country analysis. Telecommun. Policy **34**(5), 323–340 (2010)
2. Barabási, A.L.: Linked: How Everything is Connected to Everything Else and What it Means. Penguin Group, New York (2003)
3. Barnett, G.A., Park, H.W.: The structure of international internet hyperlinks and bilateral bandwidth. Ann. Telecommun. **60**(9), 1110–1127 (2005)
4. Beilock, R., Dimitrova, D.V.: An exploratory model of inter-country internet diffusion. Telecommun. Policy **27**(3), 237–252 (2003)
5. Benkler, Y.: The Wealth of Networks: How Social Production Transforms Markets and Freedom. Yale University Press, New Haven (2006)
6. Bonacich, P.: Power and centrality: a family of measures. Am. J. Sociol. **92**, 1170–1182 (1987)
7. Castells, M.: The Information Age: Economy, Society and Culture. The Power of Identity, vol. 2. Wiley, New York (2000)
8. Chadwick, A.: Internet Politics: States, Citizens, and New Communication Technologies. Oxford University Press, USA (2006)
9. Csardi, G., Nepusz, T.: The igraph software package for complex network research. Inter. J. Complex Syst., 1695 (2006)
10. Duggan, M., Smith, A.: Social media update 2013. Pew Internet and American Life Project (2013)
11. Guillén, M.F., Suárez, S.L.: Explaining the global digital divide: economic, political and sociological drivers of cross-national internet use. Soc. Forces **84**(2), 681–708 (2005)

12. Howard, P.N.: The digital origins of dictatorship and democracy: information technology and political Islam. Oxford Univ Pr on Demand (2010)
13. Mefalopulos, P.: Communication for sustainable development: Applications and challenges. Media and glocal change. Rethinking communication for development, pp. 247–260 (2005)
14. Naude, A.M., Froneman, J.D., Atwood, R.A.: The use of the internet by ten south african non-governmental organizations–a public relations perspective. Public Relat. Rev. **30**(1), 87–94 (2004)
15. R Core Team: R: A Language and Environment for Statistical Computing. R Foundation for Statistical Computing, Vienna, Austria (2013)
16. Seo, H., Thorson, S.J.: Networks of networks: changing patterns in country bandwidth and centrality in global information infrastructure, 2002–2010. J. Commun. **62**(2), 345–358 (2012)
17. Sunstein, C.R.: Infotopia: How Many Minds Produce Knowledge. Oxford University Press, New York (2006)
18. TeleGeography: Global Internet geography. Technical report, PriMetrica, Inc. (2011)

Optimal Containment of Misinformation in Social Media: A Scenario-Based Approach

Yongjia Song[1] and Thang N. Dinh[2]([✉])

[1] Department of Statistical Sciences and Operations Research, Virginia Commonwealth University, Richmond, VA 23284, USA
ysong3@vcu.edu
[2] Department of Computer Science, Virginia Commonwealth University, Richmond, VA 23284, USA
tndinh@vcu.edu

Abstract. The rapid expanding of online social networks (OSNs) in terms of sizes and user engagements have fundamentally changed the way people communicate and interact nowadays. While OSNs are very beneficial in general, the spread of misinformation or rumors in OSNs not only causes panic in general public but also leads to serious economic and political consequences. Several studies have proposed strategies to limit the spread of misinformation via modifying the topology of the diffusion networks, however, a common limit is that parameters in these diffusion models are difficult, if not impossible, to be extracted from real-world traces. In this paper, we focus on the problem of selecting optimal subset of links whose removal minimizes the spread of misinformation and rumors, relying only on *actual cascades* that happened in the network. We formulate the link removal problem as a mixed integer programming problem and provide efficient mathematical programming approaches to find exact optimal solutions.

Keywords: Social networks · Rumor blocking · Mathematical programming

1 Introduction

The rapid expanding of online social networks (OSNs) in terms of sizes and user engagements have fundamentally changed the way people communicate and interact nowadays. Billions of users are actively exchanging information on OSNs such as Facebook and Twitter. These social sites also act as the major news sources that can rapidly spread news to a vast amount of audience. Many popular events such as the election results and the death of Bin Laden were first broke out in OSNs, reaching the crowd before traditional media. While OSNs are very beneficial in general, the spread of misinformation or rumors in OSNs not only causes panic in general public but also leads to serious economic and political consequences. Examples include the endless spread of Ebola misinformation on social media [1] that hinder the fight against the disease and the widespread of

© Springer International Publishing Switzerland 2014
Z. Zhang et al. (Eds.): COCOA 2014, LNCS 8881, pp. 547–556, 2014.
DOI: 10.1007/978-3-319-12691-3_40

the false report that President Obama was killed from the hacked Fox News' Twitter account.

Several studies have proposed strategies to limit the spread of misinformation via modifying the topology of the diffusion networks [2–8]. For instance, several works study how authorities can choose to temporarily disable some links between users or even suspend accounts to limit harmful spreads. The predominant approaches are to model diffusion networks using either independent cascade or linear threshold model [9]. However, it is well-known that parameters in these models are difficult to be learned from real-world traces and in many cases the models do not fit actual data very well [10].

In this paper, we focus on the problem of selecting optimal subset of links whose removal minimizes the spread of misinformation and rumors, relying only on *actual cascades* that happened in the network. On one hand, our approach is universal as it does not assume any diffusion model. On the other hand, our approach can be integrated with any diffusion models when real-world traces are difficult to be obtained, e.g., by first simulating the cascades using the given diffusion model and then using the generated cascades to identify optimal links.

Assuming several cascade scenarios are given as the input, we formulate different formulations of the problem. We also provide efficient mathematical programming approaches to find exact optimal solutions.

Related works. Kempe et al. [9] formulated the influence maximization problem as an optimization problem. They show the problem to be NP-complete and devise an $(1 - 1/e - \epsilon)$ approximation algorithm. Later, Leskovec et al. [11] study the influence propagation in a different perspective in which they aim to find a set of nodes in networks to detect the spread of virus as soon as possible. They improve the simple greedy method with the lazy-forward heuristic (CELF), which is originally proposed to optimize submodular functions in [12]. The greedy algorithm is furthered improved by Chen et al. [13] by using an influence estimation. For the Linear Threshold model, Chen et al. [14] propose to use local directed acyclic graphs (LDAG) to approximate the influence regions of nodes. The influence maximization is also studied in other diffusion models including the majority threshold model [15] or when both positive and negative influence are considered [16].

Several studies have proposed strategies to limit the spread of misinformation via modifying the topology of the diffusion networks [2–8]. However, they often rely on diffusion models that are difficult to match with real-world traces [10].

2   Model and Problem Definition

We represent a social network using a directed graph $G = (V, E)$. Assume that we have a collection of N observed cascades. Each cascade $\tilde{G} = (V, E^x)$, also called a scenario, is a realization of the random graph obtained by removing the edge set represented by x from E. Each scenario $\tilde{G}$ specifies how information propagates in that particular cascade: a directed edge (u, v) indicates that information may propagate from u to v. Given a set of rumor $r \in R$, a set of source nodes $S(r)$

that currently hold rumor r, a set of terminal nodes $T \subsetneq V$. The objective is to choose a subset of edges in E to interdict under a budget constraint, so that the expected total damage to T caused by all the rumors is minimized over all given cascade scenarios.

3 A Scenario-Based Formulation

We assume that each rumor $r \in R$ may have a different probability model for information propagation, and we assume that each scenario k captures all these models independently for each source. Let E_r^k be the set of available edges in scenario $k \in N$, under the probabilistic information propagation from the set of source nodes $S(r), r \in R$. For each scenario $k \in N$, each terminal $t \in T$, each rumor $r \in R$, we introduce a binary variable $z_t^{k,r}$ to indicate whether node t is disconnected from the set of rumors $S(r)$ in scenario k. Let $\mathcal{P}_k^{s,t}$ be the set of all s-t paths in scenario $k \in N$ for $s \in S(r)$, $t \in T$. Our objective is to minimize the expected damage caused by the influence from various sources. Let the interdiction cost vector be $c \in \mathbb{R}_+^{|E|}$, and let the damage associated with each source $r \in S$ be $d_r > 0$, given a budget of interdiction actions B, the expected damage minimization problem from multiple sources could be formulated as:

$$\min \frac{1}{N} \sum_{k \in N} \sum_{r \in R} \sum_{t \in T} d_r z_t^{k,r} \tag{1a}$$

$$\text{s.t.} \sum_{e \in P^{s,t}} x_e \geq 1 - z_t^{k,r}, \ \forall P^{s,t} \in \mathcal{P}_k^{s,t}, \ \forall s \in S(r), \ \forall r \in R, \ \forall t \in T, \ \forall k \in N \tag{1b}$$

$$c^\top x \leq B \tag{1c}$$

$$x_e \in \{0,1\}, \forall e \in E \tag{1d}$$

$$z_t^{k,r} \in \{0,1\}, \ \forall k \in N, \ \forall s \in S(r), \ \forall r \in R, \ \forall t \in T. \tag{1e}$$

According to the problem structure, we see that (1) with relaxed z variables, i.e., $z_t^{k,r} \in [0,1]$, gives the same optimal objective value as (1), and hence is an exact formulation.

Proposition 1. *The optimal solution of (1) with continuous z variables must be an integer vector.*

Proof. Suppose $(\bar{x}, \bar{z})$ is an optimal solution to (1) with continuous z, $\bar{x} \in \{0,1\}^{|E|}$, and $\exists k \in N, s \in S(r), t \in T$ that $\bar{z}_t^{k,r} \in (0,1)$. Since $\bar{z}_t^{k,r} \geq \sum_{e \in P^{s,t}} \bar{x}_e$, $\forall P^{s,t} \in \mathcal{P}_k^{s,t}$, and the right-hand sides of these inequalities are all integers, we claim that $\sum_{e \in P^{s,t}} \bar{x}_e = 0$, $\forall P^{s,t} \in \mathcal{P}_k^{s,t}$. Therefore, $(\bar{x}, \bar{z})$ is suboptimal since we can decrease $\bar{z}_t^{k,r}$ down to 0, which causes a decrease in the objective value, and the solution remains feasible. □

z variables in (1) can be projected out easily, by recognizing that when a feasible $\hat{x} \in \{0,1\}^{|E|}$ that $c^{\top}\hat{x} \leq B$ is given, the optimal z should be: $z_{r,i}^k = \max\{\max_{p^{r,i} \in \mathcal{P}^{r,i}} 1 - \sum_{e \in p^{r,i}} \hat{x}_e, 0\}$. Although this projection enables one to solve (1) in the x space, due to the nonsmoothness of function $\max\{\max_{p^{r,i} \in \mathcal{P}^{r,i}} 1 - \sum_{e \in p^{r,i}} x_e, 0\}$, we choose to solve (1) using a branch-and-cut algorithm by adding constraints (1b) dynamically during the branch-and-bound procedure. We start solving (1) without the formulation-defining valid inequalities (1b), and during the branch-and-bound tree, we add them back to the formulation when they are necessary by solving a separation problem. For each $r \in S, i \in T, k \in N$, given a relaxation solution $\hat{x}$, the exact separation of (1b) involves solving a shortest r-i path problem in graph G_k using $\hat{x}$ as the edge weights, where a shortest r-i path is the shortest r_s-i path where r_s is one of the sources induced by rumor r.

From a slightly different perspective, we may see the connection from any rumor to any target node as a failure scenario, and we would like to control the probability that such failure scenario happens, then a variant of (1) can be formulated as a chance-constrained program. Introducing a binary variable z_k for each scenario $k \in N$ to indicate whether all the nodes in T are disconnected from r in scenario k. Let p be the number of scenarios that are allowed to fail, a scenario-based mixed integer programming formulation can be written as follows:

$$\min c^{\top}x \tag{2a}$$

$$\text{s.t.} \sum_{e \in P^i} x_e \geq z_k, \forall k \in N, P^i \in \mathcal{P}_k^i, \forall i \in T \tag{2b}$$

$$\sum_{k \in N} z_k \geq N - p \tag{2c}$$

$$x \in \{0,1\}^{|E|}, z \in [0,1]^N, \tag{2d}$$

where $\mathcal{P}_k^i$ is the set of all r-T_i paths in scenario $k \in N$ from any rumor node r. Again we do not need to enforce the integrality on z variables, since optimal z to (2) will be integral automatically. Although the scenario-based path inequalities (2b) are exponentially many, similar to (1b), they can be separated in polynomial time.

4   Strength of Valid Inequalities

In this section, we study the strength of scenario-based path inequalities (2b). For simplicity, we just assume that there is a single rumor node r. We define the feasible set Y of (2) as:

$$Y := \{(x,z) \in \{0,1\}^E \times \{0,1\}^N \mid (x,z) \text{ satisfies (2b), (2c)}\} \tag{3}$$

The dominant $Y^{\uparrow}$ of Y is defined as:

$$Y^{\uparrow} = \{(x,z) \in \mathbb{Z}_+^E \times \mathbb{Z}_+^N \mid \exists (x',z'), x' \leq x, z' \leq z, (x',z') \in Y\}, \tag{4}$$

and the partial dominant $\bar{Y}^\uparrow$ of Y is defined as:

$$\bar{Y}^\uparrow = \{(x,z) \in \mathbb{Z}_+^E \times \{0,1\}^N \mid \exists x', x' \le x, (x',z) \in Y\}, \tag{5}$$

We study the convex hull of the dominant set $Y^\uparrow$, and the partial dominant set $\bar{Y}^\uparrow$, which is justified by the fact that $\min\{c^\top x \mid (x,z) \in Y\} = \min\{c^\top x \mid (x,z) \in Y^\uparrow\} = \min\{c^\top x \mid (x,z) \in \bar{Y}^\uparrow\}$ since c vector is nonnegative.

Given a set of edges P, we define $F^P := \{k \in N \mid G_k(P)$ contains a $r - i$ path for some i $\in$ T$\}$. We also use the following notation throughout this section: for a given set Q, let $\mathbf{1}_Q := \sum_{a \in Q} e^a$, where e^a is a unit vector with a one in component a, and zero elsewhere.

Definition 1. $P \in E_k$ is called a minimal scenario-based path if P connects the source node, r, to a unique terminal node, $i \in T$.

Theorem 1. Let $k \in N$, the scenario-based path inequality, $\sum_{e \in P} x_e \ge z_k$, is facet-defining for $conv(\bar{Y}^\uparrow)$ if and only if

1. P is a minimal scenario-based path.
2. $|F^P| \le p$.
3. For any scenario $k' \in F^P$, there exists some edge $a_{k'} \in P$, that $P \backslash \{a_{k'}\}$ does not connect the source and terminal set in scenario k', i.e., consider shrinking all terminal nodes as a single node T, P does not a two-arc disjoint path from r to T.

Proof. The "only if" part of the proof is trivial, since otherwise, the inequality (2b) is dominated by: case (1), a scenario-based path inequality defined by the minimal path $P' \subseteq P$; case (2), a stronger valid inequality without z variable with the same set of edges P, case (3), inequality $\sum_{e \in P} x_e \ge z_k + z_{k'}$. Now we prove the "if" part. First, $conv(\bar{Y}^\uparrow)$ is full-dimensional, since rays $(e^i, \mathbf{0}), \forall i \in E$ are feasible directions, and points $(\mathbf{1}_E, \mathbf{1}_N - e^k), \forall k \in N$ are feasible points. Let $a_0 \in P$ be some point in P, we next prove there are $|E| + |N|$ feasible points that satisfy the inequality (2b) at equality, and are affinely independent:

- $\mathbf{x}^1 = \{(\mathbf{1}_{E \backslash P}, \mathbf{1}_{N \backslash F^P})\}$, 1 point;
- $\mathbf{x}^2 = \{(e^a + \mathbf{1}_{E \backslash P}, e^k + \mathbf{1}_{N \backslash F^P}) \mid a \in P\}$, $|P|$ points;
- $\mathbf{x}^3 = \{(e^{a_0} + e^a + \mathbf{1}_{E \backslash P}, e^k + \mathbf{1}_{N \backslash F^P}) \mid a \in E \backslash P\}$, $|E| - |P|$ points;
- $\mathbf{x}^4 = \{(e^{a_0} + e^a + \mathbf{1}_{E \backslash P}, e^k + \mathbf{1}_{N \backslash F^P} - e^{k'}) \mid k' \in N \backslash F^P\}$, $|N| - |F^P|$ points;
- $\mathbf{x}^5 = \{(e^{a_{k'}} + \mathbf{1}_{E \backslash P}, e^k + \mathbf{1}_{N \backslash F^P} + e^{k'} \mid k' \in F^P \backslash \{k\}\}$, $|F^P| - 1$ points.

By definition, P is a minimal path, so interdicting any edge in the path will disconnect r and terminal set T in graph $G_k(P)$. It is then not hard to see all these points satisfy the three requirements listed above. □

We now consider the *partial path inequality*, $\sum_{e \in P} x_e + \sum_{k \in T} z_k \ge 1$, where $T \subseteq N \backslash F^P$, and $|T| = q + 1 - |F^P|$. If $|F^P| \le q$, so that there exists such "recovery set" T, then the partial path inequality is valid by the integrality argument: if $\sum_{e \in P} x_e \ge 1$, the inequality is satisfied trivially, otherwise, we

Table 1. Random (Erdos-Reyni) network

$p/$	Time			Nodes			UB			#Cuts		
Budget	5	10	20	5	10	20	5	10	20	5	10	20
0.1	0.0	0.0	0.0	0.0	0.0	0.0	0.1	0.0	0.0	41	19	17
0.2	0.2	0.1	0.1	0.0	0.0	1.0	1.0	0.4	0.0	456	406	144
0.3	0.4	0.2	0.7	9.0	0.0	18.0	2.4	0.7	0.1	1109	891	784
0.4	1.3	0.4	2.5	21.0	0.0	66.0	4.6	1.1	0.3	2471	1486	1812
0.5	2.7	1.9	4.8	21.0	7.0	139.0	6.3	2.2	0.3	3722	3488	2881
0.6	5.1	5.3	6.7	35.0	6.0	126.0	7.6	2.5	0.3	5265	4843	4218
0.7	8.1	6.4	14.3	60.0	17.0	45.0	8.6	2.9	0.4	6700	6448	7855
0.8	14.5	9.6	16.4	46.0	27.0	114.0	9.3	3.3	0.4	8948	7339	7519
0.9	19.8	30.3	777.1	24.0	100.0	11696.0	14.3	7.2	2.1	13165	12184	14228

know at least one $z_k, k \in T$ has to be 1 to satisfy the chance constraint. Given a set of edges P, we define $\delta^P(e) = \{k \in F^P \mid G_k(P \setminus \{e\})$ disconnects s and $T\}$. We now prove the facet-inducing results for these inequalities.

Theorem 2. *Let $P \subseteq E$ be such that $|F^P| \leq p$, the corresponding partial path inequality is facet-defining for* conv$(Y^\uparrow)$ *if and only if $\delta^P(e) \neq \emptyset, \forall e \in P$.*

Proof. The "only if" part is trivial, since otherwise, the inequality is dominated by a corresponding inequality using $P \setminus \{e\}$. We next prove the "if" part. Let a_0 be a fixed edge in P, it is clear that the folliwing $|E| + |N|$ points are feasible, satisfy the inequality at equality, and are affinely independent:

- $\mathbf{x}^1 = \{(\mathbf{1}_{E \setminus P} + e^a, \mathbf{1}_{N \setminus F^P} + \mathbf{1}_{\delta^P(a)}) \mid a \in P\}$, $|P|$ points;
- $\mathbf{x}^2 = \{(\mathbf{1}_{E \setminus P} + e^{a_0} + e^{a'}, \mathbf{1}_{N \setminus F^P} + \mathbf{1}_{\delta^P(a_0)}) \mid a' \in E \setminus P\}$, $|E| - |P|$ points;
- $\mathbf{x}^3 = \{(\mathbf{1}_{E \setminus P}, \mathbf{1}_{N \setminus (F^P \cup T)} + e^k) \mid k \in T\}$, $|T|$ points;
- $\mathbf{x}^4 = \{(e^{a_0} + \mathbf{1}_{E \setminus P}, \mathbf{1}_{N \setminus (F^P \cup T)} + e^k + \mathbf{1}_{\delta^P(a_0)}) \mid k \in N \setminus T\}$, $|N| - |T|$ points; $\qquad\square$

5   Computational Experiments

In this section, we perform numerical studies on both synthesis networks of different topologies as well as real-world networks to show the performance of our proposed method.

5.1   Data Sets

We perform experiments on four synthesis networks of roughly equal sizes. The *synthesis networks* are generated with the following network models.

Table 2. Scale-free (Barabasi-Albert) network

p/	Time			Nodes			UB			#Cuts		
Budget	5	10	20	5	10	20	5	10	20	5	10	20
0.1	0.0	0.0	0.0	1.0	0.0	0.0	0.0	0.0	0.0	100	53	51
0.2	0.1	0.1	0.0	0.0	0.0	0.0	0.1	0.0	0.0	384	318	60
0.3	0.3	0.3	0.4	1.0	32.0	2.0	2.8	0.4	0.0	1391	721	446
0.4	1.1	2.3	0.4	4.0	37.0	2.0	7.9	4.6	0.0	3422	3761	1164
0.5	2.1	4.8	0.6	8.0	51.0	0.0	10.7	6.6	0.0	5177	5278	1640
0.6	3.3	5.3	0.5	0.0	49.0	0.0	9.3	3.8	0.0	6855	6285	413
0.7	4.5	13.3	0.8	1.0	75.0	0.0	11.2	5.8	0.0	9814	8639	508
0.8	9.1	48.4	0.9	5.0	148.0	0.0	12.9	8.0	0.0	11661	14564	674
0.9	51.1	38.9	25.3	220.0	139.0	252.0	7.5	3.6	0.5	13616	12911	7074

Table 3. Small-world (Watts-Strogatz) network

p/	Time			Nodes			UB			#Cuts		
Budget	5	10	20	5	10	20	5	10	20	5	10	20
0.1	0.0	0.0	0.0	0.0	0.0	0.0	0.0	0.0	0.0	33	19	19
0.2	0.1	0.1	0.0	1.0	0.0	0.0	0.3	0.0	0.0	203	131	58
0.3	0.2	0.6	0.0	0.0	8.0	0.0	1.3	0.4	0.0	528	519	131
0.4	1.2	1.3	0.1	1.0	0.0	0.0	3.2	0.9	0.0	1401	1010	449
0.5	4.6	5.4	0.1	13.0	154.0	0.0	5.8	2.0	0.0	3798	2201	595
0.6	15.2	27.1	0.3	76.0	184.0	1.0	5.6	2.3	0.0	5387	5976	926
0.7	22.8	59.9	0.9	77.0	266.0	0.0	6.6	2.8	0.0	7882	7917	2667
0.8	37.2	78.8	1.0	87.0	185.0	0.0	7.2	3.0	0.0	9610	14009	2844
0.9	29.0	159.0	0.2	42.0	388.0	0.0	7.6	3.1	0.0	11717	14787	443

- *Erdos-Reyni*: A random graph of 100 vertices and 200 edges following the Erdos-Reyni model [17].
- *Barabasi-Albert*: A power-law model using preferential attachment mechanism [18].
- *Watts–Strogatz*: A random graph which exhibit small-world phenomenon following model [19] with the dimension of the lattice 2 and the rewiring probability 0.3.

We also test our algorithm on two real social networks, namely American College Football network with 115 nodes and 613 links and scientific collaboration network with 1589 nodes and 2742 links [20].

5.2 Parameter Settings

For each network, we randomly generate the scenarios following the Independent Cascade model [9]. The edge probabilities are set to be constant values p between

0.1 and 0.9. We plan into the network three rumors. Each rumor consists of 2 % of the nodes in the networks. The number of terminal is set to 20 % of the network size. All edges have the same interdiction cost one, and the tested values of the budget are 5, 10, and 20, i.e., we seek to remove 5, 10, and 20 edges from the network.

5.3   Computational Results

We report the following measures for each network.

- *Time*: Computational time for solving (1), we use a time limit of 1 h. We use "-" to denote the case when the time limit is reached.
- *Nodes*: number of branch-and-bound nodes processed by the algorithm.
- *UB*: the best feasible solution obtained when the time limit is reached. If the algorithm finishes within the time limit, UB represents the optimal objective value.
- #Cut: Number of cuts generated.
- *OptGap*: Optimality gap when the time limit is reached. We use "-" to denote the case when the algorithm finishes within the time limit. Otherwise, the optimality gap is calculated by $(UB - LB)/UB$, where UB is the best feasible solution obtained, and LB is the best relaxation bound obtained.

The results for the above networks are shown in Tables 1, 2, and 3, respectively. All optimal solutions are found within few minutes with the exception of the Erdos-Reyni network (Table 1) when the edge probabilities are 0.9 and the budget is 20. In that case, it takes our branch and bound algorithm roughly 13 min to find the optimal solutions after generating a large number of cuts. Overall, the Erdos-Reyni network, that approximates expander graphs, is the hardest instance. The easiest network is the scale-free network, in which all

Table 4. Football network

p/	Time			Nodes			UB			OptGap		
Budget	5	10	20	5	10	20	5	10	20	5	10	20
.1	.1	.1	.1	0	1	0	1.2	.6	.1	-	-	-
.2	1.0	8.3	1.4	3	36	3	5.7	3.9	1.4	-	-	-
.3	5.0	7.7	142.1	15	67	3179	9.5	6.7	3.1	-	-	-
.4	8.2	19.8	236.2	1	72	2669	15.8	11.5	7.1	-	-	-
.5	8.2	19.8	236.2	1	72	2669	17.6	12.4	7.6	-	-	-
.6	12.5	28.3	411.1	19	52	1292	19.2	12.8	7.8	-	-	-
.7	12.6	38.6	277.1	22	94	959	20.3	12.9	7.8	-	-	-
.8	18.4	45.9	309.5	15	98	1147	21.4	13.0	7.8	-	-	-
.9	43.3	72.4	608.2	87	158	897	22.5	13.0	7.8	-	-	-

Table 5. Scientific collaboration network

p/	Time			Nodes			UB			OptGap		
Budget	5	10	20	5	10	20	5	10	20	5	10	20
.1	.1	.1	.1	0	0	0	0	0	0	-	-	-
.2	1.6	1.9	2.1	0	1	2	4.7	3.7	2.1	-	-	-
.3	3.3	0.2	0.2	1	0	0	6.0	0	0	-	-	-
.4	56	.2	4.4	93	0	0	15.8	0	0	-	-	-
.5	93.1	.3	5.3	95	0	0	24.5	0	0	-	-	-
.6	1062.3	0.5	11.1	352	0	0	31.5	0	0	-	-	-
.7	-	-	-	346	364	396	63.3	32.7	32.0	70.5 %	55.3 %	75.3 %
.8	-	-	-	170	266	291	107.8	108.2	106.0	90.1 %	94.8 %	97.8 %
.9	-	-	-	122	187	274	109.1	109.0	57.1	92.0 %	96.1 %	33.3 %

instances are solved within one minute. Note that most social networks have heavy-tailed degree distribution, i.e., scale-free networks.

Not surprisingly, when the edge probabilities increase, the solving time as well as the number of generated cuts also increase. The reason is that denser network scenarios are generated with higher edge probabilities. For example, in Small-world network (Table 3) with edge probabilities 0.9 and budget 10, 14787 cuts are generated in total.

The results for two real-world social networks are shown in Tables 4 and 5. Our proposed algorithm finds all optimal solutions in the American Football College and the first six instances of the Scientific Collaboration network, which have thousands of nodes. These are exact solutions for the largest network instances ever found. In the previous research [8] (on a different model), exact solutions only can be found for networks with dozens of vertices, which are too small for practical purposes.

6 Conclusion

We propose a new approach to contain the spread of misinformation in social networks. Our approach ultilizes historical information on cascades that happened in the network and is not constrained to any diffusion model. We provide several formulations and efficient approaches to solve for the optimal solutions. Our experiments on both synthetic networks and real-world traces suggest that optimal interdiction by removing edges can be found for social networks of moderate sizes. For large-scale networks, it is desirable to come up with scalable approximation algorithms (or heuristic) to find near-optimal solutions.

References

1. Izadi, E.: Important alert from Chinese state media: No, Ebola isnt a zombie virus, The Washington Post (2014). http://wapo.st/1mRyC09. Accessed 21 August 2014
2. Kimura, M., Saito, K., Motoda, H.: Blocking links to minimize contamination spread in a social network. ACM Trans. Knowl. Discov. Data 3(2), 9:1–9:23 (2009)

3. Budak, C., Agrawal, D., El Abbadi, A.: Limiting the spread of misinformation in social networks. In: Proceedings of the 20th International Conference on World Wide Web, WWW '11, pp. 665–674. ACM, New York (2011)
4. Schneider, C.M., Mihaljev, T., Havlin, S., Herrmann, H.J.: Suppressing epidemics with a limited amount of immunization units. Phys. Rev. E **84**, 061911 (2011)
5. He, X., Song, G., Chen, W., Jiang, Q.: Influence blocking maximization in social networks under the competitive linear threshold model. In: SDM, pp. 463–474. SIAM (2012)
6. Nguyen, N.P., Yan, G., Thai, M.T.: Analysis of misinformation containment in online social networks. Comput. Netw. **57**(10), 2133–2146 (2013)
7. Kuhlman, C., Tuli, G., Swarup, S., Marathe, M., Ravi, S.: Blocking simple and complex contagion by edge removal. In: 2013 IEEE 13th International Conference on Data Mining (ICDM), pp. 399–408, Dec 2013
8. Hemmati, M., Cole Smith, J., Thai, M.T.: A cutting-plane algorithm for solving a weighted influence interdiction problem. Comput. Optim. Appl. **57**(1), 71–104 (2014)
9. Kempe, D., Kleinberg, J., Tardos, É.: Maximizing the spread of influence through a social network. In: KDD'03, pp. 137–146. ACM, New York (2003)
10. Goyal, A., Bonchi, F., Lakshmanan, L.V.: Learning influence probabilities in social networks. In: Proceedings of the Third ACM International Conference on Web Search and Data Mining, WSDM '10, pp. 241–250. ACM, New York (2010)
11. Leskovec, J., Krause, A., Guestrin, C., Faloutsos, C., VanBriesen, J., Glance, N.: Cost-effective outbreak detection in networks. In: ACM KDD '07, pp. 420–429. ACM, New York (2007)
12. Minoux, M.: Accelerated greedy algorithms for maximizing submodular set functions. In: Stoer, J. (ed.) Optimization Techniques. Lecture Notes in Control and Information Sciences, vol. 7, pp. 234–243. Springer, Heidelberg (1978)
13. Chen, N.: On the approximability of influence in social networks. SIAM J. Discrete Math. **23**(3), 1400–1415 (2009)
14. Chen, W., Wang, C., Wang, Y.: Scalable influence maximization for prevalent viral marketing in large-scale social networks. In: ACM KDD '10, pp. 1029–1038. ACM, New York (2010)
15. Dinh, T., Zhang, H., Nguyen, D., Thai, M.: Cost-effective viral marketing for time-critical campaigns in large-scale social networks. IEEE/ACM Trans. Networking (2014)
16. Zhang, H., Dinh, T., Thai, M.: Maximizing the spread of positive influence in online social networks. In: 2013 IEEE 33rd International Conference on Distributed Computing Systems (ICDCS), pp. 317–326, July 2013
17. Erdos, P., Renyi, A.: On the evolution of random graphs. Publ. Math. Inst. Hungary. Acad. Sci. **5**, 17–61 (1960)
18. Barabasi, A., Albert, R., Jeong, H.: Scale-free characteristics of random networks: the topology of the world-wide web. Physica A **281**, 69–77 (2000)
19. Watts, D.J., Strogatz, S.H.: Collective dynamics of 'small-world' networks. Nature **393**(6684), 440–442 (1998)
20. Agarwal, G., Kempe, D.: Modularity-maximizing graph communities via mathematical programming. Eur. Phys. J. B **66**, 409–418 (2008)

Multivariate Heavy Tails in Complex Networks

Golshan Golnari$^{(\boxtimes)}$ and Zhi-Li Zhang

Department of Computer Science and Engineering,
University of Minnesota, Minneapolis, USA
{golnari,zhzhang}@cs.umn.edu

Abstract. It has been shown that many real networks are not only "scale-free" (i.e., having a power-law degree distribution), but also contain more complex structures such as "hierarchy" or "self-similarity" that cannot be captured by the preferential attachment random network model. These observations have led to a number of more sophisticated models being proposed in the literature. In this paper we advocate a *multivariate analysis* perspective based on the notion of MRVs as a unifying framework to study complex structures in networks. We demonstrate the existence of "multivariate heavy tails" in existing network models and real networks, and argue that they better capture the "hierarchical" or "self-similar" structures in these networks.

1 Introduction

Complex networks arising from natural, social, and engineered systems have been a topic of extensive studies in the past decades. A prevailing feature characterizing most of these complex networks is the *power-law* degree distribution, $P(k) \approx k^{-\gamma}$, where k represents the node degree. This gives rise to the term *scale-free* (SF) networks. Generative network models such as *preferential attachment* (PA) model have been proposed to provide a plausible explanation for the origin of the power-law degree distribution (also the "small-world" phenomenon observed in such networks, characterized by the average shortest path length in the logarithmic (log) order of the network size, $O(\log N)$). Besides the power-law degree distribution and small-world properties, many real-world complex networks also exhibit other important features, such as modularity (e.g., as characterized by a high average clustering coefficient) [1] or apparent "fractality" [5,7] that are absent in random scale-free networks generated via the PA model.

As an attempt to capture the more complex structures observed in real-world networks, many additional models (mostly *deterministic*) have been introduced in the literature, some of which yield sometimes confusing, if not contradictory, statements about the structures of complex networks. For example, the authors in [1,6] define a "hierarchical" (scale-free) network as a SF network where the (average) clustering coefficient of nodes of degree k also obeys a scaling law, $C(k) \approx k^{-\beta}$, where $\beta = 1$ in general, and provide a deterministic construction (and also a stochastic variant) for a specific family of growing *hierarchical*

© Springer International Publishing Switzerland 2014
Z. Zhang et al. (Eds.): COCOA 2014, LNCS 8881, pp. 557–570, 2014.
DOI: 10.1007/978-3-319-12691-3_41

networks satisfying these properties. The authors in [3] provide another deterministic construction for another specific family of growing networks – referred to as *pseudofractal scale-free graphs* – which also satisfy the power-law degree and cluster coefficient scaling laws. Both synthetic models are constructed as a reference system to explain, understand or "make new predictions" about the properties of scale-free complex networks; for instance, in [1,6] the authors demonstrate that the cluster coefficient scaling law exists in a number of complex networks, in particular, in metabolic networks, and use their construction to argue that such scaling law is the evidence of "hierarchical organization of modularity" in these networks. In addition, as stated by the authors, growing networks constructed using both models are *not* completely "fractal." In contrast, the authors in [4] argue the existence of "hierarchical, modular" graphs with *zero clustering coefficients*, and construct a specific family of growing networks with such properties. Whereas the authors in [5,7] introduce and define *self-similar* or *fractal* networks using the box-counting approach, and argue that many real networks, for instance, the same 43 metabolic networks considered in [6], have fractal structures. (See Sect. 2 for more discussion on these and other related studies.) These somewhat contradictory statements notwithstanding, these earlier studies demonstrate that there are more complex *dependencies* in the network structures that *cannot* be captured by the power-law degree distribution alone. These earlier studies all share a *common characteristic* in their approaches to capture more complex structures in networks: in addition to the power-law degree distribution, they introduce – and look for – the *scaling law* in another metric or form, e.g., the clustering coefficients [1], or the scaling factors between the degree distributions in renormalized networks using the box-counting approach [7].

   In this paper we bring a *multivariate analysis* perspective – in particular, the notion of *multivariate heavy tails* – to study structural properties of complex networks. As a generalization of the power law distributions in one dimension, "multivariate heavy tailed" (more precisely, *multivariate regularly varying* or *MRV* in short, see Sect. 3 for details) distributions embody more complex structures[1], and have been applied to a number of fields, e.g., multivariate time series analysis in finance to identify shared risks [2]. Like in the previous studies, this *multivariate analysis* perspective allows us to study the structural properties of complex networks using multiple metrics (that go beyond the degree distribution); but unlike the previous studies, it enables us to *explicitly and directly* examine the *dependence* structures defined by a number of different metrics, e.g., node degree and clustering coefficient, or degree-degree dependence structures, in complex networks. Such dependence structures cannot be revealed by studying each of the network statistical features in the marginal form alone. For example, intuitively "hierarchical" or "self-similar" structures introduce dependencies among sub-network structures at multiple scales and these dependencies are all explicitly built through *recursive construction* in the growing network

---

[1] For example, a multivariate distribution can have a power-law *marginal* distribution in each variable, but not jointly MRV, i.e., multivariate heavy-tailed.

models introduced in [1,3–5]. However, finding a scaling law in the marginal distribution in each of the metrics of interest (e.g., degrees and clustering coefficients) in a *real* network that matches those in the *synthetic* growing network models does not necessarily imply that the real network has the same dependence structures – at least theoretically speaking, those marginal heavy tails can occur *independently* in the network. Nonetheless, existence of MRVs in the joint distribution of these metrics provides a much stronger evidence for the "heavy-tailed" dependence structures. Hence we believe that existence of MRVs provides a better measure to capture the "hierarchical" or "self-similarity" structures in complex networks. Due to space limitation, we only provide a few examples in Sect. 4. In particular, we illustrate that the joint degree-clustering coefficient distribution in the synthetic growing networks of [1,3] is " multivariate heavy-tailed." Furthermore, we show that these two models have *distinct* structures in that the model in [3] contains a MRV joint *degree-degree* distribution, whereas that in [1] does not.

In summary, in this paper we advocate a *multivariate analysis* perspective based on the notion of MRVs as a unifying framework to study complex structures in networks. To the best of our knowledge, we believe ours is the first to apply such a framework in the study of complex networks, and demonstrate the existence of "multivariate heavy tails" in synthetic and real networks (Fig. 1).

2 Beyond the Power-Law Degree Distribution: An Overview of Existing Models

It is known that the structures of many real complex networks are not completely random: they are highly modular, and some have some "self-repeating" hierarchical patterns; these complex structures cannot be captured by the scale-free random network models such as the preferential attachment (PA) model, as shown in [1,5–7]. Several studies have attempted to capture this "self-similarity" or "hierarchy" either in the form of proposing deterministic graph models or in the form of suggesting a measure. They have common characteristic in their

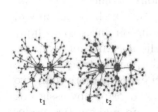

Fig. 1. Prefrential attachement model

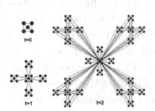

Fig. 2. Ravasaz *et al.* model.

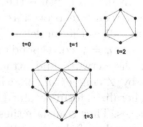

Fig. 3. Dorogovtsev *et al.* model.

approaches by finding scaling law relations in different forms and different metrics, in addition to power-law degree distribution. In the following, we present a brief overview of the existing models.

Ravasz and Barabasi [1] suggest the scaling law between the degree and clustering coefficient of the nodes $C(k) \sim k^{-\beta}$ as a quantity revealing the intrinsic hierarchy of the networks. They propose a deterministic graph model having this scaling law with an iterative construction leading to a hierarchical structure. The construction starts from a 5-vertex clique with one of the five nodes indicated as the center. It continues with replicating four copies of this cluster and connecting the center node of the central cluster to the peripheral nodes of the non-central clusters (see Fig. 2). Combining the scale-free topology with high modularity, the hierarchical modularity of the model results in the aforementioned scaling law between the node degrees and clustering coefficients. The authors believe that the hierarchy in real networks is the result of combining many small but densely connected clusters to form larger but less cohesive groups, with this process repeating recursively. They verify the suggested scaling law in a number of real networks and conclude that while this relation holds for many networks such as the Internet at the Autonomous System (AS) level and the world-wide-web (www) graphs, some "geographically organized" networks such as power grid networks and the Internet at the router level do not obey this scaling law, and therefore considered not hierarchical. The authors also suggest a stochastic variant of their deterministic network model.

Dorogovtsev et al. [3] propose a different deterministic network model which obeys the same scaling law between the node degree and clustering coefficient, in addition to the power-law degree distribution. The recursive construction of this graph starts from an edge connecting two nodes and it grows by adding a node for every edge in the network and attaching it to both ends of the edge (see Fig. 3). The random variation of this construction is creating a node per unit time and connecting it to both ends of a randomly chosen edge. Since this graph has no fixed finite fractal dimension, they call it a *pseudo-fractal* web. This network model also exhibits a strong short-range degree-degree correlation identified by $P(k, k') \sim k^{1-\gamma} k'^{-2}$.

A number of other deterministic graph models have been proposed in the literature to capture the "self-similarity property" of real networks; all resort to a recursive procedure but with different constructions. Comellas et al. [8] generalizes the model in [3] by starting from a clique (q-vertex clique for any q, where $q = 2$ corresponds to [3]) and continuing by adding one node per clique in the network and attaching it to all nodes in the clique. Another example is by Zhang et al. [9] which uses the basic Sierpinski Gasket structure or a generalized form obtained by dividing the edges of the triangles to more than two pieces. They translate these fractal geometrical structures to graphs by assigning the nodes of the graph to downward pointing triangles, and making two nodes connected if the boundaries of the corresponding triangles have touching points, considering the three sides of the outer triangle as three different nodes as well. The networks constructed thereof all have power-law degree distributions and obey the scaling law between the node degrees and clustering coefficients.

In contrast to the above deterministic network models which obey the scaling law between the node degree and clustering coefficients, Chen *et al.* [4] present a deterministic network model with a power-law degree distribution, but all nodes have zero clustering coefficient. It is recursively constructed from square-shaped elements, thus all nodes having zero clustering coefficient. They state that their network models are consistent with some real networks such as electronic circuits and the Internet at the router level which have reduced clustering coefficients. Because of its "modular" and "hierarchical" construction, the authors argue that their network models provide a counter-proof that hierarchical organization of modularity in complex networks must obey a scaling law between the node degrees and clustering coefficients as claimed in [1,6].

Several studies also attempt to directly capture the "self-similar" structures in complex networks by modifying the standard metrics or methods used in fractal geometry to defined and compute *fractal dimension*. For example, Song *et al.* [7] apply a modified version of the standard box-counting method to networks by placing nodes in "boxes" of varying sizes to obtain "coarser-versions" of the networks, and exploit the renormalization procedure to examine whether a scaling law between the degree distributions of the original networks and renormalized networks exists. If such a scaling law exists, they refer to such networks as *self-similar'* or *fractal*, otherwise they are non-fractal. The authors show that scale-free random networks generated by the PA model are non-fractal, and propose a variant of the model to produce fractal networks. They also verify this definition of fractality in a number of real networks such as www graphs and protein-protein interaction network (PIN) [6]. The authors believe that in fractal networks the growth takes place multiplicatively in a *correlated* "self-similar, modular" fashion. Examining the correlation profiles between the nodes and boxes within the networks' configurations, they find out that the fractal networks show a higher degree of *disassortativity* compared with the non-fractal networks – in other words, high-degree nodes in the networks are less likely to be connected to other high-degree nodes. They conclude that the key principle causing the fractal architecture of networks is a strong repulsion between the hubs (high degree nodes) on all length scales in a network. In addition, Kitsak *et al.* [13] apply the same definition of fractal networks in [5,7] to study the scaling law in node betweenness centrality. Through empirical data analysis, they observe that in scale-free fractal and non-fractal networks, the betweenness centrality obeys a scaling law, $P(b) \sim b^{-\gamma}$; but in fractal networks the exponent $\gamma = 2$, whereas in non-fractal networks $\gamma < 2$, with $2 - \gamma$ equal to the inverse of the fractal dimension of the network.

Lastly, s-metric is a self-similarity quantity suggested by Li *et al.* [10], which has a simple definition $s = \sum_{(i,j)\in E} k_i k_j$, calculating the multiplication of the degrees of the end points for all the edges in the network. The authors believe that the large value of s-metric is indicative of the self-similar networks. They propose a heuristic procedure to construct the s_{max} graph given a degree sequence and argue the self-similarity of this graph by coarse-graining it to a smaller graph which is also s_{max}, for the resulting truncated degree sequence.

3   Multivariate Heavy Tails: A Quick Primer

The theory of regularly varying functions is an essential analytical tool for dealing with heavy tails, long-range dependence and domains of attraction, see [2,11]. Roughly speaking, regularly varying functions are those functions which behave asymptotically like "power-law" functions. In the following, we provide a quick introduction to *multivariate regularly varying* (MRV) functions; the interested reader is referred to [2] for more details. We conclude with a proposition and a method of our own which provide a sufficient condition and a convenient tool to check for the existence of MRVs empirically.

Definition 1. [2] A measurable function $U : \mathbb{R}_+ \mapsto \mathbb{R}_+$ is *regularly varying* at ∞ with tail index $\alpha \in \mathbb{R}$ if for any $x > 0$,

$$\lim_{t \to \infty} \frac{U(tx)}{U(t)} = x^\alpha. \tag{1}$$

To check whether the distribution $F(x)$ of a single random variable X is regularly varying, the complementary cumulative distribution function (CCDF) is used in the definition above by substituting $U(x) = 1 - F(x)$ (for large values of x, $U(x)$ gives us the tail distribution). To get a better sense of Eq. (1) it is easy to check that it holds for a random variable x having the Pareto distribution, as a special case of regularly varying distributions, with parameters c and α, and cumulative distribution function $F(x) = 1 - (\frac{c}{x})^\alpha$. The equivalent form of the definition above for a regularly varying distribution with a tail index of α is as follow [11]:

$$\lim_{t \to \infty} t\mathrm{P}(X > t^{\frac{1}{\alpha}}x) = x^{-\alpha}, \text{ for } x > 0. \tag{2}$$

Empirically in data analysis the most convenient (and visual) method to verify whether a random variable has a regularly varying distribution (i.e., it is "heavy-tailed") is to plot its CCDF in a log-log scale (*cf.* the *QQ-plot*) and check its linearity. This is what we will use in this paper also. For an instance, our analysis in Fig. 5(a) shows that the degree of nodes as random variable X, has regularly varying distribution because of its linear behavior in log-log scale of CCDF plot. But random variable y in Fig. 9(b) is not regularly varying.

Generalizing **Definition** 1 to more than one random variable (or equivalently, to higher-dimensional measurable functions) gives us the notion of *multivariate regular variations* (MRVs). (We note that one cannot generalize the definition of "power-law" in Eq. (2) to more than one-dimension in a straightforward manner.) A *necessary* condition for a multivariate distribution (measurable function) to be *multivariate regularly varying* is that all of its *marginal* distributions must be regularly varying. In the following we provide the definition of MRV measurable functions/distributions.

Definition 2. Consider a random vector $\mathbf{X}$ with dimension $d(\geq 1)$ and a cone $C \subset \mathbb{R}^d$, where $\mathbf{1} = (1, \dots, 1) \in C$. We say a (non-decreasing) measurable function $U(\mathbf{x})$ defined on C, $U : C \mapsto [0, \infty)$, is *multivariate regularly varying* (MRV) on C with limit function $\lambda(\cdot)$ if $\lambda(\cdot) > 0$ and for all $\mathbf{x} \in C$ we have [2]

$$\lim_{t \to \infty} \frac{U(t\mathbf{x})}{U(t\mathbf{1})} = \lambda(\mathbf{x}). \tag{3}$$

Since in practical data analysis we often deal with *bivariate* distributions, we provide the equivalent form of the conditions for a random vector of two variables (X, Y) to be MRV as below:

(i) X has a regularly varying distribution with tail index of α,
(ii) the distribution of Y is regularly varying with tail index β,
(iii) the limit function $\mu(\cdot) > 0$ does exist for any choice of $(x, y) \in (0, \infty)$, in the following relation

$$\lim_{t \to \infty} tP(X > t^{1/\alpha}x, Y > t^{1/\beta}y) = \mu(x, y). \tag{4}$$

In empirical data analysis conditions (i) and (ii) can be checked using the standard log-log plot of the CCDF (or the Q-Q plot) of each variable. However, there is in general no easy visual tool to check for condition (iii). In the following we show that it suffices to verify condition (iii) by checking for the existence of the limit function along a particular line $(\breve{x}, r\breve{x})$ for a fixed $\breve{x}$ and $r \in (0, \infty)$ in the cone $(0, \infty)$.

Proposition. Assume that conditions (i) and (ii) hold true and that $\mu(\breve{x}, r\breve{x})$ in Eq. (4) exists for a fixed $\breve{x} > 0$ and all $r \in (0, \infty)$. Then the limit $\mu(x, y)$ in Eq. (4) exists for all possible pairs of $(x, y) \in (0, \infty)$.

Proof. First we note that the existence of the limit $\mu(x, y)$ along the X- or Y-axes $\{(x, 0), x > 0\}$, or $\{(0, y), y > 0\}$ is guaranteed by condition (i) or (ii). Hence we only need to consider the cases $x > 0, y > 0$. For a fixed $\breve{x} > 0$, we can rewrite (x, y) in the form of $(c_1\breve{x}, c_2\breve{x})$, where $c_1 := x/\breve{x} > 0$ and $c_2 := y/\breve{x} > 0$. Then we have

$$\lim_{t \to \infty} tP(X > t^{\frac{1}{\alpha}}x, Y > t^{\frac{1}{\beta}}y) = \lim_{t \to \infty} tP(X > t^{\frac{1}{\alpha}}c_1\breve{x}, Y > t^{\frac{1}{\beta}}c_2\breve{x})$$

$$= \lim_{t \to \infty} tP(X > (c_1^\alpha t)^{\frac{1}{\alpha}}\breve{x}, Y > (c_1^\alpha t)^{\frac{1}{\beta}}\frac{c_2}{c_1^{\alpha/\beta}}\breve{x})$$

$$= \frac{1}{c_1^\alpha} \lim_{t' \to \infty} t'P(X > t'^{\frac{1}{\alpha}}\breve{x}, Y > t'^{\frac{1}{\beta}}\frac{c_2}{c_1^{\alpha/\beta}}\breve{x})$$

$$= \frac{1}{c_1^\alpha}\mu(\breve{x}, r\breve{x}) > 0, \quad \text{where } r = c_2/c_1^{\alpha/\beta} > 0. \tag{5}$$

$\square$

Without loss of generality, we set $\breve{x} = 1$ throughout our analyses. Hence, this proposition reduces condition (iii) that the limit in Eq. (4) exists for the entire *quadrant* $\{(x, y) \in (0, \infty)\}$ to that along the *half-line* $\{(1, r) : r \in (0, \infty)\}$. However, checking the existence of the limit (of a bivariate function) along this half-line is still not straightforward, especially when applying to practical data practical. In the following we transform this problem into that of checking a *parameterized* family of *univariate* random variables $Z(r)$ are regularly varying. Substituting $(x, y) = (1, r)$ into Eq. (4), we have

$$\lim_{t \to \infty} t\mathrm{P}(X > t^{\frac{1}{\alpha}}, Y > t^{\frac{1}{\beta}}r) = \lim_{t \to \infty} t\mathrm{P}(X^{\alpha}r^{\beta} > tr^{\beta}, Y^{\beta} > tr^{\beta})$$

$$= \lim_{t \to \infty} t\mathrm{P}(\min\{X^{\alpha}r^{\beta}, Y^{\beta}\} > tr^{\beta})$$

$$= \lim_{t \to \infty} t\mathrm{P}(Z(r) > tz), \tag{6}$$

where the (univariate) random variable $Z(r) := \min\{X^{\alpha}r^{\beta}, Y^{\beta}\}$ and $z = r^{\beta}$. Hence if $Z(r)$ is regular varying with the tail index $\gamma = 1$ (*cf.*, Eq. (2)), i.e., $\lim_{t \to \infty} t\mathrm{P}(Z(r) > tz) = z^{-1}$ for any $z > 0$, then Eq. (6) holds for $z = r^{\beta}$.

The above results yield a convenient *direct* and *visual* tool to check for the existence of *bivariate heavy tails* when performing empirical data analysis. Using the same log-log plotting procedure described earlier for empirically checking the existence of univariate RVs, we first compute $Z(r)$ from the data for a range of different values of r, then plot the CCDF of $Z(r)$ in the log-log scale – in a sense this produces a form of (log-log) "contour" plot of the random variables $Z(r)$ – and check for the linearity of the "contours." For large $z > 0$, more linear all the contours appear, stronger the empirical evidence suggests the existence of MRVs in the data. In the next section we provide several examples as illustrations.

4   MRV in Networks

We now apply the theory of MRVs to the existing "hierarchical" and other network models proposed in the literature. In spite of their different constructions, a common characteristic these models share is that they contain *multivariate heavy tails* in one form or another. We also demonstrate that a number of real networks also contain complex structures indicative of multivariate heavy tails. This suggests the theory of MRVs as a *unifying* framework to study the more complex structures in networks.

We first consider the deterministic "hierarchical" network model proposed in [1] (see Fig. 2), where in addition to the power-law degree distribution, the node clustering coefficients also exhibits a "scaling law" as a function of the node degree. In Fig. 4(a, b), we plot the CCDF of the degree distribution $X := k_i$ and the *clustering coefficient inverse* (CCI), $Y = 1/C_i$, for all nodes i's in the log-log scale. The linearity of both plots indicates that both (marginal) distributions are indeed regularly varying (i.e., are heavy-tailed). Looking at the *joint degree-CCI* distribution, we define and compute $Z(r) := \{X^{\alpha}r^{\beta}, Y^{\beta}\}$ (where the tail indices α and β are estimated from Fig. 4(a, b)) for a range of r values (see the right bar in Fig. 4(c)). Using the method presented at the end of Sect. 3 (the plots in this section are best viewed *in color*), we plot the CCDF of Z's in the log-log scale for this range of r values. The linear behavior of CCDFs for a wide range of different ratios r indicates that the joint degree-CCI distribution is *bivariate heavy tailed*, capturing the "hierarchical" relation between the high-degree center nodes and the "modules" in the recursive construction of the network model in [1].

Applying the same MRV analysis to the "hierarchical" network model proposed in [3] (see Fig. 3), we show that the joint degree-CCI distribution of this model is also bivariate heavy-tail: Fig. 5(a, b) show the marginals are regularly

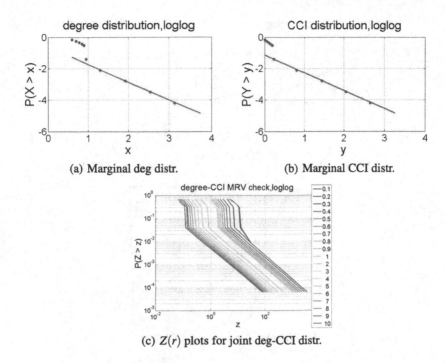

(a) Marginal deg distr.

(b) Marginal CCI distr.

(c) $Z(r)$ plots for joint deg-CCI distr.

Fig. 4. MRV Analysis for the Ravasz *et al.* model.

varying, while Fig. 5(c) verifies the existence of MRV. Comparing these two deterministic network models, besides their "hierarchy" (in the sense of [1]), the network models have very different structural properties. For instance, Dorogovtsev's model [3] grows hierarchically by "gluing" the smaller models at the high-degree nodes ("hubs") to form a larger cluster and repeating this process recursively. In contrast, Ravasz's model [1] grows hierarchically by recursively attaching the *peripheral nodes* of the non-central clusters to the *hub* of the central cluster. The distinction between these two models can be revealed when we examine and perform the MRV analysis on the *joint degree-degree* distributions[2] as shown in Fig. 6(a, b). The clear non-linearity of the "contours" in Fig. 6(a) indicates the lack of MRV for degree-degree pairs in Ravasz's model [1], whereas the joint degree-degree distribution in Dorogovtsev's model [3] is bivariate heavy-tailed.

That MRV analysis captures the *common* as well as *distinctive* structural characteristics of these two models illustrates the ability of the proposed MRV framework as a tool to help us better analyze and understand the *(hierarchical) structure* of complex networks: both networks are "hierarchical" in the sense of [1], but they differ in their "nature" of *hierarchy* (or near "self-similarity").

---

[2] Joint degree-degree pairs are defined for the edges of a network, where each degree belongs to one endpoint of an edge.

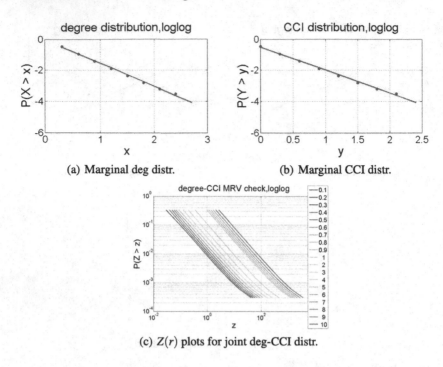

(a) Marginal deg distr.   (b) Marginal CCI distr.

(c) $Z(r)$ plots for joint deg-CCI distr.

Fig. 5. MRV Analysis for the Dorogovtsev *et al.* model.

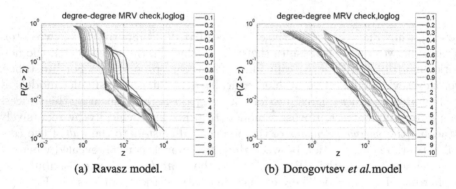

(a) Ravasz model.   (b) Dorogovtsev *et al.*model

Fig. 6. MRV Analysis for Joint Degree-Degree Distributions

In addition, we note that both networks have similar marginal features, such as power-law marginal degree and CCI distributions as well as the "small-world" property. This signifies that marginals alone cannot capture the *dependence structure* among various constituting network "modules" and thus are not as informative as when examining in a *multivariate* context.

We have also performed MRV analysis to other network models proposed in the literature. For example, we find that the deterministic network model *with zero clustering coefficients* proposed in [4] contains a bivariate heavy-tailed

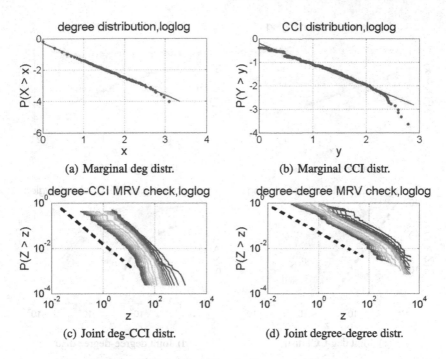

(a) Marginal deg distr.   (b) Marginal CCI distr.

(c) Joint deg-CCI distr.   (d) Joint degree-degree distr.

Fig. 7. MRV Analysis for the Internet AS dataset.

degree-degree distribution. This captures part of the "modular" structure built in the recursive construction process similar to that of [3], but using squares instead of triangles. (The authors claim that their network model is also "hierarchical" as a contradiction to the definition/claim made in [1]). In fact, we believe that instead of measuring "modularity" in terms of clustering coefficients, if one uses a generalized modularity metric (e.g., counting the number of squares a node's neighbors are in), this network model will likely exhibits a multivariate heavy tail in terms of degree and this generalized modularity metric. We also show that the "self-similar" or "fractal" networks as defined and identified by Song *et al.* in [5,7] (e.g., the PIN networks discussed below) contain MRVs in one form or another (e.g., joint degree-CCI distribution). Moreover, scale-free networks with self-similarity patterns suggested by high s-metric value [10], tend to show MRV in joint degree-degree distribution. Due to space limitation, we do not delve into details.

Last but not the least, we have also performed the MRV analysis on number of real network datasets. Due to space limitation, we provide only two examples: Internet at Autonomous System (AS) level and Protein-Protein Interaction (PIN) networks. These networks have been characterized as "hierarchical" [1,12] or "self-similar" [5,7,13]. The log-log plots of the marginal degree and CCI and joint degree-CCI as well as the joint degree-degree distribution for the AS network are shown in Fig. 7(a–d) (for the same range of r values as before,

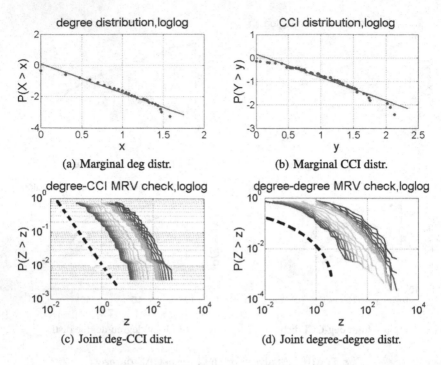

(a) Marginal deg distr.

(b) Marginal CCI distr.

(c) Joint deg-CCI distr.

(d) Joint degree-degree distr.

Fig. 8. MRV Analysis for the PIN dataset.

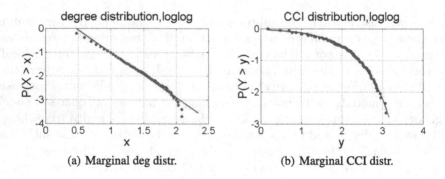

(a) Marginal deg distr.

(b) Marginal CCI distr.

Fig. 9. Marginals for the PA random network

which are not shown in Fig. 7(c, d)). Those for the PIN network are shown in
Fig. 8(a–d). We see that all marginals are (approximately) regularly varying.
The "contours" in the log-log plots for the degree-CCI distributions for both
networks are fairly linear (with the AS network showing stronger linearity in
the "tails"), suggestive of a bivariate heavy tail. On the other hand, the degree-
degree of the AS network contains a clear bivariate heavy tail, while that of the
PIN network lacks a clear bivariate heavy tail. At the end, we provide the MRV
analysis for a random network generated by PA method to emphasize that MRV

properties appear only in "hierarchical" or "self-similar" networks. Figure 9(a, b) show the marginals for degree and CCI of this network indicating the lack of even the *necessary* conditions for MRV in degree-CCI (CCI is not heavy-tailed). This network does not show MRV properties for the other metrics either.

5   Conclusion

We have advocated a *multivariate analysis* perspective based on the notion of MRVs as a unifying framework to study complex structures in networks. Applying the theory of MRVs to complex network analysis, we have demonstrated the existence of "multivariate heavy tails" in synthetic and real networks. Our analysis also poses a number of new research questions such as how to best characterize hierarchical or modular structures in complex networks. Answering these questions are part of ongoing research. To the best of our knowledge, we believe ours is the first to apply such a framework in the study of complex networks, and demonstrate the existence of "multivariate heavy tails" in synthetic and real networks.

Acknowledgment. This research was supported in part by DoD ARO MURI Award W911NF-12-1-0385, DTRA grant HDTRA1- 09-1-0050, and NSF grants CNS-10171647, CNS-1117536 and CRI-1305237.

References

1. Ravasz, E., Barabási, A.-L.: Hierarchical organization in complex networks. Phys. Rev. E **67**(2), 026112 (2003)
2. Resnick, S.: Heavy-Tail Phenomena: Probabilistic and Statistical Modeling. Springer, New York (2007)
3. Dorogovtsev, S.N., Goltsev, A., Mendes, J.F.F.: Pseudofractal scale-free web. Phys. Rev. E **65**(6), 066122 (2002)
4. Chen, L., Comellas, F., Zhang, Z.: Self-similar planar graphs as models for complex networks, arXiv preprint arXiv:0806.1258 (2008)
5. Song, C., Havlin, S., Makse, H.A.: Origins of fractality in the growth of complex networks. Nat. Phys. **2**(4), 275–281 (2006)
6. Ravasz, E., Somera, A.L., Mongru, D.A., Oltvai, Z.N., Barabási, A.-L.: Hierarchical organization of modularity in metabolic networks. Science **297**(5586), 1551–1555 (2002)
7. Song, C., Havlin, S., Makse, H.A.: Self-similarity of complex networks. Nature **433**(7024), 392–395 (2005)
8. Comellas, F., Fertin, G., Raspaud, A.: Recursive graphs with small-world scale-free properties, arXiv preprint cond-mat/0402033 (2004)
9. Zhang, Z., Zhou, S., Fang, L., Guan, J., Zhang, Y.: Maximal planar scale-free sierpinski networks with small-world effect and power law strength-degree correlation. EPL (Europhysics Letters) **79**(3), 38007 (2007)
10. Li, L., Alderson, D., Doyle, J.C., Willinger, W.: Towards a theory of scale-free graphs: Definition, properties, and implications. Internet Math. **2**(4), 431–523 (2005)

11. Resnick, S.: On the foundations of multivariate heavy-tail analysis. J. Appl. Probab. 191–212 (2004)
12. Vázquez, A., Pastor-Satorras, R., Vespignani, A.: Large-scale topological and dynamical properties of the internet. Phys. Rev. E **65**(6), 066130 (2002)
13. Kitsak, M., Havlin, S., Paul, G., Riccaboni, M., Pammolli, F., Stanley, H.E.: Betweenness centrality of fractal and nonfractal scale-free model networks and tests on real networks. Phys. Rev. E **75**(5), 056115 (2007)

Mixed Degree-Degree Correlations in Directed Social Networks

Michael Mayo[1]([✉]), Ahmed Abdelzaher[2], and Preetam Ghosh[2]

[1] Environmental Laboratory,
US Army Engineer Research and Development Center,
Vicksburg, MS, USA
michael.l.mayo@usace.army.mil
[2] Department of Computer Science,
Virginia Commonwealth University,
Richmond, VA, USA

Abstract. Many complex networks exhibit homophilic, or assortative degree mixing–the tendency for networked nodes to connect with others of similar degree. For social networks, this phenomenon is often referred to colloquially by the mantra 'your friends have more friends than you do.' We analyzed datasets for 16 directed social networks, and report that some of them exhibit both assortative (positive correlations) and disassortative (negative correlations) degree mixing across the totality of their degrees. We show that this mixed trend can be predicted based on the value of Pearson correlations computed for the directed networks. This stands in contrast to previous results reported for social networks that mark them as purely assortative. Finally, we discuss mechanisms by which these trends emerge from random models of network creation.

1 Introduction

Complex networks are graphs that link objects together into webs that exhibit nontrivial topological features, such as clustering [1], which are not generally found in canonical graphs, such as lattices. Such data are often collated from the real-world, which may include directed networks, such as transcriptional [42] or scientific co-authorship [32–34] networks, or undirected ones such as movie-actor collaboration [40] networks. Complex networks exhibit interesting topological properties, such as power-law degree distributions [4,5], short path-lengths [47], robustness to random failures [2], and resilience to rewiring [16] or adding new links [18]. For a given node, we may define its degree as the number of nodes connected to it, and the tendency of some nodes to connect with others of similar degree is termed homophily [30] or assortativity [35]. In an example from social networks, the tendency for socialites to befriend other socialites is advertised by the popular mantra, "...your friends have more friends than you do" [12]. In contrast, disassortative degree mixing refers to the phenomenon wherein nodes connect to others of dissimilar degree [35,36].

© Springer International Publishing Switzerland 2014
Z. Zhang et al. (Eds.): COCOA 2014, LNCS 8881, pp. 571–580, 2014.
DOI: 10.1007/978-3-319-12691-3_42

Assortative degree mixing has been widely recognized as a social phenomenon [30], though not exclusively [39]. Studies have shown that people more often interact with others based on similar levels of happiness [7] and even genetic similarity [14]. Assortative mixing may be an inherent and general feature of social networks, because it been observed before in the social interactions of other species, such as fish [11], which tend to interact assortatively based on body length and shoaling tendency. Similar observations have been made before in dolphins [27]. One mechanism that explains the assortativity observed in social networks relies on the existence of groups [37]: if the size of network communities vary, then the individuals mix assortatively.

While assortative mixing may emerge naturally in some social networks, it may not constitute a favorable design principle for man-made networks, which can be construed from the observation that assortative mixing appears infrequently in technological networks [35]. This begs the question of whether it may attribute unfavorable qualities to the network dynamics. Indeed, assortative degree mixing may lead to unstable fixed points [8]. However, technological networks often exhibit disassortative degree mixing [35], which has been shown, under certain conditions, to enhance synchronization [44]. Interestingly, some social networks have been shown to exhibit primarily disassortative degree mixing [17].

There is some evidence [28,38] that disassortativity in both directed and undirected networks at least partially stems from the restriction that node-pairs host at most one link–a demand that disallows any multi-link node-pairs. Intuitively, the high-degree nodes host fewer links than they normally would if unconstrained, which naturally lends more weight to the high-degree/low-degree node pairs, resulting in overall degree anticorrelations. Because this mechanism is general, Park and Newman conjectured [38] that disassortativity is the natural state of degree correlation in complex networks, and that social networks, in particular, are assortative due to great bias toward relationships that favor similar-degree pairings. However, another possibility is that an entropy-maximization principle determines whether disassortative networks are realized out of an ensemble of possible networks [19].

Here we test predictions that social networks exhibit strong assortative tendencies, by estimating the degree-degree correlation of a large number of directed social networks, ranging in size from thousands to millions of nodes. We employ Pearson [35] and Spearman [26] correlation coefficients to quantify the overall trend of the degree-degree relationships. However, by further examination of the point distribution, we demonstrate that Pearson correlation scores obscure the underlying details of the actual degree-degree relationship. In particular, we show that some networks exhibit both assortative and disassortative tendencies, wherein typically the assortative tendency can be observed in nodes with smaller degree, and anticorrelations can be generally observed in nodes of larger degree. In addition, some networks show no degree-degree correlation; according to the conjecture of Park and Newman, modest assortative tendencies are present even in degree-uncorrelated networks.

2 Methods

2.1 Network Datasets

We analyzed 16 datasets across four classes of directed social network: citation, communication, product co-purchasing, and online social networks. Each of these networks ranged in size from thousands to millions of nodes. Here we briefly describe each of these networks.

Citation networks map the relationship between scientific papers via the citation list. For example, a directed link from i to j is defined if paper i cites paper j. We used two networks extracted from the Arxiv preprint server, concerning high-energy physics phenomenology (HEP-PH) [15,23] and high-energy physics theory (HEP-TH) [15,23]. These data cover a period from January 1993 to April 2003.

We analyzed several communication networks. For the email network (Email-EU), its nodes are the individual addresses, and links between them are defined if an one address sends a message to another [24]. A snapshot of the Wikipedia page-edit history (3 January, 2008) was used to obtain another network, wherein nodes represent Wikipedia user profiles, and directed links are placed from i to j if user i edited a "Talk page" of user j. Finally, we employed two anonymized Facebook network datasets, the first mapping user-to-user links (Facebook Links [45]) and the other mapping the network of Wall posts between users (Facebook Wall [45]).

Co-purchasing networks, as reported in [20], were built using Amazon.com's "Also Bought" feature; if a product i is frequently co-purchased with product j, then a directed link is placed from i to j. These networks were created by crawling the Amazon.com website on dates: 2 March, 12 March, 5 May, and 1 June 2003, with each date resulting in a distinct, directed network.

Finally, we analyzed six online social network datasets. One dataset reflect trust relationships between users of the Advogato online community [29]. We employed two 'Slashdot Zoo' networks [25], wherein users (the nodes) may assign one another as either "friend" or "foe." These networks were obtained in November 2008 (Slashdot Zoo-2008) or February 2009 (Slashdot Zoo-2009). The Epinions network [41] is a network generated from the consumer review site Epinions.com, wherein members declare "trust relationships" that collectively form a Web of Trust. The Wikipedia Vote network was obtained from a dump of Wikipedia page-edit history (January 3 2008), from which administrator elections and vote history data were extracted [21,22]. Finally, a Youtube network (Youtube Growth) was built from a record of its users and their friendship connections [31].

2.2 Degree-Degree Correlations in Directed Networks

Whether or not a network exhibits assortative or disassortative degree mixing is determined by calculating the level of degree-degree correlation. This metric has been calculated before in two ways. First, by using the Pearson correlation, which correlates node-degrees at either end of a link, for both directed [13,39] and undirected networks [35]; or, by comparing the average neighbor degree to the degree of a given node [3,6,10,43,46]. Here, we calculate both of these metrics and compare them for a variety of directed social networks.

As shown in Fig. 1, there are four ways to define potential degree-degree correlations in a directed network. To quantify each relationship, the Pearson correlation, r, can be calculated with the equation [13]:

$$r(x,y) = \frac{L^{-1}\sum_{i=1}^{L}(x_{i,\mathrm{s}} - \langle x_\mathrm{s}\rangle)(y_{i,\mathrm{t}} - \langle y_\mathrm{t}\rangle)}{\sigma^x \sigma^y}. \tag{1}$$

Here, L denotes the total number of links in the network; $x, y \in \{R, K\}$ wherein R is a node's in-degree and K its out-degree; the subscripts s and t respectively denote the source and target nodes of the ith link. In Eq. 1, $\langle x_\mathrm{s}\rangle = L^{-1}\sum_{i=1}^{L} x_{i,\mathrm{s}}$ and $\sigma^x = \sqrt{L^{-1}\sum_{i=1}^{L}(x_{i,\mathrm{s}} - \langle x_\mathrm{s}\rangle)^2}$. The quantities $\langle y_\mathrm{t}\rangle$ and σ^y are similarly defined.

Others have measured degree-degree correlations by evaluating the curve that results when comparing node degree to its average neighbor degree [3,6, 10,43,46]. For these purposes, we label the in-degree of the ith node by R_i, its out-degree by K_i, and the associated arithmetic average of neighbor in- and out-degrees, respectively, by $\langle R^{nn}\rangle$ and $\langle K^{nn}\rangle$.

3   Results and Discussion

3.1   Mixed Positive and Negative Degree-Degree Correlations Exhibited by Social Networks

Many of the networks listed in Table 1 exhibit a qualitative "biphasic" trend in the degree-degree curves. One example is illustrated by Fig. 2, and the remaining such curves were omitted due to space constraints. How might these qualitatively different trends emerge from the underlying social interactions? We suggest that a competition between "positive" and "negative" affecters imposes these scale-dependent phenomenon. Park and Newman's result [38] that the natural state of a growing network is disassortative suggests that a tapering off of assortative social pressure for low-degree members to connect to the more popular ones may explain the non-monotonic nature of the curves (e.g. Fig. 2). Similar two-phase behavior has been observed before in real-world networks, e.g., involving Wikipedia [9].

Interestingly, all networks of Table 1 that exhibit qualitative dual correlation trends in the degree-degree curves, such as shown in Fig. 2 and denoted by asterisks in Table 1, also exhibit positive and negative values for the Pearson correlations across all four correlation types (Fig. 1). This stands in contrast to networks wherein positive and negative trends could not be identified qualitatively in their degree-degree curves, such as with the communication networks of Table 1, which host either all positive or all negative Pearson correlations.

Table 1. Structural and correlation properties of the network datasets used in this study. Pearson correlation metrics (Eq. 1) were calculated for each of the four degree-degree relationships shown in Fig. 1. Asterisks denote qualitatively non-monotonic averaged-neighbor degree vs degree curves (e.g., $\langle K^{nn} \rangle(K)$).

Network	Nodes	Links	Reference	Pearson correlations			
				in-in	in-out	out-in	out-out
Citation networks							
HEP-PH	34,546	421,578	[15, 23]	0.191	−0.0227	0.0361*	0.0636*
HEP-TH	27,770	352,807	[15, 23]	0.200	0.00474*	0.0240*	0.0811*
Communication networks							
Email-EU	265,214	420,045	[24]	−0.675	−0.170	−0.876	−0.214
Wikipedia Talk	2,394,385	5,021,410	[21, 22]	−0.0384	−0.00422	−0.0563	−0.00502
Facebook Links	63,732	1,545,686	[45]	0.160	0.161	0.161	0.162
Facebook Wall	63,892	876,993	[45]	0.438	0.441	0.472	0.454
Product co-purchasing networks							
Amazon-20030302	262,110	1,234,877	[20]	0.311*	0.0401	0.0250	0.148
Amazon-20030312	400,726	3,200,440	[20]	0.132*	0.0468*	−0.431*	0.319
Amazon-20030505	410,235	3,356,824	[20]	0.196*	0.0570	−0.378	0.222
Amazon-20030601	403,393	3,387,388	[20]	0.204*	0.0498	−0.365	0.204
Online social networks							
Advogato	3,302	32,954	[29]	−0.102	−0.0186	−0.216*	−0.0171
Slashdot Zoo-2008	77,359	905,468	[25]	−0.0488*	−0.0433*	−0.0501*	−0.0465*
Slashdot Zoo-2009	82,168	948,464	[25]	−0.0471*	−0.0419*	−0.0490*	−0.0456*
Epinions	75,887	508,837	[41]	0.102	0.0415	−0.0993*	−0.00578*
Wikipedia Vote	8,297	103,689	[21, 22]	0.00993	−0.00364	−0.0633	−0.00970
Youtube Growth	1,157,827	4,945,382	[31]	−0.0291	−0.0293	−0.0329	−0.0332

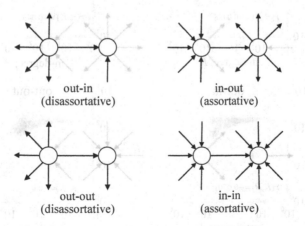

Fig. 1. All four types of degree-degree correlation in directed networks [13].

Based on the examples of Table 1, we conclude that the Pearson correlation may objectively identify networks with mixed degree-degree correlation trends, if the scores are taken holistically across all four correlation types in directed networks.

3.2 Mixed Degree-Degree Correlations and Generated Random Networks

The degree distribution cannot always be used to inform the extent of degree-degree correlation, because example networks with a given degree sequence can be provided that are free of such correlations [10]. This is problematic for network "growth" models that aim to reproduce a given set of topological network features using node-by-node attachment algorithms based purely on the node degree.

However, there is at least one growth mechanism able to explain the dual assortative/disassortative features observed in many of the social networks analyzed here in Table 1. In Reference [46], a growth model based on a random walk principle is shown to closely reproduce degree-degree correlation trends observed in the full curves (e.g. Fig 2). Whether the mixed degree-degree correlation emerges depends on values of a parameter, labeled by q_e, which is the probability a neighbor degree is added to a node of the graph [46]. For larger values of q_e, the generated networks create a peak in the averaged neighbor degree at approximately node degree = 10 [46]. This can be observed in Fig. 3, which

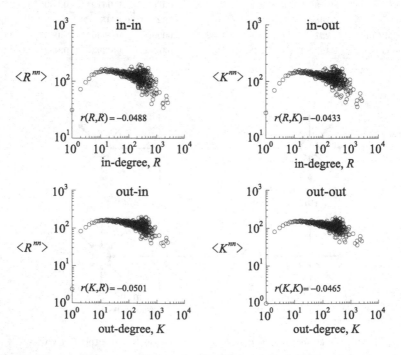

Fig. 2. Degree-degree correlations in the Slashdot Zoo-2008 social network.

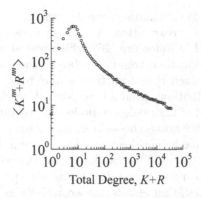

Fig. 3. Average neighbor total degree plotted against the total degree, for a random network built using the "random walk" method of Reference [46]. Figure adapted from Reference [46].

has been adapted from Reference [46], and shows the averaged neighbor degree plotted against the total degree, for a random network built using the parameter value $q_e = 0.5$.

While the authors provide no reasoning for the cross-over positioned at approximately $K + R \approx 10$, it is clearly related to the node-level growth mechanism quantified by q_e. However, given a random growth scheme and a scale-free network with negative power-law exponent, the lower degree nodes must select from an attachment pool filled with higher degree nodes, while the higher degree nodes must select from a pool filled with lower degree nodes. Thus, based on a purely random mechanism, we should expect an increasing degree-degree trend in the lower-degree nodes, and a decreasing trend in the higher-degree nodes. Indeed, this trend is observed in both Figs. 2 and 3.

4   Conclusions

We have analyzed 16 different social network datasets using two different methods often employed in the literature to assess degree-degree correlations (Table 1). The first is the Pearson correlation score, which has been recently extended to directed networks. The second is the long-standing practice of averaging nearest neighbor degrees and comparing them to the node degree, which results in a curve; e.g. $\langle K^{nn} \rangle (K)$.

We have reported that, in several social networks, this curve exhibits both increasing (positively correlated) and decreasing (negatively correlated) trends. For networks with dual non-monotone degree-degree curves, the low-degree nodes are generally positively correlated, while high-degree nodes are often negatively correlated. Based on a directed-network version of the Pearson correlation (Eq. 1), we showed that an overall non-monotone degree-degree trend, e.g. Fig. 1, could be predicted using Pearson correlation scores taken across all four types of degree correlation shown in Fig. 1.

We examined possible explanations for the emergence of this non-monotone trend in the degree-degree correlation. A random node-attachment mechanism was previously proposed in Reference [46], which was able to generate networks representative of non-monotone trends similar to those of the networks given in Table 1. We hypothesized that such trends may result from the form of the underlying degree distribution: given a power-law degree distribution with negative exponent, the pool of higher-degree nodes will be rarer than those of lower degree. Thus, lower-degree nodes choose from an attachment pool biased toward higher-degree nodes, while higher-degree nodes choose from an attachment pool awash with lower-degree nodes. That random networks generate an overall disassortative degree-degree correlation trend is in line with previous results obtained by Park and Newman [38], that claims disassortativity as the "natural" state of a complex network, social or otherwise.

Acknowledgments. This work was funded in part by the US Army's Environmental Quality and Installations 6.1 basic research program. Opinions, interpretations, conclusions, and recommendations are those of the author(s) and are not necessarily endorsed by the U.S. Army.

References

1. Albert, R., Barabasi, A.: Statistical mechanics of complex networks. Rev. Mod. Phys. **74**, 47 (2002)
2. Albert, R., Jeong, H., Barabasi, A.: Error and attack tolerance of complex networks. Nature **406**, 378–382 (2000)
3. Bagler, G., Sinha, S.: Assortative mixing in protein contact networks and protein folding kinetics. Bioinformatics **23**, 1760–1767 (2007)
4. Barabasi, A., Albert, R.: Emergence of scaling in random networks. Science **286**, 509–512 (1999)
5. Barabasi, A., Albert, R., Jeong, H.: Mean-field theory for scale-free random networks. Physica A **272**, 173–187 (1999)
6. Barrat, A., Barthelemy, M., Pastor-Satorras, R., Vespignani, A.: The architecture of complex weighted networks. Proc. Natl. Acad. Sci. U.S.A. **101**, 3747 (2004)
7. Bollen, J., Gonalvesj, B., Ruan, G., Mao, H.: Happiness is assortative in online social networks. Artif. Life **17**, 237–251 (2011)
8. Brede, M., Sinha, S.: Assortative mixing by degree makes a network more unstable. arXiv preprint cond-mat, p. 0507710 (2005)
9. Capocci, A., Servedio, V., Colaiori, F., Buriol, L., Donato, D., Leonardi, S., Caldarelli, G.: Preferential attachment in the growth of social networks: The internet encyclopedia wikipedia. Phys. Rev. E **74**, 036116 (2006)
10. Catanzaro, M., Bogu, M., Pastor-Satorras, R.: Generation of uncorrelated random scale-free networks. Phys. Rev. E **71**, 027103 (2005)
11. Croft, D., James, R., Ward, A., Botham, M., Mawdsley, D., Krause, J.: Assortative interactions and social networks in fish. Oecologia **143**, 211–219 (2005)
12. Feld, S.: Why your friends have more friends than you do. Am. J. Sociol. **96**, 1464–1477 (1991)
13. Foster, J., Foster, D., Grassberger, P., Paczuski, M.: Edge direction and the structure of networks. Proc. Natl. Acad. Sci. U.S.A **107**, 10815–10820 (2010)

14. Fowler, J., Settle, J., Christakis, N.: Correlated genotypes in friendship networks. Proc. Natl. Acad. Sci. U.S.A. **108**, 1993–1997 (2011)
15. Gehrke, J., Ginsparg, P., Kleinberg, J.M.: Overview of the 2003 kdd cup. ACM SIGKDD Newsl. **5**, 149–151 (2003)
16. Holme, P., Kim, B., Yoon, C., Han, S.: Attack vulnerability of complex networks. Phys. Rev. E **65**, 056109 (2002)
17. Hu, H., Wang, X.: Disassortative mixing in online social networks. Europhys. Lett. **86**, 18003 (2009)
18. Isalan, M., Lemerle, C., Michalodimitrakis, K., Horn, C., Beltrao, P., Raineri, E., Garriga-Canut, M., Serrano, L.: Evolvability and hierarchy in rewired bacterial gene networks. Nature **452**, 840 (2008)
19. Johnson, S., Torres, J., Marro, J., Munoz, M.: Entropic origin of disassortativity in complex networks. Phys. Rev. Lett. **104**, 108702 (2010)
20. Leskovec, J., Adamic, L., Adamic, B.: The dynamics of viral marketing. ACM Trans. Web **1**, 5 (2007)
21. Leskovec, J., Huttenlocher, D., Kleinberg, J.: Predicting positive and negative links in online social networks. In: Proceedings of the 19th International Conference on World wide web, pp. 641–650. ACM (2010)
22. Leskovec, J., Huttenlocher, D., Kleinberg, J.: Signed networks in social media. In: Proceedings of the SIGCHI Conference on Human Factors in Computing Systems, pp. 1361–1370. ACM (2010)
23. Leskovec, J., Kleinberg, J., Faloutsos, C.: Graphs over time: densification laws, shrinking diameters and possible explanations. In: Proceedings of the Eleventh ACM SIGKDD International Conference on Knowledge Discovery in Data Mining, pp. 177–187. ACM (2005)
24. Leskovec, J., Kleinberg, J., Faloutsos, C.: Graph evolution: Densification and shrinking diameters. ACM Trans. Knowl. Discov. Data **1**, 2 (2007)
25. Leskovec, J., Lang, K., Dasgupta, A., Mahoney, M.: Community structure in large networks: Natural cluster sizes and the absence of large well-defined clusters. Internet Math. **6**, 29–123 (2009)
26. Litvak, N., van der Hofstad, R.: Uncovering disassortativity in large scale-free networks. Phys. Rev. E **87**, 022801 (2013)
27. Lusseau, D., Newman, M.: Identifying the role that animals play in their social networks. P. Roy. Soc. B-Biol. Sci. **271**, S477–S481 (2004)
28. Maslov, S., Sneppen, K., Zaliznyak, A.: Detection of topological patterns in complex networks: Correlation properties of the internet. Physica A **333**, 529–540 (2004)
29. Massa, P., Salvetti, M., Tomasoni, D.: Bowling alone and trust decline in social network sites. In: Eighth IEEE International Conference on Dependable, Autonomic and Secure Computing, 2009. DASC '09, pp. 658–663. IEEE (2009)
30. McPherson, M., Smith-Lovin, L., Cook, J.: Birds of a feather: Homophily in social networks. Annu. Rev. Sociol. **27**, 415–444 (2001)
31. Mislove, A., Marcon, M., Gummadi, K., Druschel, P., Bhattacharjee, B.: Measurement and analysis of online social networks. In: Proceedings of the 7th ACM SIGCOMM Conference on Internet Measurement. pp. 29–42. ACM (2007)
32. Newman, M.: Scientific collaboration networks I. Network construction and fundamental results. Phys. Rev. E **64**, 016131 (2001)
33. Newman, M.: Scientific collaboration networks II. shortest paths, weighted networks, and centrality. Phys. Rev. E **64**, 016132 (2001)
34. Newman, M.: The structure of scientific collaboration networks. Proc. Natl. Acad. Sci. U.S.A. **98**, 404–409 (2001)

35. Newman, M.: Assortative mixing in networks. Phys. Rev. Lett. **89**, 208701 (2002)
36. Newman, M.: Mixing patterns in networks. Phys. Rev. E **67**, 026126 (2003)
37. Newman, M., Park, J.: Why social networks are different from other types of networks. Phys. Rev. E **68**, 036122 (2003)
38. Park, J., Newman, M.: Origin of degree correlations in the internet and other networks. Phys. Rev. E **68**, 026112 (2003)
39. Piraveenan, M., Prokopenko, M., Zomaya, A.: Assortative mixing in directed biological networks. IEEE/ACM Trans. Comput. Biol. Bioinform. (TCBB) **9**, 66–78 (2012)
40. Ramasco, J., Dorogovtsev, S., Pastor-Satorras, R.: Self-organization of collaboration networks. Phys. Rev. E **70**, 036106 (2004)
41. Richardson, M., Agrawal, R., Domingos, P.: Trust management for the semantic web. In: Fensel, D., Sycara, K., Mylopoulos, J. (eds.) ISWC 2003. LNCS, vol. 2870, pp. 351–368. Springer, Heidelberg (2003)
42. Salgado, H., Peralta-Gil, M., Gama-Castro, S., Santos-Zavaleta, A., Muiz-Rascado, L., Garca-Sotelo, J.S., Weiss, V., Solano-Lira, H., Martnez-Flores, I., Medina-Rivera, A., Salgado-Osorio, G., Alquicira-Hernndez, S., Alquicira-Hernndez, K., Lpez-Fuentes, A., Porrn-Sotelo, L., Huerta, A.M., Bonavides-Martnez, C., Balderas-Martnez, Y.I., Pannier, L., Olvera, M., Labastida, A., Jimnez-Jacinto, V., Vega-Alvarado, L., del Moral-Chvez, V., Hernndez-Alvarez, A., Morett, E., Collado-Vides, J.: Regulondb v8.0: omics data sets, evolutionary conservation, regulatory phrases, cross-validated gold standards and more. Nucleic Acids Res. **41**(D1), D203–D213 (2013)
43. Serrano, M., Maguitman, A., Bogu, M., Fortunato, S., Vespignani, A.: Decoding the structure of the www: A comparative analysis of web crawls. ACM Trans. Web **1**, 10 (2007)
44. Sorrentino, F., Bernardo, M.D., Garofalo, F.: Synchronizability and synchronization dynamics of weighed and unweighed scale free networks with degree mixing. Int. J. Bifurcat. Chaos **17**, 2419–2434 (2007)
45. Viswanath, B., Mislove, A., Cha, M., Gummadi, K.P.: On the evolution of user interaction in facebook. In: Proceedings of the 2nd ACM SIGCOMM Workshop on Social Networks (WOSN'09) (August 2009)
46. Vazquez, A.: Growing network with local rules: Preferential attachment, clustering hierarchy, and degree correlations. Phys. Rev. E **67**, 056104 (2003)
47. Watts, D., Strogatz, S.: Collective dynamics of 'small-world' networks. Nature **393**, 440–442 (1998)

Social and Economic Network Formation:
A Dynamic Model

Omid Atabati[1][(✉)] and Babak Farzad[2]

[1] Department of Economics, University of Calgary, Calgary, AB T2N 1N4, Canada
oatabati@ucalgary.ca
[2] Department of Mathematics, Brock University,
St Catharines, ON L2S 3A1, Canada
bfarzad@brocku.ca

Abstract. We study the dynamics of a game-theoretic network formation model that yields large-scale small-world networks. So far, mostly
stochastic frameworks have been utilized to explain the emergence of
these networks. On the other hand, it is natural to seek for game-theoretic
network formation models in which links are formed due to strategic
behaviors of individuals, rather than based on probabilities. Inspired by
Even-Dar and Kearns' model [8], we consider a more realistic framework
in which the cost of establishing each link is dynamically determined during the course of the game. Moreover, players are allowed to put transfer
payments on the formation and maintenance of links. Also, they must
pay a maintenance cost to sustain their direct links during the game. We
show that there is a small diameter of at most 4 in the general set of
equilibrium networks in our model. We achieved an economic mechanism
and its dynamic process for individuals which firstly; unlike the earlier
model, the outcomes of players' interactions or the equilibrium networks
are guaranteed to exist. Furthermore, these networks coincide with the
outcome of pairwise Nash equilibrium in network formation. Secondly; it
generates large-scale networks that have a rational and strategic micro-
foundation and demonstrate the main characterization of small degree of
separation in real-life social networks. Furthermore, we provide a network
formation simulation that generates small-world networks.

Keywords: Network formation · Linking game with transfer payments ·
Pairwise stability · Pairwise nash equilibrium · Small-world phenomenon

1 Introduction

In recent years, networks have been extensively studied mostly in terms of their
structure, but also their formation and dynamics. Structural characteristics of
various networks, which emerge from disciplines, such as economics, computer
science, sociology, biology and physics, have been investigated. Many of these
networks, in spite of their different origins, indicate large commonalities among

© Springer International Publishing Switzerland 2014
Z. Zhang et al. (Eds.): COCOA 2014, LNCS 8881, pp. 581–592, 2014.
DOI: 10.1007/978-3-319-12691-3_43

their key structural properties, such as small diameter, high clustering coefficient, and heavy-tailed degree distribution which are often quantified by power-law probability distributions. Hence, it is an exciting challenge to study network formation models capable of explaining how and why these structural commonalities both occur and evolve. The series of experiments by Milgram in the 1960s [17] were among the pioneering works that quantified the *small-world phenomenon*[1] and introduced the "six degree of separation". Recent experiments [6] showed that today's online social networks such as Facebook indicate that the degree of separation (for almost any two individuals in a given database) must be even smaller than 4.

The *small-world model* by [20] was one of the first models which generates networks with small diameter. This work followed by Kleinberg's stochastic model [16] that was located in a grid graph. It introduced a process that adds links with distance d to the grid with a probability proportional to $1/d^\alpha$. These models, however, can not be applicable when there is a strategical purpose in players' making or losing their connections. In these cases, players, which are represented by vertices, strategically establish and sever their connections to obtain an advantageous position in the social structure. Hence, we refer to a class of *game-theoretic network formation*, also known as *strategic network formation* (See [7,12] for comprehensive surveys). Models in this class are in their early efforts. They generally assume that players make connections based on a utility maximization and treat the network as the equilibrium result of the strategic interactions among players.

1.1   Our Contribution

Our game-theoretic network formation model is mainly inspired by Even-Dar and Kearns [8]. In their model, players (i.e., vertices) seek to minimize their collective distances to all other players. The network formation starts from a seed grid. Also, the cost of establishing each link in this model is considered to be the grid distance between the endpoint players of that link to the power of α, which is the parameter of the model. Hence, their model uses a *fixed link-pricing* for each link. Both link creation and link severance are considered unilateral by players. In addition, the equilibrium is defined in terms of *link stability*: no players benefit from altering a single link in their link decisions. Even-Dar and Kearns' model achieves small diameter link stable networks within the threshold of $\alpha = 2$. However, they faced an unbounded diameter that grows with the number of players, when $\alpha > 2$.

In this paper, we define three types of costs for links: (i) the link-price, (ii) the maintenance cost, and (iii) the transfer payment. The link-price p_{ij} is the price of establishing link ij. Only the initiator of connection would bear its payment. It is a one-time charge when establishing the link. We introduce a new viewpoint to this game that better echoes with reality by constructing

---

[1] The principle that individuals are all linked by short chains of connections and acquaintances.

a *dynamic link-pricing*. When characterizing the formation of a network, the involved dynamics is a crucial and determining element. We aim to effectuate the impact of this dynamics in our model with the revised link-pricing. We update the used distances of each pair of players in the related link-prices from the current network rather than sticking with the initial grid distances.

In addition, we introduce maintenance costs to make the model more real where a player can give up her payment and sever her connection. Also, it is reasonable to assume that refunding the link-prices may not be possible in lots of real-world scenarios. Hence, maintenance costs make the link severance scenario well-defined. In our model, player i is charged for all of its incident links by considering recurring maintenance costs c_{ij}. In other words, for each decision made in the game, players should take the maintenance cost of their incident links into their consideration. Lastly, we allow individuals to put transfer or side payments on their links. Transfers are a sort of communication between players for their connections. In fact, without transfer payments, many agreements on these connections would simply never exist.

In this paper, we use the myopic notion of Pairwise Stability with direct and indirect transfers (PS^t)[2] as our equilibrium notion. This notion has the advantage of being compatible with the cooperative and bilateral nature of link formation. Moreover, the pairwise stability has the desirable simplicity required for analyzing players' behaviors under this notion.[3]

On the other hand, due to the bilateral agreement for any link formation, the typical notion of Nash equilibria has some drawbacks in terms of coordination failures; e.g. an empty network is always a Nash equilibrium. In other words, Nash equilibria networks can contain some mutually beneficial link(s) that are left aside. To solve this coordination problem when employing Nash equilibria, the notion of pairwise Nash stability[4] was introduced. Pairwise Nash Stable (PNS^t) networks are at the intersection of the set of Nash equilibrium networks and the set of pairwise stable networks.

In this paper, we not only guarantee the existence of pairwise stable networks[5], but also demonstrate that, in our model, the set of pairwise stable networks coincide with the set of pairwise Nash stable networks. Finally, we show that the general set of equilibrium networks exhibits a short diameter of at most 4 as desired in social network. The rest of this paper is organized as follows. In Sect. 2, we explain the required preliminaries and provide the setup of our model. We then provide the analysis for our grid-based model with dynamic link-pricing and transfer payments in Sect. 3. In Sect. 4, we present the outcome of a network formation simulation.

---

[2] The pairwise stability is the major notion of stability that assumes myopic players and has been studied in related literature. In a linking game with transfers, it was first introduced as an extension in [15] and then developed in [2,3].

[3] Computing the best responses of players in Nash equilibria within some similar models [9,18] are proved to be NP-hard.

[4] See [2–4,11].

[5] It can be seen that the condition for ruling out the potential cycles from [13] can be adapted in our linking game with transfers.

2   Preliminaries

The network and players. Let $N = \{1, ..., n\}$ be the set of n players forming a network G. Network G is undirected and includes a list of pairs of players who are linked to each other. Link $ij \in G$ indicates that player i and player j are linked in G. Let G^N denote the complete network. The set $\mathcal{G} = \{G \subseteq G^N\}$ consists of all possible networks on N. We define network G_0 to be the starting network of the game, which is also called the *seed network*. The set of player i's neighbors in G is $\mathcal{N}_i(G) = \{j | ij \in G\}$. Similarly, $\mathcal{L}_i(G) = \{ij \in G \mid j \in \mathcal{N}_i(G)\}$ denotes the set of links, which are incident with player i in G. If l is a subset of $\mathcal{L}_i(G)$, then $G - l$ is the network resulted by removing the existing links in the set l from G. Similarly, if $l = \{ij \mid j \notin \mathcal{N}_i(G), j \neq i\}$, then the network $G + l$ is obtained by adding the links in set l to G.

The *utility* of network G for player i is given by a function $u_i : G \rightarrow \mathbb{R}^+$. Let $\mathbf{u}$ denote the vector of utility functions $\mathbf{u} = (u_1, ..., u_n)$. So, $\mathbf{u} : \mathcal{G} \rightarrow \mathbb{R}^N$. Also, the value of a network, $v(G)$, is the summation of all players' utilities in the network G; i.e., $v(G) = \sum_{i=1}^n u_i(G)$. For any network G and any subset $l_i(G) \subseteq \mathcal{L}_i(G)$, the marginal utility for a player i and a set of links $l_i(G)$ is denoted by $mu_i(G, l_i(G)) = u_i(G) - u_i(G - l_i(G))$.

Strategies; transfer payments. Each player $i \in N$ announces an action vector of transfer payment $\mathbf{t}^i \in \mathbb{R}^{n(n-1)/2}$. The entries in this vector indicate the transfer payment that player i offers (to pay) or demands (to gain) on the link jk. If $i \in \{j, k\}$, then we call it a *direct* transfer payment. Otherwise, it is called an *indirect* transfer payment. Typically, individuals can make demands (negative transfers) or offers (non-negative transfers) on their direct connections. However, they can only make offers (and not demands) on the indirect transfer payments.[6] In addition, a link jk is formed if and only if $\sum_{i \in N} t^i_{jk} \geq 0$. Thus, the profile of strategies or the announced vectors of transfer payments for all players is defined: $\mathbf{t} = (\mathbf{t}^1, ..., \mathbf{t}^n)$. Consequently, the network G, which is formed by this profile of strategies t, can be denoted as follows:

$$G(\mathbf{t}) = \{jk \mid \sum_{i \in N} t^i_{jk} \geq 0, \text{ where } j, k \in N\}.$$

The payoff function. The *distance* between a pair of players i and j in G, denoted by $d_G(i, j)$, is defined as the length of a shortest path between i and j in G. Similar to the model of Even-Dar and Kearns, players seek to minimize their total distances to all players. This benefit would be considered for each player with respect to the network G and links benefit both endpoints.[7] The link-price is defined to be $p_{ij} = d_G(i, j)^\alpha$ for $\alpha > 0$. The link-price function is non-decreasing and follows Kleinberg's stochastic model. Also, function c_{ij} denotes the maintenance cost for the link ij. The *utility function* of player i is the negative of her total distances and links expenses and is defined as follows:

---

[6] This assumption is reasonable in our framework, since the formation of other links cannot hurt the utility of non-involved players with respect to the distance-based structure of our utility function in (1).

[7] See e.g. [3,9,15] for some application instances of distance-based payoff structures.

$$u_i(G(\mathbf{t})) = -\sum_{j \in N} d_{G(\mathbf{t})}(i, j) - \sum_{j \in \mathcal{N}_i} (p_{ij} + c_{ij}) - \sum_{jk \in G(\mathbf{t})} t^i_{jk}. \tag{1}$$

The dynamic process. The following notion is stated from [13] that motivates the desired dynamics for our analysis.

Definition 1. *An improving path represents a sequence of changes from one network to another. The changes can emerge when individuals create or sever a single link based on the improvement in the resulting network relative to the current network.*

In each round of the game, one player adapts her strategy with respect to the current state of the network. We assume a random meeting mechanism for vertices (randomly choosing a pair of players), but we start with a seed network instead of an empty network [14, 19]. If two networks G and G' differ in exactly one link, they are said to be *adjacent* networks. Also, if there exists an improving path from G to G', then G' *defeats* G.

The equilibrium strategies. In every equilibrium profile of strategies $\mathbf{t}^*$, there is no excess in the offer of transfer payments. A transfer payment t^{*i}_{ij} is negative, if and only if there is a utility gap for player i in maintenance of link ij, or in other words, keeping link ij is not beneficial for i. Also, player i can only use a payment equal to her utility gap. Hence, for an equilibrium profile of strategies t^{*i}_{jk} that forms equilibrium network G,

$$G(\mathbf{t}^*) = \{jk \mid \sum_{i \in N} t^{*i}_{jk} = 0, j, k \in N\}.$$

We would like to indicate that other generalization of transfers' distribution among players are not among the main focuses of this paper.[8]

Definitions of equilibrium notions.

Definition 2. *A network G is Pairwise Stable with Transfers (PS^t) with respect to a profile of utility functions $\mathbf{u}$ and a profile of strategies $\mathbf{t}$ that creates network G if*

(a) $ij \in G \implies u_i(G) \geq u_i(G - ij)$ as well as $u_j(G) \geq u_j(G - ij)$,
(b) $ij \notin G \implies u_i(G) \geq u_i(G + ij)$ as well as $u_j(G) \geq u_j(G + ij)$.

Also, $PS^t(u)$ denotes the family of pairwise stable network with transfers.

A pure strategy profile $\mathbf{t}^* = (\mathbf{t}^{*1}, ..., \mathbf{t}^{*n})$ forms a *Nash equilibrium* in the linking game with transfers if

$$u_i(G(\mathbf{t}^i, \mathbf{t}^{*-i})) \leq u_i(G(\mathbf{t}^*))$$

---

[8] See [1, 10] for some instances of study in the case of bargaining between players on network. In fact, despite the rich literature in general for bargaining between players, bargaining on networks is in its early attempts.

holds for all $i \in N$ and all $t_i \in T_i$, where $\mathbf{t}^*_{-i}$ is the equilibrium strategy for all players other than i, and T_i is the set of all available strategies for i. We can also indicate that in the context of network formation, a network G is Nash stable iff $\forall i \in N$, and $\forall l_i(G) \subseteq \mathcal{L}_i(G)$:

$$u_i(G) \geq u_i(G - l_i(G)). \tag{2}$$

Definition 3. *A pure strategy profile* $\mathbf{t}^* = (\mathbf{t}^{*1}, ..., \mathbf{t}^{*n})$ *forms a pairwise Nash equilibrium in the linking game with transfers if*

1. *it is a Nash equilibrium, and*
2. *there does not exist any* $ij \notin G(\mathbf{t}^*)$, *and* $t \in T$ *such that*
    (a) $u_i(G(t^i_{ij}, t^j_{ij}, \mathbf{t}^*_{-ij})) \geq u_i(G(\mathbf{t}^*))$,
    (b) $u_j(G(t^i_{ij}, t^j_{ij}, \mathbf{t}^*_{-ij})) \geq u_j(G(\mathbf{t}^*))$, *and*
    (c) *at least one of (1) or (2) holds strictly,*

where $\mathbf{t}^*_{-ij}$ *includes all players' strategies in* $\mathbf{t}^*$ *except player* i.

3 Dynamic Link-Pricing Model with Transfer Payments

3.1 Existence of Pairwise Stable Network with Transfers

In all game-theoretic problems, one of the primary questions concerns the existence of equilibria or stable states. This question in the framework of network formation is translated to the existence of pairwise stable networks and have been first addressed in [13]. We show that the arguments in [5,13] can be extended and adapted in our model. As a result, we guarantee the existence of pairwise stable network with transfers in our model.

Definition 4. *A cycle C is a set of networks $(G_1, ..., G_k)$ such that for any pair of networks $G_i, G_j \in C$, there exists an improving path connecting G_i to G_j. In addition, a cycle C is a closed cycle, if for all networks $G \in C$, there does not exist an improving path leading to a network $G' \notin C$.*

While improving paths that start from a seed network may end in an equilibrium network, it is also possible to find the formation of cycles as the result of an improving path. Jackson and Watts [13] showed that in any network formation model there exists either a pairwise stable network or a closed cycle. Their argument is based on the fact that a network is pairwise stable if and only if it does not lie on an improving path to any other network. We provide the following lemma and refer to the original paper for its proof, where the exact arguments can be applied for our notion of PS^t.

Lemma 1. *In the network formation model with transfer payments, there exists either an equilibrium network from $PS^t(u)$ or a closed cycle of networks.*

Theorem 1. *In the linking game with direct and indirect transfers given the utility function in* (1),

(a) there are no cycles,
(b) there exists at least one pairwise stable network $(PS^t(u))$.

Proof. We can rule out the existence of cycles in a network formation model if we show that the following holds: If for any two networks G and G', G' defeats G if and only if $v(G') > v(G)$ and G and G' are adjacent.[9] We can briefly argue that our linking game satisfies this condition. Since the direct and indirect transfer payments between players prevent the situations, where a player's utility can get hurt by actions (link addition or deletion) of others. In fact, this is one of the main function of transfers. Therefore, the value of networks through each improving path must be increased. Conversely, if G and G' are adjacent in an improving path such that $v(G') > v(G)$, G' must defeat G, where G is a network in the cycle.

Now, since there are finitely many networks that can be reached though the dynamic process, if there is a cycle, then the exact pairwise monotonicity of our linking game implies $v(G) > v(G)$; contradiction. Ruling out the existence of cycles along with Lemma 1 guarantees the existence of at least one pairwise stable network with transfer payments. □

3.2   Convergence to Pairwise Nash Stability

Definition 5. *Let* $\alpha \geq 0$. *A utility function* $u(.)$ *is* α-*submodular in own current links on* $\mathcal{A} \subseteq \mathcal{G}$ *if* $\forall i \in N, G \in \mathcal{A}$, *and* $l_i(G) \subseteq \mathcal{L}_i(G)$, *it holds that*

$$mu_i(G, l_i(G)) \geq \alpha \sum\nolimits_{ij \in l_i(G)} mu_i(G, ij).$$

The case $\alpha = 1$ corresponds to submodularity, also called superadditivity in [2].

Lemma 2. *The utility defined in* (1) *is submodular in own current links.*

Theorem 1 in [4] shows the equivalency of pairwise stable networks and pairwise Nash stable networks, given a utility function that is α-submodular. It targets the simple observation that given a α-submodular utility function, if a player does not benefit from severing any single link, then she does not benefit from cutting any subset of links simultaneously as well. A similar argument can be adapted to our linking game with transfers as well. So, we provide the following proposition without proof.

Proposition 1. *Given a profile of utility functions* u *in* (1) *in a linking game with transfers,* $PS^t(u) = PNS^t(u)$.

---

[9] This condition is denoted as *exact pairwise monotonicity* by Jackson and Watts.

3.3   Small Diameter in Equilibrium Networks

We take a large-scale $\sqrt{n} \times \sqrt{n}$ grid as the seed network in this model. In order to prove the main result for the diameter of the equilibrium networks, we provide the following lemmas.

Let $T_{G(t)}(i,j)$ be the set of players that use link ij in their unique shortest paths to i in the network $G(t) : T_{G(t)}(i,j) = \{k \in N \mid d_{G'(t)}(i,k) > d_{G(t)}(i,k)\}$, where $G' = G - ij$.

Lemma 3. *Let $G(t)$ be an equilibrium network $(G(t) \in PS^t(u))$ and $i, j \in N$ be an arbitrary pair of players in this network. If $ij \notin G(t)$ then*

$$|T_{G(t)}(i,j)| < \frac{d_{G(t)}(i,j)^\alpha + c_{ij} + t_{ij}^i}{d_{G(t)}(i,j) - 1}.$$

Proof. Since i and j are not linked in the equilibrium network, the benefit of establishing ij has to be less than its linking costs for i and j. On the other hand, $T_{G(t)}(i,j)$ represents the set of players that creates a part of this benefit by reducing the distance $d_{G(t)}(i,j)$ between i and j to 1. Hence, we can state that paying $d_{G(t)}(i,j)^\alpha + c_{ij} + t_{ij}^i$, which is necessary for establishing ij, cannot be beneficial for player i. As a result,
$|T_{G(t)}(i,j)|(d_{G(t)}(i,j) - 1) < d_{G(t)}(i,j)^\alpha + c_{ij} + t_{ij}^i$. $\qquad \square$

Remark 1. *For any $i, j \in N$, c_{ji} can be noted as an upper bound for the transfer payment t_{ij}^i. Hence, if $c = \max_{\forall i, j \in N}(c_{jk})$, it is an upper bound for any direct transfer payment in the network.*

Lemma 4. *In any equilibrium network $G(t)$, for any player $i \in N$, let $S_i^d = \{k \in N \mid d_{G(t)}(i,k) \leq d\}$. Then,*

$$|S_i^d|(1 + \frac{d^\alpha + 2c}{d-1}) \geq n \tag{3}$$

where $c = \max_{\forall i, j \in N}(c_{ij})$.

Lemma 5 shows an upper bound for the set $|S_i^2|$ that is the next step.

Lemma 5. $|S_i^2| \leq \Delta^\alpha + 2c \Big/ k\Big(\Delta - \Big(h_1 + h_2(g_1 + 2) + h_3(2f_1 + f_2 + 3)\Big)\Big)$

where Δ is the diameter of network, and $0 \leq k, f_i, g_i, h_i \leq 1$ denote some fractions of players in set S_i^2 based on their reduced distances to i in case of forming link ij. Also, $f_1 + f_2 + f_3 = g_1 + g_2 = h_1 + h_2 + h_3 = 1$.

Theorem 2. *For a sufficiently large network, there is a small diameter of at most 4 for any equilibrium network in the dynamic link-pricing model with transfer payments.*

Proof. Based on our statements in Lemmas 4 and 5, we can imply that

$$n \leq (1 + 2^\alpha + 2c)(\Delta^\alpha + 2c) \Big/ k\Big(\Delta - \Big(h_1 + h_2(g_1 + 2) + h_3(2f_1 + f_2 + 3)\Big)\Big). \tag{4}$$

For sufficiently large network, when the diameter is greater than $\lfloor h_1 + h_2(g_1 + 2) + h_3(2f_1 + f_2 + 3) \rfloor$, it contradicts Inequality (4). Clearly we can specify that $3 \leq 2f_1 + f_2 + 3 \leq 5$ and $2 \leq g_1 + 2 \leq 3$. Thus in this case, the upper bound for the diameter is the weighted average of 1, $2f_1 + f_2 + 3$ and $g_1 + 2$ and it is surely smaller than 5. Therefore, diameter cannot be bigger than 4 for any choice of parameters. However, we cannot have the same claim for smaller diameter. □

4   Simulations

We carried out a set of simulations that improves Even-Dar and Kearns model by implementing the dynamic link-pricing and a fixed maintenance cost c. These simulations generate networks that show (i) a small diameter of at most 4, (ii) a high clustering coefficient (with respect to edge density), and (iii) a power-law degree distribution. The dynamical simulations are implemented on a grid with $n \approx 1000$. At each iteration of the dynamic process, two players i (the initiator) and j (the responder), are chosen uniformly at random. Then, player i with probability $1/2$ considers establishing a link to j and with probability $1/2$, investigates the option of giving up its link to j (if ij exists). We used the notion of link stability [8] (refer to Sect. 1.1). In this set of simulations, we aim to indicate our improvements on the earlier model in order to generate small-world networks. It is important to note that by using the dynamic link-prices, the emergence of a small diameter of at most 4 in link stable networks are directly implied similarly by our argument in Sect. 3.3.[10]

In many instances of our simulations, it can be seen that the degree distribution is a good estimation for power-law degree distributions in real-life social networks. The left panel of Fig. 1 shows the impact of parameters c and α on the degree distribution of resulted networks in vertical and horizontal moves,

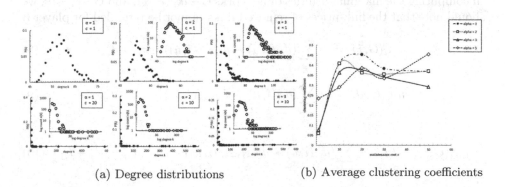

(a) Degree distributions   (b) Average clustering coefficients

Fig. 1. The structural properties of networks, achieved in the simulations.

---

[10] Note that although the existence of stable networks and convergence to the Nash outcomes would not be guaranteed in this assumption, we achieved a set of link stable networks by implementing many trials for different sets of α and c.

respectively. The right panel of Fig. 1 demonstrates the clustered structure of link stable networks: a high average clustering coefficient is present in all instances after increasing the maintenance cost from $c = 1$. The high clustering in these networks can be highlighted further by pointing out their small edge-density in the range from 0.007 for the network with $c = 50$, $\alpha = 5$ to 0.069 for the network with $c = 1$ and $\alpha = 1$. Also, the diameter in all instances was either 3 or 4 as expected.

Appendix: Omitted Proofs

Proof of Lemma 2
The proof is inspired by the arguments in [10]. First, we show the related inequality in Definition 5 holds for the case when the subset $l_i(G)$ consists of two distinct links ij and ik, which is indicated in below inequality.

$$mu_i(G, ij + ik) \geq mu_i(G, ij) + mu_i(G, ik) \tag{5}$$

If we consider any player such as $\mathbf{u}$ in network G, the distance between i and $\mathbf{u}$ $(d_G(i, u))$ contributes to the distance expenses in i's utility. It is important to note that removing any link such as ij or ik from the network G cannot decrease this distance, however if the removed link belongs to the shortest path between i and $\mathbf{u}$ in G, then the distance would be increased. This argument can be extended to removing two links such as ij and ik from G.

$$d_G(i, u) \leq d_{G-ij}(i, u) \leq d_{G-ij-ik}(i, u) \tag{6}$$

$$d_G(i, u) \leq d_{G-ik}(i, u) \leq d_{G-ij-ik}(i, u) \tag{7}$$

In computing the marginal utilities of networks $G - ik, G - ij$, and $G - ij - ik$, we should note that the link-prices of removed links cannot be refunded for player i.

$$mu_i(G, ij) = -\sum_{u \neq i}(d_G(i, u) - d_{G-ij}(i, u)) - c_{ij} - t_{ij}^i \tag{8}$$

$$mu_i(G, ik) = -\sum_{u \neq i}(d_G(i, u) - d_{G-ik}(i, u)) - c_{ik} - t_{ik}^i \tag{9}$$

$$mu_i(G, ij + ik) = -\sum_{u \neq i}(d_G(i, u) - d_{G-ij-ik}(i, u)) - c_{ij} - c_{ik} - t_{ij}^i - t_{ik}^i \tag{10}$$

According to Inequalities (6) and (7), we can simply imply the Inequality (5). Moreover, we can easily extend this argument for any subset of links $l_i(G)$.   □

Proof of Lemma 4
The set S_i^d consists of players in the neighborhood of i within a distance at most d. Furthermore, for each of these players such as k in the set S_i^d, according to

Lemma 3, we consider the set $T_{G(\mathbf{t})}(i, k)$. All players outside of this set should use one of players such as k in their shortest path to i. As a result, we can cover all players outside the set S_i^d by allocating a set $T_{G(\mathbf{t})}(i, k)$ to i for all players in set S_i^d. By doing so, an upper bound of $|T_{G(\mathbf{t})}(i, k)||S_i^d| + |S_i^d|$ for the number players in network (n) is achieved.

In order to obtain an upper bound for the set $T_{G(\mathbf{t})}(i, k)$ in wide range of different possible choices for i and k, we define c to be the maximum maintenance cost for all possible links in network. According to Remark 1, this is an upper bound for the all possible direct transfer payments in network as well, hence, $|T_{G(\mathbf{t})}(i, k)| \leq \dfrac{d^{\alpha} + 2c}{d - 1}$. By substituting the upper bounds of $T_{G(\mathbf{t})}(i, k)$ and S_i^d in $|T_{G(\mathbf{t})}(i, k)||S_i^d| + |S_i^d| \geq n$, the desired inequality can be achieved.   □

Proof of Lemma 5

Let G be an arbitrary instance from the set of equilibrium networks in our model, which are the set of pairwise stable networks with transfer $(G \in PS^t(u))$, given the utility function $u(.)$ in (1). Also, let $\mathbf{t}$ be the profile of strategies for players that forms G. Further, assume that the largest distance between any two players (or diameter) in network G exists between two players i and j. We denote Δ to be the size this distance. Note that the pair of i and j is not necessarily unique.

Based on the stable state, we can imply that creation ij is not beneficial for neither i nor j. If j wants to establish a link to i, the left side of Inequality (12) is a lower bound for the j's benefit that comes from the reduced distances to players in S_i^2. This set includes i itself and two subsets of players that are in distance 1 (type 1) and 2 (type 2) from i. First, let k represents players in S_i^2 such that their distances to j can be reduced by adding ij, as a fraction with respect to all players in $|S_i^2|$. Moreover, let h_1 represents player i itself as a fraction with respect to all players in $|S_i^2|$. By establishing ij, j's distance to i reduced by $\Delta - 1$.

Furthermore, let h_2 and h_3 represent the fractions of the number of type 1 players and type 2 players, respectively, in S_i^2. Their reduced distances for j is computed according to the initial distances of these two types of players in S_i^2 from j. Among the type 1 players, there are two subsets of players that g_1 and g_2 are their fractions with distance of $\Delta - 1$ and Δ from j, respectively. Furthermore, in type 2 players, there are three subsets of players in terms of their distance from j with fractions of f_1, f_2, f_3 that are in distance of $\Delta - 2, \Delta - 1, \Delta$ from j, respectively.

$$k|S_i^2|\Big(h_1(\Delta - 1) + h_2\big(g_1(\Delta - 3) + g_2(\Delta - 2)\big) + h_3\big(f_1(\Delta - 5) + f_2(\Delta - 4) + f_3(\Delta - 3)\big)\Big)$$

$$\leq \Delta^{\alpha} + c_{ji} + t_{ij}^j \leq \Delta^{\alpha} + 2c \tag{11}$$

$$\implies |S_i^2| \leq \Delta^{\alpha} + 2c\Big/ k\Big(\Delta - \big(h_1 + h_2(g_1 + 2) + h_3(2f_1 + f_2 + 3)\big)\Big) \tag{12}$$

where

$$0 < f_1 + f_2 + f_3 = g_1 + g_2 = h_1 + h_2 + h_3 = 1, \text{ and } 0 \leq k, f_i, g_i, h_i \leq 1. \tag{13}$$

□

References

1. Bayati, M., Borgs, C., Chayes, J., Kanoria, Y., Montanari, A.: Bargaining dynamics in exchange networks. In: ACM-SIAM Symposium on Discrete Algorithms, pp. 1518–1537 (2011)
2. Bloch, F., Jackson, M.O.: Definitions of equilibrium in network formation games. Int. J. Game Theory **34**(3), 305–318 (2006)
3. Bloch, F., Jackson, M.O.: The formation of networks with transfers among players. J. Econom. Theory **133**(1), 83–110 (2007)
4. Calvó-Armengol, A., Ilkiliç, R.: Pairwise-stability and nash equilibria in network formation. Int. J. Game Theory **38**(1), 51–79 (2009)
5. Chakrabarti, S., Gilles, R.: Network potentials. Rev. Econom. Des. **11**(1), 13–52 (2007)
6. Daraghmi, E.Y., Yuan, S.: We are so close, less than 4 degrees separating you and me!. Comput. Hum. Behav. **30**, 273–285 (2014)
7. de Martí, J., Zenou, Y.: Social networks. IFN Working Paper No. 816 (2009)
8. Even-Dar, E., Kearns, M.: A small world threshold for economic network formation. Adv. Neural Inf. Process. Syst. **19**, 385–392 (2007)
9. Fabrikant, A., Luthra, A., Maneva, E.N., Papadimitriou, C.H., Shenker, S.: On a network creation game. In: 22nd Annual ACM Symposium on Principles of Distributed Computing, pp. 347–351 (2003)
10. Gallo, E.: Essays in the economics of networks. Ph.D. Dissertation, University of Oxford (2011)
11. Hellman, T.: On the existence and uniqueness of pairwise stable networks. Int. J. Game Theory **42**, 211–237 (2012)
12. Jackson, M.O.: Social and Economic Networks. Princeton University Press, Princeton, NJ (2008)
13. Jackson, M.O., Watts, A.: The existence of pairwise stable networks. Seoul J. Econom. **14**(3), 299–321 (2001)
14. Jackson, M.O., Watts, A.: The evolution of social and economic networks. J. Econom. Theory **106**, 265–295 (2002)
15. Jackson, M.O., Wolinsky, A.: A strategic model of social and economic networks. J. Econom. Theory **71**, 44–74 (1996)
16. Kleinberg, J.: The small-world phenomenon: An algorithmic perspective. In: 32nd Annual ACM Symposium on the Theory of Computing, pp. 163–170 (2000)
17. Milgram, S.: The small world problem. Psychol. Today **1**, 61–67 (1967)
18. Myerson, R.: Game Theory: Analysis of Conflict. Harvard University Press, Cambridge, MA (1991)
19. Watts, A.: A dynamic model of network formation. Games Econom. Behav. **34**, 331–341 (2001)
20. Watts, D., Strogatz, S.: Collective dynamics of small-world networks. Nature **393**, 440–442 (1998)

A Region Growing Algorithm
for Detecting Critical Nodes

Mario Ventresca[1]($\boxtimes$) and Dionne Aleman[2]

[1] School of Industrial Engineering, Purdue University, West Lafayette, USA
mventresca@purdue.edu
[2] Department of Mechanical and Industrial Engineering,
University of Toronto, Toronto, Canada

Abstract. In this paper we apply a region growing bicriteria approximation algorithm of [5] for determining solutions to the critical node detection problem. This problem takes as input a number K and a connected, unweighted graph, and has the goal of selecting $\leq K$ vertices to remove such that the residual network has minimum pairwise connectivity. This problem has numerous applications, including those in network security, disease mitigation, marketing and antiterrorism. The algorithm achieves an $\mathcal{O}(\log n)$ approximation on the number of vertices needed to attain an $\mathcal{O}(1)$ bound on the objective function. Four random graph models and four real-world networks from different application areas are used to demonstrate that the algorithm performs within the predicted bounds.

1 Introduction

Given a graph $G = (V, E)$, critical node detection involves ascertaining a (typically small) subset of vertices, $R \subseteq V$, whose existence is somehow important, based on the context of the problem being solved. In this paper we focus on the optimization problem of critical node detection (CNDP) as formulated in 2009 by [2]. Given a connected network and upper limit K bounding the number of vertices that can be removed, the goal is to determine the vertices whose removal will leave the network with minimum pairwise connectivity. The CNDP has several applications to social networks, for instance to identify individuals to vaccinate during a pandemic [16]. Other potential applications such as viral marketing and rumor spread can be based on discovering individuals responsible for information transmission between network communities.

The main contributions in this paper revolve around an efficient algorithm for estimating solutions to the CNDP. The approach follows the algorithmic and analytical framework for rounding a fractional solution to an integral one through a process known as region growing [9,11]. While many problems have been solved using this general framework, CNDP is not among them, but we will provide one based on a straightforward adaptation of [5]. In that paper a game-theoretic analysis is conducted that requires a solution to a generalization of the sum-of-squares partitioning problem [4]. The problem under consideration differs

© Springer International Publishing Switzerland 2014
Z. Zhang et al. (Eds.): COCOA 2014, LNCS 8881, pp. 593–602, 2014.
DOI: 10.1007/978-3-319-12691-3_44

in that CNDP aims to minimize pairwise connectivity in the residual graph but the $\mathcal{O}(\log n)$ performance guarantee will still be valid.

Di Summa et al. [6] examined the restricted case of CNDP where the graph is a tree structure and proved the $\mathcal{NP}$-completeness of the problem for non-unit edge costs. A polynomial time dynamic programming algorithm for solving the problem with unit edge costs is also provided. The dynamic programming solution is applied to variants of the CNDP [1]. In [7], an integer linear programming model with a non-polynomial number of constraints is given and branch and cut algorithms are proposed. A reformulation of the CNDP is proposed in [19] that reduces the number of constraints from $\Theta(n^3)$ to $\Theta(n^2)$, for $n = |V|$, and consequently they are able to solve significantly larger problems (≈ 10 times larger and $1,000$ times faster).

Heuristics without guaranteed approximation bounds have also been studied [2]. Two stochastic search algorithms are employed in [14] in order to allow significantly larger networks to be solved within reasonable time and without significant resources. These simulated annealing and population-based incremental learning algorithms are shown to yield quality results. Both of these works lack any approximation bound on solution quality or the number of vertices selected. Two randomized rounding algorithms were proposed in [15,16], with an instance-specific approximation bound. A $\mathcal{O}(\log n \log \log n)$ pseudo-approximation algorithm was proposed in [8]. An extension of this problem, aimed to optimize connectivity and cohesiveness properties was studied in [17,18].

The remainder of this paper is organized as follows. Section 2 formally describes the critical node detection problem. The proposed region growing approach is described in Sect. 3, where its approximation abilities are also proven. A set of random graph and complex networks models is described in Sect. 4, and the results of experimentation are also given. Finally, the conclusion and directions for future work are given in Sect. 5.

2   The Critical Node Detection Problem

In this section, an integer program problem formulation for the critical node detection problem (CNDP) is outlined. We assume graph $G = (V, E)$ is undirected and unweighted. The connected components of G are sets of vertices such that all vertices in each set are mutually connected (reachable by some path), and no two vertices in different sets are reachable. Then, the CNDP can be expressed as the following integer program [3]:

$$\text{minimize} \sum_{u,v \in V} x_{uv} \tag{1}$$

$$\text{subject to } x_{uv} + y_u + y_v \geq 1, \quad \forall (u,v) \in E \tag{2}$$

$$x_{uv} + x_{vw} + x_{wu} \neq 2, \quad \forall u, v, w \in V \tag{3}$$

$$\sum_{v \in V} y_v \leq K, \tag{4}$$

$$x_{uv} \in \{0,1\}, \quad \forall u, v \in V, \tag{5}$$
$$y_u \in \{0,1\}, \quad \forall u \in V. \tag{6}$$

where

$$x_{uv} = \begin{cases} 1 & \text{if } u, v \text{ are in the same component of } G(V \setminus R) \\ 0 & \text{otherwise} \end{cases} \tag{7}$$

The optimal solution to this problem will result in a graph with minimum pair-wise connectivity. The first constraint (Eq. 2) enforces the separation of vertices into different components by allowing edge x_{uv} to be deleted if either or both vertices u and v are deleted. Equation (3) is a short-hand notation concerned with a triangle inequality. The final constraint, in Eq. (4), ensures the removal of no more than K vertices. The number of constraints is of the order $\mathcal{O}(|V|^3 + |E|)$, which greatly limits the size of the network one can consider.

3 Proposed Approach

The algorithm and analysis directly follow from [5] for the sum-of-squares partition problem, which we outline here for completeness. Let $(\mathbf{x}^*, \mathbf{y}^*)$ with objective function value z^* be the optimal solution to CNDP, and $(\hat{\mathbf{x}}, \hat{\mathbf{y}})$ with objective function value $\hat{z}$ be the optimal solution to the LP relaxation of CNDP. The rounded solution $(\bar{\mathbf{x}}, \bar{\mathbf{y}})$ with objective function value $\bar{z}$.

3.1 Region Growing

Preliminaries. Given a shortest path $\pi_{uv} = \langle u, \dots, v \rangle$ from vertex u to v, the distance between them is computed according to

$$d_{uv} = \sum_{v \in \pi_{uv} - v} \hat{y}_v, \tag{8}$$

which follows from Eq. (2) and also implies that $x_{uv} = \max(0, 1 - d_{uv} - \hat{y}_v)$.

Each d_{uv} is the shortest path distance between vertices u and v, but does not include the terminal vertex value $\hat{y}_v$. That is, compute d_{uv} as the distance from vertex u to the boundary of its target vertex v. The hemimetric d induces a topology over the space of V and allows for the definition of a ball, which defines a set of vertices a given distance from a single point.

Definition 1 (ball). *A ball $B(u, r)$ of radius $r > 0$ about a vertex u consists of all nodes v satisfying $B(u, r) = \{v | d_{uv} \le r\}$, where d_{uv} is the distance between vertices u and v.*

To each ball is assigned a volume that measures the amount of vertices it contains. This measure is different from the cardinality of the ball.

Definition 2 (volume). *The volume of a ball is the volume of $B(u,r)$, including the fractional distance of vertices not fully within its boundary:*

$$vol(B(u,r)) = \sum_{v \in B(u,r)} \phi(v) \tag{9}$$

where $\phi(v)$ determines the amount of overlap of $B(u,r)$ and v:

$$\phi(v) = \begin{cases} \hat{y}_v & \text{if } d_{uv} + \hat{y}_v \leq r \\ r - d_{uv} & \text{if } d_{uv} \leq r \leq d_{uv} + \hat{y}_v \\ 0 & \text{if } d_{uv} > r \end{cases} \tag{10}$$

As indicated in Eq. (10), a vertex may not be fully contained within a ball $B(u,r)$ if the radius is not chosen to be sufficiently large. These partially contained vertices belong to the shell of the ball.

Definition 3 (shell). *The shell of a ball $B(u,r)$, $shell(B(u,r)) = \{v | d_{uv} \leq r < d_{uv} + \hat{y}_v\}$ is the set of vertices partially overlapped by the ball. The equivalent notation $shell(u,r)$ will also be used.*

The following lemma relates the cost of a set of vertices in the shell to the volume of a ball, and provides an upper bound for the radius of a ball. This result is similar to the edge-cut solution described in [12,21], but for vertex cuts [5].

Lemma 1. *Given a feasible solution to the LP relaxation of the CNDP, for any vertex $u \in V$, a radius $r < 1/2$ exists such that*

$$w(shell(u,r)) \leq \varepsilon \times vol(B(u,r)) \tag{11}$$

where $w \colon V^m \mapsto \mathbb{R}^+$ is a user defined weight or cost function over m vertices and $\varepsilon = c\ln(n+1)$, for constant $c \geq 2$.

The weight function $w \colon V^m \mapsto \mathbb{R}^+$ is, in our context, a sum of the values of an arbitrary non-empty set ζ (of cardinality m) of vertices

$$w(\zeta) = \sum_{v \in \zeta} \hat{y}_v. \tag{12}$$

Bicriteria Algorithm. The approach is based on the region growing method of [5,9,11], which iteratively grows balls of at most some fixed radius using the fractional values $\hat{x}_{uv}$ and $\hat{y}_v$ of the LP relaxation to the CNDP. This process repeats until all nodes of the graph are contained in one of the balls. The vertices to be removed from the graph are determined as the union of all the shells of each ball.

The volume of an initial ball (i.e., ball $B(u,0)$) is equal to K/n, where n is the number of non-zero $\hat{y}_v$ values (i.e., $|\{\hat{y}_v | \hat{y}_v > 0\}|$) and $K > 0$ is the constraint limiting the number of vertices that can be removed from G. The requirement for non-zero $B(u,0)$ is for technical reasons in the derivation of Lemma 1.

Algorithm 1. The region growing algorithm.

1: Solve the LP relaxation of CNDP to
    1. determine $\hat{x}, \hat{y}$
    2. calculate d_{uv} $\forall u, v \in V$
2: $\mathcal{S} = \emptyset$
3: $H = G$
4: $\varepsilon = c \ln(n + 1)$
5: **while** $V(H) \neq \emptyset$ **do**
6:    choose some $u \in V(H)$
7:    grow the ball centered at u that minimizes $\left\{ r : \frac{w(shell(u,r))}{vol(B(u,r))} \mid shell(u, r) > 0 \right\}$
8:    $\mathcal{S} = \mathcal{S} \cup shell(u, r)$
9:    $H = H \setminus \{B(u, r) \cup shell(u, r)\}$
10: **end while**
11: **return** $\mathcal{S}$

Line 1 of Algorithm 1 solves the LP relaxation to CNDP in polynomial time, and allows for calculation of d_{uv}. The distances between vertices as shown in Eq. (8) can then be calculated for utilization by the algorithm. Set $\mathcal{S}$ stores the union of vertices belonging to the shell of each ball. At each iteration a vertex u is selected in line 6 as the initial growing region from the set of remaining vertices. For instance, u can be determined randomly, according to some other probability distribution taking local properties such as vertex degree into account, or be a predefined ordering such as according to increasing vertex degree.

A region about u is grown in line 7 according to Lemma 1. The Lemma has been rearranged as a minimization problem, which by definition will be less than $\leq \varepsilon$. The expansion process is such that $shell(u, r) \neq \emptyset$, meaning the minimum value is greater than 0. This is also a polynomial time operation, with initial radius $r = 0$ that increments to $1/c$ (or some other amount that guarantees the shell is non-empty) by sufficiently small steps. The vertices belonging to $shell(u, r)$ and $B(u, r)$ are removed from the graph H and each $shell(u, r)$ is added to set $\mathcal{S}$. The algorithm terminates when H does not contain any more vertices.

Let $\mathcal{B} = \{B_1, \ldots, B_M\}$ be the disjoint balls determined through region-growing, and let $\mathcal{S} = \{S_1, \ldots, S_M\}$ be the associated shells of each ball:

Theorem 1. *The region growing algorithm achieves a $2c \ln(n + 1)K$ approximation to the number of vertices in a solution to CNDP.*

Proof. At each iteration i, we have from Lemma 1 that

$$w(S_i) \leq c \ln(n + 1) vol(B_i) \tag{13}$$

At each i, we remove $\{B_i, S_i\}$ from G, and therefore each vertex will belong to only one ball. The total weight of the removed vertices in the rounded solution is

$$\sum_{v \in \mathcal{S}} \bar{y}_v \leq c \ln(n + 1) \sum_{B_i \in \mathcal{B}} vol(B_i) \leq c \ln(n + 1) \left(\sum_{v \in \mathcal{B}} \bar{y}_v + \sum_{B_i \in \mathcal{B}} \frac{K}{n} \right) \tag{14}$$

$$\leq c \ln(n+1)\left(K + \frac{nK}{n}\right) \tag{15}$$

$$= 2c \ln(n+1)K, \tag{16}$$

where the volume of each ball was split into its fractional and initial values.

Theorem 2. *The region growing algorithm achieves an $\mathcal{O}(1)$ approximation to the objective of CNDP.*

Proof. Recall that $\hat{x}_{uv} = \max(0, 1 - d_{uv} - \hat{y}_v)$ and $c = 1/r$. Consider, two vertices fully contained in an arbitrary ball centered at vertex a, i.e., $u, v \in B_i^*(a, r) = B_i \setminus S_i$. It follows that the distance between u, v where $r < 1/2$ must be $\hat{x}_{uv} = 1 - d_{uv} - \hat{y}_v \geq 1 - \frac{2}{c}$ for some $c > 2$; otherwise at least one of u, v would be in $shell(a, r)$. This implies $\frac{c\hat{x}_{uv}}{c-2} \geq 1$. So, for $G \setminus S$ the objective function can be bounded as

$$\sum_{u,v \in V \setminus S} \bar{x}_{uv} \leq \sum_{B_i^* \in \mathcal{B}} \sum_{u,v \in B_i^*} 1 \leq \frac{c}{c-2} \sum_{B_i^* \in \mathcal{B}} \sum_{u,v \in B_i^*} \hat{x}_{uv} = \frac{c}{c-2}\hat{z} \tag{17}$$

where $\hat{z}$ is the optimal fractional result.

The integrality gap of the LP relaxation determines the best approximation factor possible for a given problem instance I. For a minimization problem this bound is [12]

$$\sup_I \frac{z^*(I)}{\hat{z}(I)} \tag{18}$$

where $\sup(\cdot)$ is the suprememum function and $z^*(I)$ and $\hat{z}(I)$ refer to the optimal solutions to the IP and LP formulations of the problem instance I, respectively.

Theorem 3. *The integrality gap of the linear program relaxation of the CNDP is $\Omega(\log n)$ in the worst case.*

Due to space limitations the proof has been omitted, but as in [5], is based on graph expanders.

4   Experimental Results

We use four standard network structures as our benchmark data set. These models capture properties of preferential attachment, randomness and small-world behaviour.

4.1   Results

We ran 50 trials of the algorithm for each (graph, K) pair, region growing requiring negligible ($\approx$1 s) computation time. To obtain optimal solutions we utilized Gurobi version 4.61 with a maximum time limit of 2,000 s. Similarly, we allow 2,000 s to solve the LP relaxation. Experiments were conducted on a 3.4 GHz computer with 16 GB RAM. More in-depth experiments were conducted, but omitted due to space limitations.

4.2   Approximation

We examine the approximation ratio from a practical perspective by comparing against common complex network models. Typical results are shown in Fig. 1 where the top row compares the optimal and approximations in terms of objective value, and the bottom row compares the number of vertices removed. The rounding procedure may remove a number of vertices not equal to K. However, both approximations are shown to be high quality. Note that major discrepancies such as for $K = 10$ in Fig. 1(c) are a result of the initial LP not finding a solution within the given time consideration and hence the results seem worse than the approximation bound indicates (the proof assumes an LP solution exists). The graphs used were an Erdös-Renyi graph with 150 vertices, a Barabasi-Albert graph with 200 vertices, a Watts-Strogatz network with 125 vertices and a Forest Fire network with 125 vertices as well.

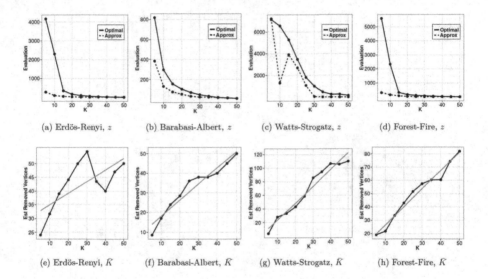

(a) Erdös-Renyi, z (b) Barabasi-Albert, z (c) Watts-Strogatz, z (d) Forest-Fire, z

(e) Erdös-Renyi, K (f) Barabasi-Albert, K (g) Watts-Strogatz, K (h) Forest-Fire, K

Fig. 1. Top: comparison of the optimal objective function value $z*$ and the SHELL-DETECT objective $\bar{z}$. Bottom: the number of removed vertices $\bar{K}$ compared to the desired limit. The dip in (c) is due to not finding a true optimal solution within the given time limit.

4.3   Real World Networks

Attaining quality result on the benchmark problems is one measure of success, but we also compare to four real-world networks where an LP solution was attainable within $10,000$ s. The first data set was originally utilized to estimate of the cover and size of human protein-protein interaction, or "interatcome", networks [13]. We consider only the largest connected component, which contains

Table 1. Network properties and solution characteristics of real world networks.

Data set	Vertices	Edges	K	z^*	$\bar{K}$	$\bar{z}$
Human	141	161	10	176	19	18
Terror	62	152	10	169	22	26
Yeast	124	422	15	87	56	18
Elegans	314	363	30	161	44	34

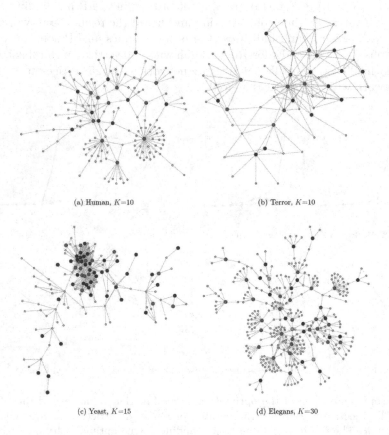

(a) Human, $K=10$ (b) Terror, $K=10$

(c) Yeast, $K=15$ (d) Elegans, $K=30$

Fig. 2. Real-world examples. Red nodes: common vertices between the optimal and proposed algorithm solutions. Blue nodes: extra vertices selected by region growing. Green nodes: optimal vertices not selected by region growing.

141 proteins and 161 interactions. The second data set is from Kreb's terrorist network based on the attack of September 11, 2001 [10] and contains 121 proteins and 190 interactions. The third network is the largest component of the yeast interatcome [20] for only unclassified proteins contains 124 proteins and 422 interactions. The final network is of *Caenorhabditis elegans* and contains 314

vertices and 363 edges [13]. The structure of these networks as well as the target number of vertices to identify is summarized in Table 1.

Figure 2 presents the results for each real-world network. In all cases, the approximation is in accordance with the approximation guarantees. However, the number of extra vertices required to maintain a constant approximation to the CNDP objective seems typically higher than desirable, but the improvement in objective value is significant.

5 Conclusion and Future Work

In this paper we proposed the use of a region-growing algorithm as proposed in [5] and applied to the CNDP. The approach detects vertex shells, instead of cut edges, in order to determine nodes to remove from the network. The approach has a logarithmic approximation bound to the number of nodes to remove, which guarantees an optimal solution within a constant factor. The integrality gap was also shown in [5] to be logarithmic, and thus our algorithm attains the asymptotically best bound. Experiments were conducted to highlight the quality of results on classic complex network models and real world problems.

Future work includes many possible directions. The algorithm attains solutions by LP rounding, but this need not be the sole manner to attain appropriate distance measures over the space of vertices. Two possible courses of action are to investigate other relaxations such as semi-definite programming, or to use a more statistical or heuristic approach. Using more efficient algorithms will also allow for significantly larger networks to be considered. The extension of the algorithms to weighted networks is also a task of high priority.

References

1. Addis, B., Di Summa, M., Grosso, A.: Complexity results and polynomial algorithms for the case of bounded treewidth. Discrete Appl. Math. **161**(16–17), 2349–2360 (2013)
2. Arulselvan, A., Commander, C.W., Elefteriadou, L., Pardalos, P.M.: Detecting critical nodes in sparse graphs. Comput. Oper. Res. **36**(7), 2193–2200 (2009)
3. Arulselvan, A., Commander, C.W., Pardalos, P.M., Shylo, O.: Managing network risk via critical node identification. In: Gulpinar, N., Rustem, B. (eds.) Risk Management in Telecommunication Networks. Springer, Heidelberg (2011)
4. Aspnes, J., Chang, K., Yampolskiy, A.: Inoculation strategies for victims of viruses and the sum-of-squares partition problem. In: Proceedings of the Sixteenth Annual ACM-SIAM Symposium on Discrete Algorithms, SODA '05, pp. 43–52. Society for Industrial and Applied Mathematics (2005)
5. Chen, P., David, M., Kempe, D.: Better vaccination strategies for better people. In: Proceedings of the 11th ACM Conference on Electronic Commerce, pp. 179–188. ACM (2010)
6. Di Summa, M., Grosso, A., Locatelli, M.: Complexity of the critical node problem over trees. Comput. Oper. Res. **38**(12), 1766–1774 (2011)
7. Di Summa, M., Grosso, A., Locatelli, M.: Branch and cut algorithms for detecting critical nodes in undirected graphs. Comput. Optim. Appl. **53**(3), 649–680 (2012)

8. Dinh, T.N., Xuan, Y., Thai, M.T., Pardalos, P.M., Znati, T.: On new approaches of assessing network vulnerability: hardness and approximation. IEEE/ACM Trans. Netw. **20**(2), 609–619 (2012)

9. Garg, N., Vazirani, V., Yannakakis, M.: Approximate max-flow min-(multi)cut theorems and their applications. SIAM J. Comput. **25**, 698–707 (1993)

10. Krebs, V.: Uncloaking terrorist networks, first monday (2001). http://www.orgnet.com/hijackers.html

11. Leighton, T., Rao, S.: An approximate max-flow min-cut theorem for uniform multicommodity flow problems with applications to approximation algorithms. In: Proceedings of the 29th Annual Symposium on Foundations of Computer Science, pp. 422–431. IEEE Computer Society (1988)

12. Vazirani, V.: Approximation Algorithms. Springer, Heidelberg (2001)

13. Venkatesan, K., Rual, J.F., Vazquez, A., Stelzl, U., Lemmens, I., Hirozane-Kishikawa, T., Hao, T., Zenkner, M., Xin, X., Goh, K., Yildirim, M.A., Simonis, N., Heinzmann, K., Gebreab, F., Sahalie, J.M., Cevik, S., Simon, C., de Smet, A., Dann, E., Smolyar, A., Vinayagam, A., Yu, H., Szeto, D., Borick, H., Dricot, A., Klitgord, N., Murray, R., Lin, C., Lalowski, M., Timm, J., Rau, K., Boone, C., Braun, P., Cusick, M., Roth, F., Hill, D., Tavernier, J., Wanker, E., Barabasi, A.L., Vidal, M.: An empirical framework for binary interactome mapping. Nat. Methods **6**(1), 83–90 (2009)

14. Ventresca, M.: Global search algorithms using a combinatorial unranking-based problem representation for the critical node detection problem. Comput. Oper. Res. **39**(11), 2763–2775 (2012)

15. Ventresca, M., Aleman, D.: A derandomized approximation algorithm for the critical node detection problem. Comput. Oper. Res. **43**, 261–270 (2014)

16. Ventresca, M., Aleman, D.: A randomized algorithm with local search for containment of pandemic disease spread. Comput. Oper. Res. **48**, 11–19 (2014)

17. Veremyev, A., Boginski, V., Pasiliao, E.L.: Exact identification of critical nodes in sparse networks via new compact formulations. Optim. Lett. **8**(4), 1245–1259 (2014)

18. Veremyev, A., Prokopyev, O.A., Pasiliao, E.L.: An integer programming framework for critical elements detection in graphs. J. Comb. Optim. **28**(1), 233–273 (2014)

19. Veremyev, A., Boginski, V., Pasiliao, E.L.: Exact identification of critical nodes in sparse networks via new compact formulations. Optim. Lett. **8**, 1245–1259 (2014)

20. von Mering, C., Krause, R., Snel, B., Cornell, M., Oliver, S., Fields, S., Bork, P.: Comparative assessment of large-scale data sets of protein-protein interactions. Nature **417**, 399–403 (2002)

21. Williamson, D.P., Shmoys, D.B.: The Design of Approximation Algorithms. Cambridge University Press, New York (2011)

A Fast Greedy Algorithm for the Critical Node Detection Problem

Mario Ventresca[1](✉) and Dionne Aleman[2]

[1] School of Industrial Engineering, Purdue University, West Lafayette, USA
mventresca@purdue.edu
[2] Department of Mechanical and Industrial Engineering,
University of Toronto, Toronto, Canada

Abstract. The critical node detection problem (CNDP) aims to fragment a graph $G = (V, E)$ by removing a set of vertices R with cardinality $|R| \leq K$ such that the residual graph has minimum pairwise connectivity. Algorithms that are capable of finding R in graphs with many thousands or millions of vertices are needed since existing approaches require significant computational cost and subsequently are useful for only very small network instances. An efficient method for evaluating the impact of removing any $v \in V$ on the CNDP objective function within reasonable time and space complexity is then necessary. In this paper we propose a depth-first search solution to this problem that requires $\mathcal{O}(|V| + |E|)$ complexity, and employ the method in a greedy algorithm for quickly identifying R in large networks. We evaluate the results using six real-world benchmark problems. The proposed algorithm can be easily extended to vertex and edge-weighted variants of the critical vertex detection problem.

1 Introduction

Detecting important or critical vertices in a graph has many important applications. Depending on the context, we may wish to promote or mitigate the diffusive process acting on the network by identifying and utilizing critical nodes. In the context of promoting a spreading process, such as advertisements or health warnings, the notion of "critical" refers to the identification of individuals who are most likely to be influential spreaders and maximally allow the information spread through the network. In such cases the selected individuals may be targeted for demonstrations, promotions or invited to public events. For mitigation contexts, such as stopping computer virus spread or for construction of stable power delivery networks, the identified vertices are those whose removal from the graph will maximally mitigate the threat, hopefully alleviating it entirely.

Specific definitions of what a critical node is have been investigated, including junctions in cell-signalling or protein-protein networks [4], highly influential individuals [12], smart grid vulnerability [17], targeted vaccination for pandemic prevention [3,23] or keys to decipher brain functionality [11]. In some contexts an accurate mathematical definition for a critical vertex, particularly for highly

© Springer International Publishing Switzerland 2014
Z. Zhang et al. (Eds.): COCOA 2014, LNCS 8881, pp. 603–612, 2014.
DOI: 10.1007/978-3-319-12691-3_45

complex systems such as the brain [20], may not yet exist. In any event, it is important to note that both promotion and mitigation can often be defined in a mathematically similar manner.

We concentrate in this paper on the critical node detection problem (CNDP) [2]. Given graph $G = (V, E)$ where $|V| = n$ and $|E| = m$, ascertain (a typically small) subset of vertices, $R \subseteq V$, $|R| \leq K$, whose removal leaves the graph with minimum pairwise connectivity, i.e.,

$$\arg\min_{R \subseteq V} \sum_{\mathcal{C}_i} \binom{|\mathcal{C}_i|}{2} \tag{1}$$

where the sum is over all connected components $\mathcal{C}_i$ of the residual graph $G \backslash R$, and $|\mathcal{C}_i|$ indicates the number of vertices in component i. The optimal network is therefore one that is maximally fragmented, and simultaneously minimizes the variance amongst the component sizes. That is, the residual network $G \backslash R$ will contain a relatively large set of connected components each containing a similar number of vertices. The problem has been shown to be $\mathcal{NP}$-hard [2,9].

1.1   Related Work

The case where G is a tree structure has been examined and proven to be $\mathcal{NP}$-complete when considering non-unit edge costs [8]. A polynomial time dynamic programming algorithm with worst-case complexity $\mathcal{O}(n^3 K^2)$ for solving the problem with unit edge costs was also provided and applied to variants of the CNDP [1]. In [7], an integer linear programming model with a non-polynomial number of constraints is given and branch-and-cut algorithms are proposed. A reformulation of the CNDP that requires $\Theta(n^2)$ constraints was recently shown and optimal solutions for small networks were ascertained [27,28].

Heuristic solutions lacking provable approximation bounds have also been investigated, but computation time for very large networks can be further improved. The CNDP work of [2] utilizes a solution to the maximum-independent set problem as a starting point for local search, repeating the process until a desired termination criteria is reached. The algorithm is tested on a limited number of network structures with promising results. Two stochastic search algorithms are employed in [22] that permit solutions to significantly larger networks (up to a few thousand vertices) to be solved within reasonable time and without significant resources. These simulated annealing and population-based incremental learning algorithms are shown to yield quality results. Randomized rounding-based algorithms have been also proposed in [24,26], but without approximation bounds (although, an instance-specific bound was derived). A $\mathcal{O}(\log n \log \log n)$ pseudo-approximation algorithm was proposed in [9].

Among the many problems that have been defined, some of the most similar include the following. The minimum contamination problem [14] aims to minimize the expected size of contamination by removing a set of edges of at most a given cardinality. A variant of the problem is also proposed with the goal of minimizing the proportion of vertices in the largest resulting network. A bi-criteria

algorithm is given that is able to achieve an $\mathcal{O}\left(1 + \epsilon, \frac{1+\epsilon}{\epsilon}(\log n)\right)$ approximation. In [6], a game-theoretic analysis is conducted that requires a solution to
a generalization of the sum-of-squares partitioning problem [3]. Exact methods
for link-based vulnerability assessment using edge-disruptors have also recently
been investigated [10].

The remainder of this paper is organized as follows. The proposed greedy
algorithm and its motivation are provided in Sect. 2. Section 3 provide brief
experimentation using six real-world networks. Conclusions are then presented
in Sect. 4.

2   The Proposed Algorithm

In this section we propose a greedy algorithm for the CNDP. To overcome computational bottlenecks for larger networks a linear-time algorithm for evaluating
the CNDP objective is given and a priority-queue based implementation to yield
significant practical performance increases is briefly described. Our $\mathcal{O}(K(n+m))$
algorithm is similar to the $\mathcal{O}(Kn^2)$ approach used for the maximum cascading
algorithm in [17]. Both algorithms are based on Tarjan [21].

For large networks with many thousands or millions of vertices (and edges) a
computationally efficient approach to minimizing the CNDP objective is required.
Selecting critical subset R from a single observation of the network is easily
deceived due to the influence of cut vertices, as indicated in Lemma 1 [2]. That
is, selecting R in a sequential fashion will allow for the discovery of a set that is
more likely to fragment the network by detecting cut vertices that are not obvious
unless a sequential approach is taken.

Lemma 1. *Let M_1 and M_2 be two sets of partitions obtained by deleting D_1
and D_2 sets of vertices from graph $G = (V, E)$, where $|D_1| = |D_2| = K$. Let L_1
and L_2 be the number of components in M_1 and M_2 respectively and $L_1 \geq L_2$. If
$C_h = C_\ell, \forall h, \ell \in M_1$, then we obtain a better objective function value by deleting
the set D_1.*

Thus, we propose a sequential greedy approach as shown in Algorithm 1. At each
iteration the vertex whose removal will have the largest decrease on the objective
function (Eq. 1) is selected for removal and added to set R, where $f()$ computes
the objective value. Computation of line 3 is a bottleneck to solving large-scale
problems. Naively, it implies removal of each $v \in V \setminus R$ and re-evaluation of the
objective function.

Here we provide an $\mathcal{O}(K(n + m))$ algorithm based on a modified depth-
first search (DFS). The iterative (versus recursive) algorithm of DFS should be
used as the framework because sufficiently large networks will quickly encounter
stack overflow errors during the search. Performing a DFS on G will construct an
equivalent representation as a DFS-tree $DFS(v)$ rooted at arbitrary v. Figure 1
provides an example of the conversion between an original graph to a DFS tree
rooted at $v = 0$. Our solution is based on the DFS-tree viewpoint.

Algorithm 1. High-level pseudocode for GREEDY-CNDP.

Require: $K > 1$ upper limit on the number of vertices to remove
1: Let $R = \emptyset$ be a set of removed vertices.
2: **repeat**
3:    select $v^* = \arg\min_{v \in V} f(G \setminus \{v^*\})$
4:    remove v^* from G, i.e., $G = G \setminus \{v^*\}$
5:    $R = R \cup \{v^*\}$
6: **until** $|R| = K$ or $|E| = 0$

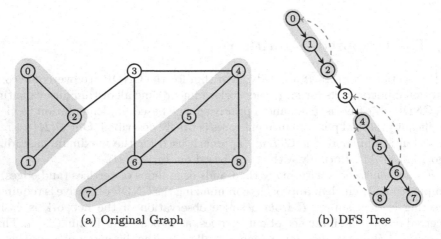

(a) Original Graph   (b) DFS Tree

Fig. 1. Equivalent original graph and DFS-tree. Back-edges are indicated as dashed arrows in DFS(0). The shaded areas correspond to resulting connected components if vertex $v = 3$ is removed from the graph. A traditional application of DFS is to detect cut vertices (i.e., articulation points), which forms the basis for our approach.

Observation 1. *Ignoring back-edges, a leaf vertex v has no children. Hence, the subtree rooted at v contains a single vertex whose deletion will not create any new connected components.*

Observation 2. *Let $\delta(v)$ be the set of children vertices of v in the DFS-tree, ignoring back edges. Then, the total number of descendant vertices can be recursively defined as*

$$s(v) = \sum_{w \in \delta(v)} \begin{cases} s(w), & \text{if } w \text{ is an internal or root vertex} \\ 1, & \text{if } w \text{ is a leaf} \end{cases} \tag{2}$$

Observation 3. *Ignoring back-edges, each internal vertex v will either be a cut vertex or not. Removing any $v \in V$ will result in an updated objective value but if v^* is a cut vertex, then the residual graph $G(v \setminus \{v^*\})$ will contain a set of new DFS subtrees $T(v)$. Otherwise, the graph will have the same number of connected components as before removing v^*, but G will contain one less vertex. This can be summarized as,*

$$\sum_{t_i \in T(v)} \binom{|t_i|}{2} \tag{3}$$

where $|t_i|$ is the number of vertices in tree t_i. This sum can be computed during the DFS-search when a vertex is found to be a cut vertex, and hence incurs only $\mathcal{O}(1)$ overhead per edge.

Observation 4. If $v \in V$ is a cut vertex it will be identified upon recursing after visiting the entire subtree of each of its children, respectively. However, the order in which vertices are visited during the DFS-based search does not guarantee that all non-descendant vertices in the graph will be explored before reaching v. Hence, the number of vertices in the ancestor DFS-tree of v can be determined by computing $|V(v)| - s(v)$, where $V(v)$ indicates the set of vertices reachable from v before it is removed. $V(v)$ can be monitored at run-time.

The above observations imply that v^* can be computed in linear time by augmenting a DFS for cut vertices to additionally calculate the impact of removing any vertex $v \in V$, which can then be used to identify v^*,

$$v^* = \arg\min_{v \in V} f(v) = \arg\min_{v \in V} \left(|V(v)| - s(v) + \sum_{t_i \in T(v)} \binom{|t_i|}{2} \right) \tag{4}$$

Since DFS has running time complexity of $O(n + m)$ and Eq. (4) can be executed in constant time per node during the search, the proposed greedy algorithm requires $O(K(n + m))$ complexity to remove K vertices from G.

Observation 5. Let $Q \subseteq V \setminus \{v^*\}$ be the subset of vertices not reachable in graph G from vertex v^*. Then, it is not necessary to recompute the impact of removing any $w \in Q$ from $G \setminus \{v^*\}$ since v^* and each w belong to different network components. That is, only vertices $u \in V(v^*)$ must be re-examined.

Observation 6. Each connected component C_i can be identified by a root vertex associated with a DFS tree. The vertex whose removal will maximally decrease the objective function can be recorded during the DFS search and returned as this root, along with $V(v)$. This requires no significant computational or memory overhead.

Observations 5 and 6 indicate that further practical improvements are possible. Specifically, a priority queue can be implemented to store the set of connected components C, which are represented and ordered by their respective root vertices. For each C_i, removing its root vertex will most significantly decrease the CNDP objective value versus any other vertex in the same component. After v^* is removed from the queue, the component that had contained it will be re-evaluated using the modified DFS-search and any newly resulting connected components will be added to the priority queue. Depending on queue implementation, maintaining proper queue ordering requires $\mathcal{O}(\log |C|)$. The per-iteration runtime significantly improves as the number of vertices in each connected component decreases. Effectively, the expected computation time will be $\approx K \left(\frac{(n+m)}{|C|} + \log |C| \right)$, although the worst case remains $\mathcal{O}(K(m + n))$.

Table 1. Benchmark networks and their properties. $|V|$ and $|E|$ are the number of vertices and edges, ρ is the global clustering coefficient, δ is the diameter and ξ is the degree assortativity.

Network	Type	$|V|$	$|E|$	ρ	δ	ξ
Conmat [15]	Collaboration	23,133	93,439	0.264	15	0.134
Ego [16]	Social	4,039	88,234	0.519	8	0.064
Flight [18]	Transportation	2,939	15,677	0.255	14	0.051
Powergrid [29]	Power grid	4,941	6,594	0.103	46	0.003
Relativity [15]	Collaboration	5,242	14,484	0.630	17	0.659
Oclinks [19]	Social	1,899	13,838	0.057	8	-0.188

3   Experimental Results

We evaluate the algorithm using six benchmark real-world problems to demonstrate the practical approximation quality and runtime. The benchmark networks and their properties are given in Table 1. All networks are simplified before use (no self-loops or multi-edges). Our experimental results compare Algorithm 1 to three centrality measures used in greedy sequential fashion such as in [25]. These are used only as a base-level comparison for the quality of the greedy approach. We consider node degree, PageRank [5] and Authority score [13] centrality attacks.

3.1   Summary of Results

To compare the quality of the greedy approach we vary K as 0.01, 0.05, 0.10, 0.15, 0.20 and 0.25 proportion of the network, respectively. The computer used for simulations was a 3.4 GHz Intel i7 processor with 16 GB RAM, running open-SUSE Linux. We compare results to other methods of network attack (degree, PageRank and Kleinberg's Authority score) in a similar greedy sequential approach. These strategies have been recognized as potentially useful for network fragmentation when considering other robustness measures such as minimizing the largest network component [25]. Due to limited space a more rigorous experimental analysis was not possible, which would include an expanded set of real-world and benchmark network instances as well as other CNDP algorithms such as [2,22]. It should be noted that betweenness and closeness centrality, which are often also employed to test network vulnerability were too computationally inefficient to be considered for these networks.

Table 2 compares the objective value for $K = \{0.10n, 0.20n\}$ of the vertices in each network, respectively. The greedy approach outperforms the alternative strategies in all cases. Figure 2 plots the same experimental results over the entire range of K values for each network where the significance of the greedy solution quality is better highlighted. The proposed algorithm is especially destructive

Table 2. Comparison of the CNDP objective value after removing 10 % and 20 % of vertices from each network in Table 1.

Problem	K	GREEDY	Degree	PageRank	Authority
Comnat	2313	58,796,393	103,398,683	87,630,163	126,804,602
	4627	83,686	90,610	92,242	7,399,785
Ego	404	2,717,347	5,339,614	3,816,109	6,320,816
	808	1,848,740	2,070,535	2,886,709	3,438,031
Flight	294	322,527	484,331	467,962	1,014,305
	588	1,457	1,698	1,715	1,567
Powergrid	494	22,182	51,508	212,369	56,815
	988	3,639	4,580	14,744	3,771
Relativity	524	224,010	1,628,337	302,309	3,382,195
	1,048	4,089	4,896	9,023	6,390
Oclinks	190	637,936	785,662	758,328	835,297
	380	218,215	258,277	246,876	306,289

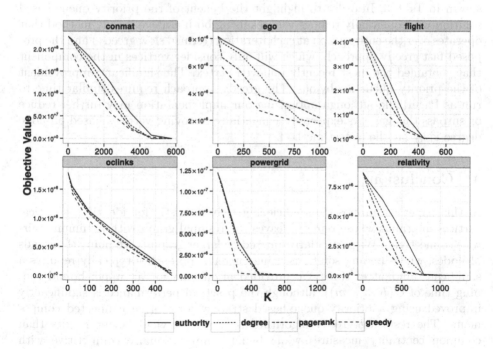

Fig. 2. Performance of each of the four strategies on the objective value. In all cases the greedy approach proposed in this paper yields the most desirable result.

Table 3. Comparing the run times (in milliseconds) of each approach. The proposed greedy approach is considered for both cases of using the queue-based strategy (fast greedy) or not (slow greedy).

Problem	Slow greedy	Fast greedy	Degree	PageRank	Authority
Conmat	1,732,841	21,229	379	16,810	26,027
Ego	7,036	2,129	92	1,480	3,750
Flight	10,199	207	25	144	231
Powergrid	29,896	252	41	288	3,470
Relativity	43,679	421	44	180	601
Oclinks	2,301	143	11	114	360

for $K < 0.15n$. All of the networks except the ego network exhibit a power-law degree distribution, which seems to be a major influence on the ability of fragmenting the networks for centrality-based approaches. The greedy algorithm significantly outperforms in these situations. Moreover, the global clustering coefficient, diameter or degree assortativity do not seem to have such an obvious impact as the degree distribution does.

Running time (in milliseconds) for these networks was also investigated and shown in Table 3. In order to highlight the benefit of the priority queue-based solution we sequentially remove vertices using both a naive greedy method that operates over the entire graph at each iteration (termed slow greedy) and the proposed fast greedy approach, which will only consider vertices in the component that contained the most recently removed vertex. The significant improvement of the priority queue is obvious. The greedy approach requires similar time to run as PageRank, although optimizing our implementation may further reduce or surpass this gap. As expected, sequentially removing vertices based on node degree is by far the fastest method.

4   Conclusion

In this paper we proposed an efficient greedy heuristic for identifying critical vertices in graphs whose removal leaves the residual graph with minimum pairwise connectivity. We particularly focused on larger graphs with many thousands of nodes, where finding solutions using current approaches typically requires a significant amount of time. We provided arguments for an upper-bound running time of $\mathcal{O}(K(n + m))$, although the practical performance is significantly improved using a priority-queue-based strategy for storing connected components. The resulting greedy algorithm is shown to yield better results than common centrality measures while being computationally competitive with degree-based greedy vertex removal.

The algorithm proposed in this paper was given without any proof of approximation quality. Future work will prove this bound. Moreover, experimentation on different network types and much larger sized, including runtime should also be conducted. Potential improvements in runtime may be attainable if within-connected component objective function evaluation was parallelized or a relationship between the nodes can be identified so that only an $\mathcal{O}(n)$ process is required to update the impact a vertex removal will have on its associated connected component. Extension to vertex and edge-weight variants of the CNDP is also possible, but not done in this paper.

References

1. Identifying critical nodes in undirected graphs: Complexity results and polynomial algorithms for the case of bounded treewidth. Discrete Appl. Math. **161**(16–17), 2349–2360 (2013)
2. Arulselvan, A., Commander, C.W., Elefteriadou, L., Pardalos, P.M.: Detecting critical nodes in sparse graphs. Comput. Oper. Res. **36**(7), 2193–2200 (2009)
3. Aspnes, J., Chang, K., Yampolskiy, A.: Inoculation strategies for victims of viruses and the sum-of-squares partition problem. In: Proceedings of the Sixteenth Annual ACM-SIAM Symposium on Discrete Algorithms, SODA '05, pp. 43–52. Society for Industrial and Applied Mathematics (2005)
4. Boginski, V., Commander, C.: Identifying critical nodes in protein-protein interaction networks. In: Clustering Challenges in Biological, Networks, pp. 153–166 (2009)
5. Brin, S., Page, L.: The anatomy of a large-scale hypertextual web search engine. Comput. Netw. ISDN Syst. **30**, 107–117 (1998)
6. Chen, P., David, M., Kempe, D.: Better vaccination strategies for better people. In: Proceedings of the 11th ACM Conference on Electronic Commerce, pp. 179–188. ACM (2010)
7. Di Summa, M., Grosso, A., Locatelli, M.: Branch and cut algorithms for detecting critical nodes in undirected graphs. Comput. Optim. Appl. **53**, 649–680 (2012)
8. Di Summa, M., Grosso, A., Locatelli, M.: Complexity of the critical node problem over trees. Comput. Oper. Res. **38**(12), 1766–1774 (2011)
9. Dinh, T.N., Xuan, Y., Thai, M.T., Pardalos, P.M., Znati, T.: On new approaches of assessing network vulnerability: Hardness and approximation. IEEE/ACM Trans. Networking **20**(2), 609–619 (2012)
10. Dinh, T.N., Thai, M.T., Nguyen, H.T.: Bound and exact methods for assessing link vulnerability in complex networks. J. Comb. Optim. **28**(1), 3–24 (2014)
11. Joyce, K.E., Laurienti, P.J., Burdette, J.H., Hayasaka, S.: A new measure of centrality for brain networks. PLoS ONE **5**(8), e12200 (2010)
12. Kempe, D., Kleinberg, J., Tardos, E.: Maximizing the spread of influence in a social network. In: Proceedings of the 9th International Conference on Knowledge Discovery and Data Mining, pp. 137–146 (2003)
13. Kleinberg, J.M.: Authoritative sources in a hyperlinked environment. J. ACM **46**(5), 604–632 (1999)
14. Anil Kumar, V.S., Rajaraman, R., Sun, Z., Sundaram, R.: Existence theorems and approximation algorithms for generalized network security games. In: Proceedings of the 2010 IEEE 30th International Conference on Distributed Computing Systems, pp. 348–357 (2010)
15. Leskovec, J., Kleinberg, J.M., Faloutsos, C.: Graph evolution: densification and shrinking diameters. ACM Trans. Knowl. Dis. Data **1**(1), 2–41 (2007)

16. McAuley, J.J., Leskovec, J.: Learning to discover social circles in ego networks. In: NIPS, pp. 548–556 (2012)
17. Nguyen, D.T., Shen, Y., Thai, M.T.: Detecting critical nodes in interdependent power networks for vulnerability assessment. IEEE Trans. Smart Grid 4(1), 151–159 (2013)
18. Opsahl, T.: Why anchorage is not (that) important: Binary ties and sample selection (2011). http://wp.me/poFcY-Vw
19. Opsahl, T., Panzarasa, P.: Clustering in weighted networks. Soc. Netw. 31(2), 155–163 (2009)
20. Sporns, O.: Networks of the Brain. The MIT Press, Cambridge (2010)
21. Tarjan, R.: Depth-first search and linear graph algorithms. SIAM J. Comput. 1(2), 146–160 (1972)
22. Ventresca, M.: Global search algorithms using a combinatorial unranking-based problem representation for the critical node detection problem. Comput. Oper. Res. 39(11), 2763–2775 (2012)
23. Ventresca, M., Aleman, D.: Evaluation of strategies to mitigate contagion spread using social network characteristics. Soc. Netw. 35(1), 75–88 (2013)
24. Ventresca, M., Aleman, D.: A derandomized approximation algorithm for the critical node detection problem. Comput. Oper. Res. 43, 261–270 (2014)
25. Ventresca, M., Aleman, D.: Network robustness versus multi-strategy sequential attack. J. Complex Netw. (2014)
26. Ventresca, M., Aleman, D.: A randomized algorithm with local search for containment of pandemic disease spread. Comput. Oper. Res. 48, 11–19 (2014)
27. Veremyev, A., Boginski, V., Pasiliao, E.L.: Exact identification of critical nodes in sparse networks via new compact formulations. Optim. Lett. 8(4), 1245–1259 (2014)
28. Veremyev, A., Prokopyev, O.A., Pasiliao, E.L.: An integer programming framework for critical elements detection in graphs. J. Comb. Optim. 28(1), 233–273 (2014)
29. Watts, D.J., Strogatz, S.H.: Collective dynamics of 'small-world' networks. Nature 393, 400–442 (1998)

Integer Programming Formulations for Minimum Spanning Forests and Connected Components in Sparse Graphs

Neng Fan[✉] and Mehdi Golari

Systems and Industrial Engineering Department,
University of Arizona, Tucson 85721, USA
{nfan,golari}@email.arizona.edu

Abstract. In this paper, we first review several integer programming formulations for the minimum spanning tree problem, and then adapt these formulations for solving the minimum spanning forest problem in sparse graphs. Some properties for the spanning forest and connected components are studied, and then we present the integer programming formulation for finding the largest connected component, which has been widely used for network vulnerability analysis.

Keywords: Minimum spanning forest · Minimum spanning tree · Connected components · Network vulnerability analysis · Integer programming

1 Introduction

Given an undirected graph $G = (V, E)$, where V is the vertex set with cardinality $|V| = n$, and E is the edge set. A graph is *connected* if there exists a path between any pair of its vertices. A maximal connected subgraph of G is called its *connected component*.

A connected graph has exactly one connected component, consisting of the whole graph. Generally, the number of connected components in graph G can be computed several approaches:

(a) It can be interpreted as the zeroth Betti number of the graph G, in topological theory (see [1]).
(b) It equals to the multiplicity of 0 as an eigenvalue of the Laplacian matrix of the graph G, in algebraic graph theory.
(c) It is also the index of the first nonzero coefficient of the chromatic polynomial of a graph.

In this paper, we assume that the number m of connected components in graph G is known and given, and the second approach will be used to compute the exact value of m for a given graph for numerical experiments purpose. The Laplacian

© Springer International Publishing Switzerland 2014
Z. Zhang et al. (Eds.): COCOA 2014, LNCS 8881, pp. 613–622, 2014.
DOI: 10.1007/978-3-319-12691-3_46

matrix for graph G is defined as $L := (l_{ij})_{n \times n}$ such that $L = D - A$, where D is the degree matrix and A is the adjacency matrix of G, or directly expressed as

$$l_{ij} = \begin{cases} deg(v_i), & \text{if } i = j, \\ -1, & \text{if } i \neq j \text{ and } v_i \text{ is adjacent to } v_j, \\ 0, & \text{otherwise,} \end{cases}$$

where $deg(v_i)$ is degree of the vertex v_i.

A *tree* is a connected graph without cycles. A subgraph of a connected graph G is its *spanning tree* T if it is a tree connecting all vertices of G. A *spanning forest* F of a disconnected graph G is a collection of spanning trees, one for each of its connected components. Given a graph G with weight matrix $W = (w_{ij})_{n \times n}$, where $w_{ij} \geq 0$ is the weight of edge $(i, j) \in E$ such that $w_{ij} = w_{ji}$ and $w_{ii} = 0$ for all $i = 1, \cdots, n$, a *minimum spanning forest (MSF)* of a weighted graph is a spanning forest such that the sum of its edge-weight is minimum. When the number of connected components in G is 1, the MSF problem reduces to the well-known *minimum spanning tree (MST)* problem.

There are two algorithms commonly used to find the MST in a graph: Prim's algorithm and Kruskal's algorithm. If the graph is not connected, the Kruskal's algorithm finds the MSF directly. Running Prim's algorithm for each connected component, we can also find the MSF. The Kruskal's algorithm has computational complexity as $O(|E| \log |V|)$ while Prim's algorithm can have $O(|E| + |V| \log |V|)$. For MSF problem, in [2,3], more efficient algorithms to find MSF that run in $O(\log |V|)$ time were designed. These are all heuristic algorithms. Recently, because of advances in high-performance computing and efficient algorithms found in commercial and open-source software, integer programming (IP) methods are more broadly used for obtaining exact solutions to many combinatorial optimization problems. For example, Pop [4] presented a survey on different integer programming formulations of the generalized MST problem.

In this paper, we study the problem of finding the minimum spanning forest in a given graph (mainly in sparse graphs, i.e., graphs with only a few edges) by integer programming. Besides the MST problem, the MSF problem also has close relations with other classic combinatorial optimization problems, such as finding the largest connected components of a graph, and interdependent power network disruptor problem for vulnerability analysis. Therefore, we also discuss these problems and their relations to network analysis.

The remainder of this paper is organized as follows. In Sect. 2, several IP formulations for MSF problem will be presented and discussed with their corresponding algorithms. Section 3 will discuss problems related to largest connected component in a network. Finally, Sect. 4 concludes this paper.

2   IP Formulations for the MSF Problem

Given an undirected graph $G = (V, E)$ with a nonnegative weight matrix W, let $x_{ij} \in \{0, 1\}$ denote whether edge $(i, j) \in E$ is selected into the forest if $x_{ij} = 1$

or not if $x_{ij} = 0$. Let x denote the vector formed by x_{ij}'s for all $(i,j) \in E$. The MSF found by optimal x^*, denoted F^*, will be a subgraph $F^* = (V, E^*)$, where $E^* = \{(i,j) \in E : x_{ij}^* = 1\}$ denotes the selected edge into the spanning forest.

2.1   IP Formulations for MST Problem

If this graph is connected, i.e., $m = 1$, the MSF problem becomes the MST problem. Recall that there are several basic properties related to tree:

Proposition 1. *A tree is an undirected simple graph T (with n vertices) that satisfies any one of the following equivalent conditions:*

(i) T has no simple cycles and has $n - 1$ edges;
(ii) T is connected and has $n - 1$ edges;

Based on each of these properties, the following IP formulations are presented correspondingly for the MST problem:

(1) Subtour elimination formulation is based on the fact that G has no simple cycles and has $n - 1$ edges (Proposition 1(i)):

$$[\text{MST1}] \quad \min_x \sum_{(i,j)\in E} w_{ij} x_{ij}$$

$$s.t. \begin{cases} \sum_{(i,j)\in E} x_{ij} = n - 1 \\ \sum_{(i,j)\in E(S)} x_{ij} \leq |S| - 1, \ \forall S \subset V, S \neq V, S \neq \emptyset \\ x_{ij} \in \{0,1\}, \ \forall (i,j) \in E \end{cases}$$

where $E(S) \subset E$ is a subset of edges with both ends in subset $S \subset V$. Constraint $\sum_{(i,j)\in E(S)} x_{ij} \leq |S| - 1$ ensures that there is no cycles in subset S.

(2) Cutset formulation based on the fact that G is connected and has $n - 1$ edges (Proposition 1(ii)):

$$[\text{MST2}] \quad \min_x \sum_{(i,j)\in E} w_{ij} x_{ij}$$

$$s.t. \begin{cases} \sum_{(i,j)\in E} x_{ij} = n - 1 \\ \sum_{(i,j)\in \delta(S)} x_{ij} \geq 1, \ \forall S \subset V, S \neq V, S \neq \emptyset \\ x_{ij} \in \{0,1\}, \ \forall (i,j) \in E \end{cases}$$

where the cutset $\delta(S) \subset E$ is a subset of edges with one end in S and the other end in $V \setminus S$. Constraints $\sum_{(i,j)\in\delta(S)} x_{ij} \geq 1$ ensures that subsets S and $V \setminus S$ are connected.

(3) Single-commodity flow formulation. Additionally, if considering the Proposition 1(ii), the single-commodity flow constraints can be used to ensure the connectivity of a tree found by x. Correspondingly, the following formulation can be presented:

$$[\textbf{MST3}] \quad \min_{x,f} \sum_{(i,j)\in E} w_{ij}x_{ij}$$

$$s.t. \begin{cases} \sum_{(i,j)\in E} x_{ij} = n-1 \\ \sum_j f_{j1} - \sum_j f_{1j} = n-1 \\ \sum_j f_{ji} - \sum_j f_{ij} = 1, \forall i \in V, i \neq 1 \\ f_{ij} \leq (n-1)x_{ij}, \; f_{ji} \leq (n-1)x_{ij}, \forall (i,j) \in E \\ f_{ij} \geq 0, \; \forall (i,j) \in E \cup E' \\ x_{ij} \in \{0,1\}, \; \forall (i,j) \in E \end{cases}$$

where the decision variable f_{ij} denotes the flow from vertex v_i to vertex v_j (f is vectors for flows on all possible pairs in two directions). The constraints in [**MST3**] to find the MST include (in order): the total number of selected edges into the tree is $n-1$; the total amount of flow from vertex v_1 is $n-1$; each vertex consumes exactly one unit of flow; the flow from v_i to v_j or from v_j to v_i may have flow if edge (i,j) is chosen into the spanning tree, and have upper bounds $n-1$; the flow is nonnegative, where E' denotes set of edges with opposite direction for edges in E; and binary choice for edge selection into the spanning tree. Through this formulation, $n-1$ edges selected by x connects all n vertices, and by Proposition 1(ii), x finds the MST.

Let P_{sub} be polyhedron formed by constraints of [**MST1**], i.e.,

$$P_{sub} = \{x \in \mathbb{R}^{|E|} : 0 \leq x_{ij} \leq 1, \forall (i,j) \in E, \sum_{(i,j)\in E} x_{ij} = n-1,$$

$$\sum_{(i,j)\in E(S)} x_{ij} \leq |S| - 1, \; \forall S \subset V, S \neq V, S \neq \emptyset\}.$$

Similarly,

$$P_{cut} = \{x \in \mathbb{R}^{|E|} : 0 \leq x_{ij} \leq 1, \forall (i,j) \in E, \sum_{(i,j)\in E} x_{ij} = n-1,$$

$$\sum_{(i,j)\in \delta(S)} x_{ij} \geq 1, \; \forall S \subset V, S \neq V, S \neq \emptyset\}$$

is the polyhedron formed by constraints of [**MST2**], and

$$P_{flow} = \{x \in \mathbb{R}^{|E|} : 0 \leq x_{ij} \leq 1, \forall (i,j) \in E,$$

all constraints in [**MST3**] except the integraty conditions}

is the polyhedron formed by constraints of [**MST3**]. There are some properties regarding these two sets: (i) The polyhedron P_{sub} has integral extreme points; (ii) $P_{sub} \subseteq P_{cut} \subseteq P_{flow}$. These properties can be used in solving IP formulations, and we refer to [5,6] for the first property for more details, and [6–8] for the second property.

Both formulations [**MST1**] and [**MST2**] have exponential number of constraints in the worst-case, while formulation [**MST3**] has polynomial number of constraints. Additionally, in [9], Martin presented a reformulation with polynomial number of constraints for the MST problem. This method was also used in [10] by Yannakakis, who attributed to [9], and was recently mentioned in [6,11].

(4) Martin's formulation:

$$[\textbf{MST4}] \quad \min_{x,y} \sum_{(i,j)\in E} w_{ij} x_{ij}$$

$$s.t. \begin{cases} \sum_{(i,j)\in E} x_{ij} = n-1 \\ y_{ij}^k + y_{ji}^k = x_{ij}, \ \forall (i,j) \in E, k \in V \\ \sum_{k\in V\setminus\{i,j\}} y_{ik}^j + x_{ij} = 1, \ \forall (i,j) \in E \\ x_{ij}, y_{ij}^k, y_{ji}^k \in \{0,1\}, \ \forall (i,j) \in E, k \in V \end{cases}$$

where the decision variable $y_{ij}^k \in \{0,1\}$ for each $(i,j) \in E, k \in V$ denotes that edge (i,j) is in the spanning tree and vertex k is on the side of j (i.e., vertex k is within the resulted part containing j after removal of (i,j) from the spanning tree) if $y_{ij}^k = 1$, and edge (i,j) is on the tree and k is not on side of j or if edge (i,j) is not in the tree if $y_{ij}^k = 0$.

Here, in this formulation, the first constraint ensures that there are exactly $n-1$ edges selected into the spanning tree. The second constraint for $(i,j) \in E, k \in V$ grantees that if $(i,j) \in E$ is selected into the tree ($x_{ij} = 1$), any vertex $k \in V$ must be either on the side of j ($y_{ij}^k = 1$) or on the side of i ($y_{ji}^k = 1$); if $(i,j) \in E$ is not in the tree ($x_{ij} = 0$), any vertex k cannot be on the side of j nor i ($y_{ij}^k = y_{ji}^k = 0$). The third constraint for $(i,j) \in E$ ensures that if $(i,j) \in E$ is in the tree ($x_{ij} = 1$), edges (i,k) who connects i are on the side of i ($y_{ik}^j = y_{ij}^k = 0, y_{ij}^k = 1$); if $(i,j) \in E$ is not in the tree ($x_{ij} = 0$), there must be an edge (i,k) such that j is on the side of k ($y_{ik}^j = 1$ for some k). These constraints with the binary requirements of all x_{ij}'s and y_{ij}^k's ensures that E^* presents a spanning tree for a connected graph, and they are proposed by Martin in [9] and discussed in [11].

2.2   IP Formulations for MSF Problem

In the following, we will first generalize these formulations for a general graph with m connected components, and therefore, a MSF can be found.

Without loss of generality, assume that the m connected components of G have vertex sets as $V_1, V_2 \cdots, V_m$, respectively. Definitely, these sets are disjoint and their union is V. Correspondingly, assume E_k ($k = 1, 2, \cdots, m$) is the edge set induced by vertices in V_i from graph G. Thus, each connected component of G can be considered as a subgraph $G_k = (V_k, E_k)$ ($k = 1, 2, \cdots, m$) of G. Therefore, a spanning forest of G consists of spanning trees of all G_k's, and one for each G_k. The following theorem is a direct result.

Theorem 1. *For the graph G with m connected components, denoted by G_1, $G_2, \cdots, G_m$, the forest F^*, consisting of spanning trees $T_1^*, T_2^*, \cdots, T_m^*$, is a minimum spanning forest of G if and only if each T_i^* is a minimum spanning tree for subgraph G_i $(i = 1, 2, \cdots, m)$. Furthermore, the number of edges in a spanning forest of G is $n - m$.*

From this theorem, if a simple graph has no simple cycles and exactly $n - m$ edges, where m is the number connected components in the graph, it is a forest. Now, we adapt subtour elimination constraints and cutset constraints to find the MSF in a graph with m connected components.

Considering $S \subset V, S \neq \emptyset, S \neq V$, there are three cases for the subtour elimination constraints and cutset constraints:

(i) if $S \subset V_i$, $\sum_{(i,j) \in E(S)} x_{ij} \leq |S| - 1$; $\sum_{i \in S, j \in V \setminus S} x_{ij} \geq 1$;
(ii) if $S \subset V_{i_1} \cup V_{i_2} \cup \cdots \cup V_{i_k}$ $(2 \leq k \leq m)$ and $S \cap V_{i_1} \neq \emptyset, \cdots, S \cap V_{i_k} \neq \emptyset$, $\sum_{(i,j) \in E(S)} x_{ij} \leq |S| - k$; $\sum_{i \in S, j \in V \setminus S} x_{ij} \geq k$;
(iii) if $S = V_{i_1} \cup V_{i_2} \cup \cdots \cup V_{i_k}$ $(1 \leq k < m)$, $\sum_{(i,j) \in E(S)} x_{ij} \leq |S| - k$; $\sum_{i \in S, j \in V \setminus S} x_{ij} \geq 0$.

The first set of constraints in case (i) for this type of subsets are enough to ensure that x_{ij}'s generate all spanning trees for all connected components, and thus, they also ensure a spanning forest for G. Therefore, for a general subset $S \subset V$, if we can make constraints in cases (ii) and (iii) be redundant, a slight modification of formulations [**MST1**] and [**MST2**] will find the MSF of G.

For subtour elimination constraints, the case (ii) can have the relations as $\sum_{(i,j) \in S} x_{ij} \leq |S| - k < |S| - 1$, and the case (iii) also has $\sum_{(i,j) \in S} x_{ij} \leq |S| - k \leq |S| - 1$. Therefore, changing them to $\sum_{(i,j) \in S} x_{ij} \leq |S| - 1$ will make them be redundant for these types of subsets S's. An IP formulation, similar to [**MST1**] for the MST problem, is presented for the MSF problem:

$$[\textbf{MSF1}] \quad \min \sum_{(i,j) \in E} w_{ij} x_{ij}$$

$$s.t. \quad \sum_{(i,j) \in E} x_{ij} = n - m$$

$$\sum_{(i,j) \in E(S)} x_{ij} \leq |S| - 1, \ \forall S \subset V, S \neq V, S \neq \emptyset$$

$$x_{ij} \in \{0, 1\}, \ \forall (i,j) \in E$$

where the first constraint ensures that there are $n - m$ edges in the spanning forest.

For the cutset constraints, using the one $\sum_{i \in S, j \in V \setminus S} x_{ij} \geq 1$ in case (i) to replace the one $\sum_{i \in S, j \in V \setminus S} x_{ij} \geq k > 1$ will make the constraints be redundant for S's in case (ii). However, the one $\sum_{i \in S, j \in V \setminus S} x_{ij} \geq 0$ in case (iii) is stronger than the one in case (i). Let $a_{ij} = 1$ if the pair $(i,j) \in E$, and $a_{ij} = 0$ if

$(i, j) \notin E$ for any $i, j \in V$ (a_{ij} is actually an element in adjacency matrix A). By introducing the variable, the case (iii) can be fixed as follows:

$$[\textbf{MSF2}] \quad \min \quad \sum_{(i,j) \in E} w_{ij} x_{ij}$$

$$s.t. \quad \sum_{(i,j) \in E} x_{ij} = n - m$$

$$\sum_{i \in S, j \in V \backslash S, (i,j) \in E} x_{ij} \geq \max_{i \in S, j \in V \backslash S} a_{ij}, \ \forall S \subset V, S \neq V, S \neq \emptyset$$

$$x_{ij} \in \{0, 1\}, \ \forall (i, j) \in E$$

where the right-hand-side $\max_{i \in S, j \in V \backslash S} a_{ij}$ of the second constraints will be 1 for S chosen as in cases (i) and (ii), and 0 for S in case (iii).

From arguments discussed above in cases (i)–(iii), both formulations [**MSF1**] and [**MSF2**] will find the minimum spanning forest directly without finding the connected components. In the following, we will propose several IP formulations to find the connected components in a graph first and then to find the minimum spanning tree of each component, which can form the minimum spanning forest as shown in Theorem 1.

Next, we present a formulation to find all components such that vertices in the same component are connected and vertices in different components are not connected.

Theorem 2. *Assume that graph G has m connected components $G_1, G_2, \cdots, G_m$. Let $z_{ik} \in \{0, 1\}$ be a decision variable such that $z_{ik} = 1$ if vertex v_i is in set V_k of component G_i, for all $v_i \in V, k = 1, 2, \cdots, m$. Additionally, let u_{ij} be a decision variable such that $u_{ij} = 1$ if vertices v_i, v_j are in the same component, and $u_{ij} = 0$ otherwise, for all $v_i, v_j \in V, i \neq j$. The formulation*

$$u_{ij} \geq a_{ij}, \ \forall i, j \in V, i \neq j \tag{1a}$$

$$\begin{cases} u_{i_1 i_2} + u_{i_2 i_3} - u_{i_1 i_3} \leq 1, \\ u_{i_1 i_2} - u_{i_2 i_3} + u_{i_1 i_3} \leq 1, \qquad \forall \ triplets \ i_1, i_2, i_3 \in V \\ -u_{i_1 i_2} + u_{i_2 i_3} + u_{i_1 i_3} \leq 1, \end{cases} \tag{1b}$$

$$u_{ij} = \sum_{k=1}^{m} z_{ik} z_{jk}, \ \forall i, j \in V, i \neq j \tag{1c}$$

$$\sum_{k=1}^{m} z_{ik} = 1, \ \forall i \in V \tag{1d}$$

$$z_{ik} \in \{0, 1\}, u_{ij} \in \{0, 1\}, \ \forall i, j \in V, k = 1, \cdots, m \tag{1e}$$

will find all m connected components in graph G, i.e., the vertex set V_k of the kth connected component can be expressed as $V_k = \{v_i \in V : z_{ik} = 1\}$, and vertices in V_k should be connected ($k = 1, 2, \cdots, m$).

Proof. Inequalities in (1a) imply that v_i, v_j are in the same component (i.e., $u_{ij} = 1$) if there exists an edge $(i, j) \in E$ (i.e., $a_{ij} = 1$) and $u_{ij} = 0$ otherwise. Inequalities (1b) imply that for all triplets $i_1, i_2, i_3 \in V$, if i_1, i_2 are in the same component and i_2, i_3 are in the same component, then i_1, i_3 are in the same component. Equalities in (1c) ensure that if i, j are partitioned into the same component, $u_{ij} = 1$ and otherwise $u_{ij} = 0$. Equalities in (1d) ensure that each vertex is partitioned into one connected component, while the last ones ensure the binary choices of z_{ik}'s and u_{ij}'s. □

In Theorem 2, we use the pairwise connectivity to find all connected components. In the formulation in Theorem 2, a feasible solution to constraints (1a)–(1b) with $u_{ij} \in \{0, 1\}$ for all $i, j \in V, i \neq j$ can decide every pair of vertices in the same component or not. By all values of u_{ij}'s from this feasible solution, constraints (1c)–(1d) with $z_{ik} \in \{0, 1\}$ for all $i \in V, k = 1, \cdots, m$ can then divide the graph into components directly.

Also, by similar method in [**MST3**], we can propose another approach by the multi-commodity flow to find all m connected components by sending m types of flows. By Martin constraints, a MST will be found for each connected component and thus a MSF can be found directly. Therefore, different from [**MSF1**] and [**MSF2**], both of which have exponential number of constraints, the following formulation [**MSF3**] has polynomial number of constraints.

$$[\textbf{MSF3}] \quad \min_{x,y,z,u} \sum_{(i,j) \in E} w_{ij} x_{ij} \tag{2a}$$

$$s.t. \text{ Constraints in (1a)–(1e)} \tag{2b}$$

$$\sum_{(i,j) \in E} x_{ij} = n - m \tag{2c}$$

$$y_{ij}^{kl} + y_{ji}^{kl} = x_{ij} z_{lk}, \quad \forall (i,j) \in E, l \in V, k = 1, \cdots, m \tag{2d}$$

$$\sum_{l \in V \setminus \{i,j\}} y_{il}^{kj} z_{lk} + x_{ij} = 1, \quad \forall (i,j) \in E, k = 1, \cdots, m \tag{2e}$$

$$y_{ij}^{kl}, y_{ji}^{kl} \leq \min\{z_{ik}, z_{jk}, z_{lk}, x_{ij}\}, \quad \forall (i,j) \in E, l \in V, k = 1, \cdots, m \tag{2f}$$

$$x_{ij}, y_{ij}^{kl}, y_{ji}^{kl} \in \{0, 1\}, \quad \forall (i,j) \in E, l \in V, k = 1, \cdots, m \tag{2g}$$

The objective (2a) is the same as those one in [**MSF1**] and [**MSF2**] to minimize the cost associated the spanning forest. Constraint (2c) ensures the number of chosen edges into the spanning forest. Constraints in (2d)–(2f) ensure that the chosen edges by x_{ij} imply a spanning tree in each connected component by Martin constraints.

Proposition 2. *In graph G, let $N(i) = \{v_j \in V : (i, j) \in E\}$ be the set of neighbors of vertex v_i. If v_i is not an isolated vertex, i.e., $N(i) \neq \emptyset$, $\sum_{j \in N(i)} x_{ij} \geq 1$ is a valid inequality for above IP formulations for the MSF problem.*

Proposition 3. *The anticycle inequalities*

$$z_{11} = 1, z_{21} + z_{22} = 1, \cdots, z_{m1} + z_{m2} + \cdots + z_{mm} = 1, \tag{3}$$

can be added to **[MSF3]** *to reduce the computational complexity.*

The m equalities in (3) denote that: the first vertex is in the 1st component; the second vertex can be either in the 1st or 2nd component; similarly, the mth vertex can be in one of the first m components.

Remark. If there exist some isolated vertices, the formulations in **[MSF1]**, **[MSF2]**, **[MSF3]** can still find the minimum spanning forests in sparse graphs.

3   Largest Connected Component

Following the method for the formulation **[MSF3]**, the connected component G_k has number of vertices as $|V_k| = \sum_{i:v_i \in V} z_{ik}$. Now based on this method, we present an IP formulation for finding the largest connected component (LCC) in a graph G with m components:

$$\textbf{[LCC]} \quad \max \sum_{k=1}^{m} \delta_k \sum_{i:v_i \in V} z_{ik} \tag{4a}$$

$$s.t. \sum_{k=1}^{m} \delta_k = 1 \tag{4b}$$

$$\text{Constraints in (1a)-(1e)} \tag{4c}$$

$$\delta_k \in \{0,1\}, \ k = 1, 2, \cdots, m \tag{4d}$$

where the objective with constraint (4b) and binary choices of δ_k's ensures that for some k, $\delta_k = 1$ and the corresponding $|V_k|$ is the largest one among all components. The one k with $\delta_k = 1$ indicates the largest connected component, and the objective $\sum_{k=1}^{m} \delta_k \sum_{i:v_i \in V} z_{ik}$ gives the size of the components as all other δ_k's are 0. Similarly, inequalities in (3) can be added to **[LCC]** to reduce the computational complexity.

The largest connected component has been used as a measurement for power network's vulnerability in case of cascading failures. In [12], the *interdependent power network disruptor* (IPND) problem was presented to find the minimum size of the largest connected component in case of node failures. Following the notations and definitions in [12], the interdependent network can be defined as a system, consisting of two graphs $G_s = (V_s, E_s)$ and $G_c = (V_c, E_c)$ and their interdependencies E_{sc}, where V_s, V_c are node sets and E_s, E_c, E_{sc} are edge set (called *intra-links*). The set E_{sc} includes *inter-links* coupling G_s and G_c, and is denoted by $E_{sc} = \{(u,v) : u \in V_s, v \in V_c\}$. The IPND problem is to find a subset $T \subseteq V_s$ of a fix size, such that the size of the largest connected component of G_s after the cascading failures caused by removal of T is minimized.

4   Conclusions

In this paper, we first reviewed several integer programming formulations for the minimum spanning tree problem, and then adapted these formulations for solving the minimum spanning forest problem in sparse graphs. Some properties for the spanning forest and connected components are studied, and we then present the integer programming formulation for finding the largest connected component, which has been widely used for network vulnerability analysis. Further research directions include studying the polyhedral properties of constraints in different integer programming formulations, and also integer programming formulations for network vulnerability analysis, especially by largest connected component, for social networks, and related combinatorial optimization problems, such as k-minimum spanning tree problem or K-cardinality tree problem, subgraph isomorphism problem, etc.

References

1. Carlsson, G.: Topology and data. Bull. Am. Math. Soc. **46**(2), 255–308 (2009)
2. Cole, R., Klein, P.N., Tarjan, R.E.: Finding minimum spanning forests in logarithmic time and linear work using random sampling. In: SPAA '96 Proceedings of the Eighth Annual ACM Symposium on Parallel Algorithms and Architectures, pp. 243–250 (1996)
3. Pettie, S., Ramachandran, V.: A randomized time-work optimal parallel algorithm for finding a minimum spanning forest. SIAM J. Comput. **31**(6), 1879–1895 (2002)
4. Pop, P.C.: A survey of different integer programming formulations of the generalized minimum spanning tree problem. Carpathian J. Math. **25**(1), 104–118 (2009)
5. Edmonds, J.: Matroids and the greedy algorithm. Math. Program. **1**, 127–136 (1971)
6. Conforti, M., Cornuéjols, G., Zambelli, G.: Extended formulations in combinatorial optimization. 4OR **8**(1), 1–48 (2010)
7. Myung, Y.S., Lee, C.H., Tcha, D.W.: On the generalized minimum spanning tree problem. Networks **26**(4), 231–241 (1995)
8. Feremans, C., Labbe, M., Laporte, G.: A comparative analysis of several formulations for the generalized minimum spanning tree problem. Networks **39**(1), 29–34 (2002)
9. Martin, R.K.: Using separation algorithms to generate mixed integer model reformulations. Oper. Res. Lett. **10**, 119–128 (1991)
10. Yannakakis, M.: Expressing combinatorial optimization problems by linear programs. J. Comput. Syst. Sci. **43**(3), 441–466 (1991)
11. Kaibel, V., Pashkovich, K., Theis, D.O.: Symmetry matters for the sizes of extended formulations. In: Eisenbrand, F., Shepherd, F.B. (eds.) IPCO 2010. LNCS, vol. 6080, pp. 135–148. Springer, Heidelberg (2010)
12. Nguyen, D.T., Shen, Y., Thai, M.T.: Detecting critical nodes in interdependent power networks for vulnerability assessment. IEEE Trans. Smart Grid **4**(1), 151–159 (2013)

Complexity, Cryptography and Game

On the Parameterized Complexity of Dynamic Problems with Connectivity Constraints

Faisal N. Abu-Khzam[1,2](✉), Judith Egan[1], Michael R. Fellows[1],
Frances A. Rosamond[1], and Peter Shaw[1]

[1] School of Engineering and Information Technology,
Charles Darwin University, Darwin, NT 0909, Australia
{Judith.Egan,Michael.Fellows,Frances.Rosamond,Peter.Shaw}@cdu.edu.au
[2] Department of Computer Science and Mathematics,
Lebanese American University Beirut, Beirut, Lebanon
faisal.abukhzam@lau.edu.lb

Abstract. In a dynamic version of a (base) problem X it is assumed that some solution to an instance of X is no longer feasible due to changes made to the original instance, and it is required that a new feasible solution be obtained from what "remained" from the original solution at a minimal cost. In the parameterized version of such a problem, the changes made to an instance are bounded by an *edit-parameter*, while the cost of reconstructing a feasible solution is bounded by some *increment-parameter*. Parameterized versions of a number of dynamic problems are studied where the solution to the base problem is assumed to be connected. We show that connectivity of solutions plays a positive role with respect to the edit-parameter by proving that the dynamic versions of CONNECTED DOMINATING SET and CONNECTED VERTEX COVER are fixed-parameter tractable with respect to the edit-parameter. On the other hand, the two problems are shown to be $W[2]$-hard with respect to the increment-parameter. We illustrate further the utility of connected solutions by proving that DYNAMIC INDEPENDENT DOMINATING SET is $W[2]$-hard with respect to the edit-parameter and we discuss some dynamic versions of maximization problems.

Keywords: Dynamic problems · Re-optimization · Dynamic connected dominating set · Parameterized complexity

1   Introduction

A dynamic version of a problem X takes as input a quintuple (I, I', S, k, r) where I and I' are instances of X and S is a solution to I, not necessarily optimal. The instances I and I' are at a given "edit distance" k, termed the *edit-parameter*. The question posed is whether a solution S' to I', also not necessarily optimal, can be obtained such that the Hamming distance between S and S' is bounded by r, which we refer to henceforth as the *increment-parameter*.

© Springer International Publishing Switzerland 2014
Z. Zhang et al. (Eds.): COCOA 2014, LNCS 8881, pp. 625–636, 2014.
DOI: 10.1007/978-3-319-12691-3_47

Dynamic problems appear in the realm of re-optimization as in the (initial) work on vehicle routing [13] and other approximation problems [3,16]. In a parameterized dynamic problem the original solution is not assumed to be optimal, and we seek a solution (of the new instance) that can be obtained at some given affordable cost. Dynamic problems have also been studied from a parameterized complexity viewpoint under the name "incremental problems" as in [9]. The term "incremental problem" has been coined in [4] in a seemingly initial work on programming languages.

Parameterized dynamic problems are formulated for the first time in [6], where DYNAMIC DOMINATING SET was shown to be fixed-parameter tractable. In this paper, we consider DYNAMIC CONNECTED DOMINATING SET as a main example of dynamic problems whose solutions are required to be connected. CONNECTED DOMINATING SET has been used as a model for virtual backbone (for message routing) of a wireless ad-hoc network (see [2,5,14,15,17]). It is natural in such applications to assume that some links between backbone nodes disappear when two nodes are no longer within the transmission range, due mainly to mobility. In this case, DYNAMIC CONNECTED DOMINATING SET (henceforth DCDS) is a better model.

We shall study the parameterized complexity of DCDS and prove that it is fixed-parameter tractable with respect to the edit-parameter, where the editing is the addition or deletion of edges. When parameterized by the increment-parameter, the problem is shown to be $W[2]$-hard via a reduction from DOMINATING SET. We also consider the DYNAMIC CONNECTED VERTEX COVER problem, which exhibits a similar behavior. In both cases, reduction to the STEINER TREE problem is used to obtain the positive results.

The rest of this paper is structured as follows. Section 2 presents some needed preliminaries. In Sects. 3 and 4 we study DYNAMIC CONNECTED DOMINATING SET and DYNAMIC CONNECTED VERTEX COVER problems (respectively). Section 5 presents an example of a domination problem where the solution is not connected. In Sect. 6, we discuss dynamic maximization problems with connectivity constraints, and we conclude in Sect. 6 with a summary and open problems.

2   Preliminaries

All graphs considered in this paper are assumed to be finite, simple and undirected. We adopt common graph theoretic terminology, but we shall define some notions that may be useful.

For a given graph $G = (V, E)$, we let $V(G) = V$ and $E(G) = E$. For $E' \subseteq E$ we denote by $V(E')$ the set of vertices that are incident on elements of E'. The set of *neighbors*, or *open neighborhood*, of a vertex v is defined as $N_G(v) = \{u \in V(G) \mid uv \in E(G)\}$ while the *closed neighborhood* of v is $N_G[v] = N_G(v) \cup \{v\}$. If $S \subseteq V(G)$, then $N_G(S) = \bigcup_{v \in S} N_G(v)$ and $N_G[S] = N_G(S) \cup S$. Moreover, the subgraph of G induced by S is denoted by $G[S]$.

If C, D are subsets of $V(G)$ satisfying $C \subseteq N_G[D]$, then we say that D *dominates* C, or equivalently, that C *is dominated by* D. When a singleton set

$\{v\}$ is dominated by D, we say that v is dominated by D. Otherwise, $N_G[v] \cap D = \emptyset$, hence v is called a *non-dominated vertex*, with respect to D. A *dominating set* for G is a set of vertices $S \subseteq V(G)$ such that $V(G)$ is dominated by S. When S is (required to be) connected, it is called a *connected dominating set* of G.

The *Hamming distance* between two subsets A, B of a superset V is the sum of elements of $A \backslash B$ and $B \backslash A$ (or the cardinality of $A \Delta B$). When A and B are sets of vertices of a graph, we denote this distance by $d_v(A, B)$. For two graphs G, G' with the same set of vertices, we denote by $d_e(G, G')$ the Hamming distance between $E(G)$ and $E(G')$.

Contracting an edge $e = uv$ of a graph G results in replacing u and v by a single vertex whose open neighborhood is $N_G(u) \cup N_G(v)$. Contracting a subgraph H of G consists of contracting all the edges of H, and results in a single vertex whose neighborhood is $N_G(V(H))$.

A set of vertices $J \subseteq V(G)$ is an *independent set* if $N_G(J) \cap J = \emptyset$. A set of vertices $C \subseteq V(G)$ is called a *vertex cover* of G if every edge of G is incident with a vertex in C. i.e. $V(G) \backslash C$ is an independent set. If $G[C]$ is connected then we say that C is a *connected vertex cover* for G.

Given a weighted graph G and a partition $\{X, Y\}$ of $V(G)$, a *Steiner tree* of G is any tree that contains all elements of Y. In this context, elements of X and Y are called *Steiner nodes* and *terminal nodes*, respectively. A *minimum Steiner tree* is one whose total weight (sum of weights of edges) is minimum. The corresponding minimization problem is called MINIMUM WEIGHTED STEINER TREE. When parameterized by $k = |Y|$, the problem can be solved in $\mathcal{O}^*(2^k)$ time and polynomial space [12].

3   Dynamic Connected Dominating Set

The parameterized DYNAMIC CONNECTED DOMINATING SET problem is defined formally as follows:

DYNAMIC CONNECTED DOMINATING SET (DCDS)

Input: A graph $G = (V, E)$ and a connected dominating set S for G;
          a graph $G' = (V, E')$ obtained from G with $d_e(G, G') \leq k$;
          positive integers k, r.
Parameter: (k, r)
Question: Does there exist $S' \subseteq V$ such that $d_v(S, S') \leq r$,
           and S' is a connected dominating set for G'?

Let (G, G', S, k, r) be a given instance of DCDS. Then we assume $G'[S]$ is not connected or S is not a dominating set of G'. We further assume the following:

1. If G' has more than one connected component then we have a NO-instance. Thus, assume that G' is connected.
2. Every vertex in G' has degree at least 2. If $deg(v) = 1$ and v is dominated by S, then remove v. If v is not dominated then put its neighbor in S', then remove v.

3. We can assume that $S \subseteq S'$. This is because we want $d_v(S, S') \leq r$ (or as small as possible). Including the elements of S in S' does not increase $d_v(S, S')$ which we may assume to be identical to $|S'| - |S|$ in this case. Note that "connecting" the elements of S (while computing S') does not incur additional cost since each such vertex must (eventually) have a neighbor in S'.
4. Adding edges is "harmless" in the sense that, in G, every vertex is dominated and S is connected. The addition of an edge cannot alter these properties.
5. The deletion of an edge is harmful only if at least one of its endpoints is in S.

3.1   A Fixed-Parameter Algorithm

We now present the main steps of an algorithm that computes S' such that $|S'| - |S|$ is minimum. We are going to use the fixed-parameter algorithm for the MINIMUM WEIGHTED STEINER TREE problem where the terminal nodes form an independent set. First let L be the list of deleted edges, and let $A \subset V(G)\backslash S$ be the set of vertices of G that are incident on deleted edges and such that A is not dominated by S in G'. That is, $A = \{v \in V(G) \mid v \in V(L)\backslash N_{G'}[S]\}$. Note that, since $|L| \leq k$, A contains at most k vertices.

Obviously $G'[S \cup A]$ is not connected (unless A is empty and $G[S]$ is connected, which is an excluded case). The required solution S' must dominate all elements of A, but maybe some of them must also be in S' to secure connectivity. Our algorithm proceeds as follows:

Enumerate (in $\mathcal{O}(2^k)$-time) all the subsets of A that could be added to the target connected dominating set (i.e., branch on all subsets of A). For each such subset A' of A we do the following:

1. Delete all elements of $A\backslash A'$ that are (now) dominated by A'. Due to the branching, we know these vertices cannot be added to the solution.
2. Delete edges between the remaining vertices of $A\backslash A'$. Again these vertices cannot dominate each other. Let $A'' = A\backslash A'$ (initially).
3. Contract every connected component of $G'[S \cup A']$ and add the resulting isolated vertices to A''. Note that A'' remains an independent set with at most $k + 1$ vertices.
4. Let H be the resulting graph and let $B = V(H)\backslash A''$. We proceed as in the algorithm described in [1] by assigning a weight of one to each edge of $H[B]$ and a weight of $n = |V(H)|$ to each edge uv where $u \in A''$ and $v \in B$.
5. At this stage, we treat $(H, |A''|)$ as an instance of MINIMUM WEIGHTED STEINER TREE where A'' is the set of (at most $k + 1$) terminals and B is the set of Steiner nodes.

Run the fixed-parameter WEIGHTED STEINER TREE algorithm of [12]. Let T be the resulting solution, and let $I(T)$ be the set of interior nodes of T. Then, due to the assigned weights, all the interior nodes are Steiner nodes (as shown in [1]).

6. If $|A' \cup I(T)| \leq r$ (in G'), then generate $S' = S \cup A' \cup I(T)$ as solution and stop.

Observe that $I(T)$ forms a connected dominating set of H. Moreover, as shown in [1] and since A'' is an independent set of H, the number of internal vertices of T is the smallest possible. Therefore, if none of the generated candidate solutions (S') satisfies $|S'| - |S| \leq r$ then we have a NO-instance. Otherwise, it would be easy to see that S' is a connected dominating set of G'.

As for the running time, and since $|A| \leq k$ and the WEIGHTED STEINER TREE algorithm runs in $\mathcal{O}(2^{|A''|})$, the total running time of the above algorithm is in $\mathcal{O}(4^k)$.

Theorem 1. *The* DYNAMIC CONNECTED DOMINATING SET *problem is fixed-parameter tractable with respect to the edit-parameter.*

3.2   Complexity of DCDS with Respect to the Increment Parameter

We next use a reduction from DOMINATING SET to show that DCDS is $W[2]$-hard with respect to the increment-parameter r only, assuming the edit-parameter k is arbitrary.

Given an instance (H, r) of DOMINATING SET, where H has n vertices, we construct two graphs G and G' as follows:

1. G consists of two n-cliques H_1 and H_2 and some edges between them as follows:
   - To each vertex v of H we have two vertices v_1 and v_2 that correspond to it so that $v_1 \in V(H_1)$ and $v_2 \in V(H_2)$. We refer to v_i as the image of v in H_i or (simply) the ith image of v.
   - For each vertex $v \in H$, we connect its first image v_1 to the vertices of H_2 that are images of all the elements of $N_H[v]$ (so v_1 and v_2 are also connected).
2. Let $S = \{v_1, v_2\}$ for some vertex v of H. Then, since H_1 and H_2 are cliques in G, S is a connected dominating set for G.
3. G' is obtained from G by deleting all the edges of H_2.

Now we can generate n instances $(G, G', S, k, r - 1)$ of DCDS (one instance for each possible set $S = \{v_1, v_2\}$). Since H_2 is edgeless, a solution of any DCDS instance consists of adding at most $r - 1$ vertices of H_1 only, in addition to the copy v_2 of v.

If H has a dominating set of size r, then at least one of the n DCDS instances would have a solution given by adding $r - 1$ vertices of H_1 to dominate all the vertices of the edgeless subgraph H_2. We have now established the following.

Theorem 2. *The* DYNAMIC CONNECTED DOMINATING SET *problem is $W[2]$-hard with respect to the increment-parameter.*

Proof. Follows from the above reduction, and the fact that DOMINATING SET is $W[2]$-hard [8]. □

4   Dynamic Connected Vertex Cover

The method used to obtain a fixed-parameter algorithm for DCDS, with respect to the edit-parameter, can be applied to the DYNAMIC CONNECTED VERTEX COVER problem.

DYNAMIC CONNECTED VERTEX COVER (DCVC)

Input: A graph $G = (V, E)$ and a connected vertex cover S for G;
            a graph $G' = (V, E')$ obtained from G with $d_e(G, G') \leq k$;
            positive integers k, r.
Parameter: (k, r)
Question: Does there exist $S' \subseteq V$ such that $d_v(S, S') \leq r$,
            and S' is a connected vertex cover for G'?

Let (G, G', S, k, r) be an instance of DYNAMIC CONNECTED VERTEX COVER (DCVC). First, we note that edge deletion could only affect the connectivity constraint while edge addition may cause some added edges not to be covered. We assume that $k = k_1 + k_2$ where we have at most k_1 added edges to cover and at most k_2 deleted edges, so at most $k_2 + 1$ connected components in $G'[S]$. To obtain a fixed-parameter algorithm we first branch on all the k_1 edges to be covered, by adding one of the two endpoints (or the other) to S', (S' is initially equal to S.) Then we contract each connected component of $G'[S']$, which again yields an instance of the WEIGHTED STEINER TREE problem, as described in the case of DCDS above.

Theorem 3. *The* DYNAMIC CONNECTED VERTEX COVER *problem is fixed-parameter tractable with respect to the edit-parameter.*

Note that a faster algorithm can be obtained if the parameter is $k + r$. simply, contract all the connected components of $G'[S]$ and add a pendant (degree-one) vertex to each of the resulting vertices, then treat the graph as an instance of CONNECTED VERTEX COVER with parameter $r + c \leq r + k$ where c is the number of components of $G'[S]$.

We prove that the problem is $W[2]$-hard with respect to the increment-parameter by reduction from the following parameterized version of the STEINER TREE problem, where the parameter is the number of Steiner nodes in the solution tree:

MINIMUM STEINER TREE (parameterized by number of Steiner nodes)

Input: A graph $G = (V, E)$, together with a partition $\{X, Y\}$ of V
            and a positive integer r.
Parameter: r
Question: Does there exist a subset Y' of Y such that $G[Y' \cup X]$
            is connected and $|Y'| \leq r$?

The above version of STEINER TREE is known to be $W[2]$-hard by a simple reduction from SET COVER, as noted in [11]. We now present the following reduction to DYNAMIC CONNECTED VERTEX COVER, parameterized by the increment cost.

Let (H, r) be an instance of the above STEINER TREE problem. We further assume that $|X| > 1$ and $H[X]$ is not connected (otherwise it would be an obvious YES-instance). We construct two graphs G and G' as follows:

Graph G is obtained from H by subdividing each and every edge that has at least one endpoint from Y. Let T be the set of the newly introduced degree-two vertices (due to the edge subdivision). We also add one vertex u that is joined to all the elements of $X \cup T$. Let $S = X \cup T \cup \{u\}$. Then S is a connected vertex cover of G.

Graph G' is obtained from G by removing all the edges between u and the elements of T and all but one of the edges between u and the elements of X (since we assumed that $|X| > 1$). Observe that $X \cup T \cup \{u\}$ remains a vertex cover in G' but, since we assumed $G[X]$ is not connected (and $|X| > 1$), $G'[X \cup T \cup \{u\}]$ is not connected in G'; $G'[X]$ is not connected and every vertex of $T \cup \{u\}$ has at most one neighbor in X.

If it exists, a solution S' to the instance $(G, G', X \cup T \cup \{u\}, k, r)$ of DYNAMIC CONNECTED VERTEX COVER consists of adding at most r elements of the set Y to the set S (to reconnect the vertex cover). It would be easy to see that $S' \backslash S$ is a solution to the initial instance of the STEINER TREE problem if and only if S' is a solution to the corresponding DYNAMIC CONNECTED VERTEX COVER instance.

Theorem 4. *The* DYNAMIC CONNECTED VERTEX COVER *problem is* $W[2]$-*hard with respect to the increment-parameter.*

5   Dynamic Independent Dominating Set

In the previous two sections, the connectivity of solutions played a role in obtaining fixed-parameter algorithms with respect to the edit-parameter. To further highlight the role of connectivity, and not domination, we give an example of a domination problem whose solution is (completely) disconnected. We show that DYNAMIC INDEPENDENT DOMINATING SET is $W[2]$-hard with respect to the edit-parameter.

DYNAMIC INDEPENDENT DOMINATING SET (DIDS)

Input: A graph $G = (V, E)$ and an independent dominating set S
           for G; a graph G' obtained from G with $d_e(G, G') \leq k$;
           positive integers k, r.
Parameter: (k, r)
Question: Does there exist S' such that $d_v(S, S') \leq r$ and S' is an
           independent dominating set for G'?

We will show a reduction from the $W[2]$-complete problem MULTICOLOR DOMINATING SET (k) [7,10].

Multicolour Dominating Set (MCDS)

Input: A graph $G = (V, E)$ and a proper vertex colouring,
            $c \mid V \rightarrow \{1, 2, \ldots, k\}$ for G;
            a positive integer k.
Parameter: k
Question: Does there exist a dominating set D for G such that
            $|D| \leq k$, and D consists of distinct coloured vertices?
            i.e. for each pair of distinct vertices $\{u, v\} \subseteq D$, $c(u) \neq c(v)$.

Theorem 5. *The problem* Dynamic Independent Dominating Set *is* $W[2]$-*hard with respect to the aggregate of the edit-parameter and the increment-parameter.*

Proof. We reduce from MCDS(k). Let $I = (H, k)$ be an instance of MCDS (k). Construct an instance $I' = (G, G', S, k' = 3k, r = 2k)$ of DIDS (k, r) as follows.

For $1 \leq i \leq k$, let the i-th color class of $V(H)$ consist of n_i vertices, $V_i = \{v_{i,1}, v_{i,2}, \ldots, v_{i,n_i}\}$.

Associated with color class i of H, the graph G has three disjoint sets of vertices of cardinality $4, n_i$ and n_i, respectively. The set $A_i = \{a_{i,1}, a_{i,2}, a_{i,3}, a_{i,4}\} \subseteq V(G)$ contains distinguished vertices which will function as a "selection gadget." The other two sets, $B_i, C_i \subseteq V(G)$, represent images of each of the vertices of V_i. For each $v_{i,j} \in V_i$, we have two corresponding vertices in G: $b_{i,j} \in B_i$ and $c_{i,j} \in C_i$, where $j \in \{1, 2, \ldots, n_i\}$.

Let $\mathcal{A} = \bigcup_{i=1}^{k} A_i$, $\mathcal{B} = \bigcup_{i=1}^{k} B_i$ and $\mathcal{C} = \bigcup_{i=1}^{k} C_i$. We specify that the induced subgraph $G[B_i]$ is an n_i-clique and that $G[C_i]$ is an independent set. Other edges of G are next described. For $i \in \{1, 2, \ldots, k\}$:

1. Vertex $a_{i,1}$ is connected to every vertex in $(A_i \backslash \{a_{i,1}\}) \cup B_i \cup C_i$.
2. Each $a_{i,j}$, for $j = 2, 3, 4$, is connected to every vertex in B_i.
3. Each $b_{i,j}$, for $j \in \{1, 2, \ldots, n_i\}$, is connected to the image in $\mathcal{C}$ of the closed neighborhood $N_H[v_{i,j}]$. So each $b_{i,j}$ is adjacent to only one vertex in C_i, namely $c_{i,j}$ (the other image of $v_{i,j}$), but could have neighbors in other components of $\mathcal{C}$ depending on the colors of the neighbors of $v_{i,j}$ in H.

This completes our description of the transformation from H to G. An illustration of the construction of G is given in Fig. 1.

Let $S = \{a_{i,1} \mid i = 1, 2, \ldots, k\}$. It should be clear from the construction of G that S is indeed an independent dominating set, as required. Let $X = \{uv \mid u, v \in A_i, 1 \leq i \leq k\}$. Then G' is obtained from G by deleting the $3k = k'$ edges in X. Hence in G', the vertices in $\mathcal{A} \backslash S$ are not dominated by S.

Claim 1: For each $i \in \{1, 2, \ldots, k\}$, domination by S' of the vertices in A_i adds at least 2 to $d_v(S, S')$. Hence, if S' is a solution of an instance then $d_v(S, S') \geq 2k$.

Proof of claim 1: We consider all possibilities for domination of A_i.

First suppose that $a_{i,1} \in S'$. Then, considering that A_i is an independent set in G' and every element of $B_i \cup C_i$ is a neighbor of $a_{i,1}$, the other three vertices

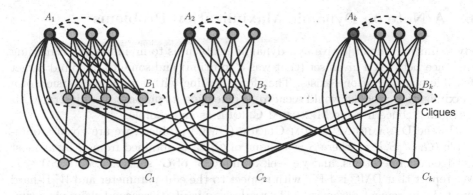

Fig. 1. The graph G constructed in the reduction from MCDS to DIDS.

of A_i (and not in S) must be dominated by themselves in G'. Hence, $a_{i,1} \in S'$ implies that $d_v(S, S')$ is increased by 3.

Next we consider the case that $a_{i,1} \notin S'$. Then $a_{i,1}$ must be dominated by a vertex v which lies in $B_i \cup C_i$. If $v \in B_i$ then v dominates all of A_i but then, since S' is an independent set, $d_v(S, S')$ is increased by 2. (Vertex $a_{i,1} \in S$ but not in S' contributes 1 to $d_v(S, S')$.) If $v \in C_i$ then $d_v(S, S')$ is increased by at least 3 as the vertices $a_{i,j}$, for $i = 2, 3, 4$, cannot be dominated by a vertex outside of $A_i \cup B_i$.

We have considered all possibilities for domination of A_i, hence the first statement of the claim is proved. The choice of i is arbitrary so summing over $i = 1, 2, \ldots, k$ shows that $d_v(S, S') \geq 2k$.

Claim 2: If I' is a YES-instance then $S' \subset \mathcal{B}$ and S' contains, for $i = 1, 2, \ldots k$, exactly one element of B_i.

Proof of claim 2: If I' is a YES-instance then $d_v(S, S') \leq r = 2k$. As shown in the details of proof of Claim 1, $d_v(S, S') \geq 2k$ and equality is achieved only in the case that there exists, for each $1 \leq i \leq k$, a vertex $v \in B_i$ such that v dominates all of A_i (in addition to all of B_i and some vertices in C.)

Now if I' is a YES-instance then S' consists of k vertices of $\mathcal{B}$ such that $N_{G'}[S'] = V(G')$. Moreover, the selection of such vertices is forced, by Claim 2, to include exactly one vertex from each of the k color classes i of H. Thus a solution $S' = \{b_{1,j_1}, b_{2,j_2}, \ldots, b_{k,j_k} \in V(G')\}$ implies that I is a YES-instance of MCDS(k) with solution $D = \{v_{i,j} \in V(H) \mid b_{i,j} \in S' \subseteq V(G')\}$ (and $|D| = k$). Conversely, if I is a YES-instance of MCDS(k) then the construction of G, G' together with the proof of Claims 1 and 2 show that I' is a YES-instance of DIDS with $d_v(S, S') = 2k \leq r$. □

6   A Note on Dynamic Maximization Problems

By definition, the objective of a dynamic problem is to minimize the Hamming distance between a given set (that was a solution) and some to-be-found target feasible solution, if one exists. Therefore, dealing with a dynamic version of a maximization problem would seem to be defeating its purpose. We illustrate this intriguing concern using MAXIMUM CLIQUE as an example.

In the DYNAMIC MAXIMUM CLIQUE (DMC) problem, we are given a quintuple (G, G', S, k, r) where S is a clique of G, G' is obtained from G by at most k edge-edit operations, and we seek a clique S' of G' such that $d_v(S, S') \leq r$. We report that DMC is FPT with respect to the edit-parameter and $W[1]$-hard w.r.t the increment-parameter. The reader can verify the second claim via a simple reduction from the MAXIMUM CLIQUE problem. We prove the first claim, as follows:

Let (G, G', S, k, r) be a given instance of DMC, and let L be the set of edges deleted from G to obtain G'. That is, $L = \{uv \in E(G) \mid uv \notin E(G')\}$. Let $A = \{v \in S \mid uv \in L\}$. Clearly, if $u, v \in S$ and uv is a deleted edge, then at least one of u and v is not in S'. So deletion of elements of S to obtain S' is unavoidable. It is straight-forward to see that minimizing the number of vertices of S which must be discarded from the new solution, i.e. minimizing $|S \backslash S'|$, is equivalent to finding a minimum sized vertex cover of the graph H where $H = \big(A, L \cap E(G[S])\big)$. Since $|L| \leq k$, computing a minimum vertex cover of H takes $\mathcal{O}(2^k)$ time.

Let c be the size of a computed minimum vertex cover, then $d_v(S, S')$ is at least c. Therefore, if $c > r$ then we have a NO-instance. Otherwise, the clique obtained by removing the c elements of the computed vertex cover (of H) is a feasible solution that meets the objective of DMC, so we have a YES-instance. In other words, DMC can be solved in $\mathcal{O}^*(f(k))$ time, which proves our claim.

The lesson learned from the above reduction to VERTEX COVER is that the increment-parameter was used only as a bound on how much smaller the new solution can be. We need not attempt to add more vertices to S'. Of course, larger cliques are preferred but we are limited by $d_v(S, S') \leq r$. Accordingly, we believe it would be more meaningful to require that $|S| - |S'| \leq r$ (instead of $d_v(S, S') \leq r$). This would force r to be an upper bound on the "decrement" in solution size and, at the same time, allows $|S| - |S'|$ to be negative, so S' can be as large as possible.

Finally, note that DYNAMIC MAXIMUM CLIQUE is another example of a dynamic problem that falls in the class FPT when parameterized by the edit-parameter. Moreover, the solution is (implicitly) connected. We pose as open problem whether all such problems are fixed-parameter tractable with respect to the edit-parameter. Other examples of maximization problems to consider are LONG CYCLE, LONG PATH and LARGEST CONNECTED COMMON SUBGRAPH.

7   Conclusion

We considered dynamic versions of parameterized graph problems whose solutions are required to induce connected subgraphs. We showed that under the editing parameter (edge-editing cost) DYNAMIC CONNECTED DOMINATING SET and DYNAMIC CONNECTED VERTEX COVER fall in the class FPT while they are both $W[2]$-hard when parameterized by the increment-parameter only.

Parameterized dynamic problems have not been well studied, so there are many open problems at his stage. For example, would the dynamic version of an FPT problem always be FPT? Our hardness result for DYNAMIC CONNECTED VERTEX COVER gives a negative answer to this question if the parameter is the increment cost. What about the edit-parameter? We pose as open problem whether, for problems whose solutions are required to be connected, the dynamic and static versions of a parameterized problem exhibit the same behavior with respect to the edit-parameter.

References

1. Abu-Khzam, F.N., Mouawad, A.E., Liedloff, M.: An exact algorithm for connected red-blue dominating set. J. Discrete Algorithms **9**(3), 252–262 (2011)
2. Bharghavan, V., Das, B.: Routing in ad hoc networks using minimum connected dominating sets. In: IEEE International Conference on Communication, pp. 376–380 (1997)
3. Böckenhauer, H.J., Hromkovic, J., Královic, R., Mömke, T., Rossmanith, P.: Reoptimization of steiner trees: changing the terminal set. Theor. Comput. Sci. **410**(36), 3428–3435 (2009)
4. Carlsson, M.: Monads for incremental computing. In: Proceedings of the Seventh ACM SIGPLAN International Conference on Functional Programming, ICFP '02, pp. 26–35. ACM, New York (2002)
5. Chen, Y.P., Liestman, A.L.: Approximating minimum size weakly-connected dominating sets for clustering mobile ad hoc networks. In: 3rd ACM International Symposium on Mobile Ad Hoc Networking and Computing (2002)
6. Downey, R.G., Egan, J., Fellows, M.R., Rosamond, F.A., Shaw, P.: Dynamic dominating set and turbo-charging greedy heuristics. J. Tsinghua Sci. Technol. **19**(4), 329–337 (2014)
7. Downey, R.G., Fellows, M.R.: Fixed parameter tractability and completeness. In: Ambos-Spies, K., Homer, S., Schöning, U. (eds.) Complexity Theory: Current Research, pp. 191–225. Cambridge University Press, Cambridge (1992)
8. Downey, R.G., Fellows, M.R.: Parameterized Complexity. Monographs in Computer Science. Springer, New York (1999)
9. Hartung, S., Niedermeier, R.: Incremental list coloring of graphs, parameterized by conservation. Theor. Comput. Sci. **494**, 86–98 (2013)
10. Lackner, M., Pfandler, A.: Fixed-parameter algorithms for finding minimal models. In: KR (2012)
11. Mölle, D., Richter, S., Rossmanith, P.: Enumerate and expand: Improved algorithms for connected vertex cover and tree cover. Theory Comput. Syst. **43**(2), 234–253 (2008)

12. Nederlof, J.: Fast polynomial-space algorithms using möbius inversion: improving on steiner tree and related problems. In: Albers, S., Marchetti-Spaccamela, A., Matias, Y., Nikoletseas, S., Thomas, W. (eds.) ICALP 2009, Part I. LNCS, vol. 5555, pp. 713–725. Springer, Heidelberg (2009)
13. Psaraftis, H.N.: Dynamic vehicle routing: Status and prospects. Ann. Oper. Res. **61**, 143–164 (1995)
14. Ramamurthy, B., Iness, J., Mukherjee, B.: Minimizing the number of optical amplifiers needed to support a multi-wavelength optical lan/man. In: IEEE International Conference on Computer Communications, pp. 261–268 (1997)
15. Salhieh, A., Weinmann, J., Kochha, M., Schwiebert, L.: Power efficient topologies for wireless sensor networks. In: International Conference on Parallel Processing, pp. 156–163 (2001)
16. Shachnai, H., Tamir, G., Tamir, T.: A theory and algorithms for combinatorial reoptimization. In: Fernández-Baca, D. (ed.) LATIN 2012. LNCS, vol. 7256, pp. 618–630. Springer, Heidelberg (2012)
17. Sivakumar, R., Das, B., Bharghavan, V.: An improved spine-based infrastructure for routing in ad hoc networks. In: IEEE Symposium on Computer and Communications (1998)

Parameterized and Subexponential-Time Complexity of Satisfiability Problems and Applications

Iyad Kanj[1]([⊠]) and Stefan Szeider[2]

[1] School of Computing, DePaul University, Chicago, USA
ikanj@cs.depaul.edu
[2] Vienna University of Technology, Vienna, Austria
stefan@szeider.net

Abstract. We study the parameterized and the subexponential-time complexity of the weighted and unweighted satisfiability problems on bounded-depth Boolean circuits. We establish relations between the subexponential-time complexity of the weighted and unweighted satisfiability problems, and use them to derive relations among the subexponential-time complexity of several NP-hard problem. For instance, we show that the weighted monotone satisfiability problem is solvable in subexponential time if and only if CNF-SAT is. The aforementioned result implies, via standard reductions, that several NP-hard problems are solvable in subexponential time if and only if CNF-SAT is. We also obtain threshold functions on structural circuit parameters including depth, number of gates, and fan-in, that lead to tight characterizations of the parameterized and the subexponential-time complexity of the circuit problems under consideration. For instance, we show that the weighted satisfiability problem is FPT on bounded-depth circuits with $O(\log n)$ gates, where n is the number of variables in the circuit, and is not FPT on bounded-depth circuits of $\omega(\log n)$ gates unless the Exponential Time Hypothesis (ETH) fails.

1 Introduction

The NP-hardness theory suggests that NP-hard problems cannot be solved efficiently—in polynomial time. This suggestion, however, has not discouraged researchers in theoretical computer science from seeking super-polynomial time algorithms for NP-hard problems due to the practical importance of such problems. While some NP-hard problems admit subexponential-time algorithms (*e.g.*, problems on planar graphs [1]), many NP-hard problems have resisted for decades all attempts to improve the brute-force exponential-time algorithms for solving them. A famous example of such NP-hard problems is CNF-SATISFIABILITY, shortly CNF-SAT, which is the canonical problem of the complexity class NP. After a tremendous amount of research work spanning several decades, the best algorithm for solving CNF-SAT remains the brute-force algorithm that enumerates every truth assignment to the variables, resulting in a $2^n m^{O(1)}$ running time

© Springer International Publishing Switzerland 2014
Z. Zhang et al. (Eds.): COCOA 2014, LNCS 8881, pp. 637–651, 2014.
DOI: 10.1007/978-3-319-12691-3_48

(n is the number of variables in the input instance, and m is the instance size). Whether CNF-SAT can be solved in time $2^{cn}m^{O(1)}$, for some constant $c < 1$, remains a longstanding open question.

If one restricts CNF-SAT to instances in which the width of each clause is at most k, where $k > 0$ is an arbitrary integer constant, one obtains the k-CNF-SAT problem, which is known to be solvable in time $O(2^{c_k n})^1$ for some constant $c_k < 1$ that depends on k (for example, see [17]). A significant amount of research has been devoted to designing algorithms for k-CNF-SAT that achieve smaller and smaller values of c_k, in particular, for $k = 3$ (see [2]). For an NP-hard problem that admits a $2^{cn}m^{O(1)}$-time algorithm (n is the search space and m is the instance size) for some $c < 1$, such as k-CNF-SAT, a natural question is whether we can improve the exponent in the exponential term of the upper bound on the running time indefinitely, that is, whether we can solve the problem in time $2^{cn}m^{O(1)}$ for $every$ $0 < c < 1$; if this is possible, we say that the problem can be solved in $subexponential\ time$.

Impagliazzo, Paturi, and Zane [18] showed that the subexponential-time complexity of the k-CNF-SAT problem ($k \geq 3$), is equivalent to the subexponential-time complexity of a large class of well-known NP-hard problems; this class is closed under subexponential-time preserving reductions, called $serf$-$reductions$. This lead to the formulation of the $Exponential\ Time\ Hypothesis$ (ETH) [18] stating that k-CNF-Sat cannot be solved in subexponential time, and the ETH has become a standard hypothesis in complexity theory. On the other hand, the subexponential-time complexity of weighted satisfiability on bounded-depth circuits is closely related to the fixed-parameter tractability of the W-hierarchy in parameterized complexity theory (see the recent survey [7]). Research on parameterized complexity discovered more subtle relations between the computational complexity of NP-hard problems and the subexponential-time complexity of satisfiability problems [3,5,6,9–11,19,21]. For example, it was shown that parameterized algorithms that are asymptotically more efficient than the exhaustive search approach for many W-hard problems are dependent of the subexponential-time complexity of various satisfiability problems (see for instance [3,5,6,9–11,19,21]). Due to the aforementioned research, it is now known that the classification of parameterized intractability corresponds to the exact complexity of satisfiability problems on various circuit families. This research of the structural connection between parameterized complexity and subexponential-time complexity resulted in new tools for deriving computational lower bounds.

To understand the difference between the complexities of CNF-SAT and k-CNF-SAT, Calabro, Implagliazzo, and Paturi [4] studied the relation between the complexities of k-CNF-SAT and that of CNF-SAT restricted to formulas with certain structural parameters, such as density (ratio of clauses to variables), and frequency of variables. They proved several results, under the assumption that the ETH holds, including a result showing that CNF-SAT restricted to instances

---

[1] For k-CNF-Sat the instance size is polynomial in n, and hence, gets absorbed in the exponential term $O(2^{c_k n})$. Throughout the paper, we will omit the polynomial factor in the instance size whenever it is polynomial in n.

of any fixed density can be solved *more efficiently* than CNF-SAT. Under the *Strong Exponential Time Hypothesis* (SETH), which states that k-CNF-SAT requires time $\Omega(2^n)$ when k goes to infinity, Cygan et al. [12] proved equivalences among the subexponential-time complexity of several natural problems. Assuming the SETH, they showed that the existence of an algorithm that runs in time $2^{cn}m^{O(1)}$, for some $c < 1$, for any of the problems in a set of problems that includes CNF-SAT, HITTING SET, and SET COVER, would imply the existence of algorithms that run in time $2^{c'n}m^{O(1)}$, for some $c' < 1$, for all the problems in this set. Patrascu and Williams [21] studied the relation between parameterized algorithms for the DOMINATING SET problem and algorithms for CNF-SAT. They showed that, for any $k \geq 3$, a parameterized algorithm for DOMINATING SET that runs in time $O(n^{k-\epsilon})$, for any $\epsilon > 0$, would lead to an algorithm for CNF-SAT that runs in time $2^{cn}m^{O(1)}$ for some $c < 1$.

In this paper we study the parameterized and the subexponential-time complexity of weighted and unweighted satisfiability problems on Boolean circuits of bounded (constant) depth. We focus on satisfiability problems because they are the canonical problems for the complexity class NP, and for the different levels of the W-hierarchy in parameterized complexity. By deriving relations among satisfiability problems with respect to their subexponential-time complexity, we are able to infer similar results for other natural NP-hard and W-hard problems. We start by establishing relations between the subexponential-time complexity of the weighted and unweighted satisfiability problems, and use them to derive relations among the subexponential-time complexity of several NP-hard problem. We show that the weighted monotone (resp. antimonotone) satisfiability problem (*i.e.*, weighted montone/antimontone CNF-SAT) is solvable in subexponential time if and only if CNF-SAT is. The aforementioned results imply, via standard reductions, that several NP-hard problems, including RED-BLUE DOMINATING SET and INDEPENDENT SET ON HYPERGRAPHS are solvable in subexponential time if and only if CNF-SAT is. We then study the parameterized and the subexponential-time complexity of circuit satisfiability problems with respect to some natural structural parameters of the underlying circuit. We obtain threshold functions on structural circuit parameters including depth, number of gates, and fan-in, that lead to tight characterizations of the parameterized and the subexponential-time complexity of the circuit satisfiability problems under consideration. For example, we show that the weighted satisfiability problem is FPT on bounded-depth circuits with $O(\log n)$ gates, and is not FPT on bounded-depth circuits of $\omega(\log n)$ gates unless the ETH fails. Our results shed light on how some natural structural parameters of the circuit dictate the complexity of circuit satisfiability problems.

2   Preliminaries

The complexity functions used in this paper are assumed to be proper complexity functions that are computable, unbounded, and nondecreasing. For a proper complexity function $f : \mathbb{N} \rightarrow \mathbb{N}$, define its inverse function f^{-1} by $f^{-1}(h) =$

$\max\{q \mid f(q) \leq h\}$. Since the function f is nondecreasing and unbounded, the function f^{-1} is also nondecreasing and unbounded, and satisfies $f(f^{-1}(h)) \leq h$.

A *circuit* is a directed acyclic graph. The vertices of indegree 0 are called the input *variables*, and are labeled either by *positive literals* x_i or by *negative literals* $\overline{x}_i$. The vertices of indegree larger than 0 are called the *gates* and are labeled with Boolean operators AND or OR. A special gate of outdegree 0 is designated as the *output* gate. We do not allow NOT gates in the circuit since, by De Morgan's laws, a general circuit can be efficiently converted into the above circuit model. A circuit is said to be *monotone* (resp. *antimonotone*) if all its input literals are positive (resp. negative). The *depth* of a circuit is the maximum distance from an input variable to the output gate of the circuit. The *size* of a circuit C, denoted $|C|$, is the size of the underlying graph (*i.e.*, number of vertices and edges). An *occurrence* of a variable in C is an edge from the variable or its negation to a gate in C. The *fan-in* of a gate is the indegree of the gate. The *fan-in* of a circuit is the maximum fan-in over all gates in the circuit.

We consider circuits whose output gate is an AND-gate and that are in the *normalized* form (see [13,16]). In the normalized form every gate has outdegree at most 1, and starting from the output AND-gate, the gates are structured into alternating levels of ORs-of-ANDs-of-ORs.... Note that this alternation can be achieved by merging gates of the same type whenever one is incoming to the other. Note also that there is no restriction on the outdegree of a literal, and literals may be incoming to gates at any level in the circuit. We denote a circuit that is in the normalized form and that is of depth at most $t \geq 2$ by a Π_t circuit. We write Π_t^+ to denote a monotone Π_t circuit, and Π_t^- to denote an antimonotone Π_t circuit. The *bottom fan-in* of the circuit is the maximum fan-in over all gates whose inputs are *only* literals. Π_t circuits naturally represent the satisfiability of t-normalized propositional formulas; that is, formulas that are products-of-sums-of-products..., including the canonical problem CNF-SAT. A circuit C is *satisfiable* if there is a truth assignment to the input variables of C that makes C evaluate to 1. The *weight* of an assignment τ is the number of variables assigned the value 1 by τ.

A *parameterized problem* Q is a subset of $\Omega^* \times \mathbb{N}$, where Ω is a fixed alphabet and $\mathbb{N}$ is the set of all non-negative integers. Each instance of the parameterized problem Q is a pair (x, k), where the second component, *i.e.*, the non-negative integer k, is called the *parameter*. We say that the parameterized problem Q is *fixed-parameter tractable* [13], shortly FPT, if there is an algorithm that decides whether an input (x, k) is a member of Q in time $f(k)|x|^{O(1)}$, where f is a computable function of k; we use *fpt-time* to denote time complexity of the form $f(k)|x|^{O(1)}$, where $|x|$ is the instance size, k is the parameter, and f is a computable function of k. Let FPT denote the class of all fixed-parameter tractable parameterized problems. (We abused the notation "FPT" above for simplicity.) A parameterized problem Q is *fpt-reducible* to a parameterized problem Q' if there is an algorithm that maps each instance (x, k) of Q to an instance $(x', g(k))$ of Q' in fpt-time, where g is a proper complexity function of k, such that $(x, k) \in Q$ if and only if $(x', g(k)) \in Q'$. Based on the notion of fpt-reducibility, a hierarchy of fixed-parameter intractability, *the W-hierarchy* $\bigcup_{t\geq 0} W[t]$, where $W[t] \subseteq W[t+1]$ for all $t \geq 0$, is defined. The 0-th level $W[0]$ of the W-hierarchy is the class FPT. The notions of hardness and completeness are defined for each level $W[i]$ of the W-hierarchy for $i \geq 1$ [13]. It is commonly believed that $W[1] \neq$ FPT (see [13]), and $W[1]$-hardness is a hypothesis for fixed-parameter intractability.

For $t \geq 2$, the *weighted Π_t-circuit satisfiability* problem, abbreviated WSAT $[t]$, is: Given a Π_t-circuit C and a parameter k, decide if C has a satisfying assignment of weight k. The *weighted monotone Π_t-circuit satisfiability* problem, abbreviated WSAT$^+[t]$, and the *weighted antimonotone Π_t-circuit satisfiability* problem, abbreviated WSAT$^-[t]$ are the WSAT$[t]$ problems on monotone circuits and antimonotone circuits, respectively. It is known that for each even integer $t \geq 2$, WSAT$^+[t]$ is $W[t]$-complete, and for each odd integer $t \geq 2$, WSAT$^-[t]$ is $W[t]$-complete; moreover, WSAT$^-[2]$ is $W[1]$-complete [13,16]. SAT$[t]$ $(t \geq 2)$ is the satisfiability problem on Π_t circuits. Note that SAT$[2]$ is CNF-SAT, and SAT$[2]$ with bottom fan-in 3 is 3-CNF-SAT.

It is clear that the WSAT$[t]$ and SAT$[t]$ problems $(t \geq 2)$ are solvable in time $2^n|C|^{O(1)}$, respectively, where C is the input circuit and n is the number of variables in C. We say that the WSAT$[t]$ (resp. SAT$[t]$) problem is solvable in *subexponential time* if there exists an algorithm that solves the problem in time $2^{o(n)}|C|^{O(1)}$. Using the results of [6,16] (see also [8]), the above definition is equivalent to the following: WSAT$[t]$ (resp. SAT$[t]$) is solvable in *subexponential time* if there exists an algorithm that takes as input an instance (C, k) of WSAT$[t]$ (resp. C of SAT$[t]$) and $\ell \in \mathbb{N}$, and there exists a computable function h, such that the algorithm solves the problem in time $h(\ell)2^{n/\ell}|C|^{O(1)}$. We will use both equivalent notions of subexponential-time solvability interchangeably depending on which one will be more convenient for deriving the results in question.

Let Q and Q' be two problems, and let μ and μ' be two parameter functions defined on instances of Q and Q', respectively. In the case of WSAT$[t]$ or SAT$[t]$, the μ and μ' under consideration will be the number of variables in the instances of these problems. A *subexponential-time reduction family* [18] (see also [16]), shortly a serf-reduction, is an algorithm A that takes as input a pair (I, ℓ) of Q and produces an instance $I' = A(I, \ell)$ of Q' such that: (1) A runs in time $f(\ell)2^{\mu(I)/\ell}|I|^{O(1)}$ for some proper complexity function f; and (2) $\mu'(I') \leq g(\ell)(\mu(I) + \log |I|)$. Intuitively, a serf-reduction from Q to Q' reduces an instance of Q in subexponential time to an instance of Q' such that the parameter of the produced instance of Q' is linear in the parameter of the instance of Q. A *subexponential-time Turing reduction family* [18] (also see [16]), shortly a Turing-serf-reduction is an algorithm A equipped with an oracle to Q' such that there are computable functions $f, g : \mathcal{N} \longrightarrow \mathcal{N}$ with: (1) given a pair (I, ℓ) of Q, where $\ell \in \mathbb{N}$, A decides I in time $f(\ell)2^{\mu(I)/\ell}|I|^{O(1)}$; and (2) for all oracle queries of the form "$I' \in Q'$" posed by A on input (I, ℓ), we have $\mu'(I') \leq g(\ell)(\mu(I) + \log |I|)$. Hence, if Q is Turing-serf-reducible to Q', and if Q' is solvable in subexponential time then Q is solvable in subexponential time.

The optimization class SNP introduced by Papadimitriou and Yannakakis [20] consists of all search problems expressible by second-order existential formulas whose first-order part is universal. Impagliazzo, Paturi, and Zane [18] introduced the notion of *completeness* for the class SNP under serf-reductions, and identified a class of problems which are complete for SNP under serf-reductions/Turing-serf reductions, such that the subexponential-time solvability for any of these problems implies the subexponential-time solvability of all problems in SNP. Many well-known NP-hard problems are proved to be complete for SNP under the serf-reduction, including 3-CNF-SAT, VERTEX COVER, and INDEPENDENT SET, for which extensive efforts have been made in the last three decades to develop subexponential-time algorithms with no success. This fact has led to the *Exponential Time Hypothesis* (ETH), which is equivalent to the statement that not all SNP problems are solvable in subexponential time:

> *Exponential-Time Hypothesis* (ETH):   The problem k-CNF-SAT, for any $k \geq 3$, cannot be solved in time $O(2^{o(n)})$, where n is the number of variables in the input formula. Therefore, there exists a constant $c > 0$ such that k-CNF-SAT cannot be solved in time $O(2^{cn})$.

The ETH is a standard hypothesis in complexity theory.

The following result is implied, using the standard technique of renaming variables, from [18, Corollary 1] and from the proof of the Sparsification Lemma [18], [15, Lemma 16.17]:

Lemma 1. k-CNF-SAT *($k \geq 3$) is solvable in $2^{o(n)}$ time if and only if k-CNF-SAT with a linear number of clauses and in which the number of occurrences of each variable is at most 3 is solvable in time $2^{o(n)}$, where n is the number of variables in the formula (note that the size of an instance of k-CNF-SAT is polynomial in n). In particular, choosing $k = 3$ we get: 3-CNF-SAT in which every variable occurs at most 3 times, denoted 3-3-SAT, is solvable in $O(2^{o(n)})$ time if and only if the ETH fails.*

Remark 1. In this paper, when we consider SAT[t] (resp. WSAT[t]) restricted to instances in which a certain parameter is $\Omega(g(n))$ (resp. $\omega(g(n))$, $O(g(n))$, $o(g(n))$), for some proper complexity function $g(n)$ of the number of variables n in the instance, we mean SAT[t] (resp. WSAT[t]) restricted to all the instances in which the parameter is *upper bounded* by a prespecified function that is $\Omega(g(n))$ (resp. $\omega(g(n))$, $O(g(n))$, $o(g(n))$). For example, when we say "SAT[t] on circuits with $o(n)$ gates is solvable in subexponential time" we mean the following: For any given proper complexity function $g(n) = o(n)$, the problem consisting of the restriction of SAT[t] to instances in which the number of gates in the circuit is at most $g(n)$ is solvable in subexponential time.

3   An Essential Lemma

We dedicate this section to the following lemma, which will be useful for proving several results in the paper pertaining to instances of SAT[t] or WSAT[t] in which the number of gates in the circuit is smaller than the number of variables. We note that this lemma is true for circuits of unbounded depth as well.

Lemma 2. *An instance of WSAT[t] on a circuit C of n variables and s gates can be decided in time $2^{O(s)}n^{O(1)}$.*

Proof. Let (C, k) be an instance of WSAT[t] such that C has n variables and s gates. We assume that no literal in C is incoming to the output AND-gate of C; otherwise, such literal must be assigned 1 in any satisfying assignment, and we can assign it 1 and update C and k accordingly. We start by enumerating all possible truth values to the gates in C in a potential satisfying truth assignment to C of weight k; there are 2^s such enumerations. Fix an enumeration to the gates in which the output gate is assigned the value 1 (we reject the other enumerations). We check the consistency of the (fixed) enumeration and reject the enumeration if it is not consistent. Checking the consistency of the enumeration can be done as follows: for each AND-gate whose inputs are all gates (*i.e.*, non-literals), we verify that all its inputs are assigned 1 by the enumeration, and for each OR-gate whose inputs are all gates, we verify that at least one of its inputs is assigned 1. After checking consistency, we remove from the circuit every gate (either AND-gate or OR-gate) that is assigned 0 (and update the circuit); we do not care if such a gate ends up evaluating to 1—different from its value in the enumeration—in the resulting assignment because C would still be satisfied. All remaining gates at this point are assigned value 1 by the enumeration.

Next, for each AND-gate that has some incoming edges from literals, we assign all its incoming literals 1 and update k accordingly; if such an assignment to the literals is inconsistent (assigns some literal and its negation the same value), we reject the enumeration. After this step, which may have assigned values to some literals, we check if all inputs to some OR-gate were assigned 0, and reject the enumeration in such case. Now we remove every AND-gate from the circuit (because all its inputs have been assigned 1, and hence, the gate is satisfied by the enumeration), and every OR-gates that has an input assigned 1. Let k' be the resulting value of the parameter k at this point. What remains of the circuit (after updating it) are OR-gates whose inputs are all literals that have not been assigned any values. Therefore, the resulting circuit can be modeled as a Boolean CNF-SAT formula F whose clauses correspond to the disjunction of the literals that are incoming to the remaining OR-gates of C. Consequently, it is sufficient to decide if F, which consists of at most s clauses, has a weight-k' satisfying assignment. We now apply the following branching algorithm to F. Choose a variable x that appears both positively and negatively in F. If all variables appear only positively or only negatively in F, then the problem can be easily reduced to a HITTING SET problem on a hypergraph of at most n vertices and s (hyper)edges (each clause corresponds to an edge), which can be solved in time $2^s n^{O(1)}$ using dynamic programming (see Theorem 1 in [14]). Therefore, we can assume that there exists a variable x that appears both positively and negatively in F. Now we can apply a folklore branching algorithm that branches on x by considering the two possible truth assignments of 0 and 1 to x; in the branch where x is assigned 1 the value of k' is reduced by 1. We also update F appropriately following each branch. Since x appears both positively and negatively in F, each branch satisfies, and hence eliminates, at least one clause in F. If at a leaf-node in the binary search tree corresponding to this branching algorithm $k' = 0$ and all the clauses in F are satisfied, we accept the instance; if no such leaf-node exists in the search tree we reject the instance.

Since the number of clauses in F is at most s, the depth of the binary search tree corresponding to the branching algorithm is at most s, and hence the running time of this algorithm is $2^{O(s)}n^{O(1)}$. Since there are at most 2^s enumerations to consider, the operations to the circuits can be implemented in polynomial time, and since the branching algorithm runs in time $2^{O(s)}n^{O(1)}$, it follows that the instance (C, k) can be decided in time $2^{O(s)}n^{O(1)}$.   □

4   Subexponential-Time Equivalences

In this section we establish relations between the subexponential-time complexity of weighted and unweighted satisfiability problems. Since satisfiability problems are canonical NP-hard problems, these relations can be employed to establish similar relations among the subexponential-time complexity of other NP-hard problems, as we illustrate in this section. We start with the following lemma:

Proposition 1. *For any $t \geq 2$, WSAT[t] $\in 2^{o(n)}m^{O(1)}$ if and only if SAT[t] $\in 2^{o(n)}m^{O(1)}$, where n is the number of variables in the instance and m is the instance size for both problems.*

Proof. Clearly, if WSAT[t] $\in 2^{o(n)}m^{O(1)}$ then SAT[t] $\in 2^{o(n)}m^{O(1)}$ because we can decide an instance of SAT[t] by invoking an algorithm for WSAT[t], for every possible weight between 0 and n of a satisfying truth assignment for SAT[t]; this gives a Turing-serf-reduction from SAT[t] to WSAT[t]. Conversely, SAT[t] $\in 2^{o(n)}m^{O(1)}$ implies that WSAT[t] $\in 2^{o(n)}m^{O(1)}$ because we can encode the weight (*i.e.*, the parameter k) of a desired satisfying assignment to WSAT[t] using a linear number of 3-CNF clauses/gates (*e.g.*, see [16]), which provides a serf-reduction from WSAT[t] to SAT[t].   □

Theorem 1. *WSAT$^+$[2] $\in 2^{o(n)}m^{O(1)}$ if and only if CNF-SAT $\in 2^{o(n)}m^{O(1)}$, where n is the number of variables and m is the instance size for both problems.*

Proof. By Proposition 1, CNF-SAT $\in 2^{o(n)}m^{O(1)}$ implies that WSAT$^+$[2] $\in 2^{o(n)}m^{O(1)}$ because WSAT$^+$[2] is a special case of WSAT[2].

To prove the converse, suppose that WSAT$^+$[2] is solvable in $2^{o(n)}m^{O(1)}$ time. Then WSAT$^+$[2] is solvable in time $2^{n/\lambda(n)}m^{O(1)}$ for some proper complexity function $\lambda(n)$. Let $\mu(n)$ be a proper complexity function satisfying $2^{\mu(n)} = O(\lambda(n))$. Given an instance C of CNF-SAT (we view the instance as a circuit), we partition the n variables of C into $r = n/\mu(n)$ blocks each of size $\mu(n)$ (we ignore ceilings and floors for simplicity). For each block B_i, $i = 1, \ldots, r$, we replace the variables in B_i with $2^{\mu(n)}$ new variables that *encode* all possible truth assignments to the variables in B_i; that is, assigning 1 to a new variable is equivalent to choosing the truth assignment for B_i — out of the $2^{\mu(n)}$ possible truth assignments to B_i—that corresponds to that particular new variable. We connect a new variable to an OR-gate in C if and only if the (partial) truth assignment corresponding to the new variable satisfies the OR-gate (*i.e.*, satisfies at least one of the original literals that were incoming to the gate). Finally, we set $k = r$, and we add an enforcement circuit to ensure that at least one new variable from the set of new variables corresponding to each block B_i is assigned 1. The enforcement circuit consists of r OR-gates, each of which feeds into the output AND-gate of C and has as inputs all the $2^{\mu(n)}$ new variables corresponding to the same block B_i, for $i = 1, \ldots, r$. Since $k = r$, the enforcement circuit ensures that exactly one new variable from the set of variables corresponding to a block B_i is assigned 1 in any satisfying assignment of weight k. Let C' be the resulting circuit. Note that the size of the enforcement circuit is linear in the number of new variables. Clearly, C is satisfiable if and only if C' has a satisfying assignment of weight k. Note that C' is monotone, and has $N = 2^{\mu(n)} \cdot n/\mu(n)$ variables and size $M = O(N + m)$. Using the (hypothetical) algorithm for WSAT$^+$[2], we can decide if C is satisfiable in time $2^{N/\lambda(N)} M^{O(1)} = 2^{O(\lambda(n)) \cdot n/(\mu(n) \cdot \lambda(N))} m^{O(1)}$ by the choice of $\mu(n)$, which is $2^{o(n)} m^{O(1)}$ since $\lambda(N) \geq \lambda(n)$ (λ is nondecreasing) and $\mu(n)$ is a proper complexity function. $\qquad\square$

Corollary 1. *WSAT$^-$[2] $\in 2^{o(n)} m^{O(1)}$ if and only if CNF-Sat $\in 2^{o(n)} m^{O(1)}$, where n is the number of variables and m is the instance size for both problems.*

Proof. For an instance (C, k) of WSAT$^+$[2] we associate an instance $(C', n - k)$ of WSAT$^-$[2], where C' is the circuit obtained from C by negating all input literals. It is easy to see that this association provides a polynomial time (serf) reduction from WSAT$^+$[2] to WSAT$^-$[2]. The same reduction serves as a serf reduction from WSAT$^-$[2] to WSAT$^+$[2] as well. The result now follows from Theorem 1. $\qquad\square$

Remark 2. Theorem 1 and Corollary 1 show that $WSAT^+[2]$ and $WSAT^-[2]$ are equivalent with respect to their subexponential-time solvability, even though they differ with respect to their parameterized complexity ($WSAT^+[2]$ is $W[2]$-complete while $WSAT^-[2]$ is $W[1]$-complete)—assuming that the W-hierarchy does not collapse. This is not surprising as we will show in the corollary below.

Corollary 2. *For every $t \geq 2$, there exists a $W[t]$-complete problem that is solvable in subexponential time if and only if the ETH fails.*

Proof. Consider the CIRCUIT SATISFIABILITY problem (unbounded depth circuits) restricted to instances in which the size of the circuit is linear in the number of variables, denoted LINEAR CIRCUIT SAT. By Imapagliazzo et al.'s results [18], one can easily show that LINEAR CIRCUIT SAT is solvable in subexponential time if and only if the ETH fails.

To prove the statement of the corollary, consider the $W[t]$-complete problem WSAT[t] with linear number of clauses (in terms of the number of variables in the circuit), denoted LINEAR WSAT[t]; this problem is clearly in $W[t]$, and its $W[t]$-hardness follows by an easy fpt-reduction (using a padding argument[2]) from the general WSAT[t] problem. A similar argument to that made in the proof of Proposition 1 shows that if LINEAR CIRCUIT SAT is solvable in subexponential time then so is LINEAR WSAT[t] (by a reduction that encodes the weight). On the other hand, consider the Turing-serf-reduction from 3-3-SAT to LINEAR WSAT[2] that iterates through all possible weights of a satisfying assignment to the instance of 3-3-SAT, and for each weight reduces the instance to an instance of LINEAR WSAT[2]. This reduction proves that if LINEAR WSAT[t], for any $t \geq 2$, is solvable in subexponential time then so is 3-3-SAT, which, in turn, implies that the ETH fails. $\qquad\square$

The RED-BLUE DOMINATING SET problem is: Given a bipartite graph whose first partition is colored red and whose second partition is colored blue, and a parameter k, decide if there is a set S of at most k red vertices such that each blue vertex is adjacent to at least one vertex in S. The following corollary follows from Theorem 1 via standard reductions from HITTING SET and RED-BLUE DOMINATING SET to WSAT$^+$[2] and vice versa that preserve the value of n (e.g., see [13]):

Corollary 3. HITTING SET *and* RED-BLUE DOMINATING SET *are solvable in time* $2^{o(n)}m^{O(1)}$ *if and only if* CNF-SAT *is solvable in time* $2^{o(n)}m^{O(1)}$, *where m is the size of the instance in each of the three problems, and n is the number of vertices in* HITTING SET, *the number of red vertices in* RED-BLUE DOMINATING SET, *and the number of variables in* CNF-SAT.

We note that the results in the above corollary can be implied from the stronger result proved in [12], albeit using more involved reductions.

The RED-BLUE NONBLOCKER problem is: Given a bipartite graph with one partition colored red and the other blue, decide if there exists a set S of k red vertices such that every blue vertex has a red neighbor not in S. The INDEPENDENT SET ON HYPERGRAPHS problem is the generalization of the INDEPENDENT SET problem to hypergraphs: Given a hypergraph H and a parameter k, decide if there exists a set S of k vertices in H such that no hyperedge of H is contained in S. The following corollary follows from Corollary 1 via standard reductions from RED-BLUE NONBLOCKER and INDEPENDENT SET ON HYPERGRAPHS to WSAT$^-$[2] and vice versa that preserve the value of n (see [13]):

Corollary 4. RED-BLUE NONBLOCKER *and* INDEPENDENT SET ON HYPER-GRAPHS *are solvable in time* $2^{o(n)}m^{O(1)}$ *if and only if* CNF-SAT *is solvable in time* $2^{o(n)}m^{O(1)}$, *where m is the size of the instance in each of the three problems,*

---

[2] A padding argument is a general tool that is used in complexity theory to extend a result to a larger class of problems. For our purpose in this paper, the padding argument works by adding/padding a "dummy" part to the instance to create an equivalent new instance in which a relation holds true between certain parameters in the new instance. We will use the padding argument a couple of times in this paper, and skip the details when it is clear how the instance can be padded.

and n is the number of vertices in INDEPENDENT SET ON HYPERGRAPHS, *the number of red vertices in* RED-BLUE NONBLOCKER, *and the number of variables in* CNF-SAT.

5   Number of Gates

In this section we study the subexponential-time complexity of SAT[t] and WSAT[t], and the parameterized complexity of WSAT[t], with respect to the number of gates in the circuit.

Proposition 2. *For any integer $t \geq 2$, SAT[t] is solvable in $2^{o(n)}$ time on circuits of $o(n)$ gates, and unless the ETH fails, SAT[t] is not solvable in $2^{o(n)}m^{O(1)}$ time on circuits of $\Omega(n)$ gates, where n is the number of variables in the circuit and m is the instance size. The same result holds for WSAT[t].*

Proof. The hardness results follow by the simple (Turing) serf-reductions from 3-3-SAT to SAT[2] and WSAT[2] (for the Turing-serf-reduction to WSAT[2], we try all possible weights) that result in instances of SAT[2] and WSAT[2] whose number of gates is linear in n.

Lemma 2 implies that WSAT[t] is solvable in $2^{o(n)}$ time on circuits of $o(n)$ gates. The same lemma also implies that SAT[t] is solvable in $2^{o(n)}$ time on circuits of $o(n)$ gates because we can iterate over all possible weights, and for each weight we invoke the algorithm for WSAT[t], thus adding only a linear factor to running time (which gets absorbed in the term $2^{o(n)}$) of an algorithm that decides WSAT[t] in $2^{o(n)}$ time on circuits of $o(n)$ gates.   □

Proposition 3. *For any integer $t \geq 2$, WSAT[t] is FPT on circuits of $O(\log n)$ gates, and unless the ETH fails, WSAT[t] is not FPT on circuits of $\omega(\log n)$ gates, where n is the number of variables in the circuit.*

Proof. Lemma 2 implies that WSAT[t] is FPT on circuits of $O(\log n)$ gates.

To prove the hardness result, let $g(n) = \omega(\log n)$ be a proper complexity function. Then $g(n) \geq \log n \cdot \lambda(n)$, for some proper complexity function $\lambda(n)$. We provide a reduction from 3-3-SAT to WSAT[2] on circuits of at most $g(n)$ gates; this reduction will imply that the ETH fails, assuming that WSAT[2] on circuits of at most $g(n)$ gates is FPT. (Note that the hardness result for WSAT[2] implies the hardness result for WSAT[t] for $t > 2$.)

Suppose that WSAT[2] is FPT on circuits of at most $g(n)$ gates, and let A be an algorithm that decides it in time $f(k)M^{O(1)}$, where f is a proper complexity function and M is the instance size. Let f^{-1} be the inverse function of f. Let C be an instance of 3-3-SAT on n variables and $m = O(n)$ clauses (viewed as a circuit). We divide the n variables in C into $f^{-1}(m)$ many blocks, and proceed as in the reduction described in the proof of Theorem 1. Let C' be the resulting circuit. C' has $N = 2^{n/f^{-1}(m)} \cdot f^{-1}(m)$ variables and $O(n) + f^{-1}(m)$ gates (the $f^{-1}(m)$ gates are for the enforcement circuit). Observing that if WSAT[2] is solvable in time $f(k)M^{O(1)}$ then it is also solvable in time $h(k)M^{O(1)}$ for any computable function $h > f$, we can choose f large enough so that $f^{-1}(m) = o(\lambda(n))$ (note that $m = O(n)$). With this choice of f, it is not difficult to verify that $O(n) + f^{-1}(m) \leq \log N \cdot \lambda(N) \leq g(N)$, and hence the number of gates of C' is at most $g(N)$. We now use that algorithm A to decide the instance $(C', k = f^{-1}(m))$, and hence C, in time $f(f^{-1}(m))(N + m)^{O(1)} = 2^{o(n)}$. This shows that if WSAT[2] is FPT on circuits of at most $g(n)$ gates then 3-3-SAT is solvable in $2^{o(n)}$ time, a consequence that implies that the ETH fails. $\square$

6   Circuit Fan-In

In this section we study the subexponential-time complexity of SAT[t] and WSAT[t], and the parameterized complexity of WSAT[t], with respect to the fan-in of the circuit. First observe that if $\mu(n)$ is the fan-in of a circuit on n variables and of depth t then $\mu(n)$ must satisfy $\mu(n) \geq n^{1/t}$ (otherwise, some variables will be disconnected).

Proposition 4. *For any $t \geq 2$ and $c > 1$, the WSAT[t] problem on circuits of fan-in $\mu(n)$, where $\mu(n) \geq cn^{1/t}$ (n is the number of variables in the circuit) is $W[t]$-hard if t is even and $W[t-1]$-hard if t is odd.*

Proof. The proof is via an fpt-reduction (padding argument) from WSAT$^+$[t], which is $W[t]$-hard if t is even and $W[t-1]$-hard if t is odd. Let (C, k) be an instance of WSAT$^+$[t], where C has n variables and n_g gates, and note that the fan-in of each gate in C is at most $n + n_g$. We construct an instance (C', k) of $WSAT$[t]. To make the fan-in in C' satisfy the upper bound stipulated by the function $\mu()$, we construct C' as follows. We start with C and introduce $(n+n_g)^{2t}$ new negative literals. We connect the negative literals using a layered circuit with $(t - 1)$ layers (excluding the new variables layer) consisting of new gates of alternating types (AND-of-OR-of-AND-..., or OR-of-AND-of-OR-..., depending on the parity of t), such that each gate has fan-in $(n + n_g)^2$; this can be achieved by a top-down construction of the layered circuit (we assume that the output gate of the circuit is at the bottom), in which each gate in a layer receives its inputs from $(n + n_g)^2$ (distinct) gates (arbitrarily chosen) in the previous layer (the top layer receives its input form the negative literals). The last layer of the layered circuit consists of $(n + n_g)^2$ gates that are incoming to the output gate of C. Clearly, the construction of (C', k) can be done in polynomial time. If C has a weight-k satisfying assignment, then C' also has weight-k satisfying assignment obtained by assigning the k positive variables in C that satisfy C the value 1 and all the remaining variables in C' the value 0. Conversely, if C' has a weight-k satisfying assignment, then by the construction of C' there must exist at most k positive variables (and also exactly k positive variables because C is monotone) that satisfy C. This proves the correctness of the fpt-reduction. Since each gate in C has fan-in at most $n + n_g$, the output gate of C' has fan-in at most $(n + n_g)^2 + (n + n_g)$, and each gate in the layered circuit has fan-in at most $(n + n_g)^2$, C' has fan-in at most $(n + n_g)^2 + (n + n_g)$. For large enough n (larger than a fixed constant that depends on t), we have

$(n + n_g)^2 + (n + n_g) \leq c(n + n_g)^2 = c((n + n_g)^{2t})^{1/t} \leq cN^{1/t}$, where $N = (n + n_g)^{2t} + n$ is the total number of variables in C'. This gives an fpt-reduction from WSAT$^+[t]$ to WSAT$[t]$ on circuits of fan-in at most $\mu(n)$, where n is the number of variables in the circuit, thus completing the proof.   □

Now we turn our attention to the subexponential-time complexity of SAT$[t]$ and WSAT$[t]$ with respect to the fan-in of the circuit.

Proposition 5. *For any integer $t \geq 2$, SAT[t] on circuits of fan-in $\mu(n)$, where $\mu(n) = o(n^{\frac{1}{t-1}})$ and n is the number of variables in the circuit, is solvable in $2^{o(n)}$ time. The same is true for WSAT[t].*

Proof. If $\mu(n) = o(n^{\frac{1}{t-1}})$ then the number of gates in the circuit is $o(n)$. By Proposition 2, it follows in this case that SAT$[t]$ and WSAT$[t]$ on circuits of fan-in $\mu(n) = o(n^{\frac{1}{t-1}})$ are solvable time $2^{o(n)}$ (note that in this case the size m of the circuit is $O(n)$).   □

7   Bottom Fan-In and Number of Occurrences

In this section we study the subexponential-time complexity of SAT$[t]$ and WSAT$[t]$, and the parameterized complexity of WSAT$[t]$, with respect to the bottom fan-in of the circuit, and with respect to the maximum number of occurrences of any variable in the circuit. We assume that every gate has fan-in at least 2 — and hence the bottom fan-in of the circuit is at least 2 — because a gate with fan-in 1 can be merged or removed. This proposition follows from Lemma 1:

Proposition 6. *For any $t \geq 2$ and $\Delta \geq 3$, unless the ETH fails, SAT[t] on circuits of bottom fan-in Δ is not solvable in $2^{o(n)} m^{O(1)}$ time, where n is the number of variables in the circuit and m is the instance size.*

Note that for $t = 2$, SAT$[t]$ on circuits of bottom fan-in 2 is the 2-CNF-SAT problem, which is solvable in polynomial time. On the other hand, if $t \geq 3$, then it is easy to see that SAT$[t]$ on circuits of bottom fan-in 2 is not solvable in $2^{o(n)} m^{O(1)}$ time unless CNF-SAT is.

The following proposition follows by the standard (polynomial-time) serf-reduction from INDEPENDENT SET, which is SNP-complete under serf-reductions, to WSAT$[2]$ of bottom fan-in 2 (see [13]):

Proposition 7. *For any $t \geq 2$ and $\Delta \geq 2$, unless the ETH fails, WSAT[t] on circuits of bottom fan-in Δ is not solvable in $2^{o(n)} m^{O(1)}$ time, where n is the number of variables in the circuit and m is the instance size.*

Observe that for $t = 2$, and for circuit of bottom fan-in 2, SAT$[t]$ is solvable in polynomial time, whereas WSAT$[t]$ is not solvable in subexponential-time (unless the ETH fails). This does not contradict Proposition 1 since the proof of the proposition (one direction) relies on the ability to encode the weight of the desired assignment using 3-CNF-SAT clauses (*i.e.*, gates of bottom fan-in 3).

The following proposition characterizes the parameterized complexity of WSAT[t] with respect to the circuit bottom fan-in. The proof of the proposition follows by the standard (polynomial-time) fpt-reduction from INDEPENDENT SET to (antimonotone) WSAT[2]of bottom fan-in 2 alluded to above ([13]).

Proposition 8. *For any integer $t \geq 2$ and $\Delta \geq 2$, WSAT[t] on circuits with bottom fan-in Δ is $W[1]$-hard.*

We now switch our attention to the maximum number of occurrences of any variable in the circuit. This proposition follows from Lemma 1:

Proposition 9. *For any $t \geq 2$ and $\Delta \geq 3$, unless the ETH fails, SAT[t] on circuits in which each variable occurs at most Δ times is not solvable in $2^{o(n)}$ time, where n is the number of variables in the circuit. The same holds true for WSAT[t].*

Consider the k-FLIP SAT problem that was proved to be $W[1]$-hard on instances in which every variable occurs at most three times [22]: Given an instance F of CNF-SAT and a truth assignment τ to F, decide if the values of k variables assigned by τ can be flipped so that F is satisfied. By a simple fpt-reduction from k-FLIP SAT to WSAT[t] (in the reduction, τ is the assignment that assigns every variable the value 0), we have:

Proposition 10. *For any $t \geq 2$ and $\Delta \geq 3$, $WSAT[t]$ on instances in which every variable occurs at most Δ times is $W[1]$-hard.*

References

1. Alber, J., Bodlaender, H., Ferneau, H., Niedermeier, R.: Fixed parameter algorithms for dominating set and related problems on planar graphs. Algorithmica **33**, 461–493 (2002)
2. Brüggemann, T., Kern, W.: An improved deterministic local search algorithm for 3SAT. Theoret. Comput. Sci. **329**, 303–313 (2004)
3. Cai, L., Juedes, D.: On the existence of subexponential parameterized algorithms. J. Comput. Syst. Sci. **67**(4), 789–807 (2003)
4. Calabro, C., Impagliazzo, R., Paturi, R.: A duality between clause width and clause density for SAT. In: IEEE Conference on Computational Complexity, pp. 252–260 (2006)
5. Chen, J., Chor, B., Fellows, M., Huang, X., Juedes, D., Kanj, I., Xia, G.: Tight lower bounds for certain parameterized NP-hard problems. Inf. Comput. **201**(2), 216–231 (2005)
6. Chen, J., Huang, X., Kanj, I., Xia, G.: Strong computational lower bounds via parameterized complexity. J. Comput. Syst. Sci. **72**(8), 1346–1367 (2006)
7. Chen, J., Kanj, I.A.: Parameterized complexity and subexponential-time computability. In: Bodlaender, H.L., Downey, R., Fomin, F.V., Marx, D. (eds.) Fellows Festschrift 2012. LNCS, vol. 7370, pp. 162–195. Springer, Heidelberg (2012)

8. Chen, J., Kanj, I., Xia, G.: On parameterized exponential time complexity. Theoret. Comput. Sci. **410**(27–29), 2641–2648 (2009)
9. Chen, Y., Flum, J.: On miniaturized problems in parameterized complexity theory. Theoret. Comput. Sci. **351**(3), 314–336 (2006)
10. Chen, Y., Flum, J.: Subexponential time and fixed-parameter tractability: exploiting the miniaturization mapping. J. Logic Comput. **19**(1), 89–122 (2009)
11. Chen, Y., Grohe, M.: An isomorphism between subexponential and parameterized complexity theory. SIAM J. Comput. **37**(4), 1228–1258 (2007)
12. Cygan, M., Dell, H., Lokshtanov, D., Marx, D., Nederlof, J., Okamoto, Y., Paturi, R., Saurabh, S., Wahlström, M.: On problems as hard as CNF-SAT. In: IEEE Conference on Computational Complexity, pp. 74–84. IEEE (2012)
13. Downey, R., Fellows, M.: Fundamentals of Parameterized Complexity. Springer, New York (2013)
14. Fernau, H.: EDGE DOMINATING SET: efficient enumeration-based exact algorithms. In: Bodlaender, H.L., Langston, M.A. (eds.) IWPEC 2006. LNCS, vol. 4169, pp. 142–153. Springer, Heidelberg (2006)
15. Flum, J., Grohe, M.: Parameterized complexity theory. In: Brauer, W., Rozenberg, G., Salomaa, A. (eds.) Texts in Theoretical Computer Science. An EATCS Series, vol. XIV. Springer, Berlin (2006)
16. Flüm, J., Grohe, M.: Parameterized Complexity Theory. Springer-verlag, Berlin (2010)
17. Impagliazzo, R., Paturi, R.: On the complexity of k-SAT. J. Comput. Syst. Sci. **62**(2), 367–375 (2001)
18. Impagliazzo, R., Paturi, R., Zane, F.: Which problems have strongly exponential complexity? J. Comput. Syst. Sci. **63**(4), 512–530 (2001)
19. Lokshtanov, D., Marx, D., Saurabh, S.: Slightly superexponential parameterized problems. In: Proceedings of the Twenty-Second Annual ACM-SIAM Symposium on Discrete Algorithms, pp. 760–776 (2011)
20. Papadimitriou, C., Yannakakis, M.: Optimization, approximation, and complexity classes. J. Comput. Syst. Sci. **43**, 425–440 (1991)
21. Patrascu, M., Williams, R.: On the possibility of faster SAT algorithms. In: Proceedings of the Twenty-First Annual ACM-SIAM Symposium on Discrete Algorithms, pp. 1065–1075 (2010)
22. Szeider, S.: The parameterized complexity of k-flip local search for SAT and MAX SAT. Discrete Optim. **8**(1), 139–145 (2011)

Kolmogorov Structure Functions for Automatic Complexity in Computational Statistics

Bjørn Kjos-Hanssen[⊠]

University of Hawai'i at Mānoa, Honolulu, HI 96822, USA
bjoernkh@hawaii.edu
http://math.hawaii.edu/wordpress/bjoern/

Abstract. For a finite word w of length n and a class of finite automata $\mathcal{A}$, we study the Kolmogorov structure function h_w for automatic complexity restricted to $\mathcal{A}$. We propose an approach to computational statistics based on the minimum p-value of $h_w(m)$ over $0 \leq m \leq n$. When $\mathcal{A}$ is the class of all finite automata we give some upper bounds for h_w. When $\mathcal{A}$ consists of automata that detect several success runs in w, we give efficient algorithms to compute h_w. When $\mathcal{A}$ consists of automata that detect one success run, we moreover give an efficient algorithm to compute the p-values.

1 Introduction

Shallit and Wang [6] introduced automatic complexity as a computable alternative to Kolmogorov complexity. They considered deterministic automata, whereas Hyde and Kjos-Hanssen [4] studied the nondeterministic case, which in some ways behaves better. Unfortunately, even nondeterministic automatic complexity is somewhat inadequate. The string 00010000 has maximal nondeterministic complexity, even though intuitively it is quite simple. One way to remedy this situation is to consider a structure function analogous to that for Kolmogorov complexity.

The latter was introduced by Kolmogorov at a 1973 meeting in Tallinn and studied by Vereshchagin and Vitányi [8] and Staiger [7].

The Kolmogorov complexity of a finite word w is roughly speaking the length of the shortest description w^* of w in a fixed formal language. The description w^* can be thought of as an optimally compressed version of w. Motivated by the non-computability of Kolmogorov complexity, Shallit and Wang studied a deterministic finite automaton analogue.

Definition 1 (Shallit and Wang *[6]*). *The* automatic complexity *of a finite binary string* $x = x_1 \dots x_n$ *is the least number* $A_D(x)$ *of states of a deterministic finite automaton* M *such that* x *is the only string of length* n *in the language accepted by* M.

This work was partially supported by a grant from the Simons Foundation (#315188 to Bjørn Kjos-Hanssen). The author also acknowledges the support of the Institute for Mathematical Sciences of the National University of Singapore during the workshop on *Algorithmic Randomness*, June 2–30, 2014.

Z. Zhang et al. (Eds.): COCOA 2014, LNCS 8881, pp. 652–665, 2014.
DOI: 10.1007/978-3-319-12691-3_49

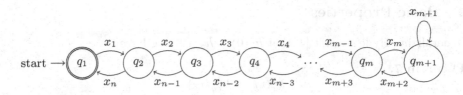

Fig. 1. A nondeterministic finite automaton that only accepts one string $x = x_1 x_2 x_3 x_4 \ldots x_n$ of length $n = 2m + 1$.

Hyde and Kjos-Hanssen [4] defined a nondeterministic analogue:

Definition 2. *The nondeterministic automatic complexity $A_N(w)$ of a word w is the minimum number of states of an NFA M, having no ϵ-transitions, accepting w such that there is only one accepting path in M of length $|w|$.*

The minimum complexity $A_N(w) = 1$ is only achieved by words of the form a^n where a is a single letter.

Definition 3. *Let $n = 2m+1$ be an odd number. A finite automaton of the form given in Fig. 1 for some choice of symbols $x_1, \ldots, x_n$ and states $q_1, \ldots, q_{m+1}$ is called a* Kayleigh graph[1].

Theorem 4 (Hyde *[3]*). *The nondeterministic automatic complexity $A_N(x)$ of a string x of length n satisfies*

$$A_N(x) \le b(n) := \lfloor n/2 \rfloor + 1.$$

Proof. If the length of n is odd, then a Kayleigh graph witnesses this inequality. If the length of n is even, a slight modification suffices, see [3]. $\square$

Definition 5. *The* complexity deficiency *of a word x of length n is*

$$D_n(x) = D(x) = b(n) - A_N(x).$$

The structure function of a string x is defined by $h_x(m) = \min\{k : \text{there is a } k\text{-state NFA } M \text{ which accepts at most } 2^m \text{ strings of length } |x| \text{ including } x\}$. In more detail:

Definition 6 (Vereshchagin, personal communication, 2014, inspired by *[8]*). *In an alphabet Σ containing b symbols,*

$$h_x(m) = \min\{k : \exists \, k\text{-state NFA } M, x \in L(M) \cap \Sigma^n, |L(M) \cap \Sigma^n| \le b^m\}.$$

We also the define the "converse" structure function

$$g_x(m) = h_x(|x| - m)$$

and its maximum

$$G_n(m) = \sup_{|x|=n} g_x(m).$$

---

[1] The terminology is a nod to the more famous Cayley graphs as well as to Kayleigh Hyde's first name.

2   Basic Properties

Generalizing Hyde's result [4] that $h_x(0) \leq h_{xy}(0)$, we have

Theorem 7.
$$h_x(m) \leq h_{xy}(m)$$

and hence for $b \in \{0,1\}$,

$$h_x(|x| - k) = h_x(n - k) \leq h_{xb}(n - k) = h_{xb}(|xb| - (k+1))$$

$$g_x(k) \leq g_{xb}(k+1)$$

hence

$$G_n(k) \leq G_{n+1}(k+1)$$

Conjecture 8. $G_n(k) \leq G_{n+1}(k)$.

Conjecture 9 (The right upper bound conjecture). $\forall k \lim_n G_n(k) = k + 1$.

We have verified Conjecture 9 for $k = 0$ and $k = 1$ by simple proofs.

Theorem 10. *There exist x and b with $h_x(|x| - 1) \nleq h_{xb}(|xb| - 1)$.*

Proof. Let $x = 0100$ and $b = 0$. Then $h_x = (3, 3, 2, 2, 1)$ and $h_{xb} = (3, 3, 2, 2, 2, 1)$. That is, the string 01000 is accepted by an automaton with 2 states accepting only 1/16th of all strings of length 5, but 0100 cannot be accepted by any such automaton for length 4. □

Theorem 11. *We have the following inequalities for all strings x and y of length n, symbols b, and $0 \leq m \leq n$.*

$$h_x(|x| - 1) + 1 \geq h_{xb}(|xb| - 2)$$

$$h_x(m) + |y| \geq h_{xy}(m)$$
$$g_x(n - m) + |y| \geq g_{xy}(n + |y| - m)$$
$$g_x(k) \leq g_{xb}(k+1) \leq g_x(k) + 1$$

Proof. Suppose there is M accepting only half the strings of length n, including x. Then there is M' accepting only a quarter of the strings of length $n + 1$, including xb, with one extra state (namely, just add a new state and one new edge labeled by b). □

Definition 12. *The entropy function $\mathcal{H} : [0, 1] \to [0, 1]$ is given by*

$$\mathcal{H}(p) = -p \log_2 p - (1 - p) \log_2(1 - p).$$

It is then fairly canonical to define the inverse entropy function $\mathcal{H}^{-1} : [0, 1] \to [0, 1/2]$ by

$$\mathcal{H}^{-1}(y) = x \iff \mathcal{H}(x) = y \quad and \quad 0 \leq x \leq 1/2.$$

Theorem 13. *For* $0 \le k \le n$,

$$\log_2 \binom{n}{k} = \mathcal{H}(k/n)n + O(\log n).$$

Proof. Let $\log = \ln = \log_e$. For $u \in \mathbb{N}$, let

$$S_u = \sum_{k=2}^{u} \log k, \quad I_u = \int_1^u \log x \, dx, \quad \text{and} \quad J_u = \int_2^{u+1} \log x \, dx.$$

Let

$$\alpha_n = \log \binom{n}{k} = S_n - S_k - S_{n-k}.$$

Note $I_u \le S_u \le J_u$ and

$$J_u - I_u = \int_u^{u+1} \log x \, dx - \int_1^2 \log x \, dx \le \log(u+1),$$

Thus up to $O(\log n)$ error terms we have

$$\alpha_n = \int_1^n \log x \, dx - \int_1^k \log x \, dx - \int_1^{n-k} \log x \, dx$$

$$= (n \log n - n) - (k \log(k) - k) - [n - k \log(n-k) - n - k]$$

$$= n \log n - k \log(k) - n - k \log(n-k)$$

$$= -k \log(k/n) - n - k \log(1 - k/n)$$

and hence

$$\log_2 \binom{n}{k} = -k \log_2(k/n) - n - k \log_2(1 - k/n) = \mathcal{H}(k/n) \cdot n.$$

$\square$

Theorem 14. *Suppose the number of* 0*s in the binary string* x *is* $p \cdot n$. *Then*

$$h_x(\mathcal{H}(p)n) \le pn + O(\log n).$$

Proof. Consider an automaton M as in Fig. 3 that has $[pn]$ many states, and that has one left-to-right arrow labeled 0 for each 0, and a loop in place labeled 1 for each consecutive string of 1s. Since M accepts exactly those strings that have $[pn]$ many 0s, the number of strings accepted by M is $\binom{n}{[p \cdot n]}$. By Theorem 13 this is $\le 2^k$ approximately when $\mathcal{H}(p)n \le k$, and we are done. $\square$

Example 15. *A string of the form* $0^a 1^{n-a}$ *satisfies* $h_x(\log_2 n) = 2$ *whereas* $h_x(0)$ *may be* $n/2$. *For instance* 0011 *has* $h_x(2) = 2$. *On the other hand* $h_x(1) = 3$ *which is why this string is more complicated than* 0110.

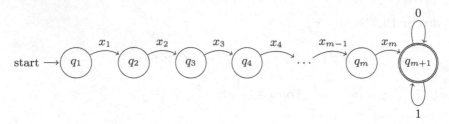

Fig. 2. An automaton illustrating the linear upper bound on the automatic structure function from Theorem 16.

Theorem 16. *For any x of length n,*

$$1 \leq h_x(m) \leq n - m + 1 \text{ for } 0 \leq m \leq n.$$

Proof. $1 \leq h_x(n - k) \leq k + 1$ because we can start out with a sequence of determined moves, after which we accept everything, as in Fig. 2. □

Remark 17. *We do not know whether $h_w(0)$, i.e., $A_N(w)$, is polynomial-time computable as a function of w. However, $g_w(m)$ is polynomial-time computable for each fixed parameter m, since there is an upper bound that only depends on m by Theorem 16.*

Theorem 16 suffices to calculate h_x for $n = 0$. For x of length 0, we have $h_x(0) = 1$.

Theorem 18. *We have $h_x(0) = A_N(x)$, the automatic complexity of x. We have $h_x(m) \geq h_x(m+1)$ for each $0 \leq m < n$.*

Theorem 19. *We have $h_x(n - k) \geq 2$ unless x is unary or $k = 0$.*

Proof. If x consists of both 0s and 1s then a 1-state automaton is useless. □

3   An Approach to Computational Statistics

We propose to study automatic sufficient statistics by looking at the p-value of an event $h_x(m) \leq q$. The m that gives the lowest p-value for a given x gives the model that we use to explain our data x. If this p-value is not less than a threshold such as Fisher's $p = 1/20$ then we just use a null hypothesis of an arbitrary binary string. Considering alphabets larger than size 2 is forced upon us in that the most complex string of length 4 is naturally 0123, and we thus get a better understanding of upper bounds. There is an automaton with 1 state accepting only binary strings. For a ternary alphabet we would redefine the structure function.

$$2^n \leq 3^m$$

when

$$n \log 2 \leq m \log 3$$

$$n\frac{\log 2}{\log 3} \le m$$

Of course, $\log 2/\log 3 = \log_3 2 = 0.63\cdots < 1$. Now the p-value for $h_x(\lceil n\log_3 2\rceil)$ ≤ 1 should be $< 1/20$ as soon as n is sufficiently large, since the only thing you can do with one state is to limit the alphabet. Thus $h_x(m) \le 1$, for $m < n$, only happens for $3\cdot 2^n - 3$ out of the 3^n strings. For $n \ge 11$,

$$\frac{3\cdot 2^n - 3}{3^n} = \frac{2^n - 1}{3^{n-1}} \le \frac{1}{20}.$$

Thus a binary string of length 11 should lead to rejection of the null hypothesis that we have a random ternary string. If we observe a binary string and are considering the null hypothesis of a quaternary alphabet, we need

$$\frac{\binom{4}{2}2^n - (\binom{4}{2} - 3) + 4}{4^n} = \frac{6\cdot 2^n - 8}{4^n} = \frac{3\cdot 2^{n-1} - 2}{4^{n-1}} < \frac{1}{20}$$

which gives $n - 1 \ge 6$, $n \ge 7$.

Of course, the probability of a binary $\{0,1\}$ string in a 4-ary alphabet is just $(2/4)^n$, but here we account for possibilities $\{i,j\} \ne \{0,1\}$.

Theorem 20. *Let x be a string of length n in an a-ary alphabet with uniform distribution.*

1. *The probability that x turns out to be binary, i.e., to be a string over some 2-element alphabet, is*
$$\frac{\binom{a}{2}2^n - a(a - 2)}{a^n}.$$
2. *The probability that x turns out to be binary over $\{0,1\}$ is $(2/a)^n$.*
3. *The probability that x turns out to be $a - 1$-ary, i.e., to be a string over some $(a-1)$-element alphabet, is*
$$a^{-n}\sum_{k=1}^{a-1}\binom{a}{k}(a-k)^n(-1)^{k+1}.$$

Proof. Part (2) is obvious and Part (3) follows from the inclusion-exclusion principle. We prove part (1). In the estimate $\binom{a}{a-1}(a-1)^n$, we are counting the a many unary strings a wrong number c_a of times, but how many times does not depend on n. To find the number c_a, we solve

$$\frac{\binom{a}{2}2^1 - c_a}{a^1} = 1 = \frac{\binom{a}{2}2^2 - c_a}{a^2}$$

which gives $c_a = a(a - 2) = 2a(a - 1) - a^2$.   □

Corollary 21. *Let x be a string in an a-ary alphabet with uniform distribution. The probability that $h_x(m) = 1$ for some $m < n$ is 0 in the limit as $n \to \infty$.*

Example 22. *In the case $n = 6$, we need*
$$\frac{\binom{5}{2}2^n - 15}{5^n} = \frac{10\cdot 2^n - 15}{5^n} < \frac{1}{20}$$
which is true for $n \ge 6$.

4 Run Complexity

4.1 Algorithms for Single-Run Complexity: p-values and Structure Function

One severe restriction that is certainly polynomial time computable is to require that each step should be from a state s to $s + 1$, except for one state that can have self-loops.

An implementation is available at [2].

Note that for a ternary alphabet the number of unary valences equals the number of binary valences. In general, we have to account for the number of valences when we assign a p-value. In a quaternary alphabet there are $\binom{4}{2}$ binary valences, but only 4 unary valences, which means that the p-values for an observed run has to be adjusted accordingly when compared with a run of a "different-ary" valence.

To specify an automaton we then only need to specify the location of the repeat, number of repeat cycles there, and labels on the edges. This is implicitly studied in Alikhani's Master's thesis [1]. In a ternary alphabet this could allow a block with a limited alphabet. This would give $n - \log_2 n$ as probabilistic upper bound on $h_x(0)$, and $h_x(m) \leq n - m + 1$.

If we have a 3-ary language with 2 self-loops at a given state, then the structure function goes up by $\lceil \log 3 / \log 2 \rceil = 2$ not 1 as we decrease m from that point. Actually it varies because it is $\lceil (m - mm) \log 3 / \log 2 \rceil$.

The structure function for a binary string will then be constant $h_x(0)$ until it hits the $n - m + 1$ curve, because the only types of automata allowed have one or two self-loops at the repeatable state. For a ternary string, there will be one more phase: Suppose as a random example $x = 1010020210$. The longest run is 00 giving $h_x(0) = n + 1 - 2 = 9$. The longest binary run is 10100 or 00202 giving $2^5 \leq 3^m$, $h_x(\lceil 5 \log 2 / \log 3 \rceil = 4) = 6$. So we get the structure function 99876654321. Now we can talk about explanatory power: this model says that the string is totally random except for one specified simple block. But what it does is identify whether a certain binary block is more surprising than another unary block, or a ternary block, which is good. In this case, having a run of two is very likely: $1 - (2/3)^{n-1} = 97.4\,\%$. The probability of having a binary run of five is at most

$$\frac{6 \cdot 2^5 \cdot 3^5}{3^{10}} = 0.79,$$

so the best explanation for this sequence, as having come from a distribution with a restricted alphabet block, is that there is a 2-letter-alphabet block of size 5.

If alphabet size is not restricted, we get a nontrivial notion.

Theorem 23. *We can compare probabilities of runs from different alphabet sizes in polynomial time.*

Proof. We can tabulate the cumulative distribution function effectively. Indeed, let R_n denote the longest run of heads in a sequence of n coin tosses. As mentioned by Alikhani [1] and Schilling [5], for $x \leq n - 1$,

$$\Pr(R_n \leq x) = \sum_{i=1}^{x+1} \Pr(R_n \leq x \mid H^{i-1}T) \Pr(H^{i-1}T)$$

$$= \sum_{i=1}^{x+1} \Pr(R_{n-i} \leq x) \Pr(H^{i-1}T)$$

$$= \sum_{i=1}^{x+1} \Pr(R_{n-i} \leq x) p^{i-1}(1-p)$$

Here H is an outcome in the restricted alphabet. So in the alphabet $\{0, 1, 2\}$ we would find the longest run from each of the following alphabets:

$$\{0\}, \{1\}, \{2\}, \{0, 1\}, \{0, 2\}, \{1, 2\}.$$

For the first three we have $p = 1/3$ and for the last three we have $p = 2/3$.   □

Example 24. *There are four nonisomorphic examples for $n = 3$: 000: unary run length 3 001: unary run length 2, binary run length 3 010: binary run length 3 012: no runs at all The most interesting is 001. Here we can use two states and get 1 accepted string, or 1 state and get 4 accepted strings. The probability of a unary run of length at least 2 is: $(1/3) + (1/3) - (1/3)(1/3) = 5/9$. The probability of a binary run of length 3 here is: $1 - (3/3)(2/3)(1/3) = 7/9$. Of course any unary run of length 2 would have to be part of a binary run of length 3, so it is not a really interesting case.*

Example 25. *For $n = 4$ we have: 0010 unary length 2, binary length 4 (interesting) (0011 is similar, with two unary length 2s) For 0010, the probability of a ternary string of length 4 being binary is $3(2/3)^4 - 3(1/3)^4 = (16 - 1)/27 = 5/9$, exactly the same as the probability of a ternary string of length 4 having a unary run of length 2.*

4.2   Algorithm for Structure Function of Multi-run Complexity for Fixed Valency

Next we could allow several repeat states but no going-back edges as in Figs. 3 and 4. It measures the presence of a collection of blocks of different kinds; in particular it detects Bernoulli distributions and even detects changes from one Bernoulli distribution to another. It does not detect things like $(01)^*$. Considering for instance $x = 0111122222$, we can imagine that it is better to use two repeat states than just one.

Theorem 26. *For multi-run complexity, there is a polynomial-time algorithm to determine whether $h_w(m) \leq q$ for a binary alphabet $\{0, 1\}$.*

Proof. We will look for automata with $\ell \leq q$ many self-loops. Let ℓ be minimal such that there is a solution to

$$x_1 + \cdots + x_\ell \geq n + 1 - q$$

consisting of lengths of disjoint runs in w. The number of solutions (x_i), $x_i \geq 0$, of $x_1 + \cdots + x_\ell = n + 1 - q$ is

$$\binom{n + 1 - q + \ell - 1}{\ell - 1}$$

and we want to know whether

$$\binom{n + 1 - q + \ell - 1}{\ell - 1} \leq b^m$$

where $b = 2$ is the alphabet size. □

We can extend this argument to the general case, but the polynomial-time algorithm will only work when all runs are required to have the same valence. When the valences can vary, we cannot simply form a decreasing sequence of all the longest runs, but have to consider arbitrary collections of disjoint runs.

Theorem 27. *For multi-run complexity in an arbitrary finite alphabet of, say, size* b, $h_w(m) \leq q$ *iff there exist disjoint runs having lengths* $x_1, \ldots, x_\ell$ *and valences* $v_1, \ldots, v_\ell$, $\ell \leq q$, *(where for instance* $v_i = \{1, 2\}$ *means that the run*

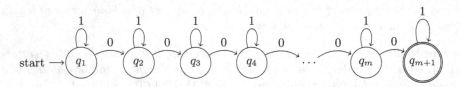

Fig. 3. An automaton illustrating multi-run complexity for a string of length n containing m many 0s, and $n - m$ many 1s.

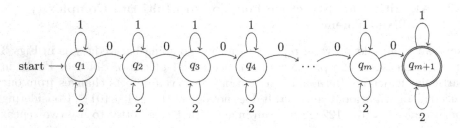

Fig. 4. An automaton illustrating multi-run complexity for a ternary string of length n containing m many 0s, and $n - m$ many 1s and 2s.

consists of 1s and 2s only; $|v_i| = 2$ *is the cardinality of* $\{1, 2\}$*) such that* $\sum x_i \geq n + 1 - q$, *and*

$$\sum \left\{ \prod_{i=1}^{\ell} |v_i|^{x_i} : \sum_{i=1}^{\ell} x_i = n + 1 - q, x_i \geq 0 \right\} \leq b^m.$$

For instance, if $|v_i|$ is a constant v, this says

$$v^{n+1-q} \left| \left\{ (x_i) : \sum_{i=1}^{\ell} x_i = n + 1 - q, x_i \geq 0 \right\} \right| \leq b^m,$$

or equivalently

$$v^{n+1-q} \binom{n + 1 - q + \ell - 1}{\ell - 1} \leq b^m.$$

If $v = b$, in other words we allow no runs at all, only reducing the number of states by the cop-out of allowing arbitrary symbols, which shows $h_w(m) \leq n + 1 - m$, then we can let $\ell = 1$ and then this says $n + 1 - q \leq m$, i.e., $h_w(m) = n + 1 - m$.

5   Upper Bounds on Structure Function for Automatic Complexity

Theorem 28. *The automatic structure function of a string* x *of length* n *is a function* $h_x : [0, n] \to [0, \lfloor n/2 \rfloor + 1]$. *Assume* x *is a binary string, so the alphabet size* $b = 2$. *Let*

$$\tilde{h} : [0, 1] \to [0, 1/2]$$

$$\tilde{h}(a) = \limsup_{n \to \infty} \max_{|x|=n} \frac{h_x([a \cdot n])}{n}$$

where $[x]$ *is the nearest integer to* x. *We make the following upper bound for* $\tilde{h}$ *(see Fig. 5):*

$$\tilde{h}(a) \leq u(a) := \begin{cases} \frac{1}{2} - \mathcal{H}^{-1}(a) & a \leq \mathcal{H}(\frac{1}{2} - \frac{\sqrt{3}}{4}) \approx 0.35 \\ \frac{2-a}{\alpha} & \mathcal{H}(\frac{1}{2} - \frac{\sqrt{3}}{4}) \leq a \leq \frac{\alpha-2}{\alpha-1} \approx 0.64 \\ 1 - a & \frac{\alpha-2}{\alpha-1} \leq a \leq 1. \end{cases}$$

where

$$\alpha = \frac{4}{\sqrt{3}} \left(2 - \mathcal{H} \left(\frac{1}{2} - \frac{\sqrt{3}}{4} \right) \right) = \mathcal{H}' \left(\frac{1}{2} - \frac{\sqrt{3}}{4} \right) \approx 0.379994.$$

and where $\mathcal{H}$ *is the entropy function (Definition 12). Note that*

$$u^{-1}(p) = \begin{cases} \mathcal{H}(\frac{1}{2} - p) & \frac{\sqrt{3}}{4} \leq p \leq \frac{1}{2}, \\ 2 - \alpha p & \frac{1}{\alpha-1} \leq p \leq \frac{\sqrt{3}}{4}, \\ 1 - p & 0 \leq p \leq \frac{1}{\alpha-1}. \end{cases}$$

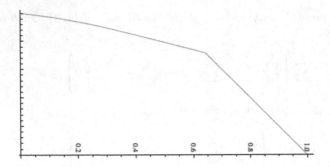

Fig. 5. Bounds for the automatic structure function for alphabet size $b = 2$; see Theorem 28.

Proof. Consider a path of length n through a Kayleigh graph with $q = pn$ many states. Let t_1 be the time spent before reaching the loop state for the first time. Let t_2 be the time spent after leaving the loop state for the last time. Let s be the number of self-loops taken by the path. Let us say that *meandering* is the process of leaving the loop state after having gone through a loop, and before again going through a loop. For fixed p let

$$\gamma(t_1, t_2, s, n) = \binom{t_1}{\frac{t_1 - pn}{2}}\binom{t_2}{\frac{t_2 - pn}{2}}\binom{n - t_1 - t_2}{s}\binom{n - t_1 - t_2 - s}{\frac{n - t_1 - t_2 - s}{2}}b^s$$

Then the number of such paths is[2]

$$N \le \sum_s \sum_{t_1} \sum_{t_2} \gamma(t_1, t_2, s, n) \tag{1}$$

since half of the meandering times must be backtrack times. Since

$$\limsup_{n \to \infty} \frac{\log_b \sum_1^n a_i}{n} \le \limsup_{n \to \infty} \frac{\log_b(n \cdot \max a_i)}{n} = \limsup_{n \to \infty} \frac{\log_b \max a_i}{n},$$

the sums can be replaced by maxima, i.e.,

$$\limsup_{n \to \infty} \frac{\log_b N}{n} \le \limsup_{n \to \infty} \frac{\log_b \gamma(t_1, t_2, s, n)}{n}, \quad (t_1, t_2, s) \in \arg \max \gamma(\cdot, \cdot, \cdot, n).$$

By Theorem 13,

$$\limsup_{n \to \infty} \frac{\gamma(t_1, t_2, s, n)}{n} \le \limsup_{n \to \infty} \frac{\delta(t_1, t_2, s, n)}{n}$$

---

[2] We can actually replace $\binom{n - t_1 - t_2}{s}$ by $\binom{(n - t_1 - t_2 + s)/2}{s}$, since the number of non-loops between loops must be even. This would give a better upper bound, but would be harder to analyze using elementary functions.

where

$$\delta = \sum_{i=1}^{2} t_i \hat{\mathcal{H}} \left(\frac{1}{2} - \frac{pn}{2t_i} \right) + (n - t_1 - t_2) \hat{\mathcal{H}} \left(\frac{s}{n - t_1 - t_2} \right) + (n - t_1 - t_2 - s) \hat{\mathcal{H}} \left(\frac{1}{2} \right) + s$$

$$= \sum_{i=1}^{2} t_i \hat{\mathcal{H}} \left(\frac{1}{2} - \frac{pn}{2t_i} \right) + (n - t_1 - t_2) \hat{\mathcal{H}} \left(\frac{s}{n - t_1 - t_2} \right) + n - t_1 - t_2 + (1 - 1/\log_2 b)s,$$

where $\hat{\mathcal{H}} = \mathcal{H}/\log_2 b$. Note that $\hat{\mathcal{H}}(1/2) = 1/\log_2 b$. Now let $\Delta(T_1, T_2, r) = \delta(T_1 n, T_2 n, rn, n)/n$ for any n. It does not matter which n, since

$$\Delta(T_1, T_2, r) = \sum_{i=1}^{2} T_i \hat{\mathcal{H}} \left(\frac{1}{2} - \frac{p}{2T_i} \right) + (1 - T_1 - T_2) \hat{\mathcal{H}} \left(\frac{r}{1 - T_1 - T_2} \right) + 1 - T_1 - T_2 + (1 - 1/\log_2 b)r.$$

Lemma 29. $\Delta(T_1, T_2, r)$ *is maximized at* $T_1 = T_2$.

Proof. Rewriting with $T = T_1 + T_2$ and $\epsilon = T_1 - T_2$, it suffices to show that with $g(x) = x\mathcal{H}(1/2 - 1/x)$, the function $f(\epsilon) = g(x + \epsilon) + g(x - \epsilon)$ is maximized at $\epsilon = 0$. This is equivalently to g being concave down, which is a routine verification.   □

In light of Lemma 29, we now let $\Delta(T, r) = \Delta(T/2, T/2, r)$, so that

$$\Delta(T, r) = T\hat{\mathcal{H}} \left(\frac{1}{2} - \frac{p}{T} \right) + (1 - T)\hat{\mathcal{H}} \left(\frac{r}{1 - T} \right) + 1 - T + (1 - \log_b 2)r.$$

Lemma 30. $\partial \Delta / \partial r = 0$ *has the solution* $r = (1 - T)\frac{b}{b+2}$.

Proof. Note that the inverse function of the derivative $\mathcal{H}'(x) = \log_2(1-x) - \log_2 x$ is $y \mapsto \frac{1}{2^y + 1}$. Thus, we calculate

$$\frac{\partial \Delta}{\partial r} = \hat{\mathcal{H}}'(r/(1 - T)) + 1 - 1/\log_2 b = 0$$

$$\mathcal{H}'(r/(1 - T)) + \log_2 b - 1 = 0$$

$$\frac{r}{1 - T} = (\mathcal{H}')^{-1}(1 - \log_2 b) = \frac{1}{2^{1 - \log_2 b} + 1} = \frac{1}{\frac{2}{b} + 1} = \frac{b}{b + 2}$$

     □

Then

$$\lim_{n \to \infty} \frac{\log_2 N}{n} \leq \varphi(T, p) := \Delta(T, (1 - T)\frac{b}{b + 2}) =$$

$$= T\hat{\mathcal{H}} \left(\frac{1}{2} - \frac{p}{T} \right) + (1 - T)c_b$$

where

$$c_b := \left(1 + \hat{\mathcal{H}} \left(\frac{b}{b + 2} \right) + (1 - 1/\log_2 b)\frac{b}{b + 2} \right)$$

Note that $c_2 = 2$.

Lemma 31. *Fix* $0 \le p \le 1/2$ *and assume* $2p \le T \le 1$. *Then we have* $0 < \frac{\partial \varphi}{\partial T}$ *iff*

$$T < T(p) := \frac{p}{\sqrt{\frac{1}{4} - \left(\frac{2}{b(2+b)}\right)^2}} = \frac{2p}{\sqrt{1 - \left(\frac{4}{b(2+b)}\right)^2}} = \frac{4p}{\sqrt{4 - \left(\frac{8}{b(2+b)}\right)^2}} = \frac{4p}{\sqrt{3}}, \quad b = 2.$$

Let

$$L_b = \frac{2p}{T(p)} = \sqrt{1 - \left(\frac{4}{b(2+b)}\right)^2} \le 1.$$

Note that $T(p) \le 1$ iff $p \le L_b/2$, and

$$\varphi(T(p), p) = T(p)\hat{\mathcal{H}}\left(\frac{1}{2} - \frac{p}{T(p)}\right) + (1 - T(p))c_b$$

$$= \frac{p}{L_b}\hat{\mathcal{H}}\left(\frac{1}{2} - L_b\right) + (1 - \frac{p}{L_b})c_b$$

$$= c_b - \left(c_b - \hat{\mathcal{H}}\left(\frac{1}{2} - L_b\right)\right)\frac{p}{L_b} =: c_b - \alpha_b p.$$

Proof. Let $\beta(T) = \frac{1}{2} - \frac{p}{T}$. We have

$$\frac{\partial \varphi}{\partial T} = \hat{\mathcal{H}}\left(\frac{1}{2} - \frac{p}{T}\right) + T\hat{\mathcal{H}}'\left(\frac{1}{2} - \frac{p}{T}\right)\left(\frac{p}{T^2}\right) - c_b = \hat{\mathcal{H}}(\beta) + T\hat{\mathcal{H}}'(\beta)\left(\frac{p}{T^2}\right) - c_b$$

$$= -\beta \log_b \beta - (1 - \beta)\log_b(1 - \beta) + (\log_b(1 - \beta) - \log_b \beta)(p/T) - c_b$$

$$= -\beta \log_b \beta - (1 - \beta)\log_b(1 - \beta) + (\log_b(1 - \beta) - \log_b \beta)(1/2 - \beta) - c_b$$

Now $b^0 < b^{\partial \varphi / \partial T}$ iff

$$1 < \beta^{-\beta}(1 - \beta)^{-(1-\beta)}((1 - \beta)/\beta)^{1/2 - \beta}b^{-c_b}$$

$$1 > \beta^\beta (1 - \beta)^{(1-\beta)}\left(\frac{\beta}{1 - \beta}\right)^{1/2 - \beta}b^{c_b} = \beta^{1/2}(1 - \beta)^{1/2}b^{c_b}$$

$$1 > \beta(1 - \beta)b^{2c_b}, \quad 0 \le \beta \le 1/2$$

giving

$$\beta < \frac{1 - \sqrt{1 - 4b^{-2c_b}}}{2} = \frac{1 - \sqrt{3/4}}{2}, \quad b = 2$$

$$p/T = 1/2 - \beta > \frac{\sqrt{1 - 4b^{-2c_b}}}{2} = \frac{\sqrt{3/4}}{2}, \quad b = 2$$

$$T < \frac{2}{\sqrt{1 - 4b^{-2c_b}}}p = \frac{4}{\sqrt{3}}p, \quad b = 2$$

Note that $b \ge 2$ and $4b^{-2c_b} = \frac{2^4}{b^2(b+2)^2} = \left(\frac{4}{b(b+2)}\right)^2$ give $1 - 4b^{-2c_b} > 0$, as required. □

Hence

$$\lim_{n\to\infty} \frac{\log_2 N}{n} \leq \psi(p) := \varphi\left(\min\{1, T(p)\}, p\right)$$

$$= \begin{cases} \varphi(1,p) = \hat{\mathcal{H}}(1/2 - p), & p \geq L_b/2; \\ \varphi(T(p),p) = c_b - \alpha_b p, & p \leq L_b/2. \end{cases}$$

Consequently $\tilde{h}(\psi(p)) \leq p$. Note that $\psi = u^{-1}$. $\square$

As Theorem 28 shows, the largest number of paths is obtained by going *fairly* straight to the loop state; spending half the time looping and half the time meandering; and then finally going equally fairly straight to the start state. The optimal value of r obtained shows that half of the time between first reaching the loop state and finally leaving the loop state should be spent looping.

References

1. Alikhani, M.: American option pricing and optimal stopping for success runs. Master's thesis, University of Hawaii at Manoa, USA (2013)
2. Kjos-Hanssen, B.: Structure function for run complexity (2014). http://kjos-hanssen.appspot.com/structure-function
3. Hyde, K.: Nondeterministic finite state complexity. Master's thesis, University of Hawaii at Manoa, USA (2013)
4. Hyde, K.K., Kjos-Hanssen, B.: Nondeterministic automatic complexity of almost square-free and strongly cube-free words. In: Cai, Z., Zelikovsky, A., Bourgeois, A. (eds.) COCOON 2014. LNCS, vol. 8591, pp. 61–70. Springer, Heidelberg (2014)
5. Schilling, M.F.: The longest run of heads. Coll. Math. J. **21**(3), 196–207 (1990)
6. Shallit, J., Wang, M.-W.: Automatic complexity of strings. J. Autom. Lang. Comb. 6(4), 537–554 (2001). 2nd Workshop on Descriptional Complexity of Automata, Grammars and Related Structures (London, ON) (2000)
7. Staiger, L.: The Kolmogorov complexity of infinite words. Theor. Comput. Sci. **383**(2–3), 187–199 (2007)
8. Vereshchagin, K.N., Vitányi, P.M.B.: Kolmogorov's structure functions and model selection. IEEE Trans. Inform. Theory **50**(12), 3265–3290 (2004)

Improved Even Order Magic Square
Construction Algorithms and Their Applications

Zhenhua Duan, Jin Liu, Jie Li, and Cong Tian[(⊠)]

ICTT and ISN Laboratory, Xidian University,
Xi'an 710071, People's Republic of China
zhhduan@mail.xidian.edu.cn, janeleexd@gmail.com,
{liujin_xd,tico_tools}@163.com

Abstract. This paper presents improved even order magic square construction algorithms, including both single even order magic square and double even order magic square construction algorithms. Further, in order to show how the algorithms work, two specific magic squares are constructed. Moreover, the analysis of the algorithms is given. Finally, the improved even order magic square construction algorithms are applied in secure communication and authentication areas for multi-user shared account in detail.

Keywords: Even order magic squares · Algorithm · Identity authentication · Cryptography

1 Introduction

An $n \times n$ magic square is an arrangement of n^2 distinct numbers, $1, 2, 3, \cdots, n^2$ (each number is used once), in a square matrix, where the sums of the numbers in each row, each column, and the forward and backward main diagonals are equal. This sum is normally called the magic constant. According to whether size n is an even or odd number, a magic square can be called an even order or odd order one. Further, the even order magic squares can be classified into double even order and single even order ones depending on whether or not n can be divisible by 4.

There are many well-known odd order magic square construction algorithms, such as de la Loubere's algorithm [1,2] and the lozenge method [3]. However, there are just a few for specific even order ones. One is based on a mathematical game called medjig, which can be used to construct even order magic squares. Though it is much difficult to construct single even magic squares, there exist several methods, including the John Horton Conway's LUX method [4] and the Strachey method [5]. Another algorithm presented in [6], makes use of two

The research is supported by the National Program on Key Basic Research Project of China (973 Program) Grant No. 2010CB328102, NSFC under Grant Nos. 61133001, 61322202 and 61420106004.

© Springer International Publishing Switzerland 2014
Z. Zhang et al. (Eds.): COCOA 2014, LNCS 8881, pp. 666–680, 2014.
DOI: 10.1007/978-3-319-12691-3_50

essential ideas: the way to fill in the matrix with consecutive positive integers and the rule to select and swap some elements. Different combinations of swapped elements can result in different magic squares in the same scale.

However, there are some problems for the existing magic square construction algorithms. For instance, they are complicated, and for a specified scale, the number of constructed magic squares is not as large as we expect. Based on the algorithm presented in [6], we improve it by introducing some new rules to select and swap elements in the forward and backward main diagonals of a matrix. Thus, our algorithms are not only easy in the filling of the magic squares, but can also generate a great number of different magic squares in the same scale. Moreover, our algorithms have low complexities of time and space, which are in fact $O(n^2)$ where n is the scale of squares.

Magic squares have fascinated humanity throughout ages. During the latter part of the 19th century, mathematicians applied magic squares to problems in probability and analysis. Today, magic squares are studied in relation to factor analysis [7], combinatorial mathematics [8], matrices [9–11], and geometry [12]. In consideration of the huge number of magic squares and the difficulty to construct them, magic squares can be applied to graph theory [8] and cryptography area, such as digital image encryption [13,14] and image authentication [15] in contemporary society. This paper applies magic square construction algorithms in secure communication [16] and authentication [17] areas for multi-user shared account, which involves electronic account key distribution and identity authentication.

The paper is organized as follows. In Sect. 2, improved even order construction algorithms are described in detail, including both single and double even order magic square construction algorithms. Section 3 gives a specific instance to show how the improved even order construction algorithms can be used in secure communication and authentication areas for multi-user shared account. Finally, we discuss related work and draw conclusions in Sect. 4.

2   Improved Algorithm

In this section, algorithms of constructing even order magic squares are presented. Based on the algorithm presented in [6], we improve it by adding some new rules to select and swap diagonal elements in the matrix, which can generate much more magic squares in the same scale. In the improved algorithms, firstly an $n \times n$ matrix with the number sequence $1, 2, 3, ...n^2$ is filled in, and then, some elements are selected to swap. The swapping rules take the case that both diagonal elements are selected into account, so that the number of combinations of selected elements is increased in each line, thus leading to a dramatically growth of the quantity of magic squares in the same scale.

2.1   Single Even Order Magic Square Construction Algorithm

Let N be the set of all positive integers and $x \leftrightarrow y$ operation for swapping x and y. The algorithm of constructing single even order magic squares is shown below.

Algorithm 1. SINGLE EVEN ORDER MAGIC SQUARE

Require: a $(4m+2) \times (4m+2)$ array A, $m \in N$;
Ensure: a $(4m+2) \times (4m+2)$ magic square array A_2;

$\{$1-1. Filling $1, 2, \cdots, (4m+2)^2$ in $A\}$

1: **for** $i = 1$ to $2m+2$ **do**
2:    **for** $j = 1$ to $4m+2$ **do**
3:       **if** $i \% 2 \neq 0$ **then**
4:          $A[i,j] = (i-1)(4m+2) + j$
5:       **else**
6:          $A[i,j] = (i-1)(4m+2) + 4m + 3 - j$
7:       **end if**
8:    **end for**
9: **end for**
10: **for** $i = 2m+3$ to $4m+2$ **do**
11:    **for** $j = 1$ to $4m+2$ **do**
12:       **if** $i \% 2 \neq 0$ **then**
13:          $A[i,j] = (i-1)(4m+2) + 4m + 3 - j$
14:       **else**
15:          $A[i,j] = (i-1)(4m+2) + j$
16:       **end if**
17:    **end for**
18: **end for**

$\{$1-2. Rules to swap elements in some rows of $A\}$

19: $A[2m+2, 2m+1] \leftrightarrow A[2m+2, 2m+2]$
20: Select one ascending row denoted as $Rise$
    $\{Rise \neq 2m+1, A[Rise, i] < A[Rise, i+1], 1 \leq i \leq 4m+1\}$
    $A[Rise, 2m+1] \leftrightarrow A[Rise, 2m+2]$
21: Select a random number from $\{1, 2, \cdots, 2m\}$ denoted as $Rand$
    $A[2m+1, Rand] \leftrightarrow A[2m+1, 4m+3-Rand]$
22: Select one descending row denoted as Dec
    $\{Dec \neq 2m+2,\ Rand,\ \text{or}\ 4m+3-Rand;\ A[Dec, i+1] < A[Dec, i],$
    $1 \leq i \leq 4m+1\}$
    $A[Dec, Rand] \leftrightarrow A[Dec, 4m+3-Rand]$
23: The current A is denoted as A_1

$\{$1-3. Rules to swap elements in some columns of $A_1\}$
$\{$1-3.1. Swapping elements of the $(2m+1)^{th}$ row and the $(2m+2)^{th}$ row$\}$

24: $A_1[2m+1, 2m+1] \leftrightarrow A_1[2m+2, 2m+1]$
25: $A_1[2m+1, 2m+2] \leftrightarrow A_1[2m+2, 2m+2]$
26: either $A_1[2m+1, Rand] \leftrightarrow A_1[2m+2, Rand]$,
    or $A_1[2m+1, 4m+3-Rand] \leftrightarrow A_1[2m+2, 4m+3-Rand]$;
27: Select $m-1$ off-diagonal elements randomly,
    for each selected element $A[2m+1, j_t], j_t \in \{1, 2, \cdots, 4m+2\}, t \in \{1, 2, \cdots, m-1\}$

$A_1[2m + 1, j_t] \leftrightarrow A_1[2m + 2, 4m + 3 - j_t]$, and
$A_1[2m + 1, 4m + 3 - j_t] \leftrightarrow A_1[2m + 2, j_t]$;

{1-3.2. Swapping elements in the $Rise^{th}$ row of A_1}

28: $A_1[Rise, 2m + 1] \leftrightarrow A_1[4m + 3 - Rise, 2m + 1]$
29: $A_1[Rise, 2m + 2] \leftrightarrow A_1[4m + 3 - Rise, 2m + 2]$
30: Either select $2m - 1$ off-diagonal elements, or 2 diagonal ones and $2m - 3$ off-diagonal ones. For each selected element $A[Rise, j_t]$, $j_t \in \{1, 2, \cdots, 4m + 2\}$, $t \in \{1, 2, \cdots, 2m - 1\}$
   $A_1[Rise, j_t] \leftrightarrow A_1[4m + 3 - Rise, j_t]$

{1-3.3. Swapping elements in the Dec^{th} row of A_1}

31: $A_1[Dec, Rand] \leftrightarrow A_1[4m + 3 - Dec, Rand]$
32: $A_1[Dec, 4m + 3 - Rand] \leftrightarrow A_1[4m + 3 - Dec, 4m + 3 - Rand]$
33: Either select $2m - 1$ off-diagonal elements, or 2 diagonal ones and $2m - 3$ off-diagonal ones. For each selected element $A[Dec, j_t]$, $j_t \in \{1, 2, \cdots, 4m + 2\}$, $t \in \{1, 2, \cdots, 2m - 1\}$
   $A_1[Dec, j_t] \leftrightarrow A_1[4m + 3 - Dec, j_t]$

{1-3.4. Swapping elements in the rows which have not been modified}

34: **for** $i = 1$ to $2m$ **do**
35:    **if** $i \neq Rise$ and $i \neq Dec$ **then**
36:       Either select $2m + 1$ off-diagonal elements, or 2 diagonal ones and $2m - 1$ off-diagonal ones. For each selected element $A[i, j_t]$, $j_t \in \{1, 2, \cdots, 4m + 2\}$, $t \in \{1, 2, \cdots, 2m + 1\}$
       $A_1[i, j_t] \leftrightarrow A_1[4m + 3 - i, j_t]$
37:    **end if**
38: **end for**
39: The current A_1 is denoted as A_2
40: **return** A_2

Now we illustrate Algorithm 1 with a specific example. Let m be 2, i.e. a 10×10 array A. Based on A, we can construct a single even order magic square by Algorithm 1 step by step as shown below.

1. Fill in A shown in Fig. 1(1).
2. Suppose $Rise = 1, Rand = 3, Dec = 4$. Then according to Rule 1-2 in Algorithm 1, swap $A[6, 5]$ and $A[6, 6]$ (Fig. 1(2)), $A[1, 5]$ and $A[1, 6]$ (Fig. 1(3)), $A[5, 3]$ and $A[5, 8]$ (Fig. 2(4)), $A[4, 3]$ and $A[4, 8]$ (Fig. 2(5)) respectively.
3.1. Swap $A[5, 5]$ and $A[6, 5]$, $A[5, 6]$ and $A[6, 6]$ first (Fig. 2(6)). As to line 26 in Algorithm 1, we swap $A[5, 3]$ and $A[6, 3]$ (Fig. 2(7)). Since $m - 1 = 2 - 1 = 1$, we choose and swap $A[5, 1]$ and $A[6, 10]$, $A[6, 1]$ and $A[5, 10]$ (Fig. 2(8)).
3.2. Modify the $Rise^{th}$ row. Swap $A[1, 5]$ and $A[10, 5]$, $A[1, 6]$ and $A[10, 6]$ first (Fig. 2(9)). Since $2m - 1 = 4 - 1 = 3$, we choose and swap $A[1, 1]$ and $A[10, 1]$, $A[1, 2]$ and $A[10, 2]$, $A[1, 10]$ and $A[10, 10]$ (Fig. 2(10)).
3.3. Modify the Dec^{th} row. Swap $A[4, 3]$ and $A[7, 3]$, $A[4, 8]$ and $A[7, 8]$ first (Fig. 2(11)). Since $2m - 1 = 3$, we choose and swap $A[4, 4]$ and $A[7, 4]$, $A[4, 5]$ and $A[7, 5]$, $A[4, 7]$ and $A[7, 7]$ (Fig. 2(12)).
3.4. Modify the rows which do not be changed, i.e. the 2nd and 3rd rows. The elements needed to be swapped can be $A[2, 4]$, $A[2, 5]$, $A[2, 6]$, $A[2, 7]$ and

$A[2,8]$ in the 2nd row (Fig. 2(13)), while $A[3,3]$, $A[3,4]$, $A[3,5]$, $A[3,8]$ and $A[3,9]$ in the 3rd row (Fig. 2(14)).

1	2	3	4	5	6	7	8	9	10
20	19	18	17	16	15	14	13	12	11
21	22	23	24	25	26	27	28	29	30
40	39	38	37	36	35	34	33	32	31
41	42	43	44	45	46	47	48	49	50
60	59	58	57	56	55	54	53	52	51
70	69	68	67	66	65	64	63	62	61
71	72	73	74	75	76	77	78	79	80
90	89	88	87	86	85	84	83	82	81
91	92	93	94	95	96	97	98	99	100

(1)

1	2	3	4	5	6	7	8	9	10
20	19	18	17	16	15	14	13	12	11
21	22	23	24	25	26	27	28	29	30
40	39	38	37	36	35	34	33	32	31
41	42	43	44	45	46	47	48	49	50
60	59	58	57	*55*	*56*	54	53	52	51
70	69	68	67	66	65	64	63	62	61
71	72	73	74	75	76	77	78	79	80
90	89	88	87	86	85	84	83	82	81
91	92	93	94	95	96	97	98	99	100

(2)

1	2	3	4	6	5	7	8	9	10
20	19	18	17	16	15	14	13	12	11
21	22	23	24	25	26	27	28	29	30
40	39	38	37	36	35	34	33	32	31
41	42	43	44	45	46	47	48	49	50
60	59	58	57	*55*	*56*	54	53	52	51
70	69	68	67	66	65	64	63	62	61
71	72	73	74	75	76	77	78	79	80
90	89	88	87	86	85	84	83	82	81
91	92	93	94	95	96	97	98	99	100

(3)

Fig. 1. An example of 10×10 magic square-1

The analysis of Algorithm 1:

Filling $1, 2, \cdots, (4m+2)^2$ in A needs $O(n^2)$ time and space, where $n = 4m+2$.

In the selecting and swapping steps, the analysis is given as follows. In the $(2m + 1)^{th}$ row, we need to select $m - 1$ elements which satisfy the condition that $1 \leq j_t \leq 2m$, where j_t is a column number. Consequently there are $2C_{2m}^{m-1}$ selections.

There are $2m$ different choices to select the $Rise^{th}$ row. For each $Rise$, there are C_{4m}^{2m-1} choices if 2 diagonal elements are not selected; otherwise, C_{4m}^{2m-3} choices. Namely, $2m(C_{4m}^{2m-1} + C_{4m}^{2m-3})$ choices in total.

There are $2m$ different choices to select the Dec^{th} row and $2m - 1$ different choices to select the $Rand^{th}$ column. For each pair of Dec and $Rand$, there are C_{4m}^{2m-1} choices if 2 diagonal elements are not selected; otherwise, C_{4m}^{2m-3} choices. That is, $2m(2m - 1)(C_{4m}^{2m-1} + C_{4m}^{2m-3})$ choices in total.

When $m = 1$, i.e. a 6×6 matrix, diagonal elements can not be swapped in each row, consequently the total number computed by Algorithm 1 is:

$$(2m)^2(2m - 1)(C_{4m}^{2m-1})^2(2C_{2m}^{m-1}) \tag{1}$$

When $m \geq 2$, and for any i, $1 \leq i \leq 2m$, $i \neq Rise$ and $i \neq Dec$, there are C_{4m}^{2m+1} if diagonal elements are not selected; otherwise, C_{4m}^{2m-1} choices. Since $2m - 2$ rows are to be adjusted, there are totally $(C_{4m}^{2m+1} + C_{4m}^{2m-1})^{2m-2}$ choices.

As a result, after the algorithm being improved the number of magic square in a $(4m + 2) \times (4m + 2)$ scale can reach up to:

$$(2m)^2(2m - 1)(C_{4m}^{2m-1} + C_{4m}^{2m-3})^2(C_{4m}^{2m+1} + C_{4m}^{2m-1})^{2m-2}(2C_{2m}^{m-1}) \tag{2}$$

Notice that: $C_{4m}^{2m+1} = C_{4m}^{2m-1}$.

According to the analysis above, a conclusion can be drawn that both the time complexity and space complexity are $O(n^2)$, where $n = 4m + 2$.

Due to the limitation of space, the detailed proof is omitted.

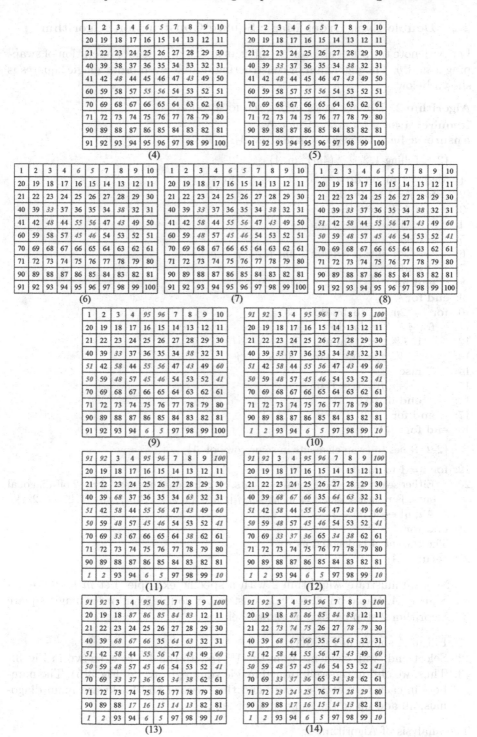

Fig. 2. An example of 10 × 10 magic square-2

2.2   Double Even Order Magic Square Construction Algorithm

Let N denote the set of all positive integers and $x \leftrightarrow y$ denote operation of swapping x and y. The algorithm of constructing double even order magic squares is shown below.

Algorithm 2. DOUBLE EVEN ORDER MAGIC SQUARE
Require: a $4m \times 4m$ array A, $m \in N$;
Ensure: a $4m \times 4m$ magic square array A_1;

   $\{$2-1. Filling $1, 2, \cdots, (4m)^2$ in $A\}$

1: **for** $i = 1$ to $2m$ **do**
2:    **for** $j = 1$ to $4m$ **do**
3:       **if** $i \% 2 \neq 0$ **then**
4:          $A[i, j] = (i - 1)4m + j$
5:       **else**
6:          $A[i, j] = (i - 1)4m + 4m + 1 - j$
7:       **end if**
8:    **end for**
9: **end for**
10: **for** $i = 2m + 1$ to $4m$ **do**
11:    **for** $j = 1$ to $4m$ **do**
12:       **if** $i \% 2 \neq 0$ **then**
13:          $A[i, j] = (i - 1)4m + 4m + 1 - j$
14:       **else**
15:          $A[i, j] = (i - 1)4m + j$
16:       **end if**
17:    **end for**
18: **end for**

   $\{$2-2. Rules to swap elements in some rows of $A\}$

19: **for** $i = 1$ to $2m$ **do**
20:    Either select $2m$ off-diagonal elements, or 2 diagonal ones and $2m-2$ off-diagonal ones. For each selected element $A[i, j_t]$, $j_t \in \{1, 2, \cdots, 4m\}$, $t \in \{1, 2, \cdots, 2m\}$
        $A[i, j_t] \leftrightarrow A[4m + 1 - i, j_t]$
21: **end for**
22: The current A is denoted as A_1
23: **return** A_1

Now we illustrate Algorithm 2 with a specific example. Let m be 2, i.e. an 8×8 array A. Based on A, we can construct a double even order magic square by Algorithm 2 step by step as shown below.

1. Fill in A shown in Fig. 3(1).
2. Select and swap elements from the 1st row to the 4th row, shown in Fig. 3.
3. Thus, we obtain the well formed magic square array, i.e. Fig. 3(5). The numbers in each row, each column, and the forward and backward main diagonals, all add up to the same number 260.

The analysis of Algorithm 2:
   Filling $1, 2, \cdots, (4m)^2$ in A needs $O(n^2)$ time and space, where $n = 4m$.

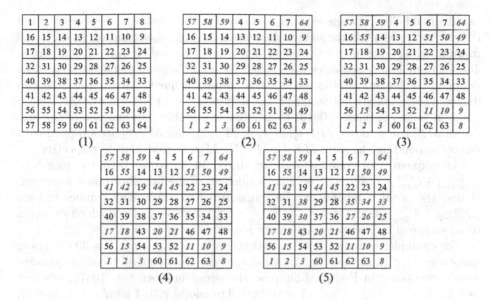

Fig. 3. An example of 8 × 8 magic square

In the selecting and swapping steps, the analysis is given as follows. If diagonal elements are not selected, there are C_{4m-2}^{2m} choices for each row; otherwise, C_{4m-2}^{2m-2} choices. Accordingly, $C_{4m-2}^{2m} + C_{4m-2}^{2m-2}$ choices for each row in total. Considering there are $2m$ rows to be adjusted, the number of magic squares with a size of $4m \times 4m$ generated by Algorithm 2 can reach up to:

$$\left(C_{4m-2}^{2m} + C_{4m-2}^{2m-2}\right)^{2m} \tag{3}$$

Notice that: $(2m) + (2m - 2) = 4m - 2$ and $C_{4m-2}^{2m} = C_{4m-2}^{2m-2}$. Based on this, we have:

$$\left(C_{4m-2}^{2m} + C_{4m-2}^{2m-2}\right)^{2m} = \left(C_{4m-2}^{2m} + C_{4m-2}^{2m}\right)^{2m} = 2^{2m}\left(C_{4m-2}^{2m}\right)^{2m} \tag{4}$$

According to the analysis above, a conclusion can be drawn that both the time complexity and space complexity are $O(n^2)$, where $n = 4m$.

Due to the limitation of space, the detailed proof is omitted.

2.3 Serial Number of Double Even Order Magic Squares

In this subsection, we present a corresponding relation between serial numbers and double even order magic squares. As for a $4m \times 4m$ matrix filled in with the consecutive positive integers according to step 1 of Algorithm 2, adjustment for each row can be partitioned into 2 cases. One is to swap arbitrary $2m$ off-diagonal elements, i.e. C_{4m-2}^{2m} choices. The other is to swap 2 diagonal elements first and then arbitrary $2m - 2$ off-diagonal elements, i.e. C_{4m-2}^{2m-2} choices. We use a $(4m - 2)$-digit binary sequence to represent each combination, where the two diagonal elements are omitted. 1 stands for the off-diagonal element in the ith row swapped with the one in the $(4m + 3 - i)^{th}$ row residing in the same column, while 0 stands for the one which is not changed. If $2m - 2$ numbers in $4m - 2$-digit binary sequence are 1, then the two diagonal elements need to be swapped. If $2m$ numbers in $4m - 2$-digit binary sequence are 1, then the two diagonal elements are not swapped. For instance, for an 8×8 matrix M, we use a 6-digit binary sequence 000101 to denote one possible combination for the first row of M. That is to say, $M[1,1]$, $M[1,5]$, $M[1,7]$, and $M[1,8]$ are swapped with the corresponding elements $M[8,1]$, $M[8,5]$, $M[8,7]$, and $M[8,8]$ respectively.

All sequences of combinations are stored in an array with the length of $(C_{4m-2}^{2m} + C_{4m-2}^{2m-2})$. Actually, only $2m$ sequences of the first $2m$ rows are enough. Hence, the serial number r can be transformed to a $2m$-digit number in base $(C_{4m-2}^{2m} + C_{4m-2}^{2m-2})$, and $0 \leq r \leq (C_{4m-2}^{2m} + C_{4m-2}^{2m-2})^{2m} - 1$. Then which elements to be swapped by accessing the binary sequence array is clear.

For example, as for an 8×8 matrix, there are $C_{8-2}^4 + C_{8-2}^2 = 30$ swapping sequences, that is, the base is 30. Define an array E storing all binary sequences, which are shown in Fig. 4(1). Suppose the serial number $r = 10379$, where r satisfies $0 \leq r \leq (C_6^4 + C_6^2)^4 - 1 = 809999$. Transform r to 4 numbers in base 30, and store them in an array $num[4]$, i.e. $num[0] = 29$, $num[1] = 15$, $num[2] = 14$ and $num[3] = 0$. Next access E according to $num[i]$, where $0 \leq i \leq 3$ and determine the off-diagonal elements to be swapped in the $(i + 1)^{th}$ row. Finally we can obtain a boolean matrix meaning which elements need to be swapped in the whole matrix, i.e. Fig. 4(2). Figure 4(3) shows the matrix filled according to step 1 of Algorithm 2. After swapping the specific elements according to Fig. 4(2), we can obtain the magic square, as shown in Fig. 4(4).

i	0	1	2	3	4	5	6	7	8	9
$E[i]$	001111	010111	011011	011101	011110	100111	101011	101101	101110	110011
i	10	11	12	13	14	15	16	17	18	19
$E[i]$	110101	110110	111001	111010	111100	000011	000101	000110	001001	001010
i	20	21	22	23	24	25	26	27	28	29
$E[i]$	001100	010001	010010	010100	011000	100001	100010	100100	101000	110000

(1) The binary sequence array E

1	2	3	4	5	6	7	8
16	15	14	13	12	11	10	9
17	18	19	20	21	22	23	24
32	31	30	29	28	27	26	25
40	39	38	37	36	35	34	33
41	42	43	44	45	46	47	48
56	55	54	53	52	51	50	49
57	58	59	60	61	62	63	64

(3) The initialized matrix

57	58	59	4	5	6	7	64
16	55	14	13	12	51	50	49
41	42	19	44	45	22	23	24
32	31	38	29	28	35	34	33
40	39	30	37	36	27	26	25
17	18	43	20	21	46	47	48
56	15	54	53	52	11	10	9
1	2	3	60	61	62	63	8

(4) The 8×8 magic square

1	1	1	0	0	0	0	1
0	1	0	0	0	1	1	1
1	1	0	1	1	0	0	0
0	0	1	0	0	1	1	1

(2) A boolean matrix

Fig. 4. Relation between serial numbers and 8×8 magic squares

3   Application

With the development of e-bank and e-commerce, online bank provides convenient service for commercial activities. The security is an utmost important issue. Fraud making use of the internet brings severe loss to users. In order to solve this problem, we propose a safe and convenient method of multi-user shared electronic account based on magic squares, including a key distribution method and an identity authentication method consisting of multi-user identity authentication method and single-user identity authentication method.

Multi-user shared electronic account is a kind of electronic account managed by multiple users together, usually 2 to 5 users, aimed at solving the trust relationship among individuals and achieving the sharing of bank savings. The identity authentication information of all users is needed if the electronic account needs to be modified, while each of users own identity authentication information is required if the account needs only to be checked.

Each electronic account in the server database consists of an account number, an $n \times n$ complementary magic square and the information of all users which includes the username and an account boolean matrix obtained by a hash operation of the account password; while each user holds the account number, his/her own username, the account password, a key password and an electronic key.

In what follows, we use an 8×8 even order magic square to distribute keys for two users, Bob and Jenny, and simulate the process of identity authentication.

3.1   Magic Square Based Key Distribution Method

With our approach, key distribution is based on magic squares. Its procedure is illustrated in detail below.

1. The server allocates an account number for an electronic account. Each user is asked to input his/her own username, account password and key password, both consisting of 6 digits.

    The information of Bob and Jenny is as follows:

    $$Bob, \quad 065432, \quad 023456$$
    $$Jenny, \quad 987654, \quad 456789$$

2. The server can obtain a random number as the magic square serial number after a hash operation according to usernames and account passwords of all users with the time to apply for the account. Here, we assume that the serial number $r = 10379$.

3. According to the serial number r, an 8×8 magic square A is constructed by Algorithm 2, as shown in Fig. 5(a).

4. For each user, we use his/her account password to construct an account boolean matrix. First, the 6-digit account password is hashed to n different numbers by n hash functions which are distributed from 0 to $2^n - 1$. Then these numbers are transformed to an n-digit binary boolean sequence

and are filled in an $n \times n$ matrix in sequential order. Thus, the construction of an $n \times n$ account boolean matrix is finished.

Here we use the following hash function:

$$Account\_Hash(k) = \lfloor 2^8 \times ((k \times 0.6180339887) mod\ 1) \rfloor, \qquad (5)$$

where k is the user's account password.

The hash function for Bob is:

$$acc\_out\_Bob[1] = \lfloor 2^8 \times ((123456 \times 0.6180339887) mod\ 1) \rfloor \qquad (6)$$

$$acc\_out\_Bob[i] = \lfloor 2^8 \times ((acc\_out\_Bob[i-1] \times 0.6180339887) mod\ 1) \rfloor \qquad (7)$$

where $2 \leq i \leq 8$. Accordingly, we can obtain 8 numbers which are distributed from 0 to 255. It is similar for Jenny. Next, it is easy to get the account boolean matrixes for Bob and Jenny, as shown in Fig. 5(b) and (c) respectively.

5. We first construct a magic square fragment for each user according to his/her account boolean matrix. The account boolean matrix, the magic square fragment and the magic square are matrixes in the same size. Then we scan the account boolean matrix from left to right and top to bottom. If the number is 1, we put the number at the same position as where 1 is in A to the corresponding position in magic square fragment; otherwise, we fill in the same position with a random number, which values from 1 to n^2. In this example, random numbers are all 5 for the convenience of explanation and understanding for observation, as shown in Fig. 5(d) and (e) respectively.

6. We add all users' magic square fragments and obtain a new $n \times n$ matrix. The n^2 elements modulo n^2 can form a matrix B. Let complementary magic square $C = A - B$. C is stored in the server database while A is deleted, as shown in Fig. 6(1). Therefore, nobody can speculate or guess the constitution of the magic square.

7. We construct a key boolean matrix for each user according to his/her key password. We choose another group of n hash functions and hash the 6-digit key password to n different numbers, which are distributed from 0 to $2^n - 1$. Then we transform these numbers to an n-digit binary boolean sequence and fill in a new $n \times n$ matrix, which is exactly the key boolean matrix. In fact, the hash functions differ from the ones in step 4. Here, we use the same hash functions for simplification.

8. We encrypt the magic square fragment with the key boolean matrix for each user. First we define a new $n \times n$ matrix and then scan the key boolean matrix from left to right and top to bottom. If the element is 1, we extract the number from the magic square fragment at the same position and store it in the new matrix in sequential order. After the first traversal, we scan the key boolean matrix for the second time. If the element is 0, we extract the above information in the similar way. Thus, the new encrypted matrix is the electronic key and is allocated to the user, as shown in Fig. 6(2).

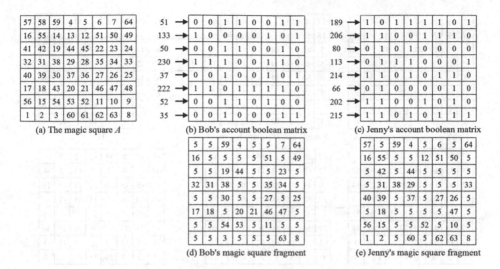

(a) The magic square A

57	58	59	4	5	6	7	64
16	55	14	13	12	51	50	49
41	42	19	44	45	22	23	24
32	31	38	29	28	35	34	33
40	39	30	37	36	27	26	25
17	18	43	20	21	46	47	48
56	15	54	53	52	11	10	9
1	2	3	60	61	62	63	8

(b) Bob's account boolean matrix

	0	0	1	1	0	0	1	1
51 →	0	0	1	1	0	0	1	1
133 →	1	0	0	0	0	1	0	1
50 →	0	0	1	1	0	0	1	0
230 →	1	1	1	0	0	0	1	0
37 →	0	0	1	0	0	1	0	1
222 →	1	1	0	1	1	1	1	0
52 →	0	0	1	1	0	1	0	0
35 →	0	0	1	0	0	0	1	1

(c) Jenny's account boolean matrix

189 →	1	0	1	1	1	1	0	1
206 →	1	1	0	0	1	1	1	0
80 →	0	1	0	1	0	0	0	0
113 →	0	1	1	1	0	0	0	1
214 →	1	1	0	1	0	1	1	0
66 →	0	1	0	0	0	0	1	0
202 →	1	1	0	0	1	0	1	0
215 →	1	1	0	1	0	1	1	1

(d) Bob's magic square fragment

5	5	59	4	5	5	7	64
16	5	5	5	5	51	5	49
5	5	19	44	5	5	23	5
32	31	38	5	5	35	34	5
5	5	30	5	5	27	5	25
17	18	5	20	21	46	47	5
5	5	54	53	5	11	5	5
5	5	3	5	5	5	63	8

(e) Jenny's magic square fragment

57	5	59	4	5	6	5	64
16	55	5	5	12	51	50	5
5	42	5	44	5	5	5	5
5	31	38	29	5	5	5	33
40	39	5	37	5	27	26	5
5	18	5	5	5	5	47	5
56	15	5	5	52	5	10	5
1	2	5	60	5	62	63	8

Fig. 5. Magic square based key distribution-1

3.2 Multi-user Identity Authentication Method

With our approach, multi-user identity authentication is based on magic squares. Its procedure is illustrated in detail below.

1. Each time to modify an account, all relevant users negotiate the specific details, such as the amount. Within a period of time, all users (Bob and Jenny) input their own account passwords, key passwords and electronic keys respectively.

$$Bob, \quad 065432, \quad 023456$$
$$Jenny, \quad 987654, \quad 456789$$

2. At each user terminal, a hash operation is made towards his/her account password and the hashed result is obtained, which is only a part of the account boolean matrix, such as the first x rows where $1 \le x < n$. Note that the hash operation and hash functions must be the same as the ones in step 4 of Sect. 3.1. Let x be 1 in order for convenience to explain.

$$Bob, \quad 00110011$$
$$Jenny, \quad 00110011$$

3. We construct a key boolean matrix for each user with his/her key password. The process must be the same as that in step 7 of Sect. 3.1. Then we decrypt the electronic key with the key boolean matrix, which is an inverse process with the encryption. We define a new $n \times n$ matrix and then scan the key boolean matrix from left to right and top to bottom. If the element is 1, we extract the current first number from the electronic key and put it to the

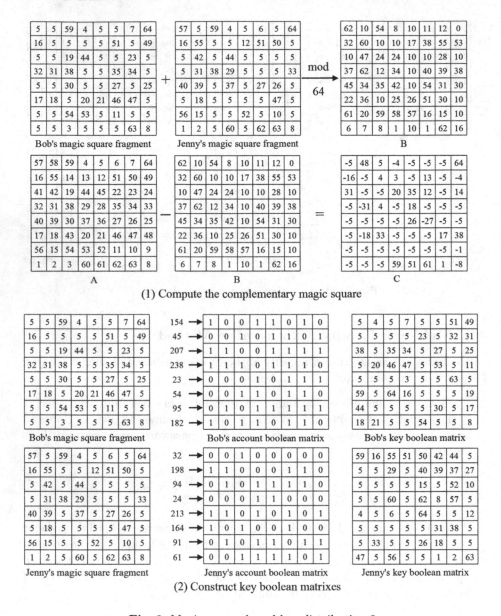

Fig. 6. Magic square based key distribution-2

same position of the new matrix. After the first traversal, we scan the key boolean matrix for the second time. If the element is 0, then we extract the above information in the similar way. Thus, the new decrypted matrix is the same as the unencrypted magic square fragment.

4. Each user sends his/her username, partial hashed result from account password and decrypted magic square fragment to the bank server via the internet.

5. The server compares the partial hashed results of all users with the account boolean matrixes stored in the database. If successfully matched, turn to step 6; otherwise, multi-user identity authentication fails and turn to step 7. In our example, we can easily know that information of Bob and Jenny matches successfully.
6. We add all users' magic square fragments and obtain a new $n \times n$ matrix. The 64 elements modulo 64 form a matrix D. Let matrix $E = C + D$. If E is a magic square, modification operations are allowed; otherwise, the failing information is reminded to each user. Here we can easily obtain that E is a magic square and $D = B$, $E = A$. Then Bob and Jenny can modify their account successfully.
7. The procedure of multi-user identity authentication is finished.

3.3 Single-User Identity Authentication Method

The procedure is similarly to that of multi-user identity authentication method. The difference is that there is only one user and the only thing needs to do for the server is to compare the received partial hashed result with stored information. No other operations are needed. If the required information is successfully matched, the user can have a view on the account information.

4 Conclusion

The construction of magic squares is an ancient mathematical problem. It is easy to construct an odd order magic square with some simple construction algorithms. For an even order one, it is usually hard to build. The algorithm presented in [6] does not take diagonal elements into account. This paper improves the original algorithm given in [6] by considering diagonal elements. There can be a great number of magic squares constructed by the new proposed algorithms. For example, there are 810000 8×8 magic squares. The number of magic squares increases dramatically as the size of the square increases. Moreover the time and space complexity are both $O(n^2)$, where n is the size of magic squares. Thus, the algorithms can be well applied in secure communication and authentication areas. In the future, we will further investigate the application of magic squares in secure communication, encryption and image authentication areas.

References

1. Weisstein, E.W.: CRC Concise Encyclopedia of Mathematics. Chapman and Hall/CRC, New York (2002)
2. Pickover, C.A.: The Zen of Magic Squares, Circles, and Stars, p. 38. Princeton University Press, United Kindom (2002)
3. Guy, R.K.: Unsolved Problems in Number Theory. Springer, New York (1997)
4. Erickson, M.: Aha! Solutions, MAA Spectrum, Mathematical Association of America (2009)

5. Rouse Ball, W.W.: Mathematical Recreations and Essays. MacMillan, London (1914)
6. Duan, Z.: An algorithm of constructing even order magic square. Microelectron. Comput. **4**, 13–16 (1990)
7. Ward III, J.E.: Vector spaces of magic squares. Math. Mag. **53**, 108–119 (1980)
8. Krishnappa, H.K., Srinath, N.K., Ramakanth Kumar, P.: Magic square construction algorithms and their applications. IUP J. Comput. Math. **3**(3), 34–50 (2010)
9. Hruska, F.E.: Magic squares, matrices planes and angles. J. Recreational Math. **23**(3), 183–189 (1991)
10. Diaconis, P., Gamburd, A., Matrices, R.: Magic squares and matching polynomials. J. Comb. **11**(4), 1–26 (2004)
11. Deng, P., Hwang, F.K., Wu, W., MacCallum, D., Wang, F., Znati, T.: Improved construction for pooling design. J. Comb. Optim. **15**(1), 123–126 (2008)
12. Truini, P., Olivieri, G., Biedenharn, L.C.: The Jordan Pair content of the magic square and the geometry of the scalars in N=2 supergravity. Lett. Math. Phys. **9**(3), 255–261 (1985)
13. Lin, K.T.: Hybrid encoding method by assembling the magic-matrix scrambling method and the binary encoding method in image hiding. Opt. Commun. **284**(7), 1778–1784 (2011)
14. Gunjal, B.L.: Wavelet based color image watermarking scheme giving high robustness and exact corelation. Int. J. Emerg. Trends Eng. Technol. (IJETET) **1**(1), 1–10 (2011)
15. Chang, C.-C., Kieu, T.D., Wang, Z.-H., Li, M.-C.: An image authentication scheme using magic square. In: 2nd IEEE International Conference on Computer Science and Information Technology (ICCSIT), pp. 1–4 (2009)
16. Du, H., Jia, X., Wang, F., Thai, M., Li, Y.: A note on optical network with non-splitting nodes. J. Comb. Optim. **10**(2), 199–202 (2005)
17. Chen, H.B., Du, D.Z., Hwang, F.K.: An unexpected meeting of four seemingly unrelated problems: graph testing, DNA complex screening, superimposed codes and secure key distribution. J. Comb. Optim. **14**(2–3), 121–129 (2007)

The Complexity of the Positive Semidefinite Zero Forcing

Shaun Fallat[1], Karen Meagher[1], and Boting Yang[2]([⊠])

[1] Department of Mathematics and Statistics, University of Regina, Regina, Canada
{Shaun.Fallat,Karen.Meagher}@uregina.ca
[2] Department of Computer Science, University of Regina, Regina, Canada
boting.yang@uregina.ca

Abstract. The positive zero forcing number of a graph is a graph parameter that arises from a non-traditional type of graph colouring, and
is related to a more conventional version of zero forcing. We establish
a relation between the zero forcing and the fast-mixed searching, which
implies some NP-completeness results for the zero forcing problem. For
chordal graphs much is understood regarding the relationships between
positive zero forcing and clique coverings. Building upon constructions
associated with optimal tree covers and forest covers, we present a linear
time algorithm for computing the positive zero forcing number of chordal
graphs. We also prove that it is NP-complete to determine if a graph has
a positive zero forcing set with an additional property.

1 Introduction

The zero forcing number of a graph was introduced in [1] and related terminology was extended in [3]. First and foremost, the interest in this parameter
has been on applying zero forcing as a bound on the maximum nullities (or,
equivalently, the minimum rank) of certain symmetric matrices associated with
graphs, although this parameter has been considered elsewhere see, for example,
[2]. Independently, physicists have studied this parameter, referring to it as the
graph infection number, in conjunction with control of quantum systems [6].

The same notion arises in computer science within the context of fast-mixed
searching [13]. At its most basic level, edge search and node search models represent two significant graph search problems [10,11]. Bienstock and Seymour [4]
introduced the mixed search problem that combines the edge search and the node
search problems. Dyer et al. [7] introduced the fast search problem. Recently, a
fast-mixed search model was introduced in an attempt to combine fast search

---

Shaun Fallat: Research supported in part by an NSERC Discovery Research Grant,
Application No.: RGPIN-2014-06036.
Karen Meagher: Research supported in part by an NSERC Discovery Research
Grant, Application No.: RGPIN-341214-2013.
Boting Yang: Research supported in part by an NSERC Discovery Research Grant,
Application No.: RGPIN-2013-261290.

Z. Zhang et al. (Eds.): COCOA 2014, LNCS 8881, pp. 681–693, 2014.
DOI: 10.1007/978-3-319-12691-3_51

and mixed search models [13]. For this model, we assume that the simple graph G contains a single *fugitive*, invisible to the searchers, that can move at any speed along a "searcher-free" path and hides on vertices or along edges. In this case, the minimum number of searchers required to capture the fugitive is called the *fast-mixed search number of* G. As we will see, the fast-mixed search number and the zero forcing number of G are indeed equal.

Suppose that G is a simple finite graph. We begin by specifying a set of initial vertices of the graph (which we say are coloured black, while all other vertices are white). Then, using a designated colour change rule applied to these vertices, we progressively change the colour of white vertices in the graph to black. Our colouring consists of only two colours (black and white) and the objective is to colour all vertices black by repeated application of the colour change rule to our initial set. In general, we want to determine the smallest set of vertices needed to be black initially, to eventually change all of the vertices to black.

The conventional zero forcing rule results in a partition of the vertices of the graph into sets, such that each such set induces a path in G. Further, each of the initial black vertices is an end point of one of these paths. More recently a refinement of the colour change rule, called the positive zero forcing colour change rule, was introduced. Using this rule, the positive semidefinite zero forcing number was defined (see, for example, [3,8,9]). When the positive zero forcing colour change rule is applied to a set of initial vertices of a graph, the vertices are then partitioned into sets, so that each such set induces a tree in G.

As mentioned above, one of the original motivations for studying these parameters is that they both provide an upper bound on the maximum nullity of both symmetric and positive semidefinite matrices associated with a graph (see [2,3]). For a given graph $G = (V, E)$, define

$$\mathcal{S}(G) = \{A = [a_{ij}] : A = A^T, \text{ for } i \neq j, a_{ij} \neq 0 \text{ iff } \{i, j\} \in E\}$$

and let $\mathcal{S}_+(G)$ denote the subset of positive semidefinite matrices in $\mathcal{S}(G)$. We use null(B) to denote the nullity of the matrix B. The *maximum nullity* of G is defined to be $\mathrm{M}(G) = \max\{\text{null}(B) : B \in \mathcal{S}(G)\}$, and, similarly, $\mathrm{M}_+(G) = \max\{\text{null}(B) : B \in \mathcal{S}_+(G)\}$, is called the *maximum positive semidefinite nullity of* G.

The zero forcing number has been studied under the alias, the fast-mixed searching number, the complexity of computing the zero forcing is generally better understood. Consequently, our focus will be on the algorithmic aspects of computing the positive semidefinite zero forcing number for graphs. As with most graph parameters, defined in terms of an optimization problem, these parameters are complicated to compute in general. However, some very interesting exceptions arise such as the focus of this paper, chordal graphs, whose positive zero forcing number can be found in linear time. However, when we consider a variant of the positive zero forcing problem, called the min-forest problem, we will show that this variant is NP-complete even for echinus graphs, which are a special type of split graphs.

2 Preliminaries

Throughout this paper, we only consider finite graphs with no loops or multiple edges. We use $G = (V, E)$ to denote a graph with vertex set V and edge set E, and we also use $V(G)$ and $E(G)$ to denote the vertex set and edge set of G respectively. We use $\{u, v\}$ to denote an edge with endpoints u and v. For $v \in V$, the vertex set $\{u : \{u, v\} \in E\}$ is the *neighbourhood* of v, denoted as $N_G(v)$. For $V' \subseteq V$, the vertex set $\{x : \{x, y\} \in E, x \in V \setminus V' \text{ and } y \in V'\}$ is the *neighbourhood* of V', denoted as $N_G(V')$. We use $G[V']$ to denote the subgraph induced by V', which consists of all vertices of V' and all of the edges that connect vertices of V' in G. We use $G - v$ to denote the subgraph induced by $V \setminus \{v\}$.

Let G be a graph in which every vertex is initially coloured either black or white. If u is a black vertex of G and u has exactly one white neighbour, say v, then we change the colour of v to black; this rule is called the *colour change rule*. In this case we say "u forces v" and denote this action by $u \to v$. Given an initial colouring of G, in which a set of the vertices is black and all other vertices are white, the *derived set* is the set of all black vertices, including the initial set of black vertices, resulting from repeatedly applying the colour change rule until no more changes are possible. If the derived set is the entire vertex set of the graph, then the set of initial black vertices is called a *zero forcing set*. The *zero forcing number* of a graph G is the size of the smallest zero forcing set of G; it is denoted by $Z(G)$. The procedure of colouring a graph using the colour rule is called a *zero forcing process* or simply a *forcing process*. A zero forcing process is called *optimal* if the initial set of black vertices is a zero forcing set of the smallest possible size.

If Z is a zero forcing set of a graph G, then we may produce a list of the forces in the order in which they are performed in the zero forcing process. This list can then be divided into paths, known as forcing chains. A *forcing chain* is a sequence of vertices $(v_1, v_2, \ldots, v_k)$ such that $v_i \to v_{i+1}$, for $i = 1, \ldots, k-1$ in the forcing process. In every step of a forcing process, each vertex can force at most one other vertex; conversely every vertex not in the zero forcing set is forced by exactly one vertex. Thus the forcing chains that correspond to a zero forcing set partition the vertices of a graph into disjoint sets, such that each set induces a path. The number of these paths is equal to the size of the zero forcing set and the elements of the zero forcing set are the initial vertices of the forcing chains and hence end points of these paths. We observe that the concept of clearing an edge by sliding in the fast-mixed search model is equivalent to the notion of a black vertex forcing a unique white neighbour.

The most widely-studied variant of the zero forcing number is called positive semidefinite zero forcing or the positive zero forcing number, and was introduced in [3], see also [8,9]. The positive zero forcing number is also based on a colour change rule similar to the zero forcing colour change rule. Let G be a graph and B a set of vertices; we will initially colour the vertices of B black and all other vertices white. Let $W_1, \ldots, W_k$ be the sets of vertices in each of the connected components of G after removing the vertices in B. If u is a vertex in B and w

is the only white neighbour of u in the graph induced by the subset of vertices $W_i \cup B$, then u can force the colour of w to black. This is the *positive colour change rule*. The definitions and terminology for positive zero forcing, such as, colouring, derived set, positive zero forcing number etc., are similar to those for zero forcing, except we use the positive colour change rule.

The size of the smallest positive zero forcing set of a graph G is denoted by $Z_+(G)$. Also for all graphs G, since a zero forcing set is also a positive zero forcing set we have that $Z_+(G) \leq Z(G)$. Moreover, in [3] it was shown that $M_+(G) \leq Z_+(G)$, for any graph G.

As noted above, applying the zero forcing colour change rule to the vertices of a graph produces a path covering of the vertices in that graph. Analogously, applying the positive colour change rule produces a set of vertex disjoint induced trees in the graph, referred to as *forcing trees*. Next we define a *positive zero forcing tree cover*. Let G be a graph and let Z be a positive zero forcing set for G. Observe that applying the colour change rule once, two or more vertices can perform forces at the same time, and a vertex can force multiple vertices from different components at the same time. For each vertex in Z, these forces determine a rooted induced tree. The root of each tree is the vertex in Z and two vertices are adjacent if one of them forces the other.

More generally, a *tree covering of a graph* is a family of induced vertex disjoint trees in the graph that cover all vertices of the graph. The minimum number of such trees that cover the vertices of a graph G is the *tree cover number* of G and is denoted by $T(G)$. Any set of zero forcing trees corresponding to an optimal positive zero forcing set is of size $Z_+(G)$. Hence, for any graph G, we have $T(G) \leq Z_+(G)$.

In this paper we will focus on *chordal graphs*. A graph is chordal if it contains no induced cycles on four or more vertices. Further, we say a vertex v is *simplicial*, if the graph induced by the neighbours of v forms a clique. If $(v_1, v_2, \ldots, v_n)$ is an ordering of the vertices of a graph G, such that the vertex v_i is simplicial in the graph $G - \{v_1, v_2, \ldots, v_{i-1}\}$, then $(v_1, v_2, \ldots, v_n)$ is called a *perfect elimination ordering*. Every chordal graph has an ordering of the vertices that is a *perfect elimination ordering*.

Let $cc(G)$ denote the fewest number of cliques needed to cover (or to include) all the edges in G. This number, $cc(G)$, is often referred to as the *clique cover number* of G. It is known for any chordal graph G, that $M_+(G) = |V(G)| - cc(G)$. From [3,5] we know that for any graph G,

$$|V(G)| - cc(G) \leq M_+(G) \leq Z_+(G). \tag{1}$$

3   The Complexity of Zero Forcing

Let G be a connected graph. In the fast-mixed search model, G initially contains no searchers and it contains only one fugitive who hides on vertices or along edges. The fugitive is invisible to searchers, and he can move at any rate and at any time from one vertex to another vertex along a searcher-free path between the two vertices. An edge (resp. a vertex) where the fugitive may hide is said to be *contaminated*, while an edge (resp. a vertex) where the fugitive cannot hide is said to be *cleared*. A vertex is said to be *occupied* if it has a searcher on it. There are two types of actions for searchers in each step of the fast-mixed search model:

1. a searcher can be placed on a contaminated vertex, or
2. a search may slide along a contaminated edge $\{u, v\}$ from u to v if v is contaminated and all edges incident on u except $\{u, v\}$ are cleared.

In the fast-mixed search model, a contaminated edge becomes cleared if both endpoints are occupied by searchers or if a searcher slides along it from one endpoint to the other. The graph G is *cleared* if all edges are cleared. The minimum number of searchers required to clear G (i.e., to capture the fugitive) is the *fast-mixed search number* of G, denoted by $\mathrm{fms}(G)$. We first show that the fast-mixed search number of a graph is equal to its zero forcing number.

Theorem 1. *For any graph G, $\mathrm{fms}(G) = Z(G)$.*

From Theorem 1 and [13, Theorems 6.3, 6.5, Corollary 6.6], respectively, we have the following results.

Corollary 1. *Given a graph G and a nonnegative integer k, it is NP-complete to determine whether G has a zero forcing process with k initial black vertices such that all initial black vertices are leaves of G. This problem remains NP-complete for planar graphs with maximum degree 3.*

Corollary 2. *Given a graph G with ℓ leaves, it is NP-complete to determine whether $Z(G) = \lceil \ell/2 \rceil$. This problem remains NP-complete for graphs with maximum degree 4.*

Corollary 3. *Given a graph G and a nonnegative integer k, it is NP-complete to determine whether $Z(G) \leq k$. This problem remains NP-complete even for biconnected graphs with maximum degree 4.*

At the end of this section, we introduce a searching model, which is an extension of the fast-mixed searching, that corresponds to positive zero forcing. This searching model, called the *parallel fast-mixed searching*, follows the same setting as the fast-mixed searching except that the graph may be split into subgraphs after each placing or sliding action, in such a way that these subgraphs may be cleared in a parallel-like fashion.

Initially, G contains no searchers, and so all vertices of G are contaminated. To begin, let $\mathcal{G} = \{G\}$. After a placing, (e.g., place a searcher on a contaminated vertex u), the subgraph $G - u$ is the graph induced by the current contaminated vertices. If $G - u$ is not connected, let $G_1, \ldots, G_j$ be all of the connected components of $G - u$. We update $\mathcal{G}$ by replacing G by subgraphs $G[V(G_1) \cup \{u\}], \ldots, G[V(G_j) \cup \{u\}]$, where u is occupied in each subgraph $G[V(G_i) \cup \{u\}]$, $1 \leq i \leq j$. Consider each subgraph $H \in \mathcal{G}$ that has not been cleared. After a placing or sliding action, let X be the set of the contaminated vertices in H. If $H[X]$ is not connected, let $X_1, \ldots, X_j$ be the vertex sets of all connected components of $H[X]$. We update $\mathcal{G}$ by replacing H by subgraphs $H[X_1 \cup N_H(X_1)\}], \ldots, H[X_j \cup N_H(X_j)]$, where $N_H(X_i)$ is occupied in each subgraph $H[X_i \cup N_H(X_i)\}], 1 \le i \le j$. We can continue this searching and branching process until all subgraphs in $\mathcal{G}$ are cleared. It is easy to observe that we can arrange the searching process so that subgraphs in $\mathcal{G}$ can be cleared in a parallel-like way.

The graph G is *cleared* if all subgraphs of $\mathcal{G}$ are cleared. The minimum number of placings required to clear G is called the *parallel fast-mixed search number* of G, and is denoted by pfms(G).

To illustrate the difference between the parallel fast-mixed searching and the fast-mixed searching, let G_k be a unicyclic graph, with $k \ge 4$, with vertex set $V = \{v_0, v_1, \ldots, v_{k-1}, v_k\}$ and edge set $E = \{\{v_0, v_i\} : i = 1, \ldots, k\} \cup \{\{v_{k-1}, v_k\}\}$. Initially $\mathcal{G} = \{G_k\}$. After we place a searcher on vertex v_0, the set $\mathcal{G}$ is updated such that it contains $k - 1$ subgraphs, i.e., the edges $\{v_0, v_1\}, \ldots, \{v_0, v_{k-2}\}$, and the 3-cycle induced by the vertices $\{v_0, v_{k-1}, v_k\}$, where v_0 is occupied by a searcher in each subgraph. Each edge $\{v_0, v_i\}$ ($1 \le i \le k - 2$) can be cleared by a sliding action. For the 3-cycle on vertices $\{v_0, v_{k-1}, v_k\}$, since no sliding action can be performed, we have to place a new searcher on a vertex, say v_{k-1}. Since the graph induced by the contaminated vertices is connected (just an isolated vertex v_k in this case), we do not need to update $\mathcal{G}$. Now we can slide the searcher on vertex v_{k-1} to vertex v_k. After the sliding action, the 3-cycle is cleared, and thus, G_k is cleared. This search strategy contains two placing actions, and it is easy to see that any strategy with only one placing action cannot cleared G_k. Thus pfms(G_k) = 2. On the other hand, we can easily show that fms(G_k) = $k - 2$.

Similar to Theorem 1, we can prove the following relation between the parallel fast-mixed searching and the positive zero forcing.

Theorem 2. *For any graph G, pfms(G) = $Z_+(G)$.*

4   Positive Zero Forcing of Chordal Graphs

In this section, we give an algorithm for finding optimal positive zero forcing sets of chordal graphs. Our algorithm is a modification of the algorithm for computing clique covers as presented in [12]. In the remainder of the paper, we suppose that all graphs G contains at least two vertices.

Algorithm. ZPLUS-CHORDAL
Input: A connected chordal graph G with $n(\ge 2)$ vertices.
Output: An optimal positive zero forcing tree cover and an optimal positive zero forcing set of G.

1. Let $(v_1, v_2, \ldots, v_n)$ be the perfect elimination ordering of G given by the lexicographic breadth-first search. Let $i \leftarrow 1$ and $G_i \leftarrow G$.
2. For the simplicial vertex v_i in G_i, let C_i be the clique whose vertex set consists of v_i and all its neighbours in G_i.

3. If there is an edge e of G_i incident to v_i that is uncoloured, then colour e black, colour all other uncoloured edges of C_i red, colour v_i white, and go to Step 5.
4. If all edges of G_i incident to v_i are coloured, then colour v_i black.
5. Set $G_{i+1} \leftarrow G_i - v_i$. If $i < |V(G)| - 1$, then set $i \leftarrow i + 1$ and go to Step 2; otherwise, colour v_{i+1} in G_{i+1} black and go to Step 6.
6. Remove all red edges from G. Let $\mathcal{T}$ be the set of all connected components of the remaining graph after all red edges are removed. Output $\mathcal{T}$ and its black vertex set and stop.

Let $V_{\text{black}}(G)$ be the set of all black vertices and $V_{\text{white}}(G)$ be the set of all white vertices from Algorithm ZPLUS-CHORDAL.

Lemma 1. *For a connected chordal graph G, $V_{\text{black}}(G)$ is a positive zero forcing set of G.*

Proof. The set $V_{\text{black}}(G)$ is the set of all vertices that are initially black in G. Let $V_{\text{white}}(G) = \{w_1, w_2, \ldots, w_m\}$ where w_i is removed before w_{i+1} in Algorithm ZPLUS-CHORDAL for $1 \leq i < m$. At the iteration when w_j is coloured white, let $e_j = \{w_j, b_j\}$ be the edge that is coloured black. In particular, for the last white vertex w_m, the black edge is $e_m = \{w_m, b_m\}$ and b_m is in the set $V_{\text{black}}(G)$. We claim that in a positive zero forcing process in G starting with $V_{\text{black}}(G)$, the vertex b_m can force w_m.

Let H_m be the connected component in the subgraph of G induced by the vertices $V(G)\backslash V_{\text{black}}(G)$ that contains the last white vertex, w_m. We will show that the only vertex in H_m that is adjacent to b_m is w_m. The vertex b_m is adjacent to w_m and assume that it is also adjacent to another vertex, say u_1 in H_m (this vertex must be part of $V_{\text{white}}(G)$). Let $\{b_m, u_1, u_2, \ldots, u_k, w_m, b_m\}$ be a cycle of minimal length with $u_1, u_2, \ldots, u_k \in H_m$. Such a cycle exists since H_m is connected.

If this cycle has length three, then u_1 and w_m are adjacent. But at the iteration when u_1 is marked white, the edge $\{w_m, b_m\}$ will be coloured red. This is a contradiction, as the edge $\{w_m, b\}$ is black.

Assume this cycle has length more than three and let u_ℓ be the vertex in the cycle that was coloured white first. At the iteration where u_ℓ is coloured white, it is a simplicial vertex. This implies that the neighbours of u_ℓ in the cycle are adjacent. But this is a contradiction with the choice of $\{b_m, u_1, u_2, \ldots, u_k, w_m, b\}$ being a cycle of minimal length with $u_1, u_2, \ldots, u_k \in H_m$.

By the positive zero forcing rule, we know that b_m can force w_m to be black. We now change the colour of vertex w_m to black, add it to $V_{\text{black}}(G)$ and delete it from $V_{\text{white}}(G)$.

Similarly, at the iteration when w_{m-1} is coloured white, let $e_{m-1} = \{w_{m-1}, b_{m-1}\}$ be the edge that is coloured black. Using the above argument, we can show that b_{m-1} can force w_{m-1}. Continuing this process, all the white vertices are forced to be black. Therefore, $V_{\text{black}}(G)$ is a positive zero forcing set for G.

For the set $V_{\text{white}}(G) = \{w_1, w_2, \ldots, w_m\}$ defined in the above proof, let C_i be the clique whose vertex set consists of w_i and all its neighbours at the point when w_i is coloured white. Define $C(G) = \{C_1, C_2, \ldots, C_m\}$. Every edge of G is coloured in Algorithm ZPLUS-CHORDAL and at the iteration when it is coloured, it must belong to some clique C_i. Thus we have the following lemma.

Lemma 2. *Let G be a chordal graph. Then $C(G)$ is a clique cover of G.*

From Lemmas 1 and 2, we can prove the correctness of Algorithm ZPLUS-CHORDAL as follows.

Theorem 3. *Let G be a chordal graph. Then $V_{\text{black}}(G)$ is an optimal positive zero forcing set of G.*

Proof. Without loss of generality, we suppose that G is a connected chordal graph. From Lemma 1 we know that $V_{\text{black}}(G)$ is a positive zero forcing set of G, so we only need to show that it is the smallest possible.

From Lemma 2 we know that $C(G)$ is a clique cover of G. Thus, $cc(G) \leq |C(G)|$, using this with (1) we have that

$$|V(G)| - |C(G)| \leq |V(G)| - cc(G) \leq Z_+(G) \leq |V_{\text{black}}(G)|.$$

The pair $V_{\text{black}}(G)$ and $V_{\text{white}}(G)$ form a partition of $V(G)$ with $|V_{\text{white}}(G)| = |C(G)|$, so $|V(G)| - |C(G)| = |V_{\text{black}}(G)|$.

Therefore, $|V(G)| - cc(G) = Z_+(G) = |V_{\text{black}}(G)|$, and $V_{\text{black}}(G)$ is an optimal positive zero forcing set for G.

As a byproduct from the proof of Theorem 3, we have the following result for chordal graphs.

Corollary 4. *For a chordal graph G,*

(1) $C(G)$ is an optimal clique cover for G, and
(2) $|V(G)| - cc(G) = Z_+(G)$.

Note that Corollary 4 (1) is proved in [12] by using primal and dual linear programming. Corollary 4 (2) can be deduced from the work in [5], where the concept of orthogonal removal is used along with an inductive proof technique.

We can easily modify Algorithm ZPLUS-CHORDAL so that it can also output the coloured graph G. In this graph, the number of black edges is equal to the number of white vertices. From the proof of Lemma 1, we know that every black edge can be used to force a white vertex to black. Define $T_{\text{black}}(G)$ to be the subgraph of G formed by taking all the edges (and their endpoints) that are coloured black. The next result gives some of the interesting properties of $T_{\text{black}}(G)$.

Theorem 4. *Let G be a chordal graph and let $T_{\text{black}}(G)$ be the subgraph formed by all the edges that are coloured black in Algorithm ZPLUS-CHORDAL. Then*

1. *the graph $T_{\text{black}}(G)$ is a forest;*
2. *all white vertices of G are contained among the vertices of $T_{\text{black}}(G)$;*
3. *each component of $T_{\text{black}}(G)$ contains exactly one black vertex;*
4. *$T_{\text{black}}(G)$ is an induced subgraph of G.*

Proof. First, we will simply denote $T_{\text{black}}(G)$ by T_{black}. Note that in each iteration of the Algorithm ZPLUS-CHORDAL when an edge is coloured black in Step 3, it is removed in Step 5 in the same iteration. Thus at any iteration i, the graph G_i does not contain any black edges in Step 2.

Suppose there is a cycle in T_{black}. Let v_i be the simplicial vertex in G_i found in Step 2, which is the first vertex to be coloured among all the vertices on the cycle in T_{black}. Assume that v_i is coloured at iteration i. Let u_i and u'_i be the two neighbours of v_i on the cycle. If v_i is coloured black, then both the edges $\{v_i, u_i\}$ and $\{v_i, u'_i\}$ must be coloured, since none of the edges of G_i can be black, they must all be red. But then these edges are not in T_{black}. This is a contradiction. If v_i is coloured white, then at most one of the edges $\{v_i, u_i\}$ and $\{v_i, u'_i\}$ is black, and contradiction again. Thus T_{black} contains no cycles, and hence T_{black} must be a forest.

Whenever a vertex of G is coloured white in Step 3, an edge incident to it is marked black in the same step. Thus T_{black} contains all the vertices that are coloured white by Algorithm ZPLUS-CHORDAL.

The vertices of G, ordered by the perfect elimination ordering, are $\{v_1, v_2, \ldots, v_{|V(G)|}\}$. For a component T in T_{black}, let $V(T) = \{v_{i_1}, v_{i_2}, \ldots, v_{i_t}\}$, where $i_1 < i_2 < \cdots < i_t$, in the perfect elimination ordering. We will show that v_{i_t} is the only black vertex in $V(T)$.

Since $v_{|V(G)|}$ is black, if $i_t = |V(G)|$ then we are done; so we will assume that $i_t < |V(G)|$. If the vertex v_{i_t} is white, then at Step 3 in the algorithm it is coloured white and there is an edge $e = \{v_{i_t}, v_j\}$ that is coloured black. The edge e must be in T (as it is black) and further, $v_j \in V(G_{i_t}) \setminus \{v_{i_t}\}$. Since the vertices are being removed in order, this implies that $j > i_t$; this is a contradiction since v_{i_t} is the last vertex in the ordering that is in T. Thus v_{i_t} is coloured black; next we will show that v_{i_t} is the only vertex in T that is coloured black.

Assume that v_{i_j} is the vertex with the smallest subscript among $\{i_1, i_2, \ldots, i_t\}$ that is coloured black in the algorithm. Unless $v_{i_j} = v_{i_t}$, the vertex v_{i_j} will be adjacent, in G, to a vertex in T with a larger subscript. At Step 5 vertex v_{i_j} is removed and all edges incident with it are removed as well. So v_{i_j} is not adjacent, in T, to any vertex with a larger subscript; similarly, no vertex with index less than i_j is adjacent in T to a vertex with index larger than i_j. This implies that T is not connected, which is a contradiction. Therefore, v_{i_t} is the only black vertex in T.

Finally, we will show that any component T in T_{black} is an induced subgraph of G. If this is not the case, then there are two vertices v_{i_p} and v_{i_q} (we will assume that $i_p < i_q$), in $V(T)$ such that $\{v_{i_p}, v_{i_q}\}$ is an edge in G but not in T. Pick the vertices v_{i_p} and v_{i_q} so that their distance in T is minimum over all pairs of vertices that are non-adjacent in T, but adjacent in G. At the i_p-th iteration of the algorithm, v_{i_p} is the simplicial vertex in G_{i_p} which is found in Step 2. Let $\{v_{i_p}, v_{i_j}\}$ be the edge marked black in Step 3. Since both v_{i_j} and v_{i_q} are adjacent to v_{i_p} in G_{i_p}, the edge $\{v_{i_j}, v_{i_q}\}$ is coloured red at this step. Thus the edge $\{v_{i_j}, v_{i_q}\}$ is not in T. Since T is a tree and $i_p < i_q$, we know that the path in T that connects v_{i_j} and v_{i_q} is shorter than the path in T that connects v_{i_p} and v_{i_q}. This is a contradiction. Hence T is an induced subgraph of G.

From Theorem 4, we have the following result.

Corollary 5. *Let G be a chordal graph. Then the output T of Algorithm ZPLUS-CHORDAL is an optimal positive zero forcing tree cover of G.*

A chordal graph is called *non-trivial* if it has at least two distinct maximal cliques; thus a trivial chordal graph is just a complete graph. Further, a simplicial vertex is called *leaf-simplicial* if none of its neighbours are simplicial. A tree with only one vertex and no edges is called a *trivial* tree. In a positive zero forcing tree cover for a graph, any tree that is trivial consists of precisely one black vertex.

Lemma 3. *Let G be a non-trivial chordal graph.*

1. *For any optimal tree cover $\mathcal{T}(G)$, each simplicial vertex of G is either a trivial tree in $\mathcal{T}(G)$ or a leaf of a non-trivial tree in $\mathcal{T}(G)$.*
2. *There is an optimal tree cover $\mathcal{T}(G)$ of G such that each leaf-simplicial vertex of G is a leaf of a non-trivial tree in $\mathcal{T}(G)$.*

Proof. Let v be a simplicial vertex of G. Let $\mathcal{T}(G)$ be an optimal tree cover of G. Let T_v be the tree in $\mathcal{T}(G)$ that contains v; this means that T_v is an induced tree in G. The neighbours of v form a clique in G, so at most one of them can be in T_v. This implies that the degree of v in T_v is less than 2; thus either T_v only contains the vertex v or v is a leaf in T_v.

Now further assume that v is a leaf-simplicial vertex and suppose that T_v is a trivial tree, this is, it only contains the vertex v. We will show that $\mathcal{T}(G)$ can be transformed into a new optimal tree covering of G in which v is a leaf of a non-trivial tree. Let C be in the clique in G that contains v and all its neighbours. We will consider two cases.

First, assume that no edge of C is also an edge of a tree in $\mathcal{T}(G)$. For any $u \in C$, let T_u be the tree in $\mathcal{T}(G)$ that contains u. Then T_v and T_u can be merged by adding the edge $\{u, v\}$. But this is a contradiction, as it implies that $\mathcal{T}(G)$ is not an optimal tree covering of G.

Second, assume that there is an edge $\{u, w\}$ in C that is also an edge of the tree T_u in $\mathcal{T}(G)$. We can split T_u into two subtrees by deleting the edge $\{u, w\}$, and then merge T_v and one subtree by adding the edge $\{v, u\}$ (the other tree will contain w, which is not simplicial, so it will not be a tree that contains only one leaf-simplicial vertex). Thus we obtain another optimal tree cover of G, in which v is a leaf in a non-trivial tree. By using the above operation, we can transform all trivial trees in T that contain a leaf-simplicial vertex so that these vertices are leaves in trees for another optimal tree cover.

Theorem 5. *Let G be a chordal graph with n vertices and m edges. Then Algorithm ZPLUS-CHORDAL can be implemented to find an optimal positive zero forcing set of G and an optimal positive zero forcing tree cover of G in $O(n + m)$ time.*

5   Minimum Forest Covers for Graphs

In this section, we consider the structure of the trees in the zero forcing tree covers of the graph. So for a given graph G and a positive integer ℓ, among all the positive zero forcing tree covers of G with size ℓ, we want to minimize the number of positive zero forcing trees that are non-trivial trees. We call this problem *Min-Forest*. If we consider the corresponding parallel fast-mixed searching model, each non-trivial positive zero forcing tree corresponds to an induced tree cleared by a "mobile" searcher and each trivial tree corresponds to an "immobile" searcher (perhaps a trap or a surveillance camera). Typically, the goal is to minimize the number of mobile searchers among parallel fast-mixed search strategies with a given number of searchers. The decision version of the Min-Forest problem is as follows.

> MIN-FOREST
> **Instance:** A graph G and positive integers k and ℓ.
> **Question:** Does G have a positive zero forcing tree cover of size ℓ in which there are at most k positive zero forcing trees that are non-trivial?

A *split graph* is a graph in which the vertices can be partitioned into two sets C and I, where C induces a clique and I induces an independent set in the graph. It is not difficult to show that a graph is split if and only if it is chordal and its complement is also chordal.

Theorem 6. MIN-FOREST *is NP-complete. The problem remains NP-complete for split graphs whose simplicial vertices all have degree 2.*

An *echinus graph* is a split graph with vertex set $\{C, I\}$, where C induces a clique and I is an independent set, such that every vertex of I has two neighbours in C and every vertex of C has three neighbours in I. It is easy to see that echinus graphs are special chordal graphs.

Corollary 6. MIN-FOREST *remains NP-complete even for echinus graphs.*

If a graph G has a positive zero forcing tree cover of size ℓ in which there are at most k non-trivial trees, then for any $n \geq \ell' > \ell$, G has a positive zero forcing tree cover of size ℓ', with at most k non-trivial trees. Note that the smallest possible value of ℓ is the positive zero forcing number. So next we consider the case when ℓ equals the positive zero forcing number. The following theorem presents a characterization of the chordal graphs for which there exists an optimal positive zero forcing tree cover that contains only one non-trivial tree.

Theorem 7. *Let G be a connected non-trivial chordal graph. There is an optimal positive zero forcing tree cover of G in which only one tree is non-trivial if and only if for every maximal clique C in G, there are two vertices $x_C, y_C \in C$ such that any other maximal clique C' in G with $V(C') \cap V(C) \neq \emptyset$ must contain exactly one of x_C and y_C.*

This result can be generalized to a family of graphs that are not chordal.

Lemma 4. *Let G be a graph and T an induced tree in G. If $|V(T)| - 1 = cc(G)$, then G has an optimal positive zero forcing set with only one non-trivial positive zero forcing tree.*

6   Cycles of Cliques

Let G be a graph and assume that $\{C_1, C_2, \ldots, C_k\}$ is a set of maximal cliques in G that covers all the edges in G. We say that G is a *cycle of cliques* if $V(C_i) \cap V(C_j) \neq \emptyset$ whenever $j = i + 1$ or $(i, j) = (k, 1)$ and $V(C_i) \cap V(C_j) = \emptyset$ otherwise. If $k = 1$ then G is a clique; we will not consider a graph that is a clique to be a cycle of cliques.

Lemma 5. *If G is a cycle of cliques $\{C_1, C_2, \ldots, C_k\}$ with $k \geq 3$, then there is a zero forcing set of size $|V(G)| - (k - 2)$ and exactly one non-trivial forcing tree.*

If G is a cycle of cliques, then the cliques $\{C_1, C_2, \ldots, C_k\}$ form a clique cover of G. So we have $|V(G)| - |cc(G)| = |V(G)| - k \leq Z_+(G) \leq Z(G) \leq |V(G)| - k + 2$. Observe that the positive zero forcing sets in the previous lemma may not always be optimal positive zero forcing sets. But, in some cases it is possible to find an optimal zero forcing set for a cycle of cliques that has exactly one non-trivial forcing tree and is also an optimal positive zero forcing set.

Lemma 6. *Assume that G is a graph that is a cycle of cliques $\{C_1, C_2, \ldots, C_k\}$ with $k \geq 3$. Further assume that there is a vertex $x \in V(C_1)$ that is in no other clique and a vertex $y \in V(C_k)$ but is not in any other clique. Then there is an optimal positive zero forcing set of size $|V(G)| - k$ with exactly one non-trivial forcing tree.*

This also gives a family for which the positive zero forcing number and the zero forcing number agree.

7   Further Work

The complexity of computing any type of graph parameter is an interesting task. For zero forcing parameters it is known that the problem of finding $Z(G)$ for a graph G is NP-complete. We also suspect that the same is true for computing $Z_+(G)$ for a general graph G. In fact, we resolve part of this conjecture, by assuming an additional property on the nature of the zero forcing tree cover that results. However, for chordal graphs G, we have verified that determining the exact value of $Z_+(G)$ can be accomplished via a linear time algorithm; the best possible situation. We also believe that it would be interesting to consider the complexity of determine $Z_+(G)$ when G is a partial 2-tree.

A solution of the Min-Forest problem describes an inner structure between maximal cliques of G. In Sect. 5, we highlight a couple of instances where upon if we restrict the number of non-trivial trees in a positive zero forcing tree cover of a given size, then conclusions concerning the complexity of computing the positive zero forcing number can be made. We are interested in exploring this notion further.

References

1. AIM Minimum Rank-Special Graphs Work Group: Zero forcing sets and the minimum rank of graphs. Linear Algebra Appl. **428**(7), 1628–1648 (2008)
2. Barioli, F., Barrett, W., Fallat, S., Hall, H.T., Hogben, L., Shader, B., van den Driessche, P., van der Holst, H.: Parameters related to tree-width, zero forcing, and maximum nullity of a graph. J. Graph Theory **72**, 146–177 (2013)
3. Barioli, F., Barrett, W., Fallat, S., Hall, H.T., Hogben, L., Shader, B., van den Driessche, P., van der Holst, H.: Zero forcing parameters and minimum rank problems. Linear Algebra Appl. **433**(2), 401–411 (2010)
4. Bienstock, D., Seymour, P.: Monotonicity in graph searching. J. Algorithms **12**, 239–245 (1991)
5. Booth, M., Hackney, P., Harris, B., Johnson, C.R., Lay, M., Mitchell, L.H., Narayan, S.K., Pascoe, A., Steinmetz, K., Sutton, B.D., Wang, W.: On the minimum rank among positive semidefinite matrices with a given graph. SIAM J. Matrix Anal. Appl. **30**, 731–740 (2008)
6. Burgarth, D., Giovannetti, V.: Full control by locally induced relaxation. Phys. Rev. Lett. **99**(10), 100501 (2007)
7. Dyer, D., Yang, B., Yaşar, Ö.: On the fast searching problem. In: Fleischer, R., Xu, J. (eds.) AAIM 2008. LNCS, vol. 5034, pp. 143–154. Springer, Heidelberg (2008)
8. Ekstrand, J., Erickson, C., Hall, H.T., Hay, D., Hogben, L., Johnson, R., Kingsley, N., Osborne, S., Peters, T., Roat, J., et al.: Positive semidefinite zero forcing. Linear Algebra Appl. **439**, 1862–1874 (2013)
9. Ekstrand, J., Erickson, C., Hay, D., Hogben, L., Roat, J.: Note on positive semidefinite maximum nullity and positive semidefinite zero forcing number of partial 2-trees. Electron. J. Linear Algebra **23**, 79–97 (2012)
10. Kirousis, L., Papadimitriou, C.: Searching and pebbling. Theoret. Comput. Sci. **47**, 205–218 (1986)
11. Megiddo, N., Hakimi, S., Garey, M., Johnson, D., Papadimitriou, C.: The complexity of searching a graph. J. ACM **35**, 18–44 (1988)
12. Scheinerman, E.R., Trenk, A.N.: On the fractional intersection number of a graph. Graphs Combin. **15**, 341–351 (1999)
13. Yang, B.: Fast-mixed searching and related problems on graphs. Theoret. Comput. Sci. **507**(7), 100–113 (2013)

A Potential Reduction Algorithm for Ergodic Two-Person Zero-Sum Limiting Average Payoff Stochastic Games

Endre Boros[1], Khaled Elbassioni[2(✉)], Vladimir Gurvich[1], and Kazuhisa Makino[3]

[1] MSIS Department of RBS and RUTCOR, Rutgers University, 100 Rockafeller Road, Piscataway, NJ 08854-8054, USA
{endre.boros,valdimir.gurvich}@rutgers.edu
[2] Masdar Institute of Science and Technology, Abu Dhabi, UAE
kelbassioni@masdar.ac.ae
[3] Research Institute for Mathematical Sciences (RIMS), Kyoto University, Kyoto 606–8502, Japan
makino@kurims.kyoto-u.ac.jp

Abstract. We suggest a new algorithm for two-person zero-sum undiscounted stochastic games focusing on stationary strategies. Given a positive real ϵ, let us call a stochastic game ϵ-ergodic, if its values from any two initial positions differ by at most ϵ. The proposed new algorithm outputs for every $\epsilon > 0$ in finite time either a pair of stationary strategies for the two players guaranteeing that the values from any initial positions are within an ϵ-range, or identifies two initial positions u and v and corresponding stationary strategies for the players proving that the game values starting from u and v are at least $\epsilon/24$ apart. In particular, the above result shows that if a stochastic game is ϵ-ergodic, then there are stationary strategies for the players proving 24ϵ-ergodicity. This result strengthens and provides a constructive version of an existential result by Vrieze (1980) claiming that if a stochastic game is 0-ergodic, then there are ϵ-optimal stationary strategies for every $\epsilon > 0$. The suggested algorithm extends the approach recently introduced for stochastic games with perfect information, and is based on the classical potential transformation technique that changes the range of local values at all positions without changing the normal form of the game.

1 Introduction

1.1 Basic Concepts and Notation

Stochastic games were introduced in 1953 by Shapley [Sha53] for the discounted case, and extended to the undiscounted case by Gillette [Gil57]. Each such game

---

Part of this research was done at the Mathematisches Forschungsinstitut Oberwolfach during a stay within the Research in Pairs Program from March 7 to March 20, 2010.

© Springer International Publishing Switzerland 2014
Z. Zhang et al. (Eds.): COCOA 2014, LNCS 8881, pp. 694–709, 2014.
DOI: 10.1007/978-3-319-12691-3_52

$\Gamma = (p_{k\ell}^{vu}, r_{k\ell}^{vu} \mid k \in K^v, \ell \in L^v, u, v \in V)$ is played by two players on a finite set V of vertices (states, or positions); K^v and L^v for $v \in V$ are finite sets of actions (pure strategies) of the two players; $p_{k\ell}^{vu} \in [0, 1]$ is the transition probability from state v to state u if players chose actions $k \in K^v$ and $\ell \in L^v$ at state $v \in V$; and $r_{k\ell}^{vu} \in \mathbb{R}$ is the reward player 1 (the maximizer) receives from player 2 (the minimizer), correpsonding to this transition. We assume that the game is non-stopping, that is, $\sum_{u \in V} p_{k\ell}^{vu} = 1$ for all $v \in V$ and $k \in K^v$, $\ell \in L^v$. To simplify later expressions, let us denote by $P^{vu} \in [0, 1]^{K^v \times L^v}$ the transition matrix, the elements of which are the probabilities $p_{k\ell}^{vu}$, and associate in Γ a *local expected reward matrix* A^v to every $v \in V$ defined by

$$(A^v)_{k\ell} = \sum_{u \in V} p_{k\ell}^{vu} r_{k\ell}^{vu}. \tag{1}$$

In the game Γ, players first agree on an initial vertex $v_0 \in V$ to start. Then, in a general step $j = 0, 1, ...$, when the game arrives to state $v_j = v \in V$, they choose mixed strategies $\alpha^v \in \Delta(K^v) := \{y \in \mathbb{R}^{K^v} \mid \sum_{i \in K^v} y_i = 1, \ y_i \geq 0 \ \text{for} \ i \in K^v\}$ and $\beta^v \in \Delta(L^v)$, player 1 receives the amount of $b_j = \alpha^v A^v \beta^v$ from player 2, and the game moves to the next state u chosen according to the transition probabilities $p_{\alpha,\beta}^{vu} = \alpha^v P^{vu} \beta^v$.

The *undiscounted limiting average (effective) payoff* is the *Cesaro average*

$$g^{v_0}(\Gamma) = \liminf_{N \to \infty} \frac{1}{N+1} \sum_{j=0}^{N} \mathbb{E}[b_j], \tag{2}$$

where the expectation is taken over all random choices made (according to mixed strategies and transition probabilities) up to step j of the play. The purpose of player 1 is to maximize $g^{v_0}(\Gamma)$, while player 2 would like to minimize it.

In 1981, Mertens and Neymann in their seminal paper [MN81] proved that every stochastic game has a value from any initial position in terms of history dependent strategies. An example (the so-called Big Match) showing that the same does not hold when restricted to stationary strategies was given in 1957 in Gillette's paper [Gil57]; see also [BF68].

In this paper we shall restrict ourselves (and the players) to the so-called *stationary* strategies, that is, the mixed strategy chosen in a position $v \in V$ can depend only on v but not on the preceding positions or moves before reaching v (i.e., not on the history of the play). We will denote by $\mathcal{K}(\Gamma)$ and $\mathcal{L}(\Gamma)$ the sets of stationary strategies of player 1 and player 2, respectively, that is,

$$\mathcal{K}(\Gamma) = \bigotimes_{v \in V} \Delta(K^v) \quad \text{and} \quad \mathcal{L}(\Gamma) = \bigotimes_{v \in V} \Delta(L^v).$$

Vrieze [Vri80, Theorem 8.3.2] showed that if a stochastic game Γ has a value $g^{v_0}(\Gamma) = m$, which is a constant, independent of the initial state $v_0 \in V$, then it has a value in ϵ-optimal stationary strategies for any $\epsilon > 0$. We call such games *ergodic* and extend their definition as follows.

Definition 1. *For $\epsilon > 0$, a stochastic game Γ is said to be ϵ-ergodic if the game values from any two initial positions differ by at most ϵ, that is, $|g^v(\Gamma) - g^u(\Gamma)| \leq \epsilon$, for all $u, v \in V$. A 0-ergodic game will be simply called ergodic.*

Our main result in this paper is an algorithm that decides, for any given stochastic game Γ and $\epsilon > 0$, whether or not Γ is ϵ-ergodic, and provides a witness for its ϵ-ergodicity/non-ergodicity. As a corollary, we get a constructive proof of the above mentioned theorem of Vrieze [Vri80]. A notion central to our algorithm is the concept of a *potential transformation* introduced in the following section.

1.2   Potential Transformations

In 1958 Gallai [Gal58] suggested the following simple transformation. Let $x : V \to \mathbb{R}$ be a mapping that assigns to each state $v \in V$ a real number x^v called the *potential* of v. For every transition (v, u) and pair of actions $k \in K^v$ and $\ell \in L^v$ let us transform the payoff $r_{k\ell}^{vu}$ as follows:

$$r_{k\ell}^{vu}(x) = r_{k\ell}^{vu} + x^v - x^u.$$

Then the one step expected payoff amount changes to $\mathbb{E}[b_j(x)] = \mathbb{E}[b_j] + \mathbb{E}[x^{v_j}] - \mathbb{E}[x^{v_{j+1}}]$, where $v_j \in V$ is the (random) position reached at step j of the play. However, as the sum of these expectations telescopes, the limiting average payoff remains the same for all finite potentials:

$$g^{v_0}(\Gamma(x)) = g^{v_0}(\Gamma) + \lim_{N \to \infty} \frac{1}{N} \mathbb{E}[x^{v_0} - x^{v_N}] = g^{v_0}(\Gamma).$$

Thus, the transformed game remains equivalent with the original one.

Using potential transformations we may be able to obtain a proof for ergodicity/non-ergodicity. This is made more precise in the following section.

1.3   Local and Global Values and Concepts of Ergodicity

Let us consider an arbitrary potential $x \in \mathbb{R}^V$, and define the *local value* $m^v(x)$ at position $v \in V$ as the value of the $|K^v| \times |L^v|$ local reward matrix game $A^v(x)$ with entries

$$a_{k\ell}^v(x) = \sum_{u \in V} p_{k\ell}^{vu}(r_{k\ell}^{vu} + x^v - x^u), \qquad \text{for all } k \in K^v, \ell \in L^v, \tag{3}$$

that is,

$$m^v(x) = \text{Val}\,(A^v(x)) := \max_{\alpha^v \in \Delta(K^v)} \min_{\beta^v \in \Delta(L^v)} \alpha^v A^v(x) \beta^v = \min_{\beta^v \in \Delta(L^v)} \max_{\alpha^v \in \Delta(K^v)} \alpha^v A^v(x) \beta^v.$$

To a pair of stationary strategies $\alpha = (\alpha^v | v \in V) \in \mathcal{K}(\Gamma)$ and $\beta = (\beta^v | v \in V) \in \mathcal{L}(\Gamma)$ we associate a Markov chain $\mathcal{M}_{\alpha,\beta}(\Gamma)$ on states in V, defined by the transition probabilities $p_{\alpha,\beta}^{vu} = \alpha^v P^{vu} \beta^v$. Then, this Markov chain has unique limiting probability distributions $(q_{\alpha,\beta}^{vu}|\ u \in V)$, where $q_{\alpha,\beta}^{vu}$ is the probability of staying in state $u \in V$ when the initial vertex is $v \in V$. With this notation, The limiting average payoff (2) starting from vertex $v \in V$ can be computed as

$$g^v(\alpha, \beta) \ = \ \sum_{u \in V} q_{\alpha,\beta}^{vu} \left(\alpha^u A^u \beta^u\right). \tag{4}$$

The game is said to be to be solvable in *uniformly optimal* stationary strategies, if there exist stationary strategies $\bar{\alpha} \in \mathcal{K}(\Gamma)$ and $\bar{\beta} \in \mathcal{L}(\Gamma)$, such that for all initial states $v \in V$

$$g^v(\bar{\alpha}, \bar{\beta}) \ = \ \max_{\alpha \in \mathcal{K}(\Gamma)} g^v(\alpha, \bar{\beta}) \ = \ \min_{\beta \in \mathcal{L}(\Gamma)} g^v(\bar{\alpha}, \beta). \tag{5}$$

This common quantity, if exists, is the value of the game with initial position $v \in V$, and will be simply denoted by $g^v = g^v(\Gamma)$.

1.4   Main Result

Given an undiscounted zero-sum stochastic game, we try to reduce the range of its local values by a potential transformation $x \in \mathbb{R}^V$. If they are equalized by some potential x, that is, $m^v(x) = m$ is a constant for all $v \in V$, we say that the game is brought to its *ergodic canonical form* [BEGM13]. In this case, one can show that the values g^v exist and are equal to m for all initial positions $v \in V$, and furthermore, locally optimal strategies are globally optimal [BEGM13]. Thus, the game is solved in uniformly optimal strategies. However, typically we are not that lucky.

To state our main theorem, we need more notation. Let us introduce $W > 0$ such that for all positions $v, u \in V$ and actions $k \in K^v$ and $\ell \in L^v$ we have either $p_{k\ell}^{vu} = 0$ or $p_{k\ell}^{vu} \geq 1/W$. Let us further define R as the smallest real such that

$$0 \leq r_{k\ell}^{vu} \leq R \tag{6}$$

for all positions $v, u \in V$ and actions $k \in K^v$ and $\ell \in L^v$ (where we assume w.l.o.g. that all rewards are non-negative), and set $N = \max_{v \in V}\{\max\{|K^v|, |L^v|\}\}$. Finally, set $n = |V|$, and let $\eta = \max\{\log_2 R, \log_2 W\}$ be the maximum bit length for the rational rewards and transition probabilities.

Theorem 1. *For every stochastic game and* $\epsilon > 0$ *we can find in* $\left(\frac{nNWR}{\epsilon}\right)^{O(2^{2n} nN)}$ *time either a potential vector* $x \in \mathbb{R}^V$ *proving that the game is locally* (24ϵ)-*ergodic, or stationary strategies for the players proving that it is not* ϵ-*ergodic.*

The proof of Theorem 1 will be given in Sect. 4. One major hurdle that we face is that the range of potentials can grow doubly exponentially as iterations proceed, leading to much worse bounds than those stated in the theorem. To deal with this issue, we use quantifier elimination techniques [BPR96, GV88, Ren92] to reduce the range of potentials after each iteration; see the discussion preceding Lemma 8. Proofs of some of the intermediate lemmas will be omitted due to lack of space.

2   Related Work

The above definition of ergodicity follows Moulin's concept of the ergodic extension of a matrix game [Mou76] (which is a very special example of a stochastic game with perfect information). Let us note that slightly different terminology is used in the Markov chain theory; see, for example, [KS63].

The following four algorithms for undiscounted stochastic games are based on stronger "ergodicity type" conditions: the strategy iteration algorithm by Hoffman and Karp [HK66] requires that for any pair of stationary strategies of the two players the obtained Markov chain has to be irreducible; two value iteration algorithms by Federgruen are based on similar but slightly weaker requirements; see [Fed80] for the definitions and more details; the recent algorithm of Chatterjee and Ibsen-Jensen [CIJ14] assumes a weaker requirement than the strong ergodicity required by Hoffman and Karp [HK66]: they call a stochastic game *almost surely ergodic* if for any pair of (not necessarily stationary) strategies of the two players, and any starting position, some strongly ergodic class (in the sense of [HK66]) is reached with probability 1.

While these restrictions apply to the structure of the game, our ergodicity definition only restricts the value. Moreover, the results in [HK66] and [CIJ14] apply to a game that already satisfies the ergodicity assumption, which seems to be hard to check. Our algorithm, on the other hand, always produces an answer, regardless whether the game is ergodic or not.

Interestingly, potentials appear in [Fed80] implicitly, as the differences of local values of positions, as well as in [HK66], as the dual variables to linear programs corresponding to the controlled Markov processes, which appear when a player optimizes his strategy against a given strategy of the opponent. Yet, the potential transformation is not considered explicitly in these papers.

We prove Theorem 1 by an algorithm that extends the approach recently obtained for stochastic games with perfect information [BEGM10]. It is also somewhat similar to the first of two value iteration algorithms suggested by Federgruen in [Fed80], though our approach has some distinct characteristics: It is assumed in [Fed80] that the values g^v exist and are equal for all v; in particular, this assumption implies the ϵ-ergodicity for every $\epsilon > 0$. For our approach we do not need such an assumption. We can verify ϵ-ergodicity for an arbitrary given $\epsilon > 0$, or provide a proof for non-ergodicity (with a small gap) in a finite time. Moreover, while the approach of [Fed80] was only shown to converge, we provide a bound in terms of the input parameters for the number of steps.

Several other algorithms for solving undiscounted zero-sum stochastic games in stationary strategies are surveyed by Raghavan and Filar; see Sects. 4 (B) and 5 in [RF91]. The only algorithmic results that we are aware of that provide bounds on the running time for approximating the value of general (undiscounted) stochastic games are those given in [CMH08,HKL+11]: in [CMH08], the authors provide an algorithm that approximates, within any factor of $\epsilon > 0$, the value of any stochastic game (in history dependent strategies) in time $(nN)^{nN}\text{poly}(\eta, \log\frac{1}{\epsilon})$. In [HKL+11], the authors give algorithms for discounted and recursive stochastic games that run in time $2^{N^{O(N^2)}} \text{poly}(\eta, \log(\frac{1}{\epsilon}))$, and claim also that similar bounds can be obtained for general stochastic games, by reducing them to the discounted version using a discount factor of $\delta = \epsilon^{\eta N^{O(n^2)}}$ (and this bound on δ is almost tight [Mil11]). These results are based on quantifier elimination techniques and yield very complicated history-dependent strategies. For almost sure ergodic games, a variant of the algorithm of Hoffman and Karp [HK66] was given in [CIJ14]; this algorithm finds ϵ-optimal stationary strategies in time (roughly) $\left(\frac{Nn^2W^n}{\epsilon}\right)^{nN} \text{poly}(N, \eta)$. This result is not comparable to ours, since the class of games they deal with are somewhat different (although both generalize the class of strongly ergodic games of [HK66]). Furthermore, the algorithm in Theorem 1 exhibits the additional feature that it either provides a solution in stationary strategies in the ergodic case, if one exists, or produces a pair of stationary strategies that witness the non-ergodicity.

3   Pumping Algorithm

We begin by describing our procedure on an abstract level. Then we specialize it to stochastic games in Sect. 4.

Given a subset $S \subseteq V$, let us denote by $e_S \in \{0, 1\}^V$ the characteristic vector of S.

Let us further assume that $m^v(x)$ for $v \in V$ are functions depending on potentials $x \in \mathbb{R}^n$ (where $n = |V|$) and satisfying the following properties for all subsets $S \subseteq V$ and reals $\delta \geq 0$:

(i) $m^v(x - \delta e_S)$ is a monotone decreasing function of δ if $v \in S$;
(ii) $m^v(x - \delta e_S)$ is a monotone increasing function of δ if $v \notin S$;
(iii) $|m^v(x) - m^v(x - \delta e_S)| \leq \delta$ for all $v \in V$.

We show in this section that under the above conditions we can change iteratively the potentials to some $x' \in \mathbb{R}^n$ such that either all values $m^v(x')$, $v \in V$, are very close to one another or we can find a decomposition of the states V into disjoint subsets proving that such convergence of the values is not possible.

Our main procedure is described in Algorithm 2 below. Given the current vector of potentials x_τ at iteration τ, the procedure partitions the set of vertices into four sets according to the local value $m^v(x)$. If either the first (top) set T_τ or forth (bottom) set B_τ is empty, the procedure terminates; otherwise, the potentials of all the vertices in the first and second sets are reduced by the same amount δ, and the computation proceeds to the next iteration.

We can show next that properties (i), (ii) and (iii) above guarantee some simple properties for the above procedure.

Lemma 1. *We have $T_{\tau+1} \subseteq T_\tau$, $B_{\tau+1} \subseteq B_\tau$ and $M_{\tau+1} \supseteq M_\tau$ for all iterations $\tau = 0, 1, \ldots$*

Algorithm 1. $\text{PUMP}(x, S)$

Input: a stochastic game Γ a subset S of states.
Output: a potential $x \in \mathbb{R}^S$.
1: Initialize $\tau := 0$, and $x_\tau := x$.
2: Set $m^+ := \max_{v \in S} m^v(x_\tau)$, $m^- := \min_{v \in S} m^v(x_\tau)$, and $\delta := (m^+ - m^-)/4$.
3: Define
$$T_\tau := \{v \in S \mid m^v(x_\tau) \geq m^- + 3\delta\}$$
$$B_\tau := \{v \in S \mid m^v(x_\tau) < m^- + \delta\}$$
$$M_\tau := S \setminus (T_\tau \cup B_\tau).$$

4: **if** $T_\tau = \emptyset$ or $B_\tau = \emptyset$ **then**
5:     **return** x_τ
6: **end if**
7: Otherwise, set $P_\tau := \{v \in S \mid m^v(x_\tau) \geq m^- + 2\delta\}$ and update

$$x_{\tau+1}^v := \begin{cases} x_\tau^v - \delta & \text{if } v \in P_\tau \\ x_\tau^v & \text{otherwise.} \end{cases}$$

8: Set $\tau := \tau + 1$ and Goto step 3.

Proof. Indeed, by (i) and (iii) we can conclude that $m^v(x_\tau) \geq m^- + \delta$ holds for all $v \in P_\tau$. Analogously, by (ii) and (iii) $m^v(x_\tau) < m^- + 3\delta$ follows for all $v \notin P_\tau$. $\square$

Lemma 2. *Either $T_\tau = \emptyset$ or $B_\tau = \emptyset$ for some finite τ, or there are nonempty disjoint subsets $I, F \subseteq S$, $I \supseteq T_\tau$, $F \supseteq B_\tau$, and a threshold τ_0, such that for every real $\Delta \geq 0$ there exists a finite index $\tau(\Delta) \geq \tau_0$ such that*

(a) $m^v(x_\tau) \geq m^- + 2\delta$ *for all $v \in I$ and $m^v(x_\tau) < m^- + 2\delta$ for all $v \in F$, and for all $\tau \geq \tau_0$;*
(b) $x_\tau^u - x_\tau^v \geq \Delta$ *for all $v \in I$ and $u \notin I$, and for all $\tau \geq \tau(\Delta)$;*
(c) $x_\tau^v - x_\tau^u \geq \Delta$ *for all $v \in F$ and $u \notin F$, and for all $\tau \geq \tau(\Delta)$.*

Proof. By Lemma 1 sets T_τ and B_τ can change only monotonically, and hence only at most $|S|$ times. Thus, if $\text{PUMP}(x, S)$ does not stop in a finite number of iterations, then after a finite number of iterations the sets T_τ and B_τ will never change and all positions in T_τ remain always pumped (that is, have their potentials reduced), while all positions in B_τ will be never pumped again.

Assuming now that the pumping algorithm $\text{PUMP}(x, S)$ does not terminate, let us define the subset $I \subseteq S$ as the set of all those positions which are always pumped with the exception of a finite number of iterations. Analogously, let F be the subset of all those positions that are never pumped with the exception of a finite number of iterations. Since I and F are finite sets, there must exist a finite τ_0 such that for all $\tau \geq \tau_0$ we have $I \subseteq P_\tau$ and $F \cap P_\tau = \emptyset$, implying (a). Note that any vertex in T_τ is always pumped by (iii) and hence $T_\tau \subseteq I$ for any $\tau \geq \tau_0$; similarly, $B_\tau \subseteq F$ for any $\tau \geq \tau_0$.

Let us next observe that all positions not in $I \cup F$ are both pumped and not pumped infinitely many times. Thus, since δ is a fixed constant, for every Δ there must exist an iteration $\tau(\Delta) \geq \tau_0$ such that all positions not in I are not pumped by at least Δ/δ many more times than those in I, and all positions not in F are pumped by at least Δ/δ many more times than those in F, implying (b) and (c). □

Let us next describe the use of $\text{PUMP}(x, S)$ for repeatedly shrinking the range of the m^v values, or to produce some evidence that this is not possible. A simplest version is the following:

Algorithm 2. REPEATEDPUMPING(ϵ)

1: Initialize $h := 0$, and $x_h := 0 \in \mathbb{R}^V$.
2: Set $m^+(h) := \max_{v \in V} m^v(x_h)$ and $m^-(h) := \min_{v \in V} m^v(x_h)$.
3: If $m^+(h) - m^-(h) \leq \epsilon$ then STOP.
4: $x_{h+1} := \text{PUMP}(x_h, V)$; $h := h + 1$.
5: Goto step 2.

Note that by our above analysis, REPEATEDPUMPING either returns a potential transformation for which all m^v, $v \in V$ values are within an ϵ-band, or returns the sets I and F as in Lemma 2 with arbitrary large potential differences from the other positions. In the next section we use a modification of these procedures for stochastic games, and show that those large potential differences can be used to prove that the game is not ϵ-ergodic.

4   Application of Pumping for Stochastic Games

We show in this section how to use REPEATEDPUMPING to find potential transformations verifying ϵ-ergodicity, or proving that the game is not ϵ-ergodic, thus establishing a proof of Theorem 1. Towards this end, we shall give some necessary and sufficient conditions for ϵ-non-ergodicity, and consider a modified version of the pumping algorithm described in the previous section which will provide a constructive proof for the above theorem.

Let us first observe that the local value function of stochastic games satisfies the properties required to run the pumping algorithm described in the previous section.

Lemma 3. *For every subset $S \subseteq V$ and $\delta \geq 0$ and for all $v \in V$ we have*

$$
\begin{aligned}
m^v(x) &\geq m^v(x - \delta e_S) \geq m^v(x) - \delta \max_{k,\ell} \textstyle\sum_{u \notin S} p_{k\ell}^{vu} \quad \text{if } v \in S, \\
m^v(x) &\leq m^v(x - \delta e_S) \leq m^v(x) + \delta \max_{k,\ell} \textstyle\sum_{u \in S} p_{k\ell}^{vu} \quad \text{if } v \notin S.
\end{aligned}
\tag{7}
$$

Furthermore, the value functions $m^v(x)$ for $v \in V$ satisfy properties (i), (ii) and (iii) stated in Sect. 3.

The above lemma implies that procedures PUMP and REPEATEDPUMPING could, in principle, be used to find a potential transformation yielding an ϵ-ergodic solution. It does not offer, however, a way to discover ϵ-non-ergodicity. Towards this end, we need to find some sufficient and algorithmically achievable conditions for ϵ-non-ergodicity.

Lemma 4. *A stochastic game Γ is ϵ-non-ergodic if there exist disjoint non-empty subsets of the positions $I, F \subseteq V$, reals a, b with $b - a \geq \epsilon$, stationary strategies α^v, $v \in I$, for player 1, and β^u, $u \in F$, for player 2, and a vector of potentials $x \in \mathbb{R}^V$, such that*

(N1) $\alpha_k^v p_{k\ell}^{vu} = 0$ *for all $v \in I$, $u \notin I$, $k \in K^v$ and $\ell \in L^v$,*
(N2) $\beta_\ell^u p_{k\ell}^{uw} = 0$ *for all $u \in F$, $w \notin F$, $\ell \in L^u$ and $k \in K^u$, and*
(N3) *for all $v \in I$ and $u \in F$:*

$$\min_{\widetilde{\beta}^v \in \Delta(L^v)} (\alpha^v)^T A^v(x) \widetilde{\beta}^v \geq b \quad and \quad \max_{\widetilde{\alpha}^u \in \Delta(K^u)} (\widetilde{\alpha}^u)^T A^u(x) \beta^u < a.$$

Let us introduce a notation for denoting upper bounds on the entries of the matrices, more precisely on the part of these entries which do not depend on negative potential differences. Specifically, define

$$\widetilde{a}_{k\ell}^v(x) = \sum_{u \in V} p_{k\ell}^{vu} r_{k\ell}^{vu} + \sum_{u \in V,\ x^u \leq x^v} p_{k\ell}^{vu}(x^v - x^u)$$

$$\widetilde{b}_{k\ell}^v(x) = m^+(x) - \sum_{u \in V} p_{k\ell}^{vu} r_{k\ell}^{vu} - \sum_{u \in V,\ x^u \geq x^v} p_{k\ell}^{vu}(x^v - x^u) \tag{8}$$

where, as before, $m^+(x) := \max_v m^v(x)$, $m^-(x) := \min_v m^v(x)$. Define further

$$R^v(x) = \max_{k \in K^v, \ell \in L^v} (\widetilde{a}_{k\ell}^v(x)) \quad \text{if } m^v(x) \geq \frac{m^+(x) + m^-(x)}{2},$$
$$R^v(x) = \max_{k \in K^v, \ell \in L^v} \left(\widetilde{b}_{k\ell}^v(x)\right) \quad \text{otherwise.} \tag{9}$$

Note that

$$m^+(x) - \widetilde{b}_{k\ell}^v(x) \leq a_{k\ell}^v(x) \leq \widetilde{a}_{k\ell}^v(x) \quad \text{for all } v \in V, \, k \in K^v, \, \ell \in L^v \text{ and } x \in \mathbb{R}^V,$$

which implies

$$\begin{aligned} m^v(x) &\leq R^v(x) && \text{if } m^v(x) \geq \frac{m^+(x)+m^-(x)}{2}, \\ m^v(x) &\geq m^+(x) - R^v(x) && \text{otherwise.} \end{aligned} \quad \text{for all } v \in V \text{ and } x \in \mathbb{R}^V,$$
$$\tag{10}$$

With this notation we can state a more constructive version of Lemma 4.

Lemma 5. *A stochastic game Γ satisfying (6) is ϵ-non-ergodic if there exist disjoint non-empty subsets $I, F \subseteq V$, a vector of potentials $x \in \mathbb{R}^V$, and reals $a', b' \in [0, m^+(x)]$ with $b' - a' \geq 3\epsilon$, $a' < \frac{m^+(x)+m^-(x)}{2}$, $b' \geq \frac{m^+(x)+m^-(x)}{2}$, such that*

(N4) $m^v(x) \geq b'$ for all $v \in I$, and $m^u(x) < a'$ for all $u \in F$;

(N5) $x^u - x^v \geq |L^v| W R^v(x)^2/\epsilon$ for all $u \notin I$, and $v \in I$;

(N6) $x^u - x^v \geq |K^v| W R^v(x)^2/\epsilon$ for all $u \in F$, and $v \notin F$.

Proof. We first show that (N4)-(N5) imply the existence of strategies α^v, for $v \in I$, satisfying (N1) and (N3). We shall then observe that a similar argument can be applied to (N4) and (N6) to show the existence of strategies β^u, for $u \in F$, such that those satisfy (N2) and (N3). Consequently, our claim will follow by Lemma 4.

Let us now fix a position $v \in I$ and denote respectively by $\bar{\alpha}^v$ and $\bar{\beta}^v$ the optimal strategies of players with respect to the payoff matrix $A^v(x)$. Denote further by $\widehat{\beta}^v = \frac{1}{|L^v|}(1, 1, \ldots, 1)$ the uniform strategy for player 2, and set $\bar{K}^v = \{k \in K^v \mid \sum_{u \notin I} \sum_{\ell \in L^v} p_{k\ell}^{vu} = 0\}$.

Let us then note that we have

$$\left(A^v(x)\widehat{\beta}^v\right)_k \leq \begin{cases} R^v(x) & \text{if } k \in \bar{K}^v, \\ R^v(x) - \frac{R^v(x)^2}{\epsilon} & \text{otherwise,} \end{cases}$$

since at least one of the entries of (N5) has at least $\frac{W}{|L^v|}$ as a coefficient in rows which are not in $\bar{K}^v$.

Note that $b' > 0$ implies by (10) that $R^v(x) > 0$. Thus by the optimality of $\bar{\alpha}$ and by the above inequalities we have

$$0 < b' \leq m^v(x) \leq \bar{\alpha}^v A^v(x)\widehat{\beta}^v \leq R^v(x) - \left(\sum_{k \notin \bar{K}^v} \bar{\alpha}_k^v\right)\frac{R^v(x)^2}{\epsilon}$$

implying that $\sum_{k \notin \bar{K}^v} \bar{\alpha}_k^v < \frac{\epsilon}{R^v(x)}$. Since by (N4) we have $0 < a'$, inequalities $\epsilon < a' + 3\epsilon \leq b' < m^v(x) \leq R^v(x)$ follow, and hence $\frac{3\epsilon}{R^v(x)} < 1$ must hold, implying that the set $\bar{K}^v$ is not empty. Let us then denote by $\widetilde{\alpha}^v$ the truncated strategy defined by

$$\widetilde{\alpha}_k^v = \begin{cases} \dfrac{\bar{\alpha}_k^v}{\sum_{k \in \bar{K}^v} \bar{\alpha}_k^v} & \text{if } k \in \bar{K}^v, \\ 0 & \text{if } k \notin \bar{K}^v. \end{cases}$$

With this we have for any $\widetilde{\beta}^v \in \Delta(L^v)$

$$b' \leq m^v(x) \leq (\bar{\alpha}^v A^v(x)\widetilde{\beta}^v$$

$$= \left(\widetilde{\alpha}^v A^v(x)\widetilde{\beta}^v\right)\left(\sum_{k \in \bar{K}^v} \bar{\alpha}_k^v\right) + \sum_{k \notin \bar{K}^v} \bar{\alpha}_k^v\left(\sum_{\ell \in L^v} a_{k\ell}^v(x)\widetilde{\beta}_\ell^v\right)$$

$$\leq \left(\widetilde{\alpha}^v A^v(x)\widetilde{\beta}^v\right) + \left(\sum_{k \notin \bar{K}^v} \bar{\alpha}_k^v\right) R^v(x)$$

$$< \left(\widetilde{\alpha}^v A^v(x)\widetilde{\beta}^v\right) + \epsilon.$$

Let us then define $\alpha^v = \widetilde{\alpha}^v$ and repeat the same for all $v \in I$. Then, these strategies satisfy (N1) and (N3) with $b = b' - \epsilon$.

Let us next note that by adding a constant to a matrix game it changes its value with exactly the same constant. Furthermore, multiplying all entries by -1 and transposing it, changes its value by a factor of -1, interchanges the roles of row and column players, but leaves otherwise optimal strategies still optimal. Thus, we can repeat the above arguments for the matrices $B^u(x) = m^+(x)E^u - A^u(x)^T$, where E is the $|L^u| \times |K^u|$-matrix of all ones, and obtain the same way strategies β^u, $u \in F$ satisfying (N2) and (N3) with $a = a' + \epsilon$. This completes the proof of the lemma.   □

To create a finite algorithm to find sets I and F and potentials satisfying (N4)-(N6) we need to do some modifications in our procedures.

First, we allow a more flexible partitioning of the m-range by allowing the m-range boundaries to be passed as parameters and replacing line 2 in procedure PUMP by

2: Set $\delta := (m^+ - m^-)/4$.

Next, Let us replace in procedure PUMP, line 7 by the following lines, where $\epsilon > 0$ is a prespecified parameter, and call the new procedure with these modifications MODIFIEDPUMP(ϵ, x, S, m_-, m_+):

7a: Otherwise set $P_\tau := \{v \in S \mid m^v(x_\tau) \geq m^- + 2\delta\}$ and compute

$$R_\tau^v := \max_{k \in K^v, \ell \in L^v} (\widetilde{a}_{k\ell}^v(x_\tau)) \qquad \text{if} \quad v \in P_\tau,$$

$$R_\tau^v := \max_{k \in K^v, \ell \in L^v} \left(\widetilde{b}_{k\ell}^v(x_\tau)\right) \qquad \text{if} \quad v \notin P_\tau,$$

where $\widetilde{a}$ and $\widetilde{b}$ are defined by (8).

7b: Create an auxiliary directed graph $G = (V, E)$ on vertex set V such that $(v, u) \in E$ iff

$$x_\tau^u - x_\tau^v < \frac{|L^v|W(R_\tau^v)^2}{\epsilon} \qquad \text{if} \quad v \in P_\tau,$$

$$x_\tau^v - x_\tau^u < \frac{|K^v|W(R_\tau^v)^2}{\epsilon} \qquad \text{if} \quad v \notin P_\tau.$$

7c: Find subsets I_τ and F_τ of V such that $T_\tau \subseteq I_\tau \subseteq P_\tau$, $B_\tau \subseteq F_\tau \subseteq V \setminus P_\tau$, and no arcs are leaving these sets in G (this can be done by a finding the strong components of G, or by the method described int he proof of Theorem 1).

7d: if such sets are found STOP and output these sets, otherwise continue with step 8.

Before starting to analyze this modified pumping algorithm, let us observe that we have for all iterations

$$m^- < m^- + \frac{\epsilon}{2} < \frac{m^- + m^+}{2} \leq m^v(x_\tau) \leq R_\tau^v \qquad \text{for all} \quad v \in P_\tau \qquad (11)$$

as long as $m^+ - m^- > \epsilon$.

Lemma 6. *Procedure* MODIFIEDPUMP(ϵ, x, S) *terminates in a finite number of steps.*

Lemma 7. *Procedure* MODIFIEDPUMP(ϵ, x, V) *either shrinks the m-range by a factor of $3/4$ or outputs potentials $x = x_\tau$ and sets $I = I_\tau$ and $F = F_\tau$ which satisfy conditions (N4)-(N6) with $a' < b'$.*

Let us observe that the bounds and strategies obtained by Lemmas 6 and 7 do not necessarily imply the ϵ-non-ergodicity of the game since those positions in I_τ and F_τ may not have enough separation in m-values (i.e. the condition $b' - a' \geq 3\epsilon$ in Lemma 5 is not satisfied). To fix this we need to make one more use of the pumping algorithm, as described in the MODIFIEDREPEATEDPUMP-ING procedure below. After each range-shrinking in this algorithm, we use a routine called REDUCEPOTENTIAL(Γ, x, m_-, m_+) which takes the current potential vector x and range $[m_-, m_+]$ and produces another potential vector y such that $\|y\|_\infty \leq 2^{\mathrm{poly}(n, N, \eta)}$. We need to this because, as the algorithm proceeds, the potentials, and hence the transformed rewards, might grow doubly-exponentially high.

The potential reduction can be done as follows. We write the following quadratic program in the variables $x \in \mathbb{R}^V$, $\alpha = (\alpha^v \mid v \in V)] \in \mathcal{K}(\Gamma)$, and $\beta = (\beta^v \mid v \in V)] \in \mathcal{L}(\Gamma)$:

$$\begin{aligned} \alpha^v A^v(x') \geq m_- \cdot \mathbf{e}, &\qquad A^v(x')\beta^v \leq m_+ \cdot \mathbf{e}, &\qquad (12)\\ \alpha^v \mathbf{e} = 1, &\qquad \mathbf{e}\beta^v = 1,\\ \alpha^v \geq 0, &\qquad \beta^v \geq 0, \end{aligned}$$

for all $v \in V$, where $\mathbf{e}$ denotes the vector of all ones of appropriate dimension. This is a quadratic system of at most $6N$ (in)equalities on at most $(2N + 1)n$ variables. Moreover the system is feasible since the original potential vector x satisfies it. Thus, a rational approximation to the solution to within an additive accuracy of δ can be computed, using quantifier elimination algorithms, in time $\mathrm{poly}(\eta, N^{O(nN)}, \log \frac{1}{\delta})$; see [BPR96, GV88, Ren92]. Note that the resulting solution will satisfy (12) but within the approximate range $[m_- - \delta, m_+ + \delta]$. By choosing δ sufficiently smaller than the desired accuracy ϵ, we can ignore the effect of such approximation.

Lemma 8. MODIFIEDREPEATEDPUMPING(ϵ) *terminates in a finite number $h \leq \log \frac{R}{24\epsilon} / \log \frac{7}{8}$, of iterations, and either provides a potential transformation proving that the game is 24ϵ-ergodic, or outputs two nonempty subsets I and F and strategies α^v, $v \in I$, for player 1 and β^v, $v \in F$, for player 2 such that conditions (N4), (N5) and (N6) hold with b', a' satisfying the condition in Lemma 5.*

To complete the proof of Theorem 1, we need to analyze the time complexity of the above procedure, in particular, bounding the number of pumping steps performed in MODIFIEDPUMP.

Algorithm 3. MODIFIEDREPEATEDPUMPING(ϵ)

1: Initialize $h := 0$, and $x_h := 0 \in \mathbb{R}^V$.
2: Set $m^+(h) := \max_{v \in V} m^v(x_h)$ and $m^-(h) := \min_{v \in V} m^v(x_h)$.
3: **if** $m^+(h) - m^-(h) \le 24\epsilon$ **then**
4:    **return** x_h.
5: **end if**
6: $x_{h+1} :=$ MODIFIEDPUMP($\epsilon, x_h, V, m_-, m_+$) and let $F_\tau, I_\tau, T_\tau, B_\tau, P_\tau$ be the sets obtained from MODIFIEDPUMP.
7: **if** $T_\tau = \emptyset$ or $B_\tau = \emptyset$ **then**
8:    $x_{h+1} :=$ REDUCEPOTENTIAL($\Gamma, x_\tau, m_-(h), m_+(h)$)
9:    Set $h := h + 1$ and Goto step 2
10: **end if**
11: Otherwise set $F = F_\tau$ and $I = I_\tau$.
12: $x_{h+1} :=$ MODIFIEDPUMP($\epsilon, x_h, I_\tau, m_-, m_+$) and let T_τ, B_τ be the sets obtained from this call of MODIFIEDPUMP.
13: **if** $T_\tau = \emptyset$ **then**
14:    $x_{h+1} :=$ REDUCEPOTENTIAL($\Gamma, x_\tau, m_-(h), m_+(h)$)
15:    Set $h := h + 1$ and Goto step 2.
16: **end if**
17: **if** $B_\tau = \emptyset$ **then**
18:    Goto step 21
19: **end if**
20: Otherwise, update $I := I_\tau$.
21: **return** x_{h+1} and the sets I and F.

Let us note that as long as $m^+ - m^- > 24\epsilon$ we pump the upper half P_τ by exactly $\delta \ge 6\epsilon$. Let $\mathcal{P}_\tau(v)$ (resp., $\mathcal{N}_\tau(v)$) denote the number of iterations, among the first τ, in which position v was pumped, that is, $v \in P_\tau$ (resp., not pumped, that is, $v \notin P_\tau$).

Let us next sort the positions $v \in V$ such that we have

$$x_\tau^{v_1} \le x_\tau^{v_2} \le \cdots \le x_\tau^{v_n},$$

and write $\Delta_j = x_\tau^{v_{j+1}} - x_\tau^{v_j}$ for $j = 1, 2, ..., n - 1$. Note that $\mathcal{P}_\tau(v_1) = \tau$ and $\mathcal{N}_\tau(v_n) = \tau$.

Let i_τ be the largest index in $\{1, 2, \ldots, n\}$, such that $v_{i_\tau} \in P_\tau$. Then, by (8) we have for $i = 0, 1, 2, \ldots, i_\tau - 1$ that

$$0 \le \widetilde{a}_{k\ell}^{v_{i+1}}(x_\tau) \le R + \sum_{j=1}^{i} \Delta_j, \tag{13}$$

where the sum over the empty sum is zero by definition. Similarly, for $i = i_\tau + 1, \ldots, n$, we have

$$-R \le \widetilde{b}_{k\ell}^{v_i}(x_\tau) \le R + \sum_{j=1}^{n-i} \Delta_{n-j}. \tag{14}$$

From (13) and (14), it follows that

$$|R_\tau^{v_{i+1}}| \leq \begin{cases} R + \sum_{j=1}^{i} \Delta_i, & \text{for } i = 0, 1, 2, \ldots, i_\tau - 1 \\ R + \sum_{j=1}^{n-i-1} \Delta_{n-j}, & \text{for } i = i_\tau, i_\tau + 1, \ldots, n - 1. \end{cases} \tag{15}$$

Let $\widetilde{i}_\tau$ be the smallest index i such that

$$\Delta_i > \frac{NW(R + \sum_{j=1}^{i-1} \Delta_j)^2}{\epsilon}, \tag{16}$$

and let $\widehat{i}_\tau$ be the largest index $i \leq n - 1$ such that

$$\Delta_i > \frac{NW(R + \sum_{j=1}^{n-i-1} \Delta_{n-j})^2}{\epsilon}. \tag{17}$$

From the definition of $\widetilde{i}_\tau$, we know that

$$\Delta_i \leq \frac{NW(R + \sum_{j=1}^{i-1} \Delta_j)^2}{\epsilon}, \quad \text{for all } i = 1, \ldots, \widetilde{i}_\tau - 1.$$

Solving this recurrence, we get

$$x_\tau^{v_{\widetilde{i}_\tau}} - x_\tau^{v_1} = \sum_{i=1}^{\widetilde{i}_\tau - 1} \Delta_i \leq \left(\frac{(\widetilde{i}_\tau - 1)NWR}{\epsilon} \right)^{2^{\widetilde{i}_\tau - 1} - 1} (\widetilde{i}_\tau - 1)^2 R \leq \left(\frac{nNWR}{\epsilon} \right)^{2^n - 1} n^2 R. \tag{18}$$

Similarly, the definition of $\widehat{i}_\tau$ gives

$$\Delta_i \leq \frac{NW(R + \sum_{j=1}^{n-i-1} \Delta_{n-i})^2}{\epsilon}, \quad \text{for all } i = \widehat{i}_\tau + 1, \ldots, n - 1,$$

from which follows

$$x_\tau^{v_n} - x_\tau^{v_{\widehat{i}_\tau + 1}} \leq \left(\frac{nNWR}{\epsilon} \right)^{2^n - 1} n^2 R. \tag{19}$$

Note that if $\widetilde{i}_\tau \leq i_\tau$ then (15) implies that taking $I_\tau = \{v_1, \ldots, v_{\widetilde{i}_\tau}\}$ would satisfy condition (N5) and guarantee that $I_\tau \subseteq P_\tau$. Similarly, having $\widehat{i}_\tau \geq i_\tau$ guarantees that taking $F_\tau = \{v_{\widehat{i}_\tau + 1}, \ldots, v_n\}$ would satisfy (N6) and $F_\tau \cap P_\tau = \emptyset$. On the other hand, if $\widetilde{i}_\tau \geq i_\tau + 1$, then (18) implies that $v_{i_\tau + 1}$ was always pumped except for at most $\kappa(R) := \left(\frac{nNWR}{\epsilon} \right)^{2^n - 1} \frac{n^2 R}{\delta}$ iterations, that is, $\mathcal{N}_\tau(v_{i_\tau + 1}) \leq \kappa(R)$. Also, since $v_{i_\tau + 1} \notin P_\tau$, then at time τ, $v_{i_\tau + 1}$ is not pumped. Similarly, if $\widehat{i}_\tau < i_\tau$, then (19) implies that v_{i_τ} was never pumped except for at most $\kappa(R)$ iterations, that is, $\mathcal{P}_\tau(v_{i_\tau}) \leq \kappa(R)$, while it is pumped at time τ. Since we have at most n candidates for each of v_{i_τ} and $v_{i_\tau + 1}$, it follows that after $\tau = 2n\kappa(R) + 1$, neither of these events ($\widetilde{i}_\tau \geq i_\tau + 1$ and $\widehat{i}_\tau < i_\tau$) can happen, which by our earlier observations implies that the algorithm constructs the sets I_τ and F_τ. We can conclude that $\mathrm{MODIFIEDPUMP}(\epsilon, x, V)$ must terminate in at most $2n\kappa(R) + 1$ iterations, either producing $m^+ - m^- \leq 24\epsilon$ or outputting the subsets I_τ and F_τ proving ϵ-non-ergodicity.

One can similarly bound the running time for the second call of $\mathrm{MODIFIED}$-PUMP (line 12), and the running time for each iteration of $\mathrm{MODIFIEDREPEAT}$-$\mathrm{EDPUMPING}(\epsilon)$ (but with R replaced by $2^{\mathrm{poly}(n,N,\eta)}$). This completes the proof of the theorem. $\square$

References

[BEGM10] Boros, E., Elbassioni, K., Gurvich, V., Makino, K.: A pumping algorithm for ergodic stochastic mean payoff games with perfect information. In: Eisenbrand, F., Shepherd, F.B. (eds.) IPCO 2010. LNCS, vol. 6080, pp. 341–354. Springer, Heidelberg (2010)

[BEGM13] Boros, E., Elbassioni, K., Gurvich, V., Makino, K.: On canonical forms for zero-sum stochastic mean payoff games. Dyn. Games Appl. 3(2), 128–161 (2013)

[BF68] Blackwell, D., Ferguson, T.S.: The big match. Ann. Math. Statist. 39(1), 159–163 (1968)

[BPR96] Basu, S., Pollack, R., Roy, M.: On the combinatorial and algebraic complexity of quantifier elimination. J. ACM 43(6), 1002–1045 (1996). Preliminary version in FOCS 1994

[CIJ14] Chatterjee, K., Ibsen-Jensen, R.: The complexity of ergodic mean-payoff games. In: Esparza, J., Fraigniaud, P., Husfeldt, T., Koutsoupias, E. (eds.) ICALP 2014, Part II. LNCS, vol. 8573, pp. 122–133. Springer, Heidelberg (2014)

[CMH08] Chatterjee, K., Majumdar, R., Henzinger, T.A.: Stochastic limit-average games are in exptime. Int. J. Game Theory 37, 219–234 (2008)

[Fed80] Federgruen, A.: Successive approximation methods in undiscounted stochastic games. Oper. Res. 1, 794–810 (1980)

[Gal58] Gallai, T.: Maximum-minimum Sätze über Graphen. Acta Mathematica Academiae Scientiarum Hungaricae 9, 395–434 (1958)

[Gil57] Gillette, D.: Stochastic games with zero stop probabilities. In: Dresher, M., Tucker, A.W., Wolfe, P. (eds) Contribution to the Theory of Games III, volume 39 of Annals of Mathematics Studies, pp. 179–187. Princeton University Press (1957)

[GV88] Grigoriev, D., Vorobjov, N.: Solving systems of polynomial inequalities in subexponential time. J. Symb. Comput. 5(1/2), 37–64 (1988)

[HK66] Hoffman, A.J., Karp, R.M.: On nonterminating stochastic games. Manag. Sci. Ser. A 12(5), 359–370 (1966)

[HKL+11] Hansen, K.A., Koucky, M., Lauritzen, N., Miltersen, P.B., Tsigaridas. E.P.: Exact algorithms for solving stochastic games: extended abstract. In: Proceedings of the 43rd Annual ACM Symposium on Theory of Computing, STOC '11, pp. 205–214. ACM, New York (2011)

[KS63] Kemeny, J.G., Snell, J.L.: Finite Markov chains. Springer, New York (1963)

[Mil11] Miltersen, P.B.: Discounted stochastic games poorly approximate undiscounted ones, manuscript. Technical report (2011)

[MN81] Mertens, J.F., Neyman, A.: Stochastic games. Int. J. Game Theory 10, 53–66 (1981)

[Mou76] Moulin, H.: Prolongement des jeux à deux joueurs de somme nulle. Bull. Soc. Math. France, Memoire, 45 (1976)

[Ren92] Renegar, J.: On the computational complexity and geometry of the first-order theory of the reals. J. Symb. Comput. **13**(3), 255–352 (1992)

[RF91] Raghavan, T.E.S., Filar, J.A.: Algorithms for stochastic games: a survey. Math. Methods Oper. Res. **35**(6), 437–472 (1991)

[Sha53] Shapley, L.S.: Stochastic games. Proc. Nat. Acad. Sci. USA **39**, 1095–1100 (1953)

[Vri80] Vrieze, O.J.: Stochastic games with finite state and action spaces. Ph.D. thesis, Centrum voor Wiskunde en Informatica, Amsterdam, The Netherlands (1980)

Miscellaneous

The Popular Matching and Condensation Problems Under Matroid Constraints

Naoyuki Kamiyama[(✉)]

Institute of Mathematics for Industry, Kyushu University, Fukuoka, Japan
kamiyama@imi.kyushu-u.ac.jp

Abstract. In this paper, we first consider a matroid generalization of the popular matching problem (without ties) introduced by Abraham, Irving, Kavitha, and Mehlhorn, and give a polynomial-time algorithm for this problem. In the second half of this paper, we consider the problem of transforming a given instance of the popular matching problem (without ties) by deleting a minimum number of applicants so that it has a popular matching under matroid constraints. This problem is a matroid generalization of the popular condensation problem proposed by Wu, Lin, Wang, and Chao. By using the results in the first half, we give a polynomial-time algorithm for this problem.

1 Introduction

The popular matching problem introduced by Abraham, Irving, Kavitha, and Mehlhorn [1] is a matching problem in which applicants have preferences to posts. Roughly speaking, a matching M is said to be popular, if there exists no other matching N such that the number of applicants that prefer N to M is larger than the number of applicants that prefer M to N. The concept of popularity was introduced by Gärdenfors [2]. The goal of the popular matching problem is to decide whether a popular matching exists in a given instance, and find such a matching if one exists. Abraham, Irving, Kavitha, and Mehlhorn [1] presented a polynomial-time algorithm for this problem. Since their seminal paper, several extensions of the popular matching problem have been investigated. For example, Manlove and Sng [3] considered its many-to-one version. Mestre [4] considered a weighted version of the popular matching problem. Sng and Manlove [5] considered a weighted many-to-one version of the popular matching problem. In this paper, we first consider a matroid generalization of the popular matching problem (without ties), and present a polynomial-time algorithm for this problem. Our algorithm can be regarded as a matroid generalization of the algorithm of [1].

Unfortunately, it is known [1] that a given instance of the popular matching problem may admit no popular matching. For coping with such instances, several alternative solutions were presented. For example, Kavitha and Nasre [6] considered the problem of deciding capacities of posts so that a given instance has a popular matching. Furthermore, Kavitha, Nasre, and Nimbhorkar [7] considered the problem of augmenting capacities of posts with minimum costs.

© Springer International Publishing Switzerland 2014
Z. Zhang et al. (Eds.): COCOA 2014, LNCS 8881, pp. 713–728, 2014.
DOI: 10.1007/978-3-319-12691-3_53

These problems are hard in general. Wu, Lin, Wang, and Chao [8] considered the problem of transforming the set of applicants so that a given instance has a popular matching. Precisely speaking, they introduced the popular condensation problem whose goal is to transform a given instance by deleting a minimum number of applicants so that it has a popular matching, and gave a polynomial-time algorithm for this problem. In the second half of this paper, we consider a matroid generalization of the popular condensation problem, and give a polynomial-time algorithm for this problem. Our algorithm can be regarded as a matroid generalization of the algorithm of [8].

A matroid generalization of the stable matching problem introduced by Fleiner [9] led to the discrete convex function generalization of the stable matching problem presented by Fujishige and Tamura [10] and the matroid approach to the stable matching problem with lower quotas presented by Fleiner and Kamiyama [11]. We hope that our abstract model helps further progress in the field of the popular matching problem.

The rest of this paper is organized as follows. In Sect. 2, we give necessary notation and several properties of matroids that will be used later. In Sect. 3, we give a characterization of the existence of a popular matching under matroid constraints. In Sect. 4, we give an algorithm for the popular matching problem under matroid constraints. In Sect. 5, we propose an algorithm for the popular condensation problem under matroid constraints.

2   Preliminaries

Let $\mathbb{Z}_+$ be the set of non-negative integers. For each set X and each element x, we define $X + x := X \cup \{x\}$ and $X - x := X \backslash \{x\}$. A pair $\mathcal{M} = (U, \mathcal{I})$ is called a *matroid*, if U is a finite set and $\mathcal{I}$ is a family of subsets of U satisfying the following conditions. (I0) $\emptyset \in \mathcal{I}$. (I1) If $I \in \mathcal{I}$ and $J \subseteq I$, then $J \in \mathcal{I}$. (I2) If $I, J \in \mathcal{I}$ and $|I| < |J|$, then $I + u \in \mathcal{I}$ for some element u in $J \backslash I$.

In this paper, we are given a finite simple bipartite graph $G = (V, E)$. We assume that V is partitioned into two subsets A and P, and each edge in E connects a vertex in A and a vertex in P. We call a vertex in A an *applicant*, and a vertex in P a *post*. We denote by (a, p) the edge in E between an applicant a in A and a post p in P. For each vertex v in V and each subset M of E, we define $M(v)$ as the set of edges in M incident to v. For each subset X of A and each subset M of E, we write $M(X)$ instead of $\cup_{a \in X} M(a)$.

In addition, we are given an injective function $\pi \colon E \to \mathbb{Z}_+$. That is, $\pi(e) \neq \pi(g)$ for every distinct edges e, g in E. Intuitively speaking, π represents preference lists of applicants. For each applicant a in A and each edges e, g in $E(a)$, if $\pi(e) > \pi(g)$, then the applicant a prefers e to g. Since π is an injective function, it represents strict preference lists of applicants.[1] As in [1], we assume that for each applicant a in A, there exists the unique *last resort* post p_a in P such that

---

[1] It should be noted that in the papers [1,3,4], the case where preference lists have ties was considered. It seems that to extend the results of this paper to the case where preference lists have ties is a non-trivial open problem.

$E(p_a) = \{(a, p_a)\}$ and $\pi(e) > \pi((a, p_a))$ for every edge e in $E(a) - (a, p_a)$, and there exists an edge e in $E(a)$ with $e \neq (a, p_a)$. For each post p in P, we are given a matroid $\mathcal{M}_p = (E(p), \mathcal{I}_p)$. We assume that for every edge (a, p) in E, we have $\{(a, p)\} \in \mathcal{I}_p$.

For each subset X of A, a subset M of E is called a *matching on X*, if it satisfies the following two conditions.

- For every applicant a in X (resp., $A \backslash X$), $|M(a)| = 1$ (resp., $M(a) = \emptyset$).
- For every post p in P, we have $M(p) \in \mathcal{I}_p$.

For each subset X of A, each matching M on X, and each applicant a in X, we denote by $\mu_M(a)$ the unique edge in $M(a)$. Let M, N be matchings on a subset X of A. Let $\mathsf{pre}_M(N)$ be the number of applicants a in X with $\pi(\mu_N(a)) > \pi(\mu_M(a))$. That is, $\mathsf{pre}_M(N)$ is the number of applicants that prefer N to M.

A matching M on a subset X of A is called a *popular matching on X*, if for every matching N on X, $\mathsf{pre}_N(M) \geq \mathsf{pre}_M(N)$. The goal of the *popular matching problem under matroid constraints* is to decide whether there exists a popular matching on A, and find such a matching if one exists.

A subset X of A is called a *popular condensation*, if there exists a popular matching on X. Notice that for every applicant a in A, $\{a\}$ is a popular condensation. The goal of the *popular condensation problem under matroid constraints* is to find a maximum-size popular condensation.

It is not difficult to see that the popular matching problem under matroid constraints contains the popular matching problem (without ties) introduced by Abraham, Irving, Kavitha, and Mehlhorn [1] and its many-to-one version introduced by Manlove and Sng [3] as special cases. Furthermore, it can represent the following variant of the popular matching problem with laminar capacity constraints. For each post p in P, we are given a laminar family $\mathcal{C}_p$ of subsets of $E(p)$, i.e., $C_1 \cap C_2 = \emptyset$, or $C_1 \subseteq C_2$, or $C_2 \subseteq C_1$ for every distinct $C_1, C_2 \in \mathcal{C}_p$. For each post p in P, we are given a function $c_p \colon \mathcal{C}_p \to \mathbb{Z}_+$. For each post p in P, we define $\mathcal{I}_p$ as the family of subsets F of $E(p)$ such that $|F \cap C| \leq c_p(C)$ for every member C in $\mathcal{C}_p$. It is not difficult to see that $\mathcal{M}_p = (E(p), \mathcal{I}_p)$ is a matroid for every post p in P. Hierarchical capacity constraints naturally arise in several practical situations (see, e.g., [12]). To the best of our knowledge, this problem has not been investigated.

2.1 Matroids

Let $\mathcal{M} = (U, \mathcal{I})$ be a matroid. A member I in $\mathcal{I}$ is called an *independent set in $\mathcal{M}$*. A subset C of U is called a *circuit in $\mathcal{M}$*, if $C \notin \mathcal{I}$, but every proper subset of C is an independent set in $\mathcal{M}$. It is known [13, Sect. 39.6] that if I is an independent set in $\mathcal{M}$ and u is an element in $U \backslash I$ with $I + u \notin \mathcal{I}$, then $I + u$ contains the unique circuit $C_\mathcal{M}(u, I)$ in $\mathcal{M}$, and we have $u \in C_\mathcal{M}(u, I)$. We call $C_\mathcal{M}(u, I)$ the *fundamental circuit of u with respect to I in $\mathcal{M}$*. It is well known that $C_\mathcal{M}(u, I)$ consists of all elements w in $I + u$ with $I + u - w \in \mathcal{I}$. For each subset X of U, a subset B of X is called a *base of X in $\mathcal{M}$*, if $B \in \mathcal{I}$ and there exists no subset Y of X such that $B \subsetneq Y$ and $Y \in \mathcal{I}$. We call a base of U in $\mathcal{M}$ a *base in* $\mathcal{M}$. The condition (I2) implies that for each subset X of U, every two bases of X in $\mathcal{M}$ have the same size, which is called the *rank of X in $\mathcal{M}$* and denoted by $r_{\mathcal{M}}(X)$. Clearly, $r_{\mathcal{M}}(\cdot)$ is monotone. That is, $r_{\mathcal{M}}(X) \leq r_{\mathcal{M}}(Y)$ for every subsets X, Y of U with $X \subseteq Y$. From the submodularity of $r_{\mathcal{M}}(\cdot)$ (see [13, Sect. 39.7]), it follows that for every subsets X, Y, Z of U such that $X \subseteq Y$ and $Z \cap Y = \emptyset$, we have

$$r_{\mathcal{M}}(Y \cup Z) - r_{\mathcal{M}}(Y) \leq r_{\mathcal{M}}(X \cup Z) - r_{\mathcal{M}}(X). \tag{1}$$

It follows from the submodularity and the non-negativity of $r_{\mathcal{M}}(\cdot)$ that

$$\forall X, Y \subseteq U \colon r_{\mathcal{M}}(X \cup Y) \leq r_{\mathcal{M}}(X) + r_{\mathcal{M}}(Y). \tag{2}$$

Let S be a subset of U. Define $\mathcal{I}|S := \{X \subseteq S \mid X \in \mathcal{I}\}$ and $\mathcal{M}|S := (S, \mathcal{I}|S)$. It is not difficult to see that $\mathcal{M}|S$ is a matroid and $r_{\mathcal{M}|S}(X) = r_{\mathcal{M}}(X)$ for every subset X of S. We define a function $\hat{r}$ on $2^{U \setminus S}$ by $\hat{r}(X) := r_{\mathcal{M}}(X \cup S) - r_{\mathcal{M}}(S)$ for each subset X of $U \setminus S$. Moreover, define $\mathcal{I}/S := \{X \subseteq U \setminus S \mid \hat{r}(X) = |X|\}$ and $\mathcal{M}/S := (U \setminus S, \mathcal{I}/S)$. It is known [13, Sect. 39.3] that $\mathcal{M}/S$ is a matroid and $r_{\mathcal{M}/S}(X) = \hat{r}(X)$ for every subset X of $U \setminus S$.

Let $\mathcal{M}_1 = (U_1, \mathcal{I}_1)$ and $\mathcal{M}_2 = (U_2, \mathcal{I}_2)$ be matroids with $U_1 \cap U_2 = \emptyset$. Define

$$\mathcal{I}_1 \oplus \mathcal{I}_2 := \{X \subseteq U_1 \cup U_2 \mid X \cap U_1 \in \mathcal{I}_1, \ X \cap U_2 \in \mathcal{I}_2\},$$
$$\mathcal{M}_1 \oplus \mathcal{M}_2 := (U_1 \cup U_2, \mathcal{I}_1 \oplus \mathcal{I}_2).$$

It is not difficult to see that $\mathcal{M}_1 \oplus \mathcal{M}_2$ is a matroid and $r_{\mathcal{M}_1 \oplus \mathcal{M}_2}(X) = r_{\mathcal{M}_1}(X \cap U_1) + r_{\mathcal{M}_2}(X \cap U_2)$ for every subset X of $U_1 \cup U_2$.

Let $\mathcal{M}_1 = (U, \mathcal{I}_1)$ and $\mathcal{M}_2 = (U, \mathcal{I}_2)$ be matroids. A subset I of U is called a *common independent set of $\mathcal{M}_1$ and $\mathcal{M}_2$*, if $I \in \mathcal{I}_1 \cap \mathcal{I}_2$.

Theorem 1 (Edmonds [14]). *For every matroids $\mathcal{M}_1 = (U, \mathcal{I}_1)$ and $\mathcal{M}_2 = (U, \mathcal{I}_2)$, the maximum size of a common independent set of $\mathcal{M}_1$ and $\mathcal{M}_2$ is equal to $\min_{X \subseteq U}(r_{\mathcal{M}_1}(X) + r_{\mathcal{M}_2}(U \setminus X))$.*

Since the following properties are well known, we omit the proof.

Lemma 1. *Let $\mathcal{M} = (U, \mathcal{I})$ and S be a matroid and a subset of U, respectively.*

1. *Let B be an arbitrary base in $\mathcal{M}|S$. For every subset Y of $U \setminus S$, Y is an independent set in $\mathcal{M}/S$ if and only if $Y \cup B$ is an independent set in $\mathcal{M}$.*
2. *For every independent set I in $\mathcal{M}$ such that $I \cap S$ is a base in $\mathcal{M}|S$, I is an independent set in $\mathcal{M}|S \oplus \mathcal{M}/S$.*
3. *For every independent set I in $\mathcal{M}|S \oplus \mathcal{M}/S$ and every element u in S with $(I \cap S) + u \in \mathcal{I}|S$, $I + u$ is an independent set in $\mathcal{M}|S \oplus \mathcal{M}/S$.*
4. *For every independent set I in $\mathcal{M}|S \oplus \mathcal{M}/S$, I is an independent set in $\mathcal{M}$.*

Lemma 2. *Let $\mathcal{M} = (U, \mathcal{I})$ be a matroid. For every subsets X, Y, Z of U such that $X \subseteq Y$ and $Z \cap Y = \emptyset$, we have $r_{\mathcal{M}/X}(Z) - r_{\mathcal{M}/Y}(Z) \leq r_{\mathcal{M}}(Y) - r_{\mathcal{M}}(X)$.*

Proof. The definitions of $\mathcal{M}/X$ and $\mathcal{M}/Y$ imply that $r_{\mathcal{M}/X}(Z) = r_{\mathcal{M}}(Z \cup X) - r_{\mathcal{M}}(X)$ and $r_{\mathcal{M}/Y}(Z) = r_{\mathcal{M}}(Z \cup Y) - r_{\mathcal{M}}(Y)$. This lemma follows from this observation and the monotonicity of $r_{\mathcal{M}}(\cdot)$. $\qquad \square$

3   Characterization

For each applicant a in A, we define the f-edge $f(a)$ as the unique maximizer of $\max\{\pi((a,p)) \mid (a,p) \in E(a)\}$. For each subset X of A and each post p in P, we denote by $\Gamma_{X,p}$ the set of edges (a,p) in $E(p)$ such that $a \in X$ and $(a,p) = f(a)$. For each subset X of A and each applicant a in X, we define the s-edge $s_X(a)$ as the unique maximizer of

$$\max\{\pi((a,p)) \mid (a,p) \in E(a) - f(a),\ \{(a,p)\} \in \mathcal{I}_p/\Gamma_{X,p}\}.$$

Notice that $s_X(a)$ is well-defined because the post p_a exists. For each subset X of A, we define the *reduced edge set* Π_X by $\Pi_X := \{f(a), s_X(a) \mid a \in X\}$.

The goal of this section is to prove the following theorem. This theorem can be regarded as a matroid generalization of Theorem 2.5 in [1] and Theorem 1 in [3], and its proof follows the line of the proofs of these known theorems.

Theorem 2. *For every subset X of A and every matching M on X, M is a popular matching on X if and only if it satisfies the following two conditions.*

(P1) *For every post p in P, $M(p) \cap \Gamma_{X,p}$ is a base in $\mathcal{M}_p|\Gamma_{X,p}$.*
(P2) *M is a subset of Π_X.*

For proving Theorem 2, we prove several lemmas. We first prove lemmas that are necessary for proving the *only if* part.

Lemma 3. *If M is a popular matching on a subset X of A, then for every post p in P, $M(p) \cap \Gamma_{X,p}$ is a base in $\mathcal{M}_p|\Gamma_{X,p}$.*

Proof. Assume that M is a popular matching on a subset X of A and p is a post in P. Since $M(p)$ is an independent set in $\mathcal{M}_p$, (I1) implies that $M(p) \cap \Gamma_{X,p}$ is an independent set in $\mathcal{M}_p|\Gamma_{X,p}$. For proving this lemma by contradiction, we assume that $M(p) \cap \Gamma_{X,p}$ is not a base in $\mathcal{M}_p|\Gamma_{X,p}$. Since $M(p) \cap \Gamma_{X,p}$ is not a base in $\mathcal{M}_p|\Gamma_{X,p}$, there exists a subset B of $\Gamma_{X,p}$ such that $B \in \mathcal{I}_p|\Gamma_{X,p}$ and $M(p) \cap \Gamma_{X,p} \subsetneq B$. Thus, it follows from (I2) that there exists an applicant a in X such that $f(a) \in \Gamma_{X,p} \setminus M(p)$ and

$$(M(p) \cap \Gamma_{X,p}) + f(a) \in \mathcal{I}_p|\Gamma_{X,p}. \tag{3}$$

Define $\mu_M(a) := (a, \ell)$. Notice that $\ell \neq p$. If $\ell = p$, then $\mu_M(a) = f(a)$, which contradicts the fact that $\mu_M(a) \in M$ and $f(a) \notin M$.

We first assume that $M(p) + f(a) \in \mathcal{I}_p$. Define $N := M + f(a) - \mu_M(a)$. It follows from $N(\ell) = M(\ell) - \mu_M(a)$ and (I1) that $N(\ell) \in \mathcal{I}_\ell$. Since $N(p) = M(p) + f(a)$, we have $N(p) \in \mathcal{I}_p$. For every post p' in $P \setminus \{\ell, p\}$, since $N(p') = M(p')$, we have $N(p') \in \mathcal{I}_{p'}$. These imply that N is a matching on X. Since $\pi(\mu_N(a)) = \pi(f(a)) > \pi(\mu_M(a))$ and $\mu_N(a') = \mu_M(a')$ for every applicant a' in $X - a$, we have $\mathsf{pre}_M(N) - \mathsf{pre}_N(M) = 1$. This contradicts the fact that M is a popular matching on X.

Next we assume that $M(p) + f(a) \notin \mathcal{I}_p$. Let C be the fundamental circuit of $f(a)$ with respect to $M(p)$ in $\mathcal{M}_p$. If C is contained in $\Gamma_{X,p}$, then C is a subset of $(M(p) \cap \Gamma_{X,p}) + f(a)$, which contradicts (3) and (I1). Thus, there exists an edge (b, p) in $C \backslash \Gamma_{X,p}$, and $M(p) + f(a) - (b, p)$ is an independent set in $\mathcal{M}_p$. Define $f(b) := (b, q)$. Notice that $q \neq p$ because $(b, p) \notin \Gamma_{X,p}$.

If $M(q) + f(b) \in \mathcal{I}_q$, then we define $N := (M \backslash \{\mu_M(a), (b, p)\}) \cup \{f(a), f(b)\}$. We can prove that N is a matching on X in the same way as above. Since we have $\pi(\mu_N(a)) > \pi(\mu_M(a))$, $\pi(\mu_N(b)) > \pi(\mu_M(b))$, and $\mu_N(a') = \mu_M(a')$ for every applicant a' in $X \backslash \{a, b\}$, we have $\mathsf{pre}_M(N) - \mathsf{pre}_N(M) = 2$, which contradicts the fact that M is a popular matching on X.

Assume that $M(q) + f(b) \notin \mathcal{I}_q$. Since $\{f(b)\}$ is an independent set in $\mathcal{M}_q$, there exists an edge (c, q) in the fundamental circuit of $f(b)$ with respect to $M(q)$ in $\mathcal{M}_q$ such that $f(b) \neq (c, q)$. Notice that $M(q) + f(b) - (c, q) \in \mathcal{I}_q$. Define

$$N := (M \backslash \{\mu_M(a), (b, p), (c, q)\}) \cup \{f(a), f(b), (c, p_c)\}.$$

We can prove that N is a matching on X in the same way as above. Since we have $\pi(\mu_N(a)) > \pi(\mu_M(a))$, $\pi(\mu_N(b)) > \pi(\mu_M(b))$, and $\mu_N(a') = \mu_M(a')$ for every applicant a' in $X \backslash \{a, b, c\}$, we have $\mathsf{pre}_M(N) - \mathsf{pre}_N(M) = 1$, which contradicts the fact that M is a popular matching on X. This completes the proof. □

Lemma 4. *If M is a popular matching on a subset X of A, then there does not exist an edge (a, p) in M with $\pi(f(a)) > \pi((a, p)) > \pi(s_X(a))$.*

Proof. Assume that M is a popular matching on a subset X of A. For proving this lemma by contradiction, we assume that there exists an edge (a, p) in M with $\pi(f(a)) > \pi((a, p)) > \pi(s_X(a))$. It follows from $M(p) \in \mathcal{I}_p$ and (I1) that

$$(M(p) \cap \Gamma_{X,p}) + (a, p) \in \mathcal{I}_p. \tag{4}$$

Since Lemma 3 implies that $M(p) \cap \Gamma_{X,p}$ is a base in $\mathcal{M}_p | \Gamma_{X,p}$, it follows from (4) and Lemma 1(1) that $\{(a, p)\} \in \mathcal{I}_p / \Gamma_{X,p}$. This observation and the definition of $s_X(\cdot)$ imply that $\pi(s_X(a)) \geq \pi((a, p))$. However, this contradicts $\pi((a, p)) > \pi(s_X(a))$, which completes the proof. □

Lemma 5. *If M is a popular matching on a subset X of A, then M is a subset of Π_X.*

Proof. Assume that M is a popular matching on a subset X of A. For proving this lemma by contradiction, we assume that there exists an applicant a in X such that $\mu_M(a)$ does not belong to Π_X. Lemma 4 implies that $\pi(\mu_M(a)) < \pi(s_X(a))$. Define $\mu_M(a) := (a, \ell)$ and $s_X(a) := (a, p)$.

We first assume that $M(p) + s_X(a) \in \mathcal{I}_p$. Define $N := M + s_X(a) - \mu_M(a)$. It follows from $N(\ell) = M(\ell) - \mu_M(a)$ and (I1) that $N(\ell) \in \mathcal{I}_\ell$. Since $N(p) = M(p) + s_X(a)$, we have $N(p) \in \mathcal{I}_p$. For every post p' in $P \backslash \{\ell, p\}$, since $N(p') = M(p')$, we have $N(p') \in \mathcal{I}_{p'}$. These imply that N is a matching on X. Since $\pi(\mu_N(a)) = \pi(s_X(a)) > \pi(\mu_M(a))$ and $\mu_N(a') = \mu_M(a')$ for every applicant a' in $X - a$, we have $\mathsf{pre}_M(N) - \mathsf{pre}_N(M) = 1$. This contradicts the fact that M is a popular matching on X.

Next we assume that $M(p) + s_X(a) \notin \mathcal{I}_p$. Let C be the fundamental circuit of $s_X(a)$ with respect to $M(p)$ in $\mathcal{M}_p$. Since $\{s_X(a)\}$ is an independent set in $\mathcal{M}_p/\Gamma_{X,p}$, Lemma 3 and Lemma 1(1) imply that

$$(M(p) \cap \Gamma_{X,p}) + s_X(a) \in \mathcal{I}_p. \tag{5}$$

If C is contained in $\Gamma_{X,p} + s_X(a)$, then C is a subset of $(M(p) \cap \Gamma_{X,p}) + s_X(a)$, which contradicts (5) and (I1). Thus, there exists an edge (b,p) in $C \backslash \Gamma_{X,p}$ such that $(b,p) \neq s_X(a)$ and $M(p) + s_X(a) - (b,p) \in \mathcal{I}_p$. Define $f(b) := (b,q)$. Notice that $p \neq q$ because $(b,p) \notin \Gamma_{X,p}$.

If $M(q) + f(b) \in \mathcal{I}_q$, then define $N := (M \backslash \{\mu_M(a), (b,p)\}) \cup \{s_X(a), f(b)\}$. We can prove that N is a matching on X in the same way as above. Since we have $\pi(\mu_N(a)) > \pi(\mu_M(a))$, $\pi(\mu_N(b)) > \pi(\mu_M(b))$, and $\mu_N(a') = \mu_M(a')$ for every applicant a' in $X \backslash \{a,b\}$, we have $\mathsf{pre}_M(N) - \mathsf{pre}_N(M) = 2$, which contradicts the fact that M is a popular matching on X.

Assume that $M(q) + f(b) \notin \mathcal{I}_q$. Since $\{f(b)\}$ is an independent set in $\mathcal{M}_q$, there exists an edge (c,q) in the fundamental circuit of $f(b)$ with respect to $M(q)$ in $\mathcal{M}_q$ such that $f(b) \neq (c,q)$. Notice that $M(q) + f(b) - (c,q) \in \mathcal{I}_q$. Define

$$N := (M \backslash \{\mu_M(a), (b,p), (c,q)\}) \cup \{s_X(a), f(b), (c, p_c)\}.$$

We can prove that N is a matching on X in the same way as above. Since we have $\pi(\mu_N(a)) > \pi(\mu_M(a))$, $\pi(\mu_N(b)) > \pi(\mu_M(b))$, and $\mu_N(a') = \mu_M(a')$ for every applicant a' in $X \backslash \{a,b,c\}$, we have $\mathsf{pre}_M(N) - \mathsf{pre}_N(M) = 1$, which contradicts the fact that M is a popular matching on X. This completes the proof. $\qquad \square$

Next we prove a lemma that is necessary for proving the *if* part. For each matching M on a subset X of A and each post p in P, we define $\mathsf{bet}_M(p)$ as the set of edges (a,p) in $M(p)$ with $\pi(f(a)) > \pi((a,p)) > \pi(s_X(a))$.

Lemma 6. *Assume that M is a matching on a subset X of A and $M(p) \cap \Gamma_{X,p}$ is a base in $\mathcal{M}_p|\Gamma_{X,p}$ for every post p in P. Then, for every matching N on X and every post p in P, we have*

$$|M(p) \cap \Gamma_{X,p}| \geq |N(p) \cap \Gamma_{X,p}| + |\mathsf{bet}_N(p)|.$$

Proof. Assume that M is a matching on a subset X of A and $M(p) \cap \Gamma_{X,p}$ is a base in $\mathcal{M}_p|\Gamma_{X,p}$ for every post p in P. For proving this lemma by contradiction, we assume that there exist a matching N on X and a post p in P with $|M(p) \cap \Gamma_{X,p}| < |N(p) \cap \Gamma_{X,p}| + |\mathsf{bet}_N(p)|$. Since it follows from (I1) that $M(p) \cap \Gamma_{X,p}$ and $(N(p) \cap \Gamma_{X,p}) \cup \mathsf{bet}_N(p)$ are independent sets in $\mathcal{M}_p$, (I2) implies that there exists an edge (a,p) in

$$[(N(p) \cap \Gamma_{X,p}) \cup \mathsf{bet}_N(p)] \backslash (M(p) \cap \Gamma_{X,p})$$

with $(M(p) \cap \Gamma_{X,p}) + (a,p) \in \mathcal{I}_p$. If (a,p) belongs to $\Gamma_{X,p}$, then this contradicts the fact that $M(p) \cap \Gamma_{X,p}$ is a base in $\mathcal{M}_p|\Gamma_{X,p}$. Thus, we can assume that $(a,p) \in \mathsf{bet}_N(p)$, which implies that $\pi((a,p)) > \pi(s_X(a))$. Since $M(p) \cap \Gamma_{X,p}$ is a base in $\mathcal{M}_p|\Gamma_{X,p}$, it follows from Lemma 1(1) that $\{(a,p)\} \in \mathcal{I}_p/\Gamma_{X,p}$. This observation and the definition of $s_X(\cdot)$ imply that $\pi(s_X(a)) \geq \pi((a,p))$, which contradicts $\pi((a,p)) > \pi(s_X(a))$. This completes the proof. $\qquad \square$

Proof (Theorem 2). Since the *only if* part follows from Lemmas 3 and 5, we prove the *if* part. Let M be a matching on a subset X of A satisfying (P1) and (P2). Let N be a matching on X. We define Δ_M and Δ_N as the sets of applicants a in X such that $\pi(\mu_M(a)) > \pi(\mu_N(a))$ and $\pi(\mu_N(a)) > \pi(\mu_M(a))$, respectively. For proving the *if* part, it suffices to prove that $|\Delta_M| \geq |\Delta_N|$. For proving this, we construct an injective function $\sigma\colon \Delta_N \to \Delta_M$ as follows. Lemma 6 implies that for each post p in P, there exists an injective function φ_p from

$$[(N(p) \cap \Gamma_{X,p}) \cup \mathsf{bet}_N(p)]\backslash(M(p) \cap \Gamma_{X,p}) \tag{6}$$

to $(M(p) \cap \Gamma_{X,p})\backslash[(N(p) \cap \Gamma_{X,p}) \cup \mathsf{bet}_N(p)]$. We construct an injective function $\sigma\colon \Delta_N \to \Delta_M$ by using the injective functions φ_p for posts p in P. Let a be an applicant in Δ_N, and define $\mu_N(a) := (a,p)$. It follows from $M \subseteq \Pi_X$ and $\pi(\mu_N(a)) > \pi(\mu_M(a))$ that $\mu_M(a) = s_X(a)$. Therefore, $\mu_N(a)$ belongs to the set of (6). Define $\varphi_p(\mu_N(a)) := (b,p)$. Since $(b,p) \in \Gamma_{X,p}$ and $(b,p) \notin N(p)$, we have $b \in \Delta_M$. Define $\sigma(a) := b$. Since φ_p is an injective function for every post p in P and $|M(a')| = 1$ for every applicant a' in X, σ is an injective function. This completes the proof.   □

4   The Popular Matching Problem

In this section, we give a polynomial-time algorithm for the popular matching problem under matroid constraints. In fact, we give an algorithm for the problem in which we are given a subset X of A, and decide whether there exists a popular matching on X, and find such a matching if one exists.

For each subset X of A, we define $\mathcal{U}_X$ as the family of subsets M of Π_X such that $|M(a)| \leq 1$ for every applicant a in X, and we define a matroid $\mathcal{A}_X$ by $\mathcal{A}_X := (\Pi_X, \mathcal{U}_X)$. For each subset X of A and each post p in P, we define

$$\mathcal{P}_{X,p} := (\mathcal{M}_p|\Gamma_{X,p} \oplus \mathcal{M}_p/\Gamma_{X,p})|\Pi_X(p).$$

For each subset X of A, we define $\mathcal{P}_X := \bigoplus_{p \in P} \mathcal{P}_{X,p}$.

Lemma 7. *If M is a popular matching on a subset X of A, then M is a common independent set of $\mathcal{A}_X$ and $\mathcal{P}_X$.*

Proof. Assume that M is a popular matching on a subset X of A. It follows from (P2) of Theorem 2 that M is a subset of Π_X. Since $|M(a)| = 1$ for every applicant a in X, M is an independent set in $\mathcal{A}_X$. In addition, it follows from (P1) of Theorem 2 and Lemma 1(2) that $M(p)$ is an independent set in $\mathcal{M}_p|\Gamma_{X,p} \oplus \mathcal{M}_p/\Gamma_{X,p}$ for every post p in P, which implies that M is an independent set in $\mathcal{P}_X$. This completes the proof.   □

Our algorithm **PMuMC** is described as follows.

Step 1. Find a maximum-size common independent set M of $\mathcal{A}_X$ and $\mathcal{P}_X$.

Step 2. If there exists an applicant a in X with $M(a) = \emptyset$, then output **null** (there exists no popular matching on X). If $|M(a)| = 1$ for every applicant a in X, then go to **Step 3**.

Step 3. Set $i := 0$, $M_i := M$, and do the following steps.

(3-a) If $M_i(p) \cap \Gamma_{X,p}$ is a base in $\mathcal{M}_p | \Gamma_{X,p}$ for every post p in P, then go to **Step 4**.

(3-b) Arbitrarily choose a post p in P such that $M_i(p) \cap \Gamma_{X,p}$ is not a base in $\mathcal{M}_p | \Gamma_{X,p}$, and find an edge (a,p) in $\Gamma_{X,p} \setminus M_i(p)$ with $(M_i(p) \cap \Gamma_{X,p}) + (a,p) \in \mathcal{I}_p | \Gamma_{X,p}$. Furthermore, set $M_{i+1} := M_i + (a,p) - \mu_{M_i}(a)$.

(3-c) Update $i := i + 1$, and go to (3-a).

Step 4. Output M_i (M_i is a popular matching on X).

The following lemma implies that the algorithm **PMuMC** is well-defined.

Lemma 8 *1. In (3-b), if M_i is a common independent set of $\mathcal{A}_X, \mathcal{P}_X$, then there is an edge (a,p) in $\Gamma_{X,p} \setminus M_i(p)$ with $(M_i(p) \cap \Gamma_{X,p}) + (a,p) \in \mathcal{I}_p | \Gamma_{X,p}$.*

2. In (3-b), if M_i is a common independent set of $\mathcal{A}_X$ and $\mathcal{P}_X$, then M_{i+1} is a common independent set of $\mathcal{A}_X$ and $\mathcal{P}_X$.

*3. The number of iterations of **Step 3** is at most $|X|$.*

Proof **1.** Since $M_i(p) \cap \Gamma_{X,p} \in \mathcal{I}_p | \Gamma_{X,p}$, this statement follows from (I2).

2. It follows from $(a,p) \in \Gamma_{X,p}$ and $M_i \subseteq \Pi_X$ that $M_{i+1} \subseteq \Pi_X$. This observation and $M_i \in \mathcal{U}_X$ imply that M_{i+1} is an independent set in $\mathcal{A}_X$. Define $\mu_{M_i}(a) := (a, \ell)$. It follows from (I1) that $M_{i+1}(\ell)$ is an independent set in $\mathcal{P}_{X,\ell}$. Lemma 1(3) implies that $M_{i+1}(p)$ is an independent set in $\mathcal{P}_{X,p}$. Furthermore, for every post p' in $P \setminus \{p, \ell\}$, since $M_i(p') = M_{i+1}(p')$, $M_{i+1}(p')$ is an independent set in $\mathcal{P}_{X,p'}$, which implies that M_{i+1} is an independent set in $\mathcal{P}_X$.

3. Since the number of applicants a in X with $\mu_{M_i}(a) = f(a)$ increases, this statement follows. $\square$

Theorem 3. *The algorithm **PMuMC** correctly solves the popular matching problem under matroid constraints.*

Proof. If there exists a popular matching M on X, then it follows from Lemma 7 and the definition of a matching on X that M is a common independent set of $\mathcal{A}_X$ and $\mathcal{P}_X$ such that $|M(a)| = 1$ for every applicant a in X. Thus, if the algorithm **PMuMC** outputs **null**, then there exists no popular matching on X. Assume that there exists a common independent set M of $\mathcal{A}_X$ and $\mathcal{P}_X$ such that $|M(a)| = 1$ for every applicant a in X. Let N be the output of the algorithm **PMuMC**. It follows from the definition of the algorithm **PMuMC** that $|N(a)| = 1$ (resp., $N(a) = \emptyset$) for every applicant a in X (resp., $A \setminus X$). Let p be a post in P. It follows from Lemma 8(2) that $N(p)$ is an independent set in $\mathcal{M}_p | \Gamma_{X,p} \oplus \mathcal{M}_p / \Gamma_{X,p}$. This observation and Lemma 1(4) imply that $N(p)$ is an independent set in $\mathcal{M}_p$, which implies that N is a matching on X. Furthermore, it follows from Lemma 8(2) that N is a subset of Π_X. In addition, it follows from the definition of the algorithm **PMuMC** that $N(p) \cap \Gamma_{X,p}$ is a base in $\mathcal{M}_p|\Gamma_{X,p}$ for every post p in P. These observations and Theorem 2 imply that M is a popular matching on X. This completes the proof.   □

The following corollary follows from the algorithm **PMuMC**.

Corollary 1. *For every subset X of A, X is a popular condensation if and only if there exists a common independent set M of $\mathcal{A}_X$ and $\mathcal{P}_X$ with $|M| = |X|$.*

Here we consider the time complexity of the algorithm **PMuMC**. We denote by γ the time complexity of deciding whether $M + e \in \mathcal{I}_p$ for a post p in P, an independent set M in $\mathcal{M}_p$, and an edge e in $E(p)\backslash M$. Define $m := |E|$. For simplicity, we assume that $\gamma = \Omega(m)$. We first consider the time complexity of **Step 1**. We can compute $f(a)$ for all applicants a in X in $O(m)$ time. To compute $s_X(a)$ for all applicants a in X, we first compute bases B_p of $\mathcal{M}_p|\Gamma_{X,p}$ for all posts p in P. This can be done in $O(m\gamma)$ time. Lemma 1(1) implies that by using the bases B_p for all posts p in P, we can compute $s_X(a)$ for all applicants a in X in $O(m\gamma)$ time. Thus, we can compute Π_X in $O(m\gamma)$ time. It is not difficult to see that we can decide in $O(m)$ time whether $M + e$ is an independent set in $\mathcal{A}_X$ for an independent set M in $\mathcal{A}_X$ and an edge e in $\Pi_X\backslash M$. Next we consider the time complexity of deciding whether $M + e$ is an independent set in $\mathcal{P}_X$ for an independent set M in $\mathcal{P}_X$ and an edge $e = (a, p)$ in $\Pi_X\backslash M$. Define $N := M + e$. For every post p' in $P - p$, since $M(p') = N(p')$, $N(p')$ is an independent set in $\mathcal{P}_{X,p'}$. We can decide in $O(\gamma)$ time whether $N(p) \cap \Gamma_{X,p}$ is an independent set in $\mathcal{M}_p|\Gamma_{X,p}$. By using the base B_p and Lemma 1(1), we can decide in $O(\gamma)$ time whether $N(p)\backslash\Gamma_{X,p}$ is an independent set in $\mathcal{M}_p/\Gamma_{X,p}$. These imply that we can decide in $O(\gamma)$ time whether $M + e$ is an independent set in $\mathcal{P}_X$. Since $m = O(\gamma)$, we can decide in $O(\gamma)$ time whether $M + e$ is an independent set in $\mathcal{A}_X$ (or $\mathcal{P}_X$) for a common independent set M of $\mathcal{A}_X, \mathcal{P}_X$ and an edge e in $\Pi_X\backslash M$. It is known (e.g., [15,16]) that by using this fact we can find a maximum-size common independent set of $\mathcal{A}_X, \mathcal{P}_X$ in $O(m^3\gamma)$ time. It is not difficult to see that **Step 2** to **Step 4** can be done in $O(m^3\gamma)$ time. Thus, the time complexity of the algorithm **PMuMC** is $O(m^3\gamma)$.

5   The Popular Condensation Problem

Our algorithm **PCuMC** for the popular condensation problem under matroid constraints is described as follows. In the sequel, for simplicity, we define $s(\cdot) := s_A(\cdot)$, $\Gamma_p := \Gamma_{A,p}$, $\Pi := \Pi_A$, $\mathcal{A} := \mathcal{A}_A$, and $\mathcal{P} := \mathcal{P}_A$. Under the same assumptions in Sect. 4, the time complexity of the algorithm **PCuMC** is $O(m^3\gamma)$.

Step 1. Find a maximum-size common independent set M of $\mathcal{A}$ and $\mathcal{P}$.
Step 2. Output the set Δ of applicants a in A with $M(a) \neq \emptyset$.

In the sequel, we will prove the correctness of the algorithm **PCuMC**.

Lemma 9. *There exists a subset D of A such that*

$$|D| + r_{\mathcal{P}}(\Pi(A\backslash D)) = \min_{F\subseteq\Pi}(r_{\mathcal{A}}(F) + r_{\mathcal{P}}(\Pi\backslash F)). \tag{7}$$

Proof. For each minimizer F of the right-hand side of (7), we define X_F as the set of applicants a in A with $F \cap \Pi(a) \neq \emptyset$.

We first consider the case where there exists a minimizer F of the right-hand side of (7) with $\Pi(X_F) = F$. Since a base of F in $\mathcal{A}$ is a subset B of F such that $|B(a)| = 1$ for every applicant a in X_F. Therefore, we have $r_{\mathcal{A}}(F) = |X_F|$. In addition, since $\Pi(X_F) = F$, we have $\Pi \backslash F = \Pi(A \backslash X_F)$. These observations imply that X_F satisfies (7).

Next we consider the case where there exists no minimizer F of the right-hand side of (7) with $\Pi(X_F) = F$. Let F be a minimizer of the right-hand side of (7) that minimizes $|\Pi(X_F) \backslash F|$. Notice that $\Pi(X_F) \backslash F \neq \emptyset$. For every edge e in $\Pi(X_F) \backslash F$, the monotonicity of $r_{\mathcal{P}}(\cdot)$ implies that

$$r_{\mathcal{A}}(F + e) = r_{\mathcal{A}}(F) \ (= |X_F|), \quad r_{\mathcal{P}}(\Pi \backslash (F + e)) \leq r_{\mathcal{P}}(\Pi \backslash F).$$

This observation implies that $F + e$ is a minimizer of the right-hand side of (7), and $|\Pi(X_{F+e}) \backslash (F + e)| = |\Pi(X_F) \backslash (F + e)| < |\Pi(X_F) \backslash F|$. This contradicts the fact that F minimizes $|\Pi(X_F) \backslash F|$. This completes the proof. $\qquad\square$

Lemma 10. *For every subset X of A, if there is a common independent set M of $\mathcal{A}_X, \mathcal{P}_X$ with $|M| = |X|$, then $|Y| \leq r_{\mathcal{P}_X}(\Pi_X(Y))$ for every subset Y of X.*

Proof. Assume that X is a subset of A and M is a common independent set of $\mathcal{A}_X, \mathcal{P}_X$ with $|M| = |X|$. It follows from Theorem 1 that

$$\forall F \subseteq \Pi_X : \ |X| \leq r_{\mathcal{A}_X}(F) + r_{\mathcal{P}_X}(\Pi_X \backslash F). \qquad (8)$$

Let Y be a subset of X. Define $F := \Pi_X(X \backslash Y)$. Since a base of F in $\mathcal{A}_X$ is a subset B of F such that $|B(a)| = 1$ for every applicant a in $X \backslash Y$. This implies that $r_{\mathcal{A}_X}(F) = |X \backslash Y|$. It follows from this observation and (8) that

$$|X| \leq r_{\mathcal{A}_X}(F) + r_{\mathcal{P}_X}(\Pi_X \backslash F) = |X \backslash Y| + r_{\mathcal{P}_X}(\Pi_X(Y)),$$

which completes the proof. $\qquad\square$

The proof of the correctness of the algorithm **PCuMC** follows the line of the proof of the algorithm of [8] for the popular condensation problem. In the sequel, let M be the maximum-size common independent set of $\mathcal{A}$ and $\mathcal{P}$ that is found in **Step 1** of the algorithm **PCuMC**, and we denote by Δ the output of the algorithm of **PCuMC**. We first prove that Δ is a popular condensation.

Lemma 11. $\Pi_\Delta = \Pi(\Delta)$.

Proof. It suffices to prove that for every applicant a in Δ, we have $s_\Delta(a) = s(a)$. For proving this, we prove that for every post p, there exists a base B in $\mathcal{M}_p | \Gamma_p$ with $B \subseteq \Gamma_{\Delta,p}$. Assume that there exists such a base B. It follows from $\Gamma_{\Delta,p} \subseteq \Gamma_p$ that B is a base in $\mathcal{M}_p | \Gamma_{\Delta,p}$. This observation and Lemma 1(1) imply that for every applicant a in Δ and every edge (a, p) in $E(a) - f(a)$, $\{(a, p)\}$ is in $\mathcal{I}_p / \Gamma_p$ if and only if $\{(a, p)\}$ is in $\mathcal{I}_p / \Gamma_{\Delta,p}$. It follows from this observation that $s_\Delta(a) = s(a)$ for every applicant a in Δ.

For proving the above statement by contradiction, we assume that there exists a post p in P such that any base in $\mathcal{M}_p|\Gamma_p$ is not a subset of $\Gamma_{\Delta,p}$. We will prove that in this case, there exists a common independent set of $\mathcal{A}$ and $\mathcal{P}$ that is larger than M, which contradicts the maximality of M.

Since $M(p)$ is an independent set in $\mathcal{P}_{A,p}$, $M(p) \cap \Gamma_p$ is an independent set in $\mathcal{M}_p|\Gamma_p$. Thus, there exists a base B in $\mathcal{M}_p|\Gamma_p$ with $M(p) \cap \Gamma_p \subseteq B$. Since any base in $\mathcal{M}_p|\Gamma_p$ is not a subset of $\Gamma_{\Delta,p}$, there exists an edge (a,p) in B with $a \notin \Delta$. Notice that $a \notin \Delta$ implies that $M(a) = \emptyset$. It follows from this observation that $(a,p) \notin M$. Define $N := M + (a,p)$. We will prove that N is a common independent set of $\mathcal{A}$ and $\mathcal{P}$ that is larger than M. Since $(a,p) \in N\backslash M$ and $M \subseteq N$, we have $|N| > |M|$. Thus, what remains is to prove that N is a common independent set of $\mathcal{A}$ and $\mathcal{P}$. Since $(a,p) \in \Gamma_p$ and $M \subseteq \Pi$, N is a subset of Π.

We first prove that N is an independent set in $\mathcal{A}$. Recall that $M(a) = \emptyset$. Since M is an independent set in $\mathcal{A}$, $|N(a')| = |M(a')| \leq 1$ for every applicant a' in $A - a$. These observations imply that $|N(a')| \leq 1$ for every applicant a' in A, which implies that N is an independent set in $\mathcal{A}$.

Next we prove that N is an independent set in $\mathcal{P}$. Since $(M(p) \cap \Gamma_p) + (a,p)$ is a subset of B and B is an independent set in $\mathcal{M}_p|\Gamma_p$, it follows from (I1) that $(M(p) \cap \Gamma_p) + (a,p)$ is an independent set in $\mathcal{M}_p|\Gamma_p$. Thus, since $M(p)$ is an independent set in $\mathcal{M}_p|\Gamma_p \oplus \mathcal{M}_p/\Gamma_p$, Lemma 1(3) implies that $N(p)$ $(= M(p) + (a,p))$ is an independent set in $\mathcal{P}_{A,p}$. In addition, $N(p') = M(p')$ for every post p' in $P - p$. These imply that N is an independent set in $\mathcal{P}$, which completes the proof. □

Lemma 12. Δ *is a popular condensation.*

Proof. Since $|M| = |\Delta|$ follows from the definition of the algorithm **PCuMC**, Corollary 1 implies that if M is a common independent set of $\mathcal{A}_\Delta$ and $\mathcal{P}_\Delta$, then the proof is done. Since $M \subseteq \Pi$ and $M(a) = \emptyset$ for every applicant a in $A\backslash\Delta$, $M \subseteq \Pi(\Delta)$. It follows from this and Lemma 11 that M is a subset of Π_Δ. Thus, since M is an independent set in $\mathcal{A}$, M is an independent set of $\mathcal{A}_\Delta$. What remains is to prove that M is an independent set of $\mathcal{P}_\Delta$.

Let p be a post of P. For proving that $M(p)$ is an independent set in $\mathcal{P}_{\Delta,p}$, we first prove that $M(p) \cap \Gamma_{\Delta,p}$ is an independent set in $\mathcal{M}_p|\Gamma_{\Delta,p}$. The definition of **Step 2** of the algorithm **PCuMC** implies that $M(p) \cap \Gamma_{\Delta,p} = M(p) \cap \Gamma_p$. Since $M(p)$ is an independent set in $\mathcal{P}_{A,p}$, $M(p) \cap \Gamma_p$ is an independent set in $\mathcal{M}_p$. These imply that $M(p) \cap \Gamma_{\Delta,p}$ is an independent set in $\mathcal{M}_p|\Gamma_{\Delta,p}$.

Next we prove that $M(p)\backslash\Gamma_{\Delta,p}$ is an independent set in $\mathcal{M}_p/\Gamma_{\Delta,p}$. In the same way as in the proof of Lemma 11, we can prove that there exists a base B in $\mathcal{M}_p|\Gamma_p$ with $B \subseteq \Gamma_{\Delta,p}$. It follows from $\Gamma_{\Delta,p} \subseteq \Gamma_p$ that B is a base in $\mathcal{M}_p|\Gamma_{\Delta,p}$. Since $M(p)\backslash\Gamma_p \in \mathcal{I}_p/\Gamma_p$, Lemma 1(1) implies that $(M(p)\backslash\Gamma_p) \cup B$ is an independent set in $\mathcal{M}_p$. Since $M(p) \cap \Gamma_{\Delta,p} = M(p) \cap \Gamma_p$, $M(p)\backslash\Gamma_{\Delta,p} = M(p)\backslash\Gamma_p$. Thus, $(M(p)\backslash\Gamma_{\Delta,p}) \cup B$ is an independent set in $\mathcal{M}_p$. Since B is a base in $\mathcal{M}_p|\Gamma_{\Delta,p}$, Lemma 1(1) implies that $M(p)\backslash\Gamma_{\Delta,p}$ is an independent set in $\mathcal{M}_p/\Gamma_{\Delta,p}$. This completes the proof. □

Let D be a subset of A satisfying (7) in Lemma 9, and define $Q := A \backslash D$.

Lemma 13. $|A \backslash \Delta| = |Q| - r_{\mathcal{P}}(\Pi(Q))$.

Proof. It follows from $|\Delta| = |M|$ and Theorem 1 that

$$|\Delta| = |M| = |D| + r_{\mathcal{P}}(\Pi(A \backslash D)) = |A| - |Q| + r_{\mathcal{P}}(\Pi(Q)),$$

which completes the proof. $\qquad \square$

Lemma 14. *For every popular condensation* Φ, $|A \backslash \Phi| \geq |Q| - r_{\mathcal{P}}(\Pi(Q))$.

Proof. Let Φ be a popular condensation, and define $\Phi_0 := A \backslash \Phi$, $\Phi_1 := Q \cap \Phi_0$, and $\Phi_2 := \Phi_0 \backslash \Phi_1$. Notice that $Q \backslash \Phi_1$ is a subset of Φ. Thus, it follows from Corollary 1 and Lemma 10 that

$$|Q \backslash \Phi_1| \leq r_{\mathcal{P}_\Phi}(\Pi_\Phi(Q \backslash \Phi_1)).$$

It follows from this that if

$$r_{\mathcal{P}_\Phi}(\Pi_\Phi(Q \backslash \Phi_1)) \leq r_{\mathcal{P}}(\Pi(Q)) + |\Phi_2|, \qquad (9)$$

then the proof is done because $|A \backslash \Phi| = |\Phi_1| + |\Phi_2|$.

Let p be a post in P. We define $F_{1,p}$ as the set of edges (a, p) in Γ_p such that $a \in Q$ or $a \in \Phi_2$. In addition, we define $F_{2,p}$ as the set of edges (a, p) in $\Pi(p)$ such that $a \in Q$ and $(a, p) = s(a)$. Notice that

$$F_{1,p} \cup F_{2,p} = (\Pi(Q) \cap \Pi(p)) \cup \{(a, p) \in \Gamma_p \mid a \in \Phi_2\}.$$

It follows from this and (2) that

$$
\begin{aligned}
r_{\mathcal{P}_{A,p}}(F_{1,p} \cup F_{2,p}) &= r_{\mathcal{P}_{A,p}}((\Pi(Q) \cap \Pi(p)) \cup \{(a, p) \in \Gamma_p \mid a \in \Phi_2\}) \\
&\leq r_{\mathcal{P}_{A,p}}(\Pi(Q) \cap \Pi(p)) + r_{\mathcal{P}_{A,p}}(\{(a, p) \in \Gamma_p \mid a \in \Phi_2\}) \\
&\leq r_{\mathcal{P}_{A,p}}(\Pi(Q) \cap \Pi(p)) + |\{(a, p) \in \Gamma_p \mid a \in \Phi_2\}| \\
&= r_{\mathcal{P}_{A,p}}(\Pi(Q) \cap \Pi(p)) + |\{a \in \Phi_2 \mid f(a) \in \Pi(p)\}|.
\end{aligned}
\qquad (10)
$$

In addition, we define $F'_{1,p}$ as the set of edges (a, p) in $\Gamma_{\Phi,p}$ with $a \in Q \backslash \Phi_1$, and we define $F'_{2,p}$ as the set of edges (a, p) in $\Pi_\Phi(p)$ such that $a \in Q \backslash \Phi_1$ and $(a, p) = s_\Phi(a)$. Notice that

$$F'_{1,p} \cup F'_{2,p} = \Pi_\Phi(Q \backslash \Phi_1) \cap \Pi_\Phi(p). \qquad (11)$$

It follows from (10) and (11) that if

$$r_{\mathcal{P}_{\Phi,p}}(F'_{1,p} \cup F'_{2,p}) \leq r_{\mathcal{P}_{A,p}}(F_{1,p} \cup F_{2,p}), \qquad (12)$$

then (9) follows and the proof is done because

$$
\begin{aligned}
r_{\mathcal{P}_\Phi}(\Pi_\Phi(Q\backslash\Phi_1)) &= \sum_{q\in P} r_{\mathcal{P}_{\Phi,q}}(\Pi_\Phi(Q\backslash\Phi_1)\cap\Pi_\Phi(q))\\
&= \sum_{q\in P} r_{\mathcal{P}_{\Phi,q}}(F'_{1,q}\cup F'_{2,q})\\
&\leq \sum_{q\in P} r_{\mathcal{P}_{A,q}}(F_{1,q}\cup F_{2,q})\\
&\leq \sum_{q\in P} r_{\mathcal{P}_{A,q}}(\Pi(Q)\cap\Pi(q)) + \sum_{q\in P}|\{a\in\Phi_2 \mid f(a)\in\Pi(q)\}|\\
&= r_{\mathcal{P}}(\Pi(Q)) + |\Phi_2|.
\end{aligned}
$$

It follows from the definitions of $\mathcal{P}_{A,p}$ and $\mathcal{P}_{\Phi,p}$ that

$$
\begin{aligned}
r_{\mathcal{P}_{A,p}}(F_{1,p}\cup F_{2,p}) &= r_{\mathcal{M}_p}(F_{1,p}) + r_{\mathcal{M}_p/\Gamma_p}(F_{2,p}),\\
r_{\mathcal{P}_{\Phi,p}}(F'_{1,p}\cup F'_{2,p}) &= r_{\mathcal{M}_p}(F'_{1,p}) + r_{\mathcal{M}_p/\Gamma_{\Phi,p}}(F'_{2,p}).
\end{aligned}
\tag{13}
$$

Here we prove that

$$
\forall(a,p)\in F'_{2,p}\backslash F_{2,p}\colon\ \pi(f(a)) > \pi((a,p)) > \pi(s(a)).
\tag{14}
$$

Let (a,p) be an edge in $F'_{2,p}\backslash F_{2,p}$. Define $s(a) := (a,q)$. It follows from $(a,p)\notin F_{2,p}$ that $(a,p)\neq s(a)$. Since $(a,p)\neq f(a)$, $\pi(f(a)) > \pi((a,p))$. For proving (14), we prove $\{s(a)\}\in\mathcal{I}_q/\Gamma_{\Phi,q}$, which implies that $\pi((a,p)) = \pi(s_\Phi(a)) \geq \pi(s(a))$. Since $(a,p)\neq s(a)$, this implies (14). Let B be a base of $\mathcal{M}_q|\Gamma_{\Phi,q}$. Since $\Gamma_{\Phi,q}\subseteq \Gamma_q$, B is an independent set in $\mathcal{M}_q|\Gamma_q$. Thus, there exists a base B' of $\mathcal{M}_q|\Gamma_q$ with $B\subseteq B'$. Since $\{s(a)\}\in\mathcal{I}_q/\Gamma_q$, Lemma 1(1) implies that $B' + s(a)\in\mathcal{I}_q$. Thus, (I1) implies that $B + s(a)\in\mathcal{I}_q$. Since B is a base of $\mathcal{M}_q|\Gamma_{\Phi,q}$, this and Lemma 1(1) imply that $\{s(a)\}\in\mathcal{I}_q/\Gamma_{\Phi,q}$. This completes the proof of (14).

Since (14) implies that $\{(a,p)\}\notin\mathcal{I}_p/\Gamma_p$ for every edge (a,p) in $F'_{2,p}\backslash F_{2,p}$,

$$
r_{\mathcal{M}_p/\Gamma_p}(F_{2,p}) = r_{\mathcal{M}_p/\Gamma_p}(F_{2,p}\cup F'_{2,p}).
\tag{15}
$$

Furthermore, it follows from the monotonicity of $r_{\mathcal{M}_p/\Gamma_{\Phi,p}}(\cdot)$ that

$$
r_{\mathcal{M}_p/\Gamma_{\Phi,p}}(F'_{2,p}) \leq r_{\mathcal{M}_p/\Gamma_{\Phi,p}}(F_{2,p}\cup F'_{2,p}).
\tag{16}
$$

It follows from $\Gamma_{\Phi,p}\subseteq\Gamma_p$, $(F_{2,p}\cup F'_{2,p})\cap\Gamma_p = \emptyset$, and Lemma 2 that

$$
r_{\mathcal{M}_p/\Gamma_{\Phi,p}}(F_{2,p}\cup F'_{2,p}) - r_{\mathcal{M}_p/\Gamma_p}(F_{2,p}\cup F'_{2,p}) \leq r_{\mathcal{M}_p}(\Gamma_p) - r_{\mathcal{M}_p}(\Gamma_{\Phi,p}).
\tag{17}
$$

Since $\Gamma_p\backslash\Gamma_{\Phi,p} = F_{1,p}\backslash F'_{1,p}$ and $F'_{1,p}\subseteq\Gamma_{\Phi,p}$, it follows from (1) that

$$
r_{\mathcal{M}_p}(\Gamma_p) - r_{\mathcal{M}_p}(\Gamma_{\Phi,p}) \leq r_{\mathcal{M}_p}(F_{1,p}) - r_{\mathcal{M}_p}(F'_{1,p}).
\tag{18}
$$

It follows from (15), (16), (17), and (18) that

$$
\begin{aligned}
&r_{\mathcal{M}_p/\Gamma_{\Phi,p}}(F'_{2,p}) - r_{\mathcal{M}_p/\Gamma_p}(F_{2,p})\\
&\leq r_{\mathcal{M}_p/\Gamma_{\Phi,p}}(F_{2,p}\cup F'_{2,p}) - r_{\mathcal{M}_p/\Gamma_p}(F_{2,p}\cup F'_{2,p})\\
&\leq r_{\mathcal{M}_p}(\Gamma_p) - r_{\mathcal{M}_p}(\Gamma_{\Phi,p})\\
&\leq r_{\mathcal{M}_p}(F_{1,p}) - r_{\mathcal{M}_p}(F'_{1,p}).
\end{aligned}
\tag{19}
$$

It follows from (13) and (19) that

$$r_{\mathcal{P}_{\Phi,p}}(F'_{1,p} \cup F'_{2,p}) - r_{\mathcal{P}_{A,p}}(F_{1,p} \cup F_{2,p})$$
$$= r_{\mathcal{M}_p}(F'_{1,p}) - r_{\mathcal{M}_p}(F_{1,p}) + r_{\mathcal{M}_p/\Gamma_{\Phi,p}}(F'_{2,p}) - r_{\mathcal{M}_p/\Gamma_p}(F_{2,p})$$
$$\leq r_{\mathcal{M}_p}(F'_{1,p}) - r_{\mathcal{M}_p}(F_{1,p}) + r_{\mathcal{M}_p}(F_{1,p}) - r_{\mathcal{M}_p}(F'_{1,p}) = 0.$$

This implies (12), which completes the proof. □

Theorem 4. *The algorithm* **PCuMC** *correctly solves the popular condensation problem under matroid constraints.*

Proof. This theorem follows from Lemmas 12, 13 and 14. □

Acknowledgements. This work was supported by JSPS KAKENHI Grant Number 25730006. The author would like to thank anonymous referees for helpful comments on an earlier version of this paper.

References

1. Abraham, D.J., Irving, R.W., Kavitha, T., Mehlhorn, K.: Popular matchings. SIAM J. Comput. **37**(4), 1030–1045 (2007)
2. Gärdenfors, P.: Match making: assignments based on bilateral preferences. Behav. Sci. **20**(3), 166–173 (1975)
3. Manlove, D.F., Sng, C.T.S.: Popular matchings in the capacitated house allocation problem. In: Azar, Y., Erlebach, T. (eds.) ESA 2006. LNCS, vol. 4168, pp. 492–503. Springer, Heidelberg (2006)
4. Mestre, J.: Weighted popular matchings. ACM Trans. Algorithms **10**(1) (2014). Article 2
5. Sng, C.T.S., Manlove, D.F.: Popular matchings in the weighted capacitated house allocation problem. J. Discrete Algorithms **8**(2), 102–116 (2010)
6. Kavitha, T., Nasre, M.: Popular matchings with variable item copies. Theor. Comput. Sci. **412**(12), 1263–1274 (2011)
7. Kavitha, T., Nasre, M., Nimbhorkar, P.: Popularity at minimum cost. J. Comb. Optim. **27**(3), 574–596 (2014)
8. Wu, Y.-W., Lin, W.-Y., Wang, H.-L., Chao, K.-M.: An optimal algorithm for the popular condensation problem. In: Lecroq, T., Mouchard, L. (eds.) IWOCA 2013. LNCS, vol. 8288, pp. 412–422. Springer, Heidelberg (2013)
9. Fleiner, T.: A fixed-point approach to stable matchings and some applications. Math. Oper. Res. **28**(1), 103–126 (2003)
10. Fujishige, S., Tamura, A.: A two-sided discrete-concave market with possibly bounded side payments: an approach by discrete convex analysis. Math. Oper. Res. **32**(1), 136–155 (2007)
11. Fleiner, T., Kamiyama, N.: A matroid approach to stable matchings with lower quotas. In: SODA, pp. 135–142 (2012)
12. Biró, P., Fleiner, T., Irving, R.W., Manlove, D.F.: The college admissions problem with lower and common quotas. Theor. Comput. Sci. **411**(34–36), 3136–3153 (2010)

13. Schrijver, A.: Combinatorial Optimization: Polyhedra and Efficiency. Springer, Berlin (2003)
14. Edmonds, J.: Submodular functions, matroids, and certain polyhedra. In: Guy, R., Hanani, H., Sauer, N., Schönheim, J. (eds.) Combinatorial Structures and Their Applications, pp. 69–87. Gordon and Breach, New York (1970)
15. Aigner, M., Dowling, T.A.: Matching theory for combinatorial geometries. Trans. Am. Math. Soc. **158**(1), 231–245 (1971)
16. Lawler, E.L.: Matroid intersection algorithms. Math. Program. **9**(1), 31–56 (1975)

Incremental Computation of Pseudo-Inverse of Laplacian

Gyan Ranjan$^{(\boxtimes)}$, Zhi-Li Zhang, and Daniel Boley

Department of Computer Science and Engineering,
University of Minnesota, Minneapolis, USA
{granjan,zhzhang,boley}@cs.umn.edu

Abstract. A divide-and-conquer based approach for computing the Moore-Penrose pseudo-inverse of the combinatorial Laplacian matrix ($\mathbf{L}^+$) of a simple, undirected graph is proposed. The nature of the underlying sub-problems is studied in detail by means of an elegant interplay between $\mathbf{L}^+$ and the effective resistance distance (Ω). Closed forms are provided for a novel *two-stage* process that helps compute the pseudo-inverse incrementally. Analogous scalar forms are obtained for the converse case, that of structural regress, which entails the breaking up of a graph into disjoint components through successive edge deletions. The scalar forms in both cases, show absolute element-wise independence at all stages, thus suggesting potential parallelizability. Analytical and experimental results are presented for dynamic (time-evolving) graphs as well as large graphs in general (representing real-world networks). An order of magnitude reduction in computational time is achieved for dynamic graphs; while in the general case, our approach performs better in practice than the standard methods, even though the worst case theoretical complexities may remain the same: an important contribution with consequences to the study of online social networks.

1 Introduction

The combinatorial Laplacian matrix of a graph finds use in various aspects of structural analysis. The eigen spectrum of the Laplacian determines significant topological characteristics of the graph, such as minimal cuts, clustering and the number of spanning trees [5,13,18]. Likewise, the Moore-Penrose pseudo-inverse and the sub-matrix inverses of the Laplacian have evoked great interest in recent years. Their applications span fields as diverse as probability and mathematical chemistry, collaborative recommendation systems and social networks, epidemiology and robustness of networks and inter-dependent networks [14–17,21,23,24,31]. A brief discussion of the specific applications is provided for reference in a subsequent section (cf. Sect. 6). Despite such versatility, the pseudo-inverse and the sub-matrix inverses of the Laplacian suffer a practical handicap. These matrices are notoriously expensive to compute. The standard matrix factorization and inversion based methods employed to compute them

© Springer International Publishing Switzerland 2014
Z. Zhang et al. (Eds.): COCOA 2014, LNCS 8881, pp. 729–749, 2014.
DOI: 10.1007/978-3-319-12691-3_54

[4,31] incur an $O(n^3)$ computational time, n being the order of the graph (number of vertices in the graph). This clearly impedes their utility particularly when the graphs are either dynamic, i.e. changing with time, or simply of large orders, i.e. have millions of nodes. Online social networks (OSN), typically represented as graphs, qualify on both counts. With time, the number of users as well as the relationships between them changes, thus requiring regular re-computations. In physical networks such as data/communication networks and power grids, dynamics in network topologies may rise from link/node failures and repairs. For OSNs and other networks with millions or more of nodes, an $O(n^3)$ computational cost is clearly prohibitive. An approach for incremental updates is imperative, particularly given that such changes, in most cases, may be local in nature.

In this work, we provide a novel divide-and-conquer based approach for computing the Moore-Penrose pseudo-inverse of the Laplacian for an undirected graph. This, in turn, determines all of its sub-matrix inverses as well. The divide operation in our approach entails determining an arbitrary *connected bi-partition* of the graph $G(V, E)$ — a cut of the graph that is made up of exactly two connected sub-graphs (say $G_1(V_1, E_1)$ and $G_2(V_2, E_2)$) — by deleting κ edges from it. As $G_1(V_1, E_1)$ and $G_2(V_2, E_2)$ are simple and connected themselves, the pseudo-inverse of their Laplacians, when computed, constitute solutions to two independent sub-problems. Better still, they can be computed in parallel (given two machines instead of one). In the conquer step, we recombine these independent solutions in an iterative manner by re-introducing the edges in the cut set one at a time to reconstruct the original graph incrementally. Clearly, this process yields a sequence of intermediate connected spanning sub-graphs of G, (say $\{G_1, G_2\} \to G_3 \to G_4 \to ... \to G_{\kappa+2}$), where $G_{\kappa+2} = G(V, E)$. The first transition $\{G_1, G_2\} \to G_3$ represents a point of *singularity* whence the disjoint components $\{G_1, G_2\}$ get connected through a *bridge* edge to yield G_3, a sub-graph with exactly one component. We call this stage the *first join* in our process. Post the first join, all intermediate sub-graphs from G_4 to $G_{\kappa+2}$ are obtained by introducing an edge in a sub-graph that is already connected. We call these atomic steps *edge firings* (details in a subsequent section).

We then show that the pseudo-inverse of the Laplacian for any intermediate sub-graph in this sequence is determined entirely in terms of the pseudo-inverse of the Laplacian for its predecessor. Our results, presented in an element-wise scalar form, reveal several interesting properties of the sub-problems. First and foremost, if n be the order of the graph $G(V, E)$, then the cost incurred at each intermediate stage is $O(n^2)$ if the solution to the sub-problems for the immediate predecessor is known. Therefore, the cost of computing the pseudo-inverse for $G(V, E)$ is $O(\kappa \cdot n^2)$, if the pseudo-inverses for $G_1(V_1, E_1)$ and $G_2(V_2, E_2)$ are known. Secondly, using these forms, each element of the pseudo-inverse for an intermediate graph can be computed independent of the other elements. Hence, given multiple machines, the overall computational time is reduced further through parallelization. Moreover, we obtain similar closed form solutions for the case of structural regress of the graph, i.e. when vertices or edges are deleted from it. A straightforward consequence is that the pseudo-inverses for dynamic time-evolving graphs can now be updated when a node joins or leaves the network or an edge (a relationship) appears/disappears in it at an $O(n^2)$ cost overall (as $\kappa << n$).

Last but not least, we use these insights to compute the pseudo-inverses of the Laplacians of large real-world networks from the domain of online social networks. Real-world networks, and social ones in particular, are reported to have some notable characteristics such as edge sparsity, *power-law* and scale-free degree distributions [3], *small-world* characteristics [30] etc. Given these properties, we note that interesting algorithms (heuristics) can be developed for fast and parallel computations for the general case based on our divide-and-conquer strategy. Thus, even though the theoretical worst case costs stay at $O(n^3)$ for general graphs, the practical gains are significant enough to warrant attention. We discuss both analytical and experimental aspects of these in detail in the subsequent sections.

The remainder of the paper is organized into the following sections: we begin by introducing the preliminaries of our work — the pseudo-inverse and the sub-matrix inverses of the Laplacian along with their properties; and the interplay of the pseudo-inverse and the effective resistance distance — in Sect. 2. In Sect. 3, we describe our divide-and-conquer strategy involving connected bi-partitions and the *two-stage* process for computing the Moore-Penrose pseudo-inverse of the Laplacian. Relevant scalar forms are presented in each case. In Sect. 4, we establish the same closed forms for a graph in regress i.e. deleting edges one at a time until the graph breaks into two. We then apply the divide-and-conquer methodology to compute the pseudo-inverses for dynamically changing graphs as well as those of real world networks in Sect. 5. We provide a brief overview of related literature, focusing on specific application scenarios of L^+ in Sect. 6. The paper is finally concluded in Sect. 7 with a summary of results and a discussion of potential future works.

2   The Laplacian, Sub-matrix Inverses and a Distance Function

In this section, we provide a brief introduction to the set of matrices studied in this work, namely, the combinatorial Laplacian of a graph ($\mathbf{L}$), its Moore-Penrose pseudo-inverse ($\mathbf{L}^+$) and the set of sub-matrix inverses of $\mathbf{L}$ (Sect. 2.1). We then demonstrate how all the sub-matrix inverses of the Laplacian can be computed in terms of the pseudo-inverse in Sect. 2.2. Finally, in Sect. 2.3 we describe the relationship between the *effective resistance distance*, a Euclidean metric, and the elements of the Moore-Penrose pseudo-inverse of the Laplacian — an equivalence that we exploit to great advantage in the rest of this work.

2.1   The Laplacian and Its Moore-Penrose Pseudo-Inverse

Let $G(V, E)$ be a simple, connected and undirected graph. We denote by $n = |V(G)|$ the number of nodes/vertices in G, also called the *order* of the graph G, and by $m = |E(G)|$ the number of links/edges. The adjacency matrix of $G(V, E)$ is defined as $\mathbf{A} \in \Re^{n \times n}$, with elements $[\mathbf{A}]_{xy} = a_{xy} = a_{yx} = [\mathbf{A}]_{yx} = w_{ij}$, if $x \neq y$ and $e_{xy} \in E(G)$ is an edge; 0 otherwise. Here, the weight of the edge w_{ij} is a measure of *affinity* between nodes i and j. Clearly, $\mathbf{A}$ is real and symmetric. The degree matrix $\mathbf{D}$, is a diagonal matrix where $[\mathbf{D}]_{xx} = d_{xx} = d(x) = \sum_{y \in V(G)} a_{xy}$, is the weighted degree of node $x \in V(G)$; the sum of all edge weights (affinities) emanating from x. Also, $vol(G) = \sum_{x \in V(G)} d(x)$, is called the *volume* of the graph G — the sum total of affinities between all pairs of vertices in G. The combinatorial Laplacian of the graph is then given by:

$$\mathbf{L} = \mathbf{D} - \mathbf{A} \tag{1}$$

It is easy to see, from the definition in (1) above, that the Laplacian $\mathbf{L}$ is a real, symmetric and doubly-centered matrix (each row/column sum is 0). More importantly, $\mathbf{L}$ admits an eigen decomposition of the form $\mathbf{L} = \mathbf{\Phi \Lambda \Phi'}$ where the columns of $\mathbf{\Phi}$ constitute the set of orthogonal eigen vectors of $\mathbf{L}$ and $\mathbf{\Lambda}$ is a diagonal matrix with $[\Lambda]_{ii} = \lambda_i : 1 \leq i \leq n$; being the n eigen values of $\mathbf{L}$. It is well established that for a connected undirected graph $G(V, E)$, $\mathbf{L}$ is positive semi-definite i.e. it has a unique smallest eigen value $\lambda_1 = 0$. The rest of the $n-1$ eigen values are all positive. Thus, $\mathbf{L}$ is rank deficient ($rank(\mathbf{L}) = n - 1 < n$) and consequently singular. Its inverse, in the usual sense, does not exist.

However, the Moore-Penrose pseudo-inverse of $\mathbf{L}$, denoted henceforth by $\mathbf{L}^+$, does exist and is unique [4]. Like $\mathbf{L}$, $\mathbf{L}^+$ is also real, symmetric, doubly centered and positive semi-definite. Moreover, the eigen decomposition of $\mathbf{L}^+$ is given by $\mathbf{L}^+ = \mathbf{\Phi \Lambda^+ \Phi'}$, with the same set of orthogonal eigen-vectors as that of $\mathbf{L}$. The set of eigen values of $\mathbf{L}^+$, given by the diagonal of the matrix $\mathbf{\Lambda}^+$, is composed of $\lambda_1^+ = 0$ and reciprocals of the positive eigen-values of $\mathbf{L}$. We denote by l_{xy}^+, the element in the xth row and y^{th} column of $\mathbf{L}^+$ (a convention followed for all matrices henceforth). We emphasize that even when the matrix $\mathbf{L}$ is sparse (which is the case with real world networks), $\mathbf{L}^+$ is always a full matrix. In fact, for a connected graph, all the elements of $\mathbf{L}^+$ are non-zero.

A straightforward approach for computing $\mathbf{L}^+$ is through the eigen-decomposition of $\mathbf{L}$, followed by an inversion of its non-zero eigen values, and finally reassembling the matrix as discussed above. In practice, however, mathematical software, such as MATLAB, use singular value decomposition to compute the pseudo-inverse of matrices (cf. *pinv* in the standard library). This general SVD based method does not exploit the special structural properties of $\mathbf{L}$ and incurs $O(n^3)$ computational time, n being the number of nodes in the graph. An alternative exists [11,31] specifically for computing $\mathbf{L}^+$ for a simple, connected, undirected graph. A $rank(1)$ perturbation of the matrix $\mathbf{L}$ makes it invertible. $\mathbf{L}^+$ can then be computed from this perturbed matrix as follows:

$$\mathbf{L}^+ = \left(\mathbf{L} + \frac{1}{n}\mathbf{J}\right)^{-1} - \frac{1}{n}\mathbf{J} \tag{2}$$

where $\mathbf{J} \in \Re^{n \times n}$ is a matrix of all $1's$. Although the theoretical cost for this method is also $O(n^3)$, in practice it works faster for graphs of arbitrary orders and edge densities than the standard *pinv* method. However, for applying such a method to *dynamically evolving* graphs in which small local modifications (e.g., adding or deleting an edge or a node), repeated computations of matrix inverse incur undue heavy computational costs. In what follows, we show that the computation of the Moore-Penrose pseudo-inverse of the Laplacian can be done in a divide-and-conquer fashion. Our method allows efficient incremental updates of $\mathbf{L}^+$ for dynamically changing graphs, without having to compute $\mathbf{L}^+$ all over again. Moreover, computing $\mathbf{L}^+$ for large graphs becomes feasible, in principle, through parallelization of (smaller) independent sub-problems over multiple machines, which can then be re-combined at $O(n^2)$ cost per edge across a division (details in a subsequent section). But first we need to establish a few more preliminary results to further motivate our study.

2.2   Sub-matrix Inverses of L

As described in the previous section, the combinatorial Laplacian $\mathbf{L}$ of a connected graph $G(V, E)$, is singular and thus non-invertible. However, given that its rank is $n - 1$, any $n - 1$ combination of columns (or rows) of $\mathbf{L}$ constitutes a linearly independent set. Hence, any $(n-1 \times n-1)$ sub-matrix of $\mathbf{L}$ is invertible. Indeed, the inverses of such $(n-1 \times n-1)$ sub-matrices are made use of in several graph analysis problems: enumerating the spanning trees and spanning forests of the graph [16], determining the random-walk betweenness of the nodes of the graph [21], to name a few. However, the cost of computing an $(n-1 \times n-1)$ sub-matrix inverse is still $O(n^3)$. To compute all such sub-matrix inverses amounts to a time complexity of $O(n^4)$. In the following, we show how they can be computed efficiently through $\mathbf{L}^+$.

Theorem 1. *Let* $\mathbf{L}(\{\overline{n}\}, \{\overline{n}\})$ *be an* $(n - 1 \times n - 1)$ *sub-matrix of* $\mathbf{L}$ *formed by removing the* n^{th} *row and* n^{th} *column of* $\mathbf{L}$. *Then* $\forall (x, y) \in V(G) \times V(G)$:

$$[\mathbf{L}(\{\overline{n}\}, \{\overline{n}\})^{-1}]_{xy} = l^+_{xy} - l^+_{xn} - l^+_{ny} + l^+_{nn} \tag{3}$$

The result in Theorem 1 above expresses, in scalar form, the general element (x^{th} row, y^{th} column) of the inverse of the sub-matrix $\mathbf{L}(\{\overline{n}\}, \{\overline{n}\})$ in terms of the elements of $\mathbf{L}^+$, as claimed. As the choice of the nth row and column is arbitrary, we can see that the result holds in general for any $(n - 1 \times n - 1)$ sub-matrix (permuting the rows and columns of $\mathbf{L}$ as per need). The cost of computing $\mathbf{L}(\{\overline{n}\}, \{\overline{n}\})^{-1}$ for a given vertex n is $O(n^2)$. Therefore, all sub-matrix inverses can be computed in $O(n^3)$ time from $\mathbf{L}^+$, which itself can be computed in $O(n^3)$ time, even if the standard methods are used. This is clearly an order of magnitude improvement. Henceforth, we focus entirely on $\mathbf{L}^+$.

2.3   The Effective Resistance Distance and L⁺

An interesting analogy exists between graphs and resistive electrical circuits [12,17]. Given a simple, connected and undirected graph $G(V, E)$, the equivalent

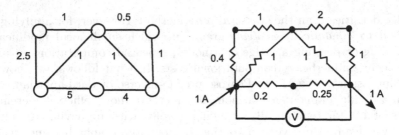

Fig. 1. A simple graph G and its EEN.

electrical network (EEN) of the graph can be formed by replacing each edge $e_{ij} \in E(G)$, of weight w_{ij} with an electrical resistance $\omega_{ij} = w_{ij}^{-1}$ ohm (cf. Fig. 1). A distance function can then be defined between any pair of nodes $(x, y) \in V(G) \times V(G)$ in the resulting EEN as follows:

Definition 1. *Effective Resistance (Ω_{xy}): The voltage developed between nodes x and y, when a unit current is injected at node x and is withdrawn at node y.*

It is well established that the square root of the effective resistance distance $(\sqrt{\Omega_{xy}})$ is a Euclidean metric with interesting applications [17]. Amongst other things, it determines the expected length of random commutes between node pairs in the graph: $C_{xy} = vol(G) \; \Omega_{xy}$, [9,29]. More importantly, Ω_{xy} can be expressed in terms of the elements of $\mathbf{L}^+$ as follows:

$$\Omega_{xy} = l_{xx}^+ + l_{yy}^+ - l_{xy}^+ - l_{yx}^+ \qquad (4)$$

We now invert the elegant form in (4) to derive an important result in the following lemma which gives us the general term of $\mathbf{L}^+$ in terms of the distance function Ω.

Lemma 1. $\forall (x, y, z) \in V(G) \times V(G) \times V(G):$

$$l_{xy}^+ = \frac{1}{2n}\left[\sum_{z=1}^{n}(\Omega_{xz} + \Omega_{zy} - \Omega_{xy})\right] - \frac{1}{2n^2}\sum_{x=1}^{n}\sum_{y=1}^{n}\Omega_{xy} \qquad (5)$$

The RHS in Lemma 1 above is composed of two terms: a triangle inequality of effective resistances [29] and a double summand over all pairwise effective resistances in the EEN. It is easy to see that the double-summand simply reduces to a scalar multiple of the trace of $\mathbf{L}^+$ $(Tr(\mathbf{L}^+) = \sum_{z=1}^{n} l_{zz}^+)$. Thus the functional half that determines the elements of $\mathbf{L}^+$, is the triangle inequality of the effective resistances, while the double summand contributes an additive constant to all the entries of $\mathbf{L}^+$. We illustrate the utility of this result, with the help of two kinds of graphs on the extremal ends of the connectedness spectrum: the star and the clique.[1]

---

[1] The graphs in these examples are assumed to be unweighted, i.e. all edges have a unit resistance/conductance.

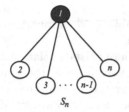

0.16	-0.04	-0.04	-0.04	-0.04
-0.04	0.76	-0.24	-0.24	-0.24
-0.04	-0.24	0.76	-0.24	-0.24
-0.04	-0.24	-0.24	0.76	-0.24
-0.04	-0.24	-0.24	-0.24	0.76

S_n $\mathbf{L}^+_{S_5}$

Fig. 2. The star graph: Pre-computed $\mathbf{L}^+_{S_n}$ for $n = 5$.

The Star. A star of order n is a tree with exactly one vertex of degree $n - 1$, referred to as the *root*, and $n-1$ pendant vertices each of degree 1, called *leaves*, (cf. Fig. 2). By definition, a singleton isolated vertex is also a degenerate star albeit with no leaves. It is easy to see that S_n, being a tree, is the most sparse connected graph of order n (with exactly $n - 1$ edges). Also, S_n is the most compact tree of its order (lowest diameter). In the following, we show how $\mathbf{L}^+_{S_n}$ can be computed using the result of Lemma 1.

Corollary 1. *For a star graph S_n of order n, with node v_1 as root and nodes $\{v_2, v_3, ..., v_n\}$ as leaves, $\mathbf{L}^+_{S_n}$ is given by:*

$$l^+_{11} = \frac{n-1}{n^2} \qquad and \qquad \forall x : 2 \leq x \leq n, \ l^+_{1x} = l^+_{x1} = -\frac{1}{n^2} \quad (6)$$

$$\forall x : 2 \leq x \leq n, \ l^+_{xx} = \frac{n^2 - n - 1}{n^2} \quad and \quad \forall x \neq y : 2 \leq x, y \leq n, \ l^+_{xy} = l^+_{yx} = -\frac{n+1}{n^2} \quad (7)$$

The Clique. On the other end of the connectedness spectrum lies the clique. A clique K_n of order n is a complete graph with $\frac{n(n-1)}{2}$ edges. Clearly, the clique is the densest possible graph of order n, as there is a direct edge between any pair of vertices in it. It is also the most compact graph of its order (lowest diameter). Then,

Corollary 2. *For a clique K_n of order n, $\mathbf{L}^+_{K_n}$ is given by:*

$$\forall x : 1 \leq x \leq n, \ l^+_{xx} = \frac{n-1}{n^2} \ and \ \forall x \neq y : 1 \leq x, y \leq n, \ l^+_{xy} = l^+_{yx} = -\frac{1}{n^2} \quad (8)$$

The results in the corollaries presented above are not just illustrative examples. They are also of interest from a computational point of view, particularly when the graph under study is an unweighted one. Both stars and cliques can occur as motif sub-graphs in any given graph. Indeed, for any non-trivial connected simple graph of order $n \geq 3$, there is at least one sub-graph that is a star. Selecting any vertex i with $d(i) \geq 2$, and conducting a one-hop breadth first search, generates a star sub-graph. Cliques, though not so universal, also occur in real world networks (e.g. citation networks). Therefore, in any divide-and-conquer methodology, both stars and cliques are likely to be found at some stage. We have already established that the cost of computing $\mathbf{L}_{S_n}^+$ and $\mathbf{L}_{K_n}^+$ is $O(1)$ (as they are determined entirely by n) and hence the solution to such a sub-problem, when found, is obtained at the lowest possible cost — a true practical gain.

To conclude, we have demonstrated that there exists a relationship between the elements of $\mathbf{L}^+$ and the pairwise effective resistances in the graph $G(V, E)$, that yields interesting closed form solutions for the pseudo-inverse for special graphs such as stars and cliques. In the subsequent sections, we demonstrate that it can be used to compute $\mathbf{L}^+$ for general graphs as well, incrementally, in a divide-and-conquer fashion.

3   From Two to One: Computing $\mathbf{L}^+$ by Partitions

In this section, we present our main result – the computation of the Moore-Penrose pseudo-inverse $\mathbf{L}^+$ of the Laplacian by means of graph bi-partitions. We first lay out a *two-stage* process — the *first join* followed by *edge firings* — that underpins our methodology. We then provide specific closed form solutions.

Note, due to space limitations, we omit the proofs from this paper. Interested readers can find the proofs for all the lemmas, theorems and corollaries in the arXiv version [25].

3.1   Connected Bi-Partitions of a Graph and the Two-Stage Process

In order to compute the Moore-Penrose pseudo-inverse of the Laplacian of a simple, connected, undirected and unweighted graph $G(V, E)$ by parts, we must first establish that the problem can be decomposed into two, or more, sub-problems that can be solved independently. The solutions to the independent sub-problems can then be combined to obtain the overall result. We start by introducing a few notations.

Definition 2. *Connected Bi-partition $(P = (G_1, G_2))$: A cut of the graph G which contains exactly two mutually exclusive and exhaustive connected sub-graphs G_1 and G_2.*

Fig. 3(a–b), shows a graph $G(V, E)$ and a connected bi-partition $P(G_1, G_2)$, obtained from the graph $G(V, E)$ by removing the set of dotted edges shown. Each partition $P(G_1, G_2)$ has certain defining characteristics in terms of the set of vertices as well the set of edges in the graph. Let, $V_1(G_1)$ and $V_2(G_2)$ be the mutually exclusive and exhaustive subsets of $V(G)$ i.e. $V_1(G_1) \cap V_2(G_2) = \phi$ and $V_1(G_1) \cup V_2(G_2) = V(G)$. Similarly, let $E_1(G_1)$ and $E_2(G_2)$ be the sets of edges in the respective sub-graphs G_1 and G_2 of P and $E(G_1, G_2)$, the set of edges that *violate* the partition $P(G_1, G_2)$ i.e. have one end in G_1 and the other in G_2. Thus, $E_1(G_1) \cap E_2(G_2) = E_1(G_1) \cap E_1(G_1, G_2) = E(G_1, G_2) \cap E_2(G_2) = \phi$ and $E_1(G_1) \cup E(G_1, G_2) \cup E_2(G_2) = E(G)$. We denote by $\mathcal{P}(G)$ the set of all such connected bi-partitions of G.

It is easy to see that for an arbitrary connected bi-partition $P(G_1, G_2) \in \mathcal{P}(G)$ both G_1 and G_2 are themselves simple, connected, undirected and unweighted

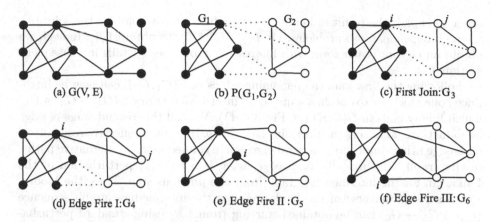

(a) G(V, E)   (b) P(G₁,G₂)   (c) First Join: G₃

(d) Edge Fire I: G₄   (e) Edge Fire II : G₅   (f) Edge Fire III: G₆

Fig. 3. Divide-and-Conquer: Connected bi-partition of a graph and the two-stage process: first join followed by three edge firings. The dotted lines represent the edges that are not part of the intermediate graph at that stage.

graphs. Hence, the discussion in Sect. 2 is applicable in its entirety to the sub-graphs G_1 and G_2 independently. Note then that $\mathbf{L}_{G_1}^+$ and $\mathbf{L}_{G_2}^+$, the Moore-Penrose pseudo-inverse of the Laplacians of the sub-graphs G_1 and G_2, must, by defin-ition, exist. The pair $\{\mathbf{L}_{G_1}^+, \mathbf{L}_{G_2}^+\}$, constitutes the solution to two independent sub-problems represented in the set $\{G_1, G_2\}$. All that remains to be shown now is that $\{\mathbf{L}_{G_1}^+, \mathbf{L}_{G_2}^+\}$ can indeed be combined to obtain $\mathbf{L}_G^+$. It is this aspect of the methodology, that we call the *two-stage* process, as explained in detail below.

The original graph $G(V, E)$ can be thought of, in some sense, as a bringing together of the disjoint spanning sub-graphs G_1 and G_2, by means of introduc-ing the edges of the set $E(G_1, G_2)$. Starting from G_1 and G_2, we iterate over the set of edges in $E(G_1, G_2)$ in the following fashion (cf. Fig. 3 for a visual reference). Let $e_{ij} \in E(G_1, G_2) : i \in V_1(G_1), j \in V_2(G_2)$, of weight w_{ij} and resistance $\omega_{ij} = w_{ij}^{-1}$ *ohm*, be an arbitrary edge chosen during the first iteration as shown in Fig. 3(c). We call this step the *first join* in our two-stage process, whereafter G_1 and G_2 come together to give an intermediate connected span-ning sub-graph (say $G_3(V_3, E_3)$). The first join represents a point of singularity in the reconstruction process, particularly from the perspective of the effective resistance distance. Note that before the first join, the effective resistance dis-tance between an arbitrary pair of nodes $(x, y) \in V(G) \times V(G)$ is infinity, if $x \in V_1(G_1)$ and $y \in V_2(G_2)$, as there is no path connecting x and y. However, once the first edge e_{ij} has been introduced during the first join, this discrep-ancy no longer exists and all pairwise effective resistances are finite. Precisely, if $\Omega^{G_1} : V_1(G_1) \times V_1(G_1) \to \Re^+$ and $\Omega^{G_2} : V_2(G_2) \times V_2(G_2) \to \Re^+$, be the pairwise effective resistances defined over the sub-graphs G_1 and G_2, the fol-lowing holds: if $x, y \in V_1(G_1)$, $\Omega_{xy}^{G_3} = \Omega_{xy}^{G_1}$; if $x, y \in V_2(G_2)$, $\Omega_{xy}^{G_3} = \Omega_{xy}^{G_2}$; and if $x \in V_1(G_1)$ & $y \in V_2(G_2)$, $\Omega_{xi}^{G_1} + \omega_{ij} + \Omega_{jy}^{G_2}$. Needless to say, this is a critical step in the process as we need finite values of effective resistances in order to exploit the result in Lemma 1. Hereafter, we can combine the solutions to the independent sub-problems, i.e. $\mathbf{L}_{G_1}^+$ and $\mathbf{L}_{G_2}^+$, to obtain $\mathbf{L}_{G_3}^+$. Indeed, we obtain an elegant scalar form with interesting properties (details in subsequent sections).

Following the first join, the remaining edges in $E(G_1, G_2)$, can now be introduced one at a time to obtain a sequence of intermediate graphs $(G_4 \rightarrow G_5 \rightarrow G_6)$ which finally ends in $G(V, E)$ (cf. Fig. 3(d–f)). We call this second stage of edge introductions, following the first join, *edge firing*. In terms of effective resistances, each edge firing simply creates parallel resistive connections, or alternative paths, in the graph. Algebraically, each edge firing is a $rank(1)$ perturbation of the Laplacian for the intermediate graph from the previous step. Thus, the Moore-Penrose pseudo-inverse of the Laplacians for the intermediate graph sequence $(G_4 \rightarrow G_5 \rightarrow G_6)$ can be obtained starting from $\mathbf{L}_{G_3}^+$ using standard perturbation methods [20] (details in subsequent sections).

To summarize, therefore, during the two-stage process we obtain a sequence of connected spanning sub-graphs of $G(V, E)$ starting from a partition $P(G_1, G_2) \in \mathcal{P}(G)$, performing the first join by arbitrarily selecting an edge $e_{ij} \in E(G_1, G_2)$, and then firing the remaining edges, one after the other, in any arbitrary order. The number of connected spanning sub-graphs of $G(V, E)$ constructed during the two-stage process is exactly $|E(G_1, G_2)|$ ($= 4$ for the example in Fig. 3). Note that, the order in which these sub-graphs are generated, is of no consequence whatsoever. Next, we use these insights to obtain $\mathbf{L}^+$ for the intermediate graphs in the sequence.

3.2   The Two-Stage Process and $\mathbf{L}^+$

We now present the closed form solutions for the Moore-Penrose pseudo-inverse of the Laplacians of the set of intermediate graphs obtained during the two-stage process.

The First Join. Given, two simple, connected, undirected graphs $G_1(V_1, E_1)$ and $G_2(V_2, E_2)$ let $\mathbf{L}_{G_1}^+$ and $\mathbf{L}_{G_2}^+$, be the respective Moore-Penrose pseudo-inverses of their Laplacians. Also, let $n_1 = |V_1(G_1)|$ and $n_2 = |V_2(G_2)|$ be the orders of the two graphs. We denote by $l_{xy}^{+(1)}$ and $l_{xy}^{+(2)}$ respectively the general terms of the matrices $\mathbf{L}_{G_1}^+$ and $\mathbf{L}_{G_2}^+$. Next, let the *first join* between G_1 and G_2 be performed by introducing an edge e_{ij} between the graphs G_1 and G_2 to obtain $G_3(V_3, E_3)$; where $i \in V_1(G_1)$ and $j \in V_2(G_2)$. Clearly, $V_3(G_3) = V_1(G_1) \cup V_2(G_2)$ and $E_3(G_3) = E_1(G_1) \cup \{e_{ij}\} \cup E_2(G_2)$. Thus, $|V_3(G_3)| = n_3 = n_1 + n_2$ and $E_3(G_3) = m_3 = m_1 + 1 + m_2$. By convention, the vertices in $V_3(G_3)$ are labeled in the following order: the first n_1 vertices $\{1, 2, ..., n_1\}$ are retained, *as is*, from $V_1(G_1)$ and the remaining n_2 vertices are labelled $\{n_1 + 1, n_1 + 2, ..., n_1 + n_2\}$ in order from $V_2(G_2)$. We denote by $\mathbf{L}_{G_3}^+$ the pseudo-inverse and $l_{xy}^{+(3)}$ its general term. Then,

Theorem 2. $\forall (x,y) \in V_3(G_3) \times V_3(G_3)$,

$$l_{xy}^{+(3)} = l_{xy}^{+(1)} - \frac{n_2 n_3 \left(l_{xi}^{+(1)} + l_{iy}^{+(1)} \right) - n_2^2 \left(l_{ii}^{+(1)} + l_{jj}^{+(2)} + \omega_{ij} \right)}{n_3^2}, \qquad if \ x,y \in V_1(G_1)$$

$$= l_{xy}^{+(2)} - \frac{n_1 n_3 \left(l_{xj}^{+(2)} + l_{jy}^{+(2)} \right) - n_1^2 \left(l_{ii}^{+(1)} + l_{jj}^{+(2)} + \omega_{ij} \right)}{n_3^2}, \qquad if \ x,y \in V_2(G_2)$$

$$= \frac{n_3 \left(n_1 l_{xi}^{+(1)} + n_2 l_{jy}^{+(2)} \right) - n_1 n_2 \left(l_{ii}^{+(1)} + l_{jj}^{+(2)} + \omega_{ij} \right)}{n_3^2}, \qquad if \ x \in V_1(G_1) \ \& \ y \in V_2(G_2)$$

The result in Theorem 2 is interesting for several reasons. First and foremost, it clearly shows that the general term of $\mathbf{L}_{G_3}^+$, is a linear combination of the elements of $\mathbf{L}_{G_1}^+$ and $\mathbf{L}_{G_2}^+$. This was indeed our principal claim. Secondly, $\forall (x,y) \in V_3(G_3) \times V_3(G_3)$, each individual $l_{xy}^{+(3)}$ can be computed independent of the others (barring symmetry, i.e. $l_{xy}^{+(3)} = l_{yx}^{+(3)}$, which we shall discuss shortly). They are determined entirely by the specific elements from the ith and j^{th} columns of the matrices $\mathbf{L}_{G_1}^+$ and $\mathbf{L}_{G_2}^+$, depending upon the membership of x and y in the disjoint graphs. This implies that all $l_{xy}^{+(3)}$ can be computed in parallel, as long as we have the relevant elements of $\mathbf{L}_{G_1}^+$ and $\mathbf{L}_{G_2}^+$.

From a cost point of view, the first join requires $O(1)$ computations per element in $\mathbf{L}_{G_3}^+$ — constant number of $\{+, -, \times, /\}$ operations — if $\{\mathbf{L}_{G_1}^+, \mathbf{L}_{G_2}^+\}$ is given *a priori*. The common term in the numerator, i.e. $(l_{ii}^{+(1)} + l_{jj}^{+(2)} + \omega_{ij})$, is an invariant for the elements of $\mathbf{L}_{G_3}^+$ and need only be computed once. This term is simply a linear multiple of the change in trace:

$$\Delta (Tr) = Tr(\mathbf{L}_{G_3}^+) - [Tr(\mathbf{L}_{G_1}^+) + Tr(\mathbf{L}_{G_2}^+)] \tag{9}$$

For details see the proof of Lemma 2 in [25]. Therefore, we achieve an overall cost of $O(n_3^2)$ for the first join. Last but not the least, we need to compute and store only the upper triangular of $\mathbf{L}_{G_3}^+$. Owing to the symmetry of $\mathbf{L}_{G_3}^+$, the lower triangular is determined automatically. As for the diagonal elements, they come without any additional cost as a result of $\mathbf{L}_{G_3}^+$ being doubly-centered (cf. Fig. 4).

Firing an Edge. We now look at the second stage that of *firing an edge* in a connected graph. Given a simple, connected, undirected graph $G_1(V_1, E_1)$, let $e_{ij} \notin E_1(G_1)$ be *fired* to obtain $G_2(V_2, E_2)$. Clearly, $V_2(G_2) = V_1(G_1)$ and $E_2(G_2) = E_1(G_1) \cup \{e_{ij}\}$. Continuing with our convention, we denote by $\mathbf{L}_{G_1}^+$ and $\mathbf{L}_{G_2}^+$ the Moore-Penrose pseudo-inverses of the respective Laplacians. Then,

Theorem 3. $\forall (x,y) \in V_2(G_2) \times V_2(G_2)$,

$$l_{xy}^{+(2)} = l_{xy}^{+(1)} - \frac{\left(l_{xi}^{+(1)} - l_{xj}^{+(1)} \right) \left(l_{iy}^{+(1)} - l_{jy}^{+(1)} \right)}{\omega_{ij} + \Omega_{ij}^{G_1}} \tag{10}$$

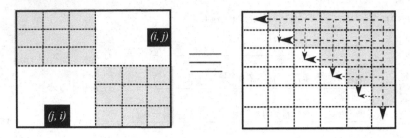

Fig. 4. The First Join: Scalar mapping $(\mathbf{L}_{G_1}^+, \mathbf{L}_{G_2}^+)$ to $\mathbf{L}_{G_3}^+$. The grey blocs represent relevant elements in $\mathbf{L}_{G_1}^+$, $\mathbf{L}_{G_2}^+$ and $\mathbf{L}_{G_3}^+$. Arrows span the elements of the upper triangular of $\mathbf{L}_{G_3}^+$ that contribute to the respective diagonal element pointed to by the arrow head: $l_{kk}^{+(3)} = -\left(\sum_{i=1}^{k-1} l_{ik}^{+(3)} + \sum_{j=k+1}^{n} l_{kj}^{+(3)} \right)$.

where $\Omega_{ij}^{G_1}$ is the effective resistance distance between nodes i and j in the graph $G_1(V_1, E_1)$ — an invariant $\forall(x, y) \in V_3(G_3) \times V_3(G_3)$ that is determined entirely by the end-points of the edge e_{ij} being fired. Once again, we observe that the general term of $\mathbf{L}_{G_2}^+$ is a linear combination of the elements of $\mathbf{L}_{G_1}^+$ and requires $O(1)$ computations per element in $\mathbf{L}_{G_2}^+$ — constant number of $\{+, -, \times, /\}$ operations — if $\mathbf{L}_{G_1}^+$ is given *a priori*. The rest of the discussion from the preceding sub-section on first join — element-wise independence and upper triangular sufficiency — holds *as is* for this stage too. However, before concluding this section, we extend the result of Theorem 3 to the pairwise effective resistances themselves in the following corollary.

Corollary 3. $\forall(x, y) \in V_2(G_2) \times V_2(G_2)$,

$$\Omega_{xy}^{G_2} = \Omega_{xy}^{G_1} - \frac{\left[\left(\Omega_{xj}^{G_1} - \Omega_{xi}^{G_1} \right) - \left(\Omega_{jy}^{G_1} - \Omega_{iy}^{G_1} \right) \right]^2}{4(\omega_{ij} + \Omega_{ij}^{G_1})} \tag{11}$$

The result above is interesting in its own right. Note that computing Ω^{G_2} when the edge density of a graph increases (or the expected commute times in random walks between nodes), is pertinent to many application scenarios [7,9,14,19,26]. Corollary 3 gives us a way of computing these distances directly without having to compute $\mathbf{L}_{G_2}^+$ first.

To conclude, therefore, we have established that the Moore-Penrose pseudo-inverses of the Laplacians of all the intermediate graphs, generated during the two-stage process, are incrementally computable from the solutions at the preceding stage, on an element-to-element basis. We shall return to specific applications of these results to dynamic (time-evolving) graphs and large graphs in general, in a subsequent section. But first, for the sake of completeness, we present the case of structural regress.

4   From One to Two: A Case of Regress

We now present analogous results in the opposite direction, that of structural regress of a graph through successive deletion of edges until the graph breaks into two. These results, similar in essence to those presented in the preceding section, are particularly significant with respect to dynamically evolving graphs that change with time (e.g. social networks). Once again, we have two cases to address with respect to edge deletions *viz.* (a) *Non-bridge edge*: an edge that upon deletion does not affect the connectedness of the graph (cf. Sect. 4.1); and (b) *Bridge-edge*: an edge that, when deleted, yields a connected bi-partition of the graph (cf. Sect. 4.2).

4.1   Deleting a Non-Bridge Edge

Given a simple, connected, undirected graph $G_1(V_1, E_1)$, let $e_{ij} \in E_1(G_1)$ be a *non-bridge* edge that is deleted to obtain $G_2(V_2, E_2)$. Clearly, $V_2(G_2) = V_1(G_1)$ and $E_2(G_2) = E_1(G_1) - \{e_{ij}\}$. Once again, we denote by $\mathbf{L}_{G_1}^+$ and $\mathbf{L}_{G_2}^+$ the Moore-Penrose pseudo-inverses of the respective Laplacians. Then,

Theorem 4. $\forall (x, y) \in V_2(G_2) \times V_2(G_2)$,

$$l_{xy}^{+(2)} = l_{xy}^{+(1)} + \frac{\left(l_{xi}^{+(1)} - l_{xj}^{+(1)}\right)\left(l_{iy}^{+(1)} - l_{jy}^{+(1)}\right)}{\omega_{ij} - \Omega_{ij}^{G_1}} \tag{12}$$

Note, as e_{ij} is a non-bridge edge, $\Omega_{ij}^{G_1} \neq 1$. In fact, given that $G_1(V_1, E_1)$ is connected, undirected and unweighted, we have: $0 < \Omega_{ij}^{G_1} < 1$. Also, as in the case of the two-stage process, we observe the same element-wise independence for $\mathbf{L}_{G_2}^+$ here as well. Once again, if the quantity of interest is Ω^{G_2} or pairwise expected commute times in random walks, we can simply use the following corollary.

Corollary 4. $\forall (x, y) \in V_2(G_2) \times V_2(G_2)$,

$$\Omega_{xy}^{G_2} = \Omega_{xy}^{G_1} + \frac{\left[\left(\Omega_{xj}^{G_1} - \Omega_{xi}^{G_1}\right) - \left(\Omega_{jy}^{G_1} - \Omega_{iy}^{G_1}\right)\right]^2}{4(\omega_{ij} - \Omega_{ij}^{G_1})} \tag{13}$$

4.2   Deleting a Bridge Edge

Finally, we deal with the case when a bridge edge is deleted from a graph, thus rendering it disconnected for the first time. This represents the point of singularity in the case of structural regress (analogous to the first join). Continuing with our convention, let $G_1(V_1, E_1)$ be a simple, connected, undirected graph with a bridge edge $e_{ij} \in E_1(G_1)$. Upon deleting e_{ij}, we obtain $G_2(V_2, E_2)$ and $G_3(V_3, E_3)$, two disjoint spanning sub-graphs of G_1. The orders of G_1, G_2 and G_3 are respectively given by n_1, n_2 and n_3, while $\mathbf{L}_{G_1}^+, \mathbf{L}_{G_2}^+$ and $\mathbf{L}_{G_3}^+$ are the respective pseudo-inverse matrices of their Laplacians. It is easy to see that $\Omega_{xy}^{G_1} = \Omega_{xy}^{G_2}$, if $x, y \in V_2(G_2)$, $\Omega_{xy}^{G_1} = \Omega_{xy}^{G_3}$, if $x, y \in V_3(G_3)$ and $\Omega_{xy}^{G_2 \times G_3} = \Omega_{xy}^{G_3 \times G_2} = \infty$, as G_1 and G_2 are disjoint. To obtain $\mathbf{L}_{G_2}^+$ and $\mathbf{L}_{G_3}^+$ from $\mathbf{L}_{G_1}^+$, we use the result in Lemma 1.

Theorem 5. $\forall(x, y) \in V_2(G_2) \times V_2(G_2)$ and $\forall(u, v) \in V_3(G_3) \times V_3(G_3)$,

$$l_{xy}^{+(2)} = l_{xy}^{+(1)} - \frac{n_2 \displaystyle\sum_{z \in V_2(G_2)} \left(l_{xz}^{+(1)} + l_{zy}^{+(1)} \right) - \displaystyle\sum_{x \in V_2(G_2)} \displaystyle\sum_{y \in V_2(G_2)} l_{xy}^{+(1)}}{n_2^2} \tag{14}$$

$$l_{uv}^{+(3)} = l_{uv}^{+(1)} - \frac{n_3 \displaystyle\sum_{w \in V_3(G_3)} \left(l_{uw}^{+(1)} + l_{wv}^{+(1)} \right) - \displaystyle\sum_{u \in V_3(G_3)} \displaystyle\sum_{v \in V_3(G_3)} l_{uv}^{+(1)}}{n_3^2} \tag{15}$$

Note also that $\mathbf{L}_{G_2}^+ \in \Re^{n_2 \times n_2}$ and $\mathbf{L}_{G_2}^+ \in \Re^{n_3 \times n_3}$. For convenience, and without loss of generality, we assume that the rows and columns of $\mathbf{L}_{G_1}^+ \in \Re^{n_1 \times n_1}$ have been pre-arranged in such a way that the first $(n_2 \times n_2)$ sub-matrix (upper-left) maps to the sub-graph G_2 and similarly the lower-right $(n_3 \times n_3)$ sub-matrix to G_3.

5   Bringing It Together: Algorithm, Complexity and Parallelization

The results obtained in Sects. 3 and 4 are summarized in Table 1. In this section, we bring together these results to bear on two important scenarios: (a) dynamic (time-evolving) graphs (cf. Sect. 5.1), and (b) real-world networks of large orders (cf. Sect. 5.2). In each case, we discuss the time complexity and parallelizability of our approach in detail.

Table 1. Summary of results: Atomic operations of the divide-and-conquer methodology.

Operation	Ω	L^+
First Join	$x, y \in G_1 : \Omega_{xy}^{G_3} = \Omega_{xy}^{G_1}$	$l_{xy}^{+(1)} - \frac{n_2 n_3 \left(l_{xi}^{+(1)} + l_{iy}^{+(1)} \right) - n_2^2 \left(l_{ii}^{+(1)} + l_{jj}^{+(2)} + \omega_{ij} \right)}{n_3^2}$
	$x, y \in G_2 : \Omega_{xy}^{G_3} = \Omega_{xy}^{G_2}$	$l_{xy}^{+(2)} - \frac{n_1 n_3 \left(l_{xj}^{+(2)} + l_{jy}^{+(2)} \right) - n_1^2 \left(l_{ii}^{+(1)} + l_{jj}^{+(2)} + \omega_{ij} \right)}{n_3^2}$
	$x \in G_1, y \in G_2 : \Omega_{xy}^{G_3} = \Omega_{xi}^{G_1} + \omega_{ij} + \Omega_{jy}^{G_1}$	$\frac{n_3 \left(n_1 l_{xi}^{+(1)} + n_2 l_{jy}^{+(2)} \right) - n_1 n_2 \left(l_{ii}^{+(1)} + l_{jj}^{+(2)} + \omega_{ij} \right)}{n_3^2}$
Edge firing	$\Omega_{xy}^{G_1} - \frac{\left[\left(\Omega_{xj}^{G_1} - \Omega_{xi}^{G_1} \right) - \left(\Omega_{jy}^{G_1} - \Omega_{iy}^{G_1} \right) \right]^2}{4\left(\omega_{ij} + \Omega_{ij}^{G_1} \right)}$	$l_{xy}^{+(1)} - \frac{\left(l_{xi}^{+(1)} - l_{xj}^{+(1)} \right) \left(l_{iy}^{+(1)} - l_{jy}^{+(1)} \right)}{\omega_{ij} + \Omega_{ij}^{G_1}}$
Non-bridge delete	$\Omega_{xy}^{G_1} + \frac{\left[\left(\Omega_{xj}^{G_1} - \Omega_{xi}^{G_1} \right) - \left(\Omega_{jy}^{G_1} - \Omega_{iy}^{G_1} \right) \right]^2}{4\left(\omega_{ij} - \Omega_{ij}^{G_1} \right)}$	$l_{xy}^{+(1)} + \frac{\left(l_{xi}^{+(1)} - l_{xj}^{+(1)} \right) \left(l_{iy}^{+(1)} - l_{jy}^{+(1)} \right)}{\omega_{ij} - \Omega_{ij}^{G_1}}$
Bridge delete	$x, y \in G_k : \Omega_{xy}^{G_k} = \Omega_{xy}^{G_1}$	$l_{xy}^{+(1)} - \frac{n_k \sum_{z \in G_k} \left(l_{xz}^{+(1)} + l_{zy}^{+(1)} \right) - \sum_{x \in G_k} \sum_{y \in G_k} l_{xy}^{+(1)}}{n_k^2}$

5.1 Dynamic Graphs: Incremental Computation for Incremental Change

Dynamic graphs are often used to represent temporally changing systems. The most intuitively accessible example of such a system is an online social network (OSN). An OSN evolve not only in terms of order, through introduction and attrition of users with time, but also in terms of the social ties (or relationships) between the users as new associations are formed, and older ones may fade off. Mathematically, we model an OSN as a dynamic graph $G_\tau(V_\tau, E_\tau)$ where the sub-index τ denotes the time parameter. We now study a widely used model for dynamic, temporally evolving, graphs called *preferential attachment* [3].

The preferential attachment model is a parametric model for network growth determined by parameters (n, κ) such that n is the desired order of the network and κ is the desired average degree per node. In its simplest form, the model proceeds in discrete time steps whereby at each time instant $1 < \tau + 1 \leq n$, a new node $v_{\tau+1}$ is introduced in the network with κ edges. This incoming node $v_{\tau+1}$, gets attached to a node $v_i : 1 \leq i \leq \tau$, through exactly one of its κ edges, with the following probability:

$$P_{\tau+1}(v_i) = \frac{d_\tau(i)}{\sum_{j=1}^{\tau} d_\tau(j)} \tag{16}$$

where $d_\tau(i)$ is the degree of node v_i at time τ. The end-points of all the edges emanating from $v_{\tau+1}$ are selected in a similar fashion. At the end of time step $\tau + 1$, we obtain $G_{\tau+1}(V_{\tau+1}, E_{\tau+1})$, and the process continues until we have a graph $G_n(V_n, E_n)$ of order n.[2]

Simplistic though it may seem, this model has been shown to account for several characteristics observed in real-world networks, including the *power law* degree distributions, the *small-world* characteristics and the logarithmic growth of network diameter with time. We return to these in detail in the next subsection while dealing with the more general case. It is easy to see that in order to study the structural evolution of dynamic networks, particularly in terms of the sub-structures like spanning trees and forests [16], or centralities of nodes and edges [21,24]; or voltage distributions in growing conducting networks [28], we require not only the final state $G_n(V_n, E_n)$, but all the intermediate states of the network. In other words, we need to compute the pseudo-inverses of the Laplacians for all the graphs in the sequence $(G_1 \rightarrow G_2 \rightarrow ... \rightarrow G_n)$. Clearly, if the standard methods are used, the cost at time step τ is $O(\tau^3)$. The overall asymptotic cost for the entire sequence is then $O\left(\sum_{\tau=1}^{n} \tau^3 = \left[\frac{n(n+1)}{2}\right]^2\right)$.

In contrast, using our incremental approach, we can accomplish this at a much lower computational cost. Note that in the case of growing networks, we do not need an explicit divide operation at all. The two sub-problems at time

---

[2] In practice, for $\kappa > 1$, the process starts with a small connected network as a base to facilitate probabilistic selection of neighbors for an incoming node. For $\kappa = 1$, we may start with a singleton node, and the resulting structure is a tree (Fig. 5).

step $\tau + 1$ are given *a priori*. We have, $G_\tau(V_\tau, E_\tau)$ and a singleton vertex graph $\{v_{\tau+1}\}$ as a pair of disjoint sub-graphs. The κ edges emanating from $\{v_{\tau+1}\}$ have end-points in G_τ as determined by (16). The conquer operation is then performed through a first join between the singleton node $\{v_{\tau+1}\}$ and the graph $G_\tau(V_\tau, E_\tau)$. We can assume that $\mathbf{L}_{G_\tau}^+$ is already known at this time step (the induction hypothesis). Also, $\mathbf{L}_{\{v_{\tau+1}\}}^+ = [0]$ and $n_2 = 1$ during the first join. Substituting in Theorem 2 we obtain the desired results. The rest of the $\kappa - 1$ edges are accounted for by edge firings (cf. the discussion in Sect. 3). Therefore, we need only $O(\kappa \cdot \tau^2)$ computations at time step τ, and hence $O\left(\kappa \cdot \sum_{\tau=1}^{n} \tau^2 = \kappa \cdot \frac{n(n+1)(2n+1)}{6} \right)$,

overall. As $\kappa << n$ in most practical cases, we have an order of magnitude lower average cost than that incurred by the standard methods. Further improvements follow from the parallelizability of our approach. Although we have not discussed it explicitly, it is evident that node and edge deletions can all be handled within this framework in the same way and at the same $O(n^2)$ cost per operation (cf. the discussion in Sect. 4).

5.2   Large Real-World Networks: A Divide-and-Conquer Approach

In order to compute $\mathbf{L}^+$ for an arbitrary graph $G(V, E)$, in a divide-and-conquer fashion, we need to first determine independent sub-graphs of G in an efficient manner. Theoretically, an optimal divide step entails determining a *balanced connected bi-partition* $P(G_1, G_2)$ of the graph G such that $|V(G_1)| \approx |V(G_2)|$ and $|E(G_1, G_2)|$, the number of edges violating the partition, is minimized. Such balanced bi-partitioning of the graph, if feasible, can then be repeated recursively until we obtain sub-graphs of relatively small orders. The solutions to these sub-problems can then be computed and the recursion unwinds to yield the final result, using our two-stage methodology in the respective conquer steps. Alas, computing such balanced bi-partitions, along with the condition of minimality of $|E(G_1, G_2)|$, belongs to the class of *NP-Complete* problems [27], and hence a polynomial time solution does not exist. We therefore need an efficient alternative to accomplish the task at hand that works well on large *real-world* networks.

Real-world networks, and particularly online social networks, have been shown to have several interesting structural properties: edge sparsity, power-law scale-free degree distributions, existence of the so called *rich club connectivity*, small-world characteristics [30] with relatively small diameters $(O(log\ n))$. Collectively, these properties amount to a simple fact: the overall connectivity between arbitrary node pairs is dependent on higher degree nodes in the network. Based on these insights, we now study two real-world online social networks — the *Epinions* and *SlashDot* networks [1] — to attain our objective of a quick and easy divide step. Table 2 gives some of the basic statistics about the two networks.[3] It is easy to see that the networks are sparse as $m = O(n) << O(n^2)$ in both cases.

---

[3] Although the networks originally have uni-directional and bi-directional links, we symmetrize the uni-directional edges to make the graphs undirected.

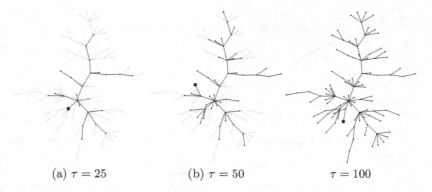

(a) $\tau = 25$ (b) $\tau = 50$ $\tau = 100$

Fig. 5. Growing a tree by preferential attachment ($n = 100$, $\kappa = 1$). The node v_τ, being added to the tree at time step τ, is emphasized (larger circle). Dotted edges at time steps $\tau = \{25, 50\}$ are a visual aid representing edges that are yet to be added in the tree until the order-limit ($n = 100$) is attained.

Table 2. Basic properties: Epinions and SlashDot networks.

$G(V, E)$	$n = |V(G)|$	$m = |E(G)|$	Leaves	Cut-off	# Comp.	$|V(GCC)|$	$|E(GCC)|$	# Cut-Edges
Epinions	75,888	405,740	35,763	4,429	30,376	37,924	61,482	102,452
SlashDot	82,168	504,230	28,499	7,012	36,311	41,084	62,225	164,719

Moreover, note that a significant fraction of nodes in the graphs are leaf/pendant nodes, i.e. nodes of degree 1 ($\approx 47\%$ for Epinions and $\approx 34\%$ for *SlashDot*). From Fig. 6(a–b), it is also evident that the node degrees indeed follow a heavy tail distribution in both cases. Thus, there are many nodes of very small degree (e.g. leaves) and relatively fewer nodes of very high degrees in these networks. Therefore, in order to break the graph into smaller sub-graphs, we adopt an incremental regress methodology of deleting high degree nodes. Ordering the nodes in decreasing order by degree, we remove them one at a time. This process divides the set of nodes into three parts at each stage:

a. **The Rich Club:** High degree nodes that have been deleted until that stage.
b. **The Giant Connected Component** (GCC): The largest connected component at that stage.
c. **Others:** All nodes that are neither in the rich club nor the GCC.

We repeat the regress, one node at a time, until the size of the GCC is less than half the size of the original graph. We call this the cut-off point. We then retain the GCC as one of our sub-graphs (one independent sub-problem) and re-combine all the non-GCC nodes together with the rich club to obtain (possibly) multiple sub-graphs (other sub-problems). This concludes the divide step.

Table 2 shows the relevant statistics at the cut-off point for the two networks. Note that the cut-off point is attained at the expense of a relatively small number of high degree nodes ($\approx 5\%$ for *Epinions* and $\approx 8\%$ for *SlashDot*). Moreover, the number of nodes in the GCC is indeed roughly half of the overall

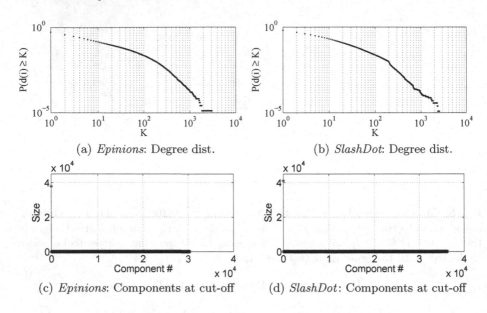

(a) *Epinions*: Degree dist.   (b) *SlashDot*: Degree dist.

(c) *Epinions*: Components at cut-off   (d) *SlashDot*: Components at cut-off

Fig. 6. Structural regress: Epinions and SlashDot networks.

order, albeit the GCC is surely sparser in terms of edge density than the over-all network ($|E(GCC)|/|V(GCC)| = 1.63$ vs. $|E(G)|/|V(G)| = 5.35$ for *Epinions* and $|E(GCC)|/|V(GCC)| = 1.51$ vs. $|E(G)|/|V(G)| = 6.13$ for *Slashdot*). Figure 6(c–d) shows the sizes (in terms of nodes) of all the connected components for the respective graphs at the cut-off point. It is easy to see that other than the GCC, the remaining components are of negligibly small orders. Recombining the non-GCC components together (including the rich club) yields an interesting result. For the *Epinions* network, we obtain two sub-graphs of orders $37,933$ and 31 respectively while for the *Slashdot* network we obtain exactly one sub-graph of order $41,084$. This clearly demonstrates that our simple divide method, yields a roughly equal partitioning of the network — and thus comparable sub-problems — in terms of nodes. The pseudo-inverses of these sub-problems can now be computed in parallel. Albeit, as in the case of all tradeoffs, this equi-table split comes at a price of roughly $\kappa = O(n)$ edges that violate the cut (cf. Table 2). This yields an $O(n^3)$ average cost for the *two-stage* process (cf. Sect. 3). However, given the element-wise parallelizability of our method, we obtain the pseudo-inverses in acceptable times of roughly 15 minutes for the *Epinions* and 18 minutes for the *SlashDot* networks.

6   Related Work

We now provide a brief review of the related work, highlighting in particular a few instances of the applications of the Moore-Penrose pseudo-inverse and the sub-matrix inverses of the Laplacian for a graph. As alluded to earlier, $\mathbf{L}^+$ is used to compute effective resistance distances between the nodes of a graph [17] as well as the one way hitting and commute times in random walks between node pairs in a graph. All these distances serve as measures of *multi-hop* dissimilarity between nodes and find applications in several graph mining contexts [9,14]. Moreover, for every connected undirected graph there is an analogous *reversible* Markov chain, L^+ finds use in the computation of relevant metrics (such as hitting time, cover time and mixing rates). L^+ is a gram-matrix. Its eigen decomposition yields an $n - dimensional$ Euclidean embedding of the graph whereby each node in the graph is represented as a point in that space. The general term l_{xy}^+ represents the inner product of the respective position vectors for the nodes x and y and thus L^+ is a valid kernel for a graph. This geometric interpretation has been used in collaborative recommendation systems [14].

In [11,32] the elements of L^+ have been given an elegant topological interpretation in terms of the *dense* spanning rooted forests and *connected bi-partitions* of the graph. In [24] the authors provide multiple interpretations of the diagonal elements of L^+ and show that they reflect both the overall positions and connectivity of nodes in a complex network. They therefore refer to L_{ii}^+ as the *topological centrality*, which provides a measure of the role of a node in the overall connectivity of a network and its robustness to random multiple failures in the network that breaks it down into two connected parts [32]. By extension, $Tr(L^+)$, also called the Kirchhoff index of a graph [17,31], is a global structural descriptor for the graph on a whole. This index is quite popular in the field of mathematical chemistry and is used to measure overall molecular strength (see, e.g., [22]). The elements of sub-matrix inverses have analogous interpretations in terms of *unrooted* spanning forests of the graph [16]. In [21], the sub-matrix inverses have been used to compute the random-walk betweenness centrality, another useful index to characterize roles of nodes in a network.

7 Conclusion and Future Work

In this work, we presented a divide-and-conquer based approach for computing the Moore-Penrose pseudo-inverse of the Laplacian (L^+) for a simple, connected, undirected graph. Our method relies on an elegant interplay between the elements of L^+ and the pairwise effective resistance distances in the graph. Exploiting this relationship, we derived closed form solutions that enable us to compute L^+ in an incremental fashion. We also extended these results to analogous cases for structural regress. Using dynamic networks and online social networks as examples, we demonstrated the efficacy of our method for computing the pseudo-inverse relatively faster than the standard methods. The insights from our work open up several interesting questions for future research. First and foremost, similar explorations can be done for the case of directed graphs (asymmetric relationships), where analogous distance functions — such as the expected commute time in random walks — are defined, albeit the Laplacians (more than one kind in literature) are no longer symmetric [6]. Secondly, matrix-distance interplays of the kind exploited in this work, also exist for a general case of the so called *forest matrix* and its distance counterpart the *forest distance* [2,10], both for undirected and directed graphs. The results presented here should find natural extensions to the forest matrix and the forest metric, at least for the undirected case. Finally, our closed forms can be used in conjunction with several interesting approaches for sparse inverse computations [8], to further expedite the pseudo-inverse computation for large generalized graphs. All these motivate ample scope for future work.

Acknowledgment. This research was supported in part by DTRA grant HDTRA1-09-1-0050, DoD ARO MURI Award W911NF-12-1-0385, and NSF grants IIS-0916750, CNS-10171647, CNS-1017092, CNS-1117536, IIS-1319749 and CRI-1305237.

References

1. http://snap.stanford.edu/data/
2. Agaev, R., Chebotarev, P.: The matrix of maximum out forests of a digraph and its applications. Autom. Remote Control **61**(9), 1424–1450 (2000)
3. Barabási, A.L., Albert, R.: Emergence of scaling in random networks. Science **286**(5439), 509–512 (1999)
4. Ben-Israel, A., Greville, T.: Generalized Inverses: Theory and Applications, 2nd edn. Springer, New York (2003)
5. Biggs, N.: Algebraic Graph Theory. Cambridge University Press, Cambridge (1993)
6. Boley, D., Ranjan, G., Zhang, Z.-L.: Commute times for a directed graph using an asymmetric laplacian. Linear Algebra Appl. **435**(2), 224–242 (2011)
7. Brand, M.: A random walks perspective on maximizing satisfaction and profit. In: Proceedings of 2005 SIAM International Conference Data Mining (2005)
8. Campbell, Y.E., Davis, T.A.: Computing the sparse inverse subset: An inverse multifrontal approach. Technical report TR-95-021, Univ. of Florida, Gainesville (1995)
9. Chandra, A.K., Raghavan, P., Ruzzo, W.L., Smolensky, R., Tiwari, P.: The electrical resistance of a graph captures its commute and cover times. In: Proceedings of Annual ACM Symposium on Theory of Computing, pp. 574–586 (1989)
10. Chebotarev, P., Shamis, E.: The matrix-forest theorem and measuring relations in small social groups. Autom. Remote Control **58**(9), 1505–1514 (1997)
11. Chebotarev, P., Shamis, E.: On proximity measures for graph vertices. Autom. Remote Control **59**(10), 1443–1459 (1998)
12. Doyle, P.G., Snell, J.L.: Random Walks and Electric Networks. The Mathematical Association of America, Washington, DC (1984)
13. Fiedler, M.: Algebraic connectivity of graphs. Czech. Math. J. **23**, 298–305 (1973)
14. Fouss, F., Pirotte, A., Renders, J.M., Saerens, M.: Random-walk computation of similarities between nodes of a graph with application to collaborative recommendation. IEEE Trans. Knowl. Data Eng. **19**, 355–369 (2007)
15. Isaacson, D., Madsen, R.: Markov Chains Theory and Applications. Wiley, New York (1976)
16. Kirkland, S.J., Neumann, M., Shader, B.L.: Distances in weighted trees and group inverse of laplacian matrices. SIAM J. Matrix Anal. Appl. **18**, 827–841 (1997)
17. Klein, D.J., Randić, M.: Resistance distance. J. Math. Chem. **12**, 81–95 (1993)

18. Luxburg, U.V.: A tutorial on spectral clustering. Stat. Comput. **17**(4), 395–416 (2007). Max Planck Institute for Biological Cybernetics. Technical report No. TR-149
19. Luxburg, U.V., Radl, A., Hein, M.: Getting lost in space: Large sample analysis of the commute distance. In: NIPS (2010)
20. Meyer, C.D.: Generalized inversion of modified matrices. SIAM J. Appl. Math. **24**(3), 315–323 (1973)
21. Newman, M.E.J.: A measure of betweenness centrality based on random walks. Soc. Netw. **27**(1), 39–54 (2005)
22. Palacios, J.L., Renom, J.M.: Bounds for the kirchhoff index of regular graphs via the spectra of their random walks. Int. J. Quant. Chem. **110**, 1637–1641 (2001)
23. Ranjan, G., Zhang, Z.-L.: How to glue a robust smart-grid: a finite network theory of inter-dependent networks (extended abstract). In: Proceedings of the 7th (2011) Cyber Security and Information Intelligence Research Workshop: THEME - Energy Infrastructure Cyber Protection (CSIIRW7) (2011)
24. Ranjan, G., Zhang, Z.-L.: Geometry of complex networks and topological centrality. Physica A: Stat. Mech. Appl. **392**(17), 3833–3845 (2013)
25. Ranjan, G., Zhang, Z.-L., Boley, D.: Incremental computation of pseudo-inverse of laplacian (2013). http://arxiv.org/abs/1304.2300
26. Sarwar, B., Karypis, G., Konstan, J., Riedl, J.: Recommender systems for large-scale e-commerce: Scalable neighborhood formation using clustering. In: Proceedings of the Fifth International Conference Computer and Information Technology (2002)
27. Sen, A., Ghosh, P., Yang, B., Vittal, V.: A new min-cut problem with application to electric power network partitioning. Eur. Trans. Electr. Power **19**(6), 778–797 (2008)
28. Tadić, B., Priezzhev, V.: Voltage distribution in growing conduction networks. Eur. Phys. J. B **30**, 143–146 (2002)
29. Tetali, P.: Random walks and effective resistance of networks. J. Theor. Probab. **4**, 101–109 (1991)
30. Watts, D.J., Strogatz, S.H.: Collective dynamics of 'small-world' networks. Nature **393**, 440–442 (1998)
31. Xiao, W., Gutman, I.: Resistance distance and laplacian spectrum. Theoret. Chem. Acc. **110**, 284–289 (2003)
32. Zhang, Z.-L., Ranjan, G.: On connected bi-partitions of a graph (manuscript under preparation 2014)

Optimal Tracking of Multiple Targets Using UAVs

David Hay[1], Shahrzad Shirazipourazad[2]([✉]), and Arunabha Sen[2]

[1] School of Computer Science and Engineering, Hebrew University, Jerusalem, Israel
dhay@cs.huji.ac.il
[2] School of Computing, Informatics and Decision System Engineering,
Arizona State University, Tempe, AZ 85281, USA
{sshiraz1,asen}@asu.edu

Abstract. Target tracking problems have been studied fairly extensively by researchers in the last few years. However, the problem of *continuous* tracking of all mobile targets using the *fewest* number of mobile trackers, *even when the trajectories of all the targets are known in advance*, has received very little attention. In this paper we study this problem, where the goal is to find the fewest number of trackers needed to track all the targets for the entire period of observation. Specifically, given a set of n targets moving in n different (known) trajectories in a two (or three) dimensional space, our objective is to find the fewest number of velocity-bounded UAVs (mobile sensors, trackers) and their trajectories, so that all the targets are tracked during the entire period of observation. We also study two other versions of the problem where not only the number of trackers but also the time during which the trackers are *active* is also taken into account. We formulate these problems as network flow problems and propose algorithms for their solution. We evaluate the performance of our algorithms through simulation and study the impact of parameters such as the speed and sensing range of the trackers.

1 Introduction

Motivated by the importance of target tracking in military and civilian environments and widespread use of UAVs in target tracking, considerable research has been conducted on target tracking problems using UAVs and mobile sensors [1–6]. There also exists a large body of research on target tracking problems using sensor networks. However, in most of these studies the sensor nodes (trackers) are static and as such the issue of path planning of trackers does not exist. The authors in [7] provide a survey of these studies. Typically, the target tracking problem using mobile trackers has two components: (*i*) estimation of target positions using sensor data, and (*ii*) mobility management of trackers (sensors). Most of the studies on target tracking using mobile sensors focus on the quality of detection of the mobile targets with a given set of UAVs (mobile sensors). In these studies one or more mobile trackers are used to track a single or multiple targets.

© Springer International Publishing Switzerland 2014
Z. Zhang et al. (Eds.): COCOA 2014, LNCS 8881, pp. 750–763, 2014.
DOI: 10.1007/978-3-319-12691-3_55

In [6] the authors study distributed mobility management of a given set of mobile sensors for better tracking of a single target. Enyang et al. in [5] study a scenario when there is one mobile sensor, a set of static sensors and a mobile sensor controller. The controller receives the data from the sensors and estimates the location of the target and direct mobility of sensors. The authors in [8] study multi-target tracking using multiple UAVs and develop a decentralized approach for target location estimation and UAV mobility management. In spite of extensive studies on target tracking problem, there exists only a handful of studies on the problem of continuous time tracking of mobile targets with an optimal number of mobile trackers.

In this paper we focus on the scenario where there are multiple mobile targets whose trajectories are known in advance. Accordingly, estimation and prediction of target location is not an issue here and the focus of this target tracking problem is on finding the minimum number of UAVs with bounded velocity and their trajectories so that every target is tracked (covered) by at least one UAV during the entire period of observation. Although it may appear that the assumption regarding complete knowledge of the trajectories of the targets makes the problem very simple (if not trivial), we show that even with this assumption, the problem of computation of minimum number of trackers and their trajectories for continuous coverage of the mobile targets remains hard, i.e., NP-complete.

In [1] the authors study energy efficiency issues related to mobile target tracking and provide a power efficient target tracking solution by adjusting the UAVs position (altitude). In [9] authors study a similar problem in which their goal is to find the smallest set of mobile backbone nodes such that mobile regular nodes are always under coverage of at least one backbone node. In these papers [1,9] it is assumed that trackers have the capability of moving very fast, i.e., infinite speed. The authors in [10] study the problem of finding the optimal set of UAVs, assuming that the UAVs have only a finite speed. They propose an algorithm in which it first computes the solution of coverage problem for each time instance (static disk coverage problem) using a greedy algorithm. Then they propose a motion assignment algorithm determining the movement of UAVs from their old positions to new positions. The greedy algorithm proposed by the authors does not guarantee optimality of the solution.

In this paper we study three different versions of the target tracking problem (TTP). The inputs to all three versions of the problem are the same and it comprises of (i) n targets with their trajectories in two or three dimensional space, (ii) speed of the trackers, and (iii) sensing range of the trackers, (iv) period of observation. We assume that a tracker can only track (sense) a target if the target is within the sensing range of the tracker. A tracker can either be in an *active* or and *inactive* state. A tracker is said to be in an active state if either (i) it is sensing a target, or (ii) moving towards a target to sense it in a future time, before the end of the period of observation. The different versions of the Target Tracking Problem studied in this paper are stated next.

TTP$_1$: The objective of TTP$_1$ is to minimize the total number of trackers needed to track all the targets during the entire period of observation.

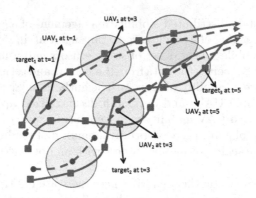

Fig. 1. Trackers and their trajectories for tracking all the three targets.

TTP$_2$: The objective of TTP$_2$ is to minimize the total active time of the trackers within the period of observation.

TTP$_{2A}$: The objective of TTP$_{2A}$ is to minimize the total active times of the trackers within the period of observation, subject to the constraint that the number of trackers do not exceed a specified number P.

TTP$_3$: The objective of TTP$_3$ is to minimize the total active times of the trackers within the period of observation, subject to the constraint that the number of trackers do not exceed the absolute minimum number of trackers necessary to track all the targets. We note that the answer to the TTP$_1$ problem provides the absolute minimum number of trackers necessary to track all the targets. Also, we note that TTP$_3$ can be solved first by computing the solution of TTP$_1$ and then computing the solution of TTP$_{2A}$ when P is set to the solution of TTP$_1$. However, in this paper we model TTP$_3$ independently.

Figure 1 depicts an example with three targets. The locations of three targets (red squares) on their trajectories (red curves) are shown at time instances 0 to 5. Blue dashed curves show the trajectories of trackers and the disks centered at the locations of trackers depict the sensing area of trackers at different time instances.

In this paper we model the target tracking problems using network flows and we propose techniques using integer linear programming to solve the three versions optimally.

The rest of the paper is organized as follows. In Sect. 2 we formulate target tracking problems. Section 3 presents our technique to solve these problems. In Sect. 4 we evaluate our technique via simulations. Section 5 concludes the paper.

2   Problems Formulation and Hardness Results

In order to formulate the three problems, we first discretize time into equal time intervals of length δ. We consider a set of n targets $\mathcal{A} = \{a_0, \ldots, a_{n-1}\}$

moving on two dimensional space over time instances $0, \ldots, T^1$. Let $p(a_i, t) = (x(a_i, t), y(a_i, t))$ be the location of target a_i at time instance t where $x(a_i, t)$ and $y(a_i, t)$ denote the x-coordinate and y-coordinate of a_i at time t. We assume that a target is covered by a tracker b_j if the distance between them is less than sensing radius r. Let $|p_1 \sim p_2|$ denote the distance between two points p_1 and p_2 on the two dimensional space.

In TTP_1 our goal is to find a smallest set of trackers $\mathcal{B}$, such that:

1. Coverage - For any target a_i and any time instance t, there is a tracker $b_j \in \mathcal{B}$ whose location in time t, denoted by $p(b_j, t)$ is in distance r from $p(a_i, t)$; i.e., $|p(a_i, t) \sim p(b_j, t)| \leq r$.
2. Mobility - For any tracker b_j with specified velocity d and any time slot $t \geq 1$, $|p(b_j, t-1) \sim p(b_j, t)| \leq d$.

Let $B(t) \subseteq \mathcal{B}$ denote the set of trackers being used at time instance t. Also, let T_j denote the number of time instances that target b_j is active. In TTP_2 the objective is to find a set of trackers $\mathcal{B}$ where the total active time of the trackers within the period of observation, i.e. $\sum_{b_j \in \mathcal{B}} T_j = \sum_{t=0}^{T} |B(t)|$ is minimized while both the Coverage and Mobility constraints of TTP_1 are satisfied. TTP_{2A} adds an additional constraint to TTP_2: the size of $\mathcal{B}$ should not exceed P. Considering both objectives of TTP_1 and TTP_2, in TTP_3 the objective is to minimize $\sum_{b_j \in \mathcal{B}} T_j = \sum_{t=0}^{T} |B(t)|$ subject to the constraint that the number of trackers do not exceed the absolute minimum number of trackers necessary to track all the targets. It may be noted that the answer to the TTP-1 problem provides the absolute minimum number of trackers necessary to track all the targets.

Theorem 1. *All three versions of TTP are NP-complete.*

Proof. Considering a special case of TTP where targets should be covered at just one time instance $t = T = 0$, all the three versions of TTP problem will be equivalent to the NP-complete Geometric Disk Cover Problem [11]; that is, given points in the plane, to identify a minimally sized set of disks (of prescribed radius) covering all points.

3 Solution

In this section we propose our techniques to compute the solution of the three versions of target tracking problem, TTP1, TTP_2 and TTP_3. In order to compute the solutions of these problems, first we model them by a flow network, by *discretizing also the space*. Next, We find the solution of the problems by computing (a modified version of) minimum flow on a directed graph $G = \langle V, E \rangle$. The flow is then splitted into paths, where each path represents movement of a tracker. Before explaining our graph construction first we give the definition of the classical *minimum flow* problem [12].

---

[1] We present the formulation in two dimensions for brevity. Extensions to higher dimensions is straightforward and is discussed in Sect. 3.

Minimum Flow Problem: Given a capacitated network $G = \langle V, E \rangle$ with a nonnegative capacity $c(i, j)$ and with a nonnegative lower bound $l(i, j)$ associated with each edge (i, j) and two special nodes, a source node S and a sink node D, a flow is defined to be a function $f : E \rightarrow \mathbb{R}^+$ satisfying the following conditions:

$$\sum_{j \in V} f(i, j) - \sum_{j \in V} f(j, i) = \begin{cases} F, & i = S \\ 0, & i \neq S, D \\ -F, & i = D \end{cases}$$

$$l(i, j) \leq f(i, j) \leq c(i, j)$$

for some $F \geq 0$ where F is the value of the flow f. The minimum flow problem is to determine a flow f for which F is minimized.

3.1   TTP Graph Construction

We construct a directed graph $G = \langle V, E \rangle$ representing the positions of the targets and the possible trackers' movements. This construction involves space discretization (in addition to the time discretization described in Sect. 2). Specifically, we consider a *grid* over the two- (or three-) dimensional space and restrict our trackers to move between points on that grid. The granularity of our grid is denoted by ε, which represents a tradeoff between the accuracy of our solution and its running time. Let $\mathcal{N}$ be the set of all points in the grid; namely, for the two-dimensional space, $\mathcal{N} = \{i \cdot \varepsilon, j \cdot \varepsilon \mid i, j \in \mathbb{Z}\}$. We note that based on the coverage constraint, for a target a_i to be covered by a tracker at a time instance t, there should be at least one tracker in the disk of radius r centered at $p(a_i, t)$. Let $D(p(a_i, t), r)$ denotes that disk. Thus, a tracker should be located in one of the points in $\mathcal{N} \cap D(p(a_i, t), r)$. For every such potential location, at any given time, we add a vertex graph, which is represented by the triplet $\langle$location, target id, timeslot$\rangle$, where *location* corresponds to the coordinates of the points on the grid (namely, the same location can be added multiple times for many targets and/or timeslots). Fig. 2(a) depicts a disk $D(p(a_i, t), r)$. Blue circles show $\mathcal{N} \cap D(p(a_i, t), r)$ and the red square represents the center of disk. In addition, we add one supersource vertex S and one supersink vertex D. In other words, the set of vertices will be

$$V = \{S, D\} \cup \bigcup_{t=0}^{T} \bigcup_{i=0}^{n-1} \langle \mathcal{N} \cap D(p(a_i, t), r), a_i, t \rangle.$$

We will have four types of edges:

1. Intra-target edges: For every target a_i and time-slot t we construct a directed ring connecting all the vertices with target id a_i and time instance t (Fig. 2(b)). We note that the order of the nodes in the ring is arbitrary.
2. Mobility edges: $\{(\langle p, a_i, t \rangle, \langle p', a_{i'}, t' \rangle) \mid i \neq i', |p' \sim p| \leq d|t' - t|\}$. Note that if $t' = t$ then $p' = p$. These edges have capacity 1 and demand 0. The direction of these edges goes from the node with lower timeslot to the node with higher timeslot, where ties are broken by nodes' target id. A mobility edge represents that a tracker can move from a location p to p' during time interval $|t' - t|$.

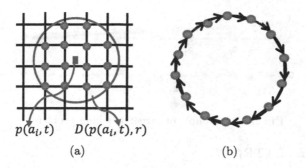

$$p(a_i, t) \qquad D(p(a_i, t), r)$$

(a)   (b)

Fig. 2. (a) A disk centered at the location of target a_i at time t (b) The ring corresponding to the discrete points in disk shown in (a).

3. Supersource edges: All vertices are connected to the supersource S with edges of capacity 1 and demand 0. These edges originate in the supersource.
4. Supersink edges: All vertices are connected to the supersink D with edges of capacity 1 and demand 0. These edges terminate in the supersink.

We note that space discretization comes with a price, as it rules out solutions in which the trackers are not at a grid point in each time-slot. As $\varepsilon \ll d$ (namely, the granularity of the grid is much finer than the maximum velocity of the tracker), these differences are negligible.

3.2 Modified Minimum Flow Problem

In Fig. 3 we illustrate an example on a line (one dimensional) with 3 targets and we show their locations on the X coordinate at two timeslots. In this example $r = 2$, $\varepsilon = 1$, and $d = 1$. Figure 4 depicts graph $G = \langle V, E \rangle$. For the sake of clarity, not all edges in E are shown. The number on each node shows the location on X dimension. In this example we note that the minimum flow value is zero while value of f of intra-target edges is 1 and all the other edges have flow zero. More specifically, all the lower bound constraints on the intra-target edges are satisfied while there is no flow starting from S. Hence, we have modified the minimum flow problem such that the lower bound constraints on the intra-target edges in graph $G = \langle V, E \rangle$ are satisfied if the flow starts from source node S. In this regard, we add additional constraints to make sure there is a path from S to every node in $V \backslash \{S, D\}$ such that the flow f on the edges of the path is 1. Let g_u be a flow function defined with respect to node $u \in V \backslash \{S, D\}$. A closer look at our construction shows that if there is a path from S to one node in a ring, there is a path to all nodes in that ring. Hence, for every ring in G, corresponding to a target a_i at a time instance t, we select an arbitrary node u (which is of form $\langle p, a_i, t \rangle$) and add the following constraints to minimum flow problem:

$$\sum_{j \in V} g_u(i, j) - \sum_{j \in V} g_u(j, i) = \begin{cases} 1, & i = S \\ 0, & i \neq S, D \\ -1, & i = D \end{cases}$$

$$g_u(i, j) \leq f(i, j)$$

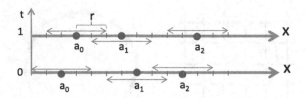

Fig. 3. An example of target tracking problem.

3.3   Solution of TTP_1

Considering that a tracker covers a target only if it is placed at a discrete point in the disk centered at the location of target at each time instance, we will prove that the minimum number of trackers required to cover the targets at all time instances is equal to the value of modified minimum flow in graph $G = \langle V, E \rangle$, and the trajectories of trackers can be computed from the solution of modified minimum flow problem.

In Fig. 4, the blue edges show the edges with flow 1 in the modified minimum flow solution. We can see that the value of modified minimum flow is 2 meaning that 2 trackers are required to cover the three targets in two time slots. The first tracker at time 0 is at location $X = 3$ ($x(a_0, 0) + \epsilon = 3$) covering target a_0 and moves to $X = 4$ at time 1 covering both a_0 and a_1 ($x(a_0, 1) + \epsilon = 4$ and $x(a_1, 1) - 2\epsilon = 4$). The second tracker at time 0 is at location $X = 9$ covering targets a_1 and a_2 where $x(a_1, 0) + 2\epsilon = 9$ and $x(a_2, 0) - \epsilon = 9$ and moves to $X = 10$, ($x(a_2, t_1) - \epsilon = 10$) covering a_2.

Lemma 1. *Any valid flow with value F on the constructed graph can be decomposed to a valid assignment of F trackers.*

The decomposition works by first assigning a tracker to a supersource edge with flow 1. Then, we get to the node $\langle p, a_i, t \rangle$ to which this supersource edge is connected. We assign the tracker b a location p at timeslot t. If the intra-edges of target a_i still have flow of size 1, then we traverse these intra-edges, and say that b covers a_i at timeslot t. After the traversal (or in case no traversal was needed), we pick one outgoing edge from $\langle p, a_i, t \rangle$ with flow 1. Such edge exists from flow conservation. We look at the edge's destination, $\langle p', a_{i'}, t' \rangle$. If $t' = t$ then $p' = p$, and then we try to take the ring of $a_{i'}$, add it to the coverage of b at time t and continue recursively with an outgoing edge from $\langle p', a_{i'}, t' \rangle$. If $t' > t$ (that the only other case), then we move b to position p' at timeslot t'. This move is legal since only mobility edges with $|p' \sim p| \leq d|t' - t|$ are added to the graph. We continue this process until we hit D. Then, we have the trajectory of b (as well as which targets it covers at each timeslot). We reduce the flow of b from the flow we got and continue the decomposition on the remaining flow (assigning more trackers). In the end of the process we will have the trajectories of all trackers, obeying their mobility restriction (movement of at most distance d at each timeslot). In addition, we ensure that every target is covered at all time instances, because the flow was valid which implies that all intra-edges had flow

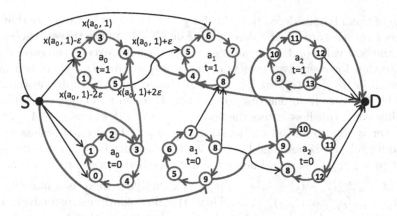

Fig. 4. The modified minimum flow solution of the example in Fig. 3

1 (meeting their demand); and therefore every target is traversed by a tracker in the decomposition of F. Since we assign one tracker to each supersource edge of flow 1, our tracker assignment will have exactly F trackers.

Lemma 2. *Any valid assignment $\mathcal{B}$ of F trackers with maximum velocity $d - \sqrt{2}\varepsilon$ and sensing radius $r - \frac{\sqrt{2}}{2}\varepsilon$ in the two-dimensional space can be represented as a valid flow with value F on this graph.*[2]

Every tracker b_j starts its trajectory at some timeslot $t_{i,j}$, at location $p_{i,j}$, covering a set of targets $A_{i,j}$. Let $p'_{i,j}$ be the closest grid point to $p_{i,j}$ (in the two-dimensional plane, $p'_{i,j}$ is at most of distance $\frac{\sqrt{2}}{2}\varepsilon$ of $p_{i,j}$). Thus, as all targets in $A_{i,j}$ are within distance $r - \frac{\sqrt{2}}{2}\varepsilon$ of $p_{i,j}$, they are within distance r of $p'_{i,j}$. Next, tracker b_j moves to a new location $p_{i',j}$ to cover a set of targets $A_{i',j}$ at timeslot $t_{i',j} > t_{i,j}$. Let $p'_{i',j}$ be the closest grid point to $p_{i',j}$. Tracker b_j will be active till it is not assigned to cover any target anymore and its trajectory finishes. We can represent the trajectory of b_j by a flow of value one in G in the following way. Corresponding to the tracker b_j, we add a flow of value one to G leaving S to the node $\langle p'_{i,j}, a_x, t_{i,j} \rangle$ where $a_x \in A_{i,j}$ and a_x has the lowest target id among other targets in $A_{i,j}$. $\langle p'_{i,j}, a_x, t_{i,j} \rangle \in V$ since a_x is within distance r of $p'_{i,j}$ and $p'_{i,j} \in \mathcal{N}$. The flow traverses the ring corresponding to a_x at $t_{i,j}$ satisfying the demand on the intra-target edges and it leaves the ring from the same node $\langle p'_{i,j}, a_x, t_{i,j} \rangle$. Next, the flow enters the node $\langle p'_{i,j}, a_y, t_{i,j} \rangle$ in the ring corresponding to the next target $a_y \in A_{i,j}$ where $y > x$. It traverses the ring and this process continues till all the rings corresponding to the targets in $A_{i,j}$ at $t_{i,j}$ are traversed by this flow (all the edges in these rings have a flow value of one). We note that if a target is covered by more than one tracker at a time instance,

---

[2] This result can be easily extended to spaces with higher dimensions. The difference between velocity d and velocity $d - \sqrt{2}\varepsilon$, or between sensing radius r and $r - \frac{\sqrt{2}}{2}\varepsilon$, is negligible for all practical purposes.

just one of them is considered to be in charge of covering the target at that time instance. In other words, just a flow of value one is selected to traverse every ring. Hence, the flow value on the intra-target edges does not exceed the capacity. We also note that for every two targets $a_x, a_y \in A_{i,j}$ at time $t_{i,j}$, there is a mobility edge from $\langle p'_{i,j}, a_x, t_{i,j} \rangle$ to $\langle p'_{i,j}, a_y, t_{i,j} \rangle$ (based on the graph construction), where the flow goes from one ring to next ring through this edge. Next, through a mobility edge, this flow enters the node $\langle p'_{i',j}, a_z, t_{i',j} \rangle$ where $a_z \in A_{i',j}$. Since, the tracker is able to move from location $p_{i,j}$ to $p_{i',j}$ between timeslots $t_{i,j}$ and $t_{i',j}$ and its velocity is bounded by $d - \sqrt{2}\varepsilon$, the distance between $p_{i,j}$ and $p_{i',j}$ is at most $(d - \sqrt{2}\varepsilon)(t'_j - t_j)$ and hence the distance between $p'_{i,j}$ to $p'_{i',j}$ is at most $\frac{\sqrt{2}}{2}\varepsilon + (d - \sqrt{2}\varepsilon)(t'_j - t_j) + \frac{\sqrt{2}}{2}\varepsilon \leq d(t'_j - t_j)$, implying there is a mobility edge $(\langle p'_{i,j}, a_y, t_{i,j} \rangle, \langle p'_{i',j}, a_z, t_{i',j} \rangle)$ in G. Then, the flow continues recursively based on the trajectory of tracker, and at the time that the tracker becomes inactive the flow goes to the supersink node D. It can be seen that the flow constraints on all the edges are satisfied. Also, since we add a flow of value one to the graph from S to D for every tracker, total flow will be exactly F.

These lemmas imply that the minimum flow yields the minimum assignment and we are done.

Extension to higher dimensions: In 2 dimensional space we use $O((r/\varepsilon)^2)$ nodes connected in a ring to represent the disk around a target. In order to extend this solution to 3 dimensions it is enough to discretize space in all 3 dimensions and use $O((r/\varepsilon)^3)$ nodes to represent the ball around a target and connect the nodes as a ring in anyway. Similarly, we can extend the model to higher dimensions.

3.4   Solution of TTP$_2$

Even though in the TTP$_1$ problem the objective is to minimize number of trackers in use during the tracking time, it may not minimize the total time that trackers are active. Fig. 5 illustrates an example in one dimensional line for two time steps. It is assumed that $r = \varepsilon = 1$. In this example minimum number of trackers needed is two. However, there can be different tracking solutions in which the two trackers are used: (1) the two trackers are active in both time steps where $p(b_i, t) = p(a_i, t), \forall i \in \{0, 1\}$ for $t = 0$ and $t = 1$. (2) at $t = 0$, $x(b_0, 0) = x(a_0, 0) + 1$ and $x(b_1, 0) = x(a_1, 0)$; at $t = 1$ just one tracker b_0 is active and $x(b_0, 1) = x(a_0, 1) + 1$. We observe that total time that trackers are active in first solution is 4 while in second solution is 3. Hence, the second solution is more efficient in using the trackers comparing to the first solution. In TTP$_2$, the objective is to minimize the total time that trackers are active in the observation period.

In order to solve TTP$_2$ we use similar flow network model $G = \langle V, E \rangle$ and ILP as in TTP$_1$. As we explained in TTP$_1$ a flow in the solution of TTP$_1$ corresponds to a tracker. In order to compute the time duration that a tracker is in use we add following costs (timestamps) to edges in E:

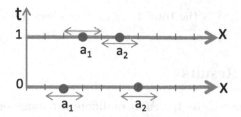

Fig. 5. An example of target tracking problem.

- Cost of supersource edges: every edge connecting the supersource to a node $\langle p, i, t \rangle \in V$ where p is location, i is target id and t is the time corresponding to this node, has time-stamp t. If a flow value on this edge is one, the time-stamp actually shows the time that the tracker corresponding to this flow begins to track.
- Cost of supersink edges: every edge connecting a node $\langle p, i, t \rangle \in V$ to the supersink node has time-stamp $t + 1$. If a flow value on this edge is one, the time-stamp actually shows the time that the tracker corresponding to this flow will no longer be in use.
- The rest of the edges have cost zero.

Let $c(u, v)$ denote the cost (time-stamp) of an edge $(u, v) \in E$. The total time that the trackers are active over observation period can be computed as $\mathcal{S} = \sum_{j \in V} c(j, D) f(j, D) - \sum_{j \in V} c(S, j) f(S, j)$. Hence, the objective of TTP$_2$ will be minimizing $\mathcal{S}$ and the constraints will be the same as TTP$_1$.

In TTP$_{2A}$ version, there is one more constraint; the number of trackers should not exceed a number P. In this case we need to add one more constraint to our flow model, i.e., $F \leq P$.

3.5 Solution of TTP$_3$

We note that solution of TTP$_2$ may not minimize total number of distinct trackers over time. For the example depicted in Fig. 5 (and explained in previous part), The objective value of TTP$_2$ is 3. A solution of TTP$_2$ with objective value 3 can include three distinct trackers where two of them are used at first time step and third tracker is located at $x(a_0, 1) + 1$ at second time step to cover both targets. The total time that trackers are active is still 3 while total number of distinct trackers is not minimized (minimum number of distinct trackers is 2). In TTP$_3$ both objectives of TTP$_1$ and TTP$_2$ are considered together. In TTP$_3$ we would like to find a tracking solution while it minimizes total time of trackers being active, the total number of distinct trackers does not exceed the solution of TTP$_1$. We note that TTP$_3$ can be solved first by computing the solution of TTP$_1$ and then computing the solution of TTP$_{2A}$ when P is set to the solution of TTP$_1$. However, we model TTP$_3$ independently as one flow network problem. In this regard, using the same flow network model in TTP$_2$, the objective of TTP$_3$ is to minimize $\mathcal{M}.F + \mathcal{S}/T$ where F is the value of flow (number of trackers), $\mathcal{S}$ is the total time that trackers are active, and $\mathcal{M}$ is a sufficient large number.

4   Simulation Results

In this section, we investigate the effect of different parameters such as coverage radius, r, and speed of trackers, d, on the number of trackers needed to cover all the targets. We also perform experiments to examine the effects of values of ε and δ on the accuracy of our technique. We use CPLEX to solve the ILP for TTP_1. In our simulations we considered 5 targets are moving on 2 dimensional area. Their trajectories over time interval $[0, 10]$, depicted in Fig. 6, are given by parametric equations.

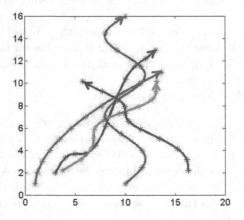

Fig. 6. Trajectories of 5 targets in time interval $[0, 10]$

In the first set of experiments, we compute the minimum number of trackers to cover the targets (Fig. 6) for different values of the tracker speed, d. In these experiments $r = 2$, $\varepsilon = 1$, $\delta = 2$ and total observation period is $[0, 10]$. Figure 7 illustrates the results of these experiments. It can be seen that when d is smaller than the speed of targets, number of trackers needed to cover all targets is more than number of targets and by increasing the value of d number of trackers decreases drastically initially and when $d \geq 2$ which is close or greater than speed of targets, three trackers can cover the targets in all time steps. We note that the speed of targets in these experiments is not constant and is not uniform since the parametric equations of their trajectories are different.

In the second set of experiments, we examine the effect of coverage radius of trackers on the solution of TTP_1. In these experiments $\varepsilon = 0.5$, $\delta = 2$ and total observation period is $[0, 10]$. Figure 8 shows the results of these experiments for two different values of d. We can see when speed of trackers is large enough (in this example $d = 2$), number of trackers needed to cover the targets will

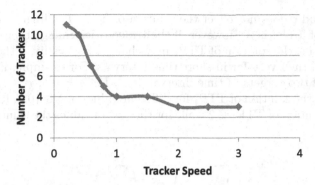

Fig. 7. Number of trackers vs. trackers speed in time interval $[0, 10]$, $r = 2$, $\varepsilon = 1$, $\delta = 2$.

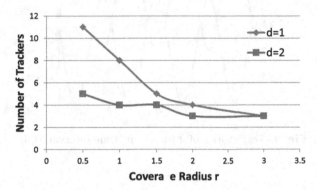

Fig. 8. Number of trackers vs. trackers coverage radius in time interval $[0, 10]$, $d = 1$, $\varepsilon = 0.5$, $\delta = 2$.

not exceed number of targets, even if tracker radius is very small. When tracker radius increases then number of trackers for both values of d is getting closer together. We note that when value of r is large enough then at least one tracker is needed to cover all the targets.

In the experiments, we examined different values for δ and ε. Generally, the smaller are ε and δ, the closer is the solution of our technique to the optimal solution of continuous scenario. We note that smaller value of δ results in lower chance that a target may not be covered by a tracker; also, more tracker may be needed to cover the targets in all time instances comparing to larger values of δ (considering that larger value of δ is an integer multiple of smaller value). On the other hand, smaller values of ε increases the possible locations that are considered for trackers and it results in smaller number of trackers in comparison to larger value of ε. While smaller values of ε and δ increase accuracy of the solution of TTP, they will increase the cost of computation. For the instance depicted in Fig. 6, we examined different values of $\delta = 1, 2, 4, 8$ and $\epsilon = 0.5, 1$ where $r = 2, d = 2$ in time interval $[0, 8]$. For this set of trajectories of targets, change of δ and ϵ does not effect the solution of TTP. In all these experiments the solution of TTP_1 is 3. However, if the trajectories change then for different values of δ and ϵ, the solution of TTP_1 may change. Especially, if the targets get very close and then very far in short time intervals. For example, Fig. 9 depicts trajectories of two targets in time interval $[0, 4]$. Let $\varepsilon = 0.5$, $r = 1$, and $d = 2$. We computed the solution of TTP_1 for different values of $\delta \in \{0.5, 1, 2, 4\}$. When $\delta = 0.5$ the solution of TTP_1 is 2 while for the rest of values the solution of TTP_1 is 1.

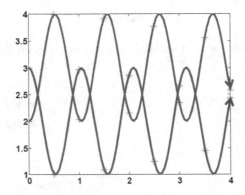

Fig. 9. Trajectories of 2 targets in time interval $[0, 4]$

5   Conclusion

In this paper, we studied the problem of continuous coverage of multiple mobile targets using mobile trackers. We considered the case that trajectories of targets are known in advance and we studied three different problems of (1) minimizing number of trackers (2) minimizing total time that trackers are active and (3) minimize the total active times of the trackers within the period of observation, subject to the constraint that the number of trackers do not exceed the absolute minimum number of trackers necessary to track all the targets. We proposed a model using flow networks to formulate these three problems. Using simulations we investigated the effect of different parameters such as coverage radius on the minimum number of trackers needed for complete coverage. In the future, we plan to present an approximation algorithm for these problems where the solution is continuous and provides all time coverage of all targets. Also, we would like to study the on-line scenario where trajectories of targets are not known in advance.

Acknowledgments. This research is supported in part by grants from the U.S. Defense Threat Reduction Agency under grant number HDTRA1-09-1-0032, the U.S. Air Force Office of Scientific Research under grant number FA9550-09-1-0120, and the Israeli Centers of Research Excellence (I-CORE) program (Center No. 4/11).

References

1. Zorbas, D., Razafindralambo, T., Luigi, D.P.P., Guerriero, F.: Energy efficient mobile target tracking using flying drones. Procedia Comput. Sci. **19**, 80–87 (2013)
2. Zhan, P., Casbeer, D., Swindlehurst, A.: A centralized control algorithm for target tracking with uavs. In: Conference Record of the Thirty-Ninth Asilomar Conference on Signals, Systems and Computers, pp. 1148–1152, October 2005
3. Wheeler, M., Schrick, B., Whitacre, W., Campbell, M., Rysdyk, R., Wise, R.: Cooperative tracking of moving targets by a team of autonomous uavs. In: IEEE/AIAA 25th Digital Avionics Systems Conference, pp. 1–9, October 2006
4. Nitinawarat, S., Atia, G., Veeravalli, V.: Efficient target tracking using mobile sensors. In: 4th IEEE International Workshop on Computational Advances in Multi-Sensor Adaptive Processing (CAMSAP), pp. 405–408, December 2011
5. Xu, E., Ding, Z., Dasgupta, S.: Target tracking and mobile sensor navigation in wireless sensor networks. IEEE Trans. Mob. Comput. **12**(1), 177–186 (2013)
6. Zou, Y., Chakrabarty, K.: Distributed mobility management for target tracking in mobile sensor networks. IEEE Trans. Mobile Comput. **6**(8), 872–887 (2007)
7. Naderan, M., Dehghan, M., Pedram, H.: Mobile object tracking techniques in wireless sensor networks. In: International Conference on Ultra Modern Telecommunications Workshops, ICUMT '09, pp. 1–8, October 2009
8. Adamey, E., Ozguner, U.: A decentralized approach for multi-UAV multitarget tracking and surveillance. In: Society of Photo-Optical Instrumentation Engineers (SPIE) Conference Series, vol. 8389. May 2012
9. Srinivas, A., Zussman, G., Modiano, E.: Construction and maintenance of wireless mobile backbone networks. IEEE/ACM Trans. Networking **17**(1), 239–252 (2009)
10. Radhakrishnan, G., Saripalli, S.: Target tracking with communication constraints: An aerial perspective. In: IEEE International Workshop on Robotic and Sensors Environments (ROSE), pp. 1–6, October 2010
11. Fowler, R.J., Paterson, M.S., Tanimoto, S.L.: Optimal packing and covering in the plane are np-complete. Inf. Process. Lett. **12**(3), 133–137 (1981)
12. Even, S.: Graph Algorithms. W. H. Freeman & Co., New York (1979)

Approximation Algorithm for the Minimum Connected k-Path Vertex Cover Problem

Xiaosong Li[1], Zhao Zhang[2]([⊠]), and Xiaohui Huang[2]

[1] College of Mathematics and System Sciences, Xinjiang University,
Urumqi 830046, Xinjiang, People's Republic of China
[2] College of Mathematics Physics and Information Engineering,
Zhejiang Normal University, Jinhua 321004, Zhejiang, People's Republic of China
hxhzz@sina.com

Abstract. A vertex subset C of a connected graph G is called a connected k-path vertex cover ($CVCP_k$) if every path of length $k - 1$ contains at least one vertex from C, and the subgraph of G induced by C is connected. This concept has its background in the field of security and supervisory control. A variation, called $CVCC_k$, asks every connected subgraph on k vertices contains at least one vertex from C. The $MCVCP_k$ (resp. $MCVCC_k$) problem is to find a $CVCP_k$ (resp. $CVCC_k$) with the minimum cardinality. In this paper, we give a k-approximation algorithm for $MCVCP_k$ under the assumption that the graph has girth at least k. Similar algorithm on $MCVCC_k$ also yields approximation ratio k, which is valid for any connected graph (without additional conditions).

Keywords: k-path vertex cover · Connected k-subgraph cover · Approximation algorithm

1 Introduction

The topology of a wireless sensor network (WSN) can be modeled as a graph, in which vertices represent sensors and edges represent communication channels between sensors. In recent years, new security protocols for WSN emerge, for example, the Canvas scheme which was first proposed by Novotny in [6]. The k-generalized Canvas scheme guarantees data integrity under the assumption that at least one node is not captured on each path of length $k - 1$, where the length of a path is the number of edges in it. Thus during the deployment and initialization of a WSN, it should be ensured that at least one protected vertex exists on each path of length $k-1$ in the communication graph. Since a protected vertex costs more, it is desirable to minimize the number of protected vertices. This problem can be formally described as follows. A path of length $k - 1$ has exactly k vertices. So, for simplicity of statement, we call such a path as a k-path.

Definition 1 (Minimum k-Path Vertex Cover). Given a graph $G = (V, E)$ and an integer $k \geq 2$, a vertex subset $C \subseteq V$ is a k-path vertex cover (VCP_k)

© Springer International Publishing Switzerland 2014
Z. Zhang et al. (Eds.): COCOA 2014, LNCS 8881, pp. 764–771, 2014.
DOI: 10.1007/978-3-319-12691-3_56

of G if each k-path in G contains at least one vertex of C. The MINIMUM k-PATH VERTEX COVER PROBLEM $(MVCP_k)$ is to find a VCP_k of the minimum cardinality.

A variation of $MVCP_k$ was proposed in [12], which is defined as follows.

Definition 2 (Minimum Connected k-Subgraph Cover). Given a graph $G = (V, E)$ and an integer $k \geq 2$, a vertex subset $C \subseteq V$ is a *connected k-subgraph cover* (VCC_k) of G if each connected subgraph in G with cardinality k contains at least one vertex of C. The MINIMUM CONNECTED k-SUBGRAPH COVER PROBLEM $(MVCC_k)$ is to find a VCC_k with the minimum cardinality.

In this paper, we study the $MVCP_k$ and $MVCC_k$ problems under the requirement that the subgraph induced by the VCP_k or VCC_k is connected. Call such problems as $MCVCP_k$ and $MCVCC_k$, respectively, where the second C represents "connected".

In [7], Liu *et al.* studied $MCVCP_k$ problem in unit disk graph (where unit disk graph is a model of a homogeneous wireless sensor network) and gave a polynomial time approximation scheme (PTAS). One basis of their algorithm is a k^2-approximation algorithm for $MCVCP_k$, which is valid for any connected graph.

In this paper give a k-approximation algorithm for $MCVCP_k$ in a general graph under the assumption that the girth of the graph is at least k, where the girth of a graph is the length of a minimum cycle in the graph. Similar algorithm can be applied to $MCVCC_k$, achieving approximation ratio k without any girth assumption on the graph. The basis of these results is a polynomial time algorithm for the corresponding problem on trees.

2   Related Works

This section focuses on related works which study algorithmic aspects of $MVCP_k$ and $MVCC_k$.

The $MVCP_k$ problem was first proposed by Novotny in [6]. Bresar *et al.* [2] gave a polynomial-time approximation-preserving reduction from the Minimum Vertex Cover problem to $MVCP_k$. So, in view of the result in [3], for any $k \geq 2$, $MVCP_k$ cannot be approximated within a factor of 1.3606 unless $\mathrm{P} = \mathrm{NP}$. Tu *et al.* [8] proved that $MVCP_3$ is NP-hard even for a cubic planar graph of girth 3. Bresar *et al.* [2] gave a linear-time algorithm for $MVCP_k$ on trees. Kardoš *et al.* [4] presented a randomized approximation algorithm for $MVCP_3$ with an expected approximation ratio 23/11. For cubic graphs, Tu *et al.* proposed a 1.57-approximation algorithm for $MVCP_3$ [8] and a 2-approximation algorithm for $MVCP_4$ [5].

All the above works are on unweighted VCP_k problem. In a more general setting, every vertex has a weight. The $MWVCP_k$ problem is to find a VCP_k of the minimum weight. Tu and Zhou [9] gave a 2-approximation for $MWVCP_3$

by using a layering method. By using a primal-dual method, they also achieved a 2-approximation [10].

In [12], we proposed the model $MWVCC_k$. Assuming that the girth of the graph is at least k, we propose a $(k-1)$-approximation algorithm for $MWVCC_k$ and showed that ratio $(k-1)$ is tight in our analysis.

The only paper studying $MVCP_k$ with a requirement of connectivity is [7], in which Liu *et al.* gave a PTAS for $MCVCP_k$ in unit disk graphs. Unit disk graph is a model of homogeneous wireless network, in which every vertex corresponds to a point on the plane, two vertices are adjacent if and only if the Euclidean distance between the corresponding points is at most one unit. A basis for the PTAS is a k^2-approximation algorithm for $MCVCP_k$ in a general graph. This k^2-approximation algorithm consists of two steps. In the first step, a VCP_k is found whose size is at most k times that of a minimum VCP_k. In the second step, more vertices are added to connect the VCP_k. The k^2 approximation ratio is a consequence of the following lemma, which will also be used in this paper. For a vertex set C of G, the subgraph of G induced by C is denoted as $G[C]$. The distance between two subgraphs G_1 and G_2 of G is the length of a shortest path in G connecting G_1 and G_2.

Lemma 1. *Suppose C is a VCP_k of a connected graph G. Then, there exist two connected components G_1 and G_2 of $G[C]$ such that $dist(G_1, G_2) \leq k$.*

In this paper, we propose an algorithm for $MCVCP_k$ which improves approximation ratio from k^2 to k in a general graph whose girth is at least k. For $MCVCC_k$, approximation ratio k is still valid and there is not girth assumption on the graph.

3   The Algorithm and Theoretical Analysis

In this section, we present a k-approximation algorithm for $MCVCP_k$ under the assumption that the girth of the graph is at least k. Similar algorithm can also be applied to $MCVCC_k$, achieving approximation ratio k without any assumption on the girth of the graph. These results are based on a polynomial time algorithm for $MCVCC_k/MCVCP_k$ on trees.

3.1   The Algorithm on Trees

We first introduce the algorithm for $MCVCP_k$ on trees.

Suppose T is a tree rooted at vertex v. For any vertex u in T, we shall use T_u to denote the subtree of T rooted at u which contains all descendants of u.

We first analyze the property of a $CVCP_k$ in a tree. Suppose C is a $CVCP_k$ in tree T and v is a vertex in C. Regard v as the root of T. Let u be a child of v. We claim that

$$\text{if } T_u \text{ contains a } k\text{-path, then } u \in C. \tag{1}$$

In fact, if T_u contains a k-path, then $V(T_u) \cap C \neq \emptyset$. Suppose $w \in V(T_u) \cap C$. Since $T[C]$ is connected, all vertices on the unique path in T connecting v and w must be in C. In particular, $u \in C$.

By the above property, a $CVCP_k$ containing given vertex v can be found in the following way: For each child u of v, if T_u has a k-path, then add u into C. Repeat this process for u. Such a recursive process is described in Algorithm 1. It can be seen from the above property that the algorithm returns the unique minimal $CVCP_k$ containing v.

Algorithm 1. Procedure $C = CVCP(T_v, k)$

Input: A tree T_v rooted at vertex v and a positive integer k.
Output: The unique minimal $CVCP_k$ of T_v containing vertex v.
  1: $C = \{v\}$.
  2: **for** each child u of v **do**
  3:      **if** T_u has a k-path **then**
  4:          $C = C \cup CVCP(T_u, k)$
  5:      **end if**
  6: **end for**
  7: Return C.

So, an optimal solution to $MCVCP_k$ on a tree can be found as long as we know one vertex in it. The following lemma shows how to locate such a vertex.

Lemma 2. *Let $P = v_1 v_2 \ldots v_t$ be a longest path in T. Suppose $t \geq k$. Regard T to be rooted at vertex v_t. Let $i_0 = \min\{i : T_{v_i}$ has a k-path$\}$. Then, there exists an optimal solution C^* such that $v_{i_0} \in C^*$.*

Proof. For any vertex set C which is a $CVCP_k$ of T, since the length of P is at least $k-1$, we have $C \cap V(P) \neq \emptyset$. Denote by $j(C) = \min\{i \in \{1, \ldots, t\} : v_i \in C\}$. Choose C^* from all optimal solutions such that $j(C^*)$ is as large as possible. Suppose $v_{i_0} \notin C^*$, we shall derive a contradiction.

Denote by $T_{v_{i_0}}^l$ (resp. $T_{v_{i_0}}^r$) the subtree of T rooted at v_{i_0} when T is regarded to be rooted at v_t (resp. v_1). See Fig. 1 for an illustration of this notation. By the definition of i_0, we have $T_{v_{i_0}}^l = T_{v_{i_0}}$ has a k-path, and thus $V(T_{v_{i_0}}^l) \cap C^* \neq \emptyset$. If $j(C^*) > i_0$, then all those vertices on the unique path of T connecting $v_{j(C^*)}$ to a vertex in $V(T_{v_{i_0}}^l) \cap C^*$ are in C^*, and thus $v_{i_0} \in C^*$, contradicting our assumption that $v_{i_0} \notin C^*$. So,

$$j(C^*) < i_0. \tag{2}$$

Then, similar argument as the above shows that $V(T_{v_{i_0}}^r) \cap C^* \neq \emptyset$ will result in the contradiction that $v_{i_0} \in C^*$. So,

$$V(T_{v_{i_0}}^r) \cap C^* = \emptyset. \tag{3}$$

By the definition of i_0, subtree $T_{v_{i_0}-1}$ does not contain k-path. Combining this with property (3), we see that $C' = \{v_{i_0}\}$ is a $CVCP_k$ of T. So, any optimal

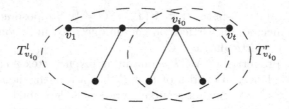

Fig. 1. An illustration for the proof of Lemma 2. For $k = 4$, the left ellipse indicates $T_{v_{i_0}}^l$ and the right ellipse indicates $T_{v_{i_0}}^r$.

solution contains exactly one element. In particular, $C^* = \{v_{j(C^*)}\}$ and C' is also an optimal solution. But $j(C') > j(C^*)$ (by (2)), contradicting to our choice of C^*. □

We now present the algorithm for $MCVCP_k$ on trees in Algorithm 2.

Algorithm 2. Algorithm for $MCVCP_k$ on Trees

Input: A tree $T = (V, E)$ and a positive integer k.
Output: A vertex set C which is a $CVCP_k$ of T.
 1: Find a longest path P in T. Suppose $P = v_1 v_2 \ldots v_t$.
 2: **if** $t \le k - 1$ **then**
 3:     Return $C = \emptyset$.
 4: **end if**
 5: Regard T to be rooted at v_t. Let $i_0 = \min\{i : T_{v_i}$ has a k-path$\}$.
 6: Return $C = CVCP(T_{v_{i_0}}, k)$.

Theorem 1. *Algorithm 2 computes an optimal solution to $MCVCP_k$ on tree T in time $O(n^2)$, where n is the number of vertices.*

Proof. The correctness of the algorithm follows from property (1) and Lemma 2. Next, we analyze the time complexity of the algorithm.

It is known that a longest path in a tree can be computed in $O(n)$ time. In fact, it was shown in [11] that for any vertex $u \in V(T)$, if v is the farthest vertex to u and w is a farthest vertex to v, then the unique path in T connecting v and w is longest in T. So, a longest path in T can be found by running Breadth First Search twice, which takes time at most $2n$ (since T is a tree).

We shall show by induction on n that Algorithm 1 runs in time at most $n^2 - 3n$ for $n \ge 3$. When $n = 3$, the tree must be a path, the time complexity is trivial. Next, suppose $n \ge 4$, and the claim is true for any smaller n. Suppose the children of v are $u_1, \ldots, u_s$, and $|V(T_{u_i})| = n_i$ for $i = 1, \ldots, s$. By induction hypothesis, the algorithm runs in time at most $\sum_{i=1}^{s} n_i^2 - 3 \sum_{i=1}^{s} n_i$. By noticing that $\sum_{i=1}^{s} n_i^2 \le (n_1 + \ldots + n_s)^2$ and $n_1 + \ldots + n_s = n - 1$, the time is at most $(n - 1)(n - 4) \le n(n - 3)$ for $n \ge 2$.

For Algorithm 2, Step 1 takes time at most $2n$. One can determine i_0 in Step 5 by checking subtrees $T_{v_1}, \ldots T_{v_t}$ sequentially to see whether the longest path in T_{v_i} has length at least $k-1$. This takes time at most $t \cdot 2n = O(n^2)$. As shown in the above paragraph, Step 6 takes time $O(n^2)$. Adding them together, the time complexity $O(n^2)$ follows.   □

Similar algorithm can be applied to $MCVCC_k$ with Line 3 of Algorithm 1 replaced by "if $|V(T_u)| \geq k$ then", and the index i_0 in Line 5 of Algorithm 2 to be determined by "$i_0 = \min\{i : |V(T_{v_i})| \geq k\}$". In fact, since determining whether $|V(T_{v_i})| \geq k$ is much easier than determining whether T_{v_i} contains a k-path, the time complexity can be reduced to $O(n)$, by modifying Algorithm 1 into a non-recursive way, and by recording some information about the number of vertices while executing the Breadth First Search.

Theorem 2. *An optimal solution to $MCVCC_k$ on a tree can be computed in time $O(n)$, where n is the number of vertices of the tree.*

3.2   The Algorithm for a General Graph

Assume that graph G is connected and has girth at least k. The algorithm for $MCVCP_k$ in G is presented in Algorithm 3.

Algorithm 3. Algorithm for $MCVCP_k$ in a General Graph with A Girth Assumption

---

Input: A positive integer k and a connected graph G with girth at least k.
Output: A $CVCP_k$ C of G.
 1: From an arbitrary vertex v, find a Depth First Search tree T of G with root v.
 2: Use Algorithm 2 to find a $CVCP_k$ C of T.
 3: **if** $v \notin C$ **then**
 4:     $C = C \cup V(P)$, where P is the shortest path from v to C.
 5: **end if**
 6: Return C.

Theorem 3. *Algorithm 3 computes in time $O(n^2)$ a $CVCP_k$ of G whose size is at most k times that of an optimal solution.*

Proof. First, we prove that the output C of Algorithm 3 is indeed a $CVCP_k$ of G. For any connected component H of $T - C$ and for any edge $e = uw \in E(G) \backslash E(T)$ with one end in $V(H)$, say $u \in V(H)$, we claim that $w \in C$. In fact, by the property of Depth First Search tree [1], vertices u and w are related, that is, one is an ancestor of the other. Notice that C induces a connected subtree of T which contains the root. So, no vertices in different connected components of $T - C$ can be related, and thus w cannot belong to another connected component of $T - C$. On the other hand, since graph G has girth at least k and H does not contain a k-path, we see that $w \notin V(H)$. Then, the claim $w \in C$ follows. As a consequence, $G - C = T - C$. Since C is a $CVCP_k$ of T, it is also a $CVCP_k$ of G.

Next, we estimate the size of output C. Denote by C_0 the $CVCP_k$ of T found by Line 2 of Algorithm 3. Then

$$|C| = |C_0| + |V(P)| - 1 \tag{4}$$

Suppose C^* is an optimal $CVCP_k$ of G. Notice that C^* is also a VCP_k of T but $T[C^*]$ might not be connected. By Lemma 1, adding at most $(k-1)$ vertices is sufficient to connect two connected components of $G[C^*]$. Recursively executing such an operation, it can be seen that adding at most $(|C^*| - 1)(k - 1)$ vertices into C^*, we obtain a $CVCP_k$ of T. It follows that

$$|C_0| \le |C^*| + (|C^*| - 1)(k - 1). \tag{5}$$

Combining inequalities (4), (5), and the observation that $|V(P)| \le k$, we have

$$|C| \le k|C^*|.$$

Hence, the approximation ratio is at most k.

The time complexity follows from Theorem 1, noticing that Step 1 and Step 3 can be executed in linear time. □

Similar algorithm can be applied to $MCVCC_k$ of a general graph G. The difference is that approximation ratio k is valid without requiring that the girth of G is at least k. In fact, by studying the proof of Theorem 3, it can be seen that the girth assumption is used to show that the other end of edge e cannot be in H itself. For the case of $MCVCC_k$, it does not matter whether edge e has both of its two ends in H: though the structure of a connected component of $T - C$ might be different from that of $G - C$, the cardinality remains the same, which is at most $k - 1$. This is sufficient to show that C is also a $CVCC_k$ of G. Notice that without girth assumption, a connected component of $T - C$ without k-path might be a connected component of $G - C$ containing a k-path, the property of a Depth First Search does not help. An example is shown in Fig. 2.

Theorem 4. *There is an $O(n)$-algorithm for $MCVCC_k$ on a connected graph with approximation ratio at most k.*

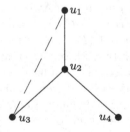

Fig. 2. An example showing that girth assumption for Theorem 3 is necessary. The solid edges form a subtree of T rooted at vertex u_1, and the dashed edge is in $E(G) \setminus E(T)$. The longest path in the subtree has length two. Adding the dashed edge (which connects a vertex to its ancestor), the graph has a 4-path $u_3 u_1 u_2 u_4$.

4   Conclusion and Future Work

In this paper, we gave a k-approximation algorithm for the $MCVCP_k$ problem under the assumption that the girth of the graph is at least k. For the $MCVCC_k$ problem, approximation ratio k can be achieved in a general graph without any girth assumption. Previous to our work, there is only one paper studying VCP_k problem under connectivity requirement [7], which achieves approximation ratio k^2 for the general graph and a PTAS for unit disk graph.

An interesting problem is to study the problems with weight.

Acknowledgments. This research is supported by NSFC (61222201), SRFDP (20126501110001), and Xingjiang Talent Youth Project (2013711011).

References

1. Bondy, J.A., Murty, U.S.R.: Graph Theory. Springer, New York (2008)
2. Bresar, B., Kardos, F., Katrenic, J., Semanisin, G.: Minimum k-path vertex cover. Discrete Appl. Math. **159**, 1189–1195 (2011)
3. Dinur, T., Safra, S.: On the hardness of approximating minimum vertex cover. Ann. Math. **162**, 439–485 (2005)
4. Kardoš, F., Katrenič, J., Schiermeyer, I.: On computing the minimum 3-path vertex cover and dissociation number of graphs. Theor. Comput. Sci. **412**, 7009–7017 (2011)
5. Li, Y., Tu, J.: A 2-approximation algorithm for the vertex cover P_4 problem in cubic graphs. Int. J. Comput. Math. (published online, 2014). doi:10.1080/00207160.2014.881476
6. Novotný, M.: Design and analysis of a generalized canvas protocol. In: Samarati, P., Tunstall, M., Posegga, J., Markantonakis, K., Sauveron, D. (eds.) WISTP 2010. LNCS, vol. 6033, pp. 106–121. Springer, Heidelberg (2010)
7. Liu, X., Lu, H., Wang, W., Wu, W.: PTAS for the minimum k-path connected vertex cover problem in unit disk graphs. J. Glob. Optim. **56**, 449–458 (2008)
8. Tu, J., Yang, F.: The vertex cover P_3 problem in cubic graphs. Inf. Process. Lett. **113**, 481–485 (2013)
9. Tu, J., Zhou, W.: A factor 2 approximation algorithm for the vertex cover P_3 problem. Inf. Process. Lett. **111**, 683–686 (2011)
10. Tu, J., Zhou, W.: A primal-dual approximation algorithm for the vertex cover P_3 problem. Theoret. Comput. Sci. **412**, 7044–7048 (2011)
11. Wu, B., Chao, K.: A note on eccentricities, diameters and radii (2004). www.csie.ntu.edu.tw/kmchao/tree04spr/diameter.pdf
12. Zhang, Y., Shi, Y., Zhang, Z.: Approximation algorithm for the minimum weight connected k-subgraph cover problem. Theoret. Comput. Sci. **535**, 54–58 (2014)

Author Index

Printed in the United States
By Bookmasters